MECHANICAL ENGINEERING THEORY AND APPLICATIONS

COMPUTATIONAL FLUID DYNAMICS THEORY, ANALYSIS AND APPLICATIONS

MECHANICAL ENGINEERING THEORY AND APPLICATIONS

Additional books in this series can be found on Nova's website under the Series tab.

Additional E-books in this series can be found on Nova's website under the E-books tab.

MECHANICAL ENGINEERING THEORY AND APPLICATIONS

COMPUTATIONAL FLUID DYNAMICS THEORY, ANALYSIS AND APPLICATIONS

ALYSSA D. MURPHY
EDITOR

Nova Science Publishers, Inc.
New York

Copyright ©2011 by Nova Science Publishers, Inc.

All rights reserved. No part of this book may be reproduced, stored in a retrieval system or transmitted in any form or by any means: electronic, electrostatic, magnetic, tape, mechanical photocopying, recording or otherwise without the written permission of the Publisher.

For permission to use material from this book please contact us:
Telephone 631-231-7269; Fax 631-231-8175
Web Site: http://www.novapublishers.com

NOTICE TO THE READER

The Publisher has taken reasonable care in the preparation of this book, but makes no expressed or implied warranty of any kind and assumes no responsibility for any errors or omissions. No liability is assumed for incidental or consequential damages in connection with or arising out of information contained in this book. The Publisher shall not be liable for any special, consequential, or exemplary damages resulting, in whole or in part, from the readers' use of, or reliance upon, this material. Any parts of this book based on government reports are so indicated and copyright is claimed for those parts to the extent applicable to compilations of such works.

Independent verification should be sought for any data, advice or recommendations contained in this book. In addition, no responsibility is assumed by the publisher for any injury and/or damage to persons or property arising from any methods, products, instructions, ideas or otherwise contained in this publication.

This publication is designed to provide accurate and authoritative information with regard to the subject matter covered herein. It is sold with the clear understanding that the Publisher is not engaged in rendering legal or any other professional services. If legal or any other expert assistance is required, the services of a competent person should be sought. FROM A DECLARATION OF PARTICIPANTS JOINTLY ADOPTED BY A COMMITTEE OF THE AMERICAN BAR ASSOCIATION AND A COMMITTEE OF PUBLISHERS.

Additional color graphics may be available in the e-book version of this book.

LIBRARY OF CONGRESS CATALOGING-IN-PUBLICATION DATA

Computational fluid dynamics : theory, analysis, and applications / editor, Alyssa D. Murphy.
    p. cm.
  Includes index.
  ISBN 978-1-61209-276-8 (hardcover)
  1. Fluid dynamics--Data processing. 2. Fluid dynamics--Mathematical models. I. Murphy, Alyssa D.
  TA357.5.D37C658 2011
  620.1'064--dc22
                            2010049663

Published by Nova Science Publishers, Inc. † New York

CONTENTS

Preface vii

Chapter 1 Computational Modeling Aspects of Polymer
Electrolyte Fuel Cell Durability 1
Maher A. R. Sadiq Al-Baghdadi

Chapter 2 A Front Tracking Method for Numerical Simulation
of Incompressible Two-Phase Flows 41
Jinsong Hua and Gretar Tryggvason

Chapter 3 Computational Fluid Dynamics in Biomedical Engineering 109
*T. A. S. Kaufmann, R. Graefe, M. Hormes,
T. Schmitz-Rode and U. Steinseifer*

Chapter 4 Prediction of Heat Transfer and Air Flow
in Solar Heated Ventilation Cavities 137
Guohui Gan

Chapter 5 A Review of Advances in Multiphase Computational
Fluid Dynamics for Microfluidics 179
Rafael M. Santos

Chapter 6 Hydrodynamics Analysis for Partially Ventilated
Cavitating Vehicle Based on Two-Fluid Model 205
M. Xiang, S. C. P. Cheung, J. Y. Tu and W. H. Zhang

Chapter 7 Computational Modelling in Tissue Engineering:
Providing Light in Times of Darkness 243
Ryan J. McCoy, Fergal J. O'Brien and Daniel J. Kelly

Chapter 8 Modelling and Simulation of an Electrostatic Precipitator 267
Shah M. E. Haque, M. G. Rasul and M. M. K. Khan

Chapter 9 Applications of Computational Fluid Dynamics
in Food Processing Operations 297
C. Anandharamakrishnan

Chapter 10	Numerical Study on Steam-Jet Vacuum Pump by Computational Fluid Dynamics Approach *Xiaodong Wang, Jiangliang Dong, Hongjian Lei and Jiyuan Tu*	**317**
Chapter 11	Computational and Raman Spectral Studies on Probing Azaindole Adsorption on Ag Nanometal Doped Sol - Gel Substrates *B. Karthikeyan, S. K. Thabasum Sheerin and M. Murugavelu*	**343**
Chapter 12	Numerical Approaches for Solving Free Surface Fluid Flows *L. Battaglia, J. D'Elía and M. Storti*	**351**
Chapter 13	Numerical Simulation of Multiphase Flow in Chemical Reactors *Chao Yang, Yumei Yong and Zai-Sha Mao*	**385**
Chapter 14	Modelling and Optimization of Microfluidic Passive Mixer using CFD Analysis *Karol Malecha and Ziemowit M. Malecha*	**475**
Chapter 15	Computer Simulation and Optimization of Stirrer Hydrodynamics at High Reynolds Numbers *O. Uğur, K. Yapici, Y. Uludağ and B. Karasözen*	**493**
Chapter 16	Fluid Structure Interaction and Galilean Invariance *Luciano Garelli, Rodrigo R. Paz, Hugo G. Castro, Mario A. Storti and Lisandro D. Dalcin*	**511**
Chapter 17	Lattice Botlzmann Method for Fluid Dynamics *Mojtaba Aghajani Delavar, Mousa Farhadi and Kurosh Sedighi*	**551**
Chapter 18	Moving Mesh Interface Tracking *Shaoping Quan*	**593**
Index		**627**

PREFACE

Computational fluid dynamics (CFD) is one of the branches of fluid mechanics that uses numerical methods and algorithms to solve and analyze problems that involve fluid flows. Computers are used to perform the millions of calculations required to simulate the interaction of liquids and gases with surfaces defined by boundary conditions. This new book presents topical research in the study of computational fluid dynamics.

Chapter 1 - Polymer electrolyte membrane (PEM) fuel cells are still undergoing intense development, and the combination of new and optimized materials, improved product development, novel architectures, more efficient transport processes, and design optimization and integration are expected to lead to major gains in performance, efficiency, durability, reliability, manufacturability and cost-effectiveness. Computational fuel cell models that allow systematic simulation, design and optimization of fuel cell systems would facilitate the integration of such advances, allow less heavy reliance on hardware prototyping, and reduce development cycles.

Durability is one of the most critical remaining issues impeding successful commercialization of broad PEM fuel cell stationary and transportation energy applications, and the durability of fuel cell stack components remains, in most cases, insufficiently understood. Lengthy required testing times, lack of understanding of most degradation mechanisms, and the difficulty of performing in-situ, non-destructive structural evaluation of key components makes the topic a difficult one. The damage mechanisms in a PEM fuel cell are accelerated by mechanical stresses arising during fuel cell assembly (bolt assembling), and the stresses arise during fuel cell running, because it consists of the materials with different thermal expansion and swelling coefficients. Therefore, in order to acquire a complete understanding of the damage mechanisms in the membrane, mechanical response under steady-state hygro-thermal stresses should be studied under real cell operating conditions and in real cell geometry (three-dimensional).

In this chapter, full three-dimensional, non-isothermal computational fluid dynamics model of a PEM fuel cell has been developed to simulate the stresses inside the PEM fuel cell, which are occurring during fuel cell assembly (bolt assembling), and the stresses arise during fuel cell running due to the changes of temperature and relative humidity. A unique feature of the present model is to incorporate the effect of hygro and thermal stresses into actual three-dimensional fuel cell model. In addition, the temperature and humidity dependent material properties are utilize in the simulation for the membrane. The model is shown to be able to understand the many interacting, complex electrochemical, transport phenomena, and stresses distribution that have limited experimental data. This model is used to study and analyse the effect of operating, design, and material parameters on the mechanical behaviour of PEM. The analysis helped identifying critical parameters and shed insight into the physical mechanisms leading to a fuel cell durability under various operating conditions..

Chapter 2 - A front-tracking method for direct numerical simulation of incompressible two-phase flows is presented along with its numerical implementation and validation. This method is based on the "one-fluid" formulation of the governing equations, in which a single set of governing equations is applied to both fluid phases, treating the different phases as one fluid with variable material properties. The unsteady Navier–Stokes equations are solved using a conventional finite volume method on a fixed grid, and the interface or front is tracked explicitly by connected marker points. The physical discontinuities across the interface are accounted for by adding the appropriate sources as delta-functions at the front separating the phases. The surface tension is computed on the front and spread onto the fixed grid. Following the motion of the front, the distribution of fluid properties such as density is updated accordingly. The front tracking method has been implemented for fully three-dimensional flow modeling, as well as for two-dimensional and axisymmetric flow systems. The proposed numerical method has been validated systematically with available experimental data, and employed widely to investigate the dynamics of multiphase flows with bubbles and drops. In addition, this numerical method has been extended to model complex multiphase interfacial flow problems involving phase change, surfactant transport and electrohydrodynamics. The extensive review of simulations that use the front tracking method to investigate various multiphase interfacial flow problems demonstrates that the front tracking method is one of the most effective numerical approaches for the interfacial multiphase flow simulations.

Chapter 3 - There are more than 1.5 million operations each year using cardiopulmonary bypass (CPB), when the function of the human heart is taken over by extracorporeal blood pumps and oxygenators. Every fourth of these patients suffers from minor or major neurologic events such as headaches, mnemonic problems, loss of neural functions or coma. This is mainly related to the blood flow to the brain. The understanding of these flow conditions is therefore of high clinical relevance. Computational Fluid Dynamics (CFD) is used to model the flow in the human vascular system and thus gain insight in flow processes invisible to experimental measurements.

To analyze these conditions, realistic models of the human vascular system are obtained using modern imaging techniques. From this data, a 3D model is reconstructed and the fluid flow can be calculated. Different blood properties (e.g. Non-Newtonian behavior) and realistic boundary conditions like systemic pressure have to be taken into account. Thereby, the flow in the cardiovascular system can be analyzed and optimized for different applications and the neurologic events during CPB operations can be reduced.

In some of these operations, clinical devices such as rotary blood pumps are implanted to support of completely take over some of the functions of the human heart, for example as a bridge-to-transplant, when a patient is waiting for a heart transplant but there is currently none available. Very common devices are Left Ventricular Assist Devices (LVADs) that support the function of the left heart. CFD plays an important role during development of these devices. For example, the hydraulic efficiency and rotor stability can be analyzed virtually. Thereby, CFD can significantly reduce the development time and cost for new prototypes.

In an extracorporeal circuit, a blood pump is often combined with an oxygenator to provide sufficient oxygen transfer via the blood. One important parameter of these devices is the gas exchange rate: oxygen O_2 diffuses from porous fibers to the blood, while carbon dioxide CO_2 is withdrawn. This process can be analyzed by CFD to support oxygenator development.

Another typical application is the design of new artificial heart valves, especially mechanical heart valves. The advantage of this approach compared to biological valves is the highly increased durability. However, patients with these valves need life-long medication to avoid thrombosis. The likelihood for thrombus formation is related to the shear rates acting on human blood. Using CFD, these shear rates as well as exposure time can be calculated. Thus, predictions of critical regions of new valves can be made.

There are many challenges when using CFD for medical applications, since interactions in the human body are very complex. Therefore, experimental as well as clinical data is needed to validate CFD. This can be done by flow visualization (e.g. MRI-flow measurements, Particle Image Velocimetry) or other experimentally obtained data (pressure, flowrate). Once validated, CFD provides many benefits for the development of biomedical devices and for the understanding of the flow conditions in the human body. In the end, it can thus increase patient survival.

Chapter 4 - Solar heated ventilation cavities are used to enhance passive cooling of buildings and building-integrated photovoltaics. This chapter is concerned with numerical prediction of buoyancy-driven heat transfer and air flow in vertical cavities for natural ventilation of buildings and inclined cavities for ventilation cooling of building-integrated photovoltaics. Computational fluid dynamics was used to predict the heat transfer and air flow rates in vertical ventilation cavities with various combinations of heat distribution ratios ranging from symmetrical to asymmetrical heating and inlet/outlet openings and also in inclined ventilation cavities with combined convective and radiative heat transfer. The natural ventilation rate and heat transfer rate in a vertical cavity have been found to vary with the total heat flux, heat distribution on the cavity wall surfaces, cavity size and opening positions. It has also been found that reducing asymmetry in heat distribution on the vertical cavity wall surfaces can increase the ventilation rate but would decrease the heat transfer rate. The air flow rate and heat transfer rate in an inclined cavity generally increase with inclination angle while the temperature of PV cells decreases with increasing inclination. General correlations between these variables in the cavities have been obtained and presented in non-dimensional forms.

Chapter 5 - Multiphase flow in microchannels has gained much interest in recent years given the numerous emerging applications of microfluidics that promise to provide technological innovations not realizable with conventional channels. Significant developments in the area of micro-scale fabrication in the last decade have allowed researchers to construct increasingly more intricate devices, generally termed Lab-on-a-chip, and used in such diverse fields as: biomedical, analytical chemistry, chemical synthesis, unit operations, among others. Moving forward microfluidics is also envisioned to allow for process intensification of industrial process, hence the production capacity is expected to span from nanoliters to cubic meters, from micrograms to tons.

The interest is derived from the fact that gas–liquid two-phase flows in microchannels often exhibit different flow behaviour than macrosized conduits, which allows for the precise control of the trajectory of fluidic particles. At the micro-scale there is increased importance and effect of surface tension forces, while gravitational forces become negligible, and inertial, shearing and drag forces have a limited effect. As a result, multiple unique flow patterns have been identified in microfluidic gas–liquid two-phase flow, including bubbly, slug, ring, churn and annular flow. Transport phenomena, such as heat and mass transfer, and reaction kinetics are also significantly improved at the micro-scale due to the increase in surface to volume ratio.

The successful design of microfluidic devices relies on the need to fully understand the flow dynamics and physics, obtained by experimental means and from numerical modelling. Large discrepancies still exist in the published data, largely as a result of difficulty and inconsistencies in experimental setup and measurement. Moreover, process properties such as heat and mass transfer, pressure drop and residence time distribution are dependent on flow parameters such as the droplet size and velocity, the void fraction, the liquid film thickness, and so forth; however these quantities often cannot be determined a priori, making the design of microfluidic devices highly dependent on empirical correlations. The accuracy and reliability of present correlations can be extensively improved if the intricacies of the flow fields are known in more detail. This has recently become possible with the advances in the field of multiphase computational fluid dynamics (CFD) and the maturity of both proprietary codes and commercial software. A number of researchers have begun exploring this tool, using techniques such as volume-of-fluid (VOF), level-set and the lattice Boltzmann method (LBM), and publications have grown extensively in recent years.

This chapter will review the recent published works in the field, focusing on the application of CFD to simulate multiphase flow in microchannels. Both works of basic research on aspects of transport phenomena, as well as applied research on the development of new chemical processes and the intensification of existing processes at the micro-scale, will be covered. Common themes and practices will be consolidated and reviewed, and potential development directions identified.

Chapter 6 - Reducing the high friction drag caused by viscosity of water has been a great challenge in developing high speed underwater vehicles. Among many proposed drag reduction techniques, ventilated cavitation has been considered as one of the promising approaches to achieve drag reduction. In this chapter, particular focus is directed to investigate the drag reduction mechanism of partial ventilated cavitation which is frequently encountered; especially for transient launch and vehicle maneuvering. Nonetheless, hydrodynamic effect of partial ventilated cavity is not only affected by dynamical effect of the deformable gas-liquid interface; but also the morphology transition from continuous cavity to dispersed gas bubbles downstream. It is therefore an extremely difficult task to gain in-depth understanding of the complex two-phase flow structure and its associated drag reduction mechanism. In this study, a numerical scheme based on Eulerian-Eulerain two-fluid framework has been proposed to analyse flow characteristics and hydrodynamics resulted from the bubbly flow downstream partially ventilated cavity. The main content can be broadly classified into four parts: (i) A detail review of basic theory and numerical approaches for ventilated cavitating underwater vehicles; (ii) Model development for the bubbly flow downstream partially ventilated cavity; (iii) Characterization and model validation of gas-liquid flow field; (iv) Hydrodynamics analysis based on the simulation results. In the first part, background theory and technical problems for ventilated cavitating underwater vehicle are introduced. Afterwards, a literature review of the historical development of numerical approaches in simulating ventilated cavitating flow is summarized. In the second part, an Eulerian–Eulerian two-fluid model incorporated with the population balance approach is introduced to predict the bubbly wake flow behind the ventilated cavity. Particular attention is devoted to establish air entrainment model at the cavity base and to improve interfacial momentum transfer models. In the third part, numerical simulation is carried out for underwater vehicle under different ventilation rate and sailing velocity. The flow field parameters including void fraction, velocity, pressure and bubble size distributions are obtained based on the simulation results. Good agreement is achieved in compared with experimental data. Finally, the drag reduction efficient for the partially ventilated vehicle is closely examined. Comparison between non-cavitating and supercavitation conditions is also presented. Flow field characteristics and meaningful conclusions for partially ventilated cavitating vehicles are summarised at the end of the article.

Chapter 7 - The human body can be considered an extremely complex machine. The demands placed on this machine over the lifetime of an individual or their natural genetic predisposition to particular conditions can result in the failure of native internal structures; for example, hips, knees, hearts or kidneys. Computational modelling has played a key role in the realisation of mechanically engineered replacements that can either be implanted internally (e.g ventricular assist devices and orthopaedic implants) or used to replace organ functionality externally (e.g kidney dialysis machines). This has enabled temporary ablation of the body's decline in some instances, but the provision of a biologically and anatomically relevant substitute capable of integrating into the host and continuing the original function unaided for the remaining lifetime of the individual, has yet to reach fruition. Interdisciplinary collaboration over the last 40 years between the engineering and scientific communities, has pursued and continues to pursue, the development of physiologically representative substitutes through the cultivation of tissue ex-vivo or through the induction of tissue growth in-vivo (tissue engineering and regenerative medicine). The intrinsic complexity of these biological systems means that experimentally prizing apart the individual mechanisms at play may not always be viable. Computational modelling tools thus have the potential to prove a very powerful addition once again; providing light where there is darkness in this emerging field and promoting furthered understanding.

This chapter endeavours to introduce the reader to computational modelling concepts undertaken within the tissue engineering and regenerative medicine fields through a series of specifically selected case studies based within the bone tissue engineering arena.

Chapter 8 - Electrostatic precipitators (ESPs) are the most common, effective and reliable particulate control devices that are mainly used in power plants and other process industries. The ESP works as a cleaning device and uses electrical forces to separate the dust particles from the flue gas. A typical ESP consists of an inlet diffuser, known as inlet evase, a rectangular collection chamber and an outlet convergent duct, known as outlet evase. Perforated plates are placed inside both the inlet and outlet evase for the purpose of flow distribution. Inside the collection chamber are placed a number of discharge electrodes (DEs) and collection electrodes (CEs). Discharge electrodes are suspended vertically between two collection electrodes. While the flue gas flows through the collection area, electrostatic precipitators accomplish particle separation by the use of an electric field in three steps. At first it imparts positive or negative charges to the particles by discharge electrodes. In the second step it attracts the charged particles to oppositely charged or grounded collection electrodes. And finally it removes the collected particles by vibrating or rapping the collection electrodes or by spraying with liquid.

This chapter presents a Computational Fluid Dynamics (CFD) model for an ESP. An ESP system consists of flow field, electrostatic field and particle dynamics. But they cannot be simultaneously applied to the modelling of an industrial ESP. The main reason for this is the lack of computational resources. A one-step algorithm incorporating all the above systems would require excessive computational memory and unacceptably long calculation time. Therefore, the ESP was modelled in two steps. Firstly, the 3D fluid (air) flow was modelled considering the detailed geometrical configuration inside the ESP. The model was then validated with the measured data obtained from the geometrically similar laboratory experiments. Numerical calculations for the gas flow were carried out by solving the Reynolds-averaged Navier-Stokes equations, and turbulence was modelled using the realizable k-ε turbulence model. In the second step, as the complete ESP system consists of an electric field and a particle phase in addition to the fluid flow field, a two dimensional ESP model was developed. An additional source term was added to the gas flow equation to capture the effect of the electric field. This additional source term was obtained by solving a coupled system of the electric field and charge transport equations. The electrostatic force was applied to the flow equations by using User Defined Functions (UDFs). A discrete phase model (DPM) was incorporated with this 2D model to study the effects of particle size, electric field and flue gas flow on the collection efficiency of particles inside the ESP.

The CFD model thus developed was successfully applied to a prototype ESP at the power plant and used to recommend options for improving the efficiency of the ESP. The aerodynamic behaviour of the flow was improved by geometrical modifications in the existing 3D numerical model. In particular, the simulation was performed to improve and optimize the flow in order to achieve uniform flow and to increase particle collection inside the ESP. The particles injected in the improved flow condition were collected with higher efficiency after increasing the electrostatic force inside the 2D model. The approach adopted in this chapter to optimize flow and electrostatic field properties is a novel approach for improving the performance of an electrostatic precipitator.

Chapter 9 - Computational fluid dynamics (CFD) has been used extensively by the scientific community worldwide for optimal design of industrial processes. CFD is a simulation tool, which uses powerful computers in combination with applied mathematics to model fluid motion. CFD is increasingly used in the design, scale-up and trouble-shooting of different unit operations in various processing industries. In recent years, a rapid development in the application of CFD in food processing operations has been witnessed. This chapter reviews the application of CFD in selected food processing operation such as spray drying, bread baking and pasteurization/sterilization processes. It also discusses the different modeling approaches and reference frames used for the CFD simulations, particles histories (temperature, velocity residence time and impact positions) during spray drying and spray-freezing, application of different radiation models (discrete transfer radiation, surface to surface and discrete ordinates) for electrical heating baking oven simulations and modeling of inactivation of enzymes during pasteurization of canned liquid food and eggs. In addition, the challenges involved in the CFD modelling of food processing, recent developments in this area and future applications are highlighted.

Chapter 10 - Steam-jet vacuum pump is one of the important equipments widely used in industry to obtain a vacuum environment for various special techniques. The primary fluid (steam) with high pressure is accelerated through a nozzle to obtain supersonic speed. The supersonic motive steam and secondary fluid mix in mixing chamber with energy and momentum exchanging. A normal shock wave is induced in throat and the flow speed suddenly drops to subsonic value. Further compression is achieved when the mixed stream passes through diffuser. The flow is complicated in the pump due to the transonic flow and difficult to be described by traditional methods. Computational fluid dynamics (CFD) can be used to investigate and predict the complicated flow problems in steam-jet pump.

A mathematic model for transonic flow was proposed to investigate the mixing flow behaviors of primary and secondary fluids in steam-jet vacuum pump. The simulation was carried out to predict the state pressure distribution among mixing chamber wall. Close agreements between the predicted results and experimental data validates the theoretical model. The velocity vectors and Mach number profiles in mixing chamber at different back pressures and the secondary fluid pathlines and mass flux profiles at different suction pressures were predicted. It is found that there are swirls separated from secondary fluid near the wall and the velocity of secondary fluid was fallen down obviously when the back pressure was bigger than critical back pressure. The above two factors lead to the entrainment ratio reduced rapidly. The flow structure in mixing chamber would be broken down and secondary flow would reverse to upstream if back pressure is in excess of the break down pressure, and the steam-jet pump lost pumping ability completely. It is also found that the mass flux increased with the increasing of suction pressures which made the entrainment ratios increased. The prediction results show that the pressure ratio is a dominant position in affecting the pump's performances.

It is found that there are spontaneously condensing phenomena in the nozzle supersonic flow process at different primary fluid initial parameters, as the primary fluid is not assumed as perfect gas. The outlet pressure of nozzle predicted is higher than that as the primary fluid assumed as perfect gas, and the outlet velocity is lower than that of general simulations, and then the efficiency of nozzle and steam-jet pump would be reduced.

Chapter 11 - Computational and Surface enhanced Raman spectral studies are used to probe the adsorption of 7-azaindole on the Ag doped sol-gel substrates. The adsorption mode and the vibrational spectral futures are discussed[1].This chapter gives a further account of application of theoretical calculations and modeling of the adsorption of azaindole on the Ag nanometal surface. The abinito and DFT level calculated Ag complexed and experimentally obtained SERS results are well agreed. It is proposed that the similar methods can be implemented in the study of surface vibrational properties of adsorbates adsorbed on nanometal embedded sol-gel substrates.

Chapter 12 - A free surface is defined as an interface between two fluids, where the lighter phase, which is usually a gas, has negligible effect over the other due to its very low values of density and viscosity. Free surface flows are common issues among several engineering disciplines, such as civil, mechanical or naval. Some typical problems are open channel flows, sloshing in tanks for storing or transporting liquids, and mold filling, among others. Different numerical methods have been developed for solving these kind of flows, being the most popular classification of techniques the one that refers to "interface tracking" and "interface capturing" methods. On one side, the interface tracking approaches are based on considering the free surface as a boundary of the domain, and defining over that boundaries some entities such as nodes or element edges of a finite element method mesh, in such a way that the fluid flow problem is solved over a single liquid phase. On the other side, interface capturing approximations are based on marking functions that indicate which part of the domain is occupied by the liquid, and which other is occupied by the gaseous phase, such that the interface position is "captured" over certain values of the marking function. In the present chapter, two finite element methods for solving free surface flows are described: an interface tracking technique developed over an arbitrary Lagrangian-Eulerian framework, and alevel set interface capturing proposal. Each approach has been considered for solving different free surface flow problems, regarding the capabilities of the methods.

Chapter 13 - More and more attentions have been paid to better understanding of the mechanisms of multiphase flow and interphase mass and heat transfer on the mesoscale and macroscale systems. The development of computational fluid dynamics (CFD) and related sciences make it possible to quantitatively describe complicated multiphase flow and solve the problems of scale-up effect and macro-control in the chemical industry processes. The current state of the art in CFD is the Reynolds-averaged Navier-Stokes method (RANS), but the marked disadvantage with this method is the modeling limited generality and unsatisfactory accuracy. Direction numerical simulation (DNS) methods completely simulate turbulent flow with high precision by a large number of grids, but up to now it is still impossible to apply it into industrial processes. The large eddy simulation (LES) method instead can be used for very general applications (geometry as well as flow fields) and the accuracy is dramatically better compared to RANS turbulence modeling. The undesirable drawback of LES is the higher computational time required for the solution.

The goal of DNS of multiphase flows is both to generate insight and understanding of the basic behavior of multiphase flow—such as the forces on a single bubble or a drop, how bubbles and drops affect the flow, and how many bubbles and drops interact in dense dispersed flows—as well as to provide data for the generation of closure models for engineering simulations of the averaged flow field. Unlike the turbulent flow of a single-phase fluid, multiphase flows generally possess a large range of scales, ranging from the sub-millimeter size of a bubble or an eddy to the size of the system under investigation. Multiphase flows, like single-phase turbulent flows, exhibit a great deal of universality and it is almost certain that re-computing small-scale behavior that is already understood is not necessary. DNS should be able to provide both the insight and the data for the modeling at the smallest scales.

Chapter 14 - This chapter includes basic information on the scaling effect in fluid mechanics. Computational fluid dynamics (CFD) is used to simulate and study the flow and mixing process of two incompressible and miscible fluids in microscale. The mathematical model of the considered flow, along with the computational procedure used to solve it, is also described. The microscale geometry of interest is a sequence of the different microconduits with a rectangular cross-section. Moreover, the influence of flow conditions and various arrangements of microconduits on mixing efficiency is also presented and discussed.

Chapter 15 - Computational fluid dynamics (CFD) is a powerful tool for solving problems associated with flow, mixing, heat and mass transfer and chemical reaction. Hence they are useful alternatives to experimental techniques that are expensive and time consuming in the simulation of mixing processes. In this chapter, computational analysis of turbulent flow field in a mixing tank and a numerical approach for the optimization of stirrer configuration parameters are presented. Velocity field and power requirement are obtained using the FASTEST, a CFD package, which employs a fully conservative finite volume method for the solution of the Navier-Stokes equations. The inheritably time-dependent geometry of stirred vessel is simulated by clicking mesh method.

In the simulation of the turbulent flow field in a mixing tank with six bladed Rushton turbine, the effects of impeller clearance and disc thickness on the power number are determined and it is found that the power number decreases with decreasing clearance and increasing disc thickness. The results are comparable with those of well established measurement techniques in terms of time-averaged velocity field, turbulent kinetic energy, dissipation rate, and power number.

The methodology for the optimization of stirrer configuration is based on a flow solver FASTEST, and a mathematical optimization tool, which are integrated into an automated procedure. Two trust region based derivative free optimization algorithms, the DFO and CONDOR are considered. Both are designed to minimize smooth functions whose evaluations are considered to be expensive and whose derivatives are not available or not desirable to approximate. An exemplary application for a standard stirrer configuration illustrates the functionality and the properties of the proposed methods.

Chapter 16 - Multidisciplinary and Multiphysics coupled problems represent nowadays a challenging field when studying or analyzing even more complex phenomena that appear in nature and in new technologies (e.g. Magneto-Hydrodynamics, Micro-Electro- Mechanics, Thermo-Mechanics, Fluid-Structure Interaction, etc.). Particularly, when dealing with Fluid-Structure Interaction problems several questions arise, namely the coupling algorithm, the mesh moving strategy, the Galilean Invariance of the scheme, the compliance with the Discrete Geometric Conservation Law (DGCL), etc. Therefore, the aim of this chapter is to give an overview of the issues involved in the numerical solution of Fluid-Structure Interaction (FSI) problems.

Regarding the coupling techniques, some results on the convergence of the strong coupling Gauss-Seidel iteration are presented. Also, the precision of different predictor schemes for the structural system and the influence of the partitioned coupling on stability are discussed.

Another key point when solving FSI problems is the use of the "Arbitrary Lagrangian Eulerian formulation" (ALE), which allows the use of moving meshes. As the ALE contributions affect the advective terms, some modifications on the stabilizing and the shock-capturing terms, are needed. Also Dirichlet constraints at slip (or non-slip) walls must be modified when the ALE scheme is used. In this chapter the presented ALE formulation is invariant under Galilean transformations.

Chapter 17 - Lattice Boltzmann method is relatively new scheme that uses microscopic models to simulate macroscopic behavior of fluid flow and dependent phenomenons. LBM can be considered as discrete version of kinetics theory. The lattice-Boltzmann method, evolved out of ideas that have been extremely investigated since 1986, when it was discovered that very simple models of discrete particles restricted to a lattice can be used to yield Navier Stokes equation to solve complicated flow problems (Frisch et al., 1986 & 1987).

The lattice method can be concerned as one of the simplest microscopic approaches for simulation of macroscopic models. It is based on the Boltzmann transport equation, which concern about the time rate of change of the particle distribution function (probability to find particles with specific velocity range at the limited position at the given time) in a particular state.

This section presents basic concepts of Lattice Boltzmann Moethod (LBM) as the basis of simulation of fluid flows. It is assumed that the reader is somewhat familiar with the physics of fluid flow and related phenomena, so here the main concentration is on to describe basic concepts of lattice Boltzmann method, in such way that it can be understood and used by reader to solve different fluid flow problems. So authors avoided prolonged mathematical analysises.

In the lattice Boltzmann method the same calculations carry out at every lattice site and only nearby particles interact with each other so it is parallel and local, which make it very suitable for programming and efficient run in parallel processing. Complex boundary conditions are included in an uncomplicated approach. The method yields a good approximation to the Navier-Stokes equations.

It is hard to shift from microscopic lattice gas model to macroscopic fluid dynamics governing equations, Navier-Stokes (NS) equations, using statistical methods for gases. After while that first model for uncompressible NS equataions was proposed by Frisch, Hasslacher and Pomeau (FHP) in 1986 (Frisch et. al, 1986), Lattice Gas Automata (LGA) has been attended as promising method to solve partial differential equations and simulation of natural phenomenons (Frisch et. al, 1987; Wolfarm 1986; Doolen 1989; Doolen, 1991).

Recently lattice Boltzmann equation has been successfully used for simulation of fluid flow and transport phenomenon. Inspite of ordinary CFD methods, LBM is based on microscopic models and mesoscopic kinetics equations, which in them behavior of collection of particle will be used to simulate continuum mechanics of system. Due to kintecis nature of LBM, it has cleared that it is appropriate for usages including interfacial dynamics, complex flows such as multi phase and multi component flows and complex boundary condition (Chen and Doolen, 1998). In addition, it must considered that modeling of the complex phenomenons like as multi phase and multi component flows, porous mediums, fluid flow in electrical and magnetic fields can be simpler in molecular view and discrete mechanics. This chapter concerns about the techniques and main concepts of LBM.

Chapter 18 - In multiphase flow simulations, there are a number of approaches that can be used to capture the interfacial dynamics as well as the surrounding fluid flow. Front Tracking, Volume of Fluid, Immersed Boundary, Lattice Boltzmann, and Boundary Integral are only a few examples. The fluid properties are usually smoothed or smeared in the above methods. In this chapter, a newly developed method, Moving Mesh Interface Tracking (MMIT) is introduced. This method treats the interface as a surface mesh, and this surface mesh connects the interior volume element into a single mesh, and then the Navier-Stokes equations are solved on this single mesh. The interface mesh moves with the fluid velocity. Thus, this method has a zero-thickness interface and naturally conserves the total mass of each phase. The jumps in fluid properties and boundary conditions across the interface are directly implemented without smoothing or smearing of the fluid properties. The interface mesh moves in a Lagrangian fashion, while the interior nodes move by a smoothing approach. This motion usually does not guarantee good mesh quality, especially for large deformation. Therefore, mesh adaptations including 3-2 and 2-3 swapping, 4-4 flipping, edge bisection, and edge contraction, are implemented to achieve good mesh quality as well as to obtain computational efficiency. Mesh separation and mesh combination are introduced to handle topological transitions such as droplet pinch-off and interface merging.

This method has been validated against a number of theoretical predictions and experimental observations. The mesh adaptation schemes are capable of dealing with large deformation as well as achieving computation efficiency. The mesh separation and mesh combination are robust in simulations of droplet pinch-off and droplets colliding. The simulations also demonstrate the great potential of this method to investigate the detailed physics for small scales such as the thin film formed during the droplet near contact motion, and the necking thread in droplet pinch-off.

In: Computational Fluid Dynamics
Editor: Alyssa D. Murphy

ISBN 978-1-61209-276-8
© 2011 Nova Science Publishers, Inc.

Chapter 1

COMPUTATIONAL MODELING ASPECTS OF POLYMER ELECTROLYTE FUEL CELL DURABILITY

Maher A. R. Sadiq Al-Baghdadi[*]

Fuel Cell Research Center,
International Energy and Environment Foundation (IEEF), Al-Najaf, Iraq

ABSTRACT

Polymer electrolyte membrane (PEM) fuel cells are still undergoing intense development, and the combination of new and optimized materials, improved product development, novel architectures, more efficient transport processes, and design optimization and integration are expected to lead to major gains in performance, efficiency, durability, reliability, manufacturability and cost-effectiveness. Computational fuel cell models that allow systematic simulation, design and optimization of fuel cell systems would facilitate the integration of such advances, allow less heavy reliance on hardware prototyping, and reduce development cycles.

Durability is one of the most critical remaining issues impeding successful commercialization of broad PEM fuel cell stationary and transportation energy applications, and the durability of fuel cell stack components remains, in most cases, insufficiently understood. Lengthy required testing times, lack of understanding of most degradation mechanisms, and the difficulty of performing in-situ, non-destructive structural evaluation of key components makes the topic a difficult one. The damage mechanisms in a PEM fuel cell are accelerated by mechanical stresses arising during fuel cell assembly (bolt assembling), and the stresses arise during fuel cell running, because it consists of the materials with different thermal expansion and swelling coefficients. Therefore, in order to acquire a complete understanding of the damage mechanisms in the membrane, mechanical response under steady-state hygro-thermal stresses should be studied under real cell operating conditions and in real cell geometry (three-dimensional).

In this chapter, full three-dimensional, non-isothermal computational fluid dynamics model of a PEM fuel cell has been developed to simulate the stresses inside the PEM fuel cell, which are occurring during fuel cell assembly (bolt assembling), and the stresses

---
[*] E-mail address: maherars@IEEFoundation.org; maherars@hotmail.com

arise during fuel cell running due to the changes of temperature and relative humidity. A unique feature of the present model is to incorporate the effect of hygro and thermal stresses into actual three-dimensional fuel cell model. In addition, the temperature and humidity dependent material properties are utilize in the simulation for the membrane. The model is shown to be able to understand the many interacting, complex electrochemical, transport phenomena, and stresses distribution that have limited experimental data. This model is used to study and analyse the effect of operating, design, and material parameters on the mechanical behaviour of PEM. The analysis helped identifying critical parameters and shed insight into the physical mechanisms leading to a fuel cell durability under various operating conditions..

1. INTRODUCTION

1.1. Background

Escalating concerns regarding the impact that conventional methods of energy conversion are having on the environment and on global economics has in recent times progressively fuelled research and development into alternative technologies. One technology that is potentially independent of fossil fuels and suited to a wide range of applications from portable through to transport and stationary systems is the Polymer Electrolyte Fuel Cell (PEFC). PEFCs have been the focus of significant research and development for over five decades. The electrochemical energy conversion device inherently mitigate the need to combust reactant gases directly which thereby prevents it from being restricted to the Carnot efficiency. PEFCs operate on a hydrogen oxidation and oxygen reduction principal to generate electrical power and water as a bi-product. Single cells themselves can be 'stacked' to meet the power demands of the target application.

Each type of PEFC application will have its own set of requirements. During the course of research and development, this has resulted in a vast multitude of materials, designs, manufacturing techniques and considerations for the different components of the cell. These variations also reflect the fact that there are indeed a multitude of factors that govern the performance of the PEFC, all of which have some element of physical design or operation associated to it that can be altered to improve an aspect of cell performance.

Variations in operating modes and general cell design according to application means that how dominant certain performance degradation and failure mechanisms are also change according to application. Automotive fuel cells, for example, are likely to operate with neat hydrogen under loadfollowing or load-levelled modes and be expected to withstand variations in environmental conditions, particularly in the context of temperature and atmospheric composition. In addition, they are also required to survive over the course of their expected operational lifetimes i.e., around 5,500 hrs, while undergoing as many as 30,000 startup/shutdown cycles. PEFCs for stationary applications would not be subjected to as many startup/shutdown cycles, however, would be expected to survive up to 10,000 - 40,000 hrs of operation whilst maintaining a tolerance to fuel impurities in the reformate feed.

An important part of the fuel cell is the electrolyte, which gives every fuel cell its name. At the core of a PEM fuel cell is the polymer electrolyte membrane that separates the anode from the cathode. The desired characteristics of PEMs are high proton conductivity, good electronic insulation, good separation of fuel in the anode side from oxygen in the cathode side, high chemical and thermal stability, and low production cost. One type of PEMs that meets most of these requirements is Nafion. This is why Nafion is the most commonly used and investigated PEM in fuel cells.

In the Polymer-Electrolyte Membrane Fuel Cell the electrolyte consists of an acidic polymeric membrane that conducts protons but repels electrons, which have to travel through the outer circuit providing the electric work. A common electrolyte material is Nafion® from DuPont™, which consists of a fluoro-carbon backbone, similar to Teflon, with attached sulfonic acid groups. The membrane is characterized by the fixed-charge concentration (the acidic groups): the higher the concentration of fixed-charges, the higher is the protonic conductivity of the membrane. Alternatively, the term "equivalent weight" is used to express the mass of electrolyte per unit charge.

For optimum fuel cell performance it is crucial to keep the membrane fully humidified at all times, since the conductivity depends directly on water content. The thickness of the membrane is also important, since a thinner membrane reduces the ohmic losses in a cell. However, if the membrane is too thin, hydrogen, which is much more diffusive than oxygen, will be allowed to cross-over to the cathode side and recombine with the oxygen without providing electrons for the external circuit.

The best catalyst material for both anode and cathode PEM fuel cell is platinum. Since the catalytic activity occurs on the surface of the platinum particles, it is desirable to maximize the surface area of the platinum particles. A common procedure for surface maximization is to deposit the platinum particles on larger carbon black particles.

Therefore, the catalyst is characterized by the surface area of platinum by mass of carbon support. The electrochemical half-cell reactions can only occur, where all the necessary reactants have access to the catalyst surface. This means that the carbon particles have to be mixed with some electrolyte material in order to ensure that the hydrogen protons can migrate towards the catalyst surface. This "coating" of electrolyte must be sufficiently thin to allow the reactant gases to dissolve and diffuse towards the catalyst surface. Since the electrons travel through the solid matrix of the electrodes, these have to be connected to the catalyst material, i.e. an isolated carbon particle with platinum surrounded by electrolyte material will not contribute to the chemical reaction.

Several methods of applying the catalyst layer to the gas diffusion electrode have been reported. These methods are spreading, spraying, and catalyst power deposition. For the spreading method, a mixture of carbon support catalyst and electrolyte is spread on the GDL surface by rolling a metal cylinder on its surface. In the spraying method, the catalyst and electrolyte mixture is repeatedly sprayed onto the GDL surface until a desired thickness is achieved. Although the catalyst layer thickness can be up to 50 μm thick, it has been found that almost all of the electrochemical reaction occurs in a 10 μm thick layer closest to the membrane.

In order to maximise the electrochemically active surface area (EASA) in the anodic and cathodic catalyst layers, the catalyst is applied as fine and widely dispersed nano-particles on the surface of a supporting particle. Typically, the catalyst is platinum or platinum alloyed with ruthenium or chromium for example, and the larger supporting particle is commonly carbon-based. Recent studies have shown that within 2000 hrs of operation, it is possible for the metal catalyst to undergo morphological change, in the form of catalytic agglomeration and/or ripening. This leads on to a gradual decrease in the EASA. Whilst agglomeration is observed for both anode and cathode catalyst layers, it is usually the cathodic particles that undergo more extensive agglomeration, where there is an increased presence of liquid water and which facilitates primary corrosion. Repetitive on/off load cycles for PEFCs can also cause platinum sintering; residual hydrogen can induce a high voltage equivalent to open-circuit voltage to the cathode, causing the sintering to occur. This can be mitigated by air-purging the anode channel. Loss of EASA due to possible agglomeration has also been observed for un-humidified PEFCs operating at a higher temperature of 150°C in PBI membranes.

Another mechanism for the loss of EASA could be attributed to the movement of platinum. When the PEFC is operated through hydrogen-air open circuit to air-air open circuit, platinum can become quite soluble and consequently liable to transportation through adjacent layers. Loss of platinum correspondingly compromises the EASA. Such phenomenon can also be accompanied by an apparent migration of platinum. Migration of metal catalyst particles in both the anode and cathode catalyst layers in PEFCs has been observed, moving towards the interface between the catalyst layer and the membrane.

1.2. Durability

Durability is one of the most critical remaining issues impeding successful commercialization of broad PEM fuel cell stationary and transportation energy applications, and the durability of fuel cell stack components remains, in most cases, insufficiently understood. Lengthy required testing times, lack of understanding of most degradation mechanisms, and the difficulty of performing in-situ, non-destructive structural evaluation of key components makes the topic a difficult one [1, 2].

The Membrane-Electrode-Assembly (MEA) is the core component of PEM fuel cell and consists of membrane with the gas-diffusion layers including the catalyst attached to each side. The fuel cell MEA durability plays a vital role in the overall lifetime achieved by a stack in field applications. Within the MEA's electrocatalyst layers are three critical interfaces that must remain properly intermingled for optimum MEA performance: platinum/carbon interface (for electron transport and catalyst support); platinum/Nafion interface (for proton transport); and Nafion/carbon interface (for high-activity catalyst dispersion and structural integrity). The MEA performance shows degradation over operating time, which is dependent upon materials, fabrication and operating conditions [3, 4].

Durability is a complicated phenomenon; linked to the chemical and mechanical interactions of the fuel cell components, i.e. electro-catalysts, membranes, gas diffusion layers, and bipolar plates, under severe environmental conditions, such as elevated temperature and low humidity [5]. In fuel cell systems, failure may occur in several ways such as chemical degradation of the ionomer membrane or mechanical failure in the PEM that results in gradual reduction of ionic conductivity, increase in the total cell resistance, and the reduction of voltage and loss of output power [6]. Mechanical degradation is often the cause of early life failures. Mechanical damage in the PEM can appear as through-the-thickness flaws or pinholes in the membrane, or delaminating between the polymer membrane and gas diffusion layers [7, 8].

Mechanical stresses which limit MEA durability have two origins. Firstly, this is the stresses arising during fuel cell assembly (bolt assembling). The bolts provide the tightness and the electrical conductivity between the contact elements. Secondly, additional mechanical stresses occur during fuel cell running because PEM fuel cell components have different thermal expansion and swelling coefficients. Thermal and humidity gradients in the fuel cell produce dilatations obstructed by tightening of the screw-bolts. Compressive stress increasing with the hygro-thermal loading can exceed the yield strength which causes the plastic deformation. The mechanical behaviour of the membrane depends strongly on hydration and temperature [9, 10].

Water management is one of the critical operation issues in PEM fuel cells. Spatially varying concentrations of water in both vapour and liquid form are expected throughout the cell because of varying rates of production and transport. Devising better water management is therefore a key issue in PEM fuel cell design, and this requires improved understanding of the parameters affecting water transport in the membrane [11, 12]. Thermal management is also required to remove the heat produced by the electrochemical reaction in order to prevent drying out of the membrane, which in turn can result not only in reduced performance but also in eventual rupture of the membrane [13, 14]. Thermal management is also essential for the control of the water evaporation or condensation rates [15]. As a result of in the changes in temperature and moisture, the PEM, gas diffusion layers (GDL), and bipolar plates will all experience expansion and contraction. Because of the different thermal expansion and swelling coefficients between these materials, hygro-thermal stresses are expected to be introduced into the unit cell during operation. In addition, the non-uniform current and reactant flow distributions in the cell can result in non-uniform temperature and moisture content of the cell, which could in turn, potentially causing localized increases in the stress magnitudes. The need for improved lifetime of PEM fuel cells necessitates that the failure mechanisms be clearly understood and life prediction models be developed, so that new designs can be introduced to improve long-term performance. Increasing of the durability is a significant challenge for the development of fuel cell technology. Membrane failure is believed to be the result of combined chemical and mechanical effects acting together [1, 2, 5]. Variations in temperature and humidity during operation cause stresses and strains (mechanical loading) in the membrane as well as the MEA and is considered to be the mechanical failure driving force in fuel cell applications [6-10]. Reactant gas cross over, hydrogen peroxide formation and movement, and cationic contaminants are all to be major factors contributing to the chemical decomposition of polymer electrolyte membranes. While chemical degradation of membranes has been investigated and reported extensively in literature [1-8], there has been little work published on mechanical degradation of the membrane. Investigating the mechanical response of the membrane subjected to change in humidity and temperature requires studying and modelling of the stress-strain behaviour of membranes and MEAs. Weber and Newman [16] developed one-dimensional model to study the stresses development in the fuel cell. They showed that hygro-thermal stresses might be an important reason for membrane failure, and the mechanical stresses might be particularly important in systems that are non-isothermal. However, their model is one-dimensional and does not include the effects of material property mismatch among PEM, GDL, and bipolar plates.

Tang et al. [17] studied the hygro and thermal stresses in the fuel cell caused by step-changes of temperature and relative humidity. Influence of membrane thickness was also studied, which shows a less significant effect. However, their model is two-dimensional, where the hygro-thermal stresses are absent in the third direction (flow direction). In addition, a simplified temperature and humidity profile with no internal heat generation ware assumed, (constant temperature for each upper and lower surfaces of the membrane was assumed).

Kusoglu et al. [18] developed two-dimensional model to investigate the mechanical response of a PEM subjected to a single hygro-thermal loading cycle, simulating a simplified single fuel cell duty cycle. A linear, uncoupled, simplified temperature and humidity profile with no internal heat generation, assuming steady-state conditions, was used for the loading and unloading conditions. Linear-elastic, perfectly plastic material response with temperature and humidity dependent material properties was used to study the plastic deformation behaviour of the membrane during the cycle. The stress evolution during a simplified operating cycle is determined for two alignments of the bipolar plates. They showed that the alternating gas channel alignment produces higher shear stresses than the aligned gas channel. Their results suggested that the in-plane residual tensile stresses after one fuel cell duty cycle developed upon unloading, may lead to the failure of the membranes due to the mechanical fatigue. They concluded that in order to acquire a complete understanding of these damage mechanisms in the membranes, mechanical response under continuous hygro-thermal cycles should be studied under realistic cell operating conditions.

Kusoglu et al. [19] investigated the mechanical response of polymer electrolyte membranes in a fuel cell assembly under humidity cycles at a constant temperature. The behaviour of the membrane under hydration and dehydration cycles was simulated by imposing a simplified humidity gradient profile from the cathode to the anode. Also, a simplified temperature profile with no internal heat generation ware assumed. Linear elastic, plastic constitutive behaviour with isotropic hardening and temperature and humidity dependent material properties were utilized in the simulations for the membrane. The evolution of the stresses and plastic deformation during the humidity cycles were determined using two-dimensional finite elements model for various levels of swelling anisotropy. They showed that the membrane response strongly depends on the swelling anisotropy where the stress amplitude decreases with increasing anisotropy. Their results suggested that it may be possible to optimize a membrane with respect to swelling anisotropy to achieve better fatigue resistance, potentially enhancing the durability of fuel cell membranes.

Solasi et al. [20] developed two-dimensional model to define and understand the basic mechanical behaviour of ionomeric membranes clamped in a rigid frame, and subjected to changes in temperature and humidification. Expansion/contraction mechanical response of the constrained membrane as a result of change in hydration and temperature was also studied in non-uniform geometry. A circular hole in the centre of the membrane can represent pinhole creation or even material degradation during fuel cell operation was considered as the extreme form of non-uniformity in this constraint configuration. Their results showed that the hydration have a bigger effect than temperature in developing mechanical stresses in the membrane. These stresses will be more critical when non-uniformity as a form of hydration profile or a physical pinhole exists across the membrane.

Bograchev et al. [21] developed a linear elastic–plastic two-dimensional model of fuel cell with hardening for analysis of mechanical stresses in MEA arising in cell assembly procedure. The model includes the main components of real fuel cell (membrane, gas diffusion layers, graphite plates, and seal joints) and clamping elements (steel plates, bolts, nuts). The stress and plastic deformation in MEA are simulated taking into account the realistic clamping conditions. Their results concluded that important variations of stresses generated during the assembling procedure can be a source of the limitation of the mechanical reliability of the system.

Suvorov et al. [22] analyzed the stress relaxation in the membrane electrode assemblies (MEA) in PEM fuel cells subjected to compressive loads using numerical simulations (finite element method). This behaviour is important because nonzero contact stress is required to maintain low electric resistivity in the fuel cell stack. In addition to the two-dimensional assumption, the temperature was kept fixed and equal to the operating temperature at all time. All properties were considered to be independent of the temperature. They showed that under applied compressive strains the contact stress in the membrane electrode assembly (MEA) will drop with time. The maximum contact stress and the rate of stress relaxation depend on the individual properties of the membrane and the gas diffusion layer.

Tang et al. [23] examined the hygro-thermo-mechanical properties and response of a class of reinforced hydrated perfluorosulfonic acid membranes (PFSA) in a fuel cell assembly under humidity cycles at a constant temperature. The load imposed keeps the membrane at elevated temperature (85 C) and linearly cycles the relative humidity between the initial (30% RH) and the hydrated state (95% RH) at the cathode side of membrane. The evolution of hygro-thermally induced mechanical stresses during the load cycles were determined for reinforced and unreinforced PFSA membranes using two-dimensional finite elements model. Their numerical simulations showed that the in-plane stresses for reinforced PFSA membrane remain compressive during the cycling. Compressive stresses are advantageous with respect to fatigue loading, since compressive in-plane stresses will significantly reduce the slow crack growth associated with fatigue failures. They showed that the reinforced PFSA membrane exhibits higher strength and lower in-plane swelling than the unreinforced PFSA membrane used as a reference, therefore, should result in higher fuel cell durability.

Bograchev et al. [24] developed two-dimensional model to study the evolution of stresses and plastic deformations in the membrane during the turn-on phase. They showed that the maximal stresses in the membrane take place during the humidification step before the temperature comes to its steady-state value. The magnitude of these stresses is sufficient for initiation of the plastic deformations in the Nafion membrane. The plastic deformations in the membrane develop during the entire humidification step. At the steady state the stresses have the highest value in the centre of the membrane; the Mises stress is equal to 2.5 MPa.

In addition to the two-dimensional assumption, the operating conditions have been taken into account by imposing the heating sources as a simplified directly related relationship between power generation and efficiency of the fuel cell. The moisture is set gradually from an initial value of 35% up to 100%. The humidity is imposed after all heat sources reach steady state. The imposed moisture is assumed to be uniformly distributed in the membrane during turn-on stage (before reaching the steady state). However, this questionable assumption leads to overestimation of the maximal stresses in the membrane during turn-on stage.

Al-Baghdadi [25] incorporated the effect of hygro and thermal stresses into non-isothermal three-dimensional CFD model of PEM fuel cell to simulate the hygro and thermal stresses in one part of the fuel cell components, which is the polymer membrane. They studied the behaviour of the membrane during the operation of a unit cell. The results showed that the displacement have the highest value in the centre of the membrane near the cathode side inlet area.

An operating fuel cell has varying local conditions of temperature, humidity, and power generation (and thereby heat generation) across the active area of the fuel cell in three-dimensions. Nevertheless, except of ref. [25], no models have yet been published to incorporate the effect of hygro-thermal stresses into actual fuel cell models to study the effect of these real conditions on the stresses developed in membrane and gas diffusion layers. In addition, as a result of the architecture of a cell, the transport phenomena in a fuel cell are inherently three-dimensional, but no models have yet been published to address the hygro-thermal stresses in PEM fuel cells with three-dimensional effect. Suvorov et al. [22] reported that the error introduced due to two-dimensional assumption is about 10%. Therefore, in order to acquire a complete understanding of the damage mechanisms in the membrane and gas diffusion layers, mechanical response under steady-state hygro-thermal stresses should be studied under real cell operating conditions and in real cell geometry (three-dimensional).

The difficult experimental environment of fuel cell systems has stimulated efforts to develop models that could simulate and predict multi-dimensional coupled transport of reactants, heat and charged species using computational fluid dynamic (CFD) methods. A comprehensive computational model should include the equations and other numerical relations needed to fully define fuel cell behaviour over the range of interest. In the present work, full three-dimensional, non-isothermal computational fluid dynamics model of a PEM fuel cell has been developed to simulate the hygro and thermal stresses in PEM fuel cell, which are occurring during the cell operation due to the changes of temperature and relative humidity. The temperature and humidity dependent material properties are utilize in the simulation for the membrane. This model is used to study the effect of operating, design, and material parameters on fuel cell performance and hygro-thermal stresses in the fuel cell MEA.

2. MODEL DESCRIPTION

The present work presents a comprehensive three–dimensional, multi–phase, non-isothermal model of a PEM fuel cell that incorporates the significant physical processes and the key parameters affecting fuel cell performance. The model accounts for both gas and liquid phase in the same computational domain, and thus allows for the implementation of phase change inside the gas diffusion layers. The model includes the transport of gaseous species, liquid water, protons, and energy. Water transport inside the porous gas diffusion layer and catalyst layer is described by two physical mechanisms: viscous drag and capillary pressure forces, and is described by advection within the gas channels. Water transport across the membrane is also described by two physical mechanisms: electro-osmotic drag and diffusion. The model features an algorithm that allows for a more realistic representation of the local activation overpotentials, which leads to improved prediction of the local current density distribution. This leads to high accuracy prediction of temperature distribution in the cell and therefore thermal stresses. This model also takes into account convection and diffusion of different species in the channels as well as in the porous gas diffusion layer, heat transfer in the solids as well as in the gases, and electrochemical reactions. The present multi-phase model is capable of identifying important parameters for the wetting behaviour of the gas diffusion layers and can be used to identify conditions that might lead to the onset of pore plugging, which has a detrimental effect of the fuel cell performance. A unique feature of the model is to incorporate the effect of hygro-thermal stresses into actual three-dimensional fuel cell model. This model is used to investigate the hygro and thermal stresses in PEM fuel cell, which developed during the cell operation due to the changes of temperature and relative humidity.

2.1. Computational Domain

A computational model of an entire cell would require very large computing resources and excessively long simulation times. The computational domain in this study is therefore limited to one straight flow channel with the land areas. The full computational domain consists of cathode and anode gas flow channels, and the membrane electrode assembly as shown in Figure 1.

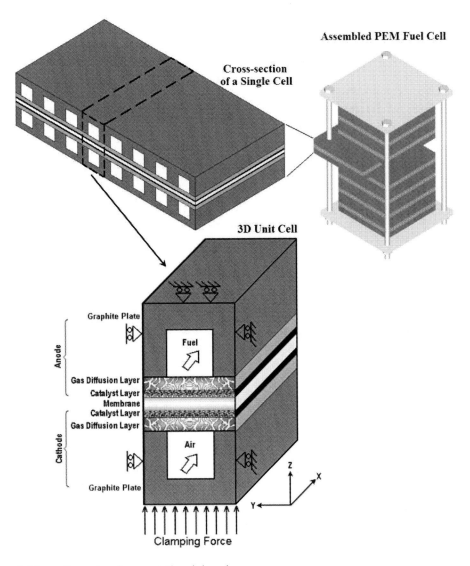

Figure 1. Three-dimensional computational domain.

2.2. Model equations

2.2.1. Gas flow channels

In the fuel cell channels, the gas-flow field is obtained by solving the steady-state Navier-Stokes equations, i.e. the continuity equation, the mass conservation equation for each phase yields the volume fraction (r) and along with the momentum equations the pressure distribution inside the channels. The continuity equation for the gas phase inside the channel is given by;

$$\nabla \cdot (r_g \rho_g \mathbf{u}_g) = 0 \tag{1}$$

and for the liquid phase inside the channel becomes;

$$\nabla \cdot (r_l \rho_l \mathbf{u}_l) = 0 \tag{2}$$

where u is velocity vector [m/s], ρ is density [kg/m^3].

Two sets of momentum equations are solved in the channels, and they share the same pressure field. Under these conditions, it can be shown that the momentum equations becomes;

$$\nabla \cdot (\rho_g \mathbf{u}_g \otimes \mathbf{u}_g - \mu_g \nabla \mathbf{u}_g) = -\nabla r_g \left(P + \frac{2}{3} \mu_g \nabla \cdot \mathbf{u}_g \right) + \nabla \cdot \left[\mu_g (\nabla \mathbf{u}_g)^T \right] \tag{3}$$

$$\nabla \cdot (\rho_l \mathbf{u}_l \otimes \mathbf{u}_l - \mu_l \nabla \mathbf{u}_l) = -\nabla r_l \left(P + \frac{2}{3} \mu_l \nabla \cdot \mathbf{u}_l \right) + \nabla \cdot \left[\mu_l (\nabla \mathbf{u}_l)^T \right] \tag{4}$$

where P is pressure (Pa), μ is viscosity [kg/(m·s)].

The mass balance is described by the divergence of the mass flux through diffusion and convection. Multiple species are considered in the gas phase only, and the species conservation equation in multi-component, multi-phase flow can be written in the following expression for species i;

$$\nabla \cdot \begin{bmatrix} -r_g \rho_g y_i \sum_{j=1}^{N} D_{ij} \frac{M}{M_j} \left[\left(\nabla y_j + y_j \frac{\nabla M}{M} \right) + (x_j - y_j) \frac{\nabla P}{P} \right] + \\ r_g \rho_g y_i \cdot \mathbf{u}_g + D_i^T \frac{\nabla T}{T} \end{bmatrix} = 0 \tag{5}$$

where T is temperature (K), y is mass fraction, x is mole fraction, D is diffusion coefficient [m^2/s]. Subscript i denotes oxygen at the cathode side and hydrogen at the anode side, and j is water vapour in both cases. Nitrogen is the third species at the cathode side.

The Maxwell-Stefan diffusion coefficients of any two species are dependent on temperature and pressure. They can be calculated according to the empirical relation based on kinetic gas theory [26];

$$D_{ij} = \frac{T^{1.75} \times 10^{-3}}{P\left[\left(\sum_k V_{ki}\right)^{1/3} + \left(\sum_k V_{kj}\right)^{1/3}\right]^2} \left[\frac{1}{M_i} + \frac{1}{M_j}\right]^{1/2} \quad (6)$$

In this equation, the pressure is in atm and the binary diffusion coefficient Dij is in [cm²/s].

The values for $\left(\sum V_{ki}\right)$ are given by Fuller et al. [26].

The temperature field is obtained by solving the convective energy equation;

$$\nabla \cdot \left(r_g \left(\rho_g Cp_g \mathbf{u}_g T - k_g \nabla T\right)\right) = 0 \quad (7)$$

where Cp is specific heat capacity [J/(kg.K)], k is gas thermal conductivity [W/(m.K)].

The gas phase and the liquid phase are assumed to be in thermodynamic equilibrium; hence, the temperature of the liquid water is the same as the gas phase temperature.

2.2.2. Gas Diffusion Layers

The physics of multiple phases through a porous medium is further complicated here with phase change and the sources and sinks associated with the electrochemical reaction. The equations used to describe transport in the gas diffusion layers are given below. Mass transfer in the form of evaporation $\left(\dot{m}_{phase} > 0\right)$ and condensation $\left(\dot{m}_{phase} < 0\right)$ is assumed, so that the mass balance equations for both phases are;

$$\nabla \cdot \left((1-sat)\rho_g \varepsilon \mathbf{u}_g\right) = \dot{m}_{phase} \quad (8)$$

$$\nabla \cdot \left(sat.\rho_l \varepsilon \mathbf{u}_l\right) = \dot{m}_{phase} \quad (9)$$

where sat is saturation, ε is porosity

The momentum equation for the gas phase reduces to Darcy's law, which is, however, based on the relative permeability for the gas phase (KP). The relative permeability accounts for the reduction in pore space available for one phase due to the existence of the second phase [27].

The momentum equation for the gas phase inside the gas diffusion layer becomes;

$$\mathbf{u}_g = -(1-sat)Kp.\nabla P/\mu_g \quad (10)$$

where KP is hydraulic permeability [m²].

Two liquid water transport mechanisms are considered; shear, which drags the liquid phase along with the gas phase in the direction of the pressure gradient, and capillary forces, which drive liquid water from high to low saturation regions [27]. Therefore, the momentum equation for the liquid phase inside the gas diffusion layer becomes;

$$\mathbf{u}_l = -\frac{KP_l}{\mu_l}\nabla P + \frac{KP_l}{\mu_l}\frac{\partial P_c}{\partial sat}\nabla sat \qquad (11)$$

where Pc is capillary pressure [Pa].

The functional variation of capillary pressure with saturation is prescribed following Leverett [27] who has shown that;

$$P_c = \tau\left(\frac{\varepsilon}{KP}\right)^{1/2}\left(1.417(1-sat) - 2.12(1-sat)^2 + 1.263(1-sat)^3\right) \qquad (12)$$

where τ is surface tension [N/m].

The liquid phase consists of pure water, while the gas phase has multi components. The transport of each species in the gas phase is governed by a general convection-diffusion equation in conjunction which the Stefan-Maxwell equations to account for multi species diffusion;

$$\nabla\cdot\begin{bmatrix}-(1-sat)\rho_g\varepsilon y_i\sum_{j=1}^{N}D_{ij}\frac{M}{M_j}\left[\left(\nabla y_j + y_j\frac{\nabla M}{M}\right) + (x_j - y_j)\frac{\nabla P}{P}\right] + \\ (1-sat)\rho_g\varepsilon y_i\cdot\mathbf{u}_g + \varepsilon D_i^T\frac{\nabla T}{T}\end{bmatrix} = \dot{m}_{phase} \qquad (13)$$

In order to account for geometric constraints of the porous media, the diffusivities are corrected using the Bruggemann correction formula [28];

$$D_{ij}^{eff} = D_{ij}\times\varepsilon^{1.5} \qquad (14)$$

The heat transfer in the gas diffusion layers is governed by the energy equation as follows;

$$\nabla\cdot\left((1-sat)\left(\rho_g\varepsilon Cp_g\mathbf{u}_gT - k_{eff,g}\varepsilon\nabla T\right)\right) = \varepsilon\beta(T_{solid}-T) - \varepsilon\dot{m}_{phase}\Delta H_{evap} \qquad (15)$$

where $keff$ is effective electrode thermal conductivity [W/(m·K)]; the term $[\varepsilon\beta(T_{solid}-T)]$, on the right hand side, accounts for the heat exchange to and from the solid matrix of the GDL. The gas phase and the liquid phase are assumed to be in thermodynamic equilibrium, i.e., the liquid water and the gas phase are at the same temperature.

The potential distribution in the gas diffusion layers is governed by;

$$\nabla \cdot (\lambda_e \nabla \phi) = 0 \qquad (16)$$

where λe is electrode electronic conductivity [S/m].

In order to account for the magnitude of phase change inside the GDL, expressions are required to relate the level of over- and undersaturation as well as the amount of liquid water present to the amount of water undergoing phase change. In the present work, the procedure of Berning and Djilali [29] was used to account for the magnitude of phase change inside the GDL.

2.2.3. Catalyst Layers

The catalyst layer is treated as a thin interface, where sink and source terms for the reactants are implemented. Due to the infinitesimal thickness, the source terms are actually implemented in the last grid cell of the porous medium. At the cathode side, the sink term for oxygen is given by;

$$S_{O_2} = -\frac{M_{O_2}}{4F} i_c \qquad (17)$$

where F is Faraday's constant (96487 [C/mole]), ic is cathode local current density [A/m^2], M is molecular weight [kg/mole].

Whereas the sink term for hydrogen is specified as;

$$S_{H_2} = -\frac{M_{H_2}}{2F} i_a \qquad (18)$$

where ia is anode local current density [A/m^2]

The production of water is modelled as a source terms, and hence can be written as;

$$S_{H_2O} = \frac{M_{H_2O}}{2F} i_c \qquad (19)$$

The generation of heat in the cell is due to entropy changes as well as irreversibility's due to the activation overpotential [30];

$$\dot{q} = \left[\frac{T(-\Delta s)}{n_e F} + \eta_{act} \right] i \qquad (20)$$

where $\dot{q}$ is heat generation [W/m^2], ne is number of electrons transfer, s is specific entropy [J/(mole.K)], ηact is activation overpotential (V).

The local current density distribution in the catalyst layers is modelled by the Butler-Volmer equation [31, 32];

$$i_c = i_{o,c}^{ref} \left(\frac{C_{O_2}}{C_{O_2}^{ref}}\right) \left[\exp\left(\frac{\alpha_a F}{RT}\eta_{act,c}\right) + \exp\left(-\frac{\alpha_c F}{RT}\eta_{act,c}\right)\right] \qquad (21)$$

$$i_a = i_{o,a}^{ref} \left(\frac{C_{H_2}}{C_{H_2}^{ref}}\right)^{1/2} \left[\exp\left(\frac{\alpha_a F}{RT}\eta_{act,a}\right) + \exp\left(-\frac{\alpha_c F}{RT}\eta_{act,a}\right)\right] \qquad (22)$$

where C_{H_2} is local hydrogen concentration [mole/m3], $C_{H_2}^{ref}$ is reference hydrogen concentration [mole/m3], C_{O_2} is local oxygen concentration [mole/m3], $C_{O_2}^{ref}$ is reference oxygen concentration [mole/m3], $i_{o,a}^{ref}$ is anode reference exchange current density, $i_{o,c}^{ref}$ is cathode reference exchange current density, R is universal gas constant (8.314 [J/(mole·K)]), αa is charge transfer coefficient, anode side, and αc is charge transfer coefficient, cathode side.

2.2.4. Membrane

The balance between the electro-osmotic drag of water from anode to cathode and back diffusion from cathode to anode yields the net water flux through the membrane;

$$N_W = n_d M_{H_2O} \frac{i}{F} - \nabla \cdot (\rho D_W \nabla c_W) \qquad (23)$$

where Nw is net water flux across the membrane [kg/(m2·s)], nd is electro-osmotic drag coefficient.

The water diffusivity in the polymer can be calculated as follow [33];

$$D_W = 1.3 \times 10^{-10} \exp\left[2416\left(\frac{1}{303} - \frac{1}{T}\right)\right] \qquad (24)$$

The variable c_W represents the number of water molecules per sulfonic acid group (i.e. mol H_2O/equivalent SO_3^{-1}). The water content in the electrolyte phase is related to water vapour activity via [34];

$$\begin{aligned} c_W &= 0.043 + 17.81a - 39.85a^2 + 36.0a^3 & (0 < a \le 1) \\ c_W &= 14.0 + 1.4(a-1) & (1 < a \le 3) \\ c_W &= 16.8 & (a \ge 3) \end{aligned} \qquad (25)$$

The water vapour activity a given by;

$$a = x_W P / P_{sat} \tag{26}$$

Heat transfer in the membrane is governed by;

$$\nabla \cdot (k_{mem} \cdot \nabla T) = 0 \tag{27}$$

where k_{mem} is membrane thermal conductivity [W/(m·K)].

The potential loss in the membrane is due to resistance to proton transport across membrane, and is governed by;

$$\nabla \cdot (\lambda_m \nabla \phi) = 0 \tag{28}$$

where λ_m is membrane ionic conductivity [S/m].

2.2.5. Hygro-Thermal Stresses in Fuel Cell

Using hygrothermoelasticity theory, the effects of temperature and moisture as well as the mechanical forces on the behaviour of elastic bodies have been addressed. An uncoupled theory is assumed, for which the additional temperature changes brought by the strain are neglected. The total strain tensor of deformation, π, is the sum;

$$\pi = \pi_e + \pi_{pl} + \pi_T + \pi_S \tag{29}$$

where, π_e is the elastic strain component, π_{pl} is the plastic strain component, and π_T, π_S are the thermal and swelling induced strains, respectively.

The thermal strains resulting from a change in temperature of an unconstrained isotropic volume are given by;

$$\pi_T = \wp (T - T_{Ref}) \tag{30}$$

where $\wp$ is thermal expansion [1/K].

The swelling strains caused by moisture change in membrane are given by;

$$\pi_S = \lambda_{mem} (\Re - \Re_{Ref}) \tag{31}$$

where λ_{mem} is membrane humidity swelling-expansion tensor and $\Re$ is the relative humidity [%]. Following the work [19], the swelling-expansion for the membrane, λ_{mem}, is expressed as a polynomial function of humidity and temperature as follows;

$$\lambda_{mem} = \sum_{i,j=1}^{4} C_{ij} T^{4-i} \Re^{4-j} \tag{32}$$

where C_{ij} is the polynomial constants, see Ref. [19].

Assuming linear response within the elastic region, the isotropic Hooke's law is used to determine the stress tensor σ.

$$\sigma = \mathbf{G}\pi \tag{33}$$

where G is the constitutive matrix.

The effective stresses according to von Mises, 'Mises stresses', are given by;

$$\sigma_v = \sqrt{\frac{(\sigma_1 - \sigma_2)^2 + (\sigma_2 - \sigma_3)^2 + (\sigma_3 - \sigma_1)^2}{2}} \tag{34}$$

where σ_1, σ_2, σ_3 are the principal stresses.

The mechanical boundary conditions are noted in Figure 1. The initial conditions corresponding to zero stress-state are defined; all components of the cell stack are set to reference temperature 20 C, and relative humidity 35% (corresponding to the assembly conditions) [17, 24, 35]. In addition, a constant pressure of (1 MPa) is applied on the surface of lower graphite plate, corresponding to a case where the fuel cell stack is equipped with springs to control the clamping force [17-19, 21, 24].

3. RESULTS AND DISCUSSION

The governing equations were discretized using a finite-volume method and solved using CFD code. Stringent numerical tests were performed to ensure that the solutions were independent of the grid size. A computational quadratic mesh consisting of a total of 64586 nodes and 350143 meshes was found to provide sufficient spatial resolution (Figure 2). The coupled set of equations was solved iteratively, and the solution was considered to be convergent when the relative error was less than 1.0×10^{-6} in each field between two consecutive iterations. The calculations presented here have all been obtained on a Pentium IV PC (3 GHz, 2 GB RAM), using Windows XP operating system.

The geometric and the base case operating conditions are listed in Table 1. Values of the electrochemical transport parameters for the base case operating conditions are taken from reference [29] and are listed in Table 2. The material properties for the fuel cell components used in this model are taken from reference [19] and are shown in Tables 3-5. The multi-phase model is validated by comparing model results to experimental data provided by Wang et al. [12]. Figure 3 shows the comparison of the polarization curves from the experimental data with the values obtained by the model at different operating fuel cell temperatures. It can be seen that the modelling results compare well with the experimental data. The importance of phase change to the accurate modelling of fuel cell performance is illustrated. Performance curves with and without phase change are also shown in Figure 3 for the base case conditions. Comparison of the two curves demonstrates that the effects of liquid water accumulation become apparent even at relatively low values of current density.

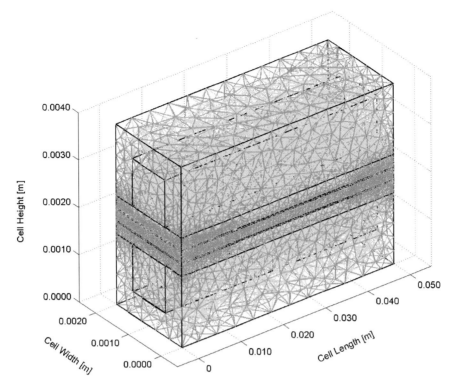

Figure 2. Computational mesh of a PEM fuel cell.

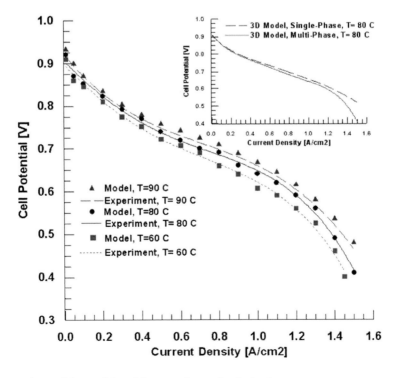

Figure 3. Comparison of the model and the experimental polarization curves.

Table 1. Geometrical and operational parameters for base case conditions

Parameter	Symbol	Value
Channel length (X-direction)	L	0.05 m
Channel width (Y-direction)	W	1e-3 m
Channel height (Z-direction)	H	1e-3 m
Land area width	W_{land}	1e-3 m
Gas diffusion layer thickness	δ_{GDL}	0.26e-3 m
Wet membrane thickness (Nafion® 117)	δ_{mem}	0.23e-3 m
Catalyst layer thickness	δ_{CL}	0.0287e-3 m
Hydrogen reference mole fraction	$x_{H_2}^{ref}$	0.84639
Oxygen reference mole fraction	$x_{O_2}^{ref}$	0.17774
Anode pressure	P_a	3e5 Pa
Cathode pressure	P_c	3e5 Pa
Inlet fuel and air temperature	T_{cell}	353.15 K
Relative humidity of inlet fuel and air	ψ	100 %
Air stoichiometric flow ratio	ξ_c	2
Fuel stoichiometric flow ratio	ξ_a	2

Table 2. Electrode and membrane parameters for base case operating conditions

Parameter	Symbol	Value
Electrode porosity	ε	0.4
Electrode electronic conductivity	λ_e	100 S/m
Membrane ionic conductivity (Nafion®117)	λ_m	17.1223 S/m
Transfer coefficient, anode side	α_a	0.5
Transfer coefficient, cathode side	α_c	1
Cathode reference exchange current density	$i_{o,c}^{ref}$	1.8081e-3 A/m^2
Anode reference exchange current density	$i_{o,a}^{ref}$	2465.598 A/m^2
Electrode thermal conductivity	k_{eff}	1.3 W/m.K
Membrane thermal conductivity	k_{mem}	0.455 W/m.K
Electrode hydraulic permeability	kp	1.76e-11 m^2
Entropy change of cathode side reaction	ΔS	-326.36 J/mol.K
Heat transfer coef. between solid and gas phase	β	4e6 W/m^3
Protonic diffusion coefficient	D_{H^+}	4.5e-9 m^2/s
Fixed-charge concentration	c_f	1200 mol/m^3
Fixed-site charge	z_f	-1
Electro-osmotic drag coefficient	n_d	2.5
Droplet diameter	D_{drop}	1e-8 m
Condensation constant	C	1e-5
Scaling parameter for evaporation	ϖ	0.01

Table 3. Material properties used in the model

Parameter	Symbol	Value
Electrode Poisson's ratio	$\Im_{GDL}$	0.25
Membrane Poisson's ratio	$\Im_{mem}$	0.25
Electrode thermal expansion	$\wp_{GDL}$	-0.8e-6 1/K
Membrane thermal expansion	$\wp_{mem}$	123e-6 1/K
Electrode Young's modulus	Ψ_{GDL}	1e10 Pa
Membrane Young's modulus	Ψ_{mem}	Table 4
Electrode density	ρ_{GDL}	400 kg/m^3
Membrane density	ρ_{mem}	2000 kg/m^3
Membrane humidity swelling-expansion tensor	λ_{mem}	from eq.(32)

Table 4. Young's modulus at various temperatures and humidities of Nafion

Young's modulus [MPa]	Relative humidity [%]			
	30	50	70	90
T=25 C	197	192	132	121
T=45 C	161	137	103	70
T=65 C	148	117	92	63
T=85 C	121	85	59	46

Table 5. Yield strength at various temperatures and humidities of Nafion

Yield stress [MPa]	Relative humidity [%]			
	30	50	70	90
T=25 C	6.60	6.14	5.59	4.14
T=45 C	6.51	5.21	4.58	3.44
T=65 C	5.65	5.00	4.16	3.07
T=85 C	4.20	3.32	2.97	2.29

Furthermore, when liquid water effects are not included in the model, the cell voltage dose not exhibit an increasingly steep drop as the cell approaches its limiting current density. This drop off in performance is clearly demonstrated by experimental data, but cannot be accurately modelled without the incorporation of phase change. By including the effects of phase change, the current model is able to more closely simulate performance, especially in the region where mass transport effects begin to dominate.

3.1. Base Case Operating Conditions

Results for the cell operates at a nominal current density of 1.2 A/cm^2 are discussed in this section. The selection of relatively high current density is due to illustrate the phase change effects, where it becomes clearly apparent between single and multi-phase model in the mass transport limited region.

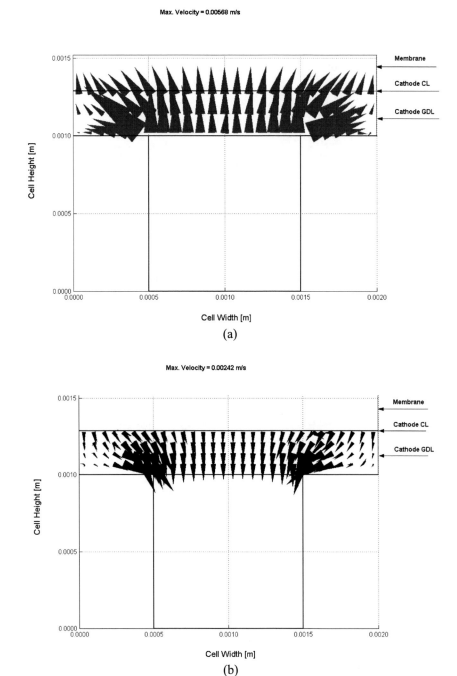

Figure 4. Gas phase velocity vectors (a) and liquid water velocity vectors (b) inside the cathode GDL.

The velocity fields inside the cathodic and anodic gas diffusion layers are shown in figures 4 and 5 for both gas and liquid phase. The pressure gradient induces bulk gas flow from the channels into the GDL. While the capillary pressure gradient drives the liquid water out of the gas diffusion layers into the flow channels. Therefore, the liquid water flux is directed from the GDL into the channel, i.e., in the opposite direction of the gas-phase velocity, where it can leave the cell.

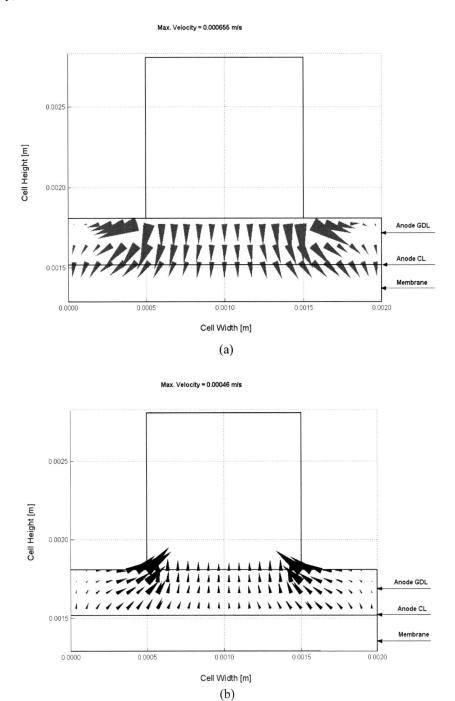

Figure 5. Gas phase velocity vectors (a) and liquid water velocity vectors (b) inside the anode GDL.

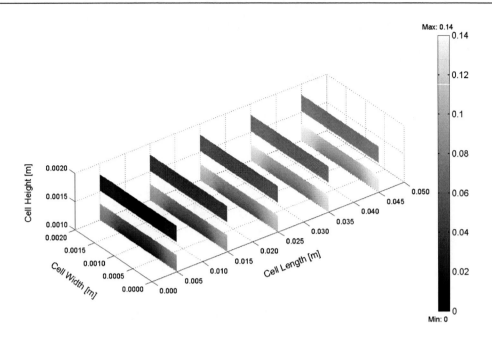

Figure 6. Liquid water saturation inside the cathode and anode GDLs.

The velocity of the liquid phase, however, is lower than for the gas phase, which is due to the higher viscosity, and the highest liquid water velocity occurs at the corners of the channel/GDL interface. The liquid water oozes out of the GDL, mainly at the corners of the GDL/channel interface.

The liquid water saturation inside the cathodic and anodic gas diffusion layers are shown in figure 6. Condensation occurs mainly in two areas inside the cathodic GDL: at the catalyst layer the molar water vapour fraction increases due to the oxygen depletion, and at the channel/GDL interface, where the oversaturated bulk flow condenses out. This term is relatively small compared to the other effects. Condensation occurs throughout the anodic GDL due to hydrogen depletion. Similar to the cathode side, the liquid water can only leave the GDL through the build-up of a capillary pressure gradient to overcome the viscous drag, because at steady state operation, all the condensed water has to leave the cell. The liquid water saturation is distributed through the entire cathodic and anodic GDL with maximum saturations found under the land areas. The reason for this is clear: once liquid water is being created by condensation, it is dragged into the GDL by the gas phase. The high spatial variation of the saturation demonstrates again the three-dimensional nature of transport processes in PEM fuel cells. Clearly, liquid water saturation depends strongly on the specified capillary pressure, and again, the permeability of the gas diffusion layer becomes the central parameter.

Several transport mechanisms in the cell affect water distribution. In the membrane, primary transport is through (i) electro-osmotic drag associated with the protonic current in the electrolyte, which results in water transport from anode to cathode; and (ii) diffusion associated with water-content gradients in the membrane. One of the main difficulties in managing water in a PEM fuel cell is the conflicting requirements of the membrane and of the catalyst gas diffusion layer. On the cathode side, excessive liquid water may block or flood the pores of the catalyst layer, the gas diffusion layer or even the gas channel, thereby inhibiting or even completely blocking oxygen mass transfer. On the anode side, as water is dragged toward the cathode via electro-osmotic transport, dehumidification of the membrane may occur, resulting in deterioration of protonic conductivity. In the extreme case of complete drying, local burnout of the membrane can result. Figure 7 shows profiles for polymer water content in the membrane for the base case conditions. The influence of electro-osmotic drag and back diffusion are readily apparent from this result.

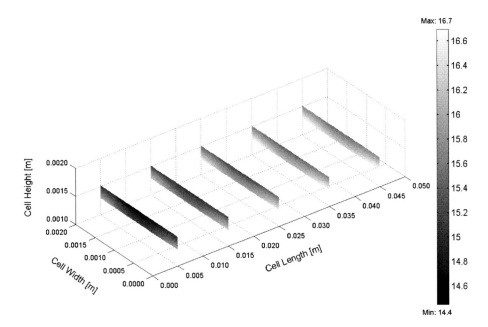

Figure 7. Water content profiles through the polymer electrolyte membrane (PEM).

Figure 8 shows the reactant gas distribution in the fuel cell cannels and the gas diffusion layers. The molar fraction of oxygen is highest under the channel, and exhibits a three-dimensional behaviour with a fairly significant drop under the land areas, and a more gradual depletion towards the outlet. The lower diffusivity of the oxygen along with the low concentration of oxygen in ambient air results in noticeable oxygen depletion under the land areas. The molar hydrogen fraction is almost constant inside the GDL due to the higher diffusivity of the hydrogen.

The water vapour molar fraction distribution in the cell is shown in figure 9. The molar water vapour fraction, however, remains almost constant throughout the gas diffusion layer in multi-phase model. In the absence of phase change, this would not be the case, as the nitrogen and water vapour fraction would increase as the oxygen fraction decreases.

The local current density distribution is shown in figure 10. The local current densities have been normalized by divided through the nominal current density (i.e. i_c/I). The local current density distribution at the catalyst layer predicted in multi-phase model has a much higher fraction of the total current, generated under the channel area. This is due to the effects of liquid water inside the GDL, which decreases its permeability to reactant gas flow (oxygen) and lead to the onset of pore plugging by liquid water. This can lead to local hot-spots inside the membrane electrode assembly, and this lead to a further drying out of the membrane, thus increasing the electric resistance, which in turn leads to more heat generation and can lead to a failure of the membrane.

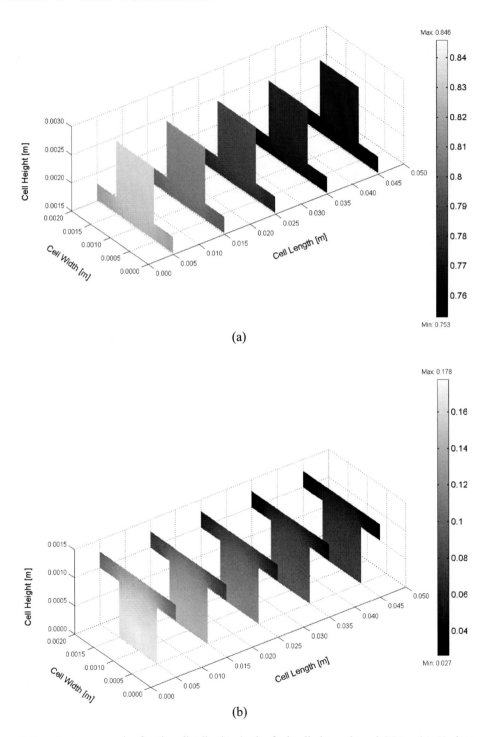

Figure 8. Reactant gases molar fraction distribution in the fuel cell channels and GDLs. (a): Hydrogen, and (b): Oxygen.

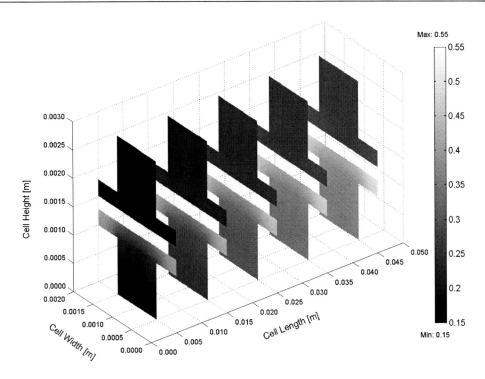

Figure 9. Water vapour molar fraction distribution in the cell.

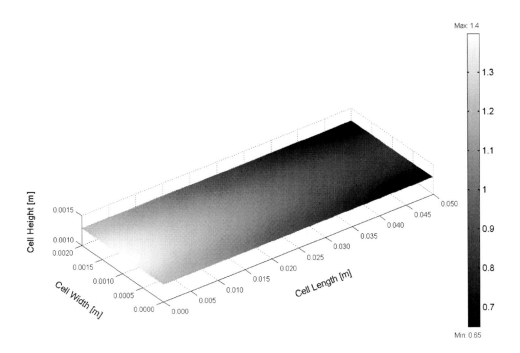

Figure 10. Dimensionless local current density distribution (i_c/I) at the cathode side catalyst layer.

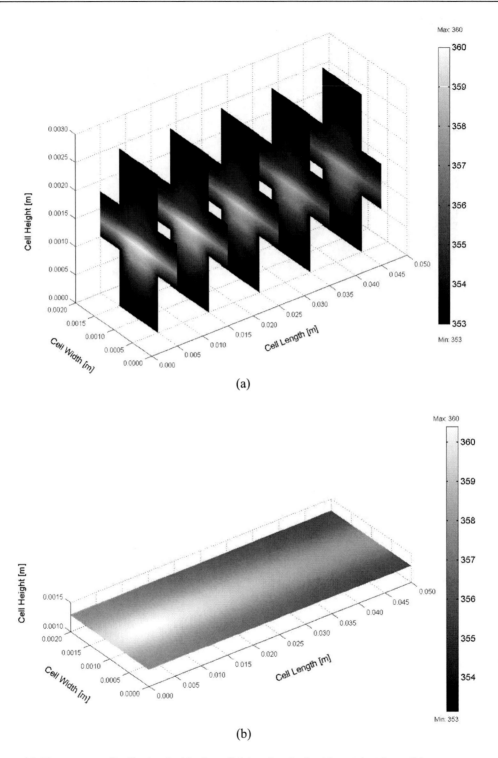

Figure 11. Temperature distribution inside the cell (a) and cathode side catalyst layer (b).

Thus, it is important to keep the current density relatively even throughout the cell. Therefore, the multi-phase model is capable of identifying important parameters for the wetting behaviour of the gas diffusion layers and can be used to identify conditions that might lead to the onset of pore plugging, which has a detrimental effect of the fuel cell performance, especially in the mass transport limited region.

The temperature distribution inside the fuel cell has important effects on nearly all transport phenomena, and knowledge of the magnitude of temperature increases due to irreversibilities might help preventing failure. Figure 11 shows the distribution of the temperature inside the cell. The figure shows that the increase in temperature can exceed several degrees Kelvin near the inlet area, where the electrochemical activity is highest. The temperature peak appears in the cathode catalyst layer, implying that major heat generation takes place in the region. In general, the temperature at the cathode side is higher than at the anode side, this is due to the reversible and irreversible entropy production.

Several sources contribute to irreversible losses in a practical fuel cell. The losses originate primarily from activation overpotential, ohmic overpotential in GDL, membrane overpotential, and diffusion overpotential. Figure 12 shows the variation of all these overpotential (losses) inside the cell. The activation overpotential profile at the cathode side CL correlates with the local current density, where the current densities are highest in the centre of the channel and coincide with the highest reactant concentrations. It can be seen that the activation overpotential is directly related to the nature of the electrochemical reactions and represents the magnitude of activation energy, when the reaction propagates at the rate demanded by the current. The magnitude of ohmic overpotential is dependent on the path of the electrons.

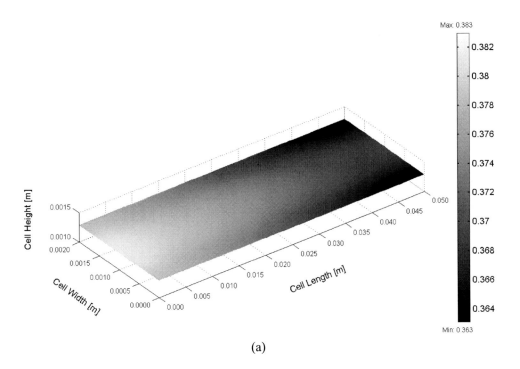

(a)

Figure 12. (Continued).

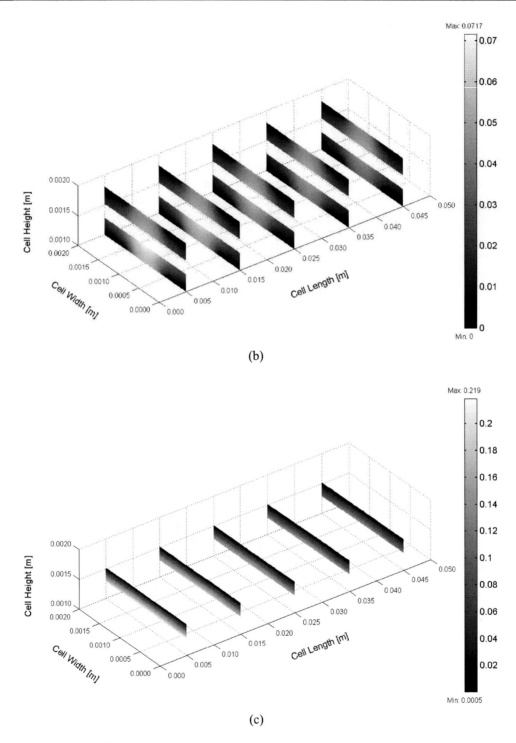

Figure 12. (Continued).

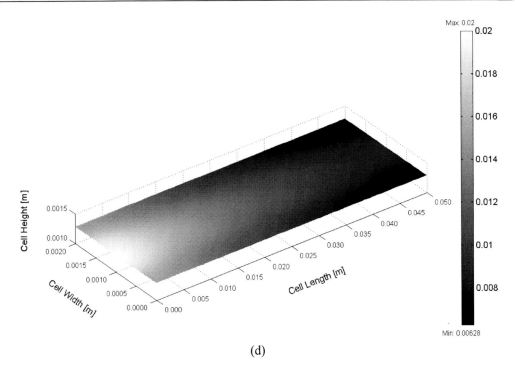

(d)

Figure 12. Variation of different loss mechanisms (overpotential) inside the cell. (a): Activation overpotential, (b) ohmic overpotential in GDLs, (c) membrane overpotential, and (d) diffusion overpotential.

The potential distributions are normal to the flow channel and the sidewalls, while there is a gradient into the land areas where electrons flow into the bipolar plate. The distributions exhibit gradients in both x and y direction due to the non-uniform local current production and show that ohmic losses are larger in the area of the catalyst layer under the flow channels. The distribution pattern of the protonic overpotential is dependent on the path travelled by the protons and the activities in the catalyst layers. It can be seen that the potential drop is more uniformly distributed across the membrane. This is because of the smaller gradient of the hydrogen concentration distribution under the channel and land areas at the anode catalyst layer due to the higher diffusivity of the hydrogen. The diffusion overpotential profile at the cathode side CL correlates with the local current density, with higher values under the channel area. This is created by the concentration gradient due to the consumption of oxygen or fuel at the electrodes.

The polymer electrolyte membrane is the core component of PEM fuel cell. It is influence by varying local conditions of temperature and humidity. Due to the varying local conditions of temperature and humidity across the PEM, the hygro and thermal stresses are introduced. Figure 13 shows total displacement distribution (contour plots) and deformation shape (scale enlarged 500 times) of the membrane. Because of the different thermal expansion and swelling coefficients between gas diffusion layers and membrane materials with non-uniform temperature distributions in the cell during operation, hygro-thermal stresses and deformation are introduced. The non-uniform distribution of stress, caused by the temperature gradient in the membrane, induces localized bending stresses, which can contribute to delaminating between the membrane and the GDLs.

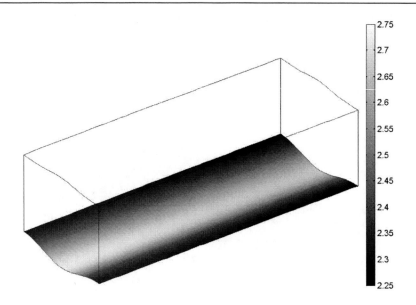

Figure 13. Total displacement distribution along the lower surface of the membrane (contour plots) and deformation shape (scale enlarged 500 times) of the membrane.

3.2. Parametric Study

The performance characteristics of the fuel cell based on a certain parameter can be obtained by varying that parameter while keeping all other parameters constant at base case conditions. Results obtained from these parametric studies will allow us to identify the critical parameters for fuel cell performance. Results with deferent operating conditions are summarized in Figure 14.

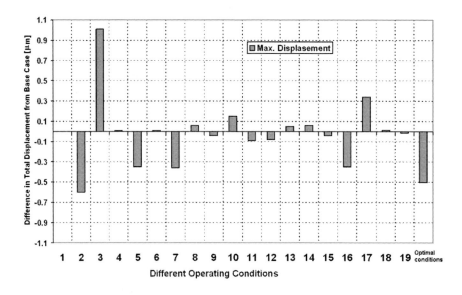

Figure 14. Change in maximum displacement with different operating conditions compared to the base case conditions.

Key:
- 1- Base case
- 2- Cell operating temperature = 60 C
- 3- Cell operating temperature = 90 C
- 4- Cell operating pressure = 1 atm
- 5- Cell operating pressure = 5 atm
- 6- Stoichiometric flow ratio = 1.5
- 7- Stoichiometric flow ratio = 3
- 8- GDL porosity = 0.3
- 9- GDL porosity = 0.5
- 10- GDL thermal conductivity = 0.5 W/m.K
- 11- GDL thermal conductivity = 2.9 W/m.K
- 12- Gas channel width = 0.8 mm
- 13- Gas channel width = 1.2 mm
- 14- GDL thickness = 0.2 mm
- 15- GDL thickness = 0.3 mm
- 16- Membrane thickness = 0.2 mm
- 17- Membrane thickness = 0.26 mm
- 18- Membrane thermal conductivity = 0.3 W/m.K
- 19- Membrane thermal conductivity = 0.6 W/m.K
- 20- Optimal conditions to achieve long cell life (Table 6)

The maximum displacement at various operating conditions is compared at a constant nominal current density of 1.2 A/cm^2. The results of parametric study are also discussed in the following subsections. In the following subsections only the parameter investigated is changed, all other parameters are at the base case conditions as outlined in Table 1 and Table 2.

3.2.1. Operating Parameters

3.2.1.1. Effect of Temperature

The temperature basically affects all the different transport phenomena inside the fuel cell. The composition of the incoming gas streams depends strongly on the temperature. Assuming the inlet gases are fully humidified, the partial pressure of water vapour entering the cell depends on the temperature only. Thus, the molar fraction of water vapour is a function of the total inlet pressure and temperature, and so the molar fraction of the incoming hydrogen and oxygen depend on the temperature and pressure as well. At an operating pressure of 1 atm the effect of the temperature on the inlet composition is much stronger than at elevated pressures. At 90 C for atmospheric pressure, almost 68% (molar) of incoming cathode side gas stream consists of water vapour and only around 6.6% is oxygen. The result is a decrease in the molar oxygen fraction at the catalyst layer. It is expected that this will affect mainly the limiting current density. The molar oxygen fraction at the catalyst layer decreases with increasing of the cell operating temperature, due to the reduction of the molar oxygen fraction in the incoming gas streams. The result is a much higher fraction of the local current density and it is generated under the channel area. The results show that the temperature distribution inside the cell show that the maximum temperature occurs at higher cell operating temperature case (90 C), and this leads to increasing the total displacement and the degree of the deformation inside the membrane, but the maximum temperature gradient appears in the lower cell operating temperature case (60 C), due to the high activation overpotential, which leads to more heat generation with drop of the cell voltage. The activation overpotential decreases with increasing of cell operating temperature. This is because of the exchange current density of the oxygen reduction reaction increases rapidly with temperature due to the enhanced reaction kinetics, which reduces activation losses. A higher temperature leads also to a higher diffusivity of the hydrogen protons in the electrolyte membrane, thereby reducing the membrane resistance and this leads to reducing the potential loss in the membrane. mass transport loss increases as the cell operating temperature increases due to the reduction of the molar oxygen fraction in the incoming gas streams and, hence, a reduction in the molar oxygen fraction at the catalyst layer.

3.2.1.2. Effect of Pressure

Similar to the temperature, the operating pressure affects numerous transport parameters that are important for the fuel cell operation. The saturation pressure of water vapour depends only on the temperature and it remains constant for a variation of the inlet pressure. A change in the operating pressure leads to a change in the inlet gas compositions, assuming the inlet gases are fully humidified. The increase in the molar oxygen fraction in the incoming gas streams is significant when the pressure is increased from atmospheric pressure up to 3 atm. A further increase in the pressure from 3 atm to 5 atm dose not lead to a significant improvement in terms of the molar oxygen fraction. The effect of the cell operating pressure on the local current distribution shows that the higher cell operating pressure results in more even distribution of the local current density due to the high oxygen concentration at the catalyst layer. This leads to the fact that for a lower cell operating pressure at a constant nominal current density, there is a much stronger distribution of current inside the cell, the maximum local current density being at the inlet under the channel area. Therefore, the maximum temperature gradient appears in the cathode side catalyst layer of the lower cell operating pressure, and this leads to increasing the total displacement and the degree of the deformation inside the membrane. The activation overpotential decreases with increasing of the cell operating pressure. This is because of the exchange current density of the oxygen reduction reaction increases with increasing of the cell operating pressure due to the enhanced reaction kinetics. To reduce mass transport loss, the cathode is usually run at high pressure. In essence, higher pressures help to force the oxygen and hydrogen into contact with the electrolyte and this leads to reducing the mass transport loss.

3.2.1.3. Effect of Stoichiometric Flow Ratio

The stoichiometric flow ratio has an important impact on the water management in the cell. The amount of incoming air determines how much water vapour can be carried out of the cell. The effect of the stoichiometric flow ratio on the local current distribution shows that the higher stoichiometric flow ratio results in more even distribution of the local current density due to the high oxygen concentration at the catalyst layer. Therefore, the maximum temperature gradient appears in the cathode side catalyst layer of the lower stoichiometric flow ratio case, and this leads to increasing the total displacement and the degree of the deformation inside the membrane. For an optimum fuel cell performance, and in order to avoid large temperature gradients inside the fuel cell, it is desirable to achieve a uniform current density distribution inside the cell.

3.2.2. Design Parameters

3.2.2.1. Effect of Gas Channels Width

A reduction in the land area width by increasing the width of the gas flow channel enhances the mass transport of the reactions to the catalyst layer that lies under the land area. The result is an increase in the molar oxygen fraction at the catalyst layer with more even distribution. It is expected that this will affect mainly the limiting current density and to a lesser degree, the voltage drop due to mass transport limitations.

The channel width has a large impact on the local current density distribution. For the narrow channel, the local current density can exceed more than 40% of the nominal current density with a sharp drop-off under the land area, where the local current density is about 40% lower than the nominal current density. The wider channel makes for a much more evenly distributed current throughout the cell. However, the temperature peak appears in the cathode side catalyst layer of the wider channel case, implying that major heat generation takes place where the electrochemical activity is the highest, and this leads to increasing the total displacement and the degree of the deformation inside the membrane. This is because of the increase in the width of the gas flow channel means that the velocity of the incoming gas has to be decreased with all remaining parameters remaining constant, and this will decrease the gases velocity in the gas diffusion layer and hence, reduced the convection heat transfer in this region.

Finally, a reduced width of the land area increases the contact resistance between the bipolar plates and the gas diffusion electrodes. Since this is an ohmic loss, it is directly correlated to the land area width.

3.2.2.2. Effect of GDL Thickness

The effect of gas diffusion layer thickness on the fuel cell performance is again mostly on the mass transport, as the ohmic losses of the electrons inside the gas diffusion layer are relatively small due to the high conductivity of the carbon fiber paper. A thinner gas diffusion layer increases the mass transport through it, and this leads to reduction the mass transport loss. The molar oxygen fraction at the catalyst layer increases with a decreasing of the gas diffusion layer thickness due to the reduced resistance to the oxygen diffusion by the thinner layer. The distribution of the local current density of the cathode side depends directly on the oxygen concentration. The thicker gas diffusion layer results in more even distribution of the local current density due to the more even distribution of the molar oxygen fraction at the catalyst layer. This leads to the fact that for a thinner gas diffusion layer at a constant current density, there is a much stronger distribution of current inside the cell, the maximum local current density being at the inlet under the channel area.

Therefore, the maximum temperature gradient appears in the cathode side catalyst layer of the thinner gas diffusion layer case, and this leads to increasing the total displacement and the degree of the deformation inside the membrane.

3.2.2.3. Effect of Membrane Thickness

The effect of membrane thickness on the fuel cell performance is mostly on the resistance of the proton transport across the membrane. The potential loss in the membrane is due to resistance to proton transport across the membrane from anode catalyst layer to cathode catalyst layer. Therefore, a reduction in the membrane thickness means that the path travelled by the protons will be decreased, thereby reducing the membrane resistance and this leads to reducing the potential loss in the membrane, which in turn leads to less heat generation in the membrane, and this leads to reducing the total displacement and the degree of the deformation inside the membrane.

These results suggested that reducing the membrane thickness played a significant role in promoting cell performance. However, there is a limitation to this reduction, due to the effect of increased gas cross-over with very thin membranes.

3.2.3. Material Parameters

3.2.3.1. Effect of GDL Porosity

The porosity of the gas diffusion layer has two comparing effects on the fuel cell performance; as the porous region provides the space for the reactants to diffuse towards the catalyst region, an increase in the porosity means that the onset of mass transport limitations occurs at higher current densities, i.e. it leads to higher limiting currents.

The adverse effect of a high porosity is increase in the contact resistance. Higher gas diffusion layer porosity improves the mass transport within the cell and this leads to reducing the mass transport loss. The molar oxygen fraction at the catalyst layer increases with more even distribution with an increasing in the porosity. This is because of a higher value of the porosity provides less resistance for the oxygen to reach the catalyst layer. A higher porosity evens out the local current density distribution. For a lower value of the porosity a much higher fraction of the total current is generated under the channel area. This can lead to local hot spots inside the membrane electrode assembly. These hot spots can lead to a further drying out of the membrane, thus increasing the electric resistance, which in turn leads to more heat generation and therefore increasing the total displacement and the degree of the deformation inside the membrane.

Thus, it is important to keep the current density relatively even throughout the cell. As mentioned above, another loss mechanism that is important when considering different gas diffusion layer porosities is the contact resistance. Contact resistance occurs at all interfaces inside the fuel cell. The magnitude of the contact resistance depends on various parameters such as the surface material and treatment and the applied stack pressure. The electrode porosity has a negative effect on electron conduction, since the solid matrix of the gas diffusion layer provide the pathways for electron transport, the higher volume porosity increases resistance to electron transport in the gas diffusion layers.

3.2.3.2. Effect of GDL Thermal Conductivity

Thermal management is required to remove the heat produced by the electrochemical reaction in order to prevent drying out of the membrane and excessive thermal stresses that may result in rupture of the membrane. The small temperature differential between the fuel cell stack and the operating environment make thermal management a challenging problem in PEM fuel cells. The maximum temperature with higher gradient appears in the cathode side catalyst layer of the lower thermal conductivity. Heat generated in the catalyst layer is primarily removed through the gas diffusion layer to the current collector rib by lateral conduction. This process is controlled by the gas diffusion later thermal conductivity. Therefore, the membrane temperature is strongly influenced by the gas diffusion layer thermal conductivity, indicating a significant role played by lateral heat conduction through the gas diffusion layer in the removal of waste heat to the ambient. Therefore, a gas diffusion layer material having higher thermal conductivity is strongly recommended for fuel cells designed to operate with high power.

3.2.3.3. Effect of Membrane Thermal Conductivity

The higher membrane conductivity results in more even distribution of the temperature inside the cell. The lower membrane conductivity means that it is likely that heat accumulates at the cathode catalyst surface during operation. Therefore, a membrane material having higher thermal conductivity is strongly recommended for fuel cells designed to operate with high power density.

3.3. Optimal Conditions to Achieve Long Cell Life

To achieve long cell life, the results show that the cell must be operate at lower cell operating temperature, higher cell operating pressure, higher stoichiometric flow ratio, and must have higher GDL porosity, higher GDL thermal conductivity, higher membrane thermal conductivity, narrower gases channels, thicker gas diffusion layers, and thinner membrane.

The parameters that achieve long cell life are presented in Table 6. These parameters have been used in the CFD model to predict the stresses, displacement, and the degree of deformation in the cell for the optimal conditions. Figure 15 shows total displacement distribution (contour plots) and deformation shape (scale enlarged 500 times) for membrane at optimal operating conditions of long cell life (Table 6) and at base case operating conditions.

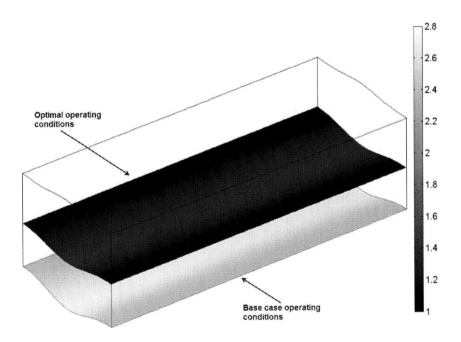

Figure 15. Total displacement [μm] distribution (contour plots) and deformation shape (scale enlarged 500 times) for membrane at optimal operating conditions of long cell life (Table 6) and at base case operating conditions.

The figure illustrates the effect of stresses on the membrane. The distribution of the total displacement is more uniform with less important displacement compared with all previous parametric results. The results show that the maximum displacement is only 1.37 micro m. This value is less than the displacement that occurs in the base case operating conditions by about 50%.

Table 6. Optimal parameters for optimum design and operating conditions to achieve long cell life

Parameter	Value
Cell operating temperature	60 C
Cell operating pressure	5 atm
Stoichiometric flow ratio	3
Gas channel width	0.8 mm
GDL porosity	0.5
GDL thickness	0.3 mm
GDL thermal conductivity	2.9 W/m.K
Membrane thermal conductivity	0.6 W/m.K
Membrane thickness	0.2 mm

All other parameters keeping constant at base case conditions. (Table 1 and 2)

CONCLUSION

A full three-dimensional, multi-phase computational fluid dynamics model of a PEM fuel cell has been developed to investigate the hygro-thermal stresses in PEM fuel cell, which developed during the cell operation due to the changes of temperature and relative humidity. A unique feature of the present model is to incorporate the effect of hygro and thermal stresses into actual three-dimensional fuel cell model. In addition, the temperature and humidity dependent material properties are utilize in the simulation for the membrane. The behaviour of the membrane during operation of a unit cell has been studied and investigated under real cell operating conditions. The results show that the non-uniform distribution of stresses, caused by the temperature gradient in the cell, induces localized bending stresses, which can contribute to delaminating between the membrane and the gas diffusion layers. These stresses may explain the occurrence of cracks and pinholes in the membrane under steady–state loading during regular cell operation. Parametric and optimization study using this model has been performed. The study quantifies the impact of operating, design, and material parameters on fuel cell performance. The model is shown to be able to: (1) understand the many interacting, complex electrochemical and transport phenomena that cannot be studied experimentally; (2) identify limiting steps and components; and (3) provide a computer-aided tool for design and optimization of future fuel cell with much higher power density, long cell life, and lower cost. To achieve long cell life, the results show that the cell must be operate at lower cell operating temperature, higher cell operating pressure, higher stoichiometric flow ratio, and must have higher GDL porosity, higher GDL thermal conductivity, higher membrane thermal conductivity, narrower gases channels, thicker gas diffusion layers, and thinner membrane.

In conclusion, the development of physically representative models that allow reliable simulation of the processes under realistic conditions is essential to the development and optimization of fuel cells, improve long-term performance and lifetime, the introduction of cheaper materials and fabrication techniques, and the design and development of novel architectures.

REFERENCES

[1] Zhang, S.; Yuan, X.; Wang, H.; Merida, W.; Zhu, H.; Shen, J.; Wu, S.; Zhang, J. A review of accelerated stress tests of MEA durability in PEM fuel cells. *Int. J. Hydrogen Energy*, 2009; 34(1): 388-404.

[2] Wu, J.; Yuan, X.Z.; Martin, J.J.; Wang, H.; Zhang, J.; Shen, J.; Wu, S.; Merida, W. A review of PEM fuel cell durability: Degradation mechanisms and mitigation strategies. *J. Power Sources*, 2008; 184(1): 104-119.

[3] Beuscher, U.; Cleghorn, S.J.C.; Johnson, W.B. Challenges for PEM fuel cell membranes. *Int. J. Energy Res.* 2005; 29(12): 1103-1112.

[4] Gode, P.; Ihonen, J.; Strandroth, A.; Ericson, H.; Lindbergh, G.; Paronen, M.; Sundholm, F.; Sundholm, G.; Walsby, N. Membrane durability in a pem fuel cell studied using PVDF based radiation grafted membranes fuel cells. *Fuel Cells* 2003; 3(1-2): 21-27.

[5] Crum, M.; Liu, W. Effective Testing Matrix for Studying Membrane Durability in PEM Fuel Cells. Part 2. Mechanical Durability and Combined Mechanical and Chemical Durability, vol. 3. Electrochemical Society Inc., Pennington, NJ 08534-2896, United States, Cancun, Mexico, 2006: 541–550.

[6] Marrony, M.; Barrera, R.; Quenet, S.; Ginocchio, S.; Montelatici, L.; Aslanides, A. Durability study and lifetime prediction of baseline proton exchange membrane fuel cell under severe operating conditions. *J. Power Sources*, 2008; 182(2): 469-475.

[7] Stanic V, Hoberech M. Mechanism of pin-hole formation in membrane electrode assemblies for PEM fuel cells. Electrode Assemblies for PEM Fuel Cells, Electrochemical Society Inc., Pennington, NJ 08534-2896, United States, Honolulu, HI, United States, 2004; p. 1891.

[8] Ramaswamy, N.; Hakim, N.; Mukerjee, S. Degradation mechanism study of perfluorinated proton exchange membrane under fuel cell operating conditions. *Electrochimica Acta* 2008; 53(8): 3279–3295.

[9] Tang Y, Karlsson AM, Santare MH, Gilbert M, Cleghorn S, Johnson WB. An experimental investigation of humidity and temperature effects on the mechanical properties of perfluorosulfonic acid membrane. *J. Mater Sci. Eng.* 2006; 425(1-2): 297-304.

[10] Xie J, Wood DL, Wayne DM, Zawodzinski T, Borup RL. Durability of polymer electrolyte fuel cells at high humidity conditions. *J. Electrochem. Soc.* 2005; 152(1): A104-A113.

[11] Kim, S.; Mench, M.M. Investigation of temperature-driven water transport in polymer electrolyte fuel cell: phase-change-induced flow. *J. Electrochem. Soc.* 2009; 156(3): B353-B362.

[12] Kim, S.; Mench, M.M. Investigation of temperature-driven water transport in polymer electrolyte fuel cell: Thermo-osmosis in membranes. *J. Membrane Sci.* 2009; 328(1-2): 113-120.

[13] Berning, T.; Lu, D.M.; Djilali, N. Three-dimensional computational analysis of transport phenomena in a PEM fuel cell. *J. Power Sources*, 2002; 106(1-2): 284-294.

[14] Berning, T.; Djilali, N. Three-dimensional computational analysis of transport phenomenon in a PEM fuel cell-a parametric study. *J. Power Sources*, 2003; 124(2): 440-452.

[15] Sivertsen, B.R.; Djilali, N. CFD based modelling of proton exchange membrane fuel cells. *J. Power Sources*, 2005; 141(1): 65-78.

[16] Webber, A.; Newman, J.A. Theoretical study of membrane constraint in polymer-electrolyte fuel cell. *AIChE J.*, 2004; 50(12): 3215–3226.

[17] Tang, Y.; Santare, M.H.; Karlsson, A.M.; Cleghorn, S.; Johnson, W.B. Stresses in proton exchange membranes due to hygro-thermal loading. *J. Fuel Cell Sci.andTech. ASME*, 2006; 3(5): 119-124.

[18] Kusoglu, A; Karlsson, A.M.; Santare, M.H.; Cleghorn, S; Johnson, W.B. Mechanical response of fuel cell membranes subjected to a hygro-thermal cycle. *J. Power Sources,* 2006; 161(2): 987-996.

[19] Kusoglu, A.; Karlsson, A.M., Santare, M.H.; Cleghorn, S.; Johnson, W.B. Mechanical behavior of fuel cell membranes under humidity cycles and effect of swelling anisotropy on the fatigue stresses. *J. Power Sources*, 2007; 170(2): 345-358.

[20] Solasi, R.; Zou, Y.; Huang, X.; Reifsnider, K.; Condit, D. On mechanical behavior and in-plane modeling of constrained PEM fuel cell membranes subjected to hydration and temperature cycles. *J. Power Sources*, 2007; 167(2): 366-377.

[21] Bograchev, D.; Gueguen, M.; Grandidier, J-C.; Martemianov, S. Stress and plastic deformation of MEA in fuel cells stresses generated during cell assembly. *J. Power Sources*, 2008; 180(2): 393-401.

[22] Suvorov, A.P.; Elter, J.; Staudt, R.; Hamm, R.; Tudryn, G.J.; Schadler, L.; Eisman, G. Stress relaxation of PBI based membrane electrode assemblies. *Int. J. Solids and Structures*, 2008; 45(24): 5987-6000.

[23] Tang, Y.; Kusoglu, A.; Karlsson, A.M.; Santare, M.H.; Cleghorn, S.; Johnson, W.B. Mechanical properties of a reinforced composite polymer electrolyte membrane and its simulated performance in PEM fuel cells. *J. Power Sources*, 2008; 175(2): 817-825.

[24] Bograchev, D.; Gueguen, M.; Grandidier, J-C.; Martemianov, S. Stress and plastic deformation of MEA in running fuel cell. *Int. J. Hydrogen Energy,* 2008; 33(20): 5703–5717.

[25] Al-Baghdadi, MARS. A CFD study of hygro-thermal stresses distribution in PEM fuel cell during regular cell operation. *Renewable Energy Journal* 2009; 34(3): 674-682.

[26] Fuller, E.N.; Schettler, P.D.; Giddings, J.C. A new methode for prediction of binary gas-phase diffusion coefficients. *Ind. Eng. Chem.*, 1966; 58(5): 18-27.

[27] Berning, T.; Djilali, N. A 3D, multi-phase, multicomponent model of the cathode and anode of a PEM fuel cell. *J. Electrochem. Soc.*, 2003; 150(12): A1589-A1598.

[28] Nguyen, P.T.; Berning, T.; Djilali, N. Computational model of a PEM fuel cell with serpentine gas flow channels. *J. Power Sources*, 2004; 130(1-2): 149-157.

[29] Al-Baghdadi, MARS.; Shahad, H. Optimization study of a PEM fuel cell performance using 3D multi-phase computational fluid dynamics model. *Journal of Zhejiang University SCIENCE-A* 2007; 8(2): 285-300.

[30] Lampinen, M.J.; Fomino, M. Analysis of free energy and entropy changes for half-cell reactions. *J. Electrochem. Soc.*, 1993; 140(12): 3537–3546.

[31] Wang, L.; Husar, A.; Zhou, T.; Liu, H. A parametric study of PEM fuel cell performances. *Int. J. Hydrogen Energy*, 2003; 28(11): 1263– 1272.

[32] Parthasarathy, A.; Srinivasan, S.; Appleby, J.A.; Martin, C.R. Pressure dependence of the oxygen reduction reaction at the platinum microelectrode/nafion interface: electrode kinetics and mass transport. *J. Electrochem. Soc.*, 1992; 139(10): 2856–2862.

[33] Siegel, N.P.; Ellis, M.W.; Nelson, D.J.; von Spakovsky M.R. A two-dimensional computational model of a PEMFC with liquid water transport. *J. Power Sources*, 2004; 128(2): 173–184.

[34] Hu, M.; Gu A.; Wang, M.; Zhu, X.; Yu, L. Three dimensional, two phase flow mathematical model for PEM fuel cell. Part I. Model development. *Energy Conversion Manage*, 2004; 45(11-12): 1861–1882.

[35] Product Information, DuPont™ Nafion® PFSA Membranes N-112, NE-1135, N-115, N-117, NE-1110 *Perfluorosulfonic Acid Polymer*. NAE101, 2005.

In: Computational Fluid Dynamics
Editor: Alyssa D. Murphy

ISBN 978-1-61209-276-8
© 2011 Nova Science Publishers, Inc.

Chapter 2

A FRONT TRACKING METHOD FOR NUMERICAL SIMULATION OF INCOMPRESSIBLE TWO-PHASE FLOWS

Jinsong Hua[*1] *and Gretar Tryggvason*[2]

[1] Institute for Energy Technology, Kjeller, Norway
[2] Department of Aerospace and Mechanical Engineering,
University of Notre Dame, Notre Dame, IN, USA

ABSTRACT

A front-tracking method for direct numerical simulation of incompressible two-phase flows is presented along with its numerical implementation and validation. This method is based on the "one-fluid" formulation of the governing equations, in which a single set of governing equations is applied to both fluid phases, treating the different phases as one fluid with variable material properties. The unsteady Navier–Stokes equations are solved using a conventional finite volume method on a fixed grid, and the interface or front is tracked explicitly by connected marker points. The physical discontinuities across the interface are accounted for by adding the appropriate sources as delta-functions at the front separating the phases. The surface tension is computed on the front and spread onto the fixed grid. Following the motion of the front, the distribution of fluid properties such as density is updated accordingly. The front tracking method has been implemented for fully three-dimensional flow modeling, as well as for two-dimensional and axisymmetric flow systems. The proposed numerical method has been validated systematically with available experimental data, and employed widely to investigate the dynamics of multiphase flows with bubbles and drops. In addition, this numerical method has been extended to model complex multiphase interfacial flow problems involving phase change, surfactant transport and electrohydrodynamics. The extensive review of simulations that use the front tracking method to investigate various multiphase interfacial flow problems demonstrates that the front tracking method is one of the most effective numerical approaches for the interfacial multiphase flow simulations.

---

[*] Email: Jinsong.Hua@ife.no

1. INTRODUCTION

Multiphase flows are numerous in both everyday life and engineering practice (Prosperetti 2004). Typical examples of multiphase flow system can be found in nature (e.g. raindrops in air and gas bubbles in water) and in many industrial processes (e.g. chemical reactions, combustion, and petroleum refining). A single gas bubble rising in an otherwise quiescent viscous liquid is one example. The understanding of the flow dynamics of such a system is of great importance in engineering applications and to the fundamental understanding of multiphase flow physics. Rising bubbles have been studied theoretically (Davies & Taylor 1950; Moore 1959), experimentally (Bhaga & Weber 1981) as well as computationally (Ryskin & Leal 1984). While all these efforts have provided us with valuable insights into the dynamics of bubbles rising in viscous liquids, there are still many questions that remain unanswered. The behavior of a single bubble is not only affected by physical properties such as the density and the viscosity of both phases (Chen et al. 1999), but also by the surface tension between the two phases and by the evolution of the bubble shape (Ohta et al. 2005; Bonometti & Magnaudet 2006). The difficulties in describing and modeling a bubble are mainly due to the strong nonlinear coupling of factors such as buoyancy, surface tension, bubble/liquid inertia, viscosity, bubble shape evolution and the bubble rise history. Therefore, most of the past theoretical work has been based on many assumptions and simplifications, and the results are only valid for certain flow regimes (Moore 1959; Taylor & Acrivos 1964). Experimental work has been limited by the available technologies to monitor, probe and sense the moving bubbles without interfering with their physics (Bhaga & Weber 1981; Shew & Pinton 2006; Tomiyama et al. 2002).

With the rapid advance of computing power and the development of robust numerical methods, direct numerical simulations promise to extend our fundamental knowledge of multiphase flow systems. However, there are still great challenges and difficulties in simulating such systems accurately. This may be attributed to the following factors: (i) the sharp interface between the gas bubble and the surrounding liquid should be tracked accurately without introducing excessive numerical smearing; (ii) the surface tension gives rise to a singular source term in the governing equations, leading to a sharp pressure jump across the interface; (iii) the discontinuity of the density and viscosity across the fluid interface may lead to numerical instability, especially when the jumps in these properties are high (for example, the density ratio of liquid to gas could be as high as 1000); (iv) the geometric complexity caused by bubble deformation and possible topological change is the main difficulty in handling the geometry of interface; a large bubble may break up into several small ones, and small bubbles may also merge to form larger bubbles; (v) the complex physics at the interface, e.g. the effects of surfactants, thin liquid film dynamics, film boiling and phase change (heat and mass transfer) and chemical reactions. Various methods for multiphase flow have been developed to address these difficulties, and each method typically has its own characteristic strengths and weaknesses. Comprehensive reviews of numerical methods for multiphase/interfacial flow simulation have been given by Scardovelli and Zaleski (1999) and van Sint Annaland et al. (2006). Most of the current numerical techniques applied in the simulation of multiphase/interfacial flows have been developed with focus on the following two aspects: (i) capturing/tracking the sharp interface, e.g. interface capturing, grid fitting, front tracking or hybrid methods; and (ii) stabilizing the flow solver to handle discontinuous fluid properties and highly singular interfacial source terms, e.g. the projection-correction method (van Sint Annaland et al. 2006) and the SIMPLE algorithm (Chen et al. 1999; Hua & Lou 2007).

Interface capturing methods include volume of fluid (Hirt & Nichols 1981; Brackbill et al. 1992), level-set (Osher & Sethian 1988; Sussman et al. 1994; Osher & Fedkiw 2001) and phase-field methods (Jacqmin 1999). In these methods the interface is captured using various volume functions defined on the grid used to solve the "one-fluid" formulation of the governing equations for multiphase flow. Since the interface can be captured using the same grid as the flow solver, these methods are relatively easy to implement. However, the accuracy of this approach is limited by the numerical diffusion in the solution of the convection equation of the volume function. Various schemes have been developed to advect, reconstruct/reinitialize the volume function to improve the accuracy in calculating the interface position. One example is the high-order shock-capturing scheme used to treat the convective terms in the governing equations (Ida 2000). Although the explicit reconstruction of the interface is circumvented, the implementation of such high-order schemes is quite complicated, and they do not work well for the sharp discontinuities encountered in multiphase/interfacial flows. In addition, a relatively fine grid is needed in the vicinity of the interface to obtain good resolution.

Grid fitting methods, on the other hand, track the moving interface by fitting the background grid points to the interface position. It can be achieved through re-meshing techniques such as deforming, moving, and adapting the background grid. This method is also well-known as the "boundary-fitting approach", and the "boundary" here refers to the interface between the fluids. The grid-fitting approach is capable of capturing the interface position accurately. Early development of this approach was done by Ryskin and Leal (1984) who used curvilinear grids to follow the motion of a bubble rising in liquid. This method is suitable for r elatively simple geometries undergoing small deformations (Kang & Leal 1987), whereas applications to complex, fully three-dimensional problems with unsteady deforming phase boundaries are very rare. This is mainly due to difficulties in maintaining the proper volume mesh quality and handling complex interface geometry such as topological change. In spite of these difficulties, the recent work by Hu et al. (2001) showed some very impressive results on 3D simulations of moving spherical particles in liquid. Quan and Schmidt (2006) presented a meshing adaptation method to track the 3D moving interface in two-phase flows.

Front tracking methods track the interface position in a Lagrangian manner using a set of interface markers and solve for the flow field on a background grid. These interface markers can be free particles without connections, or they can be logically connected elements, possibly containing accurate geometric information about the interface such as area, volume, curvature, deformation, etc. The front tracking technique was pioneered by Glimm and his coworkers (1986; 1988; 2000). They represented the front interface using a set of moving markers and solved the flow field on a separate background grid. The background grid was modified only near the front to make background grid points coincide with the front markers of the interface. In this case, some irregular grids were constructed and special finite difference stencils were created for the flow solver, increasing the complexity of the method and making it more difficult to implement. Independently, another front tracking technique was developed by Peskin and collaborators (Peskin 1977; Fogelson & Peskin 1988; Peskin & Printz 1993). In their method, the interface was represented by a connected set of particles that carry forces, either imposed externally or adjusted to achieve a specific velocity at the interface. A fixed background grid was kept unchanged even near the front, the interfacial forces were distributed onto the background grid, and the "one-fluid" formulation for the flow equations was applied to solve the flow field.

A number of combinations and improvements of these basic approaches have been proposed to enhance the capabilities in dealing with the sharp, moving interface, where complex physical phenomena and processes can occur. One of the most promising approaches is arguably the front tracking method proposed by Tryggvason and his collaborators (Unverdi & Tryggvason 1992; Tryggvason et al. 2001). Actually, this method may be viewed as a hybrid of the front capturing and the front tracking techniques: a fixed background grid is used to solve the fluid flow, while a separate interface mesh is used to track the interface position. The tracked interface carries information about jumps in the density and the viscosity and also about interfacial forces such as surface tension. Fluid properties are then distributed onto the fixed background grid according to the position of the interface. The surface tension can be calculated according to the geometry of the interface and is also distributed onto the background grid next to the interface.

Besides the numerical techniques employed to capture/track the moving interface, it is also very important to develop a stable numerical method to solve the governing equations for the flow field. Some investigators have considered simplified models such as Stokes flow (Pozrikidis 2001), where inertia is completely ignored, and inviscid potential flow (Hou et al. 2001), where viscous effects are ignored. In both cases, the motion of the deformable boundaries can be simulated with boundary integral techniques. However, when considering the transient Navier-Stokes equations for incompressible, Newtonian fluid flow, the so-called "one-fluid" formulation for multiphase flow has proved most successful (Brackbill et al. 1992; Unverdi & Tryggvason 1992; Sussman et al. 1994; Tryggvason et al. 2001). Popular modern methods that use the "one-fluid" formulation include the projection-correction method (Tomiyama et al. 2002; van Sint Annaland et al. 2006) and the SIMPLE algorithm (Chen et al. 1999; Hua & Lou 2007; Hua et al. 2008c).

Various multiphase/interfacial flow problems have been successfully simulated by the front tracking method (Tryggvason et al. 2001) with a projection-correction flow solver. Many previously reported results have been limited to flows with low Reynolds numbers (<100) and small density ratios (<100) (Bunner & Tryggvason 2002). The development of more robust methods that are applicable to a wide range of flow regimes has therefore been pursued by several investigators. Revised versions of the projection-correction method have been proposed to improve its capability in handling situations with large density and viscosity ratios (van Sint Annaland et al. 2006). Recently, Hua and Lou (2007) tested a SIMPLE-based algorithm to solve the incompressible Navier-Stokes equations. The simulation results indicate that the newly proposed method can solve the Navier-Stokes equations robustly with large density ratios up to 1000 and large viscosity ratios up to 500. Extensive simulations and model validation on a single bubble rising in a quiescent liquid were presented in Hua and Lou (2007). The simulations show good results in a wide range of flow regimes with high density and viscosity ratios, and the algorithm is promising for direct numerical simulation of multiphase flow. Unfortunately, the previous validation studies were limited to a 2D axisymmetric model (Hua & Lou 2007) where the fluid flow and bubble shapes are axisymmetric. Hence, it would be interesting to investigate the robustness of the proposed numerical approach for multiphase flow in flow regimes of higher Reynolds and Bond numbers where the bubble may no longer be axisymmetric. A fully three-dimensional modeling approach was proposed by Hua et al. (2008c). Other features such as mesh adaptation, moving reference frame and parallel processing were introduced to enhance the model capability in simulating the rise of a 3D bubble in a viscous liquid. A parallel adaptive mesh refinement (AMR) tool, PARAMESH (MacNeice et al. 2000), was integrated with the modified SIMPLE flow solver, and the governing equations were solved in a non-inertial moving reference frame attached to the rising bubble. The AMR feature allows a relatively high-resolution mesh in the vicinity of the bubble surface. The non-inertial moving reference frame technique translates the computational domain with the rising bubble, allowing the computational domain to be relatively small and always centered around the bubble. The latter feature is particularly useful for studying the path instability of a rising bubble or the interaction of multiple bubbles, which may need very long simulation periods.

The numerical algorithm proposed by Tryggvason and co-workers (Tryggvason et al. 2001; Unverdi & Tryggvason 1992) and extended further by Hua and coworkers (2007; 2008), is introduced in detail in this chapter. A single set of governing equations is used to describe the two-phase flow dynamics in the entire computational domain by treating the two fluids as one single fluid with variable fluid properties across the interface – often referred to as the "one-field" or "one-fluid" approach. The governing Navier-Stokes equations are solved on a fixed Cartesian grid, while the interface is represented by a set of explicitly tracked front markers. These markers form an adaptive triangular surface mesh that is advected with a velocity interpolated from the surrounding fluid. The interface is assumed to have a given finite thickness (normally about two to four times the background grid size) so that jumps in the fluid properties across the surface can be reconstructed smoothly by solving a Poisson equation. The details about the numerical method will be introduced in the following sections.

2. MATHEMATICAL FORMULATION FOR MULTIPHASE FLOW

A multiphase interfacial flow system consists of at least two immiscible fluids and an interface between them. It can be considered as the basic "building block" for any complex multiphase flow system. To model such a multiphase interfacial system, a mathematical model should not only describe the physics within each phase, but also the dynamics of the interface. The flow physics within each phase can be treated as a single phase flow system with an interfacial boundary which may move and involve transfer of mass and momentum. It is well known that the single phase flow system can be described well by the basic conservation equations for mass and momentum with well defined boundary conditions. Various numerical methods have been applied to solve the governing equations successfully. However, the challenges in modeling multiphase interfacial flow lie in the difficulty of defining the physical conditions at the interface such as the geometry, position, force balance and mass conservation. To obtain a mathematical formulation for multiphase interfacial flow, we will start with a short presentation of the Navier-Stokes equation for single phase flow, then discuss the physical conditions at the interface in terms of conservation laws and its mathematical description, and consequently a "one-fluid" formulation for the multiphase interfacial flow is presented. Finally, a numerical method, the front tracking method, is introduced to solve the complex governing equations.

2.1. Conservation Laws and Navier-Stokes Equation

The fluid motion in a single phase domain is governed by basic conservation laws for mass and momentum. In the flow field illustrated in Figure 2.1, the material control volume V contains a certain amount of the fluid particles. The shape and size of the control volume may deform as the fluid moves, but the mass of the fluid particles contained within it should remain constant. Therefore, mass conservation law can be expressed as

$$\frac{D}{Dt}\int_V \rho \, dV = 0, \tag{2.1}$$

where ρ stands for the fluid density. In general, the fluid density may be a function of time and space while the shape and position of the material control volume V also varies with time. This means that the above material derivative must include two effects: one related to changes in density, and the other one related to changes of the material control volume. These two effects are accounted for by the Reynolds transport theorem

$$\frac{D}{Dt}\int_V \rho \, dV = \int_V \left(\frac{\partial \rho}{\partial t} + \nabla \cdot (\rho \, \mathbf{U}) \right) dV = 0, \tag{2.2}$$

where $\mathbf{U}$ is the velocity of the fluid. Since the above integral is valid for any control volume, we obtain the following general equation for mass conservation

$$\frac{\partial \rho}{\partial t} + \nabla \cdot (\rho \, \mathbf{U}) = 0. \tag{2.3}$$

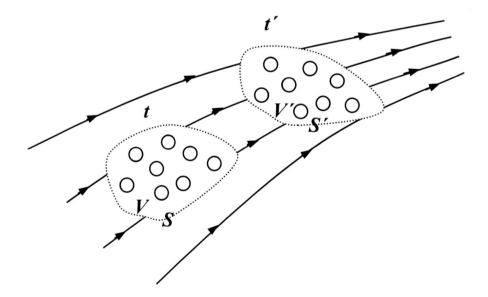

Figure 2.1. An illustration of a control volume and fluid elements.

If the density of a fluid remains constant, the flow is incompressible. The mass conservation equation (2.3) can then be reduced to $\nabla \cdot \mathbf{U} = 0$. We can change the integration over the control volume V to an integration over the bounding surface S, and obtain

$$\int_V \nabla \cdot \mathbf{U}\, dV = \int_S \mathbf{U} \cdot \mathbf{n}\, ds = 0, \tag{2.4}$$

where $\mathbf{n}$ is the unit normal vector pointing outwards on the surface of the control volume. This implies that there is no net flux through the bounding surface of the control volume and the mass contained within the control volume will remain constant.

Following the same argument as for the mass conservation, it is not difficult to get the conservation equation for the fluid momentum within the control volume

$$\frac{D}{Dt}\int_V (\rho\, \mathbf{U})\, dV = \int_S (\mathbf{J} \cdot \mathbf{n})\, ds + \int_V (\rho\, \Phi)\, dV, \tag{2.5}$$

where $\mathbf{J}$ denotes a stress on the bounding surface and Φ represents the body force within the control volume. In order to obtain a general conservation equation, we can apply the transport theorem to the left side of Equation (2.5), and the divergence theorem to the first term on the right side. Hence, the above equation can be rewritten as

$$\int_V \frac{\partial}{\partial t}(\rho\, \mathbf{U})\, dV + \int_V \nabla \cdot (\rho\, \mathbf{U}\, \mathbf{U})\, dV = \int_V (\nabla \cdot \mathbf{J})\, dV + \int_V (\rho\, \Phi)\, dV. \tag{2.6}$$

It is noted that there are only volume integrals in Equation (2.6) and the control volume can be arbitrary. Therefore, the integrand must be zero, and the general equation for momentum conservation can be expressed as

$$\frac{\partial}{\partial t}(\rho\, \mathbf{U}) + \nabla \cdot (\rho\, \mathbf{U}\, \mathbf{U}) = \nabla \cdot \mathbf{J} + \rho\, \Phi. \tag{2.7}$$

In fact, the equation of momentum conservation is an expression of Newton's second law, i.e. mass × acceleration = sum of forces. Hence, the two terms on the right side of Equation (2.7) represent the stress and body force, respectively. The conservation of angular momentum implies that the stress tensor $\mathbf{J}$ is symmetric so that $\mathbf{J} = \mathbf{J}^T$. For a Newtonian incompressible, the stress tensor ($\mathbf{T}$) can be expressed in the following form

$$\mathbf{J} = \mathbf{T} = -p\mathbf{I} + \mu(\nabla \mathbf{U} + \nabla \mathbf{U}^T), \tag{2.8}$$

in which p denotes the static pressure, $\mathbf{I}$ stands for the unit tensor, and μ is the fluid viscosity. If we substitute Equation (2.8) into Equation (2.7), the Navier-Stokes equation for Newtonian incompressible fluid flow with a gravity field $\mathbf{g}$ can be written as

$$\frac{\partial}{\partial t}(\rho \mathbf{U}) + \nabla \cdot (\rho \, \mathbf{U} \, \mathbf{U}) = -\nabla p + \nabla \cdot \mu (\nabla \mathbf{U} + \nabla \mathbf{U}^T) + \rho \mathbf{g}. \qquad (2.9)$$

This equation is always solved together with the mass conservation equation

$$\nabla \cdot \mathbf{U} = 0. \qquad (2.10)$$

So far, we have obtained the governing equations (2.9 and 2.10) for a single phase, Newtonian, incompressible fluid flow. With certain predefined boundary conditions, e.g. solid wall, inlet and outlet, we can solve the above equations numerically. The situation becomes much more complex in the case of multiphase flow. The interface between the two phases, gas-liquid or liquid-liquid, may move and deform as the fluids flow. The mass and momentum conservation equations still specify the fluid motion within each phase. But special boundary conditions are required at the interface where two different phase meet each other; these will be discussed in the following section.

2.2. Interfacial Conditions and Formulation for Multiphase Flow

As discussed in the previous section, the general conservation equation is valid in any material control volume V with bounding surface S. In the case of a multiphase flow system, there may be an interface intersecting the control volume as shown in Figure 2.2. Crossing the interface, we may move from a fluid with one set of intrinsic properties to another fluid with a different set of intrinsic properties. This implies that a step change in fluid properties, such as fluid density and viscosity, occurs across the interface within the control volume. To obtain the general conservation equations in the multiphase fluid system, special considerations are needed in deriving the boundary conditions at the interface.

The generalized transport theorem can be written as

$$\frac{D}{Dt} \int_V \varphi \, dV = \int_V \left(\frac{\partial \varphi}{\partial t} \right) dV + \int_S \varphi \, \mathbf{U} \cdot \mathbf{n} \, ds, \qquad (2.11)$$

in which V denotes a control volume whose closed boundary S is moving at velocity $\mathbf{U}$, $\mathbf{n}$ is the unit surface normal vector, and φ is the physical variable to be considered in the analysis. For multiphase flow systems, we may divide the control volume into two distinct regions V_1 and V_2, which are occupied by the two respective fluids. The interface between the two fluid phases within the control volume is denoted by Γ. The bounding surface of the control volume V can also be divided into two parts S_1 and S_2, according to the different fluid phases. We apply the transport theorem to the two fluid regions V_1 and V_2,

$$\frac{D}{Dt} \int_{V_1} \varphi \, dV = \int_{V_1} \left(\frac{\partial \varphi}{\partial t} \right) dV + \int_{S_1} \varphi \, \mathbf{U} \cdot \mathbf{n} \, ds + \int_\Gamma \varphi_1 \, \mathbf{U}_i \cdot \mathbf{n}_1 \, ds, \qquad (2.12)$$

$$\frac{D}{Dt}\int_{V_2} \varphi\, dV = \int_{V_2}\left(\frac{\partial \varphi}{\partial t}\right)dV + \int_{S_2} \varphi\, \mathbf{U}\cdot\mathbf{n}\, ds - \int_{\Gamma} \varphi_2\, \mathbf{U}_i \cdot \mathbf{n}_1\, ds. \qquad (2.13)$$

in which $\mathbf{U}_i$ is the velocity of the interface Γ, and $\mathbf{n}_1$ denotes the outward normal unit vector on the interface referring to the fluid phase 1. φ_1 and φ_2 are the values of the physical quantity at the interface on the side of the two respective fluid regions V_1 and V_2.

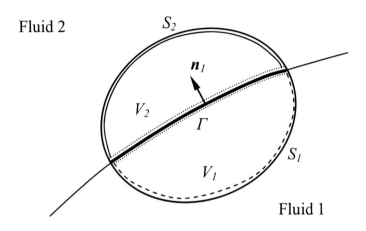

Figure 2.2. A schematic diagram showing an interface between two phases intersecting with a material control volume.

By summing Equations (2.12) and (2.13), we obtain the general transport equation for a multiphase fluid system

$$\begin{aligned}
\frac{D}{Dt}\int_V \varphi\, dV &= \frac{D}{Dt}\int_{V1} \varphi\, dV + \frac{D}{Dt}\int_{V_2} \varphi\, dV \\
&= \int_{V_1}\left(\frac{\partial \varphi}{\partial t}\right)dV + \int_{S_1} \varphi\, \mathbf{U}\cdot\mathbf{n}\, ds + \int_{\Gamma} \varphi_1\, \mathbf{U}_i\cdot\mathbf{n}_1\, ds \\
&\quad \int_{V_2}\left(\frac{\partial \varphi}{\partial t}\right)dV + \int_{S_2} \varphi\, \mathbf{U}\cdot\mathbf{n}\, ds - \int_{\Gamma} \varphi_2\, \mathbf{U}_i\cdot\mathbf{n}_1\, ds \\
&= \int_V\left(\frac{\partial \varphi}{\partial t}\right)dV + \int_S \varphi\, \mathbf{U}\cdot\mathbf{n}\, ds + \int_{\Gamma} (\varphi_1 - \varphi_2)\, \mathbf{U}_i\cdot\mathbf{n}_1\, ds
\end{aligned} \qquad (2.14)$$

The above equation can be simplified as

$$\frac{D}{Dt}\int_V \varphi\, dV = \int_V\left(\frac{\partial \varphi}{\partial t}\right)dV + \int_S \varphi\, \mathbf{U}\cdot\mathbf{n}\, ds - \int_{\Gamma} [\varphi]\, \mathbf{U}_i\cdot\mathbf{n}_1\, ds, \qquad (2.15)$$

in which $[\varphi] = \varphi_2 - \varphi_1$ refers to the jump of the quantity φ across the singular interface Γ. Applying the general conservation principle to the control volume, we obtain

$$\frac{D}{Dt}\int_V \varphi \, dV = \int_S \mathbf{J} \cdot \mathbf{n} \, ds + \int_V \phi \, dV .\tag{2.16}$$

in which **J** refers to the flux and ϕ is the source of the physical variable φ. Combining equations (2.16) and (2.15), we obtain the following equation

$$\int_V \left(\frac{\partial \varphi}{\partial t} - \phi\right) dV + \int_S (\varphi \mathbf{U} - \mathbf{J}) \cdot \mathbf{n} \, ds - \int_\Gamma [\varphi] \mathbf{U}_i \cdot \mathbf{n}_1 \, ds = 0 .\tag{2.17}$$

To derive the jump conditions across the interface, we apply the above equation to a control volume which shrinks and collapses onto the interface. For this purpose, we design a pill-box shaped control volume as shown in Figure 2.3. The pill-box has a thickness of δ. If we let the control volume thickness go to zero, its volume is going to reduce to zero as well. In addition, the bounding surfaces S_1 and S_2 will collapse onto the interface Γ. Hence, Equation (2.17) can be reduced to

$$\int_\Gamma [\varphi \mathbf{U} - \mathbf{J}] \cdot \mathbf{n}_1 \, ds - \int_\Gamma [\varphi] \mathbf{U}_i \cdot \mathbf{n}_1 \, ds = 0 .\tag{2.18}$$

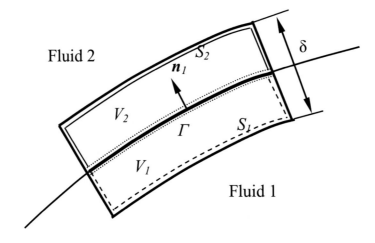

Figure 2.3. A pill-box shaped control volume with interface multiphase fluid system.

Since the control volume V with interface intersection Γ is arbitrary, the integrand in Equation (2.18) must be zero. This then yields the general jump condition across the interface in a multiphase fluid system

$$[\varphi(\mathbf{U} - \mathbf{U}_i) - \mathbf{J}] \cdot \mathbf{n}_1 = 0 .\tag{2.19}$$

In the presence of surface source $\mathbf{m}_i$ on the interface Γ, the jump condition can be adjusted to be (Slattery 1999)

$$[\varphi(\mathbf{U}-\mathbf{U}_I)-\mathbf{J}]\cdot\mathbf{n}_1 = \mathbf{m}_I. \tag{2.20}$$

The mass conservation in case of no phase change ($\varphi = \rho$, $\mathbf{J} = 0$, $\mathbf{m}_I = 0$) reduces to

$$[\rho(\mathbf{U}-\mathbf{U}_I)]\cdot\mathbf{n}_1 = 0. \tag{2.21}$$

This jump condition means that the mass flow (per unit area) leaving phase 1 must equal the mass flow entering phase 2. For the case of Newtonian, incompressible multiphase fluid flow without phase change, the fluid velocity on interface equals the interface moving velocity ($\mathbf{U} = \mathbf{U}_I$) and $[\mathbf{U}]\cdot\mathbf{n}_1 = 0$. In addition, it should be noted that the mass conservation equation (2.21) does not yield any information about the velocity tangent to the interface. For viscous fluid, an additional relation called the no-slip condition is needed to specify the velocity tangent to the interface

$$\mathbf{U}_1\cdot\mathbf{t}_1 = \mathbf{U}_2\cdot\mathbf{t}_1 \text{ or } [\mathbf{U}]\cdot\mathbf{t}_1 = 0 \tag{2.22}$$

in which $\mathbf{t}_1$ denotes a unit vector tangential to the interface. This condition states that the velocity tangent to the interface should be continuous across the interface separating the two viscous fluids.

For the momentum conservation, the forces at the interface need to be included. The surface tension can be one of the major interfacial momentum source terms, and is expressed as

$$\mathbf{m}_I = 2\kappa\sigma\mathbf{n}_1 + \nabla_s\sigma. \tag{2.23}$$

where κ denotes the mean curvature of the interface Γ, σ is the surface tension coefficient, and ∇_s is the gradient operator acting along the interface Γ. For the momentum conservation, the general jump condition across interface can be rewritten using $\varphi = \rho\mathbf{U}$, $\mathbf{J} = \mathbf{T}$ and $\mathbf{m}_I = \kappa\sigma\mathbf{n}_1 + \nabla_s\sigma$, and be given as

$$[\rho\mathbf{U}(\mathbf{U}-\mathbf{U}_I)-\mathbf{T}]\cdot\mathbf{n}_1 = \kappa\sigma\mathbf{n}_1 + \nabla_s\sigma. \tag{2.24}$$

When the fluid velocity matches the interface velocity ($\mathbf{U} = \mathbf{U}_I$) and the surface tension coefficient is uniform on the interface ($\nabla_s\sigma = 0$), the jump condition can be simplified for incompressible multiphase flow without phase change at the interface,

$$[p\mathbf{I}-\mu(\nabla\mathbf{U}+\nabla\mathbf{U}^T)]\cdot\mathbf{n}_1 = \kappa\sigma\mathbf{n}_1. \tag{2.25}$$

From the above derivation of the jump conditions across interface from the basic laws of mass and momentum conservation, it can be concluded that the jump condition, Equations (2.21) and (2.22), are included in the mass continuity equation for incompressible flows

$$\int_V (\nabla \cdot \mathbf{U}) dV = 0. \tag{2.26}$$

and that the jump condition Equation (2.25) can be embedded in the following momentum conservation equation in the following way

$$\int_V \left(\frac{\partial}{\partial t}(\rho \mathbf{U})\right) dV + \int_V (\nabla \cdot (\rho \mathbf{U}\mathbf{U})) dV = \int_V (-\nabla p) dV + \int_V \left(\nabla \cdot [\mu (\nabla \mathbf{U} + \nabla \mathbf{U}^T)]\right) dV$$
$$+ \int_\Gamma \sigma \kappa \mathbf{n}_i ds + \int_V (\rho \mathbf{g}) dV \tag{2.27}$$

And the integral Equations (2.26) and (2.17) can rewritten in the differential form as

$$\nabla \cdot \mathbf{U} = 0, \tag{2.28}$$

$$\frac{\partial}{\partial t}(\rho \mathbf{U}) + \nabla \cdot (\rho \mathbf{U}\mathbf{U}) = -\nabla p + \nabla \cdot [\mu (\nabla \mathbf{U} + \nabla \mathbf{U}^T)]$$
$$+ \int_\Gamma \sigma \kappa \mathbf{n}_i \delta(\mathbf{x} - \mathbf{x}_i) ds_i + \rho \mathbf{g} \tag{2.29}$$

where $\delta(\mathbf{x} - \mathbf{x}_i)$ is a Dirac-delta function that is zero everywhere except on the interface. A comparison of the governing equations for multiphase flow (2.28 and 2.29) and those for single phase flow (2.9 and 2.10) shows that the only difference is the term due to surface tension. In addition, we should note that the fluid property field for multiphase flow varies from one phase to the other phase in the simulation domain while it is constant for the single phase flow. If we want to solve the multiphase flow governing equation using the methods that have been successfully applied to single phase flow, we should pay special attention to the surface tension term and to the variation of fluid properties.

2.3. Front Tracking Approach

The "one-fluid" formulation as expressed in the governing equations (2.28) and (2.29) can be applied to model multiphase interfacial flow problems. We use one set of governing equations for the entire flow domain, so we treat the different fluids as a single fluid with material properties varying across the interface. The jump conditions across the interface can be embedded in the solution procedure for the governing equations. The interface is convected with the fluid flow. To develop a solution approach for these equations, it is necessary to determine the following issues for practical implementation,

- how the interface is represented;
- how the distribution of fluid properties (e.g. density and viscosity) across the interface is to be determined;
- how the surface tension is calculated;

- how the flow field is solved;
- how the interface is advected.

Following these ideas, a numerical solution procedure, called the Front Tracking Method, is introduced in this chapter. The basic concept of the front tracking method is illustrated in Figure 2.4.

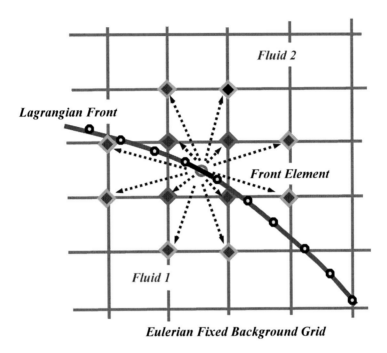

Figure 2.4. Basic concept of the front tracking method and the computational grids.

A fixed regular uniform Eulerian mesh is used to discretize the computational domain which consists of two fluid phases. The interface between the two phases is explicitly represented by a set of markers or elements. The change of the fluid properties is denoted by a smooth indicator function, which is constructed using a Dirac-delta function. The indicator function is zero in one fluid and one in the other fluid with a smooth transition over several Eulerian meshes in the interface vicinity. With the explicit shape of the interface, it is not difficult to calculate the surface tension using different methods. However, the curvature calculation on the interface normally involves high order derivatives, which may lead to inaccuracy. In this chapter, a force balance approach that calculates the net surface tension force on the interface elements will be introduced. This method is very robust and easy for numerical implementation. After we get the surface tension on the interface element, the surface force is distributed to the Eulerian grid through a smoothing process as shown by the arrows in Figure 2.4.

With the proper distributions of fluid properties and surface tension force on the Eulerian grid, the transient Navier-Stokes equation can be solved using various methods, e.g. the projection-correction method or the SIMPLE algorithm, with proper initial and boundary conditions. Therefore, the flow field in the whole domain can be obtained.

With the flow field given, the velocity of the element nodes can be obtained through an interpolation process, and the interface is advected explicitly. As the interface mesh moves, the element size may change, which may lead to a poor distribution of the elements on the interface and poor mesh quality and resolution. Hence, a mesh adaptation process is necessary to maintain the mesh quality on the interface. Another possible consequence of interface motion is that interfaces may collide with each other and result in a topology change. Special attention is needed to handle the topology change properly so that the numerical calculation can be continued in a physically sensible way.

Once we have updated the interface position, corrected the interface topology and improved the interface mesh quality, the simulation procedure can be repeated by reconstructing the indicator function, calculating the distribution of fluid properties, computing the surface tension and its volume distribution, solving the Navier-Stokes equation on the Eulerian background grid, interpolating the velocity to the nodes of the interface mesh elements, advecting the interface position, adjusting interface topology, and updating the interface mesh quality. More detailed descriptions of the implementation of the above steps are presented in the following sections.

3. NUMERICAL IMPLEMENTATION OF FRONT TRACKING

3.1. Geometrical Construction of the Fluid Interface

One of the significant advantages of the front tracking method is that it tracks the interface explicitly. In most of our simulations, we use a front structure that consists of geometry nodes (or points) connected by elements. For two-dimensional simulations, the interface is normally represented by a set of connected line elements with two end nodes. For three-dimension simulations, the interface has to be represented by a set of connected triangular elements in 3D space. The typical structures of the interface are shown in Figure 3.1

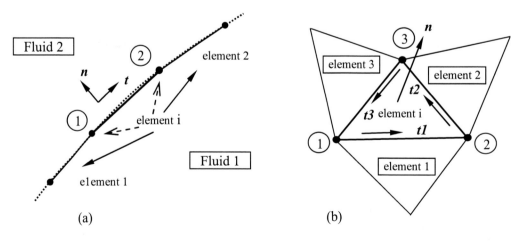

Figure 3.1. Basic structure of the front grid, (a) line elements for two-dimensional interface; and (b) triangular elements for three-dimensional interface.

Both the nodes and elements (the front objects) are stored in a linked list data-structure that contains the pointers to the previous object and the next object in the list. The order of element objects in the list can be completely arbitrary and has no relation to the actual order on the interface. Hence, the use of a linked list makes it convenient to add and remove objects. For each node object, the information stored is just its coordinates and the moving velocity used to advect the interface position. On the other hand, the element object contains most of the front information. Each element should know about the nodes that it is connected to and their order. For example, in two dimensional situations, a line element should have two nodes. The tangential direction (t) can be defined as the vector pointing from the first node to the second node. The normal direction (n) can be defined by rotating 90 degrees counterclockwise from the tangential direction (t). Hence, the normal direction always points from fluid 1 (inner fluid) to fluid 2 (outer fluid). For three dimensional situations, three nodes are connected to the triangular element in counter-clockwise order. The tangential directions on the three edges form a closed loop, and the normal direction of the element is also always pointing from fluid 1 (inner fluid) to fluid 2 (outer fluid). In other words, the nodes and edges of each element should be numbered counterclockwise when viewing the element from the outside. In addition, the element should also store the information about the neighboring elements which share a node for the 2D situation and an edge for the 3D situation.

Besides the information about geometry and connectivity, the front element should also store other physical information: the surface tension coefficient and surface tension; the jump of fluid properties such as density and viscosity; and any other surface quantities that are needed for a particular simulation.

Since the computer language Fortran 90 is used to implement the numerical method, the linked list can be easily constructed using derived data types and pointers. Hence, we define the derived data types of front, element, edge and node. The front type includes the pointers to the first object of the linked-lists of elements, edges and nodes. The element type includes the pointers to edges, nodes and the values of physical variables such as density jump and surface tension coefficient. The edge type includes the pointers to two nodes. The node type includes the coordinates and the interface velocity. Any operation involving the front is done by starting with the first object, and then using its pointer to the next object in the linked list to move to the next object and so on, until all objects have been visited.

When information is interpolated between the front and the fixed background grid, it is important to always go from the front to the grid and not the other way around. Since the fixed grid is structured and regular, it is very simple to determine the point on the fixed grid that is closest to a given front position. If we denote the total number of grid points in one direction on the mesh block by n_x, and the block coordinate in this direction starts from a_x and ends at b_x, and we can assume the grid index i starts from 0, then the grid point index i corresponds to the front position x is given by $i = \text{int}(x \cdot n_x /(b_x - a_x))$ in Fortran. If the fixed grid point index starts with 1 instead of 0, or if the grids are staggered, small modifications are obviously needed. For the reverse operation, finding the front point closest to a given grid point is a much more complex and computing intensive operation, which we should try to avoid.

If a front insects a wall boundary, we usually use the intersection point on the wall and a ghost point outside the wall. The ghost node is connected to the interaction point on the wall by a ghost element. The ghost nodes or elements are not included in the linked list for the front that describes the interface boundary, but the front element connected to the wall intersection point treats the ghost element as its neighbor. The position of the ghost point is adjusted in such as way that the tangent of the front at the intersection point on the wall has a desired value according to the boundary conditions or physical requirements. For example, for a full slip wall, we normally assume that the front tangent should be normal to the boundary. For a symmetry boundary, we normally set the ghost point to be the mirror point of the point nearest to the wall. The ghost node can also be adjusted to set the contact angle at the wall. In conclusion, the ghost node approach is quite effective to treat the interaction between the front and a wall boundary for two-dimensional problems. It is still quite challenging for three-dimensional problems.

3.2. Geometrical Restructuring of the Phase Interface

As the front nodes move with the velocity interpolated from the flow field on the fixed Eulerian grid, the front may deform and the front nodes become unevenly distributed on the front. As a result, some parts become crowded with front elements while the resolution of other parts becomes inadequate. When the element size becomes too large, additional elements must be added to maintain the calculation accuracy. When the element size becomes too small, it must be removed to maintain the calculation efficiency. In addition to reducing the total number of elements used to represent the front, element removal usually also prevents the formation of "wiggles" much smaller than the fixed grid size. Hence, a geometric restructuring process is very important in the front tracking method so that we can maintain the quality and uniformity of the front elements.

The requirement of front restructuring makes the codes that use explicit tracking more complex than other front capturing codes such as VOF or level set method. Many of the operations needed for front restructuring can be made relatively straightforward by the use of a suitable data structure, e.g. a linked-list. Figure 3.2 shows two typical restructuring operations schematically for a two-dimensional front (a) front element split and (b) front element removal.

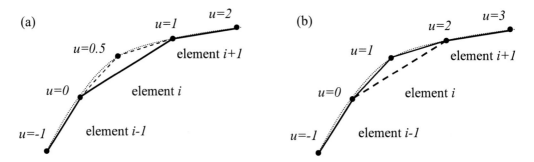

Figure 3.2. Restructuring operations for two-dimensional front: (a) front element split; (b) front element removal.

In the traditional practice of mesh refinement, we can split a large element by inserting a new point simply at the mid point between the two old end-points using a linear interpolation.

However, such a linear splitting approach is usually not suitable for the front interface element refinement because a 2D curved front interface is represented by the end-nodes linked by elements. The mid point between the end-nodes of a front element is not generally located on the interface. Hence, a higher order interpolation is needed to account for the curvature of the interface. This is particularly important when surface tension is large and non-smooth parts of the front may lead to large pressure fluctuations. In this chapter, we will introduce the Legendre interpolation, which is simple and easy for implementation of 2D front restructuring.

By numbering the points as shown in Figure 3.2, any point on the interface can be interpolated from the nodes of the three adjacent front elements,

$$f(u) = -\frac{1}{6}u(2-u)(1-u)f_{-1} + \frac{1}{2}(1-u)(1+u)(2-u)f_0 \\ + \frac{1}{2}u(1+u)(2-u)f_1 + \frac{1}{6}u(1+u)(u-1)f_2$$
(3.1)

where f is either the x or the y coordinate, u is an interpolation parameter that takes the values shown in the figure, and f_{-1}, f_0, f_1 and f_2 are the coordinates on the nodes where $u = -1, 0, 1$ and 2 respectively. The interpolation formulation (3.1) gives the exact coordinates value at the existing end nodes of the front elements:

$$f(-1) = f_{-1}; \quad f(0) = f_0; \quad f(1) = f_1; \quad f(2) = f_2.$$
(3.2)

When we split the front element by adding a new point, it is simply inserted at $u = 0.5$, resulting in

$$f(0.5) = \frac{1}{16}(-f_{-1} + 9f_0 + 9f_1 - f_2).$$
(3.3)

After inserting the new node ($u = 0.5$) in the node linked-list, two new front elements (indicated by dashed line elements) are created by connecting the newly inserted node with the two existing nodes, and should be added to the element linked-list. The old element (element i) should be removed from the linked-list.

It is more straightforward to implement the front element removal operation as shown in Figure 3.2. We need to delete the old node ($u = 1$) and the two old front elements connected to this node. A new element is created to connect the existing nodes ($u = 0$ and $u = 2$).

For three dimensional situations, the front restructuring operations are much more complicated. Not only are there several different ways to add and delete points, but other aspects of the front such as the shape and the connectivity of the elements must also be considered. The restructuring of the surface grid can be accomplished by adding or deleting elements. In some cases, reconnecting the points to form "better shaped" elements is also necessary.

There are a variety of ways to add and delete elements on the interface grid. Figure 3.3 shows the common operations based on an element edge: edge split, edge deletion and edge swap. If an edge is too long (as shown in Figure 3.3(a)), we have to split the long edge by inserting a new node (n_0) on the interface, which can be interpolated from the middle position of the long edge (e_0). Hence, the long edge is replaced with two shorter edges, and two new edges are created. The newly created four edges connect the newly inserted node (n_0) with the four existing node (n_1, n_2, n_3 and n_4), respectively. Finally, four new elements (NE_1, NE_2, NE_3 and NE_4) are formed by the new edges and existing edges. The old two elements (E_1 and E_2) sharing the long edge should be deleted from the element linked-list.

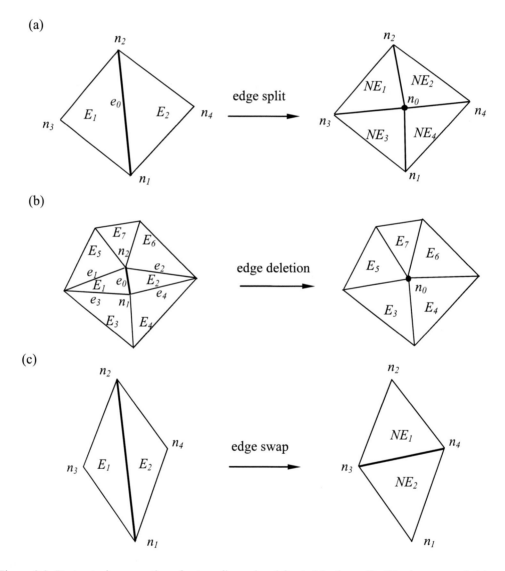

Figure 3.3. Restructuring operations for two-dimensional front: (a) edge split; (b) edge removal; (c) edge swap.

If an edge's length is too short (as shown in Figure 3.3(b)), we should remove it from the edge linked-list. Firstly, we insert a new node (n_0) on the interface, which can be interpolated from the middle position of the short edge (e_0). Secondly, we collapse both end nodes (n_1 and n_2) of the short edge (e_0) to the newly inserted node (n_0). At the same time, the two edges (e_1, e_3) of the old element (E_1) collapse, so do the two edges (e_2, e_4) of the old element (E_2). Finally, the old elements (E_1, E_2) are removed from the element linked list. The connection for remaining old elements should be updated accordingly.

Figure 3.3(c) shows the edge swap operations to improve the element shape. We delete the old elements (E_1, E_2) and the shared edge connecting the nodes n_1 and n_2. We created a new edge connecting the nodes n_3 and n_4, and two new elements (NE_1, NE_2). After the swap operation, the newly created elements should have a better shape than the old ones. The element shape can be quantified by its aspect ratio, which is defined as the length of the perimeter squared divided by the area of the element, normalized by the corresponding ratio for an equilateral triangle.

In the operations of edge split and edge deletion, we need to interpolate a new point on the interface located at the middle point of the edge. Barycentric coordinates (u, v, w) defined in Figure 3.4 are used for the interpolation. The coordinates are not independent and we must have

$$u + v + w = 1. \tag{3.4}$$

All the six nodes of the neighboring elements who have shared edges with element i can be mapped with the Barycentric coordinate system. Any interpolated coordinate on the interface can be given by

$$\begin{aligned} f(u,v,w) = & \frac{1}{2}(1-u)\left[-u f_5 + (1-v) f_3 + (1-w) f_2\right] \\ & + \frac{1}{2}(1-v)\left[(1-u) f_3 - v f_6 + (1-w) f_1\right] \\ & + \frac{1}{2}(1-w)\left[(1-u) f_2 + (1-v) f_1 - w f_4\right]. \end{aligned} \tag{3.5}$$

The interpolation formula (3.5) gives the exact coordinate values for the six nodes of the front element and its neighbors,

$$\begin{aligned} & f(1,0,0) = f_1; \quad f(0,1,0) = f_2; \quad f(0,0,1) = f_3 \\ & f(1,1,-1) = f_4; \quad f(-1,1,1) = f_5; \quad f(1,-1,1) = f_6 \end{aligned} \tag{3.6}$$

The central point (node 0) between node 1 and node 2 in the Barycentric coordinate system can be used to interpolate the new node for edge split or deletion,

$$f(0.5, 0.5, 0) = \frac{1}{2}(f_1 + f_2) + \frac{1}{4} f_3 - \frac{1}{8}(f_5 + f_6). \tag{3.7}$$

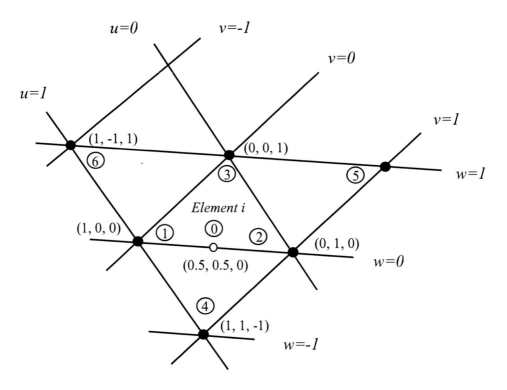

Figure 3.4. Barycentric coordinates for a surface element in three dimensional space.

To determine when it is necessary to perform the interface grid restructuring operations such as edge split or deletion, we usually define a minimum and a maximum edge length, and the operations take effect if the edge length exceeds these limits. These limits are selected such that the resolution of the front is comparable to the fixed grid. As a rule of thumb, we found that 2-4 elements per grid mesh is good for two-dimensional simulations. For three dimensional simulations, we usually examine the edge length of each element as well as its aspect ratio. The edge split operation is adopted when the edge length is longer than the preset maximum edge length, and the edge deletion operation is adopted when the edge length is shorter than the preset minimum edge length. The edge swap operation is used when the element aspect ratio is too high. If the size of the fixed background grid is h, it is found that a maximum edge length of h, a minimum edge length $0.3h$, and a maximum aspect ratio of 1.5 work well for most cases. With the proper front element restructuring operations, the front elements will maintain reasonable size and good shape, which results in better computing accuracy and efficiency.

3.3. Construction of the Phase Indicator Function and Fluid Property Distribution

Since the Navier-Stokes equations are solved on the fixed background grid while the surface tension and fluid property discontinuity are carried with the moving front, it is necessary to distribute the physical quantities (e.g. surface force, fluid density gradient) that exist on the moving front to the fixed grid. Mathematically, the distribution process from the interface front to the volume grid can be represented by a δ-function. Hence, the distribution operation corresponds to the construction of a smooth approximation to this δ-function on the fixed grid. This "smoothing" can done in several different ways, but it is always necessary to ensure that the transferred quantity is conserved. The interface quantity ϕ_s is usually expressed in units per unit area (or length in two-dimensions), but it should be given in terms of units per unit volume (ϕ_f) on the fixed grid. To ensure that the total value is conserved in the smoothing process, we must therefore require

$$\int_{\Delta S} \phi_s(s)ds = \int_{\Delta v} \phi_v(\mathbf{x})dv \text{ or } \int_{\Delta S} \phi_s(s)\delta(\mathbf{x}-\mathbf{x}')ds = \phi_v(\mathbf{x}) \tag{3.8}$$

By converting the above equation into a discretized form, we obtain the formulation for the three dimensional transfer

$$\phi_{v,\,ijk} = \sum_l \phi_{s,\,l} w_{ijk}^l \frac{\Delta s_l}{h^3}, \tag{3.9}$$

where $\phi_{v,\,ijk}$ is the value on the fixed grid with index (i,j,k), $\phi_{s,\,l}$ is the value on the front element l. Δs_l is the area of front element l and h^3 is the cell volume of the fixed grid. w_{ijk}^l is the interpolation weight factor between the front element l and the fixed grid (i,j,k), or a discretised form of the Dirac-delta function. To ensure the conservation of physical quantities, the weight factors must satisfy

$$\sum_{i,\,j,\,k} w_{ijk}^l / h^3 = 1 \tag{3.10}$$

Since the weight factors have a finite support, there is a relatively small number of front elements that contribute to the value at each point of the fixed grid. The number of grid points used in the transfer depends on the particular weighting function selected. In the actual implementation of the distribution of quantities from the front to the fixed grid, we usually loop over the interface elements and do the distribution to the grid points that are near the front element. Normally, this interpolation process has high efficiency.

Usually, the weighing function is expressed as a product of one-dimensional functions. In three-dimensions, the interpolation weight factor between the fixed grid point (i,j,k) and the front point $\mathbf{x}_f = (x_f, y_f, z_f)$ is written as

$$w_{ijk}(\mathbf{x}_f) = w_{ijk}^l / h = d(x_f - ih) \cdot d(y_f - jh) \cdot d(z_f - kh). \tag{3.11}$$

where h is the grid spacing. For two-dimensional interpolation, the third term is set to unity. There are several ways to define the one-dimensional weighting function $d(r)$, where r referrers to the distance between the points from the front and the fixed grid along the coordinate axis. The simplest interpolation is the area (volume) weighting, which is interpolated over two fixed grid points in each direction,

$$d(r) = \begin{cases} (h-|r|)/h^2, & |r| < h \\ 0, & |r| \geq h \end{cases}. \tag{3.12}$$

Peskin (1977) suggested the following interpolation scheme, which can be implemented over four grid points in each direction,

$$d(r) = \begin{cases} \dfrac{1}{4h}\left(1+\cos(\dfrac{\pi r}{2h})\right), & |r| < 2h \\ 0, & |r| \geq 2h \end{cases}. \tag{3.13}$$

In principal, it is desirable to have a grid approximation that is as compact as possible. However, a very narrow support, using only a few grid points close to the front, usually results in increased grid effects. It is found that the area weighting works well in most cases, while the function proposed by Peskin (1977) is obviously smoother.

One novelty of the front method described here is the way that the fluid property field (density and viscosity) is updated with the moving interface. Due to the discontinuity of fluid properties across the interface, we may expect either excessive numerical diffusion or problems with oscillation around the jump if they are convected using a traditional approach. To avoid these problems, the front tracking method introduces an additional computational element – the interface grid – that explicitly marks the position of the interface. Hence, an indicator function $I(\mathbf{x})$, that is one for the inner fluid and 0 for the outer fluid, is constructed from the known position of the interface. For incompressible flow, we assume that the fluid properties such as density and viscosity are constant within each fluid, the indicator function allows us to evaluate the distribution of fluid property variables at each grid point over the whole field by

$$\begin{aligned}\rho(\mathbf{x}) &= \rho_o + (\rho_i - \rho_o)\, I(\mathbf{x}) \\ \mu(\mathbf{x}) &= \mu_o + (\mu_i - \mu_o)\, I(\mathbf{x}) \end{aligned}, \tag{3.14}$$

where subscripts i and o represent the inner and outer fluid respectively, and

$$I(\mathbf{x}) = \begin{cases} 1, & \mathbf{x} \in \text{innerfluid} \\ 0, & \mathbf{x} \in \text{outerfluid} \end{cases}. \tag{3.15}$$

Actually, the indicator function can be expressed by an integral of a Dirac-delta function over the domain (Ω) enclosed by the interface (Γ),

$$I(\mathbf{x}) = \int_\Omega \delta(\mathbf{x}-\mathbf{x}')d\mathbf{x}'. \tag{3.16}$$

Taking the gradient of the indicator function and applying Stokes' theorem, we get

$$\mathbf{G}(\mathbf{x}) = \nabla I(\mathbf{x}) = \int_{\Gamma} \mathbf{n}\delta(\mathbf{x} - \mathbf{x}')ds, \qquad (3.17)$$

where $\mathbf{n}$ is the outward unit normal vector on the interface. And the discrete form of the above can be expressed as

$$\mathbf{G}(\mathbf{x}) = \sum_{l}\sum_{i,j,k} \mathbf{n}_l w_{ijk}^l \Delta s_l / h^3 = \sum_{l}\sum_{i,j,k} \mathbf{n}_l w_{ijk} \Delta s_l. \qquad (3.18)$$

Since the interface position is explicitly described by a set of front elements, it is quite straightforward to get the outward normal vector ($\mathbf{n}_l$), the area of front element (Δs_l), and the weighing factors between the front element and the fixed grid (w_{ijk}). Hence, the gradient field of the indicator function ($\mathbf{G}(x)$) can be constructed using (3.18).

The indicator function can be reconstructed from the gradient field. There are several ways to achieve this. The simplest approach is to integrate the indicator function from a point where the fluid property is known. However, this approach can produce an indicator function field that depends slightly upon the direction in which the integration is done, and the integration process also allows errors in the gradient field of the indicator function to propagate away from the front. In most implementations of the front tracking method, we take the numerical divergence of the gradient field, resulting in a numerical approximation to the Laplacian

$$\nabla^2 I(\mathbf{x}) = \nabla \cdot \mathbf{G}(\mathbf{x}). \qquad (3.19)$$

The above Poisson equation can be solved numerically on the fixed grid. The left hand side can be discretised using a standard central difference scheme for the Laplacian, and the right hand side, divergence of the gradient field, can be calculated using standard finite difference method. With appropriate boundary conditions, the above Poisson equation can be solved for the indicator function for the whole simulation domain. Even though it is a quite robust procedure to restructure the indicator function, there are still two types of possible errors. The indicator function away from the interface may not be exactly equal to what it should be (0 or 1), and small over- or undershoots are occasionally found near the interface. Small density variation in the field away from the interface may cause unphysical buoyancy flow. Undershoots may lead to negative density or viscosity, which may cause problems in the flow solver. To avoid such problems, we often solve the Poisson equation by iterating only on the points close to the interface, leaving the points away from the interface unchanged. In addition, the small unphysical over- and undershoots can be easily removed by simple filtering.

In summary, with the explicit interface position and proper front elements, the corresponding indicator function can be restructured by the procedures introduced in this section. Then, the fluid property distribution in the simulation domain can be estimated for the solution of the Navier-Stokes equation for the flow field.

3.4. Treatment of Surface Tension on the Interface

The accurate computation of the interface force is perhaps one of the most critical elements in the modeling of multiphase interfacial flows, in which the motion of the interface between two immiscible fluids should be tracked for a long time. In the front tracking method, the front is explicitly represented by a set of discrete points and elements, which makes the computation of surface tension force much more straight forward comparing to the approaches used by the VOF and level set methods. In most of our simulations, we calculate the surface tension force on the front element directly rather than from the interface curvature. This makes the computation simple and robust.

The surface force on a short segment of the front can be expressed as

$$F_e^{ST} = \int_\Gamma \sigma \kappa \mathbf{n}\, ds, \qquad (3.20)$$

where σ is the surface tension coefficient, κ is the curvature of the interface, $\mathbf{n}$ is the outward normal unit vector, and Γ is the front element.

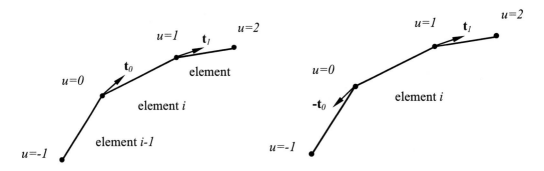

Figure 3.5. Computation of surface tension force on a 2D interface (a) tangents on end nodes; (b) tension on the front element.

For a two-dimensional interface, the curvature can be defined through

$$\kappa \mathbf{n} = \frac{\partial \mathbf{t}}{\partial s}. \qquad (3.21)$$

where $\mathbf{t}$ is the unit tangent vector on the interface. If the surface tension coefficient σ is constant, the surface tension force can be expressed using the tangents at the end points ($\mathbf{t}_0$, $\mathbf{t}_1$) as shown in Figure 3.5,

$$F_e^{ST} = \sigma \int_\Gamma \frac{\partial \mathbf{t}}{\partial s}\, ds = \sigma (\mathbf{t}_1 - \mathbf{t}_0). \qquad (3.22)$$

Hence, instead of having to find the curvature, we only need to find the tangents of the end points for the surface tension calculation. In addition to simplifying the computation, this ensures that the total force on any closed surface is zero. This is because the surface tension force on the end nodes shared by two adjacent front elements has exactly the same value but opposite direction for each front element. This conservation property not only complies with the physical definition of surface tension, but also is particularly important for long time computations where even a small error in the surface tension computation can lead to an unphysical net force on a front that can accumulate over time. The above formulation can also easily be extended to situations with variable surface tension. A simple expression of the partial derivative shows that

$$\frac{\partial \mathbf{t} \sigma}{\partial s} = \sigma \frac{\partial \mathbf{t}}{\partial s} + \mathbf{t} \frac{\partial \sigma}{\partial s} = \sigma \kappa \, \mathbf{n} + \frac{\partial \sigma}{\partial s} \mathbf{t}, \quad (3.23)$$

which is the usual expression accounting for both the normal and the tangential forces. The general form of surface tension force calculation can be rewritten as

$$F_e^{ST} = \int_\Gamma \frac{\partial \sigma \mathbf{t}}{\partial s} ds = (\sigma_1 \mathbf{t}_1 - \sigma_0 \mathbf{t}_0). \quad (3.24)$$

Therefore, the surface force on each element is computed by simply subtracting the product of the surface tension coefficient and the tangent at the end nodes of each front element for both constant and variable surface tension. The accuracy and efficiency of the computations depends on the calculation of the unit tangent vectors at the end nodes on the front.

In most two-dimensional computations, we compute the tangent directly from the Legendre polynomial fit through the end nodes of several adjacent front elements as shown in Figure 3.5 The tangent to the curve is given by

$$\mathbf{t} = \frac{\partial (f_x, f_y)}{\partial u} \Big/ \left| \frac{\partial (f_x, f_y)}{\partial u} \right|, \quad (3.25)$$

where (f_x, f_y) are the coordinates on the front using the Legendre polynomial fit (3.1). The derivatives at the end nodes $u = 0$ and $u = 1$ are then given by

$$\left. \frac{\partial f}{\partial u} \right|_{u=0} = \frac{1}{6} \left[-2f_{-1} - 3f_0 + 6f_1 - f_2 \right],$$

$$\left. \frac{\partial f}{\partial u} \right|_{u=1} = \frac{1}{6} \left[-f_{-1} - 6f_0 + 3f_1 + 2f_2 \right]. \quad (3.26)$$

For the calculation of surface tension on a three-dimensional interface, we can extend the concept developed for the two dimensional interface. The surface tension force on a surface element is

$$F_e^{ST} = \oint_\Gamma \sigma \mathbf{t} \times \mathbf{n} \, ds, \quad (3.27)$$

where **t** is the vector of the element edge and **n** is the unit outward normal vector on the interface.

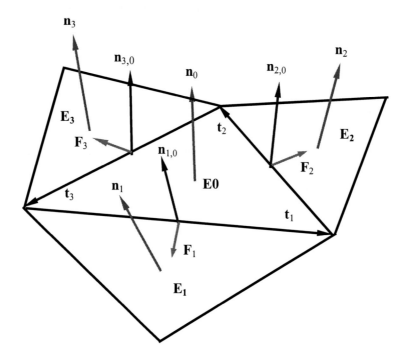

Figure 3.6. Computation of surface tension force on an interface in three dimensional space.

Figure 3.6 shows the surface forces (F_1, F_2 and F_3) exerted on a central surface element (E_0) by its neighboring elements (E_1, E_2 and E_3). The surface force acting on an edge shared between the central element and a neighboring element can be calculated by

$$\mathbf{F}_i = \sigma(\mathbf{t}_i \times \mathbf{n}_{i,0}). \tag{3.28}$$

where $\mathbf{t}_i$ is the vector of edge i pointing from one end-node to the other end-node in a counterclockwise direction, and $\mathbf{n}_{i,0}$ is the outward unit vector on the edge shared by the central element and the neighboring element i. The vector $\mathbf{n}_{i,0}$ can be estimated by averaging the unit element normal vectors of E_0 and E_i. This ensures that the surface tension forces exerting on the common edge shared by two elements are same in magnitude but opposite in direction, and that the net force on a closed surface is zero. Hence, the net surface tension force on the central element (E_0) can be expressed in discrete form as

$$\mathbf{F}_e^{ST} = \sum_{i=1}^{3} \mathbf{F}_i = \sum_{i=1}^{3} \sigma(\mathbf{t}_i \times \mathbf{n}_{i,0}). \tag{3.29}$$

It should be pointed out that there are several other ways to calculate the tangents on the interface (van Sint Annaland et al. 2005; Shin & Juric 2002; Sousa et al. 2004). We did not give a detail description here due to the limitation on the chapter length.

After the surface force is calculated on each front element, it is distributed to the fixed grid for the solution of the flow field,

$$F_v^{ST} = \sum_l F_e^{ST} w_{ijk}^l . \tag{3.30}$$

4. NUMERICAL SOLVERS AND INTERFACE ADVECTION

4.1. Numerical Solvers for the "One-Fluid" Formulation

With the methods introduced in the previous section, the interface position and shape can be tracked explicitly with reasonable resolution, and the distribution of fluid properties (density and viscosity) and surface tension on the fixed grid can also be calculated. Hence, the "one-fluid" formulation of the governing equations for multiphase interfacial flow (2.28 & 2.29) can be solved. In principle, this set of governing equations can be solved by most used numerical methods for the Navier-Stokes equations for single phase impressible flows. However, the special characteristics of multiphase interfacial flow also pose some special requirements on the flow solver. For example, the flow solver has to be robust in handling the large density and viscosity differences between the two fluid phases, and the strong local source terms in the momentum equation which may be caused by the surface tension or buoyancy. In this chapter, we present two popular numerical methods: the Projection Method and the SIMPLE algorithm, for multiphase interfacial flow simulations.

The Projection Method is also well known as the split method. Based on the flow field at the current time step n (Γ^n, $\mathbf{u}^n$, ρ^n, μ^n), the front position can be advected to the new position (Γ^{n+1}) for the next time step ($n+1$) with a time step size of Δt, and then the new distributions of fluid properties such as density and viscosity (ρ^{n+1}, μ^{n+1}) can be updated accordingly. Once the density is updated, the velocity field can be computed. In the projection method, the velocity update is split into two steps. The first is a prediction step where the effect of the pressure is ignored,

$$\frac{\rho^{n+1}\mathbf{u}^* - \rho^n\mathbf{u}^n}{\Delta t} = -\nabla \cdot \rho^h \mathbf{u}^n \mathbf{u}^n + \nabla \cdot [\mu^n(\nabla \mathbf{u}^n + \nabla^T \mathbf{u}^n)] + \mathbf{F}^{ST} + \rho^n \mathbf{g} \tag{4.1}$$

The second is a correction step, where the pressure gradient is added,

$$\frac{\rho^{n+1}\mathbf{u}^{n+1} - \rho^{n+1}\mathbf{u}^*}{\Delta t} = -\nabla p . \tag{4.2}$$

Due to the constraints of the incompressibility condition, the velocity at the new time step is divergence free,

$$\nabla \cdot \mathbf{u}^{n+1} = 0. \tag{4.3}$$

Hence, we can use Equation (4.3) to eliminate $\mathbf{u}^{n+1}$ from Equation (4.2), resulting in

$$\nabla \frac{1}{\rho^{n+1}} \cdot \nabla p = \frac{1}{\Delta t} \nabla \cdot \mathbf{u}^*. \tag{4.4}$$

Since the density is not constant within the computational domain, this equation cannot be solved by traditional fast Poisson solvers designed for separable elliptic equations. The solution of the pressure equation is usually the most time-consuming part of the computation and must therefore de done efficiently. Considerable progress is currently taking place in development of efficient methods for the solution of nonseparable elliptic equations. With the calculated pressure field, Equation (4.2) can be used to obtain the new flow field. The projection method presented here is the conventional first-order scheme, but it can easily be made second order.

To compute the momentum advection, the viscous force and the pressure term, any type of standard discretization schemes can be used. In most of our computations, a fixed, regular, staggered grid is used, and the momentum equation is discretised using a conservative, second-order centered difference scheme for spatial variables and an explicit first or second order method for time integration. In the original method, centered differencing was used for all spatial variables, simple averaging is applied to the points where the variables are not defined, and time integration is done by the simple explicit second-order projection method. When the numerical robustness is important such as at high Reynolds number, some authors (Tryggvason et al. 2001) have used QUICK scheme and other high order up-wind schemes. For the viscous terms, standard second-order centered differences can be used with simple averages for the viscosity at points where it is not defined. However, this is not the only possibility, and the geometric average can also be used. To achieve second order accuracy in time integration, an Adams-Bashford integration scheme can be used. Alternatively, a simple predictor-corrector scheme can be employed, where the first order solution at $n+1$ serves as a predictor that is then corrected using the trapezoidal rule. The latter method is particularly simple to implement since we can simply take two first-order Euler steps and then average the final solution and the one at the beginning of the time step.

Various advanced schemes with some improvements, modifications and higher order approaches have been developed for various purpose. For example, the projection method with the first-order scheme as presented in the previous paragraphs is not good at handling multiphase flow systems with large density difference, e.g. a typical air/water system. Rudman (1998) proposed a pressure-correction version of the projection method to solve the transient incompressible Navier-Stokes equations in fluid flow with large density variation. The basic elements of this scheme can be described as follows. In the projection step, the effect of the pressure is also considered,

$$\frac{\rho^{n+1}\mathbf{u}^* - \rho^n \mathbf{u}^n}{\Delta t} = -\nabla \cdot \rho^h \mathbf{u}^n \mathbf{u}^n - \nabla p^n + \nabla \cdot [\mu^n (\nabla \mathbf{u}^n + \nabla^T \mathbf{u}^n)] + \mathbf{F}^{ST} + \rho^n \mathbf{g}. \tag{4.5}$$

Then, the pressure correction term, δp, is calculated by the following equation,

$$\nabla \frac{1}{\rho^{n+1}} \cdot \nabla \delta p = \frac{1}{\Delta t} \nabla \cdot \mathbf{u}^* . \tag{4.6}$$

And finally, the velocity and pressure field can corrected as

$$\mathbf{u}^{n+1} = \mathbf{u}^* - \frac{\Delta t}{\rho^{n+1}} \nabla \delta p , \tag{4.7}$$

$$p^{n+1} = p^n + \nabla \delta p . \tag{4.8}$$

Although the above scheme is presented in its first order form, it is not difficult to implement the above algorithm in higher order, which generally produces better results.

In addition to explicit approaches, the implicit method should be a good alternative to solve the Navier-Stokes equations for multiphase flows with large density variation. Hua and Lou (2007) tried the Semi-Implicit Method for Pressure-Linked Equations (SIMPLE) algorithm. The results indicated that the SIMPLE algorithm is robust in solving the governing equations for the multiphase flow system with large density difference. The governing equations to be solved can be expressed as

$$\frac{\rho^{n+1}\mathbf{u}^{n+1} - \rho^n \mathbf{u}^n}{\Delta t} + \nabla \cdot \rho^{n+1}\mathbf{u}^{n+1}\mathbf{u}^{n+1} = -\nabla p^{n+1} + \nabla \cdot [\mu^{n+1}(\nabla \mathbf{u}^{n+1} + \nabla^T \mathbf{u}^{n+1})] \\ + \mathbf{F}_{ST}^{n+1} + \rho^{n+1} \mathbf{g} \tag{4.9}$$

$$\nabla \cdot \mathbf{u}^{n+1} = 0 . \tag{4.10}$$

Similar to the conventional SIMPLE approach, the above equation can be solved through iteration. In a multi-fluid system, due to the density jump across the interface, the mass flux conservation is not valid in control volumes crossing the front interface. Instead, the volume flux conservation is adopted here as a modification of the traditional SIMPLE algorithm. The divergence of the velocity field over the whole solution domain will be kept at zero, as long as both liquid and gas can be reasonably considered to be incompressible fluids. Based on this assumption, the SIMPLE algorithm is used to calculate the correction values of the pressure and velocity after solving the momentum equation. In discretized form, the momentum equation can be expressed as,

$$a_p \mathbf{u}_p^* = \sum a_{n,p} \mathbf{u}_{n,p}^* + S_p - B \nabla P_p^* . \tag{4.11}$$

in which n indicates the neighboring points surrounding the centre point p, and the coefficient $a_{n,p}$ involves the flow properties of convection and diffusion and geometrical properties of the control volume. S_p refers to the source term and B the coefficient for the pressure gradient term. Details about these coefficients can be obtained in the work of Patankar (1980).

An improved pressure field (p_p^{**}) and velocity field ($\mathbf{u}_p^{**}$) can be obtained by adding the correction terms (p_p', $\mathbf{u}_p'$) to the values used in Equation (4.11) based on calculation of $\mathbf{u}_p^*$ or assumption of p_p^*,

$$\mathbf{u}_p^{**} = \mathbf{u}_p^* + \mathbf{u}_p' \; ; \; p_p^{**} = p_p^* + p_p' \; . \tag{4.12}$$

Substituting the above equations into Equation (4.11), we have a relationship between the correction velocity ($\mathbf{u}_p'$) and the correction pressure (p_p') as follows,

$$\mathbf{u}_p' = \sum a_{a,p} \mathbf{u}_{n,p}' / a_p - B/a_p \cdot \nabla p_p' \; . \tag{4.13}$$

Applying incompressible fluid condition to the improved velocity field ($\nabla \cdot \mathbf{u}_p^{**} = 0$), the velocity correction ($\mathbf{u}_p'$) should satisfy the following condition,

$$\nabla \cdot \mathbf{u}' = -\nabla \cdot \mathbf{u}^* \; . \tag{4.14}$$

By taking divergence of both sides of Equation (4.13), ignoring the first term (with high order) on the right hand side of Equation (4.13) and substituting it into Eq. (4.14), the pressure correction can be obtained by solving the following equation,

$$\nabla \cdot \left[(B/a_p) \nabla p' \right] = \nabla \cdot \mathbf{u}^* \; . \tag{4.15}$$

Based on the pressure correction, the velocity correction can also be derived using Equation (4.13). The updated velocity and pressure are then used as the guessed field for the next iteration for solving the momentum Equation (4.12). Such iterations will be repeated until both the momentum and continuity equations have converged. Compared with the projection method, the SIMPLE algorithm avoids directly solving the pressure equation, which enhances the numerical robustness when density/viscosity jumps across the two-fluid interface are large. Most of the simulation results presented in the later sections of this chapter are actually based on the SIMPLE algorithm.

4.3. Interfacial Velocity Interpolation and Advection of the Front

As discussed in Section 2, one of the conditions on the interface is that it should move with the same velocity as the surrounding liquid when there is no mass transfer across the interface. After the velocity field is calculated on the fixed background grid, the velocity of the front needs to be interpolated from the velocity at the fixed grid. In Section 3, the distribution function used to spread the fluid property jump to the fixed grids near the interface was discussed. This function can similarly be used to interpolate field variables from the background grid to the end nodes of the front elements using the following equation,

$$\mathbf{u}_f = \sum_{i,j,k} w_{ijk} \mathbf{u}. \qquad (4.16)$$

where the summation is over the points on the fixed grid that are close to the front points. It is generally desirable that the interpolated front velocity should be bounded by the fixed grid values, and the front value is the same as the grid value if a front point coincides with a fixed grid point.

Once the velocity of each front point has been found, the new front position can been found easily by integration, A simple first-order explicit Euler integration gives

$$\mathbf{x}_f^{n+1} = \mathbf{x}_f^n + \mathbf{u}_f^n \cdot \Delta t. \qquad (4.17)$$

When the Navier-Stokes equations are solved by a higher-order time integration method, the same scheme can also be used to advect the front points.

After the front point is advected to the new position, the mesh size and quality of the front may have deteriorated due to the deformation of the front. For example, some element points may have moved closer resulting in very small elements, and some elements may have become larger. The element resolution on the interface has a strong effect on the information exchange between the front and the fixed background grids that in turn affects the accuracy of the simulation results. Hence, front mesh adaptation or restructuring as described in Section 2 has to be performed to maintain the front element mesh quality.

While the momentum equation is usually solved in conservative form, the explicit advection of the front is not conservative. In addition, the conservation of bubble volume may be sacrificed due to mesh refinement and coarsening. Therefore, both accurate advection of the front points and higher order calculation of the points inserted on the interface for mesh refinement and coarsening are essential to enhance the mass conservation of each fluid phase. In some cases, particularly for very long runs with many bubbles or drops where the resolution of each particle is relatively low, we may encounter unacceptably high changes in the mass of the particles. In these cases, we may correct the particle sizes every few time steps to conserver the mass. The front position is corrected by the following equation:

$$\mathbf{x}_f = \mathbf{x}_f^* + (V - V^*)/S^* \cdot \mathbf{n}^*. \qquad (4.17)$$

where $\mathbf{x}_f$, $\mathbf{x}_f^*$ are the final corrected front position and the current front position respectively. V is the theoretical/ initial bubble volume. V^*, S^* and n^* are the bubble volume, surface area and outward unit normal vector before the correction, respectively.

4.4. Change of Interface Topology

In general, numerical simulation of multiphase flow must account for changes of the topology of the interface boundary when bubbles or drops coalesce and breakup. In the front tracking method, the interface is explicitly tracked by connected points, and topology changes can be accounted by modifying the front node connections in a appropriate way. The complexity of this operation is often cited as the greatest disadvantage of the front tracking method. In some other methods (e.g. VOF or Level Set method), automatic topology change (coalescence or breakup) can be very convenient, particularly if the topology change does not need to be treated accurately. This is also a serious weakness of such methods. Physically, coalescence is usually strongly dependent on how quickly the fluid between the coalescing parts drains. Simply connecting the close parts of the interface without taking into account the coalescence physics may give an incorrect solution.

Visually, the interface topology changes whenever two interfaces or different parts of the same interface come closer than one grid spacing. In physics, topology changes in multiphase flows can be divided into two broad classes (Tryggvason et al. 2001): (1) film rupture. If a large drop approaches another drop or a flat surface, the fluid between must be "squeezed" out before the drop are sufficiently close so that the film become unstable to attractive forces and rupture; (2) thread break. A long and thin cylinder of one fluid will generally break up by interface instability where one part of the cylinder become sufficiently thin so that surface tension "pinches" it into two. The exact mechanism of how fluid threads snap and break is still under active research and investigation. There are good reasons to believe that threads can become infinitely thin in a finite time and that their breaking is "almost" described by the Navier-Stokes equation. Films, on the other hand, are generally believe to rupture as a result of short-range attractive forces, once they are a few hundred angstroms thick. These forces are usually not included in the continuum description. In order to account for draining of films prior to rupture, resolution of very small length scales is required. It seems unlikely to be practical in most cases and limited by the current available technology and known knowledge. Therefore, it is still quite a challenge to model the detailed physics related to interfacial topology changes.

In the current front tracking code, topological change is accomplished by a two-step process. First, the part of the interface that should undergo topological change must be identified, and then the actual change of the front markers must be done. In general, film rupture occurs when two parts of the interface approach each other, and thread breakup takes place when the thread is further stretched. Hence, it is necessary to search the whole front to locate where two fronts or two parts of the same front are approaching each other to form a film or a thread where topological change may occur. The simplest, but least efficient, way to conduct this search is to compute distance between the centroids of all front elements. This is an $O(N^2)$ operation, and can be very time consuming when the interface consists of a large number of front elements. Alternatively, by dividing the computational domain into small sub-regions and sorting the front elements according to location, this operation can be made reasonably efficient. This does, however, increase the complexity of the code. Once the close elements have been identified, topological changes can be made by removing or reconnecting the relevant front points. The linked-list data-structure for representing the front brings great convenience for implementing these operations in a computer code.

5 VALIDATION OF FRONT TRACKING METHOD

The "one-fluid" formulation for multiphase interfacial flows is a completely rigorous reformulation of the Navier-Stokes equations with special accounting for the jump conditions over the interface, which includes large variation of fluid properties as well as the surface tension force. The front tracking method provides a comprehensive numerical approach to track the position and shape of the interface, to calculate the fluid property distribution and the surface tension force, and to solve the governing equations. The accuracy of this numerical scheme should be established so that it can potentially be used as a numerical tool to investigate the complex phenomena and physics in a much wider range of multiphase interfacial flows. To demonstrate the accuracy of one specific implementation of the front tracking method, we simulated the rising of a single bubble in viscous liquid and the pinching off of a gas bubble from a nozzle immersed in viscous liquid, where detailed experimental observations and measurements are available. The reasonable agreement between the simulations and experiments demonstrates the accuracy of the current implementation of the front tracking method.

5.1. Single Bubble Rising in Viscous Liquid

A sound understanding of the physics of a rising bubble in viscous liquid is essential in a variety of practical applications ranging from steam bubbles in boiler tubes to gas bubbles in oil wells. It has been investigated in a wide range of experiments (Clift et al. 1978; Bhaga & Weber 1981), theoretical analysis (Davies & Taylor 1950; Moore 1959) and numerical simulations (Chen et al. 1999; van Sint Annaland et al. 2005, 2006). To some extent, bubble rising behavior in certain flow regimes has been well investigated and understood. Hence, a single bubble rising in viscous liquid due to buoyancy is an important benchmark to validate numerical methods for multiphase interfacial flows, such as in the work of Raymond and Rosant (2000), Chen et al. (1999), Ohta et al. (2005). However, most of the previous simulations have been performed under conditions of moderate density difference, and the simulations results have been validated against experimental observations of bubble shape only under some typical flow regimes. There are very few quantitative comparisons on the bubble terminal velocity as in the work of Koebe et al. (2002). In this section, front tracking methods for both 2D axisymmetric and 3D models are used to systematically investigate a single air bubble rising in a quiescent liquid solutions under a wide range of flow regimes. The simulation results are compared with experimental observations in terms of terminal bubble shape, terminal rising velocity and wake flow pattern.

For a bubble rising in a quiescent viscous liquid, it is reasonable to treat both liquid and gas phases as incompressible fluids. We scale Equations (2.28) and (2.29) by introducing dimensionless characteristic variables as follows:

$$\mathbf{x}^* = \frac{\mathbf{x}}{D}, \; \mathbf{u}^* = \frac{\mathbf{u}}{\sqrt{gD}}, \; \tau^* = \sqrt{\frac{g}{D}}t, \; \rho^* = \frac{\rho}{\rho_l}, \; p^* = \frac{p}{\rho_l gD}, \; \mu^* = \frac{\mu}{\mu_l}, \; \kappa^* = D\kappa, \; \mathbf{g}^* = \frac{\mathbf{g}}{g},$$

where D is the diameter of a sphere with the same volume as the bubble, and $g = |\mathbf{g}|$. Thus we may express the dimensionless equations as,

$$\nabla \cdot \mathbf{u}^* = 0 \tag{5.1}$$

$$\frac{\partial(\rho^* \mathbf{u}^*)}{\partial \tau^*} + \nabla \cdot \rho^* \mathbf{u}^* \mathbf{u}^* = -\nabla p^* + \frac{1}{Ar} \nabla \cdot [\mu^*(\nabla \mathbf{u}^* + (\nabla \mathbf{u}^*)^T)]$$
$$+ \frac{1}{Bo} \int_\Gamma \kappa^*_f \mathbf{n}_f \, \delta(\mathbf{x}^* - \mathbf{x}^*_f) ds + \rho^* \mathbf{g}^* \tag{5.2}$$

The non-dimensional Archimedes and Bond numbers (also known as Eotvos number) used here are defined as

$$Ar = \frac{\rho_l g^{1/2} D^{3/2}}{\mu_l} \text{ and } Bo = \frac{\rho_l g D^2}{\sigma}.$$

The Archimedes number was also used in previous work (Bonometti & Magnaudet 2006) to characterize the rise of a bubble in liquid due to buoyancy, reflecting the ratio of buoyancy to viscous forces. By studying the non-dimensional formulation, it can be noticed that the flow can be entirely characterized by the following four dimensionless parameters: the density and viscosity ratios of the fluids, the Archimedes number and the Bond number. In experimental work it is common to use a different set of dimensionless numbers (Bhaga & Weber 1981) – the most important one being the bubble Reynolds number defined as $\text{Re} = \rho_l D U_\infty / \mu_l$, where U_∞ is the experimentally measured terminal velocity of the rising bubble. Another dimensionless number one may encounter in this setting is the Froude number, $Fr = U_\infty/(gD)^{1/2}$. We thus have the following relationship between the Archimedes number of our formulation and the Reynolds number used in experiments: $\text{Re} = Ar \cdot Fr$. Finally, we mention two other dimensionless numbers frequently used by experimentalists, namely the Eotvos number, which is exactly the same as the above-defined Bond number, and the Morton number $M = g\mu_l^4/(\rho_l \sigma^3)$.

The terminal shape of a single rising bubble under a range of Reynolds and Bond numbers was observed and reported in the experimental work by Bhaga and Weber (1981) as shown in Figure 51. Generally, small bubbles, which have low Reynolds or Bond number ($\text{Re} < 1$ or $Bo < 1$), rise in a steady fashion and maintain a spherical shape. The shape of larger bubbles, with intermediate Reynolds and Bond numbers ($1 < \text{Re} < 100$ and $1 < Bo < 100$), are affected significantly by the flow conditions. Various bubble shapes (oblate ellipsoid, disk-like, oblate ellipsoidal cap, skirt bubble, and spherical-cap) have been found in various flow regimes. In spite of the difference in shapes, the bubbles rise steadily in the liquid along a straight path. With further increase of the Reynolds number ($100 < \text{Re} < 500$), the bubble shape may become toroidal in the high Bond number ($100 < Bo < 500$) regime; spherical-cap in intermediate Bond number regime ($30 < Bo < 100$) and oblate ellipsoid in the low Bond number regime ($1 < Bo < 30$). As the bubble size increases further, a turbulent wake develops behind the bubble that leads to unsteady bubble motion. The bubble may rise in a wobbly path, oscillate about a mean shape and even break up or coalesce. In general, the rising bubbles have axisymmetric shapes when the Reynolds and Bond numbers are not too higher (Re < 200, Bo < 200) within the range indicated by the red dot lines in Figure 5.1.

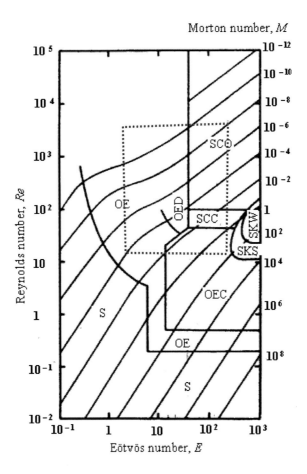

Figure 5.1. The regime map of experimentally observed shape fro a bubble rising in a liquid: S, spherical; OE, oblate ellipsoid; OED, oblate ellipsoidal (disk-like and wobbling); OEC, oblate ellipsoidal cap; SCC, Spherical cap with closed, steady wake; SCO, spherical cap with open, unsteady wake; SKS, skirted cap with smooth, steady skirt; SKW, skirted cap with wavy, unsteady skirt. Reprint from Bhaga & Weber (1981) with permission.

Based on the experimental observations, a single bubble rising in quiescent liquid may be simulated using the front tracking method using an axisymmetric co-ordinate system within certain flow regimes that will produce axisymmetric bubbles. The solution domain size in the radial direction should be large enough so that the boundary effects on bubble rising can be ignored, and the bubble can be assumed to be rising in an infinite quiescent liquid, which matches experimental conditions. The simulations reported by Hua & Lou (2007) clearly show that the boundary effect is negligible when the radial size of the solution domain is about four times the bubble diameter, which is used in this study. The solution domain size in the axial direction is about twelve times the bubble diameter. The ideal spherical bubble is located initially on the symmetry axis at two bubble diameters above the bottom. Both liquid and bubble are assumed to be stationary in the initial state, and the rising of the bubble in the simulation domain due to buoyancy is to be simulated.

Test Case	Experiments		Simulations	
	Test condition	Observed bubble terminal shape	Predicted bubble Terminal Shape	Modelling conditions
A1	E=8.67 M=711 Re=0.078	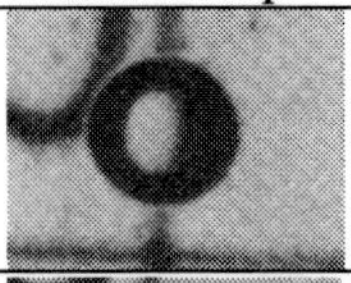		Bo = 8.67 Ar = 0.979 Fr = 0.069
A2	E=17.7 M=711 Re=0.232	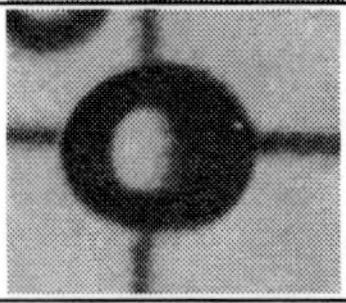	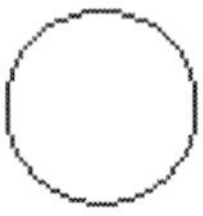	Bo = 17.7 Ar = 1.671 Fr = 0.116
A3	E=32.2 M=8.2×10^{-4} Re=55.3	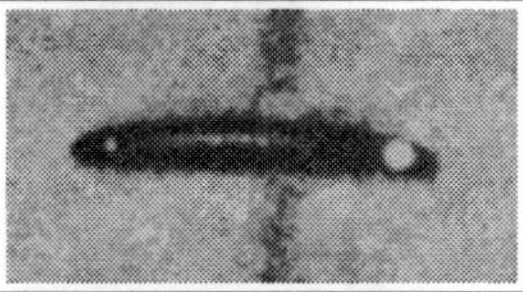		Bo = 32.2 Re = 79.88 Fr = 0.663
A4	E=243 M=266 Re=7.77	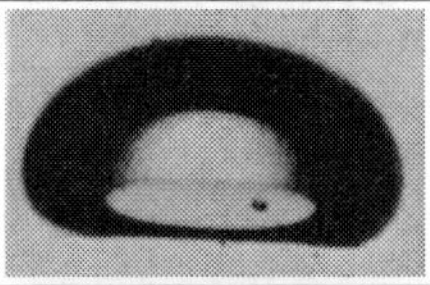		Bo = 243 Ar = 15.24 Fr = 0.551
A5	E=115 M=4.63×10^{-3} Re=94.0	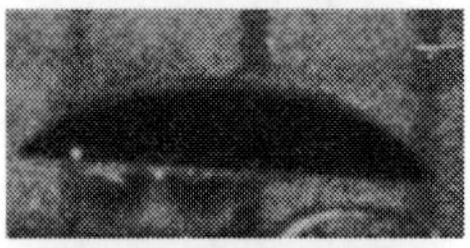		Bo = 115 Ar = 134.6 Fr = 0.659
A6	E=237 M=8.2×10^{-4} Re=259	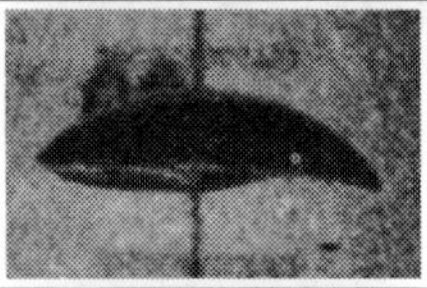		Bo = 237 Ar = 357 Fr = Unstable
A7	E=339 M=43.1 Re=18.3	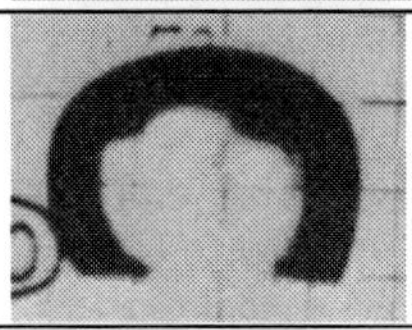		Bo = 339 Ar = 30.83 Fr = 0.581
A8	E=641 M=43.1 Re=30.3			Bo = 641 Re = 49.72 Fr = 6.02×E-1

Figure 5.2. Comparison of terminal bubble shapes observed in experiments (Bhaga & Weber 1981) and predicted in simulations under different conditions (A1-A8) of various Reynolds, Morton and Bond numbers. Reprint from Hua and Lou (2007) with permission.

Test Case	Experiments		Simulations	
	Test conditions	Observed terminal bubble shapes	Predicted terminal bubble shapes	Modelling conditions
B1	E=116 M=848 Re=2.47	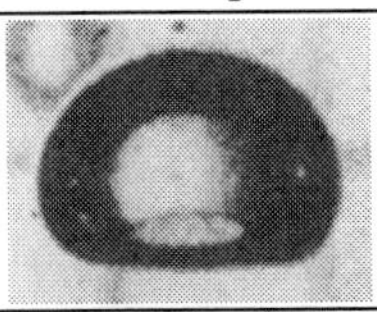		Bo = 116 Ar = 6.546 Fr = 0.354
B2	E=116 M=266 Re=3.57	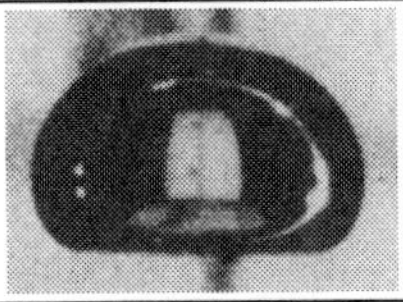		Bo = 116 Ar = 8.748 Fr = 0.414
B3	E=116 M=41.1 Re=7.16	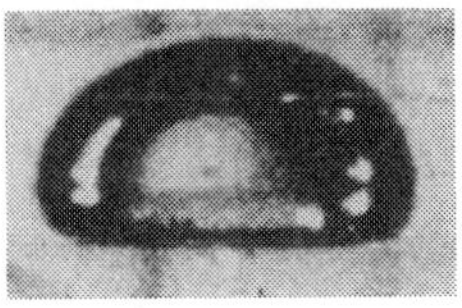		Bo = 116 Ar = 13.95 Fr = 0.502
B4	E=116 M=5.51 Re=13.3	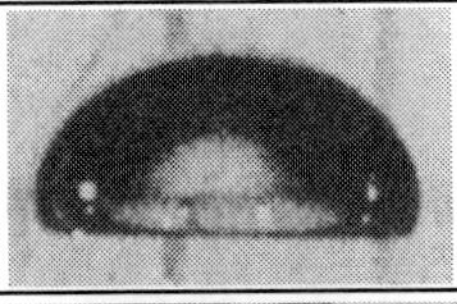		Bo = 116 Ar = 23.06 Fr = 0.571
B5	E=116 M=1.31 Re=20.4	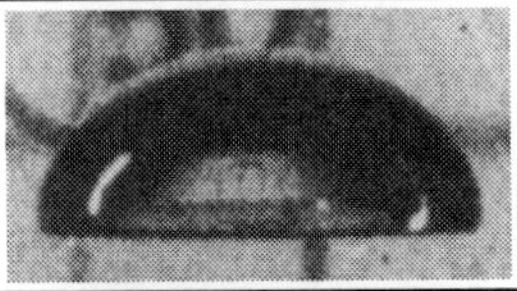		Bo = 116 Ar = 33.02 Fr = 0.602
B6	E=116 M=0.103 Re=42.2	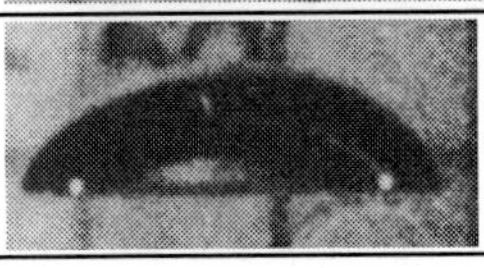		Bo =116 Ar =62.36 Fr = 0.634
B7	E=116 M=4.63×10^{-3} Re=94.0			Bo = 116 Ar = 135.4 Fr = 0.660
B8	E=116 M=8.60×10^{-4} Re=151	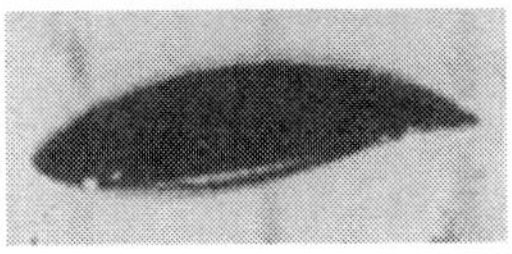		Bo = 116 Ar = 206.3 Fr = Unstable

Figure 5.3. Comparison of terminal bubble shapes observed in experiments (Bhaga & Weber 1981) and predicted in simulations under different conditions (B1-B8) of various Reynolds and Bond numbers. Reprint from Hua and Lou (2007) with permission.

Test Case	Experiments		Simulations	
	Test conditions	Observed terminal bubble shapes	Predicted terminal bubble shapes	Modelling conditions
W1	E=96.2 M=0.962 Re=18.2	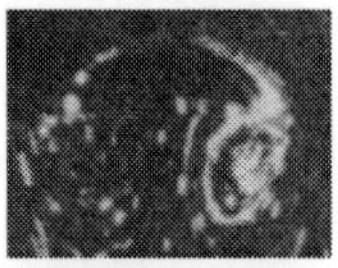	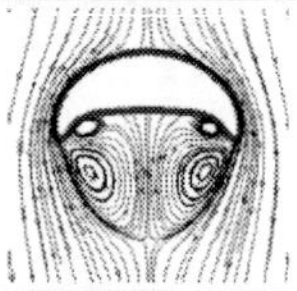	Bo=96.2 Ar=31.0
W2	E=116 M=0.962 Re=22.0	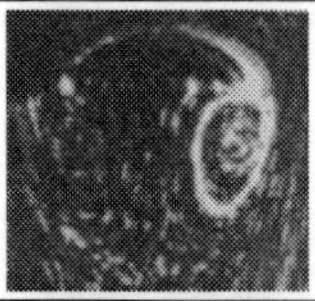	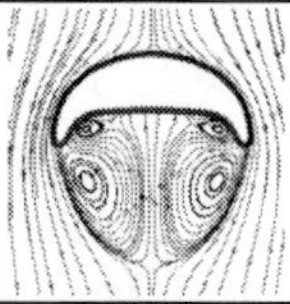	Bo=116 Ar=35.69
W3	E=95.6 M=1.44×10^{-1} Re=32.4	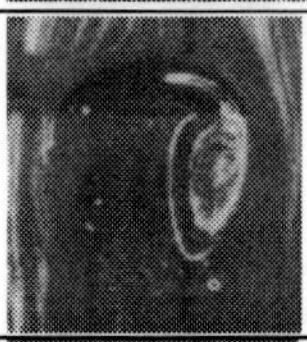	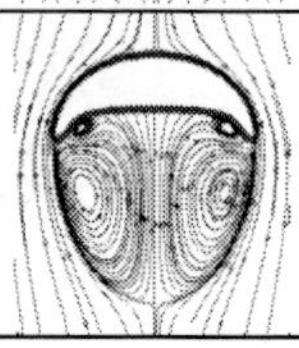	Bo=95.6 Ar=49.63
W4	E=151 M=1.44×10^{-1} Re=37.4		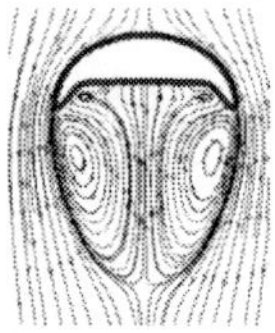	Bo=151 Ar=69.92
W5	E=94.3 M=4.85×10^{-3} Re=77.9	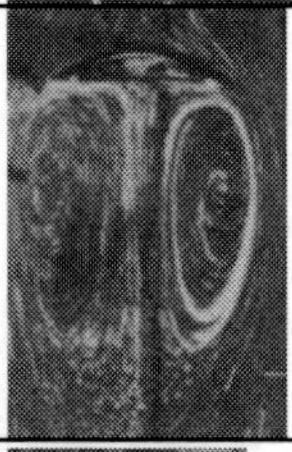	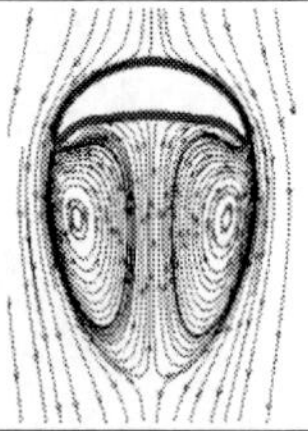	Bo=94.3 Ar=116.3
W6	E=114 M=4.85×10^{-3} Re=91.6	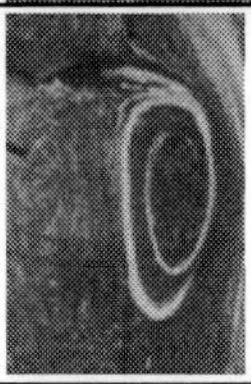	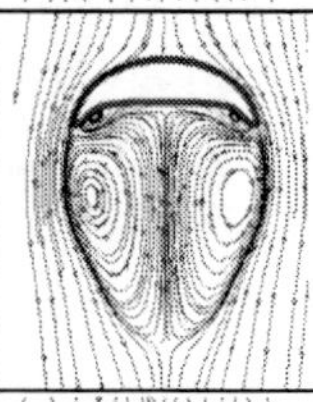	Bo=114 Ar=134.11
W8	E=292 M=26.7 Re=22.1	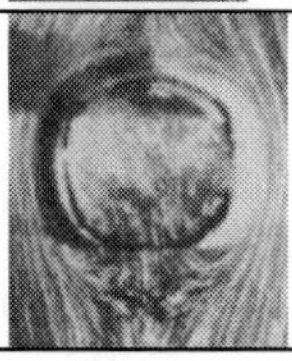	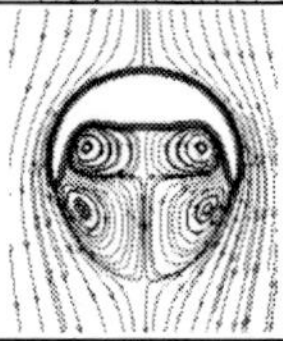	Bo=292 Ar=31.07

Figure 5.4. Comparison of terminal bubble wake observed in experiments (Bhaga & Weber 1981) and predicted in simulations under different conditions (W1-W8) of various Reynolds, Morton and Bond numbers. Reprint from Hua and Lou (2007) with permission.

A uniform fine background mesh was adopted to solve the Navier-Stokes equation. The simulation accuracy is also affected by the background mesh size. The effects of mesh size on the prediction of bubble rising velocity and terminal bubble shape were analyzed by Hua and

Lou (2007). The simulations give stable predictions when the bubble diameter is meshed with more than twenty-five grid points.

Bhaga & Weber (1981) conducted extensive experiments to investigate the behavior of a single bubble rising in viscous liquid. They examined the bubble shape, terminal velocity and wake flow pattern under a wide range of flow regimes. This provides us with many test cases to validate the accuracy of the simulation results; a generic simulation tool should be able to make reasonable predictions for different test cases under different flow regimes. Figures 5.2 and 5.3 compare the terminal bubble shapes obtained from experiments and simulations under a number of experimental test conditions, and Figure 5.4 compares the bubble wake flow patterns.

Table 5.1. Comparison of bubble terminal velocity (in terms of Reynolds number) observed in experiments (Bhaga & Weber 1981) and predicted in simulations (Hua and Lou 2007)

Test Cases	Experiment Re	Simulation			$\frac{|Re - Re,s|}{Re} \times 100$
		Ar	Fr	$Re, s = Ar \cdot Fr^*$	
A1	0.078	0.979	0.072	0.0705	9.63
A2	0.232	1.671	0.126	0.211	9.05
A3	55.300	79.880	0.663	52.960	4.23
A4	7.770	15.240	0.551	8.397	8.07
A5	94.000	134.6	0.659	88.701	5.63
A7	18.300	30.830	0.581	17.912	2.12
A8	30.300	49.72	0.602	29.931	1.22
B1	2.470	6.546	0.354	2.317	6.18
B2	3.570	8.748	0.414	3.621	1.45
B3	7.160	13.95	0.502	7.002	2.19
B4	13.300	23.06	0.571	13.167	0.99
B5	20.400	33.02	0.602	19.878	2.56
B6	42.200	62.36	0.634	39.536	6.31
B7	94.000	135.4	0.66	89.364	4.93

In general, the simulated terminal bubble shapes agree well with experimental observations (Bhaga & Weber 1981) for most of the cases shown in Figures 5.2 and 5.3 under different flow regimes as shown in Figure 5.1. However, the predicted bubble shapes for cases A6 and B8 are totally different from those observed in experiment (Bhaga & Weber 1981). In the experiment a spherical cap bubble with an open wake was observed, while a toroidal bubble was predicted in the simulation. This difference is believed to be caused by the assumption of an initially spherical bubble shape in the simulation as discussed in Hua and Lou (2007).

The existence of a closed toroidal wake has been observed in experiments by Bhaga and Weber (1981) through flow visualization using H_2 tracers. The wake flow circulation patterns predicted by the simulation agree well with the observations in experiments as shown in Figure 5.4. The wake circulation within the bubble base indentation is clearly shown in Case W3 by the photograph, where the trace track disappears behind the rim of the bubble. Similar wake circulation pattern was revealed in the simulation. In fact, the simulation show that a secondary wake circulation occurs just behind the bubble rim. This may be the reason why the bubble images for Cases W1, W2 and W3 show bright spots at the lower outside of the bubble rim. This secondary circulation in the skirted bubble wake becomes much more obvious in the simulation case W8. The image of the skirt bubble wake W8 shows a large bright spot just underneath the bubble. As the bubble size increases (with the Reynolds number increasing), the wake volume increases as well, and the wake seems to be torn away from the bubble itself. In this case, the secondary wake circulation disappears, and the bubble base indentation becomes smaller.

Test Case	Experiments		Simulations	
	Test Conditions	Observed terminal bubble shape	Predicted terminal bubble shape	Modeling conditions
A1	E = 17.7 M = 711 Re = 0.232	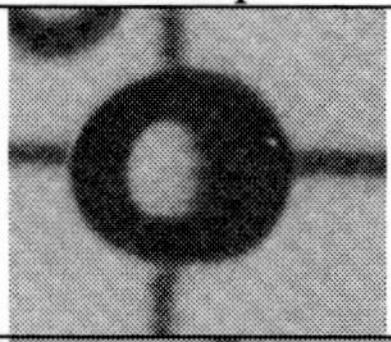	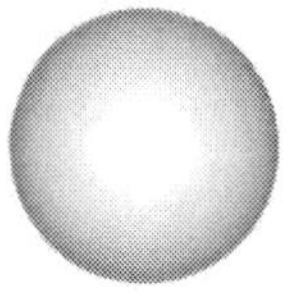	Bo = 17.7 Ar = 1.671 Fr = 0.109
A2	E = 32.2 M = 8.2·10^{-4} Re = 55.3	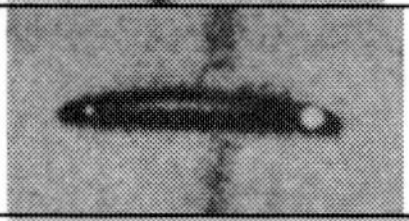	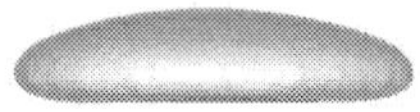	Bo = 32.2 Ar = 79.88 Fr = 0.686
A3	E = 243 M = 266 Re = 7.77		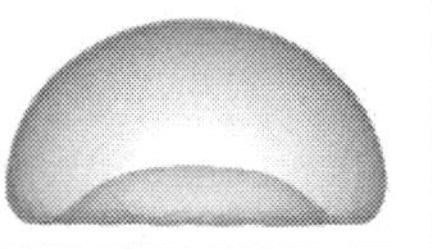	Bo = 243 Ar = 15.24 Fr = 0.499
A4	E = 115 M = 4.63·10^{-3} Re = 94.0		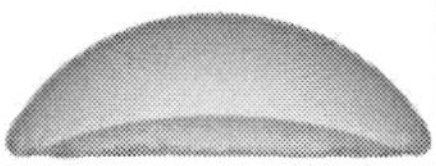	Bo = 115 Ar = 134.6 Fr = 0.666
A5	E = 339 M = 43.1 Re = 18.3	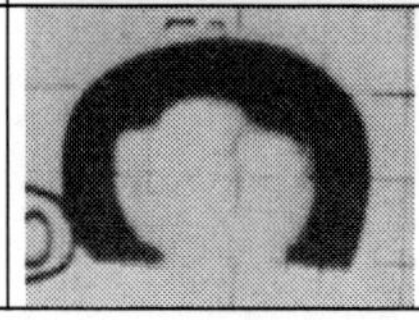	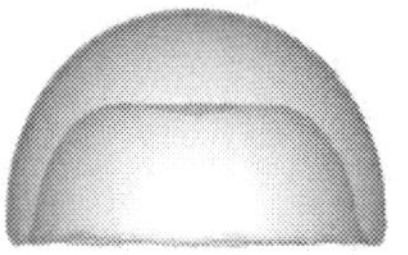	Bo = 339 Ar = 30.83 Fr = 0.576

Figure 5.5. Comparison of terminal bubble shapes observed in experiments (Bhaga & Weber 1981) and predicted by numerical simulations. Reprint from Hua et al. (2008c) with permission.

The terminal bubble velocity is another important indicator to quantitatively evaluate the difference between the experimental data and simulation results. Based on the parameters given in the experiments, the characteristic parameters (Ar, Bo, ρ_l/ρ_b, μ_l/μ_b) are derived for the simulations, which the predicted terminal bubble rise velocity can be expressed as a function of Froude number ($Fr = U_\infty/(gD)^{1/2}$). Hence, the predicted terminal bubble velocity U_∞ can be calculated as $Fr \cdot gD^{1/2}$, and the Reynolds number (Re, s) predicted in the simulation can be calculated as $Re, s = Ar \cdot Fr$. The comparison of the Reynolds numbers for the experiment and simulation cases is shown in Table 5.1. The results from simulation predictions agree with those of experiments very well within 10% difference.

We also applied the front tacking method for three dimensional simulations of single bubble rising in viscous liquid, and compared the results with experiments (Bhaga & Weber 1981). In Figure 5.5, we compare the observed and predicted terminal bubble shapes for a range of Reynolds and Bond numbers, and the results agree very well. Table 5.2 shows a comparison of the associated terminal rise velocities, and again there is reasonable agreement between experiments and our numerical predictions in most cases. However, it is noted that the relative deviation in Case A1 is a little bit high as the numerically predicted Reynolds number is about 20% lower than that observed in experiments. A possible reason is that the hindrance effect due to the limited size of the domain becomes more significant at low Reynolds numbers (Hua et al. 2008c).

Table 5.2. Comparison of terminal rise velocities found in experiment (Bhaga & Weber 1981) and predicted by the three-dimensional numerical simulations (Hua et al. 2008c)

Test Cases	Experiment Re	Simulation Ar	Fr	$Re_s = Fr \cdot Ar$	Deviation $|Re - Re_s|/Re \cdot 100$
A1	0.232	1.671	0.109	0.182	21.49
A2	55.3	79.88	0.686	54.798	0.91
A3	7.77	15.24	0.499	7.605	2.13
A4	94.0	134.6	0.666	89.644	4.63
A5	18.3	30.83	0.576	17.758	2.96

5.2. Bubble Pinch-off in Viscous Liquid

Numerical predictions of the terminal shape, speed and wake flow pattern of a single rising bubble in viscous liquid using the front tracking method have been compared in detail with corresponding experiments for model validation purpose. However, this validation exercise for the terminal steady state is not enough to prove the prediction accuracy in the time domain. The pinch-off of a gas bubble from a tiny vertical nozzle immersed in quiescent viscous liquid due to buoyancy has been investigated for decades. New technology - high speed and high resolution cameras -- makes it possible to record the details of the bubble pinch-off process, which include the bubble growth and pinch off in both domains of time and space (Thoroddsen et al. 2007). It provides another ideal case to validate the predictions of the current front tracking method.

Extensive experimental studies by Thoroddsen et al. (2007) reported the dynamics and the shape of the neck region while an air bubble pinches off slowly in various glycerine/water mixtures driven by buoyancy, using an ultrahigh-speed video camera with up to one million frames per second. The physical problem of bubble pinch-off from a capillary nozzle submerged in a viscous liquid due to buoyancy is sketched in Figure 5.6. The capillary nozzle has a radius of $r_o = 1.35 \times 10^{-3}$ m. The simulation domain for the experiment is indicated by the gray zone shown in Figure 5.6, which has a radius of $3r_o$ and a height of $12r_o$. The densities of the bubble phase and the outer liquid are ρ_g and ρ_l, respectively, and the viscosities of the two phases are μ_g and μ_l, respectively. The density ratio is defined as $\eta = \rho_g/\rho_l$, and the viscosity ratio as $\lambda = \mu_g/\mu_l$. The two fluids are assumed to be immiscible and incompressible, and the problem is assumed to be axisymmetric as the bubble grows at a very slow rate. The initial flow fields for both liquid and gas phase are assumed to be stationary. The initial position of the gas-liquid interface is assumed to be flat at the capillary nozzle outlet. A constant pressure condition is applied to the outer boundary of the computational domain. As the flow is driven by buoyancy, the gas flow rate through the nozzle is kept low to minimize any inertial effects.

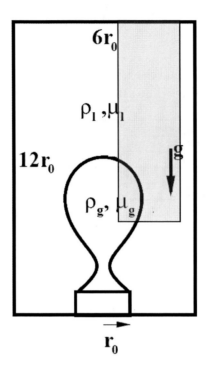

Figure 5.6. Schematic diagram of a bubble pinch-off from a capillary tube ($r_o = 1.35 \times 10^{-3}$ m) submerged in a viscous liquid due to buoyancy. Reprint from Quan and Hua (2008) with permission.

The governing equations for the fluid flow system are similar to those discussed in Section 5.1. We scale the equations by introducing the following dimensionless characteristic variables,

$$\mathbf{x}^* = \frac{\mathbf{x}}{r_o}, \ \mathbf{u}^* = \frac{\mathbf{u}}{\sqrt{gr_o}}, \ t^* = \sqrt{\frac{g}{r_o}}t, \ \rho^* = \frac{\rho}{\rho_l}, \ p^* = \frac{p}{\rho_l g r_o}, \ \mu^* = \frac{\mu}{\mu_l}, \ \kappa^* = r_o\kappa, \ \mathbf{g}^* = \frac{\mathbf{g}}{|g|}$$

where r_o is the radius of the capillary nozzle which is used as the characteristic length. The Archimedes number (Ar) and Bond number (Bo) in the governing equations are thus re-defined as,

$$Ar = \frac{\rho_l g^{1/2} r_o^{3/2}}{\mu_l} \text{ and } Bo = \frac{\rho_l g r_o^2}{\sigma}.$$

In the work of Quan and Hua (2008), the bubble pinch-off was predicted by solving the full Navier-Stokes equations for both the bubble and the liquid phases. Figure 5.7(a) shows the air bubble pinch-off process experimentally observed in water by Thoroddsen et al. (2007). The bubble shape variation during the pinch-off process predicted by the current simulation is shown in Figure 5.7(b), with a direct comparison with the experiments. The simulation and experiment were performed under the same conditions. It can be concluded the agreement between the numerical predictions and experimental observations is quite good in both spatial and time domains.

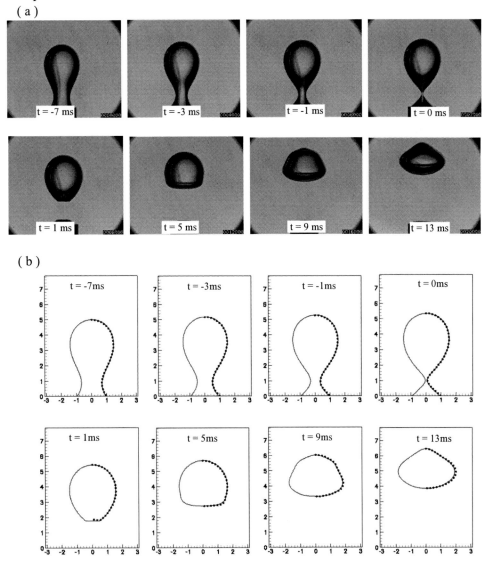

Figure 5.7. (a) Snapshots of the experimentally observed air bubble pinch-off process in water; (b) Comparison of the bubble shapes predicted by simulations (dotted line) and observed in experiments (solid line). Both experiment and simulation were performed under the conditions: $r_o = 1.35 \times 10^{-3}$ m ; $\rho_g = 1.005$ kg/m^3 ; $\rho_l = 1.0 \times 10^3$ kg/m^3 ; $\mu_g = 0.0142$ cP and $\mu_l = 1.48$ cP ; $\sigma = 6.5 \times 10^{-2}$ N/m and $g = 9.8$ m/s^2. Reprint from Quan and Hua (2008) with permission.

In order to test the accuracy and robustness of the numerical method, numerical predictions of bubble pinch-off were validated against the experiments over a range of liquid viscosities, rather than just for one specific condition. A number of simulations have been conducted in this study using the exact experimental conditions as reported by Thoroddsen et al. (2007) to investigate the effects of liquid viscosity on bubble pinch-off behavior. Figure

5.8 shows the comparison of the variation of bubble shapes before pinch-off predicted by simulations (dotted line) and observed in experiments (dashed line) under conditions of different mixture compositions, (a) 75% glycerin and 25% water ($\mu_l = 1.4\,\text{cP}$); (b) 84% glycerin and 16% water ($\mu_l = 68\,\text{cP}$); and (c) 99% glycerin and 1% water ($\mu_l = 3400\,\text{cP}$). It can be seen from Figure 5.8 that the simulated bubble shapes agree well with the experimental observations across a large range of fluid viscosity. This comparison further validates the accuracy of our numerical method. The numerical method has also been applied to investigate the effects of the viscosity, the surface tension coefficient, and the gas density on the bubble pinch-off dynamics as reported in Quan and Hua (2008).

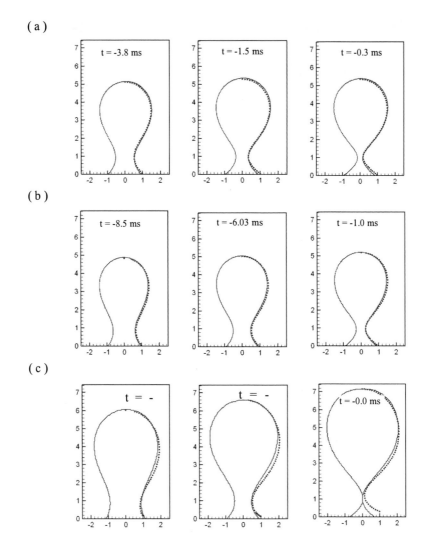

Figure 5.8. Comparison of the temporal variations of bubble shapes before bubble pinch-off predicted by simulations (dotted line) and observed in experiments (solid line) under the conditions of different glycerin/ water compositions, (a) 75% glycerin and 25% water ($\mu_l = 26\,\text{cP}$); (b) 84% glycerin and 16% water ($\mu_l = 68\,\text{cP}$); and (c) 99% glycerin and 1%water ($\mu_l = 3400\,\text{cP}$). Reprint from Quan and Hua (2008) with permission.

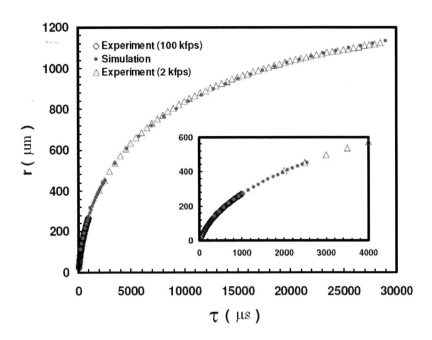

Figure 5.9. Detailed comparison of the necking radius between the experiments and the numerical simulation. $\sigma = 6.5 \times 10^{-2}$ N/m, $\rho_l = 1200 \, kg/m^3$, $\rho_g = 1.0 \, kg/m^3$, $\mu_g = 0.01 cP$, $\mu_l = 68 cP$ and $r_o = 1.35 \times 10^{-3}$ m. Reprint from Quan and Hua (2008) with permission.

In addition, a detailed comparison between the experimental results and the numerical simulations on the necking history of the neck radius prior to pinch-off is shown in Figure 5.9, with an inset showing a zoomed-in view for the period when the pinch-off process occurs during $0 \leq \tau \leq 4000 \mu s$. Here, τ is defied as $(t_{pich \, off} - t)$, the time prior to pinch off. The parameters for the case are $\sigma = 6.5 \times 10^{-2}$ N/m, $\rho_l = 1200 kg/m^3$, $\rho_g = 1.0 kg/m^3$, $\mu_g = 0.01 cP, \mu_l = 68 cP$. In experiments, the bubble pinch-off process was recorded twice using different video capturing speeds, 2,000 fames per second (fps) for the overall process and 100,000 frames per second (fps) for the detailed process at the pinch-off point. As shown in Figure 5.9, the agreement for the neck radius greater than 110 μm is excellent. This further underscores the good accuracy of the numerical method.

6. APPLICATION TO TWO-PHASE FLOW PROBLEMS WITH DROPLETS AND BUBBLES

6.1. A Single Air Bubble Rising in Water at High Reynolds Number

Two phase flow systems such as air bubbles rising in water may be observed in many industrial processes. For the design of an efficient two-phase flow system, detailed knowledge of bubble sizes and shapes, slip velocities, internal circulation, swarm behavior, bubble induced turbulence and mixing, and bubble size distribution (including coalescence and breakup) is of fundamental importance. In such industrial applications, bubbles often have non-spherical and even dynamic shapes as well as asymmetric wake structures. Extensive experimental work has been carried out to study the behavior of an air bubble rising in water (Clift et al. 1978; Tomiyama et al. 2002). Their measurement of the terminal velocity of air bubbles rising in water as a function of bubble size are presented in Figure 6.1. It is found that the measurements of the bubble terminal velocity vary significantly (or bifurcate) when the bubble size is greater than 0.5 mm and smaller than 10 mm. Traditionally this variation has been explained by the presence of surfactants (Clift et al. 1978), but more recently both Wu and Gharib (2002) and Tomiyama et al. (2002) attributed this variation to the manner in which the bubbles were generated. The phenomena continue to be a matter of discussion, as reported by Yang and Prosperetti (2003).

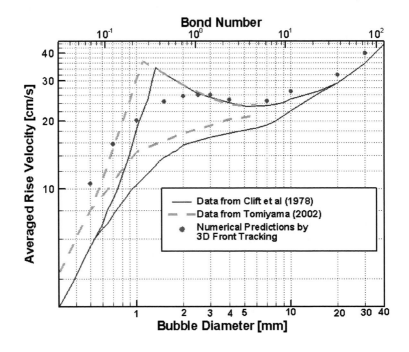

Figure 6.1. Comparison of the numerically predicted and experimentally measured terminal rise velocities of a single air bubble rising in water as a function of the bubble diameter. Reprint from Hua et al. (2008c) with permission.

Hua et al. (2008c) used the front tracking method with mesh adaptation and a moving reference frame, allowing a finer mesh in the region of the bubble surface. Consequently, better accuracy was obtained in the simulations. We simulated a single air bubble rising in initially quiescent pure water with bubble diameters ranging from 0.5 mm to 30 mm. The numerically predicted bubble rising velocities agree well with the upper bound of the experimental measurements by Tomiyama et al. (2002) within the whole range of different bubble sizes as shown in Figure 6.1. The variation of terminal shape of the rising bubble with a wide range of bubble size from 0.5mm ~ 30 mm is shown in Figure 6.2, which has reasonable agreement with experimental observations (Clift et al. 1978). When the bubble diameter is in the range from 2 mm to 10 mm, oscillation of the bubble rise velocity and the bubble shape is also predicted in the simulations. The terminal bubble rise velocity is calculated through averaging the instantaneous rise velocity over a period of time. Since the initial bubble shape was assumed to be spherical and the surface tension coefficient to be constant, the bifurcation of the bubble rise velocity found experimentally is not observed in the numerical predictions. This is as expected because the two probable causes of this bifurcation are not present in our numerical simulations: (i) There are no surfactants present in the water; and (ii) We do not simulate the bubble generation process, but rather assume the existence of a perfectly spherical bubble at the start of our simulations, eliminating any initial shape disturbances and their effects. However, including such effects numerically with the aim of reproducing the bifurcation in the rise velocities would be a most interesting research topic.

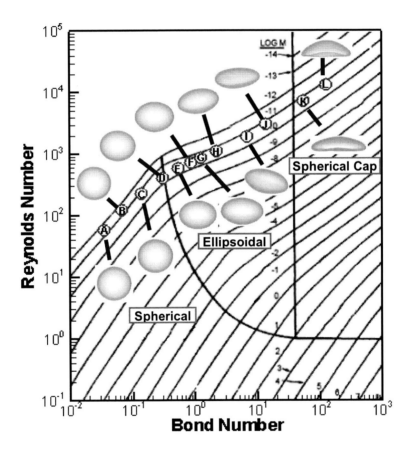

Figure 6.2. Numerically predicted terminal shapes of air bubbles rising in water for bubble diameters ranging from 0.5 ~ 30mm. The points A to L correspond to the bubble size of 0.5, 0.7, 1.0, 1.5, 2.0, 2.5, 3.0, 4.0, 7.0, 10.0, 20.0 and 30.0mm, respectively. Reprint from Hua et al. (2008c) with permission.

Many experiments (Wu & Gharib 2002; Magnaudet & Eames 2000) have demonstrated that millimetre-sized bubbles rising in low viscosity liquids do not generally follow a straight trajectory. In pure water, the transition from a straight rise path to zigzag occurs when the equivalent diameter of the bubble exceeds 1.8 mm. In this regime, the bubbles exhibit approximately oblate spheroidal shapes, and they either rise in zigzag path within a vertical plane or spiral around a vertical axis. In Hua et al. (2008b), we applied the front tracking method to simulate the rise behaviour of a single, initially stationary and spherical bubble in a quiescent viscous liquid. Both zigzag and spiral rise patterns were revealed using the current simulation algorithm.

Figure 6.3 shows that the numerically predicted rise path of the bubble (bubble mass centre) is a zigzag when the simulation parameters are set to be $Bo = 20.0$, $Ar = 2000$, $\rho_l / \rho_b = 1000$ and $\mu_l / \mu_b = 100$. A three-dimensional view of the trajectory of the bubble mass center is shown in Figure 6.3(a). It clearly indicates that the bubble starts to oscillate after it rises vertically in some distance and reaches certain velocity. When the velocity of the bubble becomes high enough, the bubble starts following a zigzag path. A projection of the trajectory into the xy-plane is shown in Figure 6.3(b), revealing that the zigzag path lies almost entirely in a single vertical plane.

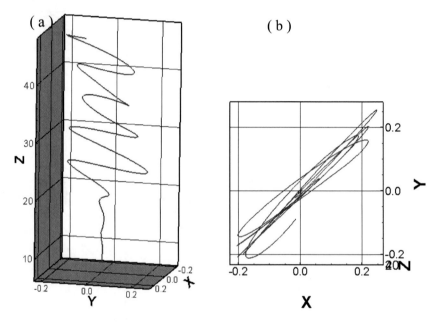

Figure 6.3. The trajectory of the mass centre of a bubble rising in a zigzag path, (a) three dimensional view and (b) XY plane view.

The mechanisms that determine the bubble rising behaviour are closely related to the wake structure created by the rising bubble (Mougin & Magnaudet 2006). A careful study of the wake structure of a rising bubble is illustrated in Figure 6.4, which shows the flow stream line paths around the bubble and a pressure contour in the bubble wake. When the bubble rising speed is low, a bubble wake with symmetric, closed recirculation rings is formed, and a low pressure zone is generated at the recirculation centre. As the rise velocity increases, the flow instability is amplified and the bubble wake starts to detach from one side of the bubble bottom. Due to the asymmetric wake structure, the drag and buoyancy forces acting on the bubble will also become unbalanced, and the bubble is tilted by the lift forces as shown in Figure 6.4(a). As the bubble speed increases further, the bubble wake becomes more asymmetric and the bubble tilting becomes more pronounced. As a result, the recirculation ring of the bubble wake is broken on one side as shown in Figure 6.4(b), and the other side of the recirculation ring attaches itself to one side of the oblate bubble. Consequently, two open recirculation rings attached to one side of the bubble are finally formed, resulting in tilting and lateral movement of the bubble. The lateral movement makes the open recirculation rings in the bubble wake switch from one side of the bubble to the other side alternatively. The oscillation of the two open wake recirculation rings causes the bubble to rise in a zigzag path.

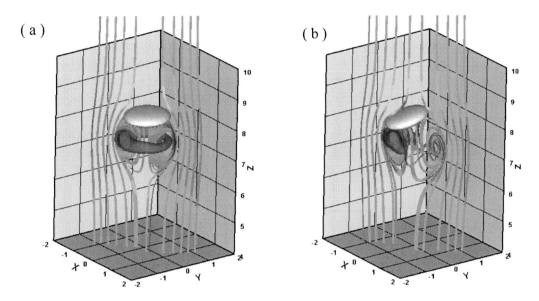

Figure 6.4. Variations of the bubble wake structure for a bubble rising in a zigzag path.

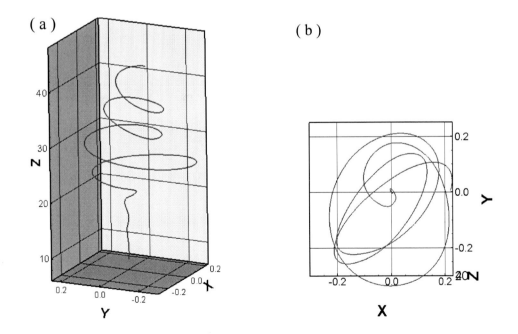

Figure 6.5. The trajectory of the mass centre of a bubble rising in a spiral path, (a) three-dimensional view and (b) XY plane view.

Figure 6.5 shows a spiral pattern for the bubble rise path as predicted by simulations when the parameters are set to be $Bo = 20.0$, $Ar = 1500$, $\rho_l/\rho_b = 1000$ and $\mu_l/\mu_b = 100$. A three dimensional view of the trajectory of the mass center of the bubble is shown in Figure 6.5(a). It clearly indicates that the bubble rises in a spiral with increasing radius until an almost constant radius is obtained. The top view of the bubble rise trajectory shown in Figure 6.5(b) indeed reveals an almost constant spiral radius as the bubble rises. A careful study of the bubble wake structure is shown in Figure 6.6. It is found that the wake structure for a spiraling bubble is totally different from the one observed for a zigzagging bubble. For the spiraling bubble, only one strong, open recirculation ring is attached to one side of the bubble and tilts the bubble. The point of attachment being the lowest point of the bubble, which is due to the high deformability of the spiraling bubble with higher Bond number. The point of attachment of the open recirculation ring moves along the bottom side edge of the oblate bubble and causes the bubble to follow a spiral trajectory.

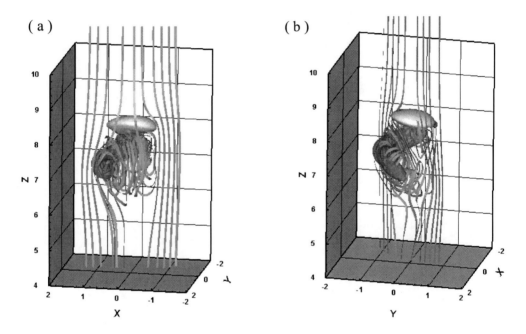

Figure 6.6. Variations of the bubble wake structure for a bubble rising in a spiral path.

6.2. Multiple Bubble Interaction and Bubbly Flow

The problem of a single bubble rising in a viscous liquid is an ideal case for numerical model validation. However, the final goal for developing a numerical model for multiphase flow is investigating not only the flow behavior of single rising bubble in viscous liquids, but also multi-fluid systems with multiple bubbles. With the confidence gained from validating the current model for a single bubble rising in a viscous liquid, we now extend the model to explore the complex interaction between two bubbles rising in a liquid. In the present study, the bubble coalescence criterion is simplified, and only geometrical constraints are considered. This means that bubble coalescence occurs when two points on the bubble surface are close enough, e.g. one fifth of the size of the bubble surface mesh. The topology change associated with the coalescence is implemented from a geometrical point of view as discussed in Section 4.4. A similar modeling strategy can also been found in the work by Shin and Juric (2002).

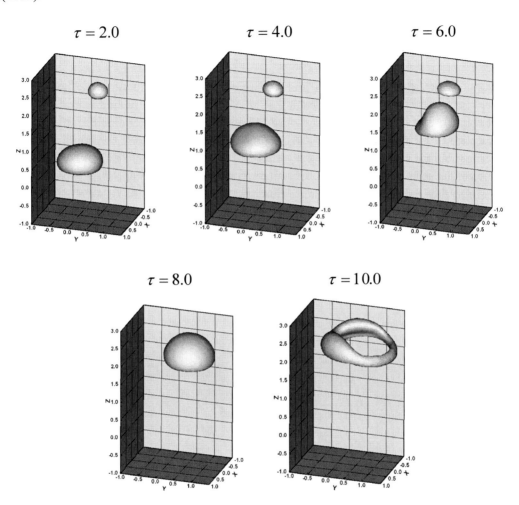

Figure 6.7. Numerical simulation of the interaction of two rising bubbles of different size. The flow conditions are: $Ar = 134.6$, $Bo = 115.0$ for the larger bubble and $Ar = 47.6$, $Bo = 28.75$ for the smaller bubble while $\rho_l/\rho_b = 1181$, $\mu_l/\mu_b = 5000$ for both bubbles. The initial diameter of the larger bubble is D=1.0, twice that of the smaller bubble. The smaller bubble is initially located 2.5D higher in vertical direction and 0.5D axis-off in the Y direction relative to the larger bubble. Reprint from Hua et al. (2008b) with permission.

Figure 6.7 illustrates the simulation of the interaction of two initially spherical bubbles rising in a quiescent liquid due to buoyancy. The smaller bubble is initially located 2.5D above the bigger bubble in vertical Z-direction, and 0.5D axis-off from the bigger bubble in the horizontal Y-direction. Here, D represents the effective diameter of the bigger bubble. The diameter of the smaller bubble is half that of the bigger bubble. The flow conditions for the bigger bubble are as follows: $Ar = 134.6$, $Bo = 115.0$, $\rho_l/\rho_b = 1181$ and $\mu_l/\mu_b = 5000$. As the bigger bubble has a higher rise velocity, it will catch up with the smaller bubble (Figure

6.7 ($\tau = 4.0$)). When they are close enough, the trailing bigger bubble is significantly affected by the low-pressure zone in the wake of the leading smaller bubble. The trailing bubble undergoes large deformations and moves towards the bottom wake zone of the leading bubble (Figure 6.7 ($\tau = 6.0$)). Finally, the trailing big bubble merges with the leading smaller bubble, and a toroidal bubble ring is formed (Figure 6.7 ($\tau = 10.0$)). Similar bubble shape evolution patterns have been predicted in other numerical studies (van Sint Annaland et al. 2005). Figure 6.8 shows the temporal variation of the bubble positions in both vertical and horizontal directions. It can be seen from Figure 6.8(a) that the trailing bigger bubble has a higher rise speed than the smaller leading bubble. The interesting finding is that, when the two bubbles are close enough, the rise speed of both bubbles increases significantly. After the coalescence of the two bubbles, the resulting merged bubble returns to the familiar situation of a single bubble rising. The lateral movement of the trailing bubble caused by the leading bubble can be seen in Figure 6.8(b). Even though initially the leading bubble moves slightly away from the trailing bubble laterally, this distance is quite small. However, the trailing bubble, despite its larger size, is significantly affected by the leading bubble and moves towards it.

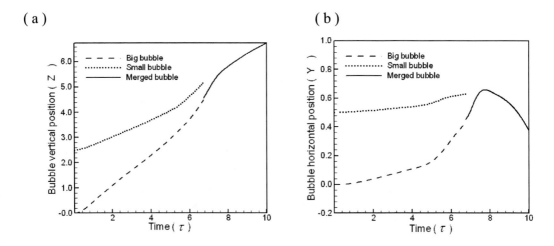

Figure 6.8. Simulation predicted (a) vertical and (b) lateral movement in Y direction of two rising bubbles and the merged bubble under the same conditions as those in Figure 6.7. Reprint from Hua et al. (2008b) with permission.

Although the simulations described above suggest that we are gaining considerable understanding of a single bubble rising behavior and bubble interaction mechanism, in many practical situations always there are many bubbles in the multiphase flow system with considerably high Reynolds number. It is well known that bubbles at high enough Reynolds number rise unsteadily, either wobbling as they rise or rising along a spiral path. Researchers are currently examining the buoyancy driven motion of many bubbles at high Reynolds numbers, so that moderately deformable bubbles show wobbly motion. In Figure 6.9, Tryggvason et al. (2006) show one frame from a simulation of 14 wobbly bubbles at about 6% void fraction, using a 128^3 grid. The Eotvos number is 4.0 and the Morton number is about 10^{-6}, giving a rise Reynolds number of 75. Initially, the bubble are placed randomly in a fully periodic domain but as they rise, they interact freely. The shape and position of bubbles as well as the fluid velocity in a plane cutting through the middle of the domain in a relatively late time are shown in Figure 6.9. While the bubbles occasionally clump together, on the average they remain essentially uniformly distributed. Examining the properties of a homogenous distribution of wobbly bubbles is relevant for realistic systems. Wobbly bubbles generally rise considerably more slowly than their non-wobbly counterpart and velocity fluctuations are generally much larger. It is noted that bubble deformation promotes unsteady rise, and that air bubbles in water do not start to wobble until the rise Reynolds number becomes relatively high.

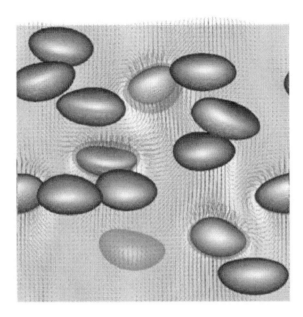

Figure 6.9. A simulation of 14 wobbly bubbles in a fully periodic domain. The average rise Reynolds number is about 75 and the domain is resolved by a 128^3 grid. The shape and position of bubbles as well as the velocity (short line segments) in a plane cutting through the middle of the domain are presented. Reprint from Tryggvason et al. (2006) with permission.

Studies of homogenous bubbly flows can yield considerable insight into many aspects of the interactions between freely rising bubbles, while there are obviously many aspects of realistic flows that depend critically on the presence of walls. Figure 6.10 shows a simulation of 16 bubbles in the so-called "minimum turbulent channel" of Jimenez and Moin (1991). The dimensions of the channel are $\pi \times \pi/2 \times 2$ units in the streamwise, spanwise, and wall normal direction, respectively. The channel is bounded by walls at the top and bottom and has periodic spanwise and streamwise boundaries. The wall Reynolds number is $\mathrm{Re}^+ = \rho u^+ h/\mu = 135$. Here, h is the half-height of the channel and $u^+ = \sqrt{\tau_w/\rho}$ is the friction velocity based on the average wall shear τ_w without bubbles. The computation were done using a grid of $256 \times 128 \times 192$ points, uniformly spaced in the streamwise and the spanwise direction but unevenly spaced in the wall normal direction. The initial velocity field was taken from spectral simulations of turbulent channel flows to avoid having to simulate the transition, and the volume flux was kept constant by adjusting the pressure gradient. The turbulence was first evolved without bubbles to ensure that the finite volume method could correctly simulate the single phase flow. The results show that slightly deformable bubbles can lead to significant reduction of wall drag by sliding over streamwise vortices and forcing them toward the wall where they are cancelled by the wall-bounded vorticity of the opposite sign. Spherical bubbles, on the other hand, often reach into the viscous sublayer where they are slowed down and lead to a large increase in drag. For more details about the simulation work, please refer to Lu et al. (2005). For other work on bubbles in a turbulent channel, the work of Kanai and Miyata (2001) and Kawamura and Kodama (2002) can be referred.

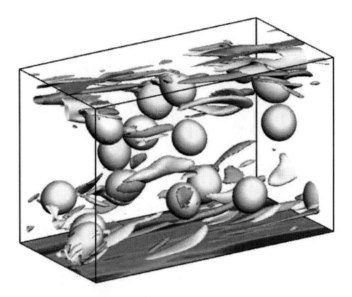

Figure 6.10. A simulation of 16 bubbles in the "minimum turbulent channel." The flow is from left to right. In addition to the bubbles, iso-contour of streamwise vorticity are shown. The vorticity froms elongated structures near the walls. Dark and light shades indicate opposite signs of vorticity. Reprint from Tryggvason et al. (2006) with permission.

6.3. Breakup of a Liquid Jet

A liquid jet is formed when liquid is injected into another immiscible liquid through a nozzle. The breakup of the liquid jet into droplets is of fundamental importance in many industrial processes. The jet breakup into droplets increases the interfacial area, enhancing heat and/or mass transfer, and sometimes chemical reactions. Thus simulation of jet formation and breakup has been used extensively to predict the jet length and droplet size. The front tracking method has also been applied to investigate the formation mechanism of liquid droplets or gas bubbles through the jet breakup mechanism.

Homma et al. (2006) investigated the breakup of an axisymmetric liquid jet, injected vertically upward from a nozzle into another quiescent immiscible liquid, into droplets using a front tracking method. Three jet breakup modes, dripping, jetting with uniform droplets and jetting with non-uniform droplets, were predicted through the simulations as shown in Figure 6.11. Under certain conditions, the jet can become quite long before it breaks up.

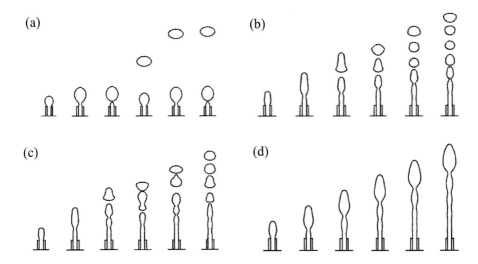

Figure 6.11. Different liquid jet breakup modes predicted by simulation (a) dripping ($Re = 377$, $We = 1.28$, $Fr = 8.18$ and $\mu_c/\mu_j = 6.67$); (b) jetting with uniform droplet ($Re = 402$, $We = 2.11$, $Fr = 27.5$ and $\mu_c/\mu_j = 1.00$); (c) jetting with uniform droplet ($Re = 228$, $We = 3.24$, $Fr = 52.9$ and $\mu_c/\mu_j = 1.00$) (d) jetting without breakup ($Re = 425$, $We = 1.70$, $Fr = 27.5$ and $\mu_c/\mu_j = 100$). Reprint from Homma et al. (2006) with permission.

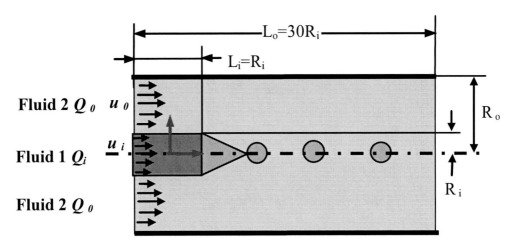

Figure 6.12. Schematic diagram of droplet formation from a capillary nozzle in a co-flowing viscous liquid (not to scale). Reprint from Hua et al. (2007) with permission.

Cramer et al (2004) studied experimentally drop formation from a liquid jet in a co-flowing ambient immiscible fluid. A typical co-flowing system of two viscous immiscible liquids is illustrated in Figure 6.12. The dispersed liquid (Fluid 1) is injected into the system through a capillary nozzle into the co-flowing liquid (Fluid 2) in a coaxial cylindrical tube. Both the dispersed phase and the continuous phase are injected into the system continuously at constant flow rates Q_i and Q_o, respectively. As the two fluid phases are immiscible with low flow rate, an axis-symmetric interface is formed between two the liquid streams. The liquid jet develops waves and eventually breaks up into droplets.

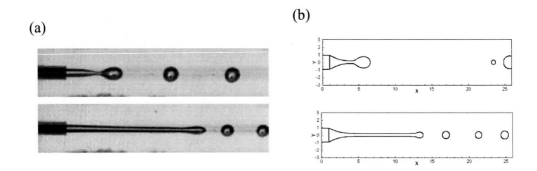

Figure 6.13. The dripping and jetting modes of droplet formation in the co-flowing liquid system, (a) observed in the experiments by Cramer et al (2004) and (b) predicted in the simulation.

Experiments (Cramer et al. 2004) revealed that there were two modes of droplet formation, as shown in Figure 6.13, depending upon the flow speed ratio of the continuous phase to the disperse phase and fluid properties such as viscosity. Numerical simulations (Hua et al. 2007) were performed to investigate the droplet formation mechanism in a micro-channel with a similar set up. The droplet formation modes, i.e. dripping and jetting, are well predicted in the simulations as shown in Figure 6.13, which agrees with experimental observation.

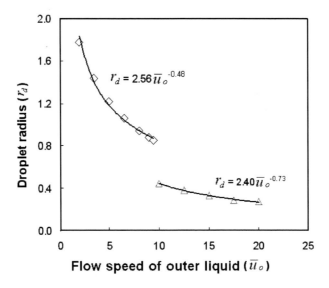

Figure 6.14. Effect of the flow speed of the continuous liquid phase on the droplet size. Reprint from Hua et al. (2007) with permission.

The effect of the continuous phase liquid flow speed on the size of the droplets formed in the micro-channel by the co-flowing liquids is shown in Figure 6.14. The simulation results indicate that the droplet generation mode changes from dripping to jetting when the flow velocity of the continuous phase is increased. Under the same droplet generation mode, the droplet size varies smoothly with the continuous change of the outer liquid flow speed, with power law correlations. However, the droplet size shows a sudden change when the transition of droplet generation mode occurs within a very small region of the outer liquid phase speed at the critical condition. A similar phenomenon has been observed in a number of experimental works (Cramer et al 2004; Utada et al. 2005). The abrupt change of the droplet generation mode and droplet size formed for a small variation of the outer flow speed may create practical challenges for the design/operation of micro-fluidic devices to produce mono-dispersed droplets.

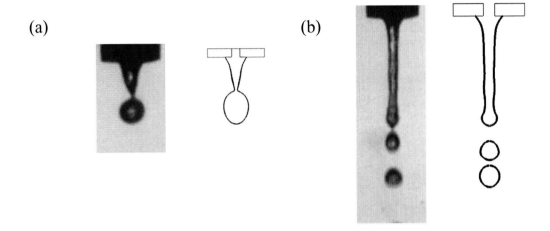

Figure 6.15. Comparison of the droplet formation modes (a) dripping and (b) jetting from a liquid jet in a solvent –antisolvent system observed in experiments and predicted by simulations. Reprint from Lee et al. (2008) with permission.

Lee et al. (2008) examined both experimentally and numerically the mechanism of a liquid jet breakup and droplet formation in near critical conditions for a solvent-antisolvent system (carbon dioxide-dichloromethane system). The liquid jet formation and breakup was examined using a high-speed camera. The size of droplets formed at different jet breakup regimes was measured with global sizing velocimetry using the shadow sizing method. A front tracking method was employed to simulate the jet formation and breakup process under the same conditions as the experiments. A comparison of the experimental observation and simulations is illustrated in Figure 6.15 for both dripping and jetting modes. They are in reasonable agreement. The simulations provided many flow details which are still difficult to measure using current techniques.

7. EXTENSION TO COMPLEX TWO-PHASE FLOW PROBLEMS

Bubbles and drops are the most common subjects in the multiphase / interfacial flows, and the aim of developing the front tracking method is to build our modeling capability, to improve our fundamental understanding of the flow physics, and to tackle the relevant problems in industrial applications. In addition, there are many physical processes that are strongly coupled with the multiphase interfacial flows. These physical processes may include phase change, surfactant accumulation and distribution, and electrohydrodynamics. In this section, we will discuss the possible extension of the methodology presented previously to these complex physical problems.

7.1. Phase Change – Boiling or Evaporation

The combination of phase change with fluid flow and heat transfer determines the efficiency of heat transfer equipment. The power generation industry takes advantage of the high heat transfer rates associated with phase change in boiling to extract energy from other sources such as solar, fossil and nuclear fuels. Hence, phase change modeling should account not only for the fluid flow governed by the Navier-Stokes equations, but also the energy and mass transfer from one phase to the other phase. At the interface, the jump conditions for mass, momentum and energy can be written as

$$\rho_l(\mathbf{u}_f - \mathbf{u}_l) \cdot \mathbf{n}_f = \rho_v(\mathbf{u}_v - \mathbf{u}_f) \cdot \mathbf{n}_f = \dot{m}, \tag{7.1}$$

$$\dot{m}(\mathbf{u}_v - \mathbf{u}_l) = (\tau_v - \tau_l) \cdot \mathbf{n}_f - (p_v - p_l)\mathbf{n}_f + \sigma \kappa \mathbf{n}_f, \tag{7.2}$$

$$\dot{m} h_{fg} = \dot{q}_f = k_v \frac{\partial T}{\partial n}\bigg|_v - k_l \frac{\partial T}{\partial n}\bigg|_l. \tag{7.3}$$

Here, $\mathbf{u}_l$ and $\mathbf{u}_v$ are the fluid velocity at the liquid and vapor side of the interface, $\mathbf{u}_f$ is the interface velocity, and $\dot{m}$ is the evaporation rate at the interface. T and k stand for temperature and thermal conductivity, respectively. h_{fg} is the latent heat due to phase change, and $\dot{q}_f$ is the heat flux across the interface due to phase change. In the derivation of equation (7.3), we have assumed that the interface temperature T_f is the same as the saturation temperature at the system pressure, i.e. $T_f = T_{sat}(p_{sys})$.

Applying the concept of the "one-fluid" formulation introduced in Section 2, we can obtain the following governing equations for the continuity, momentum and thermal energy, respectively.

$$\nabla \cdot \mathbf{u} = \int_\Gamma (\mathbf{u}_v - \mathbf{u}_f) \cdot \mathbf{n}_f \delta(\mathbf{x} - \mathbf{x}_f) ds_f, \tag{7.4}$$

$$\frac{\partial \rho \mathbf{u}}{\partial t} + \nabla \cdot \rho \mathbf{u}\mathbf{u} = -\nabla p + \nabla \cdot \mu(\nabla \mathbf{u} + \nabla \mathbf{u}^T) + \rho \mathbf{g} \\ + \int_\Gamma \sigma \kappa_f \delta(\mathbf{x} - \mathbf{x}_f) \mathbf{n}_f ds_f, \tag{7.5}$$

$$\frac{\partial \rho c T}{\partial t} + \nabla \cdot \rho c \mathbf{u} T = \nabla \cdot k \nabla T$$
$$+ \left[1 - (c_v - c_l) \frac{T_{sat}}{h_{fg}}\right] \int_\Gamma \delta(\mathbf{x} - \mathbf{x}_f) \dot{q}_f ds_f \,, \qquad (7.6)$$

Where c is the specific heat capacity of the fluid.

The velocity difference between the liquid and vapor can be related to the evaporation rate by elimination of $\mathbf{u}_f$ in Equation (7.1).

$$(\mathbf{u}_v - \mathbf{u}_l) \cdot \mathbf{n}_f = \dot{m}(1/\rho_v - 1/\rho_l), \qquad (7.7)$$

Applying the relationship between the evaporation rate and the flux $\dot{m} = \dot{q}_f / h_{fg}$ to Equation (7.7) and combining with equation (7.4), the following mass conservation equation is obtained

$$\nabla \cdot \mathbf{u} = \frac{1}{h_{fg}} \left(\frac{1}{\rho_v} - \frac{1}{\rho_l}\right) \int_\Gamma \dot{q}_f \delta(\mathbf{x} - \mathbf{x}_f) ds_f \,, \qquad (7.8)$$

Hence, we can obtain the governing equations (7.8), (7.5), (7.6) for the multiphase flow problem with phase change. To close this set of governing equations, a correlation for heat flux across the interface ($\dot{q}_f$) is needed

$$\dot{q}_f = [k_v(T_v - T_{sat}) - k_l(T_{sat} - T_l)] / |\Delta_{nf}|, \qquad (7.9)$$

where T_l and T_v are the temperature of the liquid side ($\mathbf{x}_l = \mathbf{x}_f - \Delta_{nf}$) and vapor side ($\mathbf{x}_v = \mathbf{x}_f + \Delta_{nf}$) near the phase interface. T_{sat} is the interface temperature. Δ_{nf} is the normal distance of the sampling points from the interface. T_l, T_v and T_{sat} are interpolated from the temperature field at the sampling points $\mathbf{x}_l$, $\mathbf{x}_v$ and $\mathbf{x}_f$, respectively. Hence, the flow field can be solved accounting for the phase change across the interface.

According to equation (7.1), the normal velocity of the interface can be calculated by

$$\mathbf{u}_f \cdot \mathbf{n}_f = (u_l + u_v)/2 - \frac{\dot{q}_f}{2h_{fg}} \left(\frac{1}{\rho_l} + \frac{1}{\rho_v}\right). \qquad (7.10)$$

Hence, the normal velocity of the interface consist of two components as shown in (7.10): one due to fluid advection and another one due to phase change. Once the velocity is known, the position of the interface can be advected along the normal direction,

$$\frac{d\mathbf{x}_f}{dt} = \mathbf{u}_f \cdot \mathbf{n}_f. \tag{7.11}$$

The above numerical approach has been developed and employed by Esmaeeli & Tryggvason (2004a, b) to investigate film boiling. This numerical method has been tested by simulating the film boiling in a quiescent liquid from a flat heating surface. Figure 7.1 shows the evolution of the liquid/vapor interface predicted by the simulations. In practical situations, the boiling more frequently takes place where the liquid is flowing and the geometry is complex. Esmaeeli & Tryggvason (2004c) took an important step to study such flows computationally by introducing a method that allows for incorporation of complex geometries into the flow field. They examined film boiling of an initially quiescent fluid on a horizontal cylinder and multiple cylinders and focused on heat transfer rates. Interface merging and break up is an inherent characteristic of boiling flows. Modelling the complex topology changes in boiling flow is still one of remaining challenges in simulations using the front tracking methods.

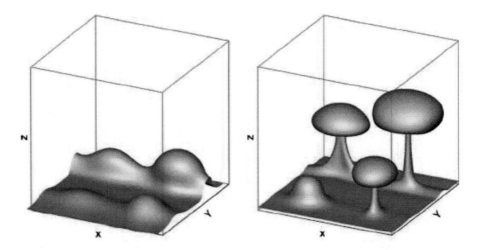

Figure 7.1. Evolution of a liquid/vapor interface in the process of film boiling. Reprint from Esmaeeli and Tryggvason (2004b) with permission.

7.2. Effect of Surfactant on the Interface

The dynamics of bubbles or drops are strongly affected by the accumulation and distribution of surfactant on the interface. Surfactants are also used in the oil/gas industry to change the rheological property of oil/water/gas mixtures for flow assurance, separation or mixing. The interfacial surface tension is generally a function of the surfactant concentration according to an equation of state. Non-uniform surfactant concentration may lead to non-uniform normal (capillary) and tangential (Maragoni) stress, which may significantly affect the motion and deformation of drops and bubbles. Distributions of surfactant on the interface result from advection and diffusion both at the interface and in bulk fluid by the motion of the fluid and by molecular mechanism, respectively. Hence, to consider the effect of surfactant on the bubble or drop dynamics, the governing equations for fluid flow must be complemented by an equation for the convection and diffusion of surfactants in the bulk liquid

$$\frac{\partial C}{\partial t} + \nabla \cdot (\mathbf{u}\, C) = \nabla \cdot (D \nabla C). \qquad (7.12)$$

Here C is the surfactant concentrations in the bulk fluid and on the interface, respectively. D is the corresponding diffusion coefficients in the bulk liquid.

In the front tracking method, we track the interface position in a Lagrangian manner. The evolution equation for surfactant concentration at the interface can be solved on the Lagrangian grid with the following equation.

$$\frac{DA\alpha}{Dt} = AD_s \nabla_s^2 \alpha + AS_\alpha . \qquad (7.13)$$

where A is the area of the interfacial element. α is the surfactant concentrations at the interface. D_s is the corresponding diffusion coefficients at the interface. S_α is the source term of surfactant absorption rate on the interface, which is related to the bulk concentration by the following formula

$$S_\alpha = -D \left(\mathbf{n} \cdot \nabla C |_{\text{interface}} \right). \qquad (7.14)$$

The above equation (7.15) can be solved explicitly while tracking the motion of the interface (Muradoglu and Tryggvason 2008). A similar equation was solved implicitly by Lai et al. (2008).

The surface tension coefficient can be expressed as a function of surfactant concentration $\sigma = f(\alpha)$. The approach to solve the flow field equations (2.28 and 2.29) with variable surface coefficient has been introduced in the previous section. The governing equation (7.12) for the surfactant concentration in the bulk fluid can be solved straightforwardly using a standard numerical method. The governing equation for surfactant transport at the interface (7.13) is solved on the Lagrange grid, which is used for tracking the interface shape and position in the front tracking method.

The above numerical modeling approach (Muradoglu and Tryggvason 2008) was developed for computations of multiphase interfacial flows with soluble surfactants. This method can be used to solve the evolution equations of the interfacial and bulk surfactant concentrations together with the incompressible Navier-Stokes equations. A non-linear equation of state was used to relate the interfacial surface tension to surfactant concentration on the interface. This method was used to model the effects of soluble surfactants on the motion of a buoyancy-driven bubble in a circular tube. Figure 7.2 shows the predicted temporal variation of the bubble shape as well as the surfactant concentration in the bulk liquid and on the interface, when a bubble rises in contaminated liquid with constant surfactant concentration. The simulation results agree with the experimental observations (Clift et al. 1978).

Figure 7.2.Contour plots of the surfactant concentration in the bulk fluid (left) and distribution of the surfactant concentration on the interface (right) for a rising contaminated bubble. Time goes from left to right. Reprint from Muradoglu and Tryggvason (2008) with permission.

7.3. Electrohydrodynamics Simulation

Electric fields have be used extensively to produce and manipulate liquid drops in many industries for atomization, inkjet printing, enhanced coalescence, emulsion breakup, etc. Under the influence of an externally applied electric field, a drop suspended in a viscous liquid may experience complex behavior. Due to the dielectric mismatch between the fluids, the applied electric field may induce a stress at the interface, which may have strong effects on the drop deformation and motion in the bulk liquid. The processes such as the drop deformation and translation, the electric charge accumulation and convection, and the induced viscous flow inside and outside drop are all coupled, i.e., one influence the others. The complex flow physics can be described by the standard conservation equations (2.28 and 2.29) by adding the electric field, given by

$$\mathbf{F}_{ES} = -\frac{1}{2}\mathbf{E} \cdot \mathbf{E} \nabla \varepsilon + q^v \mathbf{E} + \nabla \left(\frac{1}{2} \mathbf{E} \cdot \mathbf{E} \frac{\partial \varepsilon}{\partial \rho} \rho \right), \tag{7.15}$$

where $\mathbf{E}$ is the electric field strength, ε is the permittivity of fluid and q^v is the volume charge density near the interface. The first term on the right-hand side of (7.15) is due to the polarization stress, and it acts normal to the interface as a result of the term $\nabla \varepsilon$. The second term is due to the interaction of the electric charges with the electric field, acting along the direction of the electric field. The last term results from the changes in material density, usually called electrorestriction force density. This term is neglected in this study as the fluid is assumed to be incompressible. As the electric charges are located on the interface, both the polarization electric stress and the charge-field interaction electric stress exert a force on the interface. In order to calculate the electric force, the electric field strength ($\mathbf{E}$) and volume charge density (q^v) in Equation (7.15) are estimated using various electric field models: Leaky dielectric model, perfect dielectric model and surface charged model. Detailed discussion of these models for electric field calculations can be found in Hua et al. (2008a).

The experimental data of Torza et al. (1971) are chosen as the basis for the validation study. The fluid inside the drop is silicon oil and the outside fluid is oxidized castor oil. The droplet deformation increases with the electric field from left to right. Since the fluids used in the experiment are conductive, the leaky dielectric model is used to simulate the electric field effect. In addition to normal forces that deform the drops, tangential forces can induce a fluid motion either from the poles of the drops to their equator or from the equator to the poles. As shown in Figure 7.3, the simulation predictions agree well with the experimental observations on the drop deformation under the effect of externally applied electric fields.

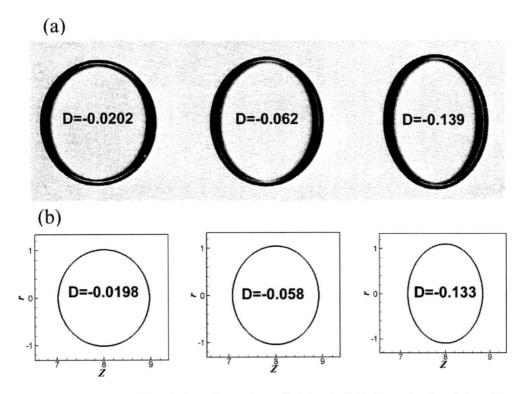

Figure 7.3. Comparison of drop deformation under applied electric fields increasing from left to right (a) observed in experiments of Torza et al. (1971) and (b) predicted in the simulation using the leaky dielectric model. D refers to the drop deformation factor. Reprint from Hua et al. (2008a) with permission.

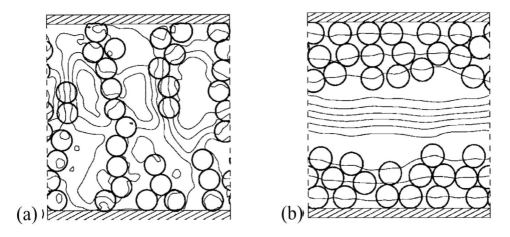

Figure 7.4. Steady-state drop distribution and streamlines for a suspension of 36 drops (a) at low flow rate (b) at high flow rate. The liquid flow in the channel is from the left to right. Reprint from Fernandez et al. (2005) with permission.

A front tracking method was also employed by Fernandez et al. (2005) to investigate the effects of an electric field applied to an emulsion of drops flowing through a channel on the distribution of drops across the cross section of the channel. Both the dielectrophoretic and the electrohydrodynamic interactions of the drops work together to "fibrate" the emulsion by lining the drops up into columns parallel to the electric field. Figure 7.4 shows snapshots of the drop distribution in the channel cross section under different channel flow rates. For low flow rate, the drops form fibers that span the channel and block the flow. At higher flow rates short fibers form but the flow breaks up longer fibers and the drops accumulate at the walls, forming an essentially stationary layer. If the flow rate is high enough, the drop interactions are dominated by the fluid shear, and drops remain in suspension.

Conclusion

Attempts to simulate multiphase flows numerically started in the early days of computational fluid dynamics. Although a few successful simulations can be found in the early literature, major progress has been made only in the past few years. One area of significant progress made in simulating two-phase flows, including the motion of many bubbles and drops, over the last decade has relied nearly exclusively on the use of the so-called "one-fluid" approach where a single set of governing equations is used to represent all the different fluids that are present. This approach has been adopted in methods such as the VOF method, level-set method, phase field method and the front tracking method. The key difference among these methods lies in how to track the interface front position and treat the physical discontinuities across the interface.

In this chapter, a front-tracking method for direct numerical simulation of incompressible multiphase interfacial flows has been presented as well as its numerical implementation. This method is based on the "one-fluid" formulation concept, in which one set of governing equations is applied to both fluid phases, treating the different phases as one fluid with variable material properties. The unsteady Navier–Stokes equations are solved using a conventional finite volume method on a fixed grid, and the interfacial front is tracked explicitly by connected marker points. The physical discontinuities across the interface are accounted for by adding the appropriate sources as delta-functions at the interfacial front separating the phases. Interfacial source terms such as surface tension are computed on the front and transferred to the fixed grid through interpolation. Advection of fluid properties such as density is done by following the motion of the front. The method has been implemented for two-dimensional and axisymmetric flow simulations, as well as for fully three-dimensional flow modeling.

Even though explicit front tracking method is generally more complex than the advection of a marker function used in other methods, e.g. VOF method, we believe that the increased accuracy and robustness are well worth the effort (Tryggvason et al. 2001). The explicit tracking of the interface not only reduces errors associated with the advection of a marker function and surface tension computations, but the flexibility inherent in the explicit tracking approach should also be valuable for application to problems where complex interface physics must be accounted for. Hence, we also introduced some extensions of the traditional front tracking method to model complex interfacial multiphase flow problems involving phase change, surfactant transport and electrohydrodynamics. Recent progress in incorporating elastic membranes and solid boundaries into the methods based on the front tracking or immersed boundary methods will extend its utility further.

The numerical method described in this chapter has been validated systematically with available experimental data for single bubble rising in viscous liquid and bubble pinch-off in liquid, and it has been widely employed in investigating the dynamics of multiphase flows with bubbles and drops. Extensive reviews of the simulation work, which adopt the front tracking method to investigate various multiphase interfacial flow problems, demonstrate that the front tracking method is one of the most effective numerical approaches for multiphase / interfacial flows.

ACKNOWLEDGMENTS

We acknowledge the fruitful discussions with Professor Christopher Lawrence, and we are grateful for his careful reading of an earlier version of this chapter.

REFERENCES

Bhaga, D. & Weber, M.E. (1981). Bubbles in viscous liquids: shapes, wakes and velocities. *J. Fluid Mech.*. 105, 61.

Bonometti, T. & Magnaudet, J. (2006). Transition from spherical cap to toroidal bubbles. *Phys. Fluids.* 18, 052102.

Brackbill, J.U., Kothe, D.B. & Zenmach, C. (1992). A continuum method for modeling surface tension. *J. Comput. Phys.* 100, 335.

Bunner, B. & Tryggvason, G. (2002). Dynamics of homogenous bubbly flows. Part I: Rise velocity and microstructure of the bubbles. *J. Fluid Mech.* 466, 17.

Chen, L., Garimella, S.V., Reizes, J.A. & Leonardi, E. (1999). The development of a bubble rising in a viscous liquid. *J. Fluid Mech.* 387, 61.

Clift, R., Grace, J.R. & Weber, M.E. (1978). *Bubbles, Drops, and Particles*. San Diego: Academic Press.

Cramer, C., Fischer, P. & Windhab, E.J. (2004). Drop formation in a coflowing ambient fluid. *Chem. Eng. Sci.* 59, 3045.

Davies, R.M., & Taylor, F.I. (1950). The mechanism of large bubbles rising through extended liquids and through liquids in tubes. *Proc. R. Soc. Lond.* A 200, 375.

Esmaeeli, A. & Tryggvason, G. (2004a). Computations of film boiling. Part I: Numerical method. *Int. J. Heat Mass Transfer* 47, 5451.

Esmaeeli, A. & Tryggvason, G. (2004b). Computations of film boiling. Part II: Multi-mode film boiling. *Int. J. Heat Mass Transfer* 47, 5463.

Esmaeeli, A. & Tryggvason, G. (2004c). A front tracking method for computations of boiling in complex geometries. *Int. J. Multiphase Flow* 24, 1037.

Fernandez, A., Tryggvason, G., Che, J. & Ceccio, S.L. (2005). The effects of electrostatic forces on the distribution of drops in a channel flow: Two-dimensional oblate drops. *Phys. Fluids* 17, 093302.

Fogelson, A.L. & Peskin, C.S. (1988). A fast numerical method for solving the three-dimensional Stokes equations in the presence of suspended particles. *J. Comput. Phys.* 79, 50.

Glimm, J., McBryan, O., Menikoff R., & Sharp, D.H. (1986). Front tracking applied to Rayleigh–Taylor instability. *SIAM J. Sci. Statist. Comput.* 7, 230.

Glimm, J., Grove, J.W., Lindquist, B., McBryan O. & Tryggvason G. (1988). The bifurcation of tracked scalar waves. *SIAM J. Sci. Stat. Comput.* 9, 61.

Glimm, J., Grove, J.W., Li, X.L. & Tan, D.C. (2000). Robust computational algorithms for dynamic interface tracking in three dimensions. *SIAM J. Sci. Comput.* 21, 2240.

Hirt, C.W. & Nichols, B.D. (1981). The volume of fluid (VOF) method for the dynamics of free boundaries. *J. Comput. Phys.* 39, 201.

Homma, S., Koga, J., Matsumoto, S., Song, M. & Tryggvason, G. (2006). Breakup mode of an axisymmetric liquid jet injection into another immiscible liquid. *Chem. Eng. Sci.* 61, 3986.

Hou, T.Y., Lowengrub, J.S. and Shelley, M.J. (2001). Boundary integral methods for multicomponent fluids and multiphase materials. *J. Comput. Phys.* 169, 302.

Hu, H.H., Patankar, N.A. & Zhu, M.Y. (2001). Direct numerical simulation of fluid-solid system using the arbitrary Lagrangian-Eularian technique, *J. Comput. Phys.* 169, 427.

Hua, J. & Lou, J. (2007). Numerical simulation of bubble rising in viscous liquid. *J. Comput. Phys.* 222, 769.

Hua, J., Lim L.K. & Wang, C.H. (2008a). Numerical simulation of deformation / motion of a drop suspended in viscous liquids under influence of steady electric fields. *Phys. Fluids* 20, 113302.

Hua J., Lin, P. & Stene J.F. (2008b). Numerical simulation of gas bubbles rising in viscous liquids at high Reynolds number. *Contemporary Mathematics*, 446, 17.

Hua, J., Stene, J.F. & Lin, P. (2008c) . Numerical simulation of 3D bubbles rising in viscous liquids using a front tracking method. *J. Comput. Phys.* 227, 3358.

Hua, J., Zhang, B. & Lou, J. (2007). Numerical simulation of microdroplet formation in coflowing immiscible liquids. *AIChE J.* 53, 2534.

Ida, M. (2000). An improved unified solver for compressible and incompressible fluids involving free surfaces. Part I. Convection. *Comput. Phys. Commun.* 132, 44.

Jacqmin, D. (1999). Calculation of two-phase Navier-Stokes flows using phase-field modeling. *J. Comput. Phys.* 155, 96.

Jimenez, J. & Moin, P. (1991). The minimal flow unit in near wall turbulence. *J. Fluid Mech.* 225, 213.

Kanai, A. & Miyata, H. (2001). Direct numerical simulation of wall turbulent flows with microbubbles. *Int. J. Numer. Meth.. Fluids* 35, 593.

Kang, I.S. & Leal, L.G. (1987). Numerical solution of axisymmetric, unsteady free-boundary problem at finite Reynolds number. I. Finite difference scheme and its application to the deformation of a bubble in a uniaxial straining flow. *Phys. Fluids* 30, 1929.

Kawamura, T. & Kodama, Y. (2002). Numerical simulation method to resolve interaction between bubbles and turbulence. *Int. J. Heat Fluid Flow* 23, 627.

Koebe, M., Bothe, D., Pruess, J. & Warnecke, H.-J. (2002). 3D direct numerical simulation of air bubbles in water at high Reynolds number, In Proceedings of the 2002 ASME Fluids Engineering Division Summer Meeting (FEDSM2002-31143), Montreal, Quebec, Canada, July 14-18.

Lai, M.C., Tseng, Y.H. & Huang, H. (2008). An immersed boundary method for interfacial flows with insoluble surfactant. *J. Comput. Phys.* 227, 7279.

Lee, L.Y., Lim, L.K., Hua, J. & Wang, C.H. (2008). Jet breakup and droplet formation in near-critical regime of carbon dioxide-dichloromethane system. *Chem. Eng. Sci.* 63, 3366.

Lu, J., Fernandez, A. & Tryggvason, G. (2005). The effect of bubbles on the wall drag in a turbulent channel flow. *Phys. Fluids* 17, 095102.

MacNeice, P., Olson, K.M., Mobarry, C., deFainchtein, R. & Packer, C. (2000). PARAMESH: A parallel adaptive mesh refinement community toolkit. *Comput. Phys. Commun.* 126, 330.

Magaudet, J. & Eames, I., (2000). The motion of high-Reynolds-number bubbles in inhomogeneous flows. *Annu. Rev. Fluid Mech.* 32, 659.

Moore, D.W. (1959). The rise of a gas bubble in viscous liquid. *J. Fluid Mech.* 6, 113.

Mougin G. & Magnaudet, J. (2006). Wake-induced forces and torques on a zigzagging / spiraling bubble. *J. Fluid Mech.* 567, 185.

Muradoglu, M. & Tryggvason, G. (2008). A front tracking method for computation of interfacial flows with soluble surfactants. *J. Comput. Phys.* 227, 2238.

Ohta, M., Imura, T., Yoshida, Y. & Sussman, M. (2005). A computational study of the effect of initial bubble conditions on the motion of a gas bubble rising in viscous liquids. *Int. J. Multiphase Flow* 31, 223.

Osher, S. & Sethian J.A. (1988). Fronts propagating with curvature dependent speed: Algorithms based on Hamilton-Jacobi formulations. *J. Comput. Phys.* 79, 12.

Osher, S. & Fedkiw, R.P. (2001). Level set methods: An overview and some recent results. *J. Comput. Phys.* 169, 463.

Patankar, S.V. (1980). *Numerical Heat Transfer and Fluid Flow*. New York:Hemisphere.

Peskin, C.S. (1977). Numerical analysis of blood flow in the heart. *J. Comput. Phys.* 25, 220.

Peskin C.S. & Printz, B.F. (1993). Improved volume conservation in the computation of flows with immersed boundaries. *J. Comput. Phys.* 105, 33.

Pozrikidis C. (2001). Interfacial dynamics for Stokes flow. *J. Comput. Phys.* 169, 250.

Prosperetti, A. (2004). Bubbles. *Phys. Fluids* 16, 1852.

Quan, S. & Schmidt, D.P. (2006). A moving mesh interface tracking method for 3D incompressible two-phase flows. *J. Comput. Phys.* 221, 761.

Quan, S. & Hua, J. (2008). Numerical studies of bubble necking in viscous liquids. *Phys. Rev. E.* 77, 066303.

Raymond, F. & Rosant, J.-M. (2000). A numerical and experimental study of the terminal velocity and shape of bubbles in viscous liquids. *Chem. Eng. Sci.* 55, 943.

Rudman, M. (1998). A volume-tracking method for incompressible multifluid flows with large density variations. *Int. J. Numer. Meth. Fluids,* 28, 357.

Ryskin, G. & Leal, L.G. (1984). Numerical simulation of free-boundary problems in fluid mechanics. Part 2. Buoyancy-driven motion of a gas bubble through a quiescent liquid. *J. Fluid Mech.* 148, 19.

Scardovelli, R. & Zaleski, S. (1999). Direct numerical simulation of free-surface and interfacial flow. *Annu.Rev. Fluid. Mech.* 31, 567.

Shew, W.L. & Pinton, J.-F. (2006). Viscoelastic effects on the dynamics of a rising bubble. *J. Stat. Mech.* P01009.

Shin, S. & Juric, D. (2002). Modelling three-dimensional multiphase flow using a level contour reconstruction method for front tracking without connectivity. *J. Comput. Phys.* 180, 427.

Slattery, J.C. (1999). *Advanced Transport Phenomena.* Cambridge University Press.

Sousa, F.S., Mangiavacchi, N., Nonato, L.G., Castelo, A., Tome, M.F., Ferreira, V.G., Cuminato, J.A., McKee, S. (2004). A front-tracking/front-capturing method for the simulation of 3D multi-fluid flows with free surfaces. *J. Comput. Phys.* 198, 469.

Sussman, M., Smereka, P. & Osher, S. (1994). A level set approach for computing solutions to incompressible two-phase flows. *J. Comput. Phys.* 114, 146.

Taylor, T.D. & Acrivos, A. (1964). On the deformation and drag of a falling viscous drop at low Reynolds number, *J. Fluid Mech.* 18, 466.

Thoroddsen, S. T., Etoh, T. G. & Takehara, K. (2007). Experiments on bubble pinch-off. *Phys. Fluids* 19, 042101.

Tomiyama, A., Celata, G.P., Hosokawa, S. & Yoshida, S. (2002). Terminal velocity of single bubbles in surface tension force dominant regime. *Int. J. Multiphase Flow* 28, 1497.

Torza, S., Cox, R.G., & Mason, S.G. (1971). Electrohydrodynamic deformation and burst of liquid drops. *Philos. Trans. R. Soc. London, Ser. A* 269, 295.

Tryggvason, G., Bunner, B.B., Esmaeeli, A., Juric, D., Al-Rawahi, N., Tauber, W., Han, J., Nas, S. & Jan, Y.J. (2001). A front-tracking method for the computations of multiphase flow. *J. Comput. Phys.* 169, 708.

Tryggvason, G., Esmaeeli, A., Lu, J., & Biswas, S. (2006). Direct numerical simulations of gas /liquid multiphase flows. *Fluid Dynamics Research* 38, 660.

Unverdi, S.O. & Tryggvason, G. (1992). A front-tracking method for viscous, incompressible, multi-fluid flows. *J. Comput. Phys.* 100, 25.

Utada, A.S., Lorenceau, E., Link, D.R., Kaplan, P.D., Stone, H.A. & Weitz, D.A. (2005). Monodisperse double emulsions generated from a micro capillary device. *Science,* 308, 537.

van Sint Annaland, M., Deen, N.G. & Kuipers, J.A.M. (2005). Numerical simulation of gas bubbles behaviour using a three-dimensional volume of fluid method. *Chem. Eng. Sci.* 60, 2999.

van Sint Annaland, M., Dijkhuizen, W., Deen, N.G. & Kuipers, J.A.M. (2006). Numerical simulation of behaviour of gas bubbles using a 3-D front-tracking method. *AIChE J.* 52, 99.

Wu, M.M. & Gharib, M. (2002). Experimental studies on the shape and path of small air bubbles rising in clean water. *Phys. Fluids* 14, 49.

Yang, B., Prosperetti, A. & Takagi, S. (2003). The transient rise of a bubble subject to shape or volume changes, *Phys. Fluids* 15, 2640.

Chapter 3

COMPUTATIONAL FLUID DYNAMICS IN BIOMEDICAL ENGINEERING

T. A. S. Kaufmann, R. Graefe, M. Hormes, T. Schmitz-Rode and U. Steinseifer

Department of Cardiovascular Engineering,
Institute of Applied Medical Engineering, Helmholtz Institute,
RWTH Aachen University and University Hospital Aachen, Germany

ABSTRACT

There are more than 1.5 million operations each year using cardiopulmonary bypass (CPB), when the function of the human heart is taken over by extracorporeal blood pumps and oxygenators. Every fourth of these patients suffers from minor or major neurologic events such as headaches, mnemonic problems, loss of neural functions or coma. This is mainly related to the blood flow to the brain. The understanding of these flow conditions is therefore of high clinical relevance. Computational Fluid Dynamics (CFD) is used to model the flow in the human vascular system and thus gain insight in flow processes invisible to experimental measurements.

To analyze these conditions, realistic models of the human vascular system are obtained using modern imaging techniques. From this data, a 3D model is reconstructed and the fluid flow can be calculated. Different blood properties (e.g. Non-Newtonian behavior) and realistic boundary conditions like systemic pressure have to be taken into account. Thereby, the flow in the cardiovascular system can be analyzed and optimized for different applications and the neurologic events during CPB operations can be reduced.

In some of these operations, clinical devices such as rotary blood pumps are implanted to support of completely take over some of the functions of the human heart, for example as a bridge-to-transplant, when a patient is waiting for a heart transplant but there is currently none available. Very common devices are Left Ventricular Assist Devices (LVADs) that support the function of the left heart. CFD plays an important role during development of these devices. For example, the hydraulic efficiency and rotor stability can be analyzed virtually. Thereby, CFD can significantly reduce the development time and cost for new prototypes.

In an extracorporeal circuit, a blood pump is often combined with an oxygenator to provide sufficient oxygen transfer via the blood. One important parameter of these devices is the gas exchange rate: oxygen O_2 diffuses from porous fibers to the blood, while carbon dioxide CO_2 is withdrawn. This process can be analyzed by CFD to support oxygenator development.

Another typical application is the design of new artificial heart valves, especially mechanical heart valves. The advantage of this approach compared to biological valves is the highly increased durability. However, patients with these valves need life-long medication to avoid thrombosis. The likelihood for thrombus formation is related to the shear rates acting on human blood. Using CFD, these shear rates as well as exposure time can be calculated. Thus, predictions of critical regions of new valves can be made.

There are many challenges when using CFD for medical applications, since interactions in the human body are very complex. Therefore, experimental as well as clinical data is needed to validate CFD. This can be done by flow visualization (e.g. MRI-flow measurements, Particle Image Velocimetry) or other experimentally obtained data (pressure, flowrate). Once validated, CFD provides many benefits for the development of biomedical devices and for the understanding of the flow conditions in the human body. In the end, it can thus increase patient survival.

ABOUT COMPUTATIONAL FLUID DYNAMICS IN MEDICAL TECHNOLOGY

Computational Fluid Dynamics is not the gold standard in medical technology. It is used in several research processes, especially to save research / development time and costs. However, for the development of medical devices, CFD can only be a small part of the developing process from idea to commercialization. This process includes in-vitro studies and in-vivo studies. Finally, clinical studies are needed in order to proof the usability and the long term effects of these devices. Additionally, these clinical trials increase the credibility, making the devices more applicable for physicians. But in the early stages of device development, CFD is a commonly used tool. It can help to gain insight in many processes that are invisible to experimental approaches.

Besides device development, understanding of the human body and the intrinsic processes is very important. The flow in the vascular system is only one example out of many. Yet, many of the interactions are still unknown, and sometimes they go down to even molecular level. Simulating of these interactions is a huge challenge if it is possible at all. So for numerical flow simulation in the human body it is crucial to make many simplifications and assumptions.

During these studies, the link between a relatively simple CFD model and the very complex reality of the clinic has to be found. Only by creating useful additional information that can be derived from common clinical data will a method be adapted by physicians.

Ultimately, the patient is the most important factor in any medical technology. The main goals are minimizing the risk for complications and thus increasing the patient survival rate as well as to increase the patient well being. Reducing the costs is another important factor, as it is for any commercialized approach.

The following part is divided in four chapters. The first and the last chapter will give two examples of successful applications of CFD for biomedical problems, the simulation of cardiopulmonary bypass conditions and the analysis of a mechanical heart valve prosthesis, especially in terms of thrombosis. The second and third chapter provide an overview over two main fields where CFD is used to improve the research in the biomedical field, the employment of CFD in the development of rotary blood pumps and the simulation of mass transfer in oxygenators.

Finally, some validation methods are discusses, first for general use and later for the projects and research fields discusses earlier. Of course, these chapters are not enough to cover the whole range of CFD in the field of biomedical engineering. They are meant to give an overview over some projects and applications. Some references related to the presented methods are included for the interested reader. For further information, a wide range of literature is available online.

I. SIMULATION OF CARDIOPULMONARY BYPASS CONDITIONS

Introduction

Cardiopulmonary Bypass (CPB) describes a clinical operation during which the function of the heart is taken over by an extracorporeal circuit.

This circuit consists of an inflow cannula that withdraws the blood out of the vena cava, an oxygenator to oxygenate the blood and eliminate carbon dioxide (CO_2), a heat exchanger to temperate the blood, a blood pump that pumps the blood and an outflow cannula that returns the blood. Worldwide, more than 1.5 million CPB operations are done each year. In these applications, the tissue perfusion can be decreased by 20% [1-4]. According to some surgeons, every fourth of these patients suffers from minor or major neurologic complications, ranging from headaches and mnemonic problems to loss of neural functions, coma or even death. These complications have been widely studied in recent years [5-10]. There are basically two reasons for those phenomena: The cerebral blood flow (CBF), so the bloodflow to the brain, might be insufficient, especially in the right part of the brain. Additionally, CPB patients have a higher risk for strokes [15, 16]. Both phenomena are linked to the flow in the cardiovascular system, mainly in the aortic arch and the vessels leading to the brain, namely the left and right carotid and vertebral arteries. This flow is dependent on the cannulation type and technique. Device cannulation to the cardiovascular system is an important consideration for CPB and cardiac assist, and the development of new cannulation methods and cannulae is an important field of study [19, 20].

There are basically two ways to return blood during CPB. The most common technique is cannula positioning in the ascending aorta. Some surgeons however prefer cannula placement in the right subclavian artery [11]. This has also received attention for other applications like Ventricular Assist Devices (see Chapter B-II), due to the potential for a minimally invasive implantation approach [12, 13]. Stenosis of the subclavian artery is a contra-indicator for this method [14].

There are experimental ways of analyzing the fluid dynamics in the aorta, for example ultrasound and MRI imaging. However, they have a limited spatial resolution and are not applicable in the case of CPB. This counts especially for MRI, due to the high magnetic fields. However, both techniques can be simulated using Computational Fluid Dynamics.

One example for such a study is described in on the following pages [17, 18]. It compares different positions of the outflow cannula in the ascending aorta and the right subclavian artery in terms of cerebral blood flow.

Methods

The first step is the model creation. It is necessary to generate a realistic model of the human vascular system. Therefore, a Computer Aided Design (CAD) model of the aorta and outgoing vessels is constructed from Magnetic Resonance Imaging (MRI) or Computed Tomography (CT) data. The vessel structure can be extracted by Hounsfield units. The images are recorded in single slices. So the structures need to be extracted separately for each slice and are then added to a 3D model. Mathematical as well as manual corrections are performed in order to generate a smooth yet realistic model of the human vascular system. This model is shown in figure 1.

Depending on the application, smaller vessels need to be implemented. The final model for this example consists of the aorta, subclavian arteries, carotid arteries and vertebral arteries.

In this model, different types of outflow cannulas are virtually placed in different locations inside the CV system. Thereby, both cannulation techniques can be analyzed. For cannulation of the aorta, the cannula positions are spread around the brachiocephalic trunk in feasible distance for surgeons, representing different approaches for cannulation found in the clinic. For cannulation of the right subclavian artery, the cannula positions are varied in dependency on the distance of the cannula tip to the branch of the vertebral artery.

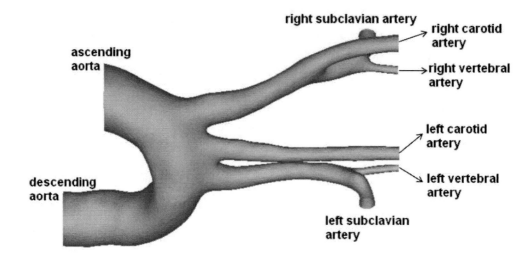

Figure 1. Model of the cardiovascular system.

The next step is the mesh generation. It has to be created separately for each different cannula position. Since the vascular system is very complex, tetrahedral elements are used as discrete the geometry. Additionally, several layers of prismatic elements are generated at the wall boundaries to resolve the boundary layers of the blood flow. Depending on the application, between 1 and 4 million elements are created to represent the geometry.

Boundary conditions:

Setting correct boundary conditions is one of the major challenges for CFD in the field of biomedical engineering. For CPB however, the boundary conditions are well defined. Since the flow is provides by an external rotary or roller pump, it can be assumed to be continuous. Thereby, the pressure in the vessels is constant.

The CPB conditions can therefore be modeled as follows:

- Inlet of the cannula: 5 l/min continuous flow (4.5 - 6.5 l/min)
- Outlet of the arteries: 55 mmHg constant static pressure (50 - 60 mmHg)
- Wall: no slip condition*

*During physiological beating of the heart, the aorta is expanded about 15-20% due to the elasticity of the vessel walls. This is called the aortic Windkessel function. Thus, CFD needs to be coupled with numerical simulations of the vessel walls. However, CPB is a constant pressure system. Thus, the Windkessel function can be neglected.

The flow in the CV system is laminar, except for high systolic flow of the beating heart. During CPB however, the jet of the cannula acts as a FREISTRAHL. Therefore, a Turbulence Model is implemented in the simulations** (SST) (**follow-up studies showed no influence on the results, so that both approaches are valid). However, the shear rates are very low, thus the non-Newtonian behavior of blood cannot be neglected. Several non-Newtonian models for the blood viscosity are available in the literature.

Steady state simulations were performed, and the convergence criterion was set to 1e-4 for the maximal residuals. Physical convergence was checked in terms of massflow to the cerebral vessels.

Results and Conclusions

For analysis of the different cannulation techniques, several flow variables as velocity, pressure and wall shear stress are compared. In particular, the CBF is of importance.

In terms of CBF, cannulation of the aorta is sensitive to location and angle. In few cases, a vortex that is generated close to the cannula jet due to the Venturi effect is located directly under the branch of the brachiocephalic trunk. Thereby, no more blood flows to that vessel. Instead, blood is withdrawn out of the trunk and thus the brain. This is depicted in figure 2. However, cerebral autoregulation, so the body's intrinsic ability to provide sufficient cerebral blood flow despite changes in blood perfusion pressure, was neglected. So in reality, the effect of reversed blood flow direction could be counteracted by autoregulation. However, it is likely that the cannula perfusion with the decreased perfusion pressure would also result in a non-optimal situation in terms of CBF.

Additionally, the level of autoregulation differs for different patients. It decreases significantly with age for example. So the effect of the decreased CBF due to a non-optimal cannula position has still some effect in the clinic.

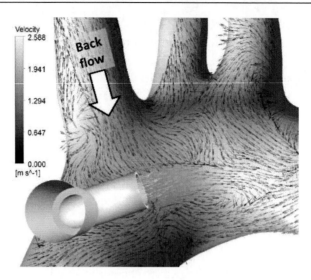

Figure 2. Cannula position in the ascending aorta with backflow from the brachiocephalic trunk.

A lateral cannula position (figure 3, alternative position) appears to provide a better overall CBF. The average flow from that position is +15% higher compared to central cannulation. This value could decrease if autoregulation would also be included.

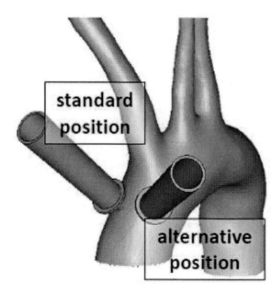

Figure 3. Two different cannulation approaches in the ascending aorta.

Cannulation to the right subclavian artery showed sufficient blood flow to all distal vessels for a cannula position several cm away from the vertebral artery. The results are shown in table 1. The flow to the right vertebral artery was quadrupled. If the cannula tip gets too close to the branch of the vertebral artery, the same effect may occur as for aortic cannulation. Thereby, blood is again withdrawn out of the brain.

Using these results, general rules for surgeons regarding cannula placement can be stated. Additionally, different types of cannulas or procedures can be analyzed by transferring this method to other applications. Since the model creation is based on realtime MRI or CT records, different patient-specific diseases can be taken into account as well. Thereby, CFD can help to decrease the risk for patient, stress for surgeons and costs for hospitals.

Table 1. Flow distribution for cannulation of the right subclavian artery during cardiopulmonary bypass

Flow distribution for cannulation to subclavian artery	
Vessel	Blood flow in % of total flow
Aorta	67
Right Subclavian Artery	0 (pinched)
Left Subclavian Artery	5.8
Right Carotid Artery	8.6
Left Carotid Artery	8.0
Right Vertebral Artery	8.6
Left Vertebral Artery	2.0

II. EMPLOYMENT OF COMPUTATIONAL FLUID DYNAMICS IN THE FIELD OF ROTARY BLOOD PUMPS

Introduction and Medical Background

Heart Failure- Organ donor shortage: Since the incidence of advanced heart failure (HF) continues to double every 10 years, it is now the most common cause of hospitalization in the western world. Whilst medical and surgical advances continue to occur, a larger number of patients are still listed for cardiac transplantation each year. However, the number of donor organs remains limited. In the United States alone, 550,000 new cases of heart failure are diagnosed each year, while less than one percent (2125) of patients were in receipt of a heart transplant in 2005 [21]. As the average life expectancy increases, so does the incidence of severe end stage heart failure.

Mechanical Assistance- A proven alternative: The most effective form of treatment for end stage heart failure is heart transplantation; however the shortage of donor organs has necessitated the implementation of mechanical cardiac support as an alternative. Mechanical circulatory support has been proven to produce more favorable outcomes than optimally medically treated heart failure patients [22]. However the uptake of such technology has been slow. The demand for mechanical circulatory assistance is nevertheless expected to increase significantly, with the indications widening to include those patients not previously considered transplant candidates. These treatment strategies may be used as a bridge to heart transplant, bridge to recovery, or indeed as a destination alternative. The latter support strategy requires a device with increased mechanical durability/lifetime.

History- Improvements in technology: Medical device technology has improved over the last few decades, enhancing the durability of bearings systems used to support moving parts. 1st generation pulsatile devices included contacting components, which limits their predicted mechanical lifetime to below 3 years. The reduced size of 2nd generation non-pulsatile rotary impeller devices accelerated them to the forefront of current VAD research. However, initial techniques for impeller support also imposed significant limitations on device lifetime, as they required a shaft, seals and bearings [23, 24]. Subsequent improvements resulted in 2nd generation devices that rely on blood immersed pivot support [25, 26]; however predicted service life is still below 5 years, and risks for thrombus/heat generated hemolysis at the contact point are elevated. Several techniques have since been developed to improve device lifetime, ranging from complete magnetic suspension [27, 28] to passive hydrodynamic suspension [29, 30]. These 3rd generation devices eliminate contact wear and reduce the number of moving components, potentially increasing device lifetime to beyond 10 years.

CFD- Fields of application: Computational fluid dynamics (CFD) has become a widely used development tool for turbomachinery and also during the last years in the field of Rotary Blood Pumps (RBPs).

Application fields can be divided into:

- Performance and efficiency
- Forces and moments on the impeller; impeller stability
- Prediction of hemolysis and thrombus formation

These major fields have not changed since the first upcoming of publications dealing with design improvements of RBPs using CFD in the early 1990s. Naturally, these fields have to be considered in a coupled manner during the design of a RBP. CFD can be a tool to balance often conflicting design requirements of a hydraulic components and blood compatibility considerations [31].

Fields of Application

Performance and efficiency: A RBP as a VAD is expected to never run at a single operational point in contrast to a larger number of cases of general industrial turbomachinery applications. Due to a residual hydraulic cardiac output of the natural heart, the pump encounters a transient inflow condition represented by systole and diastole. Moreover, the outlet boundary condition depends on the state of the supported patient. The human body possesses an autoregulation mechanism to change the flow resistance locally according to need. This affects the outlet pressure condition of the pump. As a consequence, the knowledge of the performance or pump characteristic curve (H-Q) is of significant importance to optimally support a failing heart. By means of steady-state or transient CFD, the performance of a RBP can be determined and improved in combination with design changes.

The device efficiency of a RBP is defined by the relation of output hydraulic power to electrical input power. Therefore, the device efficiency is a multiplication of motor efficiency and hydraulic efficiency. Although efficiency is not as critical as other design aspects like general blood compatibility, losses should be minimized. A low efficiency could lead to a necessity for inconveniently large batteries. Furthermore, due to the conservation of energy, energetic losses are transformed into heat that could cause temperature induced hemolysis. By means of CFD, the hydraulic efficiency can be determined and improved in combination with design changes.

A common example of CFD use to increase the hydraulic efficiency is described by Steinbrecher [32]. By means of steady-state simulations and making use of periodicity, a significant parameter study was conducted, compare figure 4. Changes to the design affected the inlet and outlet blade angles, the number of blades of the impeller, the blade height.

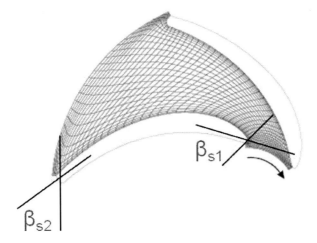

Figure 4. Mesh.

A structured mesh was employed with boundary layer refinements. An extract of the results are listed in table 2. Although the parameter combination in line 5 showed the highest efficiency, the parameter combination in line 6 was chosen for the further development. A lower blood surface contact area due to a lower blade number was preferred compared to a marginally higher hydraulic impeller efficiency.

Table 2. Results

Blade number	β_{s1}	β_{s2}	Blade height [mm]	Impeller efficiency [%]
6	40.1	49.9	2	95.5
6	31.2	49.8	2	95.7
5	40.1	42.1	2	93.9
7	40.1	42.1	2	97.3
7	40.1	42.1	1.5	98.2
6	40.1	42.1	2	98.1
6	40.1	42.1	1.5	98.0

Forces and moments on the impeller: RBPs especially as VADs face highest requirements concerning reliability and increasingly also lifetime. Depending mostly on size, an implantation, replacement or removal of a VAD can be a very traumatizing operation. RBPs as a part of an extracorporeal circuit do not face these restrictions concerning minimization and lifetime in the same magnitude. In the course of the current 3rd generation of VADs, generally there remains only one moving part which has to be supported by a bearing system. In order to maximize lifetime and minimize wear, the rotating impeller is supported by contactless bearings. Alternatives are magnetic bearings, hydrodynamic bearings or a combination thereof. The bearing system must be capable to balance forces in the axial and radial directions and tilting moments. Forces on the impeller can be divided into hydraulic forces, gravitational forces, inertia/shock forces, magnetic forces and potential hydrodynamic forces. These forces are the essential input to any bearing design procedure. By means of CFD, hydraulic and hydrodynamic forces as well as tilting moments are predicted with either steady-state or transient simulations.

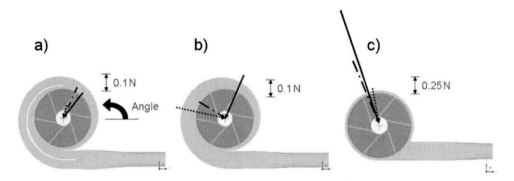

Figure 5. Radial forces for the double (a), single (b) and circular volute (c). Force vectors are shown for 3 l/min (dotted line), 5 l/min (dot-dash line) and for 7 l/min (solid line).

A recent example of CFD use to predict hydraulic forces tried to optimize a design with respect to the employment of a radial hydrodynamic bearing [33]. The results for the hydraulic forces for three different volute designs in a centrifugal RBP are shown in figure 5.

The results were generated by steady-state simulations while changing the operational flow rate to reflect the residual cardiac output.

Prediction of hemolysis and thrombus formation: Blood compatibility is an inevitable quality of RBPs. Hemolysis or blood damage can be looked upon as damage or rupture of the membrane of red blood cells (RBCs). It is important to note that hemolysis cannot be obviated completely but must be minimized nevertheless. Blood compatibility means also to avoid thrombus formation. The coagulation cascade is a complex matter and there are several means to activate thrombocytes and to trigger clot formation. The prediction of hemolysis and thrombus formation is a very active field of application for CFD. Hemolysis in a RBP can be caused by shear stress, foreign surface contact or elevated temperatures. Shear stress based hemolysis is often referred to as mechanically or flow induced hemolysis. This part of the overall hemolysis is often predicted by CFD using a number of approaches. Thrombus formation is mostly looked upon by CFD users as caused by the occurrence of blood stagnation areas. This implies that a good design for a RBP has two thresholds for the shear stress magnitude. The lower threshold is required to obviate stagnation areas and the upper threshold eliminates the creation of excessive shear stress.

The range of approaches to predict flow induced hemolysis lasts from a simple qualitative evaluation of wall shear stress magnitudes to tensor- and strain-based methods accounting for damage accumulation. Between those approaches are models with different levels of complexity which in most cases are individually validated by comparison to in-vitro blood tests. The validation might only hold true for specific pump configurations and an application of the model to a different device might lead to inaccurate hemolysis predictions. A reasonable distinction can be established by dividing models into Lagrange, Euler or hybrid models. Euler approaches have in common, that the flow field is evaluated by a global observer. A Lagrange model is based on particles representing RBCs. The exposure to stress and the exposure time are recorded as a RBC travels through the device [34]. Hybrid models consider results from both approaches and attempt to find a correlation that fits best to experimental results. A breakthrough that facilitates all approaches was established in 1995 when Bludszuweit introduced the idea to model the three dimensional stresses including normal and shear stress acting on a fluid particle by a scalar stress [35, 36]. Common results with reasonable effort in order to account for hemolysis early in the development process look like figure 6.

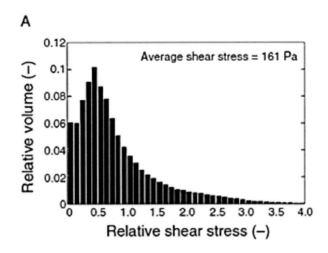

Figure 6. Volume fractions exposed to scalar stress levels [37].

A distribution like this enables a designer to quickly compare different designs and thus reduce the danger of developing non blood compatible devices.

Methods

Blood: Blood is a suspension containing cells. Because of the high concentration, the hematocrit is around 45 %, blood flow in general does not behave like a Newtonian fluid, see figure 7.

Because of the fact that the viscosity becomes reasonably constant with increasing shear rate, RBPs are in fact mostly modeled assuming a Newtonian behavior and a dynamic viscosity of the limiting value around 3.6 mPas. It has been shown that a vast majority of the volume within such a device is exposed to a shear rate above 100 s^{-1} [38].

Multiple frames of reference: Most recent RBPs are composed of stationary parts and one rotating part. To model the acceleration of the fluid due to the impeller, the employment of multiple frames of reference is the most widely used approach. The solvers that are mostly used in the field (ANSYS CFX, FLUENT) offer the use of a frozen rotor interface that effectively connects a rotating domain and a stationary domain. In doing so, a steady-state can be simulated that reduces computational load. With increasing computational resources, transient simulations become more and more feasible that enable a time dependant flow field prediction and an actual rotation of the impeller is modeled.

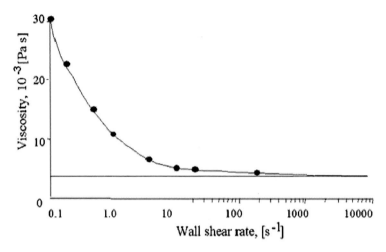

Figure 7. Blood dynamic viscosity for different shear rates.

Automated Optimization: In conjunction with analysis of a given design or component is usually the desire to improve according to different objectives. Burgreen showed in 2001 first approaches to optimize designs automatically in the field of RBPs together with CFD and coupling to an external optimization algorithm [31], see figure 8. Following this lead, other groups try to automate numerical design procedures [39, 40].

The expressed goal in the final design steps was to evaluate designs quantitatively, modify designs formally with an external optimization algorithm and modify geometries automatically including remeshing. The goal to formally optimize devices and machines and further decrease development time and costs with numerical simulation is thus naturally also present in the field of RBPs. Up to today, no integrated formal optimization algorithm internally connected to a flow solver has been used in the field of RBPs.

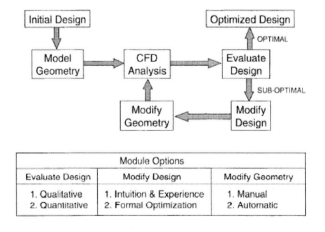

Figure 8. CFD assisted design process [31].

Results and Conclusions

The use of CFD in the field of RBPs has been taken up by a large number of research groups. The necessity to validate and always question numerical results is established and common practice. The fields of application have not changed over the years but there is a trend towards an automated design optimization process. The employment of CFD was reported to have shortened development times and costs. Aside from this, the possibility to determine, analyze and optimize flow parameters not or not easily quantifiable with experimental measurements seems to be the biggest advantage of CFD in the field.

III. COMPUTATIONAL FLUID DYNAMICS IN THE ANALYSIS AND DEVELOPMENT OF OXYGENATORS

Introduction and Medical Background

Oxygenators are used in clinical routine if the natural gas exchange of Oxygen (O_2) and Carbondioxide (CO_2) is disturbed and has to be replaced artificially. Extra corporeal oxygenators are often used to support or fully take over the limited breathing function of patients. Another common application is cardiopulmonary bypass. CPB describes a position during which the whole cardiac system including lung function is replaced (see also Chapter B-I). During the beginning of artificial oxygenation, blood was brought in direct contact with the gas (bubble-, disc- and filter-oxygenator). The advantage was that a very effective gas exchange rate could be realized. On the other hand, blood damage through direct contact with the gas phase was a strong disadvantage. Nowadays, mainly membrane oxygenators are used clinically to provide the necessary gas exchange. In these devices the gas and fluid phase are divided by a membrane, so that the blood does not get in direct contact with the gas. In modern membrane oxygenators, semi-permeable, microporous hollow fiber membranes are used. The gas flows inside these fibers while the blood flows around them. Depending on the oxygenator size, about 10.000 single hollow fiber membranes are linked to one oxygenator bundle, so that an exchange area of 0.2-2 m^2 is achieved. The natural lung however has an exchange area of about 80 m^2. So compared to physiological conditions, the exchange area of membrane oxygenators is very low.

For extracorporeal membrane oxygenation (ECMO) and CPB, oxygenators are sometimes used in combination with a rotary blood pump. An optimization of the flow path and the fiber position and orientation is necessary to provide a most effective gas exchange in these systems. An experimental measurement of the complex fluid dynamics within the oxygenators is hardly possible. Analysis of flow and transport processes using Computational Fluid Dynamics provides an easier and more cost effective alternative. However, it is crucial to choose a suitable mathematical model. In contradiction to experimental studies when only global changes of mass transport and pressure differences are defined, numerical studies are able to analyze local material values. The understanding and analysis of local flow and mass transport processes is the baseline to achieve essential improvements that are listed in the following:

- Avoidance of "dead-water-areas"
- Optimization of the flow through the fibrebundle.
- Optimization of the inlet profile of the fibrebundle
- Local analysis of mass transport

Due to the compact construction of fiber bundles within oxygenators (fiber-distance ~150μm) and the (partially) high gradients of velocity and gas concentration, a complete analysis of fiber bundles can only be realized with very large computational effort. Therefore, most analyses are performed under the following simplifications:

- 2D analysis
- Representation of fiber bundles by porous media
- Calculation of a small part of the fiber bundles, extrapolation onto the whole system

A 2D analysis is only valid and feasible for very simple geometries and problems. If instationary fluid states should be analyzed, this kind of simplification is not feasible for oxygenator systems, since the flow conditions normal to the 2D-plane are not included in the analysis. Additionally, the impact of a 3D-fiber constellation cannot be represented using a 2D analysis.

If the fiber bundle is represented by a porous media, the single fibers and the flow through them are neglected. Instead, a substitute is created, for which the impact of the fibers is considered via the porosity of a specific media. This porosity is dependent on the packing density and thus dependent on the numbers of single hollow fiber membranes. Besides porosity, the permeability of the bundle is set.

However, depending on the assembly, the permeability is not constant for all directions and has to be defined for each different flow direction separately. Small changes in fiber position and geometry that cause a different flow though the fibers, and thus the mass transport can hardly if at all be implemented in this method. One big advantage of this method is that a representation of the whole oxygenator domain is possible. On the other hand, local fluid conditions and local concentrations in between the fibers are not included in the analysis. If the flow through single fibers and thereby the local fluid conditions have to be considered, only a small part of the whole fiber bundle can be analyzed due to high computational effort. To analyze the whole geometry in terms of oxygenation power, the data has to be interpolated onto the whole system [42]. Due to the non-uniform flow through the fiber bundles of commercial systems and the thereby changed gas increase or decrease, this interpolation can cause uncertainties in the analysis. By direct numerical simulation of the flow around the fibers, this method can be applied to basic studies about the impact of different fiber assemblies and fiber geometries, for which no interpolation is needed. Also, the effect of different types of flow (constant or pulsatile flow) on the mass transport can be simulated. Mazahari and Ahmadi [43] studied the impact of the simulation by a direct modeling of fibers and the simplification using a porous media in a 2D-simulation. By analyzing pressure differences and passage times, a big discrepancy of both methods could be shown.

Modelling of Blood and Mass Transport

The fluid that flows through oxygenators is blood. The modeling of blood however is a great challenge for both computational fluid dynamics and mass transport simulation. While water is a Newtonian fluid, blood has non-Newtonian properties, so the viscosity varies in dependency on the shear rate of the fluid. This effect increases especially for small shear rates, when rouleaux formation of the erythrocytes occurs. For many applications like rotary blood pumps (Chapter B-II), this effect can be neglected. However, this simplification is not valid for oxygenators due to low flow velocities and thus low shear rates. Still, some authors accept this simplification in their studies [44-47]. Depending on the fiber assembly, up to 50% of the fluid volume are below the critical shear rate of 100 s^{-1}. Different mathematical descriptions for the non-Newtonian behavior of blood are available in the literature. They can be implemented in the numerical solver and thus the method.

Blood is a suspension of a fluid (plasma) and corpuscular parts (Hematocrit: erythrocytes, leucocytes, proteins). Due to the high corpuscular fraction (~45%), a resolution of the particles in the numerical simulation is not possible with current solver methods, since the particle solver are only valid for low particle concentrations. Therefore, blood is usually modeled as a homogenous, incompressible media with uniformly dispersed corpuscular parts (for the chemical binding of erythrocytes).

But the special properties of blood affect not only the flow. The mass transport of O_2 and CO_2 in blood is mainly dependent on two mechanisms. Firstly, the gases are dissolved in the blood plasma for the transport. Secondly, the biggest part is chemically bound to the hemoglobin within the erythrocytes. Both mechanisms have to be included in the solver. Different approaches to simulate mass transport for oxygenators following the mentioned simplifications are described in the literature. Zhang [47] describes the validation of an advection-diffusion-model for the mass transport with the assumption that the continuity approach for the fiber bundle is given by a porous media. Gage [48] showed that the simulated gas transport is significantly different for modeling of single fibers compared to modeling of the whole fiber bundle as a porous media.

Different methods can be found not only regarding the fiber bundle. For modeling of the mass transport itself, different models are used in the literature. Matsuda et al. [49] described the mass transport using Darcy's law which defines the blood flow through the fiber bundle at low Reynolds numbers. Based on key characteristics, the mass transfer coefficient which describes the mass transport from the gas phase to the fluid phase and is dependent on the flow in the fiber bundle is defined for a global solution [50-55].

For this, the Reynolds number, Sherwood number and Schmit number are used. However, small changes (e.g. the angle between the fibers) cannot be included for an estimation of the flow conditions within the fiber bundle using the Reynolds number, since these changes of the flow condition have no significant effect on the Reynolds number itself, but the altered flow around the fibers may influence the mass transport. For the Sherwood number, a different definition that has been experimentally measured is given by different authors for different fiber configurations [51]. For the simulation, a source is defined at the fibers which describes the gas volume flow into the blood in dependency on the mass transfer coefficient and the different partial pressures in the gas and fluid phase [47, 56, 57].

Vaslef [58] on the other hand invented a model that has been specially developed for hollow fiber membranes. In this model, the mass transport and the chemical binding of O_2 to the intracellular hemoglobin are described by an effective diffusivity in which an adsorption term is added to the molecular diffusivity. Svitek et al. [47] continued this idea and invented also an effective diffusivity for CO_2. The flow conditions are approximated by porous media in both cases.

Results and Conclusions

Numerical simulation of oxygenators can nowadays only be done under the restriction of many simplifications, mainly due to computational costs. The simulation of a whole fiber bundle including the single fibers is not possible. Therefore, the simplifications have to be adjusted to the corresponding problem.

Using a porous media, the whole oxygenator can be simulated and thus the global parameters can be calculated. However, analysis of the impact of different fiber parameters like angle and geometry is not possible. Additionally, the directional permeabilities of the fiber bundle need to be estimated in experiments prior to the simulation. A simulation of the flow around single fibers is limited to a small area of the whole system. But an extrapolation onto the whole oxygenator system can lead to errors in the results.

However, since the accuracy of models for mass transport in blood is constantly increasing and new methods for validation of these techniques are being developed, CFD can help to significantly increase the exchange power of oxygenators during the development process.

IV. COMPUTATIONAL FLUID DYNAMICS OF A MECHANICAL HEART VALVE PROSTHESIS

Introduction and Medical Background

If blood gets in contact with artificial materials it coagulates and forms so called thrombi. If these blood clots are swept along by the blood flow, they can cause embolism. These thromboembolic complications are one of the main problems for artificial organs with blood contact such as heart valve prostheses, especially mechanical valves [59]. 2-4% of patients per year suffer from thromboemblic complications after heart valve replacement [60]. A total of 75% of all complications of heart valve patients are related to thrombosis or anticoagulant-related bleedings [61]. Besides thrombogenic potential, these valves change the hydrodynamic performance of the heart. The analysis of the fluid dynamics of these devices is therefore of high clinical relevance during their development process.

This is done by experimental and numerical approaches [62, 63]. Experimentally, the formation of thrombi can be simulated in a specially designed valve tester [64]. But these experiments are limited in their spatial resolution, and an analysis of fluid dynamics can only be done regarding few parameters like pressure and flow curves. Additionally, some methods like Particle Image Velocimetry are not applicable due to non-transparent materials. Therefore, the experimental results are coupled to numerical data in order to gain insight in flow processes invisible to experimental approaches.

Another goal is to compare experimental and numerical results in order to find parameters to predict thrombogenesis. Thus, experimental data such as leaflet kinematics or pressure profiles are used as boundary conditions for the numerical studies. Besides flow patterns, the shear rates in pivot regions are analyzed. In particular, the opening and closing procedure of these valves are important due to cavitation and squeezeflow [65-69]. Thus, transient simulations are necessary to analyze mechanical heart valve prostheses.

In this chapter a study is presented of a numerical flow simulation of the opening and closing process of a mechanical heart valve, which is analyzed for the application of a valve tester consisting of a conical inlet, the valve, bulbs and aorta. Experimentally obtained boundary conditions like flow and pressure curves as well as the experimentally determined valve movement were used to create the numerical model. The pressure loss due to the valve provides experimental validation of the simulation.

Methods

The CAD model of the valve tester [64] consists of a conical inlet, the valve including ring and leaflets, the bulbs and a cylindrical outlet. The tested Triflo valve has three leaflets made of pyrolytic carbon and a ring of titanium alloy. Since the geometry is symmetric, $1/6^{th}$ can be used for numerical studies, whereby the computational time is significantly decreased.

During one cycle, the leaflets get in contact with the ring at different positions due to a complex bearing and guidance system. Thus, the pivot regions occur at different positions for different points in time. This is a challenge for the mesh generation, since the critical areas change over the time. Analysis of these regions however is very important, because experimental measurements in this area are not possible, but thrombus formation is very likely [70, 71].

The movement of the leaflets is described by mesh deformation, so as movement of nodes and thereby distortion of elements. To do so, mesh elements have to be created in each point in the geometry, and contact between the leaflets and the ring has to be neglected. Thus, the leaflets have to be shrinked to 99% of their original size in order to avoid contact with the bearings. This simplification is discussed later.

Using mesh deformation, the quality of the elements is decreased. A sufficient quality (e.g. angles of the elements) is necessary for feasible analysis. In order to simulate a whole cycle of systole and diastole (48° leaflet rotation in both directions), the movement is split up in small simulations of 3° each. The results from one simulation are used as initial values for the next one, so that 32 simulations complete a full heart cycle.

The shape of the leaflets is very complex. Therefore, tetrahedral elements are used to represent the geometry. These elements become very small in the pivot regions. A sufficient gradient in the mesh size has to be implemented in order to provide feasible results. Additionally, layers of prismatic elements are created around the leaflets in order to support the mesh deformation. After a mesh independency study, 900.000 elements are created to represent $1/6^{th}$ of the geometry.

The numerical studies must have the same boundary conditions as the corresponding experiments. This is done by using experimental measurements as input data for the simulations. For this application [64], the following settings are used:

- Cycle time: 0.8 s
- Systolic time: 0.28 s
- Beats per minute: 75
- Pressure profile: 140/70 mmHg

Since the test fluid of the valve tester [64] is Newtonian water-glycerine, a Newtonian blood model with a constant dynamic viscosity of 3.6 mPa*s and a density of 1.059 kg/m^3 is assumed.

Shear Stress Transport (SST) is used as turbulence model and the time step size is set in order to provide a temporal resolution of 33 ms.

To get the best possible adjustment of the boundary conditions given in the experiments, a flow field is set as inlet profile during systole. This flow field during systole is simulated separately for the tester geometry without valve, assuming that the valve itself has no effect on the inlet profile during that part of the cycle. For diastole, the corresponding pressure curves at the in- and outlet are set as opening pressures in the simulation.

The valve movement is measured with a high-speed video (2000 fps). The distance of the leaflet tips and the central point is then calculated by self-written software based on LabView which detects the leaflet position by brightness comparison. Thus, the leaflet opening angle is known for each point in time. This experimentally routed motion is then transferred to the simulations and fit to the pressure and flow curves.

Results and Conclusions

During systole, a vortex which stabilizes the flow into the aorta is generated behind the valve. This vortex moves to the bulbs during diastole and channels fluid onto the backside of the leaflets, supporting the closing procedure. High gradients occur mainly in the pivot areas, where also the highest shear is observed. The experimental validation is discussed in Chapter C.

These transient simulations of valve behavior provide the possibility to analyze flow patterns and effects invisible to experimental approaches. They can thereby support the developing process of devices and methods in the field of artificial heart valves. A coupling of experimental and numerical data has proven to be a helpful methodology to understand and describe complex setups.

However, the process of thrombus formation itself cannot be studies with rather simple numerical methods, since the interactions are very complex and partly occur on a molecular level. Thus, new methods need to be worked out to describe thrombus formation in numerical studies.

A first step is particle tracking [72]. Tracking of the pathways of single particles in the blood flow allows analysis of the integrated shear exposure of platelets, which is crucial for thrombus formation. Another possibility is to simulate the adhesion of particles. Such complex numerical models are currently developed in several institutions all over the world. However, they have not reached

VALIDATION OF COMPUTATIONAL FLUID DYNAMICS IN THE FIELD OF BIOMEDICAL ENGINEERING

Background and Necessity

Simulating the human body is a huge challenge. The body itself is a very complex system with interactions that are based on various fields of science, ranging from medicine and biology to chemistry and physics. Many of these phenomena take place on a molecular level. Numerical simulations however describe phenomena based on differential equations. But since many interactions in the human body are still unknown and the level of complexity that is necessary to describe a phenomenon varies for all applications, it is not always possible to include these phenomena in the equations. Instead, many simplifications and assumptions are necessary for these Computational Fluid Dynamics studies. That is why, in the end, only animal or clinical trials can provide an established validation for biomedical engineering and are therefore necessary before any device can go into clinical use.

There are however several experimental methods that need to be done beforehand in order to validate simulations in an in-vitro environment. Such a validation can only be performed if the boundary conditions are the same for simulation and experiment. So setting the boundary conditions precisely is a necessary step for any validation. But nevertheless it has to be kept in mind that all these experimental methods themselves are subject to errors and uncertainties. So the analysis of experimental data is by itself challenging and has to be done with care. Calibration and statistical analysis are only two of many requirements. A detailed discussion of these obstacles is not part of the following Chapter.

Some Common Methods

Flow and Pressure Probes

The basic method of comparing numerical results for biomedical applications is via flow and pressure probes. In the simulations, the flow and pressure can be read out at any point in the geometry. The probes however can only be places at accessible positions that vary for each application. Given that the exact same boundary conditions are applied for both the simulation and the experiment, the first important step is to choose a position that is relevant for analysis while accessible for the experiment and robust for the simulations. Once this is done, experimental and numerical data can be compared. It is assumed that if at these specific points the results are identical, that applies to the rest of the geometry as well. Of course, each experimental measurement has errors. So identical results are only achieved in theory. Differences of up to 20% are a common range for comparison of numerical and experimental results in the field of biomedical engineering.

Mock Circulation Loops

Mock circulation loops (MCL) are a hydraulic analogy of the human circulatory system. They are used to mimic the behavior of the human body, especially the vascular system. Typically, an MCL features both a pulmonary and systemic circuit as well as the four chambers of the heart. Not all of these components are necessary for each application, so partial MCLs exist as well. In general, MCLs are used to test devices like blood pumps or total artificial hearts in terms of physiological response. Thereby, they provide the possibility to validate numerical studies.

Particle Image Velocimetry

Particle Image Velocimetry or PIV is a method to visualize flow patterns. It uses small particles (~10μm) which are dispersed in the fluid. These particles are illuminated by lasers and their reflections are captured with high-speed cameras. The displacement of these particles after a short time interval is then used to calculate the fluid movement.

Since the particles need to be accessible by the lasers and cameras at the same time, only a sheet at a time is analyzed. Analysis of whole flow patterns is done "sheet by sheet". One other aspect is the necessity of a fully transparent geometry. Already a small optical distortion results in high losses of accuracy. Therefore, transparent models need to be crafted in order to use PIV as a validation method. Besides these challenges, PIV provides the possibility of comparing complex flow patterns for a whole range of applications.

Other Methods

The examples above are only few of the many possibilities of experimental validation of numerical studies. Basically, every part of a CFD study that can be reproduced in an experiment can be used as experimental validation, as long as it is relevant for the specific application. Of course, it has always to be considered if validation of one variable at a specific position or working point also verifies other values. This is often assumed, even though it is questionable.

Cardiopulmonary Bypass

There are several challenges for validation of the flow simulation in the cardiovascular system. Firstly, the geometry needs to be the same for both applications. In particular, the results are very sensitive to the cannula position, which needs to be adjusted carefully. Secondly, the boundary conditions need to be identical.

For physiological conditions, this can be done by MRI-flow measurements, which is done by some groups over the world. But the spatial resolution of this method is limited, and so is the level accuracy of the validation. Additionally, it is not applicable for CPB, since the operations cannot be performed within a CT or MRI scanner. Therefore, different methods are needed in order to provide a quantitative validation of the CFD results. Taking into account necessity of identical boundary conditions and cannula positioning, a combination of Particle Image Velocimetry and flow measurements is used.

The first step is the model creation, which is based on the same CAD model as the simulations. From this model, a 3-dimensional, physical model is crafted by rapid prototyping. Using silicone potting, a silicone block is generated around this model. Afterwards the inner core is removed. Thus, a transparent silicone negative of the cardiovascular system is created that has the exact same geometry as the CFD mesh. In this model the cannula is placed at a predefined position. To assure an identical position, this model is then placed in a micro-CT. The resulting data set is compared to the previous CAD model in order to have the same cannula position for the CFD analysis.

The pressure level during the PIV measurements is set via a big reservoir, and a rotary pump provides the inlet flow.

Another crucial point is the fluid viscosity. The PIV test fluid must have the same refractive index as the silicone in order to provide full optical access. Thus, it has a Newtonian viscosity of 7.6 mPa*s. In order to validate the CFD results, additional simulations with the same fluid viscosity need to be performed.

To validate the studies presented in chapter B-I, two validation steps were performed: one for validation of subclavian cannulation, one for validation of aortic cannulation. For cannulation to the subclavian artery, the flow direction is the same for all vessels. For cannulation of the aorta, a flow-splitting occurs where the jet hits the aortic arch. Thereby, two vortices are generated in the left

Rotary Blood Pumps

Validation of numerical results is commonly executed by comparison of pump characteristic curves (H-Q) from CFD and experiments with a physical model. An outstanding example of actually validating flow patterns within the impeller passages of an axial flow pump was published in 2006 by Triep [41], see figure 9. By comparison of flow patterns obtained with flow visualization experiments in a physical and transparent model to numerical results, one can be sure of a more coherent validation of the flow behavior.

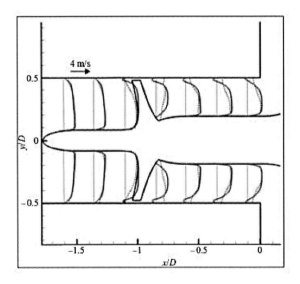

Figure 9. Comparison of PIV and CFD [41].

Oxygenators

A comparison of experimental data and numerical studies is used for validation of the numerical studies. The hydraulic assumptions are validated by measurement of the pressure difference over the fiber bundle. However, a validation of the flow path in the fiber bundle is not possible with this method. Validation of the mass transport is done by experimental measurements and the comparison of O_2 and CO_2 partial pressure or the percentage of blood saturation. Zhang et al. [47] validated their transport model by taking blood probes at different positions along a fiber bundle.

A comparison of experimental and numerical blood saturation showed good agreement. Fill et al. [56] analyzed the exchanged gas volume flow compared to the whole blood volume flow. However, the numerical and experimental results differed. A validation of the local gas concentrations within the fiber bundle is currently not possible. This can only be calculated and visualized numerically.

Heart Valves

There are several ways of validating the simulation of a mechanical valve. A qualitative validation of the flow patterns behind the valve has been done in a PIV study. In general, it shows the same flow patterns as the simulation. In particular, the vortex behind the valve that stabilizes the flow during systole and supports the closure during diastole is found.

Another way of validation is comparison of pressure curves. Thereby, differences of 2.5% during systole occur. But during diastole, this increases up to 25%. One reason for this is the assumption of 1% shrinkage of the leaflet. Thereby, the leakage flow through the valve during diastole is increased in the simulation, whereby the pressure difference is decreased.

Validation via thrombus formation is very challenging. Once a small thrombus is generated, it grows more or less randomly. However, most of the thrombi in the valve tester [64] occur in the areas where the highest shear rates occur in the simulation, so the numerical results can be validated qualitatively.

CONCLUSION

Biomedical engineering is in the main focus of interest in the modern society and is thus a constantly growing field. While some methodologies are already very sophisticated, there are still many challenges for the future. Beside very complex models for the interactions in or with the human body, saving costs for new therapies and devices is very important. For many applications in the clinic, Computational Fluid Dynamics is a powerful tool do save development time and costs.

However, the credibility of these studies has to be increased. A failure in a medical device can lead to immediate patient death. Thus, careful validation and interpretation of the results is absolutely necessary.

Additionally, Computational Fluid Dynamics can only be part of a whole range of methods including experiments, animal trials and clinical trials that need to be performed before any new development will actually be used in clinical routine.

This overview over some projects and research areas is only a tiny pinhole compared to the huge variety of research in many institutions worldwide. It can merely clarify the importance of Computational Fluid Dynamics and explain some of the essential steps to take for a successful development of medical devices or methodologies. Following these guidelines, Computational Fluid Dynamics can help to increase patient's wellbeing, safety and chances of survival.

REFERENCES

[1] Hornick P., Taylor K.; *J. Cardiothorac. Vasc. Anesth.* 1997 May;11(3):310-315: Pulsatile and nonpulsatile perfusion: the continuing controversy.

[2] Ji B., Undar A.; *ASAIO J.* 2006 Jul-Aug;52(4):357-361: An evaluation of the benefits of pulsatile perfusion during cardiopulmonary bypass procedures in pediatric and adult cardiac patients.

[3] Tarcan O., Ozatik M.A., Kale A., Akgül A., Kocakulak M., Balci M., Undar A., Kucukaksu D.S., Sener E., Tasdemir O.; *Med. Sci. Monit.* 2004 Jul;10(7):CR294-299. Epub 2004 Jun 29: Comparison of pulsatile and non-pulsatile cardiopulmonary bypass in patients with chronic obstructive pulmonary disease.

[4] Undar A., Ji B., Kunselman A.R., Myers J.L.; *ASAIO J.* 2007 Nov-Dec; 53(6): 725-729: Detection and classification of gaseous microemboli during pulsatile and nonpulsatile perfusion in a simulated neonatal CPB model.

[5] Fukae K., Tominaga R., Tokunaga S., Kawachi Y., Imaizumi T., Yasui H.; *J. Thorac. Cardiovasc. Surg.* 1996 Feb;111(2):478-484: The efects of pulsatile and nonpulsatile systemic perfusion on renal sympathetic nerve activity in anesthetized dogs.

[6] Kapetanakis E.I., Stamou S.C., Dullum M.K., Hill P.C., Halle E., Boyce S.W., Bafi A.S., Petro K.R., Corso P.J.; *Ann. Thorac. Surg.* 2004 Nov;78(5):1564-1571: The impact of aortic manipulation on neurologic outcomes after coronary artery bypass surgery: a risk-adjusted study.

[7] Nakamura K., Harasaki H., Fukumura F., Fukamachi K., Whalen R.; *Scand. Cardiovasc. J.* 2004 Mar;38(1):59-63: Comparison of pulsatile and non-pulsatile cardiopulmonary bypass on regional renal blood flow in sheep.

[8] Scarborough J.E., White W., Derilus F.E. Mathew J.P., Newman M.F., Landolfo K.P.; *Semin. Thorac. Cardiovasc. Surg.* 2003 Jan;15(1):52-62: Neurological outcomes after coronary artery bypass grafting with and without cardiopulmonary bypass.

[9] Undar A., Masai T., Frazier O.H., Fraser C.D. Jr.; *ASAIO J.* 1999 Nov-Dec;45(6):610-614: Pulsatile and nonpulsatile flows can be quantified in terms of energy equivalent pressure during cardiopulmonary bypass for direct comparisons.

[10] Verdonck P.R., Siller U., De Wachter D.S., De Somer F., Van Nooten G.; *Int. J. Artif. Organs.* 1998 Nov;21(11):705-713: Hydrodynamic comparison of aortic arch cannulae.

[11] Grossi E.A., Kanchuger M.S., Schwartz D.S., McLoughlin D.E., LeBoutillier M 3rd, Ribakove G.H., Marschall K.E., Galloway, A.C., Colvin S.B.; Ann Thorac Surg. 1995 Mar;59(3):710-712: Effect of cannula lenght on aortic arch flow: protection of the atheromatous aortic arch.

[12] Orime Y., Shiono M., Hata H., Yagi S., Tsukamoto S., Okumura H., Nakata K., Kimura S., Hata M., Sezai A., Sezai Y.; *Artif. Organs.* 1999 Jun;23(6):508-512: Cytokine and endothelial damage in pulsatile and nonpulsatile cardiopulmonary bypass.

[13] Albert A.A., Beller C.J., Arnrich B., Walter J.A., Rosendahl U.P., Hetzel A., Priss H., Ennker J., Perfusion 2002; 17; 451: Is there any impact of the shape of aortic end-hole cannula on stroke occurrence? Clinical evaluation of straight and bent-tip aortic cannulae.

[14] Joubert-Huebner E., Gerdes A., Sievers H-H., Perfusion 2000; 15; 69: An in vitro evaluation of a new cannula tip design compared with two clinically established cannula-tip designs regarding aortic arch vessel perfusion characteristics.

[15] Gerdes, A., Joubert-Huebner, E., Esders K., Sievers H-H., *Annals of Thoracic Surgery* 2000; 69; 1425-1430: Hydrodynamics of Aortic Arch Vessels During Perfusion Through the Right Subclavian Artery.

[16] Meyns B, Verbeken E, Kerkhoffs W, et al.; *The Journal of Heart and Lung Transplantation* 2006;25(2s):s123-s124: Partial Support For Chronic Heart Failure With A Subcutaneous Pump.

[17] Meyns , F . Rega , W . Droogne , A . Simon , S . Klotz , D . Burkhoff; *Journal of Cardiac Failure* , Volume 14 , Issue 6 , Pages S45 - S46 B: Initial Clinical Experience with the Circulite Synergyâ„¢ Device for Partial Circulatory Support.

[18] Blauth C.I., Arnold J.V., Schulenberg W.E., McCartney A.C., Taylor K.M.; *J. Thorac. Cardiovasc. Surg.* 1988 Apr;95(4):668-676: Cerebral microembolism during cardiopulmonary bypass. Retinal microvascular studies in vivo with fluorescein angiography.

[19] Kaufmann T.A., Hormes M., Laumen M., Timms D.L., Linde T., Schmitz-Rode T., Moritz A., Dzemali O., Steinseifer U. The impact of aortic/subclavian outflow cannulation for cardiopulmonary bypass and cardiac support: a computational fluid dynamics study. *Artif Organs*, 33(9):727–732, Sep 2009.

[20] Kaufmann T.A., Hormes M., Laumen M., Timms D.L., Linde T., Schmitz-Rode T., Moritz A., Dzemali O., Steinseifer U. Flow distribution during cardiopulmonary bypass in dependency on the outflow cannula positioning. *Artif Organs*, 33(11):988–992, Nov 2009.

[21] Rosamond W, Flegal K, Friday G, Furie K, Go A, Greenlund K, et al. Heart disease and stroke statistics–2007 update: a report from the American Heart Association Statistics Committee and Stroke Statistics Subcommittee. *Circulation.* 2007 Feb;115(5):e69–171.

[22] Rose EA, Moskowitz AJ, Packer M, Sollano JA, Williams DL, Tierney AR, et al. The REMATCH trial: rationale, design, and end points. Randomized Evaluation of Mechanical Assistance for the Treatment of Congestive Heart Failure. *Ann Thorac Surg.* 1999 Mar;67(3):723–730.

[23] Wampler R, Lancisi D, Indravudh V, Gauthier R, Fine R. A sealless centrifugal blood pump with passive magnetic and hydrodynamic bearings. *Artif. Organs.* 1999 Aug;23(8):780–784.

[24] Göbel C, Arvand A, Eilers R, Marseille O, Bals C, Meyns B, et al. Development of the MEDOS/HIA DeltaStream extracorporeal rotary blood pump. *Artif. Organs.* 2001 May;25(5):358–365.

[25] Ohara Y, Sakuma I, Makinouchi K, Damm G, Glueck J, Mizuguchi K, et al. Baylor Gyro Pump: a completely seal-less centrifugal pump aiming for long-term circulatory support. *Artif. Organs.* 1993 Jul;17(7):599–604.

[26] Jarvik RK. System considerations favoring rotary artificial hearts with blood-immersed bearings. *Artif. Organs.* 1995 Jul;19(7):565–570.

[27] Akamatsu T, Nakazeki T, Itoh H. Centrifugal blood pump with a magnetically suspended impeller. *Artif. Organs.* 1992 Jun;16(3):305–308.

[28] Masuzawa T, Ezoe S, Kato T, Okada Y. Magnetically suspended centrifugal blood pump with an axially levitated motor. *Artif. Organs.* 2003 Jul;27(7):631–638.

[29] Golding LA, Smith WA. Cleveland clinic rotodynamic pump. Ann Thorac Surg. 1996 Jan;61(1):457–462. Available from: http://dx.doi.org/10.1016/0003-4975(95)00944-2.

[30] Watterson PA, Woodard JC, Ramsden VS, Reizes JA. VentrAssist hydrodynamically suspended, open, centrifugal blood pump. *Artif. Organs*. 2000 Jun;24(6):475–477.

[31] Burgreen GW, Antaki JF, Wu ZJ, Holmes AJ. Computational fluid dynamics as a development tool for rotary blood pumps. *Artif. Organs*. 2001 May;25(5):336–340.

[32] Steinbrecher C. Numerische Simulation eines berührungsfrei gelagerten Rotors für eine Blutpumpe [dissertation]. TU Muenchen; 2004.

[33] Graefe R, Timms D, Boehning F, Schmitz-Rode T, Steinseifer U. Investigation of the Influence of Volute Design on Journal Bearing Bias Force using CFD. *Artif. Organs*. 2010;Forthcoming.

[34] Apel J, Paul R, Klaus S, Siess T, Reul H. Assessment of hemolysis related quantities in a microaxial blood pump by computational fluid dynamics. *Artif. Organs*. 2001 May;25(5):341–347.

[35] Bludszuweit C. Model for a general mechanical blood damage prediction. *Artif. Organs*. 1995 Jul;19(7):583–589.

[36] Bludszuweit C. Three-dimensional numerical prediction of stress loading of blood particles in a centrifugal pump. *Artif. Organs*. 1995 Jul;19(7):590–596.

[37] Nishida M, Maruyama O, Kosaka R, Yamane T, Kogure H, Kawamura H, et al. Hemocompatibility evaluation with experimental and computational fluid dynamic analyses for a monopivot circulatory assist pump. *Artif. Organs*. 2009 Apr;33(4):378–386. Available from: http://dx.doi.org/10.1111/j.1525-1594.2009.00730.x.

[38] Apel J. Numerische Simulation der Strömung in Miniaturkreiselpumpen zur Blutförderung [dissertation]. RWTH Aachen; 2002.

[39] Graefe R, Borchardt R, Arens J, Schlanstein P, Schmitz-Rode T, Steinseifer U. Improving oxygenator performance using computational simulation and flow field based parameters. *Artif. Organs*. 2010;Forthcoming.

[40] Zhu L, Zhang X, Yao Z. Shape optimization of the diffuser blade of an axial blood pump by computational fluid dynamics. *Artif. Organs*. 2010 Mar;34(3):185–192. Available from: http://dx.doi.org/10.1111/j.1525-1594.2009.00799.x.

[41] Triep M, Brücker C, Schröder W, Siess T. Computational fluid dynamics and digital particle image velocimetry study of the flow through an optimized micro-axial blood pump. *Artif. Organs*. 2006 May;30(5):384–391. Available from: http://dx.doi.org/-10.1111/j.1525-1594.2006.00230.x.

[42] P.W. Dierickx, D.S. De Wachter, P.R. Verdonck, Two-dimensional Finite Element Model for Oxygen Transfer in Cross Flow Hollow Fiber Membrane Artificial Organs, *The International Journal of Artificial Organs*, 2001

[43] A.R. Mazaheri, G. Ahmadi, Uniformity of the Fluid Flow Velocities Within Hollow Fiber Membranes of Blood Oxygenation Devices, *Artificial Organs*, 30(1), 10-15, 2006.

[44] Asakawa Y., Funakubo A., Fukunaga K., Taga I., Higami T., Kawamura T., Fukui Y, Development of an Implantable Oxygenator with cross flow pump, *ASAIO Journal*, 2006.

[45] I.N. Zinovik, W.J. Federspiel, Modelling of blood flow in a ballon-pulsed intravascular respiratory catheter, *ASAIO Journal* 2007, 464-468, 2007.

[46] M.J. Gartner, C.R. Wilhelm, K.L. Gage, M.C. Fabrizio, W.R. Wagner, Modelling flow effects on thrombotic deposition in a membrane oxygenator, *Artificial Organs*, 24(1), 29-36, 2000.

[47] J. Zhang, T.D.C. Nolan, T. Zhang, B.P. Griffith, Z.J. Wu, Characterization of membrane blood oxygenator devices using computational fluid dynamcs, *Journal of Membrane Science*, 288, 268-279, 2007.

[48] K.L. Gage, M.J. Gartner, G.W. Burgreen, W.R. Wagner, Predicting Membrane Oxygenator Pressure Drop Using Computational Fluid Dynamics, International Society for *Artificial Organs*, 26(7), 600-607, 2002.

[49] N. Matsuda, M. Nakamura, K. Sakai, K. Kuwana, K. Tahara, Theoretical and Experimental Evaluation for Blood Pressure Drop and Oxygen Transfer Rate in Outside Blood Flow Membrane Oxygenator, *Journal of Chemical Engineering of Japan*, 32 No. 6, 752-759, 1999.

[50] A.R. Goerke, J. Leung, S.R. Wickramasinghe, Mass and momentum transfer in blood oxygenators, *Chemical Engineering Science*, 57, 2035-2046, 2002.

[51] Gabelman, S.T. Hwang, Hollow fiber membrane contactors, *Journal of Membrane Science, 159*, 61-106, 1999.

[52] N. Matsuda, K. Sakai, Blood flow and oxygen transfer rate of an outside blood flow membrane oxygenator, *Journal of Membrane Science*, 170, 153-158, 2000.

[53] T.J. Hewwitt, B.G. Hattler, W.J. Federspiel, A Mathematical Model of Gas Exchange in an Intravenous Membrane Oxygenator, *Annals of Biomedical Engineering*, 26, 166-178, 1998.

[54] W.J. Federspiel, T.J. Hewitt, B.G. Hattler, Experimental Evaluation of a Model for Oxygen Exchange in a Pulsating Intravascular Artificial Lung, *Annals of Biomedical Engineering,* 28, 160-167, 2000.

[55] A.M. Guzman, R.A. Escobar, C.H. Amon, Methodology for Predicting Oxygen Transport on an Intravenous Membrane Oxygenator Combining Computational and Analytical Models, *Journal of Biomedical Engineering*, 127, 1127-1140, 2005.

[56] B. Fill, M. Gartner, G. Johnson, M. Horner, J. Ma, Computational fluid flow and mass transfer of a functionally integrated pediatric pump-oxygenator configuration, *ASAIO Journal,* , 214-219, 2008.

[57] R.G. Svitek, W.J. Federspiel, A Mathematical Model to Predict CO2 Removal in Hollow Fiber Membrane Oxygenators, *Annals of Biomedical Engineering*, 36 (6), 992-1003, 2008.

[58] S.N. Vaslef, L.F. Mockros, R.W. Anderson, R.J. Leonard, Use of a mathematical model to predict oxygen transfer rates in hollow fiber membrane oxygenators, *ASAIO,* 40(4), 990-996, 1994.

[59] Paul R, Marseille O, Hintze E, Huber L, Schima H, Reul H, et al. In vitro thrombogenicity testing of artificial organs. *Int. J. Artif. Organs.* 1998 Sep;21(9):548–552.

[60] Cannegieter SC, Rosendaal FR, Briët E. Thromboembolic and bleeding complications in patients with mechanical heart valve prostheses. *Circulation.* 1994 Feb;89(2):635–641.

[61] Bonow RO, Carabello BA, Chatterjee K, de Leon AC, Faxon DP, Freed MD, et al. ACC/AHA 2006 guidelines for the management of patients with valvular heart disease: a report of the American College of Cardiology/American Heart Association Task

Force on Practice Guidelines (writing Committee to Revise the 1998 guidelines for the management of patients with valvular heart disease) developed in collaboration with the Society of Cardiovascular Anesthesiologists endorsed by the Society for Cardiovascular Angiography and Interventions and the Society of Thoracic Surgeons. *J Am Coll Cardiol.* 2006 Aug;48(3):e1–148.

[62] Shi Y, Zhao Y, Yeo TJH, Hwang NHC. Numerical simulation of opening process in a bileaflet mechanical heart valve under pulsatile flow condition. *J. Heart Valve. Dis.* 2003 Mar;12(2):245–255.

[63] Redaelli A, Bothorel H, Votta E, Soncini M, Morbiducci U, Gaudio CD, et al. 3-D simulation of the St. Jude Medical bileaflet valve opening process: fluid-structure interaction study and experimental validation. *J. Heart Valve. Dis.* 2004 Sep;13(5):804–813.

[64] Maegdefessel L, Linde T, Michel T, Hamilton K, Steinseifer U, Friedrich I, Schubert S, Hauroeder B, Raaz U, Buerke M, Werdan K, Schlitt A. Argatroban and bivalirudin compared to unfractionated heparin in preventing thrombus formation on mechanical heart valves. Results of an in-vitro study. *Thromb Haemost.* 2009 Jun;101(6):1163-9.

[65] Lu PC, Liu JS, Huang RH, Lo CW, Lai HC, Hwang NHC. The closing behavior of mechanical aortic heart valve prostheses. *ASAIO J.* 2004;50(4):294–300.

[66] Graf T, Reul H, Dietz W, Wilmes R, Rau G. Cavitation of mechanical heart valves under physiologic conditions. *J. Heart Valve. Dis.* 1992 Sept;1:131-141.

[67] Graf T, Reul H, Detlefs C, Wilmes R, Rau G. Causes and formation of cavitation in mechanical heart valves. *J. Heart Valve. Dis.* 1994 April; 3(1):49-64.

[68] Lee C S, Aluri S, Chandran K B. Effect of valve holder flexibility on cavitation initiation with mechanical heart valve prostheses: an in vitro study. *J. Heart Valve. Dis.* 1996; 5(1):104-113.

[69] Chandran K B, Aluri S. Mechanical valve closing dynamics: relationship between velocity of closing, pressure transients, and cavitation initiation. *Ann Biomed Eng.* 1997; 25(6):926-938.

[70] Scharfschwerdt M, Thomschke M, Sievers HH. In-vitro localization of initial flow-induced thrombus formation in bileaflet mechanical heart valves. *ASAIO J.* 2009;55(1):19–23.

[71] Meuris B, Verbeken E, Flameng W. Mechanical valve thrombosis in a chronic animal model: differences between monoleaflet and bileaflet valves. *J. Heart Valve. Dis.* 2005 Jan;14(1):96–104.

[72] Alemu Y, Bluestein D. Flow-induced platelet activation and damage accumulation in a mechanical heart valve: numerical studies. *Artif. Organs.* 2007 Sep;31(9):677–688.

In: Computational Fluid Dynamics
Editor: Alyssa D. Murphy

ISBN 978-1-61209-276-8
© 2011 Nova Science Publishers, Inc.

Chapter 4

PREDICTION OF HEAT TRANSFER AND AIR FLOW IN SOLAR HEATED VENTILATION CAVITIES

Guohui Gan[*]

Department of Architecture and Built Environment,
University of Nottingham, University Park, Nottingham NG7 2RD, UK

ABSTRACT

Solar heated ventilation cavities are used to enhance passive cooling of buildings and building-integrated photovoltaics. This chapter is concerned with numerical prediction of buoyancy-driven heat transfer and air flow in vertical cavities for natural ventilation of buildings and inclined cavities for ventilation cooling of building-integrated photovoltaics. Computational fluid dynamics was used to predict the heat transfer and air flow rates in vertical ventilation cavities with various combinations of heat distribution ratios ranging from symmetrical to asymmetrical heating and inlet/outlet openings and also in inclined ventilation cavities with combined convective and radiative heat transfer. The natural ventilation rate and heat transfer rate in a vertical cavity have been found to vary with the total heat flux, heat distribution on the cavity wall surfaces, cavity size and opening positions. It has also been found that reducing asymmetry in heat distribution on the vertical cavity wall surfaces can increase the ventilation rate but would decrease the heat transfer rate. The air flow rate and heat transfer rate in an inclined cavity generally increase with inclination angle while the temperature of PV cells decreases with increasing inclination. General correlations between these variables in the cavities have been obtained and presented in non-dimensional forms.

INTRODUCTION

Solar heated ventilation cavity structures including solar chimneys, Trombe walls and double facades make use of thermal buoyancy for passive heating and natural ventilation of buildings. The exterior skin of a traditional Trombe wall or solar chimney is made of glass to

---
[*] Tel: 44 115 9514876, Fax: 44 115 9513159, Email: guohui.gan@nottingham.ac.uk

allow solar heat transmission and the interior wall collects and stores solar energy to preheat incoming air in cold seasons and/or enhance natural ventilation in summer. In such a ventilation structure, air flow in the cavity is mainly driven by buoyancy along the storage wall due to negligible absorption of solar heat by the exterior skin. Recent years have witnessed the rapid development of building-integrated photovoltaics as a renewable energy system. Photovoltaic (PV) devices can be mounted on a building roof or directly integrated into a building envelope including replacement of the exterior skin of double facades [1, 2] and solar chimneys [3]. PV devices can absorb a significant proportion of solar heat and thus the exterior skin of such a structure can also function like the interior storage wall under bright sunshine in terms of the driving force for air flow in the cavity. The heat transfer and air flow rates through a cavity with two heated walls would differ from those with one heated wall.

A solar chimney or similar structure can also be installed on a pitched roof to form an inclined ventilation cavity. A more widely deployed type of inclined cavity is however represented by the air gap behind a PV device mounted on a building with pitched roof. The most commonly used material for a PV device is based on crystalline silicon. The efficiency of electricity conversion of a crystalline PV module decreases with increasing temperature, therefore it is vital to provide an adequate air gap between the module and building envelope to assist buoyancy-driven ventilation cooling.

The simplest form of (two-dimensional) ventilation cavity comprises of two parallel walls/plates and heat transfer through such a cavity has been numerically studied previously [4 - 6]. Baskaya, et al. [7] investigated the effects of plate separation and inclination on natural convection between asymmetrically heated vertical and inclined parallel plates and found that the heat transfer increased with the Rayleigh number but decreased with the increase in the inclination angle from the vertical position (i.e., the heat transfer would increase with inclination angle from the horizontal position, as defined for the inclination in this chapter). Heat transfer and air flow in a ventilation cavity are influenced by the heat distribution on the wall surfaces as well as the total heat input. Rodrigues, et al. [8] found that the flow rate in a simulated vertical solar collector with asymmetrical heating at a fixed heat flux ratio of 1/5 between the cold wall and hot wall increased with channel width and total heat flux. Nguyen, et al. [9] developed a correlation between Nusselt number and Rayleigh number from the measurement of the wall temperature along a vertical channel asymmetrically heated at percentage heat distribution ratios from close to zero to nearly 50% for aspect ratios between 12 and 40.

In addition to the heat distribution and cavity size, another major factor that affects the heat transfer and air flow in a ventilation cavity is the flow resistance offered by the entrance and exit openings. For a vertical cavity, either of inlet or outlet opening can be horizontal or vertical. A horizontal inlet opening on a vertical wall of the cavity is commonly associated with natural ventilation of buildings whereas a vertical one at the bottom of the cavity represents operation of ventilation cooling of a double facade in a multistory building. The top of a solar chimney can be open to the ambient to form a vertical outlet or covered with a rain guard to form horizontal outlets allowing air to flow out of the chimney in any horizontal direction. The cavity top may also form part of a building structure such as the ventilated double facade or Trombe wall. In most previous numerical studies of buoyancy-driven cavity flow, inlet and outlet boundaries were specified with uniform properties such as air pressure and temperature right at the openings using a computational domain the same size as the cavity geometry but use of a uniform pressure at the inlet would result in a nearly uniform inlet velocity as indicated by Baskaya, et al. [7] and Rodrigues, et al. [8]. Morrone, et al. [10] optimised the plate separation of a channel formed by two parallel plates with symmetrical uniform heat flux for natural convection using an I-shaped computational domain to represent two sub-domains - the actual physical domain between the plates, and two large rectangular reservoirs placed upstream of the entrance and downstream of the exit. The author for this chapter [11, 12] has recently analysed the effect of computational domain on the prediction of buoyancy-driven natural ventilation through vertical cavities of various sizes and heat distribution ratios and the results have shown the imperative for accurate prediction of natural ventilation using an extended computational domain larger than the physical size of the cavity such that flow patterns at inlet and outlet openings are established naturally from the flow within the cavity and without.

This chapter is concerned with numerical prediction of buoyant air flow in solar heated vertical ventilation cavities for a wide range of Rayleigh numbers with both horizontal and vertical inlet/outlet openings for natural ventilation of buildings and in inclined cavities for building-integrated photovoltaics. Computational fluid dynamics (CFD) was used to predict the effects of heat distribution, cavity width and height, inlet and outlet positions and inclination angle as well as computational domain on the air flow and heat transfer rates.

METHODOLOGY

A commercial CFD package [13] was used for prediction of steady state air flow and heat transfer through ventilation cavities. Air moving through a solar heated ventilation cavity would likely involve both laminar and turbulent flow and the RNG k-ε turbulence model [14] was employed for modelling of air turbulence.

Model Equations

For an incompressible steady-state flow, the time-averaged air flow and heat transfer equations in tensor notation are as follows.

Continuity equation

$$\frac{\partial(\rho U_i)}{\partial x_i} = 0 \qquad (1)$$

where U_i is the mean air velocity component in x_i direction (m/s) and ρ is the air density (kg/m^3).

Momentum equation

$$\frac{\partial(\rho U_j U_i)}{\partial x_j} - \frac{\partial}{\partial x_j}\left[\mu_e\left(\frac{\partial U_i}{\partial x_j} + \frac{\partial U_j}{\partial x_i}\right)\right] = \rho g_i - \frac{\partial}{\partial x_j}\left[\left(P_s + \frac{2}{3}k\right)\delta_{ij}\right] \quad (2)$$

where μ_e is the effective viscosity (kg/m-s), g_i is the gravitational acceleration in i direction (m/s^2), P_s is the static pressure (Pa), k is the turbulent kinetic energy (m^2/s^2) and δ_{ij} is the Kronecker delta (δ_{ij} = 1 if i = j and δ_{ij} = 0 if i $\neq$ j).

The effective viscosity is obtained from

$$d\left(\frac{\rho^2 k}{\sqrt{\varepsilon\mu}}\right) = 1.72 \frac{(\mu_e/\mu)}{\sqrt{(\mu_e/\mu)^3 - 1 + C_v}} d(\mu_e/\mu) \quad (3)$$

where ε is the turbulent dissipation rate (m^2/s^3), μ is the laminar dynamic viscosity (kg/m-s) and $C_v \approx 100$.

Turbulent kinetic energy

$$\frac{\partial(\rho U_i k)}{\partial x_i} - \frac{\partial}{\partial x_i}\left(\frac{\mu_e}{\sigma_k}\frac{\partial k}{\partial x_i}\right) = \mu_t S^2 - \rho\varepsilon - \frac{g_i}{\rho}\frac{\mu_t}{\sigma_t}\frac{\partial \rho}{\partial x_i} \quad (4)$$

where σ_t is the turbulent Prandtl number, σ_k is the Prandtl number for turbulent kinetic energy, μ_t is the turbulent viscosity (kg/m-s) and S is the modulus of rate-of-strain tensor

$$S = \sqrt{\frac{1}{2}\left(\frac{\partial U_i}{\partial x_j} + \frac{\partial U_j}{\partial x_i}\right)\left(\frac{\partial U_i}{\partial x_j} + \frac{\partial U_j}{\partial x_i}\right)}.$$

Turbulent dissipation rate

$$\frac{\partial(\rho U_i \varepsilon)}{\partial x_i} - \frac{\partial}{\partial x_i}\left(\frac{\mu_e}{\sigma_\varepsilon}\frac{\partial \varepsilon}{\partial x_i}\right) = C_1 \mu_t S^2 \frac{\varepsilon}{k} - C_2 \rho \frac{\varepsilon^2}{k} - C_3 \frac{g_i}{\rho}\frac{\mu_t}{\sigma_t}\frac{\partial \rho}{\partial x_i}\frac{\varepsilon}{k} - Sr \quad (5)$$

where σ_ε is the Prandtl number for turbulent dissipation rate, C_1 = 1.42, C_2 = 1.68, C_3 = tanh(V_v/V_h) with V_v and V_h being the vertical and horizontal mean velocity components, respectively, and Sr is the rate of strain given by

$$Sr = \frac{C_\mu \rho \eta^3 (1 - \eta/\eta_0)}{1 + \beta\eta^3} \frac{\varepsilon^2}{k} \quad (6)$$

where $\beta = 0.012$, $\eta_0 = 4.38$, $\eta = S\,k/\varepsilon$.

σ_k, σ_t and σ_ε in Equations (4) and (5) are computed via

$$\left|\frac{\alpha - 1.3929}{\alpha_0 - 1.3929}\right|^{0.6321} \left|\frac{\alpha + 1.3929}{\alpha_0 + 1.3929}\right|^{0.3679} = \frac{\mu}{\mu_e} \quad (7)$$

$\sigma_t = 1/\alpha$ with α_0 being the laminar inverse Prandtl number ($\alpha_0 = 1/\sigma$) and $\sigma_k = \sigma_\varepsilon = 1/\alpha$ with $\alpha_0 = 1.0$.

Energy equation

$$\frac{\partial(\rho U_j H_i)}{\partial x_j} - \frac{\partial}{\partial x_j}\left[\left(\frac{\mu}{\sigma} + \frac{\mu_t}{\sigma_t}\right)\left(\frac{\partial H_i}{\partial x_j}\right)\right] = q \quad (8)$$

where H_i is the specific enthalpy in i direction (J/kg) and for dry air with negligible pressure work and kinetic energy in incompressible flow $H_i = C_p T$, C_p is the specific heat of air at a constant pressure (J/kgK), T is the absolute air temperature (K) and q is the volumetric heat production/dissipation rate (W/m^3).

The density of dry air for the calculation of thermal buoyancy effect is given by the following ideal gas law

$$\rho = \frac{P}{RT} \quad (9)$$

where P is the absolute pressure (Pa) and R is the gas constant (J/kgK).

Boundary Conditions and Solution Method

Prediction of flow in a ventilation cavity was carried out using a larger computational domain than the physical size of a cavity (defined as Domain L) but comparison was made late using a domain of the same size as the cavity (Domain S). Domain L is extended from Domain S and the minimum size of Domain L is such that further extension would not affect the flow in the cavity. This would require an extension of about five to 10 times the cavity width for symmetrically to asymmetrically heated cavities, respectively [11]. Figure 1 shows the schematic diagram of the two domains for a ventilation cavity. The cavity could be vertical or inclined with heat input from any of the two surfaces of a wall. Uniform heat fluxes on the inner surfaces of the left and right walls were equal to q_1 and q_2, respectively, and those on the outer surfaces of the left and right walls were q_3 and q_4, respectively. The top or bottom side of the cavity could be partly or fully open.

Two types of boundary were defined in the prediction of cavity flow – non-slip wall for solid wall surfaces and pressure inlet/outlet for openings. The heat flux was used as the thermal boundary condition for a wall surface and uniform pressure and temperature were prescribed as conditions at the inlet/outlet either as cavity openings (for Domain S) or the domain boundary (for Domain L). The ambient air temperature was fixed at 20°C as the thermal boundary condition for the openings. As an illustration of likely flow patterns in a cavity in Figure 1, the heat flux on either surface of the right wall was much larger than the corresponding value for the left wall, i.e., $q_2 \gg q_1$ and $q_4 \gg q_3$, such that air would flow out of the wide cavity along the right side but reverse flow might occur near the top on the left side.

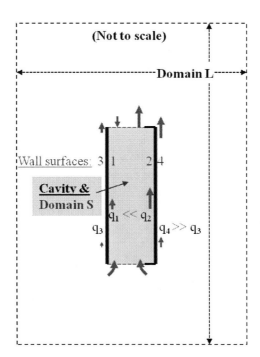

Figure 1. Schematic of two types of computational domain for prediction of cavity flow.

Near a solid wall, because the flow was not fully turbulent, the enhanced wall treatment was used to resolve the flow in the near-wall cells or control volumes. In this treatment, the two layer model (viscous-affected inner region and fully turbulent outer region) was combined with velocity, turbulence and thermal wall functions [13]. The two layer model was used to specify the turbulent viscosity and dissipation rate for the near wall control volume that lay in the viscous-affected inner region and fully turbulent outer region and the wall functions to calculate the mean velocity and temperature and derive the turbulent kinetic energy based on its local equilibrium between production and dissipation.

The above equations were solved using the control volume method and the SIMPLE algorithm. Further details of the methodology can be found in references [13, 15].

Model Validation

The model had previously been validated for buoyancy-driven air flow in a vertical ventilation cavity of constant and straight cross section [11]. Here further validation was performed by comparing the predictions with the experimental measurements of natural ventilation through an L-shaped tall cavity by Moshfegh and Sandberg [16] and natural convection in a C-shaped cavity by La Pica, et al. [17]. The L-shaped cavity consisted of a horizontal inlet section of 0.3 m long and 0.5 m high and a vertical section of 0.23 m wide and 6.5 m tall. The vertical wall on the inlet side was uniformly heated at 20 to 300 W/m^2. The C-shaped cavity of gap widths varying from 0.075 m, 0.125 m to 0.17 m consisted of a 3 m tall hot wall heated from 48 to 317 W/m^2, a cold wall made of 6 mm thick glass and silvered to minimise radiative heat transfer from the heated wall, and two 0.2 m long horizontal inlet and outlet ducts. The dimension in the third direction of the cavities was much larger than the cavity width (1.54 m compared with the width of 0.23 m for the L-shaped cavity and 1.2 m compared with the maximum width of 0.17 m for the C-shaped cavity) and heat transfer along the direction was negligible. Therefore, predictions were carried out for two dimensional flow using larger computational domains than the cavities. Figure 2 shows

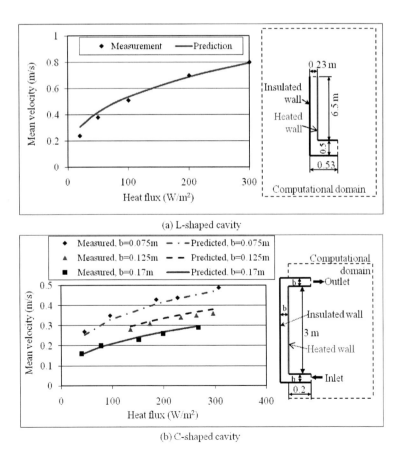

(a) L-shaped cavity

(b) C-shaped cavity

Figure 2. Comparison of predicted and measured mean air velocities in the test cavities.

good agreement between the predicted and measured mean air velocities in the two cavities. The average difference between the predicted and measured velocities in the L-shaped cavity for heat fluxes at and above 100 W/m^2 was less than 3% and that in the C-shaped cavity was below 5% for all the three cavity widths. The validated CFD model was applied to predicting heat transfer and air flow through vertical and inclined two-dimensional ventilation cavities.

VERTICAL CAVITIES

For vertical ventilation cavities, it was assumed that heat was uniformly distributed on one or both of the inner surfaces of two walls while the outer surfaces were insulated. Predictions were carried out for cavities of different heights ranging from 1 m to 6 m, widths from 0.05 m to 0.6 m, total heat fluxes from 100 to 1000 W/m^2, heat distribution ratios from 0 to 100% and different inlet/outlet positions. The inlet and outlet positions were used to characterise the type of a cavity. The most basic type of a vertical rectangular cavity (Type 1 cavity) consists of two parallel walls and two openings at bottom and top forming the vertical inlet and vertical outlet, respectively. A horizontal inlet (Type L cavity) is formed when the bottom opening is rotated to the vertical direction and positioned through or below the right wall, or below the left wall alternatively. When the outlet opening is set above one or both vertical walls, a number of more cavity configurations are obtained as shown in Figure 3 in combination with Table 1. With a vertical inlet, the original Type 1 cavity can be transformed into Type Γ or 7 cavity with one horizontal outlet or Type T cavity with two horizontal outlets over the top of two walls. For the cavity with a horizontal inlet, the original cavity with a vertical outlet, Type L, can be developed into Type Z or C cavity with a horizontal outlet in the same or opposite direction in relation to the inflow or into Type τ cavity with two horizontal outlets.

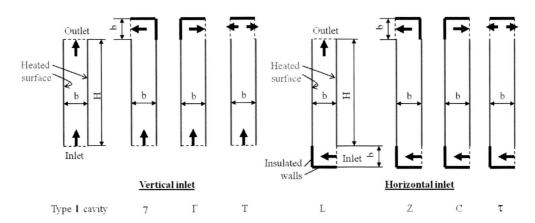

Figure 3. Types of cavity according to the inlet and outlet positions.

For the inlet or outlet positioned horizontally through either of the walls, the inlet/outlet height was set to be the same as the cavity width. To compare the effect of heat distribution, heated sections of two opposing wall surfaces had the same height and the inlet opening was below a heated vertical wall (likewise outlet above vertical wall). The horizontal bottom wall and part of the wall opposite to and with the same height as the horizontal inlet opening were insulated, so were the top wall and part of the vertical wall opposite to the outlet for a cavity with a horizontal outlet. For such cases, the two walls would not be symmetrical along the centre of the cavity width in terms of variation of a parameter with heat distribution.

Table 1. Types of two-dimensional ventilation cavity according to the inlet and outlet positions

Cavity type	I	Γ or 7	T	L	Z	C	τ
Inlet direction	Vertical	Vertical	Vertical	Horizontal	Horizontal	Horizontal	Horizontal
Outlet direction	Vertical	Horizontal	Two horizontal	Vertical	Horizontal*	Horizontal**	Two horizontal
Application	Ventilated double facade (middle floor)	Ventilated double facade (top floor)	Solar chimney	Solar chimney	Trombe wall for summer ventilation	Trombe wall for winter heating	Solar chimney

* Inflow and outflow in the same direction.
** Inflow and outflow in opposite direction.

For convenience, the wall with the (horizontal) inlet (right wall) is referred to as the inlet wall (with heat flux q_2 on the inner surface) and the other vertical wall (heated section) as the opposite wall (with heat flux q_1 on the inner surface). In this chapter, heat distribution refers to the relative values of uniform heat fluxes from, or heat uniformly distributed on, two inner wall surfaces and the heat distribution ratio (q_r) for a ventilation cavity is the ratio of heat flux from the right wall surface (q_2 in W/m^2) to the total heat flux from both (left and right) wall surfaces ($q_1 + q_2$), i.e.,

$$q_r = \frac{q_2}{q_1 + q_2} \qquad (10)$$

Thus, if heat flows from the left inner wall surface only, $q_2 = 0$ and $q_r = 0\%$; if heat flows from the right inner wall surface only, $q_1 = 0$ and $q_r = 100\%$; and if heat is equally distributed on both inner wall surfaces $q_1 = q_2$ and $q_r = 50\%$.

Large extended computational domains were employed for predicting the heat transfer and air flow in and around ventilation cavities for different combinations of inlet and outlet openings. The effect of domain type was also discussed separately. Results are presented graphically for predictions with a total heat gain of 100 W/m^2 by the two inner wall surfaces of a cavity, unless indicated otherwise, but the effect of total heat flux and all the rest of results are analysed.

The predicted heat transfer and air flow in a ventilation cavity are analysed using dimensionless numbers. The air flow rate is represented by the Reynolds number, Re. For two-dimensional flow, it is defined as the ratio of the flow rate to the kinematic viscosity of air [18]:

$$\mathrm{Re} = \frac{Vb}{v} = \frac{Q}{v} \qquad (11)$$

where V is the mean velocity of air through the cavity (m/s), b is the cavity width (m), Q is the flow rate (m³/s-m) and v is the kinematic viscosity (m²/s).

The heat transfer rate in a cavity is assessed in terms of Nusselt number, Nu. The Nusselt number is defined as

$$Nu = \frac{h_c b}{k} = \frac{q_w b}{(t_w - t_a)k} \qquad (12)$$

where h_c is the convective heat transfer coefficient (W/m²K), k is the thermal conductivity of air (W/mK), q_w [= ½ (q_1 + q_2)] is the average heat flux on the two inner wall surfaces (W/m²), t_w is the average temperature of the two opposing wall surfaces (°C) and t_a is the ambient air temperature (°C). The thermal properties of air are based on the average temperatures of inner wall surfaces and ambient air, i.e., ½ (t_w + t_a).

Cavities with Vertical Inlet and Vertical Outlet

Prediction was first performed for a cavity comprised of two vertical walls and two vertical openings to model natural ventilation through a cavity with the inlet at the bottom and outlet at the top.

Air Flow and Heat Transfer

Figure 4 shows the predicted air flow patterns around the inlet and outlet of a 3 m tall and 0.6 m wide cavity with two levels of asymmetrical heating – one with 35 and 65 W/m² heat fluxes on the left and right wall surfaces, respectively, and the other with 100 W/m² fixed on the right wall surface only. The air flow was asymmetrical in the symmetrical cavity under asymmetrical heating as a result of larger heat transfer from the right wall surface. The level of flow asymmetry increased with that in the level of asymmetric heating; the maximum flow asymmetry occurred when only one of the walls was heated. It can be seen that air flowing into the inlet formed a small recirculation or reverse flow zone near the bottom of the less heated surface. When only one wall was heated, reverse flow also occurred near the top of the cavity and away from the heated wall. The velocity of downward flow varied across the outlet and the area of reverse flow extended for about one half of the cavity width. The degree of reverse flow would however depend on the cavity width. The flow patterns for the 0.6 m wide cavity with only one wall heated suggested that the thickness of the buoyancy-driven velocity boundary layer was less than one half of the cavity width but entrainment of air from the surroundings would prevent reverse flow at the top when both walls were heated at, e.g., a heat distribution ratio of 65%.

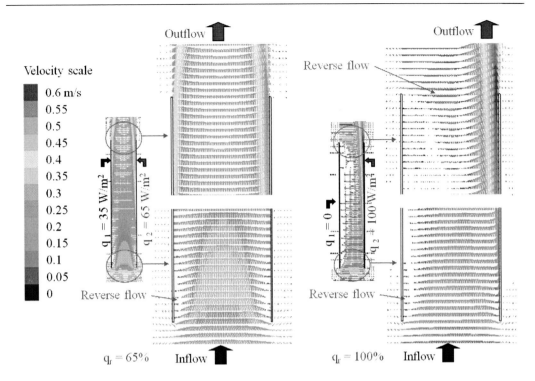

Figure 4. Predicted air flow patterns around the inlet and outlet of 3 m tall and 0.6 m wide cavity with 65% and 100% heat distribution ratios.

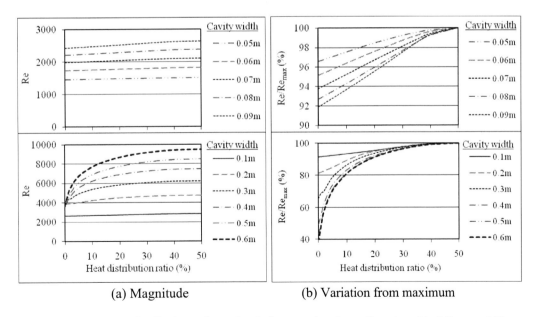

Figure 5. Effect of heat distribution ratio on the air flow rate in a 3 m tall cavity with different widths.

The effect of heat distribution on the predicted air flow rate in terms of Re through the 3 m tall cavity of different widths is shown in Figure 5. Because the cavity with a vertical inlet and outlet was symmetrical along the centre of the width, the variations with heat distribution ratio of air flow rate and heat transfer rate would be symmetrical along the point of $q_r = 50\%$. For example, the air flow rate at $q_r = 80\%$ would be the same as that at $q_r = 20\%$. Figure 5 shows that the air flow rate through a cavity increased from asymmetrical to symmetrical heating. The increase was most significant for heat distribution ratios between 0% and 10% for cavity widths equal to or greater than 0.3 m. In other words, to maximise the ventilation rate through a tall cavity with an aspect ratio (ratio of cavity height to width) less than 10, the heat flux from either of the two wall surfaces should be at least 10% of the total heat. This requirement may not be met at all times for a single-glazed ventilation cavity in cold seasons and so use of double glazing for the exterior skin of the cavity would be desirable [19] where the ventilation cavity would form a critical component of a ventilation system.

It is also seen that the air flow rate increased with cavity width if the heat distribution ratio was not less than 2%. For example, at a heat distribution ratio of 50%, the flow rate increased by about 50% and 100%, respectively, when the cavity width increased from 0.2 m to 0.4 m and to 0.6 m. However, at a heat distribution ratio less than 2%, the flow rate would increase up to a cavity width of 0.5 m. The flow rate then decreased slightly as the cavity width increased further due to excessive reverse flow near the top of the cavity. For example, at 0% heat distribution ratio, the flow rate increased by 28% when the cavity width increased from 0.2 m to 0.3 m but decreased by 11% when the cavity width increased further from 0.5 m to 0.6 m. If a ventilation cavity such as solar chimney is designed with negligible heat loss or gain through glazing, e.g. when making use of heat in its storage wall for ventilation cooling in summer, an optimum cavity width can be defined as the width at which the buoyancy-driven ventilation rate reaches a maximum and it is approximately 0.5 m for this case with free air flow in and out of the 3 m tall cavity. However, during the period of bright sunshine, the glazing of a solar chimney would likely absorb more than 2% of insolation and so the width of a 3 m tall cavity could be larger than 0.5 m for maximising natural ventilation.

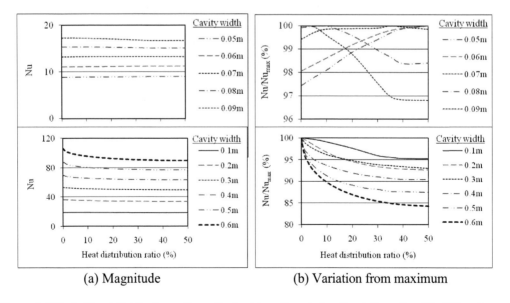

(a) Magnitude (b) Variation from maximum

Figure 6. Effect of heat distribution ratio on the heat transfer rate in a 3 m tall cavity with different widths.

Besides, the effect of heat distribution on the variation in air flow rate increased with cavity width. For example, the minimum flow rate at 0% heat distribution ratio was 9% and 20% lower than the maximum flow rate at 50% heat distribution ratio for 0.1 m and 0.2 m wide cavities, respectively, but the minimum was 37% lower than the maximum for a 0.3 m wide cavity and 65% lower for a 0.6 m wide cavity.

Figure 6 shows the variation with heat distribution ratio and cavity width of the predicted heat transfer rate in the 3 m tall cavity.

Unlike the air flow rate which increased with heat distribution ratio, the heat transfer rate in cavities wider than 0.07 m decreased with the increase in the ratio particularly for heat distribution ratios between 0 and 20%. The effect of heat distribution on the variation in heat transfer rate also increased with cavity width (Figure 6b) but the effect was not as much as that on the air flow rate. For example, the variation of minimum from maximum heat transfer rates was about 5% for a 0.1 m wide cavity and increased only to 16% for a 0.6 m wide cavity.

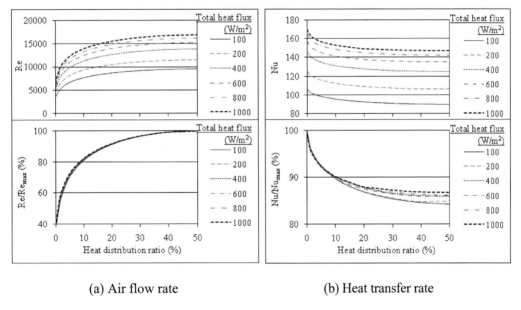

(a) Air flow rate   (b) Heat transfer rate

Figure 7. Effect of total heat flux on the variations in air flow rate and heat transfer rate with heat distribution ratio in a 3 m tall and 0.6 m wide cavity.

A similar effect of the heat distribution ratio on the air flow and heat transfer rates can be observed from Figure 7 for different total heat fluxes varying from 100 to 1000 W/m^2 on both inner wall surfaces of a 3 m tall and 0.6 wide cavity. The air flow and heat transfer rates increased with the heat flux at a similar proportion for the entire range of heat distribution ratios.

By varying both the height and width, the effect of height or aspect ratio of a cavity can be assessed. Figure 8 shows the dependence of the air flow and heat transfer rates on the height for cavities with an aspect ratio of 10 and under a total heat flux of 1000 W/m^2. The air flow and heat transfer rates increased with the height of the cavity for the same aspect ratio. Again, the trends of variations were similar for different cavity heights.

Correlations for Air Flow Rate and Heat Transfer Rate

The predicted air flow rate and heat transfer rate in terms of Re and Nu, respectively, for the vertical cavities have been shown to vary with cavity size, total heat flux and heat distribution ratio. The effects of cavity size and heat flux can be combined into a single parameter represented by the Rayleigh number, Ra.:

$$Ra = \frac{g\beta q_w b^4}{v\alpha k} \frac{b}{H} \qquad (13)$$

where β is the thermal expansion coefficient of air (1/K), α is the thermal diffusivity of air (m^2/s) and H is the height of cavity (heated walls) (m).

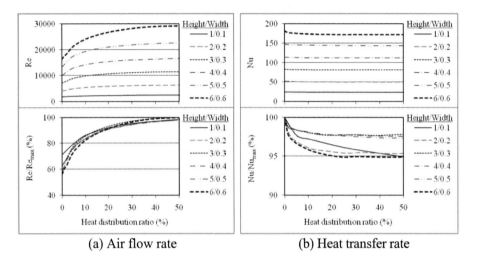

(a) Air flow rate (b) Heat transfer rate

Figure 8. Effect of cavity height on the variations in air flow rate and heat transfer rate with heat distribution ratio in cavities with an aspect ratio of 10 and total heat flux of 1000 W/m^2.

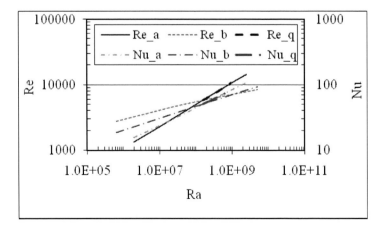

Figure 9. Variations of Reynolds number and Nusselt number with Rayleigh number based on averaged values for the cavities with vertical inlet and outlet.

The heat transfer rate is also influenced by the air flow rate. Therefore, Nu can be correlated with Ra and Re using either data obtained for all individual heat distribution ratios or values averaged over the entire range of heat distribution ratios.

Correlations based on average values

As a first step, to simply the analysis, values of Ra, Re and Nu for the entire range of heat distribution ratios were averaged to obtain one set of data for each combination of cavity width, cavity height and total heat flux. There existed log linear relations between the averaged Nu, Re and Ra as shown in Figure 9 where subscript a, b or q represents data for Ra, Nu and Re with varying one of the three parameters - the cavity aspect ratio, width or total heat flux, respectively - but fixing other one or two parameters. For example, the line representing Re_b against Ra_b was obtained with values of Re and Ra for a 3 m tall cavity with a total heat flux on both inner wall surfaces of 100 W/m² but with varying widths and the best fittings between two of the three groups are as follows:

$$Nu = 1.69\ Ra^{0.18} \qquad (R^2=1) \qquad (14)$$

$$Re = 540\ Ra^{0.124} \qquad (R^2=0.997) \qquad (15)$$

$$Nu = 0.000194\ Re^{1.44} \qquad (R^2=0.997) \qquad (16)$$

From Equations (12) through (14), $Nu \sim b^{0.9}$ and $h_c \sim b^{-0.1}$. Therefore, in terms of trend of variation, the effect of cavity width on the convective heat transfer coefficient would differ from that on the Nusselt number, with Nu increasing while h_c decreasing with the increase in cavity width as shown in Figure 10.

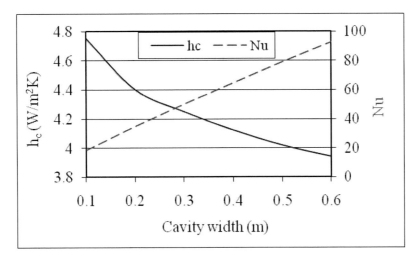

Figure 10. Effect of cavity width on the heat transfer coefficient and Nusselt number for a 3 m tall cavity with vertical inlet and outlet and with a total heat flux of 100 W/m².

To take account of all the effects of total heat flux and cavity size, the relationship between Nu and Ra and in combination with Re is represented by the following expressions:

$$Nu = c_1 \left[Ra \left(\frac{H}{b} \right)^{3/2} \right]^m \qquad (17)$$

and

$$Nu = c_2 \left(Ra^{1/3} \, \text{Re} \right)^n \qquad (18)$$

where c_1 and c_2 are constants and m and n are exponents.

From Figure 11, one obtains $c_1 = 0.137$, $m = 0.265$, $c_2 = 0.129$ and $n = 0.4$.

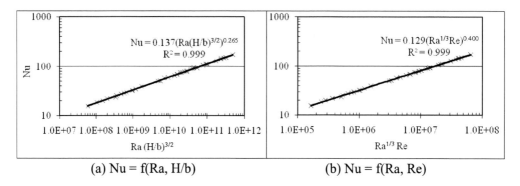

Figure 11. Effect of Rayleigh number, aspect ratio and Reynolds number on Nusselt number based on averaged values for cavities with vertical inlet and outlet.

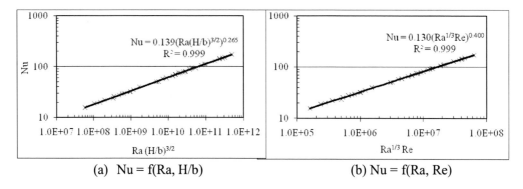

Figure 12. Effect of Rayleigh number and Reynolds number on Nusselt number for cavities with vertical inlet and outlet and 15% heat distribution ratio.

Effect of heat distribution ratio

The above correlations would be useful for estimating the heat transfer rate and air flow rate for buoyancy-driven cavity flow, if the variation with heat distribution ratio could be ignored as in circumstances where heat fluxes on two inner wall surfaces are at a similar magnitude. The effect of heat distribution ratio can be significant particularly when reverse flow near the outlet of a wide cavity occurs. Then, individual data of Ra, Re and Nu which vary with heat distribution ratio as well as with total heat flux, cavity width and height should be used and these data can be correlated for each heat distribution ratio. A typical set of data and correlations are shown in Figure 12 for a heat distribution ratio of 15%. General correlations similar to Equations (17) and (18) can then be obtained for all the data points with the regression coefficients (and constants) being related to the heat distribution ratio. However, it was found that the exponent for Ra in Equation (18) should also vary with heat distribution ratio for extremely asymmetrical heating in order to obtain better curve fitting as illustrated in Figure 13 for q_r = 0% or 100%. Therefore, general expressions should be Equation (17) together with Equation (19) below:

$$Nu = c_2 \left(Ra^{a_2} Re \right)^n \tag{19}$$

with coefficients c_1, m, c_2, a_2 and n given in Table 2.

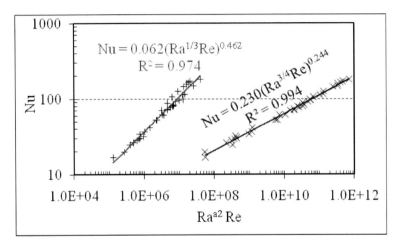

Figure 13. Effect of Rayleigh number and Reynolds number on Nusselt number for cavities with vertical inlet and outlet and 0% heat distribution ratio.

Table 2. Coefficients and constants for correlations of heat transfer rate and air flow rate in cavities with vertical inlet and outlet

Correlation	Coefficient / constant	Heat distribution ratio, q_r (%)		
		0 - 5	5 - 20	20 - 50
$Nu = c_1 \left[Ra \left(\dfrac{H}{b} \right)^{3/2} \right]^m$	c_1	$0.122 + 0.007\, q_r$ $- 0.00082\, q_r^2$	0.137	
	m	$0.271 - 0.0011\, q_r$	0.265	
$Nu = c_2 \left(Ra^{a_2} Re \right)^n$	c_2	$0.227 - 0.024\, q_r$ $+ 0.0014\, q_r^2$	$0.107 + 0.00149\, q_r$	0.135
	a_2	$0.748 - 0.163\, q_r$ $+ 0.0186\, q_r^2$	1/3	
	n	$0.25 + 0.034\, q_r$ $- 0.0018\, q_r^2$	$0.418 - 0.00113\, q_r$	0.395

For q_r > 50%, replace q_r with (1 - q_r).

Equation (17) can be used to calculate the heat transfer rate for given cavity size, heat flux and distribution on cavity wall surfaces. The air flow rate can then be obtained from Equation (19).

Cavities with Horizontal Inlet and Vertical Outlet

Results of prediction are presented for cavities with a horizontal inlet opening through the bottom of either of the walls and a vertical outlet at the cavity top (Type L cavity).

Air Flow and Heat Transfer Rates

The flow patterns of air through the horizontal inlet into a 3 m tall and 0.6 m wide cavity with heat distribution ratios of 65% and 100% are shown in Figure 14. Air flowed along the opposite wall after turning from horizontal at the inlet opening to vertical direction. Two large recirculation zones can be observed – one in the corner between the opposite wall and bottom wall and another near the inlet wall above the inlet. Compared with the cavity with vertical inlet in Figure 4, the flow patterns near the outlet were similar but, because of the deflection of incoming air towards the opposite wall, reverse flow at the outlet for the cavity with horizontal inlet was less and consequently the air flow rate was higher for the same heat flux distribution and cavity size, even though the flow resistance at the horizontal inlet opening would be larger.

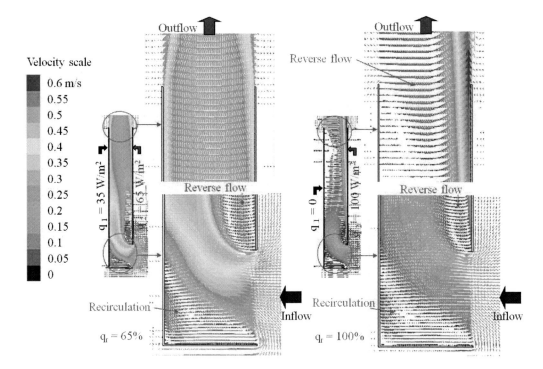

Figure 14. Predicted air flow patterns around the inlet and outlet of 3 m tall and 0.6 m wide cavity with horizontal inlet and vertical outlet and with 65% and 100% heat distribution ratios.

Figure 15 shows the variation with heat distribution ratio of the air flow rates in the 3 m tall cavity with different widths. By comparison with Figure 5, it is seen that trends of variations in air flow rate with cavity width and heat distribution ratio were similar for the two inlet positions. However, the flow rate through the cavity with horizontal inlet was not symmetrical along the equal heat flux value (q_r = 50%). The air flow rate was slightly higher when more heat flowed from the inlet wall surface than when heat flowed from the opposing wall surface and the highest flow rate through the cavity with the horizontal inlet opening was obtained when the heat distribution ratio was between 60% and 65% rather than 50% for the cavity with the vertical inlet. It is also seen that when the inlet wall was heated with a 2% or more of the total heat flux or the opposite wall with 10% or more of the total heat flux, the air flow rate increased with cavity width. When only one wall was heated, the flow rate in the cavity with the horizontal inlet also increased with cavity width up to 0.5 m and the effect of heat distribution on the variation in flow rate increased with cavity width.

Figure 16 shows that the air flow rate through a cavity with horizontal inlet, Re_h, was generally less than that in a cavity with vertical inlet, Re_v, for heat distribution ratios between 5% and 80% due to larger flow resistance through the horizontal inlet where flow had to turn at a right angle from the ambient to the cavity. For very low (< 5%) or high (> 80%) heat distribution ratios, due to reduced reverse flow at the outlet, the air flow rate in a wide cavity with the horizontal inlet was generally higher than that with the vertical inlet. However, for a very wide cavity (0.6 m wide and 3 m tall), the air flow rate through the horizontal inlet was as much as 20% lower than that through the vertical inlet at a heat distribution ratio of 1%.

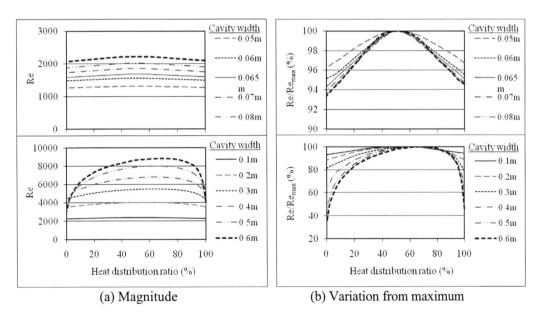

(a) Magnitude   (b) Variation from maximum

Figure 15. Effect of heat distribution ratio on the air flow rate in a 3 m tall cavity with horizontal inlet and vertical outlet and with different widths.

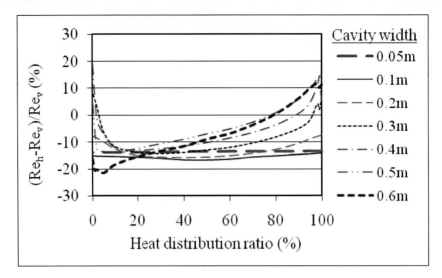

Figure 16. Comparison between two inlet positions of the flow rates through a 3 m tall cavity with a vertical outlet and with different widths.

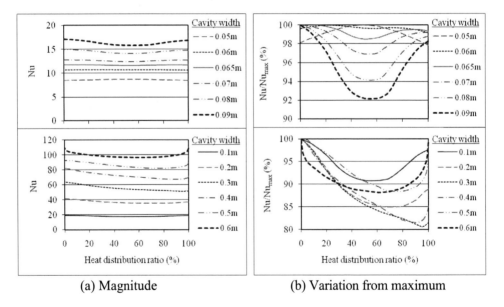

(a) Magnitude (b) Variation from maximum

Figure 17. Effect of heat distribution ratio on the heat transfer rate in a 3 m tall cavity with horizontal inlet and vertical outlet and with different widths.

The variation of the heat transfer rate with heat distribution ratio for the 3 m tall cavity is shown in Figure 17 for different cavity widths. The heat transfer rate was generally higher for cavities wider than 0.06 m when heat flowed from one wall surface only and the highest heat transfer rate was obtained when heat flowed only from the opposite wall surface whereas the lowest value was obtained where more heat flowed from the inlet wall surface than from the opposite wall surface.

The effect of heat distribution on the variation in heat transfer rate and the lowest heat transfer rate depended on the cavity width. For very wide (0.6 m) or narrow (≤ 0.1 m) cavities, the variation was small and the slowest heat transfer would take place at a heat distribution ratio of about 60%. The largest variation in heat transfer rate was 19% for a 0.3 m wide cavity at a 98% heat distribution ratio.

The effect of inlet position on the predicted heat transfer rate for the 3 m tall cavity with different heat distribution ratios is shown in Figure 18 for different cavity widths. The predicted heat transfer rate in cavities wider than 0.1 m with the horizontal inlet (Nu_h) was generally higher than that with the vertical inlet (Nu_v). The largest difference in heat transfer rate between two inlet positions was 22% for the 0.3 m wide cavity at a 5% heat distribution ratio.

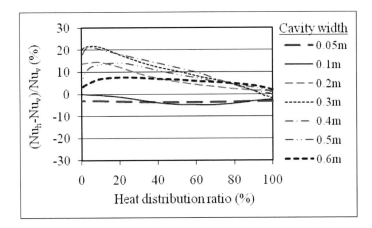

Figure 18. Comparison between two inlet positions of the heat transfer rate in the 3 m tall cavity with vertical outlet and with different widths.

Correlations

Similar correlations between the Nusselt number, Reynolds number and Rayleigh number can be obtained for cavities with a horizontal inlet below a heated vertical wall based on the individual or averaged values of Nu, Re and Ra for the entire range of heat distribution ratios. The relationships between these variables are illustrated in Figure 19 for a heat distribution ratio of 40%.

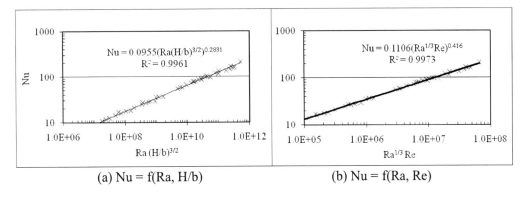

Figure 19. Relationships between Rayleigh number, aspect ratio, Reynolds number and Nusselt number for cavities with horizontal inlet and vertical outlet at 40% heat distribution ratio.

For both sets of data, the following correlations between Nu, Re and Ra are obtained for cavities with horizontal inlet and vertical outlet:

$$Nu \approx 0.1\, (Ra\, (H/b)^{3/2})^{0.28} \qquad (R^2=0.994 \text{ to } 0.996) \qquad (20)$$

and

$$Nu \approx 0.11\, (Ra^{1/3}\, Re)^{0.42} \qquad (R^2=0.987 \text{ to } 0.997) \qquad (21)$$

To obtain better correlations (with $R^2 \geq 0.998$) requires the use of more general expressions that take account of the effect of heat distribution ratio in the coefficients and exponents, Equations (17) and (18).

Table 3 lists the coefficients and exponents in Equations (17) and (18) for ventilation cavities with the horizontal inlet and vertical outlet.

Table 3. Coefficients and exponents in $Nu = c_1 \left[Ra\left(\dfrac{H}{b}\right)^{3/2} \right]^m$ **and** $Nu = c_2 \left(Ra^{1/3}\, Re \right)^n$ **for cavities with horizontal inlet and vertical outlet**

Coefficient / constant	Heat distribution ratio, q_r (%)		
	0 - 30	30-95	95-100
c_1	$0.894 - 0.03 q_r + 0.00038 q_r^2 - 0.0000015 q_r^3$	0.102	
m	$-0.111 + 0.015 q_r - 0.00019 q_r^2 + 0.00000076 q_r^3$	0.28	
c_2	$4.43 - 0.16 q_r + 0.002 q_r^2 - 0.0000084 q_r^3$	0.12	$0.067 + 0.0081 q_r - 0.0008 q_r^2$
n	$-2.96 + 0.13 q_r - 0.0016 q_r^2 + 0.0000065 q_r^3$	0.41	$0.46 - 0.0094 q_r + 0.00096 q_r^2$

The exponent for the aspect ratio in Equation (17) as well as that for the Rayleigh number was also found to depend on the heat distribution ratio as shown in Figure 20. Therefore, better correlations with simpler relations between the coefficient/exponent and heat distribution ratio can be obtained using the following expressions in which the exponent for H/b and Ra is also taken as a function of the heat distribution ratio, i.e.

$$Nu = c_1 \left[Ra\left(\dfrac{H}{b}\right)^{a_1} \right]^m \qquad (22)$$

and

$$Nu = c_2 \left(Ra^{a_2}\, Re \right)^n \qquad (23)$$

Again, Equations (22) and (23) will allow the calculation of the heat transfer rate and air flow rate in a cavity with horizontal inlet and vertical outlet for given cavity size, heat flux and distribution on the cavity wall surfaces.

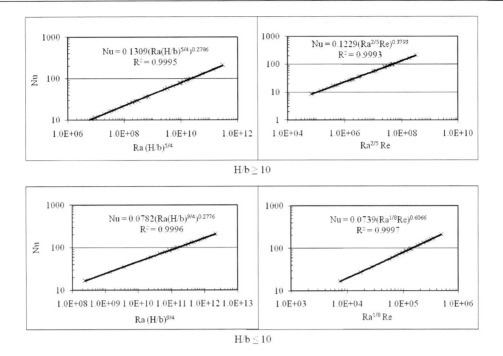

Figure 20. Effect of aspect ratio on the relationships between Rayleigh, Reynolds and Nusselt numbers for cavities with horizontal inlet and vertical outlet at 40% heat distribution ratio.

Effect of Outlet Position

The flow outlet for the cavities discussed above was set as the top horizontal side. Predictions were also carried out for cavities with the outlet opening set above one or both vertical walls.

Figure 21 shows the variations of the predicted air flow rate with heat distribution ratio for cavities of 3 m tall and 0.3 m wide with different positions of inlet and outlet illustrated in Figure 3. It is seen that for cavities with a vertical inlet, a symmetrical cavity (with one vertical outlet too), Type 1 cavity, generally had the highest air flow rate and an asymmetrical cavity with one horizontal outlet (Type Γ or 7) had the lowest flow rate due to the additional flow resistance resulting from the flow turning at the top. The air flow rate for an asymmetrical cavity decreased with increasing proportion of heat on the inner surface of the wall with the outlet up to about 80% heat distribution ratio. For example, for Type 7 cavity, the air flow rate at 0% heat distribution ratio (heat from the wall surface with the outlet only) was nearly a third less than that at 80% heat distribution ratio. The air flow rate for a symmetrical cavity with two horizontal outlets (Type T) was lower than that for the symmetrical cavity with one outlet (Type 1) due to the flow resistance by the horizontal cover above the cavity and vertical walls but was higher than that for the asymmetrical cavities because of the reduced additional flow resistance of the outlets with a larger total area. When the size of the two outlets of Type T cavity was halved (denoted as Type T1 cavity) such that the total outlet area was the same as that of the inlet of the cavity or one outlet for asymmetrical cavities, the predicted air flow rate varied less with the heat distribution ratio and consequently the air flow rate was lower than that for the asymmetrical cavities with similar levels of heating on both wall surfaces but was still higher than that for the cavities with heating mostly on the surface of the wall with the outlet opening.

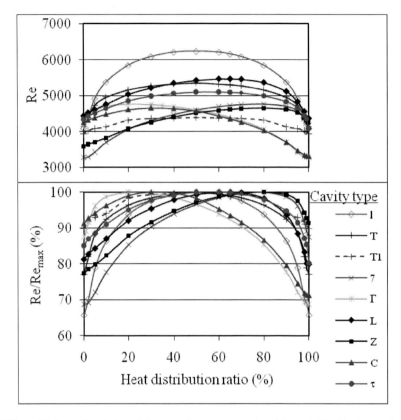

Figure 21. Effect of inlet and outlet positions on the air flow rate with heat distribution ratio for 3 m tall and 0.3 m wide cavities.

For the cavities with a horizontal inlet, the air flow rate was also highest when the outlet was vertical (Type L) and lowest for the cavities with one horizontal outlet (Types Z and C). Again, the air flow rate for the cavity with two horizontal outlets (Type τ) was slightly lower than that with one vertical outlet but higher than that with one horizontal outlet. The variation pattern of the air flow rate with heat distribution ratio for Type C cavity nearly mirrored that for Type Z cavity. That is, the air flow rate for Type Z cavity increased with heat distribution ratio from 20% to 100% but for Type C cavity it decreased with the ratio between 0% and 80%. Put it another way, the trend of increase was in opposite direction of outflow. It was noticed that the variation pattern for Type 7 cavity was similar to that for Type Z cavity or Type Γ to Type C. Both had one horizontal outlet and the air flow rate for both increased (or decreased) with heat distribution ratio, suggesting that the inlet position had less influence than the outlet in opposite flow direction according to the air flow rate. The difference in the variation magnitude from the maximum flow rate was again larger for cavities with vertical inlet than those with horizontal inlet for a given outlet position. For example, the air flow rate through Type Z cavity at a heat distribution ratio of 0% was 23% less than the maximum flow rate that occurred at a heat distribution ratio of about 60% whereas the maximum difference between the flow rates at heat distribution ratios 80% and 0% through Type 7 cavity was just over 31%.

Figure 22 compares the variations of predicted heat transfer rate with heat distribution ratio for the cavities. In general, the heat transfer rate for a cavity with vertical inlet was lower than that with horizontal inlet. The lowest heat transfer rate was given by the cavity with a vertical inlet and horizontal outlet (Type Γ or 7) and the highest was obtained for the cavity with horizontal outlet and vertical outlet (Type L). Among the cavities with vertical inlet, the one with vertical outlet had the highest heat transfer rate. In comparison, for cavities with horizontal inlet, the lowest heat transfer rate resulted from the one with one horizontal outlet in the opposite flow direction (Type C cavity at $q_r \approx 70\%$).

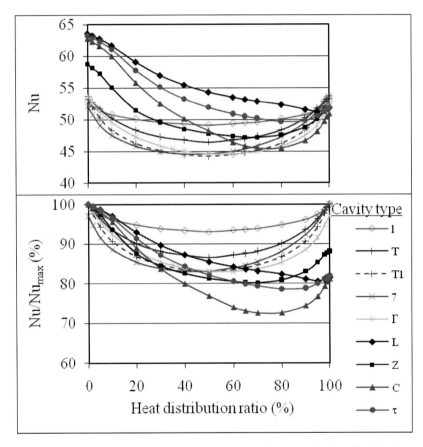

Figure 22. Effect of inlet and outlet positions on the heat transfer rate with heat distribution ratio for 3 m tall and 0.3 m wide cavities.

The maximum heat transfer rate occurred when only one of the wall surfaces was heated. For an asymmetrical cavity with vertical inlet, the maximum occurred when the wall with outlet was not heated. For the cavities with horizontal inlet, the maximum occurred when only the wall with inlet was heated regardless of the outlet position. The variation from the maximum was also larger for cavities with horizontal inlet than those with vertical inlet. The smallest variation was given by the cavity with vertical inlet and outlet (Type 1) and the largest by the cavity with horizontal inlet and outlet in the opposite flow direction (Type C).

Effect of Computational Domain

In the above predictions, larger computational domains than the cavities (Domain L) were employed. Predictions were also performed for these cavities with the computational domain of the same size as the cavity geometry (Domain S). The importance of using a large extended domain has been demonstrated by the author [11, 12]. Here the effect of the type of domain on the predicted heat transfer and air flow results is illustrated for two types of cavity with a vertical outlet.

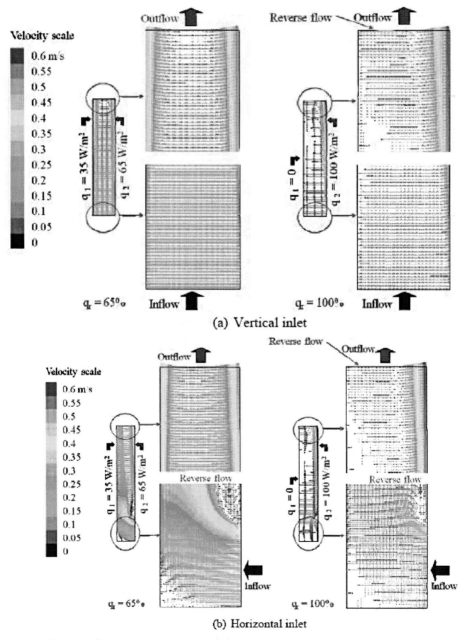

Figure 23. Predicted air flow patterns around the inlets and outlets of 3 m tall and 0.6 m wide cavities with 65% and 100% heat distribution ratios using Domain S.

Figure 23 shows the predicted air flow patterns around the openings of asymmetrically heated 3 m tall and 0.6 m wide cavities with a vertical inlet and a horizontal inlet using Domain S. The predicted air velocity was almost uniform at the vertical inlet or increased along the height of the horizontal inlet, in contrast with non-uniform velocity distribution when using the larger domain (Figures 4 and 14). After entering the cavity with vertical inlet, air flowed along the two vertical walls upwards and no recirculation near the inlet could be predicted. For cavities with a horizontal inlet, horizontal incoming air flowed along part of the bottom wall and then turned upwards along the unheated wall surface without a clear recirculation zone in the corner that was observed using the larger domain. Nevertheless, a large recirculation zone could be seen above the inlet and near the heated wall surface. Reverse flow also occurred near the top of the cavity and away from the heated wall surface but with a rather uniform velocity, similar to the inlet in terms of flow pattern. For both types of cavity, the total reverse flow at the outlet was larger with a nearly uniform velocity than with varying velocities and, as a result, using the small domain could under-predict the overall buoyancy effect in the cavities.

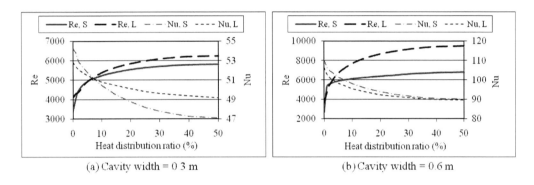

Figure 24. Comparison of the air flow rate and heat transfer rate in 3 m tall cavities with vertical inlet and outlet using Domain L or Domain S.

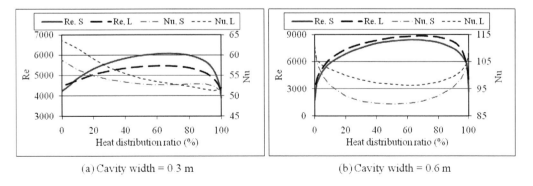

Figure 25. Comparison of the air flow rate and heat transfer rate in 3 m tall cavities with horizontal inlet and vertical outlet using Domain L or Domain S.

The effect of heat distribution ratio on the air flow rate and heat transfer rate in the 3 m tall cavities is shown in Figure 24 and Figure 25 for two cavity widths – 0.3 m and 0.6 m. Using either type of domain, the predicted air flow rate increased from the minimum when heat flowed from one wall surface only to the maximum when heat was equally distributed on both inner wall surfaces ($q_r = 50\%$) of the cavity with a vertical inlet or when the heat distribution ratio was between 60% and 65% for the cavity with a horizontal inlet. When downward flow was predicted for a wide cavity, the air flow rate was under-predicted using Domain S due to the over-prediction of reverse flow at the cavity top. The level of under-prediction in the air flow rate increased with the width of a cavity with vertical inlet but decreased for a cavity with horizontal inlet. The heat transfer rate was also under-predicted using Domain S in general but could be over-predicted in a narrow cavity with asymmetrical heating.

INCLINED CAVITIES

The heat transfer and air flow rates were also predicted in an inclined cavity representing the air gap behind a building-integrated PV module. The size of the required air gap can be determined using the numerical method [20, 21].

Predictions were carried out for one type of commercial module [22] installed in parallel onto the roof or beside the sun-facing facade of a building with an air gap of 0.1 m. The PV module was 1209 mm long, 537 mm wide and 10 mm thick but with a 50 mm thick frame. PV cells covered an area of 1125 mm by 500 mm of the module. Due to the protruding frame, the entrance and exit openings of the cavity was 0.06 m for the 0.1 m air gap between the module and building envelope. The configuration of the air gap resembled a straight ventilatiion cavity with partly open inlet and outlet openings. Two different sizes of domain – Domains S and L - were used. Figure 26a illustrates the two types of domain for the prediction of air flow and heat transfer in the air gap between the module and (part of) building envelope. Domain S could only be used for the prediction of flow in the air gap between the back side of the PV module and part of the roof surface whose length was the same as the module length. A larger domain would allow prediction of flow in the air gap and in front of the module but here Domain L was used to predict the flow generated by the heat transfer in the air gap only. To predict the flow generated by the heat transfer from both sides of the module, i.e., with heat fluxes on both sides of the module, a domain of the same size as Domain L was used but denoted as Domain L2. The magnitude of heat flux on the back side for predictions using Domains S and L was taken to be the same as that from the same side of the PV module for the prediction using Domain L2.

The air flow and heat transfer rates behind a PV module and the cell temperature would be influenced by the inclination angle or the roof pitch on which the PV module was mounted and so predictions were performed for different inclinations ranging from nearly flat roof to vertical facade installation. The incident solar radiation was assumed at 1000 W/m^2. The insolation level was taken to be constant regardless of the inclination angle. This would not be realistic for a given PV installation site at a given time but would be reasonable for investigation into the effect of the inclination of a cavity. Because of the high insolation, radaition heat transfer would play an important role in the overal heat transfer and air flow around the module. The importance of including radiation heat transfer was demonstrated by comparison between predictions of heat transfer and resulting air flow by convection only and by combined convection and radiation. A discrete transfer radiation model [23] was used for modelling the radiation heat transfer from the module to surroundings. Radiation heat transfer from the module surface to the roof surface would transform the cavity with principally one surface heated to a cavity with both opposing wall surfaces heated at a heat distribution ratio that would generally depend on the cavity size, inclination and surface properties. The predicted heat distribution ratio for the module installation varied from 59% to 67% for inclinations from 15 to 90 degrees which was close to the ratio where the maximum air flow rate was obtained for a vertical cavity with horizontal inlet and vertical outlet discussed before. Two different sizes of roof length were used for the predictions – one being the same as the module length and another longer.

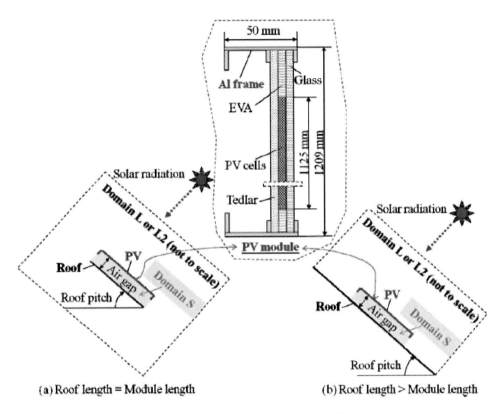

Figure 26. Schematic diagram of two computational domains for predicting air flow around a PV module.

Module Length = Roof Length

Figure 27 shows the predicted air flow patterns around the PV module installed on a roof with a 45° pitch and with the same length as the module. The principal prediction involved combined convection and radiation heat transfer. The results using a large extended domain showed that air flowed around the module frame at the entrance of the air gap and the velocity at the inlet would vary considerably. The incoming air then flowed along backside of the module to generate convection heat transfer from the hot module to cool air. Approximately 35% of the solar heat gain on the back side of the module under this installation was transferred to the roof by radiation and consequently natural convection also took place along the roof surface. In contrast, using the small domain, the predicted air velocity at the entrance of the air gap was nearly uniform.

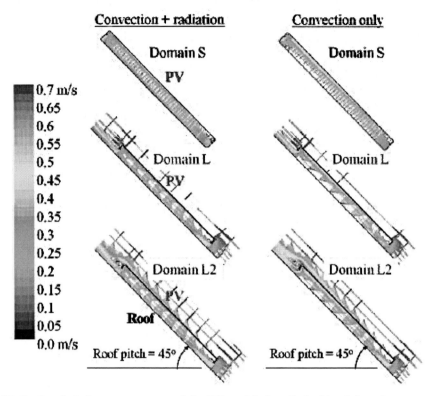

Figure 27. Predicted air flow patterns around the PV module installed with a 0.1 m air gap on a roof with a 45° pitch.

Immediately after entering the air gap, air flowed along the roof rather than the module. The overall flow in the air gap was more uniform and smooth but less efficient in removing heat from the module. As a result, at the same air flow rate, smaller heat transfer rate was predicted using the small domain than using a large domain. When heat transfer from both the front and back sides was taken into account, the buoyant flow along the front side entrained the flow from the air gap. The flow entrainment could either increase the air flow rate through the gap if the two air streams could move forward more or less vertically along a roof with a large pitch or decrease it as a result of increased flow resistance if the outgoing flow streams had to change from a nearly horizontal direction along a roof with a small pitch to the vertical direction.

The predicted air flow and heat transfer rates through the air gap in terms of Re and Nu respectively and the mean cell/PV temperature of the module are shown in Figure 28 for the 'roof' with the same length as that of module at different inclination angles. For a given inclination, the predicted air flow rate using Domain L was smaller than using Domain S because the additional flow resistances at the entrance and exit for the air gap were taken into account. The predicted PV temperature using Domain S for heat transfer through the back side was higher due to less efficient heat transfer from the hot module to cool air. When the prediction involved heat transfer from both sides of the module using Domain L2, the outgoing flow from the air gap for the module installed on a roof with a pitch greater than about 40 degrees would be assisted by buoyant flow from the front side and so the air flow rate resulting from heat transfer from both sides of the module was higher than that for heat transfer through one side only.

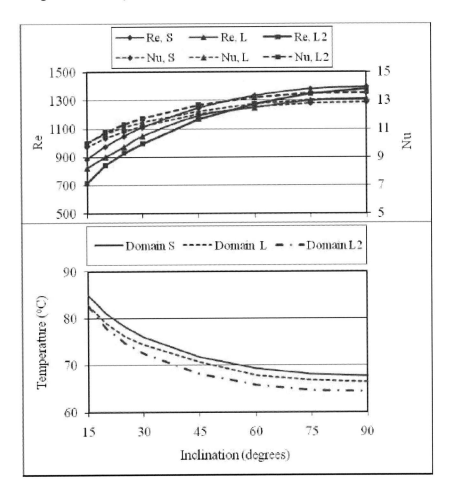

Figure 28. Predicted air flow and heat transfer rates behind the module and PV temperature using Domains S, L and L2.

For roof pitches less than 40 degrees, the entrainment by the flow in front forced the outgoing air upwards and increased the flow resistance. The net effect between the entrainment and increased flow resistance was reducing the flow rate through the air gap. Despite the reduced flow rate, the flow entrainment forced cool incoming air towards the module and increased heat transfer from the module. This led to a lower PV temperature predicted for heat transfer from both sides than that for heat transfer through one side only using a domain of the same size or the smaller domain. In other words, neglecting the heat transfer and buoyant air flow from the front side of the module would over predict the PV temperature.

Figure 29 shows the relationships between Reynolds, Nusselt and Rayleigh numbers calculated from the predicted air flow and heat transfer rates. In calculating the Rayleigh number for Equation (13), the cavity height was fixed as the module length rather than varying with inclination and hence, for the same cavity size and heat flux, the Rayleigh number would only indirectly depend on the resulting wall surface temperatures that also influence the thermal properties of air.

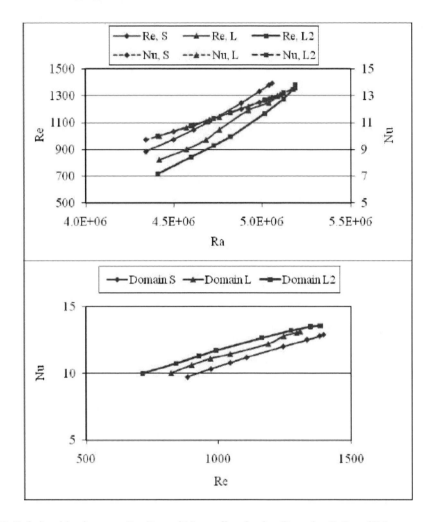

Figure 29. Relationships between Ra, Re and Nu predicted using Domains S, L and L2.

The predicted air flow and heat transfer rates increased with Rayleigh number for any type of computational domain. Similar to the effect of inclination angle, for a given Rayleigh number, the air flow rate in terms of Re was highest using Domain S but it was lower using Domain L2 than Domain L for the full range of Rayleigh number because of the dependence of the thermal properties of air on the wall surface temperatures according to Equation (13) – lower temperatures predicted using Domain L2 leading to lower thermal properties and higher Rayleigh number. However, there was no difference in the heat transfer rate, Nu, using different types of domain. The three sets of data for Nusselt number using three types of domain varied with Rayleigh number along the same line represented by the following approximation:

$$Nu = 0.00000466\, Ra - 10.68 \qquad (R^2 = 0.998) \qquad (24)$$

Figure 29 also shows that for a given air flow rate the predicted heat transfer rate was lower using Domain S than using Domain L which was itself lower than using Domain L2. This is opposite to the relationship of the heat transfer rate with the inclination angle or Rayleigh number. The heat transfer rate linearly increased with air flow rate for any type of domain. Their relationships are given by

Using Domain S,
Nu = 0.0061 Re – 4.4   (R^2 = 0.999)   (25a)

Using Domain L,
Nu = 0.0062 Re – 5.0   (R^2 = 0.995)   (25b)

Using Domain L2,
Nu = 0.0054 Re – 6.2   (R^2 = 0.994)   (25c)

Effect of Radiation Heat Transfer

When the radiation heat transfer was neglected, air flow in the air gap was asymmetric, predominantly along the module (see Figure 27).

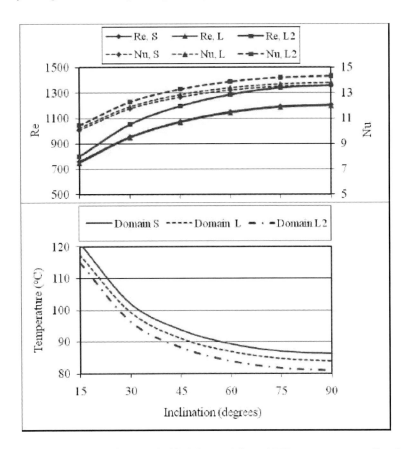

Figure 30. Air flow and heat transfer rates behind the module and PV temperature predicted with convection heat transfer only using Domains S, L and L2.

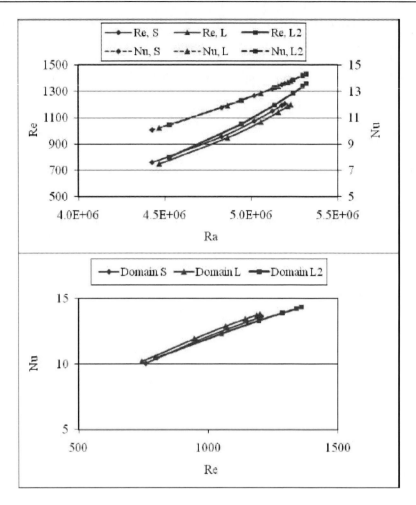

Figure 31. Relationships between Ra, Re and Nu predicted with convection heat transfer only using Domains S, L and L2.

Neglecting the radiation heat transfer also resulted in similar predicted air flow rates using Domains S and L. Consequently, as in the case of a vertical cavity where the air flow rate decreased with increasing level of asymmetry in heat distribution, the predicted air flow rate decreased by about 14% using Domain S and only 8% to 10% using Domain L, but the flow rate using Domain L2 increased by nearly 12% at an inclination of 15 degrees although the effect of radiation heat transfer diminished at larger inclination angles (Figure 30 compared with Figure 28).

The predicted heat transfer rate without radiation heat transfer increased slightly by 2% to 6% using any type of domain and this would have lead to a lower PV temperature for the same mode of heat transfer. However, because of neglecting radiation heat transfer, the PV temperature was over-predicted considerably by 17 to 36 K as seen from Figure 30 in comparison with Figure 28. Therefore, radiation heat transfer cannot be neglected in predicting PV temperatures under operating conditions, i.e. with bright sunshine. On the other hand, as shown in Figure 31, the difference in the relationships between Re, Nu and Ra using different types of domain decreased due to the increased influence of convection heat transfer to compensate for the additional resistances at the entrance and exit of the air gap.

Module Length < Roof Length

In the above predictions, the air gap was assumed as a simple ventilation cavity with two walls of the same height. In practice, however, the wall opposing the module is the roof or facade of a building and is longer than a single module. To simulate the realistic PV installation, predictions were carried out for the air gap with the wall representing the roof/facade much longer than the module such that it formed one side of the computational domain (Domain L or L2) (see Figure 26b). The direct solar heat gain by the two sections of the roof/facade – one below the entrance and the other above the exit opening of the air gap – was neglected but the indirect radiation from the module was taken into account.

Form the prediction with combined convection and radiation heat transfer, incoming air is seen from Figure 32 to flow along the roof immediately after the entrance while outgoing air moved along the roof due to the Coanda effect. The straight flow of air above the exit opening enhanced the air flow and heat transfer in the air gap. Consequently, the predicted air flow and heat transfer rates using Domain L were much higher than those using Domain S and using Domain L2 resulted in even higher rates as shown in Figure 33. However, the heat transfer from the module to air was less efficient because the increased air flow and heat transfer rates resulted from the flow along the roof as much as that along the backside of the module. As a result, the predicted PV temperature was similar using Domain L and Domain S. In other words, the temperature of PV installed along a long roof/facade would be higher than that installed along a panel of the same length as the module. When the prediction involved heat transfer from both sides of the module using Domain L2, the increased outgoing flow from the air gap would be more from the module surfaces than the roof surface and the module temperature was lower than predicted for heat transfer through one side only. However, the predicted temperature was still slightly higher than the prediction for the module of the same length as the roof except for the nearly flat roof (of 15-degree pitch) where the temperature of the module mounted on a long roof would be much lower due to the much higher air flow rate.

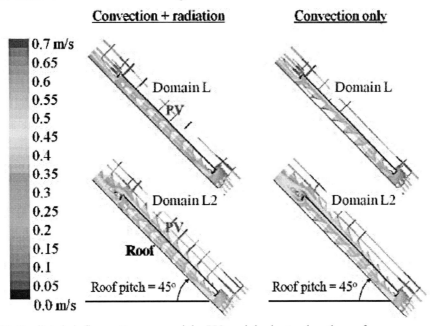

Figure 32. Predicted air flow patterns around the PV module shorter than the roof.

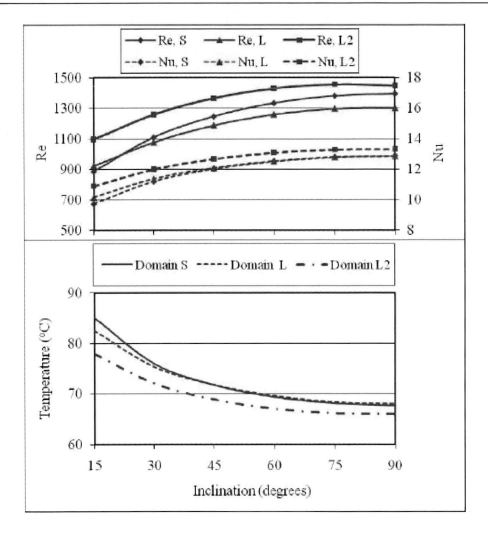

Figure 33. Predicted air flow and heat transfer rates behind the module shorter than the roof and PV temperature using Domains S, L and L2.

The variation patterns of air flow rate and heat transfer rate with inclination or Rayleigh number or the heat transfer rate against air flow rate were similar for two different lengths of roof as seen from Figure 34 and Figure 29. The air flow and heat transfer rates also increased with Rayleigh number for any type of computational domain or roof length. Similar to the effect of inclination angle of a roof longer than the module, for a given Rayleigh number, the air flow rate was highest using Domain L2. The difference between using Domains S and L was clearer because of the higher air flow rate predicted using Domain S than Domain L at the same Rayleigh number. However, there was no difference in Nusselt number in relation to Rayleigh number using different types of domain. The three sets of data for Nusselt number varied with Rayleigh number along the following relationship:

$$Nu = 0.00000472 \, Ra - 10.95 \qquad (R^2 = 0.998) \qquad (26)$$

Without taking account of radiation heat transfer, heat transfer and air flow in the air gap would again take place predominantly along the module (see Figure 32). By comparing Figure 35 with Figure 33, it is seen that the predicted air flow rate for heat transfer from the module backside only using Domain L decreased by 12% to 13% for inclinations of 30 degrees and above but by as much as 40% for an inclination of 15 degrees.

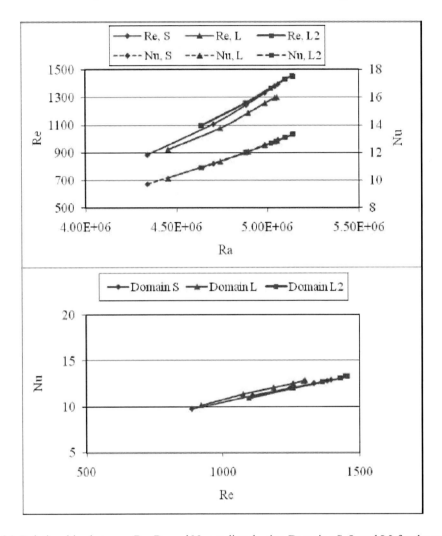

Figure 34. Relationships between Ra, Re and Nu predicted using Domains S, L and L2 for the module shorter than the roof.

By contrast, the air flow rate predicted for heat transfer by convection only from both sides of the module using Domain L2 increased by about 7% at the 15-degree inclination and the difference between the air flow rates with and without radiation heat transfer increased with inclination by 1% to 7% at larger inclinations. The effect of radiation on the predicted heat transfer rate was small but the effect on the PV temperature remained to be considerable. Using any domain for prediction, neglecting radiation heat transfer would increase the heat transfer rate by up to 6% but increase the PV temperature by 19 to 39 K, larger increase for smaller inclination.

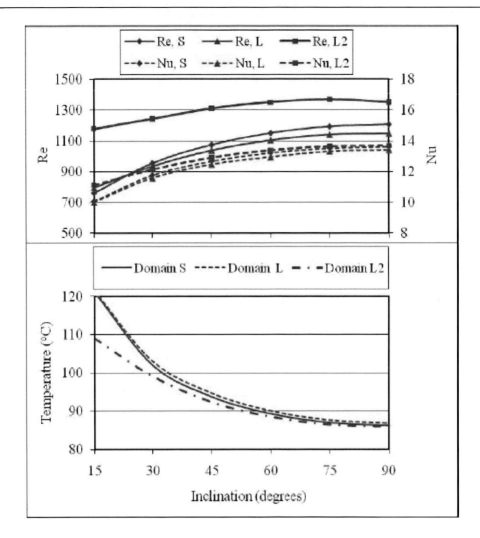

Figure 35. Air flow and heat transfer rates behind the module shorter than the roof and PV temperature predicted with convection heat transfer only using Domains S, L and L2.

Figure 36 shows that the difference in the relationships between Re, Nu and Ra using Domains S and L was negligible but the predicted air flow rate Re in relation to Rayleigh number or Nusselt number using Domain L2 was much larger than using Domain S or L.

Of the ten sets of predictions according to the domain type, heat transfer mode and roof length, the prediction with combined convection and radiation heat transfer from both sides of the module installed on a long roof/facade is considered more realistic for air flow around a building-integrated PV system.

One implication from the predicted PV temperature is that PV modules with their total length less than the roof length should be installed preferably close to the eave of a building to the middle of its roof, where feasible, to facilitate ventilation cooling. Such an installation would reduce the PV temperature not only by means of buoyancy-driven flow but also taking advantage of potential larger wind effects.

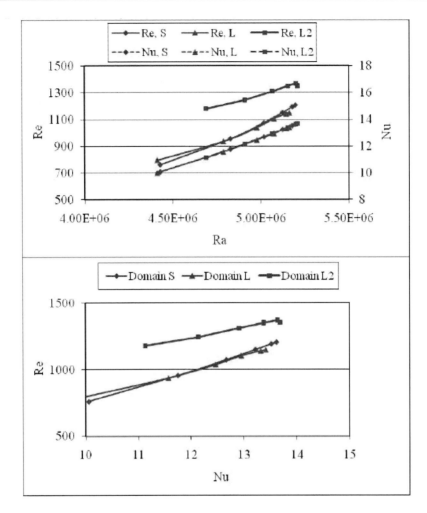

Figure 36. Relationships between Ra, Re and Nu predicted with convection heat transfer only using Domains S, L and L2 for the module shorter than the roof.

CONCLUSION

Work in this chapter has shown that the air flow and heat transfer rates generated by buoyancy in solar heated ventilation cavities vary with cavity size and inclination, total heat flux and heat distribution ratio on the two wall surfaces. Within a wide range of heat distribution ratios, both air flow rate and heat transfer rate increase with cavity width and height and total heat input. In general, the more evenly heat is distributed on the two opposing wall surfaces, the larger the air flow rate but smaller heat transfer rate. However, the buoyancy-driven flow rate would decrease if the aspect ratio of the cavity is so small, e.g., for a 3 m tall cavity wider than 0.5 m, that the total width of velocity boundary layers at the top of heated walls is smaller than the cavity width, resulting in significant reserve flow.

The air flow and heat transfer rates are also affected by the position of inlet and outlet openings. The air flow rate in a vertical ventilation cavity with a horizontal inlet (and with a vertical outlet) would be less than that for a cavity with vertical inlet but the heat transfer rate in the ventilation cavity with horizontal inlet could be higher than that with vertical inlet. Also, the position of a horizontal outlet could have a larger influence on the air flow rate in a ventilation cavity than the position of inlet.

For vertical ventilation cavities that have low or moderate levels of asymmetry in wall heating, the effects of the total heat flux, heat distribution ratio, cavity width and height on the heat transfer rate and air flow rate can be represented by the following expressions:

$$Nu = c_1 \left[Ra \left(\frac{H}{b} \right)^{3/2} \right]^m$$

and

$$Nu = c_2 \left(Ra^{1/3} \, \text{Re} \right)^n$$

For ventilation cavities with possible reverse flow (those with a low aspect ratio and highly asymmetrical heating), a more accurate representation of the relationship between Nu, Re and Ra is to employ the exponent of Ra larger than 1/3, i.e., $a_2 > 1/3$ in the following expression:

$$Nu = c_2 \left(Ra^{a_2} \, \text{Re} \right)^n$$

For ventilation cavities with a horizontal inlet, the exponent for the aspect ratio would also depend on the heat distribution ratio and a more general expression is

$$Nu = c_1 \left[Ra \left(\frac{H}{b} \right)^{a_1} \right]^m$$

For an inclined ventilation cavity representing an air gap between a PV module and a building envelope, the air flow rate and heat transfer rate generally increase with inclination angle while the temperature of PV cells decreases with increasing inclination. Building-integrated PV modules should be installed close to the eave of a building rather than on the middle of the roof to facilitate ventilation cooling.

The accuracy of predicting buoyancy-driven ventilation cooling of building-integrated PV devices would depend on the type of computational domain, the mode of heat transfer involved and the size of building envelope as well as the PV configuration and air gap. For accurate prediction of the cell temperature of building-integrated PV modules, consideration should be given to the following:

- Details of the module structure including the frame and cell distribution,
- Use of a larger computational domain than the air gap,

- Use of the roof or facade of a building as either part of the boundary of the computational domain or within an even larger domain for predicting buoyancy and/or wind effects for example,
- Convection and radiation heat transfer on both front and back sides of PV modules.

REFERENCES

[1] Eicker, U., Fux, V., Infield, D., Mei, L. and Vollmer, K., Thermal performance of building integrated ventilated PV facades, *Proceedings of International Solar Energy Conference*, Jerusalem, Israel, 4-9 July 1999, pp. 613-616.

[2] Aschehoug, Ø., PV system integrated in a double façade, *CADDET Renewable Energy Newsletter*, December 2000, pp. 26-28.

[3] Pedersen, P. V. and Rasmussen, A., Low energy ventilation system using PV modules, www.ecobuilding.dk.

[4] Aung, W., Fletcher, L. S. and Sernas, V., Developing laminar free convection between vertical flat plates with asymmetric heating, *International Journal of Heat and Mass Transfer*, 1972, 15, 2293–2308.

[5] Badr, H. M., Habib, M. A., Anwar, S., Ben-Mansour, R. and Said, S. A. M., Turbulent natural convection in vertical parallel-plate channels, *Heat and Mass Transfer*, 2006, 43, 73–84.

[6] Fedorov, A. G. and Viskanta, R., Turbulent natural convection heat transfer in an asymmetrically heated, vertical parallel-plate channel, *International Journal of Heat and Mass Transfer*, 1997, 40(16), 3849-3860.

[7] Baskaya, S., Aktas, M. K. and Onur, N., Numerical simulation of the effects of plate separation and inclination on heat transfer in buoyancy driven open channels, *Heat and Mass Transfer*, 1999, 35, 273-280.

[8] Rodrigues, A. M., Canha da Piedade, A., Lahellec, A. and Grandpeix, J. Y., Modelling natural convection in a heated vertical channel for room ventilation, *Building and Environment*, 2000, 35, 455-469.

[9] Nguyen, P. H., Yerro, P., Bernard, J. L., Sellitto, P. and Dupont, M., Wall temperature experimental investigation in a thermally driven channel, *Renewable Energy*, 2000, 19, 443-456.

[10] Morrone, B., Campo, A. and Manca, O., Optimum plate separation in vertical parallel plate channels for natural convective flows: incorporation of large spaces at the channel extremes, *International Journal of Heat and Mass Transfer*, 1997, 40(5), 993-1000.

[11] Gan, G., Impact of computational domain on the prediction of buoyancy-driven ventilation cooling, *Building and Environment*, 2010, 45(5), 1173-1183.

[12] Gan, G., Simulation of buoyancy-driven natural ventilation of building - Impact of computational domain, *Energy and Buildings*, 2010, 42(5), 1290-1300.

[13] FLUENT User's Guide. *Fluent Inc.*, USA. 2005.

[14] Yakhot, V. and Orzsag, S. A., Renormalization group analysis of turbulence: I. basic theory, *Journal of Scientific Computing*, 1986, 1, 1-51.

[15] Gan, G., Prediction of turbulent buoyant flow using an RNG k-ε model, *Numerical Heat Transfer, Part A: Applications*, 1998, 33(2), 169-189.

[16] Moshfegh, B. and Sandberg, M., Flow and heat transfer in the air gap behind photovoltaic panels, *Renewable and Sustainable Energy Reviews*, 1998, 2, 287-301.
[17] La Pica, A., Rodono, G. and Volpes, R., An experimental investigation on natural convection of air in a vertical channel, *International Journal of Heat and Mass Transfer*, 1993, **36,** 611–616
[18] Raithby, G. D. and Hollands, K. G. T., Natural Convection, in Rohsenow W. M., Hartnett J. R, Cho Y. I. (eds.), Handbook of Heat transfer, 3rd ed. McGraw-Hill, New York, 1998, pp. 4.1–4.99.
[19] Gan, G. and Riffat, S. B., A numerical study of solar chimney for natural ventilation of buildings with heat recovery, *Applied Thermal Engineering*, 1998, 18(12), 1171-1187.
[20] Gan, G., Numerical determination of adequate air gaps for building-integrated photovoltaics, *Solar Energy*, 2009, 83(8), 1253-1273.
[21] Gan, G., Effect of air gap on the performance of building-integrated photovoltaics, *Energy*, 2009, 34(7), 912-921.
[22] BP Solar, 2008. Module 485. www.bpsolar.com.
[23] Carvalho, M. G., Farias, T. and Fontes, P., Predicting radiative heat transfer in absorbing, emitting and scattering media using the discrete transfer method, *Fundamentals of Radiation Heat Transfer*, ASME FED, 1991, 160, 17-26.

In: Computational Fluid Dynamics
Editor: Alyssa D. Murphy

ISBN 978-1-61209-276-8
© 2011 Nova Science Publishers, Inc.

Chapter 5

A REVIEW OF ADVANCES IN MULTIPHASE COMPUTATIONAL FLUID DYNAMICS FOR MICROFLUIDICS

Rafael M. Santos[*]
Katholieke Universiteit Leuven, Leuven, Belgium

ABSTRACT

Multiphase flow in microchannels has gained much interest in recent years given the numerous emerging applications of microfluidics that promise to provide technological innovations not realizable with conventional channels. Significant developments in the area of micro-scale fabrication in the last decade have allowed researchers to construct increasingly more intricate devices, generally termed Lab-on-a-chip, and used in such diverse fields as: biomedical, analytical chemistry, chemical synthesis, unit operations, among others. Moving forward microfluidics is also envisioned to allow for process intensification of industrial process, hence the production capacity is expected to span from nanoliters to cubic meters, from micrograms to tons.

The interest is derived from the fact that gas–liquid two-phase flows in microchannels often exhibit different flow behaviour than macrosized conduits, which allows for the precise control of the trajectory of fluidic particles. At the micro-scale there is increased importance and effect of surface tension forces, while gravitational forces become negligible, and inertial, shearing and drag forces have a limited effect. As a result, multiple unique flow patterns have been identified in microfluidic gas–liquid two-phase flow, including bubbly, slug, ring, churn and annular flow. Transport phenomena, such as heat and mass transfer, and reaction kinetics are also significantly improved at the micro-scale due to the increase in surface to volume ratio.

The successful design of microfluidic devices relies on the need to fully understand the flow dynamics and physics, obtained by experimental means and from numerical modelling. Large discrepancies still exist in the published data, largely as a result of difficulty and inconsistencies in experimental setup and measurement. Moreover, process properties such as heat and mass transfer, pressure drop and residence time distribution are dependent on flow parameters such as the droplet size and velocity, the void fraction,

---

[*] Email: Rafael.Santos@alumni.utoronto.ca

the liquid film thickness, and so forth; however these quantities often cannot be determined a priori, making the design of microfluidic devices highly dependent on empirical correlations. The accuracy and reliability of present correlations can be extensively improved if the intricacies of the flow fields are known in more detail. This has recently become possible with the advances in the field of multiphase computational fluid dynamics (CFD) and the maturity of both proprietary codes and commercial software. A number of researchers have begun exploring this tool, using techniques such as volume-of-fluid (VOF), level-set and the lattice Boltzmann method (LBM), and publications have grown extensively in recent years.

This chapter will review the recent published works in the field, focusing on the application of CFD to simulate multiphase flow in microchannels. Both works of basic research on aspects of transport phenomena, as well as applied research on the development of new chemical processes and the intensification of existing processes at the micro-scale, will be covered. Common themes and practices will be consolidated and reviewed, and potential development directions identified.

INTRODUCTION

A large number of researchers have recently performed studies to identify multiphase flow behaviour in micrometer-scale channels. The principal reason for the increasing attention on such devices is the numerous emerging applications of microfluidics that promise to provide technological innovations not realizable with conventional channels. In particular, gas-liquid two-phase flows in microchannels often exhibit different flow behaviour than macro-sized conduits and the trajectory of fluidic particles in microchannels can be controlled precisely. There have also been significant developments in the area of micro-scale fabrication, which have allowed researchers to construct increasingly more intricate devices. In order to successfully design microfluidic devices it is crucial to fully understand the flow dynamics and physics; this knowledge is generally obtained mainly by performing laboratory experimentation and in a more limited form from numerical modelling. However, large discrepancies still exist in the published data, largely as a result of difficulty and inconsistencies in experimental setup and measurement.

Numerous correlations have been produced that rely on data regression, dimensional analysis or empirical correlations; however they are for the most part based on limited, experimentally accessible information such as the flow pattern observations and various assumptions. It could be possible to extensively improve the accuracy and reliability of such models if the intricacies of the flow fields were known in more detail, which has recently become possible with the advances in the field of numerical analysis, and more specifically multiphase computational fluid dynamics (CFD). With CFD, information about local parameters such as velocity profiles, volumetric mass fractions and interface configurations are easily obtained, whereas in experimental work such data are nearly, if not entirely, impossible to measure.

While experimental work on gas-liquid two-phase flow in microchannels is abundant, that is not the case for numerical modelling of the same process. Numerical modelling of multiphase flow phenomena is one of the most complex areas of the field of computational fluid dynamics, and it is a field that is still in its infancy and in continuous development. However, in recent years the first models that permit interface tracking have become available, in part due to the progress in the field of computing that now allow models that once required highly advanced computing machines to be solved in common personal computers in acceptable periods of time. As result, in recent years the first numerical results have been published on slug formation and flow in microfluidic geometries. Preference has typically been given to two-dimensional analysis due to its considerably lower computational cost; however the trend has recently been shifting as researchers realize that surface tension physics require full three-dimensionality for correct implementation. Nonetheless, still in both 2D and 3D cases the size of the computational geometry must be limited to several channel diameters in length, not enabling modelling of the full length of a microchannel or complex microfluidic networks.

The basis of the discussion is this chapter will be on the work of Santos and Kawaji (2010), followed by a review of the most interesting aspects with regards to the application of multiphase computational fluid dynamics of work by other authors. The work of Santos and Kawaji (2010) consisted in studying Taylor slug flow and comparing experimental results with numerical modelling. The experiments consisted in producing two-phase flow at the appropriate flow conditions to result in the formation of Taylor slugs in a microfluidic geometry composed of nearly square microchannels (with each side wall in the order of 100 µm in length) forming a T-junction, whereby liquid entered the main channel, parallel to the outlet flow, while the gas phase entered the side channel, perpendicular to the main and outlet channels. Numerical modelling, using the commercial computational fluid dynamics software FLUENT, which employs the Volume-of-Fluid (VOF) multiphase modelling technique, was performed of the same channel geometry using flow parameters identical to the experimental values. Modelling was entirely done in three-dimensional fashion, in order to fully capture the interfacial tension forces that are so important in the slug break-up process. The computational geometry was also long enough to allow the formation of a few air slugs in order to better capture the flow and pressure fields in the two-phase region. Within the study several slug flow parameters were measured from numerical and experimental runs that include Taylor slug size and velocity, liquid slug length and velocity, Taylor slug formation frequency, void fraction, cross-sectional air slug coverage and two-phase flow patterns. Results were used to construct correlations to predict these parameters based solely on a priori variables, namely the flow rates. This study was unique at the time in that both experiments and fully three-dimensional numerical simulations have been performed on the same microfluidic process, something which is not commonly found in the published literature up to the publication date of this chapter.

MICROFLUDICS

The different flow behaviour in microchannels is often attributed to the increased importance and effect of surface tension forces at micro-scales, while gravitational forces become negligible, and inertial, shearing and drag forces have a limited effect (Brauner and Moalem-Maron 1992, Akbar et al. 2003, Garstecki et al. 2006). The wetting properties of the fluids relative to the channel walls, more specifically the contact angle, have also been shown to affect the two-phase flow patterns in microchannels (Rosengarten et al. 2006).

Flow patterns that have been identified in microfluidic gas-liquid two-phase flow include bubbly, slug, ring, churn and annular flow (Kawahara et al. 2002). In some cases a thin film of liquid has been observed between the bubble and the channel wall (Irandoust et al. 1989, Rosengarten et al. 2006, Fukagata et al. 2006, Taha and Cui 2006). Over a specific range of operating conditions and using an appropriate combination of channel geometry and inlet conditions, it is possible to restrict the two-phase flow pattern in microchannels to the regime called Taylor slug flow, where the gas phase takes the form of elongated bubbles of characteristic capsular or bullet shapes, which occupy the entire cross-sectional area of the channel and are separated from each other by liquid plugs.

Compared to single phase laminar flow, Taylor slugs have been shown to increase transverse mass and heat transfer because of recirculation within the liquid plugs and the reduction of axial mixing between liquid plugs (Irandoust et al. 1989). Literature results indicate that the mass transfer, pressure drop and resident time distribution are dependent on the slug length and liquid film thickness; however these two quantities cannot be determined a priori, making the operation of microfluidic devices highly dependent on empirical correlations and modelling results (Qian and Lawal 2006).

Several techniques have been developed to produce gas slugs using different microchannel geometries. Gañán-Calvo and Gordillo (2001), Cubaud and Ho (2004) and Cubaud et al. (2005) described a method using capillary hydrodynamic flow focusing, Yasuno et al. (2004) proposed a microchannel emulsification method, Kawahara et al. (2002), Serizawa et al. (2002) and Xiong and Chung. (2007) used millimeter-sized orifice mixers to produce a pre-mixed two-phase flow that later entered a microchannel capillary, Xu et al. (2006) used a cross-flowing shear-rupturing method, and Xiong et al. (2007) used co-flowing channels. The T-junction gas slug break-up method, whereby channels of equal hydraulic diameter carry the gas and liquid phases that meet in a perpendicular junction, has become widely popular in recent studies (Thorsen et al. 2001, Okushima et al. 2004, Günther et al. 2004, Garstecki et al. 2006, Ide et al. 2007).

NUMERICAL MODELLING

Numerical Modelling Theory

The use of computational fluid dynamics (CFD) modelling techniques has become widespread for solving a variety of engineering and scientific problems that before the arrival of CFD had to be overly simplified to be solved analytically. CFD has made it possible to model complex transport phenomena and processes that involve a multitude of governing partial differential equations, which often times are highly non-linear. The backbone of transport phenomena models are the Navier-Stokes equations and these can be fully solved in 1D, 2D or 3D by CFD software packages that perform millions of calculations per numerical run to predict the interaction between fluids, gases and solids. A number of companies have developed CFD software and some of the most popular ones (used widely in both industry and academia) are: FLUENT from Fluent Inc., CFX from Ansys, Star-CD from CD-Adapco, Phoenics from CHAM, Comsol Multiphysics from Comsol, and TransAT from Ascomp. A number of researchers also choose to build their own models, often times because the source codes are not fully accessible in commercial codes and their line of research requires solution to specific models that are not available commercially. This type of endeavour usually requires an extensive amount of time and also extensive knowledge in the subject of CFD as well as computer programming. Self-built models can also lack accuracy or contain mistakes or errors, which make the validation of their predictions questionable or uncertain. As a result, commercial CFD software are often more useful as their codes have been built by large companies with extensive resources and skilled personnel, and have been extensively tested and verified by a number of sources; hence the results obtained can be reliably trusted (taking note of simplifications and limitations of the models). In this section, the discussion centers on the application of FLUENT 6.2 software, as utilized in Santos and Kawaji (2010), to model two-phase gas-liquid slug formation in a microfluidic T-junction.

In order to transform a continuous set of partial differential equations into form that can be evaluated numerically by a computer the equations must be discretized (i.e. represent and evaluate them as algebraic equations). FLUENT uses the finite volume method, whereby values are calculated at discrete locations on a meshed geometry, and the finite volume refers to the small spatial volume that surrounds a nodal point on the mesh. In this method, volume integrals containing divergence terms are converted to surface integrals and are then evaluated as fluxes at the surfaces of the finite volume. Given that the flux that enters a finite volume is equal to the flux that leaves the adjacent volume this method is termed to be conservative. The finite volume method is also suitable for models that use unstructured meshes (Fluent, 2006).

A more specialized area of computational fluid dynamics is modelling of multiphase flow phenomena, which is one of the most complex applications of CFD techniques, resulting in computationally expensive and more difficult to converge models. FLUENT offers a number of multiphase flow modelling techniques such as the Discrete Phase Model (DFM), Mixture Model, Eulerian Multiphase Flow Model, and Volume of Fluid Model (VOF). Of these, only the VOF model enables capturing and tracking the precise location of the interface between fluids. The Volume of Fluid method operates under the principles that the two or more fluids (or phases) are not interpenetrating. For each qth fluid phase in the system a new variable is introduced called the volume fraction (Ω_q). For each computational control volume the sum of all volume fractions must equal unity. All variables and properties in any computational cell are volume-averaged values, such that they are either representative of a single pure phase (when $\Omega_q = 0$ or $\Omega_q = 1$), or are representative of a mixture of the phases (when $0 > \Omega_q > 1$). For example, the density (ρ) and viscosity (μ) of a gas-liquid system are evaluated as follows:

$$\rho = \Omega_L \rho_L + \Omega_G \rho_G \tag{1}$$

$$\mu = \Omega_L \mu_L + \Omega_G \mu_G \tag{2}$$

A single continuity equation (3) and the momentum equation (4) are solved continuously across the computational domain, as would be done in the case of single phase flow. The VOF method accomplishes interface tracking by solving an additional continuity-like equation (5) for the volume fraction of the primary phase (gas), which yields the value of Ω_G. Ω_L is computed as $1 - \Omega_G$. (Fluent, 2006)

$$\frac{\partial \rho}{\partial t} + \nabla \cdot \left(\rho \vec{v} \right) = 0 \tag{3}$$

$$\frac{\partial \left(\rho \vec{v} \right)}{\partial t} + \nabla \cdot \left(\rho \vec{v} \vec{v} \right) = -\nabla p + \nabla \cdot \left[\mu \left(\nabla \vec{v} + \nabla \vec{v}^T \right) \right] + \rho \vec{g} + \vec{F} \tag{4}$$

$$\frac{\partial \Omega_G}{\partial t} + \vec{v} \cdot \nabla \Omega_G = 0 \tag{5}$$

The body force term (F) in Equation (4) is responsible for taking into account the surface tension (σ) and contact angle (θ) effects. Constant static contact angles are inputted into the model while surface tension values can be either be constant or variable (to account for Marangoni convection). These effects are computed in FLUENT by use of the continuum surface force (CSF) model developed by Brackbill et al. (1992). This model, rather than imposing the contact angle effect as a boundary condition at the wall, uses the contact angle value to adjust the interface normal in cells near the wall. As a result the curvature of the interface near the wall is adjusted, and the curvature is used to adjust the body force term. This procedure produces what is called a dynamic boundary condition. The curvature (κ) is computed as the divergence of the unit normal (6), which is computed in two manners: away from the wall (7) or adjacent to the wall (8), where the contact angle is taken into account. The surface normal (n) is simply the gradient of Ω_q (9). The surface tension causes a pressure jump across the interface, which is expressed as a volume force as given by Equation (10).

$$\kappa = \nabla \cdot \hat{n} \tag{6}$$

$$\hat{n}_{bulk} = \frac{n}{|n|} \tag{7}$$

$$\hat{n}_{near-wall} = \hat{n}_{wall} \cos(\theta) + \hat{t}_{wall} \sin(\theta) \tag{8}$$

$$n = \nabla \Omega_q \tag{9}$$

$$\vec{F} = \sigma \frac{\rho \kappa_G \nabla \Omega_G}{\frac{1}{2}(\rho_G + \rho_L)} \tag{10}$$

Numerical Modelling Procedure

In this section the numerical modelling approaches used in the FLUENT CFD software by Santos and Kawaji (2010) for solution of the multiphase VOF model are outlined. It is intended to provide a guideline for future researchers in the field. The following are the principal steps in setting up, running and analyzing the model:

Grid: The computational mesh was prepared in the Gambit software, also developed by Fluent Inc. The mesh was read into the FLUENT software, which was operated using the 3DDP mode (3-dimensional double precision), and was scaled by a factor of 10^{-6} (to convert meter lengths, which are standard, into micrometers).

Solver: A segregated axi-symmetric time-dependent unsteady solver was used. The VOF model was chosen as the multiphase flow model and the geometric interpolation scheme (which is the finest and most precise) was used for interface interpolation, along with the implicit body force formulation and a Courant number of 0.25. The Courant number was used to ensure that the time-step for cells near the interface was sufficiently small. For example, for a value of 0.25, the time-step must be at least one fourth of the characteristic transit time of a fluid element across a control volume. For discretization the PRESTO! (pressure staggering options) scheme was used for pressure interpolation, the PISO (pressure-implicit with splitting of operators) scheme was used for pressure-velocity coupling, and the second-order up-wind differencing scheme (most accurate) was used for the momentum equation. Under-relaxation factors were set to values of 1 for pressure, density and body forces, and 0.7 for momentum.

Materials and Phases: From the FLUENT materials database air and water-liquid were chosen as the fluids. The density of air can be adjusted as necessary. Air was designated as the primary phase and water was the secondary phase. Wall adhesion was turned on so that contact angles could be prescribed and a constant surface tension value of 73.5 dyn/cm was inputted.

User-defined functions: When using user-defined functions (UDF), these were compiled prior to setting boundary conditions and beginning numerical iteration.

Boundary Conditions: For the inlets, the velocity inlet boundary condition was used, while pressure outlet boundary condition was used for the outlet and the symmetry boundary condition was assigned to the symmetry face. For the inlet boundaries inlet velocities were entered (either a constant value or a user-defined function) and the water volume fractions at the boundaries were set to 1 for the main inlet and 0 for the side inlet. For the outlet boundary the backflow volume fraction was set to 1 and the gauge pressure was set to zero Pascal. Hence in the numerical calculation the static pressure is always calculated upstream of the outlet and decays to zero at the outlet. For the wall boundary condition the contact angle value was inputted and no user input was required for the symmetry boundary condition.

Initialization: The flow field was initialized, setting all velocities, pressures and volume fractions to zero. This caused the entire channel geometry to be filled with air ($\Omega = 0$) initially. In order not to have to model water filling the entire main channel, the flow field in the main channel was modified using the patching technique. The computational cells in the main channel were marked and the water volume fraction was set to a value of 1 for all cells.

Iteration: The residuals of the continuity equation and the x, y and z momentum equations were monitored during numerical computation and the convergence criterion for the model to move on to the next time step was chosen as 0.0001 for each equation. The fixed time stepping method was chosen and a time step value of 1.6×10^{-6} s/step was used.

This time step were small enough to ensure the resulting slug formation was independent of the time step sizes. The maximum number of iterations per time step was set to 30 (most numerical runs converged within approximately 15 iterations or less). The number of time steps was selected according to the flow rates in order to form as many slugs as possible before the first slug reached the outlet boundary, and typically ranged from 2,000 to 10,000. Data files were saved every 250 time steps to allow for visualization and data analysis of interim flow patterns.

Data analysis: After each numerical run was completed the data was analyzed to produce useful qualitative and quantitative results. FLUENT provides convenient forms of visualizing the flow field using contour plots. Plots were produced to show the volume fraction distribution (to visualize where air and water phases were located in the channel), the static pressure distribution and the velocity magnitude distribution. Surfaces were also created to cut across the geometry for visualization of the cross-section of the channel and air slugs. For some runs a point-type surface was created prior to the numerical iteration and it was used to monitor the value of the static pressure at that point as a function of time; this data was recorded in a text file in a form that could be read into a spreadsheet. In order to quantitatively identify flow parameters such as the gas slug velocity, the gas slug length and volume, the distance between gas slugs and the location of the centroid of gas slugs, a user-defined function was developed and used. This UDF was written in C programming language in a text file and was compiled and executed to read the numerical data after each run. The velocity of gas slugs (v_S) was calculated by averaging the axial velocity (x-direction) of the gas phase contained within the gas slug according to Equation (11). The volume of gas slugs (V_S) was calculated by the summation of the gas phase volume contained in the range of computation (cellrange) according to Equation (12).

$$v_S = \frac{\sum_{i=1}^{cellrange} \Omega_{G,i} \cdot V_{cell,i} \cdot v_{x,i}}{\sum_{i=1}^{cellrange} \Omega_{G,i} \cdot V_{cell,i}} \tag{11}$$

$$V_S = \sum_{i=1}^{cellrange} \Omega_{G,i} \cdot V_{cell,i} \tag{12}$$

A unique approach was taken by Santos and Kawaji (2010b) for the definition of the inlet velocity boundary conditions. FLUENT uses the velocity given to calculate the mass flow rate into the computational domain according to Equation (13). The velocity can be inputted as a single value that will be applied to all nodes on the boundary face, or it can be passed on to FLUENT by user-defined functions (UDF) that are capable of supplying values of velocity as a function of coordinate location, making it possible for a velocity profile to be attributed to the inlet boundary. The advantage of using the UDF approach is that the computational geometry can be simplified. Rather than making inlet channels long enough to allow time for the flow field to reach a fully developed laminar profile, it is possible to significantly shorten the inlet channels if the fully developed profile is given as a boundary condition.

$$\dot{m} = \int \rho \vec{v} \cdot d\vec{A} \tag{13}$$

The calculation procedure used by Santos and Kawaji (2010b) to analytically obtain the laminar velocity profiles for rectangular microchannels was developed by Spiga and Morini. (1994).

They used a finite Fourier transform to obtain an analytical solution of the two-dimensional velocity distribution symmetric with respect to rectangular Cartesian coordinates x and y. The set of equations is presented below.

$$W_{(x,y)} = \frac{16\psi}{\pi^4} \sum_{n...odd}^{\infty} \sum_{j...odd}^{\infty} \frac{\sin\left(n\pi\frac{z}{a}\right)\sin\left(j\pi\frac{y}{b}\right)}{nj(\psi^2 n^2 + j^2)}; n,j = 1,3,5... \tag{14}$$

$$\psi = a/b \tag{15}$$

$$v_{(x,y)} = W_{(x,y)} \times \xi \tag{16}$$

$$\xi = \frac{a^2}{\mu}\left(-\frac{\partial P}{\partial x}\right) \tag{17}$$

$$v_{max} = 1.5 \begin{pmatrix} 1 + 0.546688\psi + 1.552013\psi^2 - 4.059427\psi^3 \\ + 3.214927\psi^4 - 0.857313\psi^5 \end{pmatrix} U_{inlet} \tag{18}$$

Equations (14), (15) and (16) are the governing equations of the model and are solved by FLUENT via the UDF. However, before inputting the UDF into FLUENT, it is necessary to find the value of the variable ξ for each of the fluid inlet superficial velocity (U_{inlet}) desired. Therefore the governing equations are solved first in a spreadsheet on Microsoft Excel, with aid of Equation (18). For a given value of U_{inlet}, the maximum fluid velocity at the center of the rectangular inlet, v_{max}, is calculated using Equation (18). The value obtained from Equation (18) is then compared to the value predicted using the full set of governing equations. The value of the derivative of P with respect to x in Equation (17) is then changed until the two v_{max} values are equal. Then the final value of ξ is written into the UDF. This procedure is followed for each fluid inlet. Examples of the resulting velocity profile are illustrated in Figure 1. The contour plots are shown without node smoothing (individual facet values are shown).

To verify that the mass flow rates (m) of the runs that used the UDF are equal to the values obtained without the use of the UDF, comparable numerical runs were performed. The resulting flow profile is shown in Figure 2. It is possible to see that the flow is very similar for both cases, and that the volume of gas emerging from the T-junction is identical.

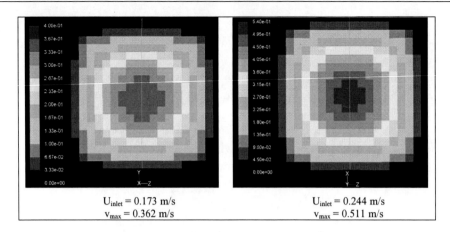

Figure 1. Fully developed laminar inlet velocity profiles.

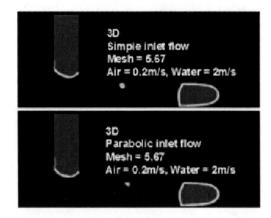

Figure 2. Effect of inlet boundary condition on numerical runs.

2D versus 3D Modelling

A large portion of computational fluid dynamics (CFD) modelling in the literature has been performed with two-dimensional models, leaving only a fraction of problems being solved with full three-dimensionality. The principal reason for this is that three-dimensional models increase the computational costs dramatically, resulting in the need for larger computer memories and in longer computational times. For example, take a 2D computational domain consisting of 100 computational nodes across each lattice, resulting in a total of 100x100, or 10,000, computational nodes. Now if this geometry is extended to the third dimension to form a cubic domain, the number of computational nodes becomes 100x100x100, or 1,000,000 nodes, which is exactly 100 times more nodes than the previous case. This means the computer memory needs to be 100 times larger, and the simulation will take 100 times longer (for example, instead of 1 hour, it will take more than 4 days). Another reason two-dimensional models are often used is because many transport phenomena problems can be simplified due to symmetry or absence of significant changes or effects of the third dimension. This may be the case for problems such as flow in pipes, diffusion through a membrane, casting of molten metal sheets, etc.

When it comes to multiphase problems, the third dimension in many cases cannot, or at least should not, be disregarded. This is especially true when interface tracking is to be accomplished. Extensive work has been done recently on bubble formation, oscillation and interaction, and jet instability and break-up, which show that fluid behaviour observed in 3D models can significantly depart from 2D counterparts. Many times 2D models are used to gain insight into industrial problems, such as plunging of steel jets, gas plume injection in molten metal, or tapping in a ferro-alloy furnace (Johansen 2002), however the over-simplicity of these models must be recognized. Esmaeeli et al. (1999) studied the gravity-driven rise of bubbles in a liquid column using both 2D and 3D models. They found that 2D models provided well converged results of a greater number of bubbles than could be modelled in 3D; however the usefulness of those results were not surprisingly mixed. While predicting correct trends, 2D results tended to over-predict bubble interaction and distribution fluctuation.

The aim of Santos and Kawaji (2010) was to track the formation of air slugs, one at a time, in a microfluidic T-junction. For a slug to form, a portion of air volume must break-off from the air inlet stream. The slug breaks off when the fluid region between the slug and the air inlet becomes very narrow, that is, having a width that is smaller than the distance between two computational nodes (in the order of a few micrometers). The narrowing process is caused by a combination of several factors, of which some may be more important than others depending on the flow dynamics. In the break-up of a two-dimensional system, such as a liquid sheet, the dominant factors contributing to break up are the growth of instabilities, viscous and kinetic effects. In three-dimensional systems such as liquid jets, the surface tension becomes a much more important factor. As necking of the jet occurs, the surface tension forces are concentrated into a single point in the fluid, all pointing in the same inward direction from all radial coordinates. In the 2D case, the surface tension is much less concentrated since the curvature only exists in one dimension, and the surface tension vector components of the interfaces on either side of the fluid point in the opposite direction, stabilizing the oscillation effect. This is because on one side of the sheet the air phase is on the convex side of the curvature and on the other it is on the concave side. The surface tension force always points towards the concave side of the interface (such as the interior of a bubble). Only in the case of multiple oscillation harmonics will liquid sheets be destabilized by the surface tension effect, since then the forces would not balance on either side of the sheet symmetrically. During the oscillation of a liquid jet the surface tension is always pinching the jet in the radial direction, leading to the growth of the oscillation amplitude and the eventual pinching of the jet and formation of a drop or bubble (Figure 3). In 2D, a computer model does not differentiate between a sheet and a jet (a jet in 2D is essentially the cross-section of a sheet), hence the jet dynamics are not properly captured. Therefore, while the modelling of liquid sheets in 2D does not suffer any consequence from the absence of the third dimension, in the case of jets and bubble break-up the third dimension is required to fully capture the pinching effect.

Another example is a drop or bubble emerging from an orifice. As the fluid stretches out of the orifice away from the bulk phase, the surface area of the fluid increases, raising the surface energy of the system. To minimize this surface energy, the surface of fluid will pull the interface back in an effort to reduce the surface area. Once the volume of the drop or bubble emerging from the bulk fluid becomes large enough, it will attempt to minimize its own surface area, acquiring a spherical shape.

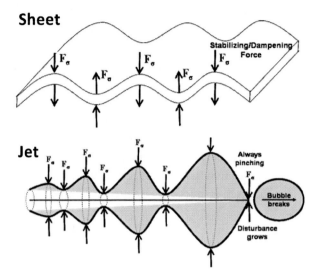

Figure 3. Effect of surface tension on the oscillation and break-up of liquid sheets and jets.

At this point, the fluid stream connecting the growing drop or bubble and the bulk fluid will begin to thin. Once the curvature near the thinning fluid becomes excessive, the surface tension will no longer be strong enough to hold the two fluid regions together, and the interface will rapidly be ripped apart and divide. At this point the interfaces of the bubble or drop and the bulk phase will retract and both will acquire a new spherically-shaped curvature. A complete solution of these phenomena was published by Wilkes et al. (1999).

To capture the true behaviour of interface break-up, Santos and Kawaji (2010) used the three-dimensional model available from FLUENT 6.2 CFD software, developed by Fluent Incorporated, which uses the volume of fluid (VOF) multiphase modelling technique to track the interface. This allowed for accurately capturing the slug formation at a microfluidic T-junction. Previous works that have used 2D models to capture similar fluid behaviour include those of Fukagata et al. (2007) and Qian and Lawal. (2006). The latter claimed that some 3D runs were performed in 2D as well, and the difference in the results, more specifically the air slug lengths, was small. They do mention, however, that the Laplace pressure (caused by interface curvature and surface tension) was twice as large in 3D results. As a result, it seems reasonable to assume that the slug break-off from the interface may be significantly affected by the different treatment of the surface tension.

To test the effect of the third-dimension, numerical runs were performed that used the same flow parameters (fluid velocities, contact angle, surface tension and mesh resolution) for both a 2D and 3D model. The resulting phase contour plots are shown in Figure 4. It can be seen that the gas slug formed in the 3D case is much more representative of typical microfluidic results than the 2D case. It can be noted that the gas slug in the 2D case does not acquire a spherical shape, likely due to the lack of Laplace pressure within the air phase, which would cause the interface to become more rigid and spherical. In the 3D case, the advancing interface acquires a typical nose shape, while the rear end becomes flattened due to the high velocity of the gas slug in this numerical run (in the order of 2.2 m/s). It is concluded, therefore, that in order to produce valuable numerical results that can be compared to experimental data, numerical simulation of this type of microfluidic process should only be performed with three-dimensional models.

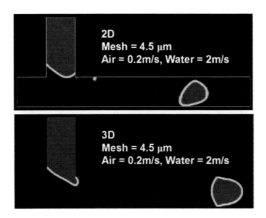

Figure 4. Comparison of 2D and 3D numerical modelling of slug formation at a microchannel T-junction.

REVIEW OF MULTIPHASE MICROFLUIDIC CFD

In microfluidic multiphase flow, fluidic particles are constantly within short distances from other fluidic particles and the channel walls. As a result, phenomena such as drop or bubble coalescence and drop or bubble attachment or detachment from solid surfaces are important to be predicted correctly. The precise understanding of such fluid physics phenomena is still progressing, and the latest models may not yet be present in commercial software or be in wide spread use. However useful works are available in literature and an extensive review detailed here leads to the discovery of interesting approaches.

Other authors that have used FLUENT to model microfluidic multiphase flow and are of note include Qian and Lawal (2006) and Yamaguchi et al. (2004). Qian and Lawal (2006) modelled a microfluidic T-junction to study gas-liquid slug formation. They compared 2D and 3D models, and noticed little difference in the results, choosing to perform the majority of experiment in 2D to save on computational time. However, their channel was 250 μm in diameter, and fluid velocities were in the range of 20 mm/s, significantly different from the dimensions and conditions tested by Santos and Kawaji (2010), where there were significant differences between 2D and 3D simulations. Yamaguchi et al. (2004) however, used FLUENT's 3D VOF model to simulated co-flowing liquids in a curved meandering microchannel of 200 μm diameter. They validated their simulation by comparison with experimental tests using confocal fluorescence microscopy to image the flow pattern.

As discussed previously on the comparison of 2D and 3D simulation of slug formation in microchannel flow as described in Santos and Kawaji (2010), the jet break-up mechanism is analogous in the effect of the surface tension forces. Hardt et al. (2005) provides an interesting study on the instabilities of a liquid jet focussed into a converging microchannel. Of most importance, they compare CFD simulation using the commercial software CFX4 with the VOF and CSF methodologies, to finite element (FE) results achieved by Ashgriz and

Mashayek (1995). The main difference between the two approaches is that the VOF model uses a fixed grid and accounts for the fluid dynamics of the external phase, whereas the FE approach utilizes a moving grid that is updated to accommodate the change in shape of the liquid jet. Furthermore in the FE model there is no outer fluid, as such the Laplace pressure due to surface curvature is the only contributor to the stress tensor on the free surface of the jet. The simulation performed with both methods used an initial perturbation to destabilize the jet, however whereas the FE technique is sensitive enough to propagate an initial perturbation of only 5% of the unperturbed jet radius, the VOF approach requires at least 15% perturbation due to the its diminished capability to resolve fine deformations of the fluid interface. However, a disadvantage of the FE moving grid approach is that it cannot predict the droplet break-up event. Ashgriz and Mashayek (1995) address this by considering arbitrarily that droplet break-up occurs when the radius of the jet is reduced to 1% of the original radius.

Hardt et al. (2005) found qualitatively good results in comparison to those of Ashgriz and Mashayek (1995). Specifically, the dynamics of the jet oscillation were very similar, though the VOF method predicted earlier jet break-up due to lack of grid resolution. An adaptive grid methodology available in more recent version of commercial software could probably address this difference satisfactorily to an extent. With regards to the microfluidic converging channels simulations, comparison was also made to analytical linear stability theory. It was found that the CFD results predicted the characteristic beads-on-a-string shape, but with a more stable jet. Analysis of the results pointed to a stabilizing effect provided by the channel walls due to shear forces created by the outer fluid phase. At the jet bulges, the outer fluid is accelerated and produces maxima of shear forces on the jet, slowing down the progression of bulge growth and consequently string thinning. The authors conclude that for simulation of the contraction geometry studied, CFD is a superior method to a semi-analytical approach, which should only be used when the jet radius is small compared to the channel radius.

Krishnan et al. (2010) provide insight into the various numerical simulation schemes that can be chosen within the commercial software FLUENT. The simulation work is done on rising bubble and flow in minichannels, however the conclusions can well be extended to modelling of microchannels as well since the focus of the work is on interface stability. It was found that the First Order Upwind (FOU) scheme and power law scheme for the momentum discretization perform better than the Second Order Upwind (SOU) scheme and the QUICK scheme, due to their first order upwind bias. Gradient calculations were captured better with the node-based scheme compared to the cell-based scheme. For capturing thin liquid films it was found the FOU performs best compared to the other schemes. The results were validated by comparing flow regimes with experimental flow maps.

The effect of wetting properties on microfluidic multiphase flow has been studied by Santos and Kawaji (2010b) as well as other authors. Rosengarten et al. (2006) were concerned with studying the influence of the contact angle on microfluidic two-phase flow. They used the commercial software CFD-ACE, utilizing the standard VOF finite volume method in combination with the CSF model of Brackbill et al. (1992) to account for surface tension effects. Their geometry consisted in a microfluidic contraction, whereby a gas bubble position in a larger inlet compartment entered the contraction as a result of drag from the continuous liquid phase. The gas bubble was generated large enough so that it would guarantee contact with the channel walls during entrance into the contraction, this way ensuring the contact angle would have an effect on the fluid flow. As a result, their modelling runs predicted significant effect of the contact angle on gas bubble flow in the microfluidic channel. They concluded that wetting properties of microchannels, when combined with proper two-phase flow introduction and slug formation, can be used for design of microfluidic droplet manipulation systems.

Another work from some of the same authors was later published by Harvie et al. (2007), also dealing with a microfluidic contraction but in this case with liquid-liquid systems. The innovation in this case was the study of shear thinning fluids, which were chosen to be either the dispersed phase or the continuous phase. In this study, the contact angle was maintained in the regime were the dispersed phase was non-wetting, so that droplet extension within the microchannel could be observed. The same Volume of Fluid methodology was used in this study. The results lead to better understanding on the characterization of viscosity ranges of the non-newtonian fluids, however no validation with experimental or theoretical results was presented.

Saha and Mitra (2009) present unique results on the use of dynamic contact angles in conjunction with VOF multiphase modelling in microfluidic devices. They incorporated a series of dynamics contact angle (θ_d) models from various authors into the open source CFD software OpenFOAM (Open Field Operation and Manipulation) 1.5 from OpenCFD Ltd.

Jiang et al. (1979)

$$\cos\theta_d = \cos\theta_e - (1+\cos\theta_e)\tanh(4.96 Ca^{0.702})$$

Shikhmurzaev (2008)

$$\cos\theta_d = \cos\theta_e - \frac{25\frac{\mu v}{\sigma}\left[(1+0.46(\cos\theta_e - 0.07))+0.54\frac{\sin\theta_d - \theta_d \cos\theta_d}{\sin\theta_d \cos\theta_d - \theta_d}\right]}{0.46\left\{\left[(1+0.46(\cos\theta_e - 0.07))+\left(12.5\frac{\mu v}{\sigma}\right)^2\right]^{1/2} + 12.5\frac{\mu v}{\sigma}\right\}}$$

Kalliadasis and Chang (1993)

$$|\tan\theta_d| = 7.48 Ca^{1/3} - 0.47863 Ca^{0.293}$$

Sikalo et al. (2005)

$$\cos\theta_d = 1 - 2\tanh\left(5.16\left[\frac{Ca + fHI}{1+1.31(Ca+fHI)^{0.99}}\right]^{0.706}\right)$$

$$fHI = \frac{1}{5.16}\left[1+1.31(Ca+fHI)^{0.99}\right]^{0.706}\tanh^{-1}\left[\frac{1-\cos\theta_e}{2}\right]$$

Fries and Dreyder (2008)
$$\cos\theta_d = \cos\theta_e - 2(1+\cos\theta_e)Ca^{0.5}$$

Cox (1986) $\qquad \theta_d^3 = \theta_e^3 + 144 Ca$

Newman (1968) $\qquad \cos\theta_d = \cos\theta_e \left(1 - e^{-\sigma t/\mu M}\right)$

The models most suitable for microfluidic flow, based on good agreement with analytical results, were found to be those of Kalliadasis and Chang (1993), Sikalo et al. (2005) and Fries and Dreyder (2008). The model of Kalliadasis and Chang (1993) however is independent of the specified static contact angle (θ_e) and depends only on the Capillary number (Ca), hence such model is only valid for fully hydrophilic cases, when the static contact angle is close to zero. The models of Sikalo et al. (2005) and Fries and Dreyder (2008) were found to be more suited to problems with partially wetting surfaces.

Gupta et al. (2009) reviewed many microfluidic multiphase modelling works and concluded that most authors do not pay particular attention to ensuring appropriate meshing near the channels walls to accurately capture the existence of liquid film. They theorized this is the reason certain authors claim the contact angle has an effect of producing gas slugs that directly contact the microchannel walls under Taylor slug flow. To address this hypothesis, they modelled Taylor slug flow with highly defined meshing close to the channel walls, in a flow configuration consisting of gas-core liquid ring feed inlet and 2D axi-symmetric computational domain. FLUENT VOF was used for the simulation. They concluded the Taylor gas slugs invariably are surrounded by a liquid film, except under the particular conditions where the gas-liquid flow is in a transient regime, where gas and liquid flow enter the channel from different streams, where the channel cross-section in non-circular, or where the homogeneous void fraction is high. They furthermore pointed that the contact angle plays a small role in microfluidic flow since the three-phase contact line does not exist except in the flow formation region. While these arguments are sustained by the simulation performed by the authors, it should be noted that slug formation is frequently studied in literature, and channel cross-sections are often non-circular due to difficulties in fabrication techniques or interest of researchers in different channel geometries for practical applications. For example Santos and Kawaji (2010) studied slug formation at a microchannel T-junction with rectangular channels. Hence the conclusions of Gupta et al. (2009) should be taken with caution, though the mesh refinement approach certainly is of merit in any circumstance if computational expense is not too great. Another useful contribution from the authors include the observation that the CSF model generates spurious currents near the gas-liquid interface, such that accurate gradient calculations and the use of grid elements of unity aspect ratio are important to minimize errors in surface tension modelling. Furthermore they published guidelines based on the Capillary number to create wall mesh grids that most ideally capture the liquid film.

Subsequent to the discussion of Gupta et al. (2009) on the modelling of microfluidic two-phase flow taking consideration of the presence of a thin liquid film around gas slugs in microchannels, Kashid et al. (2010) published a similar work but including the slug formation from a T-type microchannel mixer, rather than simply core-annular feed inlet microchannel flow. The inclusion of the T-junction ensures that both phases at one point in the channel touch the wall, hence for liquid film formation surrounding the entire gas dispersed phase, the bubble must detach from the wall in the two-phase flow region. This was confirmed in their work using the VOF CFD methodology in FLUENT. To address the need for a very fine mesh size to ensure capturing of the liquid film, the simulation was done in planar 2D. The mesh size was changed from 0.5 to 2 μm for grid independence studies.

The presence of the liquid film surrounding gas slugs was confirmed. However, it should be noted that the process conditions used may promote slug detachment and liquid film formation. The main channel in the T-mixer has a diameter of 1000 μm and the side channel had a smaller diameter of 200 μm. These dimensions are significantly larger than the channels in the order of 100 μm used by Santos and Kawaji (2010b) and may produce flow regimes that depart from true microchannel behaviour as described by Kawahara et al. (2002). Furthermore, the smaller side channel used may produce elongated gas slugs that do not reach the required volume to fill the entire channel cross-section prior to break-up. Lastly, the simulation was performed in 2D, which does not capture the correct surface tension forces. The slug shapes reported appear reasonable, however the flow velocities were of low magnitude (in the range of 27-67 mm/s) and thus shear effects that could distort the slug shape and break-op mechanism were attenuated. Indications of unstable interface dynamics can be seen by the appearance of waves along the thin liquid films reported.

Another work of note from the same group as in the discussion on Hardt et al. (2005) is that of Hardt (2005) concerning the interaction of droplets within a microfluidic device. The purpose of this work is the modification of multiphase microfluidic CFD models to account for short range forces that may explain why in some studies droplets under close contact do not proceed to coalescence. It is theorized that molecules, such as surfactants, may adsorb on fluid interfaces causing a stabilizing effect. This is contrary to the expectation from standard fluid dynamics that suggest the tendency for coalescence should be high at short distances, where the only mechanism available to slow down coalescence is viscous dissipation, which temporarily holds together the thin liquid bridges that separated the dispersed phase elements. Essentially, Hardt (2005) suggest that such pure models overestimate the rate at which fluid interfaces merge, mainly due to the dominant effect that surface tension takes in reducing the curvature of the cusps that form when two interfaces contact at a random position.

To amend these models, forces are introduced into the VOF equations within FLUENT to account for the development of an electric double layer (EDL) effect. Such repulsive force (P) take the form of $P(D) \alpha D^{-2}$, where D is the distance between the interfaces. Given that a high degree of accuracy of the short range distances is crucially important for the correct modelling of the interactions, grid refinement close to the interfaces in required and is performed using an adaptive grid-refinement method and hanging nodes. However, EDL forces act on a range of only a few nanometers, and even the most detailed refinement may not serve accurate enough in some circumstances. As such, a less rigorous approach is taken, whereby rather than modelling the details for the interface interaction, the model should simply predict two outcomes: the merging of fluid volumes if the hydrodynamic interactions are strong enough to overcome the repulsive force; or the separation of the entities if the repulsive forces are stronger.

The grid refinement method used subdivides each computational cell into four smaller cells, or 16 if necessary, based on the hanging-node scheme. This methodology is useful for achieving well-localized grid refinement compared to more global techniques. The criterion for grid refinement (e_i) is taken as the gradient of the volume fraction field (f) (19). The distribution of e_i in the whole computational domain is normalized by dividing by $(e_i)_{max}$ and a refinement interval ($e_i \in [\Gamma_r,1]$) and coarsening interval ($e_i \in [0,\Gamma_c]$) are defined by two thresholds (Γ_r, Γ_c).

$$e_i = \left|\vec{\nabla} f\right| \tag{19}$$

Fukagata et al. (2007) used the Level Set method to model microfluidic multiphase flow. The Level Set model has been popular with authors using their own in-house code, however it has not been implemented in commercial software as of yet, the exception being the TransAT software of ASCOMP GmbH of Switzerland. Fukagata et al. (2007) modelled a steady-state slug train, using the moving wall technique to impose a fluid velocity on stationary slugs. The Level Set method is used for capturing the two-phase fluid interface, while the continuity (20) and momentum (21) equation take the typical shape of the Navier-Stokes equations. Surface tension (σ) is taken into account by the last term in Equation (21), where κ, δ and $\vec{n}$ denote the curvature, the Dirac delta function, and the unit normal to the interface respectively.

$$\nabla \cdot \vec{u} = 0 \tag{20}$$

$$\frac{\partial(\rho\vec{u})}{\partial t} + \vec{u} \cdot \nabla(\rho\vec{u}) = -\nabla p + \nabla \cdot \left[\mu\left(\nabla\vec{u} + (\vec{u})^t\right)\right] - \sigma\kappa\delta\vec{n} \tag{21}$$

The interface is captured by a smooth function F as in Equation (22), where the gas-liquid interface is defined as $F = 0$, while the gas phase resides in the negative region ($F < 0$) and the liquid in the positive region ($F > 0$). Physical properties (ϕ) of the fluids are interpolated according to Equation (23), where a smoothed Heaviside function is used as an interpolating function (H_ε) in Equation (24). The surface tension term in Equation (21) is calculated using the CSF model of Brackbill et al. (1992). The curvature, Dirac delta function and unit normal vector are defined in Equations (25), (26) and (27).

$$\frac{\partial F}{\partial t} + \vec{u} \cdot \nabla F = 0 \tag{22}$$

$$\phi = \left(\frac{1}{2} + H_\varepsilon\right)\phi_L + \left(\frac{1}{2} - H_\varepsilon\right)\phi_G \tag{23}$$

$$H_\varepsilon = \begin{cases} -\frac{1}{2} & F \leq -\varepsilon \\ \frac{1}{2}\left[\frac{F}{\varepsilon} + \frac{1}{\pi}\sin\left(\frac{\pi F}{\varepsilon}\right)\right] & -\varepsilon < F < \varepsilon \\ \frac{1}{2} & F \geq \varepsilon \end{cases} \tag{24}$$

$$\delta_\varepsilon = \frac{dH_\varepsilon}{dF} = \begin{cases} \frac{1}{2}\left[\frac{1}{\varepsilon} + \frac{1}{\varepsilon}\cos\left(\frac{\pi F}{\varepsilon}\right)\right] & |F| < \varepsilon \\ 0 & |F| \geq \varepsilon \end{cases} \qquad (25)$$

$$\kappa = \nabla \cdot \vec{n} \qquad (26)$$

$$\vec{n} = \frac{\nabla F}{|\nabla F|} \qquad (27)$$

The Level Set based commercial software TransAT was tested against FLUENT by Gupta et al. (2010). Modelling with TransAT was performed using the QUICK scheme for discretization of the Level Set advection term. The re-initialization equation was solved using the WENO scheme. Pressure-velocity coupling was achieved with the SIMPLE algorithm, and the second order bounded HLPA scheme was used to discretize the advection terms. Time marching is done by the Runge-Kutta third-order explicit scheme. Because of the used of an adaptive time step method, a very small time-step had to be used ($O(10^{-7}s)$), and residuals had to be reduced by an order of 10^{-6}, increasing drastically the computational time.

Simulation of slug formation in a straight microchannel by annular gas-core liquid-ring flow inlet was performed in axi-symmetric 2D. It was found that volumes of gas and liquid slugs produced either software were very similar, the only differences being caused by slightly different gas slug detachment times. The VOF model in FLUENT captures the interface by geometrical reconstruction, while the Level Set method explicitly tracks the interface, giving rise to differences in determination of interface location, shape and surface tension forces. As such, the TransAT is expected to be more reliable than FLUENT, though no comparison to experimental data or benchmarking was presented by the authors. Also, wall shear stresses and local Nusselt numbers were found to be very similar using the two codes. Slight differences again can be explained by small differences in the calculated liquid film thickness. The authors concluded that the similarity of the two codes gives confidence in the veracity of the computational results. While TransAT uses methods that more accurately describe the fluid physics, the computational expense is higher. FLUENT provides nearly identical results in much faster computational times.

Another approach to modelling microfluidic multiphase flow that has emerged recently and is gaining popularity is a method called the Phase-Field Method. He and Kasagi (2008) describe the method and its advantages. The main difference to the more common techniques is it being based on fluid free energy instead of force. The merging and break up of fluid volumes is computed effortlessly, without extra coding, and the contact line movement is inherently simulated. The methodology is proposed specifically as an evolution upon the more widely used continuum surface force (CSF) model of Brackbill et al. (1992). The CSF model treats surface tension as a body force distributed within a transition region near the interface, and calculates the discontinuous change with a continuous treatment. It has been found that the CSF model generates spurious currents or parasitic flow in the neighbourhood of the interface, and these may cause instabilities and interface destruction when modelled within microfluidic conduits. The two methods use surface tension expressions (F_S) as extra force terms in the Navier-Stokes equations, however they take two different forms in CSF

(28) and Phase-Field (29), where µ denotes the free energy. Essentially the Phase-Field method assumes that any change in kinetic energy results in an opposite change in free energy.

$$F_S = -\sigma\kappa(\phi)\delta(\phi)\nabla\phi \tag{28}$$

$$F_S = -F\nabla\mu \tag{29}$$

Numerical simulation of air–water two-phase flow conducted by He and Kasagi (2008) show that the CSF method produces unbalanced surface tension forces that accelerate the fluid due to parasitic flow. Meanwhile the Phase-Field model reduces the parasitic flow to the level of the truncation error because the energy is correctly transferred between kinetic and surface tension components. The Phase-Field method was further validated by simulation of bubbly and slug flows in a micro tube of 600 µm in diameter. The bubble shape and flow patterns found were in good agreement with comparable experimental and analytical results.

Yet another CFD method that has been tested at the microfluidic scale is the Lattice Boltzmann Method (LBM). Two studies are covered in this section, that of Li and Chen (2005) and that of Yu et al. (2007). The first used the LBM to model a chaotic micromixer. The latter modelled a flow focussing microchannel junction to track Taylor slug formation and flow. The main difference between LBM and other CFD techniques is that rather than making use of the Navier-Stokes equations, it applies mesoscopic kinetic theory of fluids. As a consequence, interface tracking of multiphase flow is also different, such that rather than tracking the interface, in LBM phase separation takes place spontaneously. Furthermore fluid-solid and wetting phenomena are also easy to implement in LBM. The main disadvantage at this point is that LBM still has been constricted to simple simulations rather than real problems with more complex boundary conditions.

The study of Yu et al. (2007) uses a two component model where the repulsive interaction between the two components keeps them in two different phases. Yu et al. used two approaches to define the two phases. In one case both components are assumed to have same density and follow the ideal gas equation of state (EOS), and the separation of the two components comes from their repulsive interactions. Another more complex case involved assigning a non-ideal EOS to one component and an ideal gas EOS to the other, creating large density and viscosity difference between the two phases, so as to more realistically capture the actual properties of a gas-liquid system. Yu et al. (2007) used the former method more often for simplicity purpose. The distribution function of each component is governed by the discretized Boltzmann equation with single relaxation time for the collision term as given in Equation (30). Moreover number density and momentum of each component is calculated from Equation (31) and Equation (32). The discrete velocity associated with each ith direction is c_i, relaxation time is τ, and σ denotes each component. The total interacting force acting on each component (F^σ) in Equation (33) is derived from the sum of the fluid-fluid interaction force (36) and the fluid solid interaction force (37).

$$f_i^\sigma(x+c_i,t+1) - f_i^\sigma(x,t) = -\frac{1}{\tau^\sigma}\left(f_i^\sigma(x,t) - f_i^{\sigma(eq)}(n^\sigma,v^\sigma)\right) \tag{30}$$

$$n^\sigma(\bar{x},t) = \sum_i f_i^\sigma(\bar{x},t) \tag{31}$$

$$n^\sigma u^\sigma(\bar{x},t) = \sum_i c_i f_i^\sigma(\bar{x},t) \tag{33}$$

where,

$$v^\sigma = u + \frac{\tau^\sigma}{\rho^\sigma} F^\sigma \tag{34}$$

$$u = \frac{\sum_\sigma \rho^\sigma / \tau^\sigma u^\sigma}{\sum_\sigma \rho^\sigma / \tau^\sigma} \tag{35}$$

$$F_f^\sigma = -\psi^\sigma(x) \sum_\sigma G_{\sigma\bar{\sigma}} \sum_i T_i \psi^{\bar{\sigma}}(x + c_i) c_i \tag{36}$$

$$F_s^\sigma = -\psi^\sigma(x) \sum_i G_{\sigma s} T_i s(x + c_i) c_i \tag{37}$$

The modelling of Yu et al. (2007) was limited to a 2D projection of the experimental microchannel junction, which contained 125 µm rectangular channels. The inlet conditions used were velocities at the inlet and constant pressure at the outlet. A bounce back condition was implemented at the fluid-solid boundary to simulate non-slip condition for fluid velocity. Reasonable agreement to experimental results were found, however the authors point that discrepancies are likely related to erroneous settings for the wetting conditions and the lack of three-dimensional modelling.

Chung et al. (2008) propose a finite element-front tracking method (FE-FTM) to investigate the effect of viscoelasticity on drop deformation. The front tracking method is found to provide robust solutions due to a large number of front particles on the interface being advected, thereby allowing for tracking the interfaces into the sub-grid regions. Compared to other implementations of this method using finite-difference models, the author also suggest their finite element model is more flexible since it does not impose any geometrical constraints. In front tracking, the interface is comprised of successive elements, where each element has two front particles. The position of front particle on the interface (x_p) is convected with fluid velocity according to Equation (38), where u_p is the particle velocity at position x_p. The new position of the particle, x_p^{n+1}, is updated from the previous step x_p^n using the Runge-Kutta second-order method (RK2). As the interface moves the distance between two successive front nodes, which is initially set at zero, may grow. However, the distances must be kept within a certain limiting range to conserve the accuracy of the calculation, for example 20 to 50% of the diagonal size of the smallest mesh in the computational domain. This is accomplished by Chung et al. (2008) by dividing elements into two equally sized elements and inserting a new front tracking particle, when the distance grows beyond the 50%

limit, or by deleting one small element and connecting two neighbouring elements with a single front tracking particle at its centre.

$$\frac{dx_p}{dt} = u_p \qquad (38)$$

CFD is very useful in capturing very detailed flow phenomena in microchannels, however at times researchers are interested in simplified models that can be developed based on analytical techniques once the flow physics has been fully understood, or at least captured in empirical fashion. One particular work of note is that of Gleichmann et al. (2008). The authors developed a tool toolkit for system simulations of segmented flow-based microfluidic networks using functional nodes interconnected by virtual connectors that handle droplet sequences and reagent properties, rather than the more typical finite element approach. Local rules at the functional nodes are established based on experimental data or results from CFD modelling. A preliminary example is provided where self-controlled 1:1 fusion of the droplets from two droplet sequences at a Y-junction is performed and compared to CFD results. Gleichmann et al. (2008) report that while a 2D version of the simulation required 14 days of running time, the same model in the toolkit ran in only 20 minutes. The authors propose that this approach may be useful for application in the design and optimization of fluidic networks such as segmented-flow-based lab-on-a-chip systems.

Zhu et al. (2008) and Zhu et al. (2010) report consecutive works on a less typical form of multiphase flow in microchannels: formation of liquid droplets from a pore, into gas flowing in a main channel. Both works use the standard FLUENT VOF methodology in full 3D modelling. The studies focus both on the effect of channel geometry, and more interestingly from a fluidics perspective on the effect of wetting properties of the individual channel walls in the main microchannel. For instance, it is observed that the droplet alters its flow behaviour dramatically when it touches a hydrophilic wall. By setting the bottom wall where the pore is located to hydrophobic, but the other three to hydrophilic, and by changing the aspect ratio of the channel, the wall with which the droplet first interacts differently also changes. In the case of a narrow channel (high aspect ratio), interaction occurs first with the side walls, but for a flatter channel (low aspect ratio), this occurs at the top wall first. This work is further example of how surface wetting properties can be used to tune flow at the micro-scale.

Zhuan and Wang (2010) discuss modelling on nucleate boiling in microchannels. Besides the standard set of CFD equation describing flow and heat conduction and convection, the challenge with this type of technique is in obtaining the correct two-phase flow regime resulting from the nucleation of vapour bubbles at the correct channel locations, their growth and detachment. The authors performed this by using a critical temperature needed at the channel wall to trigger the onset of nucleation. Furthermore, it is found that the heat transfer mechanism through the emerging bubble surface to the bulk phase is different at the micro-scale in that surface tension suppresses bubble growth and hence Marangoni convection dominates the interface heat transfer. At a later stage the wall superheat temperature is higher than a transitional point, and the heat flux becomes large enough to overcome the surface tension, leading to rapid bubble growth, lift off and detachment from the wall. Pressure spikes are also found to emanate from the nucleation sites due to the surface tension forces, and these perturbations lead to flow disruption. Lastly, Zhuan and Wang (2010) propose that the different flow regime of nucleate boiling within microchannels, in that vapour bubbles occupy the entire channel cross-section, compared to bubbly flow typically observed in larger channels, can be a criteria for determination of the transition of macro-to-micro two-phase flow and heat transfer behaviour.

CONCLUSIONS

This review was not meant to be exhaustive; there are several other works of quality already published that were not covered here. This review selected works that provide novel findings in line with the field of study and research efforts carried out by Santos and Kawaji (2010).

It is found that there is ample agreement that computational fluid dynamics provide unique opportunities for study of microfluidic multiphase flow. It is also found that the state of present CFD technology is sufficient to provide good agreement and validation in various studies, representing well fluid phenomena such as drop elongation, drop break-up and mixing.

Less certainty is obtained with regards to modelling effects such as void fraction distributions, two-phase frictional pressure drop and liquid film occurrence. The reasons may be lack of understanding of the fundamental physics, lack of detail of computational meshes, simplicity of numerical schemes, or misrepresentation of experimental conditions (such as inlet conditions, wetting properties, etc.).

For instance, future numerical modeling could implement an advanced model for tracking the moving contact line at the wall, perhaps based on nanofluidic techniques, to improve accuracy with respect to the prediction of the interface dynamics. Rosengarten et al. (2006) have observed that because the contact line must move along the wall, which uses a no-slip boundary condition for fluid flow, the VOF code cannot be truly grid independent. In addition, current published works consistently use incompressible flow models, which may be an oversimplification of the fluid physics in microchannels, potentially contributing to the difficulties in correct simulation of the lower void fraction distributions and velocity slip measured in experimental work. Professor Masahiro Kawaji of the University of Toronto has found that numerical simulation using incompressible flow and VOF models are incompatible and cannot achieve convergence due to the way the interface movement is predicted by the advection equation, in that it is not directly dependent on the pressures of the fluid phases. Further work is warranted in this area to obtain full and accurate description of microfluidic multiphase flow.

ACKNOWLEDGMENTS

I thank Professor Masahiro Kawaji for guiding me through the microfluidics field of research since my undergraduate studies at the University of Toronto and for supervising my Master's Thesis work during 2006-2007. I also thank the financial support received from the Natural Sciences and Engineering Research Council of Canada (NSERC) for both my Master's and Doctoral studies.

REFERENCES

Akbar, M. K.; Plummer, D. A.; Ghiaasiaan, S. M. *Int. J. Multiphase Flow* 2003, 29, 855-865.
Ashgriz, N.; Mashayek, F. *J. Fluid Mech.* 1995, 291, 163-190.
Brackbill, J. U.; Kothe, D. B.; Zemach, C. *J. Comput. Phys.* 1992. 100, 335-354.
Brauner, N.; Moalem-Maron, D. Int. Commun. *Heat Mass Transfer*. 1992, 19, 29-39.
Chung, C.; Hulsen, M. A.; Kim, J. M.; Ahn, K. H.; Lee, S. J. J. Non-Newtonian Fluid Mech. 2008, 155, 80-93.
Cox, R. G. *J. Fluid Mech.* 1986, 168, 169-194.
Cubaud, T.; and Ho, C. -M. Phys. Fluids 2004, 16, 4575-4585.
Cubaud, T.; Tatineni, M.; Zhong, X.; and Ho, C. -M. *Phys. Rev. E.* 2005, 72, 037302.
Esmaeeli, A.; Tryggvason, G. *J. Fluid Mech.* 1999, 385, 325-358.
Fluent. FLUENT 6.2 documentation. Fluent Incorporated, Lebanon, NH, 2006.
Fries, N.; Dreyer, M. *J. Colloid Interface Sci.* 2008, 327, 125-128.
Fukagata, K.; Kasagi, N.; Ua-arayaporn, P.; Himeno, T. *Int. J. Heat Fluid Flow*. 2007, 28, 72-82.
Gañán-Calvo, A. M.; Gordillo, *J. M. Phys. Rev. Lett.* 2001, 87, 274501.
Garstecki, P.; Fuerstman, M. J.; Stone, H. A.; Whitesides, G. M. *Lab Chip*. 2006, 6, 437-446.
Gleichmann, N.; Malsch, D.; Kielpinski, M.; Rossak, W.; Mayera, G.; Henkel, T. *Chem. Eng. J.* 2008, 135S, S210-S218.
Günther, A.; Khan, S. A.; Thalmann, M.; Trachsel, F.; Jensen, K. F. *Lab Chip*. 2004, 4, 278-286.
Gupta, R.; Fletcher, D. F.; Haynes, B. S. *Chem. Eng. Sci.* 2009, 64, 2941-2950.
Gupta, R.; Fletcher, D. F.; Haynes, B. S. *Chem. Eng. Sci.* 2010, 65, 2094-2107.
Hardt, S. *Phys Fluids*. 2005, 17, 100601.
Hardt, S.; Jiang, F.; Schönfeld, F. *Int. J. Multiphase Flow*. 2005, 31, 739-756.
Harvie, D. J. E.; Davidson, M. R.; Cooper-White, J. J.; Rudman, M. *Int. J. Multiphase Flow*. 2007, 33, 545-556.
He, Q.; Kasagi, N. *Fluid Dyn. Res.* 2008, 40, 497-509.
Ide, H.; Kimura, R.; Kawaji, M. Proceedings of the 5th ASME ICNMM. 2007.
Irandoust, S.; Andersson, B. *Ind. Eng. Chem. Res.* 1989, 28, 1684-1688.
Jiang, T. S.; Oh, S. G.; Slattery, J. C.; *J. Colloid Interface Sci.* 1979, 69, 74-77.
Johansen, S. T. *Exp. Therm. Fluid Sci.* 2002, 26, 739-745.
Kalliadasis, S.; Chang, H. C.; *Phys. Fluids*. 1993, 6, 12-23.
Kashid, M. N.; Renken, A.; Kiwi-Minsker, L. *Chem. Eng. Res. Des.* 2010, 88, 362-368.
Kawahara, A.; Chung, P. M. -Y.; Kawaji, M. *Int. J. Multiphase Flow*. 2002, 28, 1411-1435.
Krishnan, R. N.; Vivek, S.; Chatterjee, D.; Das, S. K. *Chem. Eng. Sci.* 2010, 65, 5117-5136.
Li, C.; Chen, T. *Sens. Actuators, B.* 2005, 106, 871-877.
Newman, S. *J. Colloid Interface Sci.* 1968, 26, 209-213.
Okushima, S.; Nisisako, T.; Torii, T.; Higuchi, T. *Langmuir*. 2004, 20, 9905-9908.
Qian, D.; Lawal, A. *Chem. Eng. Sci.* 2006, 61, 7609-7625.
Rosengarten, G.; Harvie, D. J. E.; Cooper-White, *J. Appl. Math. Modell.* 2006, 30, 1033-1042.
Saha, A. A.; Mitra, S. K. *J. Colloid Interface Sci.* 2009, 339, 461-480.
Santos, R. M.; Kawaji, M. *Int. J. Multiphase Flow*. 2010, 36, 314-323.

Santos, R. M.; Kawaji, M. Proceedings of CHISA2010/ECCE-7. 2010b, I4.3, 0673.
Serizawa, A.; Feng, Z.; Kawara, Z. *Exp. Therm. Fluid Sci.* 2002, 26, 703-714.
Shikhmurzaev, Y. D. Capillary Flows with Forming Interfaces; Chapman and Hall/CRC: Boca Raton, FL, 2008.
Sikalo, S.; Wilhelm, H. D.; Roisman, I. V.; Jakirlic, S.; Tropea, C. *Phys. Fluids.* 2005, 17, 1-13.
Spiga, M.; Morini, G. L. Int. Commun. *Heat Mass Transfer.* 1994, 21, 469-475.
Taha, T.; Cui, Z. F. *Chem. Eng. Sci.* 2006, 61, 665-675.
Thorsen, T.; Roberts, R. W.; Arnold, F. H.; Quake, S. R. *Phys. Rev. Lett.* 2001, 86, 4163-4166.
Wilkes, E. D.; Phillips, S. D.; Basaran, O. A. *Phys. Fluids.* 1999, 11, 3577-3598.
Xiong, R.; Bai, M.; Chung, J. N. *J. Micromech. Microeng.* 2007, 17, 1002-1011.
Xiong, R.; Chung, J. N. *Phys. Fluids.* 2007, 19, 033301.
Xu, J. H.; Li, S. W.; Wang, Y. J.; Luo, G. S. *Appl. Phys. Lett.* 2006, 88, 133506.
Yamaguchi, Y.; Takagi, F.; Watari, T.; Yamashita, K.; Nakamura, H.; Shimizu, H.; Maeda, H. *Chem. Eng. J.* 2004, 101, 367-372.
Yasuno, M.; Sugiura, S.; Iwamoto, S.; Nakajima, M. *AIChE J.* 2004, 50, 3227-3233.
Yu, Z.; Hemminger, O.; Fan, L. -S. *Chem. Eng. Sci.* 2007, 7172-7183.
Zhu, X.; Liao, Q.; Sui, P. C.; Djilali, N. *J. Power Sources.* 2010, 195, 801-812.
Zhu, X.; Sui, P. C.; Djilali, N. *J. Power Sources.* 2008, 181, 101-115.
Zhuan, R.; Wang, W. *Int. J. Heat Mass Transfer.* 2010, 53, 502-512.

Chapter 6

HYDRODYNAMICS ANALYSIS FOR PARTIALLY VENTILATED CAVITATING VEHICLE BASED ON TWO-FLUID MODEL

M. Xiang[1,2], S. C. P. Cheung[2], J. Y. Tu[*2] and W. H. Zhang[1]

[1]Institute of Aerospace and Material Engineering, National University of Defence Technology, Changsha 410073, P. R. China
[2]School of Aerospace, Mechanical and Manufacturing Engineering, RMIT University, Victoria 3083, Australia

ABSTRACT

Reducing the high friction drag caused by viscosity of water has been a great challenge in developing high speed underwater vehicles. Among many proposed drag reduction techniques, ventilated cavitation has been considered as one of the promising approaches to achieve drag reduction. In this chapter, particular focus is directed to investigate the drag reduction mechanism of partial ventilated cavitation which is frequently encountered; especially for transient launch and vehicle maneuvering. Nonetheless, hydrodynamic effect of partial ventilated cavity is not only affected by dynamical effect of the deformable gas-liquid interface; but also the morphology transition from continuous cavity to dispersed gas bubbles downstream. It is therefore an extremely difficult task to gain in-depth understanding of the complex two-phase flow structure and its associated drag reduction mechanism. In this study, a numerical scheme based on Eulerian-Eulerain two-fluid framework has been proposed to analyse flow characteristics and hydrodynamics resulted from the bubbly flow downstream partially ventilated cavity. The main content can be broadly classified into four parts: (i) A detail review of basic theory and numerical approaches for ventilated cavitating underwater vehicles; (ii) Model development for the bubbly flow downstream partially ventilated cavity; (iii) Characterization and model validation of gas-liquid flow field; (iv) Hydrodynamics analysis based on the simulation results. In the first part, background theory and technical problems for ventilated cavitating underwater vehicle are

---

[*] Corresponding Author: Prof. Jiyuan Tu. SAMME, RMIT University, Bundoora, Melbourne, Victoria 3083, Australia. Email: jiyuan.tu@rmit.edu.au; Phone no: +61-3-9925 6191; Fax no.: +61-3-9925 6108

introduced. Afterwards, a literature review of the historical development of numerical approaches in simulating ventilated cavitating flow is summarized. In the second part, an Eulerian–Eulerian two-fluid model incorporated with the population balance approach is introduced to predict the bubbly wake flow behind the ventilated cavity. Particular attention is devoted to establish air entrainment model at the cavity base and to improve interfacial momentum transfer models. In the third part, numerical simulation is carried out for underwater vehicle under different ventilation rate and sailing velocity. The flow field parameters including void fraction, velocity, pressure and bubble size distributions are obtained based on the simulation results. Good agreement is achieved in compared with experimental data. Finally, the drag reduction efficient for the partially ventilated vehicle is closely examined. Comparison between non-cavitating and supercavitation conditions is also presented. Flow field characteristics and meaningful conclusions for partially ventilated cavitating vehicles are summarised at the end of the article.

1. INTRODUCTION

Aiming to top up the speed of underwater vehicles, substantial amount of research works had been devoted in the past decades. The high friction drag caused by high viscosity of water has been an inevitable problem impeding the development of high speed vehicles. Although it is impractical to alter the viscosity of water, flow near the surface can be replaced by a layer of gas, creating a large reduction in near-wall density. This provides an attractive prospect of friction drag reduction for external flows. Cavitation is considered to be one of the effective ways to form a stable gas layer at the surface.

Natural cavitation occurs in liquid when liquid local pressure is lower or near its vapor pressure. By means of natural evaporation, low pressure region are then filled with vapor bubbles. For cavitation analysis, dimensionless Cavitation number, a special form of Euler number, is commonly used which is given by:

$$\sigma_v = \frac{p_\infty - p_v}{1/2 \rho_l u_\infty^2} \tag{1}$$

where p_∞ is the pressure at infinity; p_v is the water vapour pressure; ρ_l is the water density; u_∞ is the mainstream velocity.

Based on the definition, lower value of the cavitation number represents more vapor bubbles will be generated in cavitating zone. In some cases with sufficiently low cavitation number (e.g. $\sigma_v < 0.1$), large amount vapor bubbles will be generated and merged together forming a continuous cavity which covers the entire underwater object. This particular cavitation phenomenon is commonly known as: Supercavitation. Owning to its great potential to reduce drag forces on underwater vehicles, much research efforts had been dedicated to investigate the complex mechanism of supercavitation [1-3]. In general, there are two commonly adopted methods to achieve supercavitation. A simple approach is to increase vehicle maneuvering speed, generally $u_\infty > 50 m/s$ for water, to achieve large liquid dynamic pressure. As a result, static pressure of liquid is significantly reduced under the saturated vapour pressure which generates bubbles through liquid evaporation. Due to the fact that majority of the bubbles are water vapour, supercavities produced by this method are also referred as vapor supercavities. Although this method can naturally achieve supercavition, apparently, it is a challenging task to maintain such high speed due to high friction of liquid.

Another approach is to initialize the cavitation by injecting gas bubbles into the rarefaction zone at non-streamlined locations. This method, commonly referred as ventilated cavitation, has recently been considered as one of the promising approaches due to its high efficiency and low energy requirement. For ventilated cavitation, the cavitation number is redefined using the cavity pressure p_c:

$$\sigma_c = \frac{p_\infty - p_c}{1/2 \rho_l u_\infty^2} \qquad (2)$$

Corresponding to different cavitation numbers, different flow conditions and ventilation parameters result in completely diverse flow structure and cavitation characteristics, including cloud cavitation, partial cavitation and supercavitation. Among all these, partial cavitation is one of the most important phenomenon which is frequently encountered; especially for transient launch and vehicle maneuvering[4]. Partial cavitation has also been successfully applied to shipbuilding [5, 6] and hydrofoils [7,8] as one of the most effective drag reduction techniques. In this book chapter, particular attention is therefore directed to discuss the flow characteristics of partial ventilated cavitation where the continuous cavity is only partially cover the vehicle surface as shown in Figure 1.

As depicted, due to the high shear stress across the gas-liquid interface, the vortical flow structure shatters the continuous cavity and exposes part of the vehicle body to liquid phase. Such cavity breakage mechanism is primarily governed by the highly nonlinear toroidal vortex structure and re-circulation motions of the re-entrained jets. In addition, the sheared off cavities are then broken up into numerous small bubbles due to the turbulent impact; exhibiting a typical bubbly flow characteristics in the wake region and significantly affect the hydrodynamic drag on the vehicle surface. In particular for the case when bubbles being entrained into the boundary layer, substantial drag reduction could be achieved through the bubble induced changes of near-wall turbulence structure and density ratio. This mechanism is known as microbubble drag reduction (MBDR) which is proved to be capable of achieving drag reduction by 20%~80% [9-11].

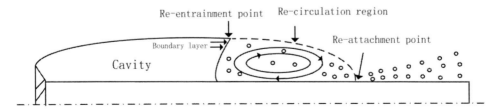

Figure 1. Schematic of toroidal vortex shedding structure and re-entrained jet of a partially ventilated cavitation.

Nevertheless, the performance of MBDR is strong governed by a number of important parameters including bubble injection rate, bubble size, bubble breakage or coalescence and vehicle geometry. In order to better control the partially ventilated cavity and optimize the vehicle performance, it is necessary to predict the features of the two-phase flow and vehicle hydrodynamics created by ventilated cavity accurately. To accurately predict its embedded physical phenomena, computational modelling of cavitation has received considerable attentions over the past decades. Aiming to simplify the computational scheme, potential flow theory was commonly adopted in early studies [1,12,13]. Nevertheless, interfacial viscous force and dynamical effect of the free-surface were ignored in their simulation. Furthermore, the method is also heavily relied on the empirical models for cavity closure; where gas leakage processes were not solved explicitly.

Recently, a more sophisticated modelling approach based on Eulerian-Eulerian framework received more and more attention among researchers [14-19]. In this approach, computational domain is assumed to be filled with the mixture of liquid and gas phases where flow field is solved homogeneously based on one set of continuity and momentum equations. Density of the mixture was evaluated based a continuous function expressed in terms of gas void fraction which is solved by an additional transport equation.

By considering the interfacial viscous force in the simulation, this method enables more detail investigations of the flow structures and its associated drag reduction characteristics. Another advantage of this model comes from solving void fraction explicitly with transport equation; allowing a more precise prediction in capturing the shape gas-liquid interface. However, noticeable prediction errors were reported for the predictions in the wake region where continuous cavity was shattered into dispersed gas bubbles. It remains an extreme challenge to comprehend the exact mechanism not only in regard to the presence of discrete interfaces with large density ratio; but also the morphology transition from continuous cavity to dispersed gas bubbles. Considering the limit knowledge on the drag reduction of partial ventilated cavity and its great influence on the vehicle hydrodynamics, particular emphasis is directed towards a better understanding via computational fluid dynamics investigations on the complex interactions between the gas bubbles emanating from the rim of the cavity and the dynamical bubble migration and mass transfer caused by breakage and coalescence events downstream of the cavity. A numerical framework based on Eulerian–Eulerian two-fluid model is adopted to model the dispersion of gas bubble and its embedded complex flow structures downstream. Following the success in previous studies on microbubble drag reduction[20,21], population balance approach based on the MUltiple-SIze-Group (MUSIG) model[22-25] is adopted to capture the evolution of the bubble size distribution due to coalescence and breakage. The contents of this chapter have been arranged as follows.

Firstly, air entrainment model at the tail of partial ventilated cavity is established to give accurate boundary condition for the bubbly flow downstream.

Secondly, details of the numerical method will be presented to elucidate the implementation of population balance in conjunction with CFD techniques.

Thirdly, to assess the performance of various turbulence models in predicting the wake behind the vehicle, sensitivity study on three turbulence models is presented by and compared against experimental data. Detail numerical results are then presented and compared with measurement to discuss the gas void fraction distribution along the vehicle and the associated flow features created by dispersed bubbles. Finally, drag reduction caused by the dispersed micro-bubbles and density effect is investigated in order to obtain meaningful conclusions to assist the vehicle design.

2. AIR ENTRAINMENT MODEL

As discussed in the previous section, for partial ventilated, gas bubbles are sheared off and ejected periodically at the base of the continuous cavity under the action of the pressure gradient and shear stress from re-entrained jet. This phenomenon, commonly refers as air entrainment, results in a continuous gas loss of the ventilated cavity; forming a typical bubbly flow structure downstream when gas bubbles further break down into small bubbles due to turbulent impact.

Theoretical speaking, air entrainment involves air and water exchange process which is governed by the microscopic counteraction between turbulent shear stress and surface tension. With limited mesh resolution, microscopic processes of air entrainment are therefore modelled by an empirical mechanistic model.

According to Chanson [26], air entrainment of ventilated cavity is analogy to the plunging jet mechanism where the tail of the cavity undergoes a periodic cycle consists of: pinches off, elongates, collapses and entrainment four stages. The air entrainment rate per unit width of the cavity can be approximated as [27]:

$$q \sim U_c \Delta \qquad (3)$$

where U_c is the flow velocity at the cavity boundary and Δ is the mean thickness of the air layer that forms the sheared-off pocket.

Assuming the gas is entrained along the boundary layer of the high velocity cavity wall [28], as shown in Figure 2, the volumetric rate of air entrainment is obtained:

$$\dot{Q} = 2\pi \int_0^\delta (R-y)u\,dy \qquad (4)$$

where δ and R refer to the boundary layer thickness and cavity radius respectively.
The auxiliary conditions are:

$$\begin{aligned} y = 0, \quad u &= U_c \\ y = \delta, \quad u &= 0 \end{aligned} \qquad (5)$$

With the assumption of $\delta/R \ll 1$, equation (4) can be rewritten as

$$\dot{Q} = 2\pi R_i \int_0^\delta u\,dy \qquad (6)$$

Introducing the displacement thickness $\delta 1$ and $\delta 2$:

$$\delta_1 U_c = \int_0^\delta u\,dy \tag{7}$$

$$\delta_2 U_c^2 = \int_0^\delta u^2\,dy \tag{8}$$

The air entrainment rate is expressed as:

$$Q = 2\pi U_c \frac{\delta_1}{\delta_2} \delta_2 R_l \tag{9}$$

With the application of the momentum balance in x-direction, the following equation is obtained:

$$\tau_w 2\pi R\,dx = \frac{d}{dx}(2\pi R \int_0^\delta \rho u^2\,dy)dx \tag{10}$$

$$\frac{\tau_w R}{\rho U_c^2} = \frac{d}{dx}(\delta_2 R) \tag{11}$$

where τw represents the interfacial shear stress and the local friction coefficient cf is defined as:

$$c_f = \frac{\tau_w}{\rho/2 U_c^2} \tag{12}$$

Substituting in Equation (11) and integrating along the cavity boundary until the entrainment point gives:

$$\delta_2 R_l = \int_0^l R c_f / 2\,dx \tag{13}$$

where l refers to the cavity length before the re-entrained position. Substituting in Equation (9) gives:

$$\dot{Q} = 2\pi U_c \frac{\delta_1}{\delta_2} \int_0^l R c_f / 2\,dx \tag{14}$$

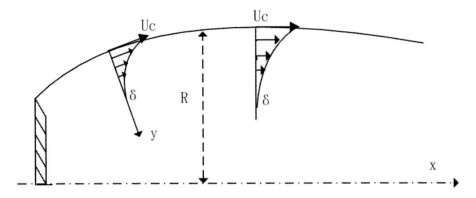

Figure 2. Sketch of air entrained along the cavity boundary layer.

The specification of the boundary layer profile is required for further progress. In practice, ventilated cavitating flows are normally subject to high Reynolds number flows, so the turbulent velocity profile over a fixed solid wall composed of the universal law of the wall and a wake function is chosen:

$$\frac{U_c - u}{u_*} = f(\frac{yu_*}{v}) + \frac{\Pi}{\chi} W(y/\delta) \tag{15}$$

where u_* is the friction velocity, defined as $u_* = \sqrt{\tau_w/\rho}$, and v the kinematic viscosity. An often used form of the wake function is:

$$W(y/\delta) = 2\sin^2(\frac{\pi y}{2\delta}) \tag{16}$$

For the law of the wall only the logarithmic part is considered:

$$f(\frac{yu_*}{v}) = \frac{1}{\chi} \ln(\frac{yu_*}{v}) + B \tag{17}$$

κ and B is the Karman constants. At the boundary $y = \delta$,

$$\frac{U_c}{u_*} = f(\frac{\delta u_*}{v}) + \frac{2\Pi}{\chi} \tag{18}$$

Equations (7), (8), (13), (15) and (18) with appropriate initial condition determines δ, δ_1, δ_2, u, c_f completely. An empirical law for c_f which is a fit to the numerical solutions of the systems of equations outlined above is applied in this axisymmetric boundary layer flow.

$$c_f = 0.024(\text{Re}_d \frac{x}{D_n})^{-1/7} \tag{19}$$

where Red is the Reynolds number based on the diameter of the cavitator. After integrating equations (7) and (8), the following relation is obtained:

$$\frac{\delta_1}{\delta_2} = 0.3688\sqrt{\frac{2}{c_f}}\chi \tag{20}$$

In accordance with the air entrainment mechanism of plunging jet, the final air entrainment rate is expressed as:

$$Q = 2\pi R_l U_c \Delta \tag{21}$$

$$\Delta = 0.3688\kappa \frac{1}{R_l}\sqrt{\frac{2}{c_f}}\int_0^l R\frac{c_f}{2}dx \tag{22}$$

For the cavity contour, the empirical formula proposed by Savchenko [29] is adopted:

$$R = \frac{1}{2}D_n\left(2\sqrt{\frac{C_d}{\ln\frac{1}{\sigma_c}}\frac{2x}{D_n} - \frac{\sigma_c}{\ln\frac{1}{\sigma_c}}\frac{4x^2}{D_n^2}}\right)^{1/2} \tag{23}$$

The final entrainment rate varies along with the re-entrainment position. For the partial ventilated cavity in this study, the cavity is assumed to be broken up by the re-entrained jet from the point where the maximum diameter of the cavity achieves due to the high local shear stress. This assumption is validated through comparison between the calculated gas loss rates with the experimental ventilation rates under different cavitation numbers as shown in Figure 3. The ventilation rate is balanced by the leakage rate, illustrating the entrainment model reasonable.

As information on the size of the inception bubbles is limited, the entrained bubble size is taken to be the maximum stable bubble diameter in the mixing zone given by Thorpe [30]:

$$r_{max} = \frac{3}{2}\sqrt[3]{\left(\frac{We_c \sigma R_l^4}{128\rho_l U_c^2}\right)\frac{1}{(R_l - r_{max})^2}} \tag{24}$$

where We_c is the critical Weber number for breakage. A value of 4.7 is adopted for We_c. σ refers to the surface tension coefficient and K is a constant taken to have a value of 3.8.

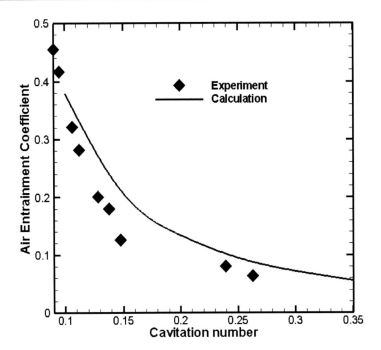

Figure 3. Air entrainment rate vs. cavitation number (Fr=29).

3. MULTIPHASE MODELLING

3.1. Two-Fluid Model

The two-fluid model based on the Eulerian–Eulerian framework solves the ensemble-averaged of mass, momentum and energy whereby the liquid is considered as the continuum phase and the gas is treated as disperse phase. Two sets of governing equations are solved for each phase. Interactions between phases are effected via interfacial transfer terms for heat, mass and momentum exchange. Since there is no interfacial mass or heat transfer between the phases in the present study, the energy equation is not needed to be solved.

In the absence of interfacial mass transfer, the continuity equation of the two-phases can be written as:

$$\frac{\partial(\rho_i \alpha_i)}{\partial t} + \nabla \cdot (\rho_i \alpha_i \vec{u}_i) = 0 \tag{25}$$

where α, ρ and $\vec{u}$ are the void fraction, density and velocity vector of each phase. Subscripts i = l or g denote the liquid and gas phase respectively. The momentum equation can be expressed as:

$$\frac{\partial(\rho_i \alpha_i \vec{u}_i)}{\partial t} + \nabla \cdot (\rho_i \alpha_i \vec{u}_i \vec{u}_i) = -\alpha_i \nabla P + \rho_i \alpha_i \vec{g} + \nabla \cdot [\alpha_i \mu_i^e (\nabla \vec{u}_i + (\nabla \vec{u}_i)^T)] + F_i \tag{26}$$

where Fi represents the interfacial forces for the interfacial momentum transfer and $\bar{g}$ is the gravity acceleration vector. It is noted that the interfacial forces appearing in the momentum equation strongly govern the distribution of the liquid and gas phases within the flow volume. Details of the different forces acting within the two-phase flow are described in the next section.

3.2. Interfacial Forces

The interfacial forces can be categorized into drag, lift, wall lubrication and turbulence dispersion. The total interfacial forces for the interfacial momentum transfer in Eq. (26) are thus:

$$F_i = F_{lg} = F_{lg}^{drag} + F_{lg}^{lift} + F_{lg}^{wall\ lubrication} + F_{lg}^{dispersion} = -F_{gl} \tag{27}$$

In Eq. (9), F_{lg} denotes the momentum transfer from the gas phase to the liquid phase and vice versa for F_{gl}.

3.2.1. Drag Force

The interfacial momentum transfer between gas and liquid due to the drag force results from shear and form drag which is modelled according to Ishii and Zuber [32] as:

$$F_{lg}^{drag} = \frac{1}{8} C_D a_{if} \rho_l |\bar{u}_g - \bar{u}_l|(\bar{u}_g - \bar{u}_l) = -F_{gl}^{drag} \tag{28}$$

where C_D is the drag coefficient which can be evaluated by correlation of several distinct Reynolds number regions for individual bubbles. The model also takes into account different bubble shape regimes, such as spherical particle regime, distorted particle regime and spherical cap regime.

Spherical Particle Regime:

$$C_D(sphere) = \max\left(\frac{24}{Re_B}(1 + 0.15 Re_B^{0.687}), 0.44\right) \tag{29}$$

where ReB is the bubble Reynolds number defined using the Sauter mean bubble diameter D_s:

$$Re_B = \frac{\rho_l |u_l - u_g| D_s}{\mu_l} \tag{30}$$

Distorted Particle Regime:

$$C_D(ellipse) = \frac{2}{3}\sqrt{Eo} \tag{31}$$

Where Eo represents the Eötvos number which is defined by:

$$Eo = \frac{g(\rho_l - \rho_g)D_s^2}{\sigma} \tag{32}$$

Spherical Cap Regime:

$$C_D(cap) = \frac{8}{3} \tag{33}$$

As implemented within ANSYS CFX 11 the regime selection is based on:

$$C_D = \begin{cases} C_D(sphere) & \text{if } C_D(sphere) \geq C_D(ellipse) \\ \min(C_D(ellipse), C_D(cap)) & \text{if } C_D(sphere) \leq C_D(ellipse) \end{cases} \tag{34}$$

In addition to bubble Reynolds number and deformation, locally high gas void fraction can influence the drag dramatically. For uniformly disperse flows, an increase drag coefficient is appropriate, and for flows where gas structures are streamlined a reduced drag coefficient is appropriate[21].

The application of the standard drag force model in the high void fraction region right behind the cavity base will give rise to too much local drag, thereby underpredicting the vortex size of the sheared off bubbles. Simonnet et al. [33] concluded that the neighbouring effect of close packed bubbles becomes significant if the void fraction exceeds 15%.

An empirical "cluster" drag model was also proposed to correlate the drag coefficient with the local gas void fraction.

$$C_D = C_{D\infty}(1-\alpha_g)[(1-\alpha_g)^m + (4.8\frac{\alpha_g}{1-\alpha_g})^m]^{-\frac{2}{m}} \tag{35}$$

where $C_{D\infty}$ is the drag coefficient of an isolated bubble in an infinite medium, which can be obtained through the balance of buoyancy, drag and gravitational as

$$C_{D\infty} = \frac{4}{3}\frac{\rho_l - \rho_g}{\rho_l}gD_s\frac{1}{u_\infty^2} \tag{36}$$

In the above equation, u_∞ represents the velocity of an isolated bubble in a quiescent liquid which can be calculated using the correlation of Jamialahmadi et al.[34]:

$$u_\infty = \frac{u_{b1}u_{b2}}{\sqrt{u_{b1}^2 + u_{b2}^2}} \tag{37}$$

where

$$u_{b1} = \frac{1}{18}\frac{\rho_l - \rho_g}{\mu_l}gD_s^2\frac{3\mu_g + 3\mu_l}{3\mu_g + 2\mu_l}$$

$$u_{b2} = \sqrt{\frac{2\sigma}{D_s(\rho_l - \rho_g)} + \frac{gD_s}{2}} \tag{38}$$

Similar observations were also found in the numerical research of microbubble drag reduction by Kunz et.al[21]. A model proposed by Johansen and Boysan[35] has been adapted to an Eulerian framework to take into account the high void fraction effect on the bubble drag force:

$$C_d = C_{d0}(1 - 1.54[\min(0.5157, \alpha_g)]^{2/3}) \tag{39}$$

where Cd0 is the original drag coefficient and the Min function is provided to ensure that the corrected drag coefficient does not drop to below 1% of the uncorrected value.

In this chapter, the Kunz model is adopted as the "cluster" drag model due to the comparable void fraction and bubble size within this research.

3.2.2. Lift Force

Owing to the velocity gradients in radial and azimuthal directions, bubbles rising in a liquid are subjected to a lateral lift force. Inclusion of the lift force is important to obtain the correct spreading of the gas phase. The force can be correlated to the relative velocity and the local liquid vorticity from Drew and Lahey [36] as:

$$F_{lg}^{lift} = \alpha_g C_L \rho_l \left(\bar{u}_g - \bar{u}_l\right)\left(\nabla \times \bar{u}_l\right) = -F_{gl}^{lift} \tag{40}$$

For the lift coefficient, CL, the Eötvos number dependent correlation proposed by Tomiyama [37] is adopted and the lift coefficient can be expressed as:

$$C_L = \begin{cases} \min[0.288\tanh(0.121\operatorname{Re}_B), f(Eo_d)] & Eo < 4 \\ f(Eo_d) = 0.00105 Eo_d^3 - 0.0159 Eo_d^2 \\ \qquad -0.020 Eo_d + 0.474 & 4 \le Eo \le 10 \\ -0.29 & Eo > 10 \end{cases} \tag{41}$$

The modified Eötvos number is:

$$Eo = \frac{g(\rho_l - \rho_g)D_h^2}{\sigma} \tag{42}$$

The variable D_h is the maximum bubble horizontal dimension that can be evaluated by using the empirical correlation of Wellek et al:

$$D_h = D_s(1 + 0.163 Eo^{0.757})^{1/3} \tag{43}$$

3.2.3. Wall Lubrication Force

This force constitutes another lateral force due to the surface tension thereby allowing the bubble to concentrate in a region close to the wall but not immediately adjacent to the wall. According to Antal et al. [38], this force can be modelled as

$$F_{lg}^{lubrication} = -\left(C_{w1} + C_{w2}\frac{D_s}{y_w}\right)\bar{n}_w \frac{\alpha_g \rho_l \left[|(\bar{u}_g - \bar{u}_l) - ((\bar{u}_g - \bar{u}_l)\cdot \bar{n}_w)\bar{n}_w|\right]^2}{D_s}$$
$$= -F_{gl}^{lubrication} \tag{44}$$

where y_w is the distance from the wall boundary and $\bar{n}_w$ is the outward vector normal to the wall. The wall lubrication constants determined through numerical experimentation for a sphere are: $C_{w1} = -0.01$ and $C_{w2} = 0.05$. To avoid the emergence of attraction force, the force is set to zero for large y_w, viz.,

$$F_{lg}^{lubrication} = -F_{gl}^{lubrication} = 0 \quad \text{if} \quad y_w > \frac{C_{w2}}{C_{w1}} D_s \tag{45}$$

3.2.4. Turbulent Dispersion Force

In order to account for the random (dispersive) influence of the turbulent eddies on the gas bubbles, the concept of the turbulent dispersion force is introduced. The effect of the turbulent dispersion force, just like that of the lift force, is also important to obtain correct dispersion of the gas. The turbulent dispersion force expression in terms of Farve-averaged variables proposed by Burns et al. [39] is adopted:

$$F_{lg}^{dispersion} = C_{TD}C_D \frac{v_{t,g}}{\sigma_{t,g}}\left(\frac{\nabla \alpha_l}{\alpha_l} - \frac{\nabla \alpha_g}{\alpha_g}\right) = -F_{gl}^{dispersion} \tag{46}$$

where CTD, CD, $v_{t,g}$ and $\sigma_{t,g}$ represent the turbulent dispersion coefficient, drag force coefficient, turbulent kinematic viscosity for the gas phase and the turbulent Schmidt number of the gas phase, respectively.

3.3. Population Balance Model

The population balance of any system is a record for the number of particles, which may be solid particles, liquid nuclei, bubbles or, variables (in mathematical terms) whose presence or occurrence governs the overall behaviour of the system under consideration. In most of the systems under concern, the record of these particles is dynamically depended on the "birth" and "death" processes that terminate existing particles and create new particles within a finite or defined space. The foundation development of the population balance equation stems from the consideration of the Boltzman equation, where such an equation is generally expressed as an integrodifferential form of the particle distribution function.

Owing to the complex phenomenological nature of events, numerous efforts have been put into developing numerical techniques for solving the population balance equaltions (PBEs). Among the many available numerical techniques, the method of discrete classes has received particular interest due to its rather straightforward implementation within CFD program. Several studies based on MUSIG model [40-42] typified the application of this particular category of population balance approach. Therefore, the MUSIG model has been adopted in this study to investigate the bubble size distribution downstream ventilated cavity. The analysis is simplified by considering bubble sizes change due to breakage and coalescence events only, i.e. nucleation, growth and/or dissolution of bubbles due to absorption (desorption) or boiling are not considered.

In the MUSIG model, the continuous particle size distribution (PSD) function is approximated by M number size fractions. The mass conversation of each size fraction is balanced by the inter-fraction mass transfer due to the mechanisms of particle coalescence/agglomeration and breakage processes. The overall population evolution can then be explicitly resolved via source terms within the transport equations. Assuming that each bubble class travel at the same mean algebraic velocity, the number density equation based on Kumar and Ramkrishna [43] can be expressed as:

$$\frac{\partial n_i}{\partial t} + \nabla \cdot (\bar{u}_g n_i) = B_C + B_B - D_C - D_B \tag{47}$$

where ni is the average bubble number density of the ith bubble class, the source terms BC, BB, DC and DB are, respectively, the birth rates due to coalescence and break-up and the death rate due coalescence and break-up of bubbles. \In a form consistent with the variables used in two-fluid model, the transport equation in terms of size fraction becomes:

$$\frac{\partial (\rho_g \alpha_g f_i)}{\partial t} + \nabla \cdot (\rho_g \bar{u}_g \alpha_g f_i) = B_C + B_B - D_C - D_B \tag{48}$$

For the MUSIG model, which employs multiple discrete bubble size groups to represent the population balance of bubbles, the birth and death rates can be written in terms of the size fraction according to

$$B_C = (\rho_g \alpha_g)^2 \frac{1}{2} \sum_{j=1}^{i} \sum_{k=1}^{i} f_j f_k \frac{M_j + M_k}{M_j M_k} a(M_j, M_k) X_{jki} \tag{49a}$$

$$D_C = (\rho_g \alpha_g)^2 \sum_{j=1}^{N} f_i f_j \frac{1}{M_j} a(M_i, M_j) \tag{49b}$$

$$B_B = \rho_g \alpha_g \sum_{j=i+1}^{N} r(M_j, M_i) f_j \tag{49c}$$

$$D_B = \rho_g \alpha_g f_i \sum_{j=1}^{N} r(M_i, M_j) \tag{49d}$$

where Mj refers to the mass of the jth bubble group. The coalescence mass matrix X_{jk} is defined as:

$$X_{jki} = \begin{cases} \dfrac{(M_j + M_k) - M_{i-1}}{M_i - M_{i-1}} & \text{if } M_{i-1} < M_j + M_k < M_i \\ \dfrac{M_{i+1} - (M_j + M_k)}{M_{i+1} - M_i} & \text{if } M_i < M_j + M_k < M_{i+1} \\ 0 & \text{otherwise} \end{cases}$$

For the merging of bubbles, the coalescence of two bubbles is assumed to occur in three stages. The first stage involves the bubbles colliding thereby trapping a small amount of liquid between them. This liquid filmthen drains until it reaches a critical thickness. The last stage features the rupturing of the liquid film subsequently causing the bubbles to coalesce. The collisions between bubbles may be caused by turbulence, buoyancy and laminar shear.

Only random collisions driven by turbulence are usually considered for bubbly flow conditions. The coalescence kernel considering turbulentcollision taken from Prince and Blanch [44] can be written as the product of the collision rate and coalescence efficiency:

$$a(M_i, M_j) = \theta_{ij} P_C(d_i, d_j) \tag{50}$$

where the collision rate can be expressed as the product of the collision cross-sectional area A and characteristic velocity u' as

$$\theta_{ij} = \underbrace{\frac{\pi}{4}\left[d_i + d_j\right]^2}_{A} \underbrace{\left(u_{ti}^2 + u_{tj}^2\right)^{1/2}}_{u'} \tag{51}$$

while the coalescence efficiency is

$$P_C(d_i, d_j) = \exp\left(-\frac{t_{ij}}{\tau_{ij}}\right) \tag{52}$$

The contact time τ_{ij} when two bubbles come together is

$$\tau_{ij} = \frac{(d_{ij}/2)^{2/3}}{(\varepsilon^c)^{1/3}} \tag{53}$$

while the time required for two bubbles to coalesce t_{ij} can be estimated to be

$$t_{ij} = \left(\frac{\rho_l (d_{ij}/2)^3}{16\sigma}\right)^{1/2} \ln\left(\frac{h_0}{h_f}\right) \tag{54}$$

The equivalent diameter dij appearing in Eqs. (52) and (53) is calculated as suggested by Chesters and Hoffman [45]:

$$d_{ij} = \left(\frac{2}{d_i} + \frac{2}{d_j}\right)^{-1} \tag{55}$$

where the diameters di and dj are evaluated based on the spherical bubble assumption as

$$d_i = \left(\frac{6}{\pi}\frac{1}{\rho_g} M_i\right)^{1/3} \qquad d_j = \left(\frac{6}{\pi}\frac{1}{\rho_g} M_j\right)^{1/3} \tag{56}$$

For air-water systems, experiments have determined h0, initial film thickness and, hf, critical film thickness at which rupture occurs to be 1×10^{-4} m and 1×10^{-8} m respectively. The turbulent velocity ut in the inertial sub-range of isotropic turbulence by Rotta [46] is given by $u_t = 1.4\varepsilon^{1/2} d^{1/3}$.

For the consideration of the breakage of bubbles, the model developed by Luo and Svendsen [47] is employed for the break-up of bubbles in turbulent dispersions. In this model, binary break-up of the bubbles is assumed and the model is based on surface energy criterion and isotropic turbulence. The break-up frequency for binary break-up of bubbles is:

$$r(M_i, M_j) = C(1-\alpha_g)\left(\frac{\varepsilon}{d_i^2}\right)^{1/3} \int_{\zeta_{min}}^{1} \frac{(1+\zeta)^2}{\zeta^{11/3}} \exp\left(-\frac{12c_f\sigma}{\beta\rho_l\varepsilon^{2/3}d_i^{5/3}\zeta^{11/3}}\right)d\zeta \quad (57)$$

where $\zeta = \lambda/d_j$ is the size ratio between an eddy and a particle in the inertial sub-range and consequently $\zeta_{min} = \lambda_{min}/d_{min}$. The lower limit of the integration is given by

$$d_{min} = \left(\frac{6}{\pi}\frac{1}{\rho_g}M_{min}\right)^{1/3} \quad \lambda_{min} = 11.4\eta \quad (58)$$

In which $\eta = (v_l^3/\varepsilon)^{1/4}$ where v_l denotes the kinematic viscosity of the continuous phase. The constants C and β are determined respectively from fundamental consideration of drops or bubbles breakage in turbulent dispersion systems to be 0.923 and 2.0.

The variable cf denotes the increase coefficient of surface area:

$$c_f = \left[f_{BV}^{2/3} + (1-f_{BV})^{2/3} - 1\right]$$ where f_{BV} is the breakage fraction: M_j/M_i.

3.4. Turbulence Modeling

Turbulence model is required to close the terms of effective viscosity in the Navier-Stokes equations. However, turbulent states are generally very rich, complex and varied, therefore no single turbulence model can thus far be readily employed to span these states since none is expected to be universally valid for all types of single-phase or multiphase flows.

For multiphase flow, separate turbulence model is adopted for the gas and liquid phases. The effective viscosity of the liquid phase is considered as being composed of the molecular viscosity μl, the turbulent viscosity μt,l and the bubble induced turbulence μt,b

$$\mu_l^e = \mu_l + \mu_{t,l} + \mu_{t,b} \quad (59)$$

For the liquid turbulent viscosity, two equation k-ε model has been widely employed for its simplicity and encouraging results.

$$\mu_{t,l} = \rho_l C_\mu \frac{k^2}{\varepsilon} \quad (60)$$

where the turbulent kinetic energy k and its dissipation rate ε in Eq. (59) of the continuous liquid phase are determined from transport equations which are straightforward extensions of the standard k-ε model [49,50]. Another common turbulence model applied in multiphase flow is the Shear Stress Transport (SST) model for its realistic prediction of turbulent dissipation close to wall. The SST model is a hybrid version of the k-ε and k-ω models with a specific blending function. Instead of using empirical wall function to bridge the wall and the far-away turbulent flow, it solves the two turbulence scalars (i.e. k and ω) explicitly down to the wall boundary.

Similar to the k-ε model, the multiphase versions of the Reynolds Stress Model and SST model are merely straightforward generalizations of their respective single-phase equations. For wall-bounded multiphase flows, it may be highly preferable and makes perfect sense that the modified SST model is applied since it dispenses with the need of prescribing the wall functions and solves the fluid flow within the viscous sub-layer adjacent to a wall as well as the bulk flow away from the wall through the consideration of fundamental transport equations. Effect of turbulent in the liquid phase on turbulence in the gas phase may be modelled by setting the viscosity to be proportional to the liquid turbulent viscosity:

$$\mu_g^e = \frac{\rho_g}{\rho_l} \mu_l^e \tag{28}$$

The extra bubble induced turbulent viscosity in Eq. (27) is evaluated according to the model by Sato et al. [48]:

$$\mu_{t,b} = \rho_l C_{\mu b} \alpha_g D_s |\bar{u}_g - \bar{u}_l| \quad \text{with} \quad C_{\mu b} = 0.6 \tag{29}$$

3.5. Numerical Details

Since the experimental water tunnel and test body are symmetric, numerical simulations are only performed on a half portion of the water tunnel with symmetry boundary conditions imposed at both sides. The front part of the test body is covered by the cavity where gas-liquid interface is assumed to be steady and evaluated according to empirical correlations of Savchenko [29]. The cavity is assumed to be ruptured at the highest interfacial shear stress point where the cavity has its maximum diameter. The final outline of test body and the interface of the cavity for the three selected cases are depicted in Figure 4.

At the bubble inlet, the gas leakage rate is specified through a velocity inlet boundary located at the circumference of the cavity base (see also in Figure 4). Gas bubbles with uniform diameter and velocity are specified at the inlet with a thickness of Δ. Detail parameters are given in section 2. The cavity interface is set up as free-slip boundary while the test body is modelled as no-slip for both liquid and gas phase.

The generic CFD code ANSYS CFX11 was utilized to investigate the liquid flow around the ventilated cavity and the two-phase flow behavior downstream of the cavity. Figure 7 illustrates the computational mesh corresponding to the schematic geometry illustrated in

Figure 4. Taking the cavitator as the reference datum, the inlet boundary was located at a distance 0.2 m upstream while the outlet boundary was placed at a distance of 0.8 m downstream, which resulted in a total tunnel length of 1 m. An outlay mesh of 200,000 hexahedral elements was constructed for the entire flow domain. Mesh sensitivity on the numerical predictions was performed by doubling the number of elements. Comparing the predicted results between the two meshes, small discrepancies were observed. The maximum differences between these two mesh levels were less than 5%. The transport equations of the two-fluid and population balance models were discretized by the finite volume approach. The convection terms were approximated by a high order resolution scheme while the diffusion terms were approximated by the second-order central difference scheme.

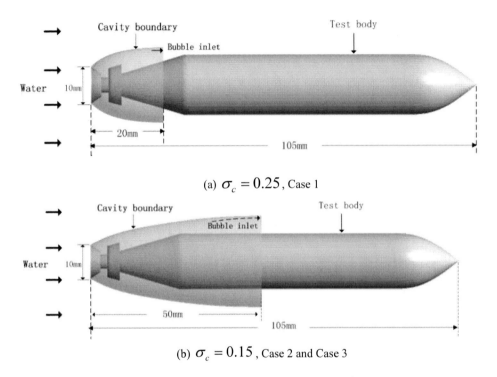

(a) $\sigma_c = 0.25$, Case 1

(b) $\sigma_c = 0.15$, Case 2 and Case 3

Figure 4. Schematic of the test body with cavity interface evaluated based on Savchenko [29].

Convergence was achieved within 2500 iterations when the RMS (root mean square) residual dropped below 10-5. A fixed physical time scale of 0.001 s was adopted to achieve steady state solutions. For the MUSIG model, bubbles ranging from 0 mm to 1 mm diameter were equally divided into 10 size groups.

4. EXPERIMENTAL DETAILS

Numerical predictions are validated against the experimental measurements in the high-speed water tunnel with a test section of 0.19m (W) x 0.19m (H) x 1m (L) at St. Anthony Falls Laboratory of the University of Minnesota [51, 52]. To study the re-entrained jet leakage mechanism of ventilated cavitation, a test body with 10mm cavitator installed at the front (Figure 5) was placed at the middle of the test section.

These experiments display what is referred to as a Gilbarg-Efros type closure[1], where the cavity terminates with a re-entrant jet onto the body. The clear cavity boundary in the front and the bubbly flow downstream were observed in the snapshot shown in Figure 6. Velocity distribution of gas bubbles were measured using Particle Image Velocimetry (PIV) technique. Photographic luminous conversion based the normalized grey-scale value was adopted to extract the time-averaged gas bubble void fraction profiles. Measurement was taken in the bubbly wake region downstream of the test body as shown also in Figure 5.

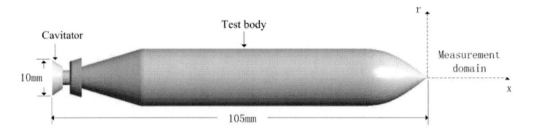

Figure 5. Schematic of the test body.

Figure 6. Snapshot of ventilated cavitation and gas leakage behaviour by Wosnik et al. (2003).

Different flow conditions are expressed in terms of Froude number, dimensionless ventilation rate and cavitation number. Definitions of Froude number Fr and the ventilation rate $\overline{Q_g}$ are given:

$$F_r = \frac{u_\infty}{\sqrt{gD_n}}, \quad \overline{Q_g} = \frac{\dot{Q}}{u_\infty D_n^2} \tag{61}$$

where u_∞ refers to the mainstream velocity, D_n is the cavitator diameter and $\dot{Q}$ is the volumetric gas ventilation rate.

Three cases shown in Table 1 are selected to compare with numerical results, including cases (Case 1 and Case 3) with the same Froude Number and cases (Case 2 and case 3) with the same cavitation number.

Table 1. Flow parameters of the three selected flow conditions for validations

	Froude Number, F_r	Ventilation rate, $\overline{Q_g}$	Cavitation Number, σ_c
Case 1	29	0.08	0.25
Case 2	20	0.18	0.15
Case 3	29	0.16	0.15

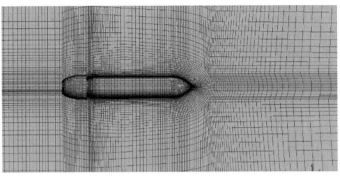

(a) $\sigma_c = 0.25$, Case 1

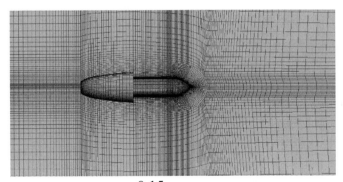

(b) $\sigma_c = 0.15$, Case 2 and Case 3

Figure 7. Mesh distribution for the computational models.

5. RESULTS AND DISSCUTION

5.1. Sensitivity Study on Turbulence Modeling

To assess the performance of various turbulence closures in predicting turbulent wake behind bluff body, three different turbulence models (i.e. standard k~ε model, RNG k~ε model and the k~ω based Shear Stress Transport (SST) model) are selected to compare based on the single phase simulation results. The predicted radial velocity distributions in the wake region behind the test body were compared with experimental data.

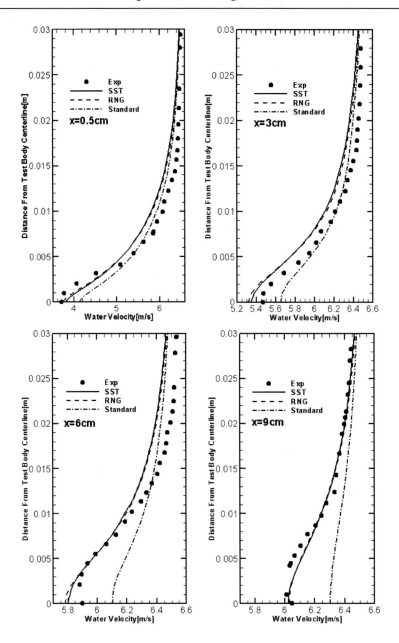

Figure 8. Predicted radial velocity distribution in the wake of the test body using different turbulence model (*Fr=20*).

As shown in Figure 8, velocity magnitude within the wake gradually reaches its freestream value as the downstream distance increases. In general, the asymptotic behavior of the axisymmetric wake is accurately captured by all turbulence models. The standard k~ε model over-predicted the velocity magnitude; while predictions of the SST model were in good agreement with measurements. As a result, all numerical predictions presented hereafter were obtained from the SST turbulence model.

5.2. Characteristic Analysis for the Multiphase Flow Field

5.2.1 Gas Phase Distribution

The predicted void fraction and streamline distribution in corresponding to the three selected cases are shown in Figure 9.

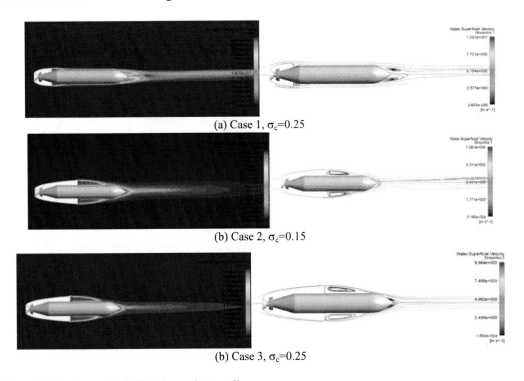

(a) Case 1, $\sigma_c=0.25$

(b) Case 2, $\sigma_c=0.15$

(b) Case 3, $\sigma_c=0.25$

Figure 9. Contours of void fraction and streamline.

As depicted by the swirling streamlines and shade of void fraction, a strong toroidal vortex structures were observed right behind the cavity base.

This re-circulation motion held up most of the leakage bubble forming a high void fraction bulk within the wake region. The mainstream of liquid flow was found re-attached to the test body. Such re-attachment of liquid flow could impose additional pressure on the test body causing unfavourable structural vibration. Furthermore, at the end of the test body, another vortex region occurs also formed behind the bluff body; causing another high void fraction area.

Table 2 presents the vortex region sizes of the selected three cases, where Lw1 and Lw2 refer to the length of the first and second vortex region respectively. As the cavitation number decreases, the first vortex region becomes bigger in volume due to larger cavity diameter. On the other hand, the volume of second vortex region is found to be decreased. This is because the liquid velocity is directed along the boundary of the former vortex region. The liquid streamline is made smoothly attaching to the test aft body with increasing cavity diameter. Therefore the disturbance made by aft body is reduced. With the same cavitation number, Case 2 and Case 3 presents similar void fraction and streamline distribution. The liquid re-attachment points were almost identical in Case 2 and Case 3, indicating the vortex region size is mainly governed by the cavitation number instead of the flow velocity.

Table 2. Wake region sizes for the selected cases

	L_{w1}/D_n	L_{w2}/D_n
Case 1	2.0	2.7
Case 2	3.3	1.1
Case 3	3.2	1.2

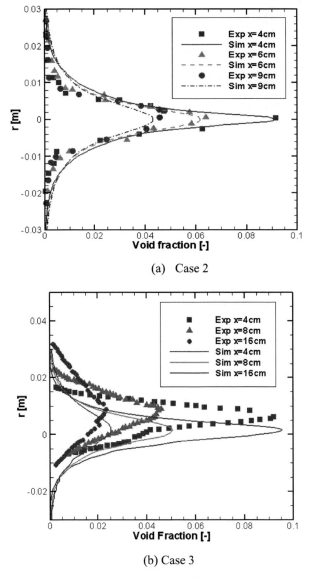

(a) Case 2

(b) Case 3

Figure 10. Void fraction distribution in the bubbly wake.

Moreover, due the gravity effect, the volume of the lower wake region was found slightly larger than upper wake region. This also contributed to the extra pressure force acting on the lower part of the test body. Figure 10 illustrates the predicted radial void fraction distribution at different axial positions which originates from the end of the test body in compared to the experimental data. In the axial direction, the void fraction reduces gradually while in the radial direction the bubbles are dispersing rapidly. Driven by the gravitation force, the peak of void fraction profiles gradually shifted upper away from the centreline of the test body. In general, as shown in the figure, magnitude and trends of void fraction profile were successfully captured by the model. Nevertheless, the locations of void fraction peak were considerably under-predicted. This could be caused by the uniform gas leakage rate imposed at the inlet boundary condition. As discussed before, gas leakage mechanism is modelled through the gas injection at the inlet boundary. Gas injection is uniformly distributed at the inlet boundary in regardless of the gravity effect. However, as confirmed by the experimental observation [51], noticeable less amount of gas leakage was observed at the lower part of the cavity. This could be caused by the gravity effect which slightly shifted the cavity interface upward and reduced the shearing-off of bubbles from the lower cavity. The uniformly distribution gas injection could be in turn over-estimated the actually gas leakage rate from the lower side, causing over-prediction of void fraction in Figure 10.

However, a qualitative agreement in the magnitude and trend is obtained between the simulation and experiment data, indicating that the numerical model, especially the "cluster" drag force model, can capture the essential physics of the complex bubbly flow downstream of ventilated cavity.

5.2.2. Velocity and Pressure Distribution

To understand the detail flow structure, a closer examination of the velocity and pressure distribution in Case 3 is presented in this section. As shown in Figure 11, the velocity is observed to be reduced in the vortex regions due to the recirculation momentum. As in the single phase flow, a wake region is formed behind the aft body when the flow passes the bluff body. Two dimensionless parameters are adopted to quantify this wake region: momentum thickness θ and wake width δ^*. Detail definitions are given as follows:

$$\delta_*^2 = \lim_{R \to \infty} \frac{1}{u_0} \int_0^R (u_\infty - u) r dr \tag{62}$$

$$\theta^2 = \lim_{R \to \infty} \frac{1}{u_\infty^2} \int_0^R u(u_\infty - u) r dr \tag{63}$$

where u0 represents the deficit velocity on the centerline which is defined as u0=u∞-ucl. Figure 13 shows the radial water velocity distribution normalized by δu which refers to the deficit velocity defined as δu=u∞-u. In this figure, x refers to distance from the end of the test body. All curves in the figure are observed to converge to one single curve, indicating the velocity in the wake abides by the same power law as in single phase flow [53].

Corresponding to the velocity and void fraction distributions, as shown in Figure 12, the pressure distribution exhibits a rapid variation within the vortex region. High pressure zone is formed at the stagnation point, while low pressure region is found right behind the cavity in the wake region. Further downstream, another low pressure region is formed in the second vortex region, which in turn produces additional pressure drag force to the test body.

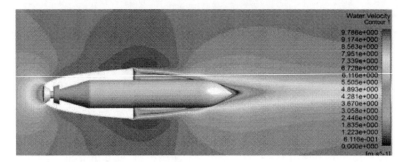

Figure 11. Water velocity contours (Case 3).

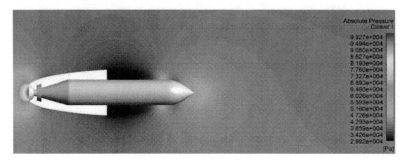

Figure 12. Pressure contours (Case 3).

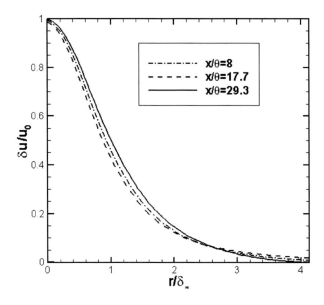

Figure 13. Dimensionless water velocity profiles (Case 3).

Figure 14 shows the pressure coefficients of Case 1 and Case 3 along the test body surface in compared with the non-cavitating case, where x refers to the axial distance from the cavitator and L is the length of the test body.

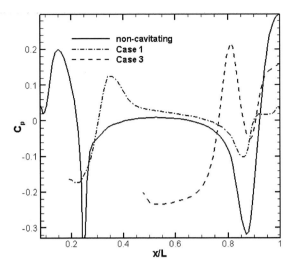

Figure 14. Pressure coefficient distribution.

According to the definition of cavitation number, the cavity pressure for supercavity increases as the cavitation number decreases. However, for the partial cavity focused on in this research, the pressure in the bubble dominated region behind the cavity base for case 1 is higher than that for case 3. This is because the pressure downstream of the cavity base is greatly increased by the high pressure acting at the reattachment point for the short cavity. Compared with the non-cavitating case, the pressure along the aft body is lower for the cases with partial cavity, indicating additional form drag induced by the ventilated cavity. However, as the streamline getting smoothly attaching to the body curve, which is represented by Case 3, the additional form drag is depressed.

5.2.3 Bubble Size Evolution Due to Breakage and Coalescence

Aiming to gain more in-depth understanding of bubble interactions behind the ventilated cavity, the bubble size evolution resulted from bubble breakage and coalescence processes is analysed. According to the applied bubble breakage model of Luo and Svendsen [47], bubble breakage rate is significantly affected by the turbulent dissipation rate. As shown in Figure 15, the predicted maximum turbulent eddy dissipation rate is located at the boundary of the vortex region. Meanwhile, large amount of bubbles could also agitate the bubble coalescence processes. As a result, vicious bubble interactions occur within the wake region behind the cavity base and the bubble size is mostly governing by the countering effect of both coalescence and breakage processes.

Figure 16 presents variations of the Sauter mean bubble diameter along with the void fraction distribution near the test body surface. As mentioned in section 5.2.1, the void fraction is of high value due to the re-circulation in the first vortex region. Moving towards the re-attachment point (defined as in Figure 1), it can be observed that void fraction gradually decreases to zero and becomes a bubble-free region near the body surface; which indicates the beginning of the transitional region. As the flow continue to develop downstream, substantial amount of bubbles disperse into boundary layer causing in a rapid increase of the void fraction at the exit of the transitional region.

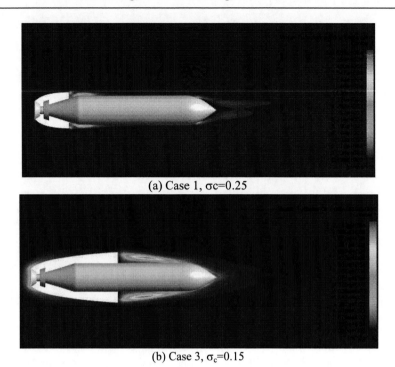

(a) Case 1, σc=0.25

(b) Case 3, σ_c=0.15

Figure 15 Distribution of turbulence dissipation rate.

Thereafter, for Case 1 the near-wall void fraction decreases due to the dispersion of the bubbles until reaching the second vortex region behind the test body. For Case 3, the bubbles are directly dispersed into the second vortex region after the transitional region.

Bubble size is greatly affected by the void fraction distribution. In case 1, coalescence processes is found to dominant at the cone section of the body; causing an increased bubble size compared with the entrained bubbles with 0.15mm diameter at the bubble inlet. After then the bubbles are broken up into small ones due to the high turbulence.

Downstream the transitional region, the near-wall bubble size is slightly increasing which is in accordance with the experimental results of microbubble drag reduction for an axisymmetric body[54]. The near-wall bigger bubbles were floated up by buoyancy, resulting in a decreasing near-wall void fraction mentioned above. For case 3, the larger first vortex region promotes stronger turbulence and eventually enhances the breakage processes of bubbles. The buoyance effect is not observed on the body surface but further affects the wake region behind the test body as shown in Figure 10.

In conclusion, the entrained bubbles were mostly disintegrated into the smaller bubbles due to the high turbulence in the vortex region (i.e. x/L≤0.33 for case 1 and x/L≤0.78 for case 3). Subsequently, the coalescence effect became slightly dominant in the bubbly flow regions leading to merging of the small bubbles. As the flow continues to develop downstream, the coalescence and breakage are almost maintained balance, resulting in a stable bubble size in the wake region away from the test body. Based on the predicted bubble sizes obtained through the current numerical model, the breakage and coalescence processes in the different flow regions downstream the ventilated cavity was successfully predicted via the population balance approach based on the MUSIG model.

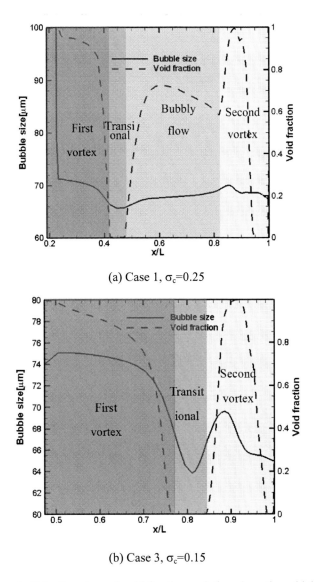

Figure 16. Sauter mean bubble diameter and void fraction variation along the vehicle body.

5.3. Microbubble Drag Reduction

5.3.1. Microbubble Distribution in the Boundary Layer

In order to get deeper understanding of the microbubble drag reduction mechanism, details of the flow field in the boundary layer is discussion hereafter. Figure 17 presents the radial void fraction corresponding to various sections: section crossing the eye of the first vortex (represented by x/L=0.24 for Case 1 and x/L=0.56 for Case 3), section at the exit of the first vortex (represented by x/L=0.43 for Case 1 and x/L=0.79 for Case 3), and sections in the bubbly flow region. As depicted, the void fraction profile exhibits totally different patterns in different regions. In the vortex eye section, extremely high void fraction is found across the section area and decreases dramatically beyond the vortex region. At the exit section of the vortex region, void fraction becomes almost zero near the wall surface (y+<20) and then gradually increases to its peak value in the buffer layer (i.e. y+=100).

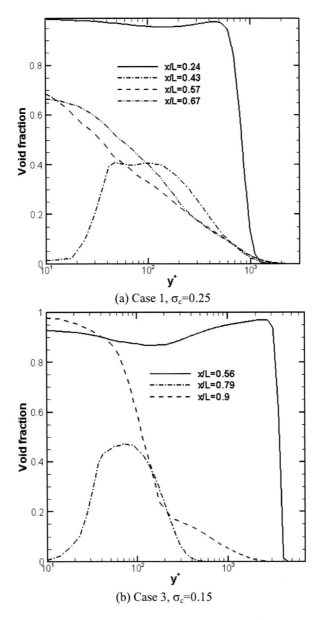

Figure 17. Radial void fraction distribution corresponding to sections of eye vortex, end of vortex region, bubbly flow region.

The void fraction is then gradually decreased due to the dispersion and washing down effect of liquid phase. As the bubbles entrained into the bubbly flow region, the peak value occurs at a distance lower than y+=10. The maximum void fraction slightly reduces as the flow develops further downstream. It can be revealed that for case 1 the void fraction profiles converges in the outer layer after y+=1000, which signifies the boundary of the no-bubble domain and extends until the maximum height of the computational domain. For case 3, the height of the bubble domain which extends to y+>5000 varies due to the curvature of the aft body.

5.3.2. Skin Friction Reduction Analysis

Density ratio, a parameter strongly depends on the near-wall void fraction, is commonly believed as one the most dominant mechanisms for microbubble drag reduction [9,10]. The reduction of local friction factor downstream of the vortex region is displayed in Figure 18, as well as the density variation near the body surface.

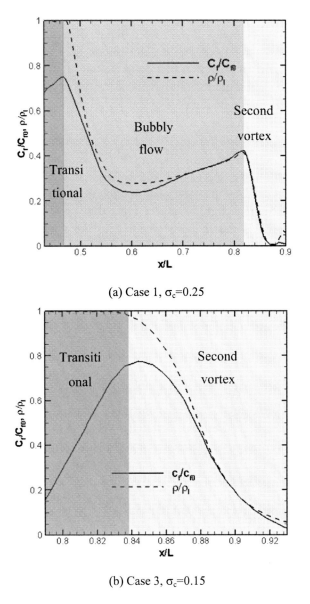

(a) Case 1, $\sigma_c=0.25$

(b) Case 3, $\sigma_c=0.15$

Figure 18. Friction coefficient reduction vs. density reduction along the test body.

Here, c_{f0} refers to the local friction factor for the test body without cavitation and ρl is the liquid density. In the transitional region, the flow density equals to the liquid density due to the bubble free layer near wall. However, the local skin friction is reduced because of the low liquid velocity induced by the flow reattachment. Then the friction coefficient increases owning to the growing liquid velocity. Thereafter, for Case 1, the friction factor decreases steeply as the bubbles enters into the bubbly region, approaching a local friction reduction as much as 70%. Subsequently, a loss of MBDR persistence is observed which is mainly contributed by the lift force. The relative velocity gives rise to a lift force away from the wall, which at the scales of interest, gradually starves the near wall boundary layer of bubbles resulting in recovery of the boundary layer toward higher skin friction values. When bubbles approach the second recirculation region, the skin friction falls again. For Case 3, the larger first vortex region gives rise to lower skin friction in the transitional region. In conclusion, the variation of the friction factor tightly follows the density curve, confirming the dominant effect of density acting on drag reduction.

5.3.3. Total Drag Reduction Analysis

In order to optimize the cavitating vehicle performance, the variation of drag reduction along with the cavitation number has to be investigated. In addition to the friction drag force which will be greatly reduced by the cavity and microbubbles, the total drag of any body is also significantly affected by the form resistance created by pressure. Especially for the bluff body, the total drag force is much more sensitive to the variation of form drag which dominates in the total drag force. For the ventilated cavitating vehicle, the form resistance is composed of three parts which is induced by the cavitator, the conical part and the body base respectively. Under the same Froude number, the form resistance induced by the cavitator is almost unchanged. Therefore, only the latter two parts are investigated. Assuming the pressure in the cavity is uniformly distributed as pc:

$$p_c = p_\infty - \frac{1}{2}\sigma_c \rho_l u_\infty^2 \tag{64}$$

The form resistance induced by the conical part is affected by the cavity pressure and the wet area (i.e. area of test body exposed to liquid). When the cavity terminates before the cone-cylinder intersection point, the variation trend of the form drag is determined by the combined effects mentioned above. When the cavity terminates on the cylinder, this part of form resistance decreases steadily with a subsequence decrease of cavitation number. The base drag is caused by the second vortex region formed behind the bluff body. As shown in Figure 9, a bigger recirculation is formed for case 1, resulting in higher base form drag. With the friction force in the front cavity ignored, the total drag force is estimated under different working conditions. The final result is presented in Table 3, where Cf, Cp, CD refer to the friction drag coefficient, form drag coefficient and total drag coefficient respectively. The last case in the table with a cavitation number of 0.1 represents the case of supercavitation which covers the whole body surface. Due to the combined effect of gas layer and microbubble drag reduction, the vehicle with partial cavity achieves friction drag reduction over 70% even at small ventilation rate. For Case 3, the friction force has been reduced by 90%. At the same time, the flow streamline smoothly attaches to the aft body which greatly depress the form drag, leading the total drag force reduced by 60%. The main effect of supercavitation is to reduce the friction force to a great degree. However, when the form drag dominates in the total drag, the drag reduction effect will be compromised because of the additional form drag. Therefore, for the vehicle with large form drag, it is more effective to achieve the desired drag reduction using partial ventilated cavity with the vehicle body reasonably designed.

Table 3. Results for drag reduction ratio

	σ_c	C_f/C_{f0}	C_p/C_{p0}	C_D/C_{D0}
Case 1	0.25	0.27	0.61	0.55
Case 3	0.15	0.09	0.39	0.34
Supercavitation	0.1	0	0.83	0.68

CONCLUSION

An Eulerian-Eulerian two-fluid model, coupled with the population balance approach, is presented in this article to handle the bubbly wake flow downstream partially ventilated cavity. To simplify the simulation, gas-liquid interface of the ventilated cavity was specified in according to empirical equations of Savchenko [29]. The air entrainment model at the cavity base was introduced to model the gas loss rate due to shearing-off bubbles. A sensitivity study on three turbulence modelling closures has demonstrated that the shear stress transport (SST) turbulence model is the best candidate to simulate the re-circulation motions within the wake region behind the test body. Furthermore, to consider closely packed bubble in high gas void fraction region, a "cluster" drag force model is adopted to take into account the neighbouring effect of bubbles. In general, numerical predictions at various axial locations downstream of the test body were in satisfactory agreement with experimental data. The vortex regions and treads of the radial void fraction profiles were successfully captured. Nonetheless, the gravity effect on the void fraction distribution was slightly under-predicted. Numerical predictions also revealed that the liquid re-attachment lengths were almost identical in cavitating cases with the same cavitation number.

The Sauter mean bubble diameter distributions predicted by the population balance approach based on the MUSIG model at different axial locations confirmed the dominance of break-up due to the high turbulent intensity behind the ventilated cavity generating more small micro-bubbles in the high void fraction region. This phenomenon is opposite to the common observations of typical bubbly flows where bubbles tend to coalesce to form elongated bubbles at high void fraction. Moving beyond the first vortex region, coalescence mechanism became slightly dominant in forming large bubbles downstream. These bubbles were then subjected to the positive lift force; floating away from the test body surface resulting a gradually decreasing trend for the near-wall gas void fraction distribution.

Based on the predicted results, the drag reduction effect caused by the micro-bubbles was investigated. At the exit of the first vortex region, the near-wall void fraction peaks at a distance around y+=100 with a free-bubble layer (y+<20) observed. As the bubbles dispersed into the bubbly flow region, the peak point occurs at a distance lower than y+=10 with the peaking value decreasing. This is mainly caused by the positive lift force which drive bubbles away from the wall, subsequently contributing to a gently loss of MBDR persistence downstream of the transitional region.

Finally, a thorough comparison between non-cavitating and supercavitating cases in terms of total drag reduction effect was presented. Benefit from the combined effect of gas layer and microbubble drag reduction, the partial cavitation achieves over 60% total drag reduction but a considerably low additional form drag. It can be therefore concluded that it is more effective to achieve desired drag reduction for the vehicle with high form drag using partial ventilated cavity.

More importantly, this study also presented a preliminary effort to incorporate population balance method into the two-fluid model to investigate the bubbly flow downstream partially ventilated cavitating vehicle. Further effort should be also focused on combing free-surface model with the present population balance approach to investigate the complex vortex structure and interaction between continuous cavity and discrete dispersed leakage bubbles.

REFERENCES

[1] Semenenko, V.N., 2001. Aritficial supercavitation: physics and calculation. , in RTO AVT/VKI special course: supercavitating flows. Belgium.

[2] David, R. S., 2001. Basic research into high-speed supercavitating bodies. *ONR Project* N00014-97-1-0126.

[3] Hargrove, J., 2004. Supercavitation and aerospace technology in the development of high-speed underwater vehicles. *AIAA* 2004-130.

[4] Abraham, N. V., 2003. High-speed Bodies in Partially Cavitating Axisymmetric Flow, in Fifth International Symposium on Cavitation, Osaka, Japan.

[5] Ivanov, A. N., and Kalyzhny, V. G., 1996. The Perspectives of Applications of Ventilated Cavities on Naval Combat Ships," *Intl. Conf. on 300th Ann. Of Russian Navy,* St. Petersburg, A2, 41–46.

[6] Latorre, R., 1997. Ship Hull Drag Reduction Using Bottom Air Injection, *Ocean Engineering* 24, 161-175.

[7] Kopriva, J., Amromin, E. L., and Arndt, R. E. A., 2008, Improvement of Hydrofoil Performance by Partial Ventilated Cavitation in Steady Flow and Periodic Gusts, *ASME J. Fluids Eng* 130, 031301.

[8] Eduard Amromin., 2010. Microbubble Drag Reduction Downstream of Ventilated Partial Cavity. *J. Fluids Eng*. 132, 051302.

[9] Sanders, W.C. and Winkel, E.S., 2006. Bubble friction drag reduction in a high-Reynolds-number flat-plate turbulent boundary layer. *J. Fluid Mech* 552, 353-380.

[10] Ortiz-Villafuerte, J. and Hassan, Y.A., 2006. Investigation of Microbubble Boundary Layer Using Particle Tracking Velocimetry. *Journal of Fluids Engineering* 128, 507-519.

[11] Shen, X., Ceccio, S.L. and Perlin, M., 2006. Influence of bubble size on micro-bubble drag reduction. *Experiments in Fluids* 41, 415-424.

[12] Kirschner, I.N., 2001.Numerical modeling of supercavitating flows, in *RTO AVT/VKI special course: supercavitating flows*. Belgium.

[13] Young, Y.L. and Kinnas S.A., 2003. Analysis of Supercavitating and Surface-piercing Propeller Flows via BEM. *Computational Mechanics* 32, 4-6.

[14] Kinzel, M.P. and Lindau, J.W., 2007. Computational Investigations of Air Entrainment, Hysteresis, and Loading for Large-Scale, Buoyant Cavities, in HPCMP USERS GROUP CONFERENCE.

[15] Owis, F.M. and Nayfeh A.H., 2002. A Compressible Multi-Phase Flow Solver for the Computation of the Super-Cavitation over High-Speed Torpedo, in AIAA-2002-0875.

[16] Qin, Q. and Song C.C.S., 2003. A Numerical Study of an Unsteady Turbulent Wake Behind a Cavitating Hydrofoil, in *Fifth International Symposium on Cavitation* (Cav2003).

[17] Kunz, R.F. and Boger, D.A., 2000. A Preconditioned Navier-Stokes Method for Two-phase Flows with Application to Cavitation Predication. *Computers and Fluids* 29, 849-875.

[18] Lindau, J.W. and Kunz, R.F., 2005. Development and Application of Turbulent, Multiphase CFD to Supercavitation, in 2nd International Symposium on Seawater Drag Reduction. Busan, Korea.

[19] Kinzel, M.P. and Lindau, J.W., 2008. Free-Surface Proximity Effects in Developed and Super-Cavitation, in DoD HPCMP Users Group Conference.

[20] Mohanarangam, K., S.C.P. Cheung and J.Y. Tu, Numerical simulation of micro-bubble drag reduction using population balance model. *Ocean Engineering*, 2009. 36, 863-872.

[21] Kunz, R.F., Gibeling, H.J., and Maxey, M.R., 2007. Validation of two-fluid Eulerian CFD Modeling for microbubble drag reduction across a wide range of reynolds numbers. *Journal of Fluids Engineering* 129, 66-79.

[22] Olmos, E., Gentric, C., Vial, Ch., Wild, G., Midoux, N., 2001. Numerical simulation of multiphase flow in bubble column. Influence of bubble coalescence and break-up. *Chemical Engineering Science* 56, 6359–6365.

[23] Yeoh, G.H., Tu, J.Y., 2004. Population balance modelling for bubbly flows with heat and mass transfer. *Chemical Engineering Science* 59, 3125–3139. 89.

[24] Yeoh, G.H., Tu, J.Y., 2005. Thermal-hydrodynamic modelling of bubbly flows with heat and mass transfer. *A.I.Ch.E. Journal* 51, 8–27.

[25] Frank, T., Zwart, P.J., Shi, J., Krepper, E., Lucas, D., Rohde, U., 2005. Inhomogeneous MUSIG model—a population balance approach for polydispersed bubbly flow. In: *Proceeding of International Conference for Nuclear Energy for New Europe*, Bled, Slovenia.

[26] Chanson, H., 1996. *Air Bubble Entrainment in Free-Surface Turbulent Shear Flows*.

[27] Ma, J., Oberai, A.A., Drew, D.A., Lahey Jr, R.T., Moraga, F.J., 2010. A quantitative sub-grid air entrainment model for bubbly flows – plunging jets. *Computers and Fluids* 39, 77-86.

[28] Spurk, J.H., 2002, On the gas loss from ventilated supercavities. *Acta Mechanica*, 125-135.

[29] Savchenko, Y. N., 1998. Investigation of high-speed supercavitating underwater motion of bodies. In: High-speed motion in water. *AGARD Report* 827, 20-1-20-12.

[30] Thorpe, R. B, Evans, G.M., 2001. Liquid recirculation and bubble breakage beneath ventilated gas cavities. *Chemical Engineering Science* 56, 6399-6409.

[31] Goncalves, E. and Patella, R.F., 2009. Numerical Simulation of Cavitating Flows with Homogeneous Models. *Computers and Fluids* 38, 1682-1696.

[32] Ishii, M., Zuber, N., 1979. Drag coefficient and relative velocity in bubbly, droplet or particulate flows. *A.I.Ch.E. Journal* 5, 843–855.

[33] Simonnet, M., Gentric, C., Olmos, E., Midoux, N., 2007. Experimental determination of the drag coefficient in a swarm of bubbles, *Chemical Engineering Science* 62, 858–866

[34] Jamialahmadi, M., Branch, C., Ullerstein-Hagen, H., 1994. Terminal bubble rise velocity in liquids, *Trans. Ins. Chem. Eng.* 72, 119-122.

[35] Johansen, S. T., and Boysan, F., 1988, Fluid Dynamics in Bubble Stirred Ladles: Part II. Mathematical Modeling. *Metall. Trans.* B, 19b, 755–764.

[36] Drew, D.A., Lahey Jr., R.T., 1979. Application of general constitutive principles to the derivation of multidimensional two-phase flow equation. *International Journal of Multiphase Flow* 5, 243–264.

[37] Tomiyama, A., 1998. Struggle with computational bubble dynamics. In: Proceeding of the Third International Conference on Multiphase Flow. Lyon, France.

[38] Antal, S.P., Lahey Jr., R.T., Flaherty, J.E., 1991. Analysis of phase distribution in fully developed laminar bubbly two-phase flow. *International Journal of Multiphase Flow* 17, 635–652.

[39] Burns, A.D., Frank, T., Hamill, I., Shi, J.-M., 2004. The Favre averaged drag model for turbulenbce dispersion in Eulerian multi-phase flows. In: *Proceedings of the Fifth International Conference on Multiphase Flow*, ICMF, Yokohama, Japan.

[40] Cheung, S.C.P., Yeoh, G.H., Tu, J.Y., 2007. On the numerical study of isothermal vertical bubbly flow using two population balance approaches, *Chemical Engineering Science*. 62, 4659-4674.

[41] Yeoh, G.H., Tu, J.Y., 2004. Population balance modelling for bubbly flows with heat and mass transfer. *Chemical Engineering Science* 59, 3125–3139.

[42] Frank, T., Zwart, P.J., Shi, J., Krepper, E., Lucas, D., Rohde, U., 2005. Inhomogeneous MUSIG model—a population balance approach for polydispersed bubbly flow. In: *Proceeding of International Conference for Nuclear Energy for New Europe*, Bled, Slovenia.

[43] Kumar, S., Ramkrishna, D., 1996. On the solution of population balance equations by discretisation—I. A fixed pivot technique. *Chemical Engineering Science* 51, 1311–1332.

[44] Prince, M.J., Blanch, H.W., 1990. Bubble coalescence and break-up in air sparged bubble columns. *A.I.Ch.E. Journal* 36, 1485–1499.

[45] Chesters, A.K., Hoffman, G., 1982. Bubble coalescence in pure liquids. *Applied Scientific Research* 38, 353–361.

[46] Rotta, J.C., 1972. Turbulente Stromungen. B.G. Teubner, Stuttgart.

[47] Luo, H., Svendsen, H., 1996. Theoretical model for drop and bubble break-up in turbulent dispersions. *A.I.Ch.E. Journal* 42, 1225–1233.

[48] Sato, Y., Sadatomi, M., Sekoguchi, K., 1981. Momentum and heat transfer in two-phase bubbly flow—I. *International Journal of Multiphase Flow* 7, 167–178.

[49] Lopez de Bertodano, M., Lahey Jr., R.T., and Jones, O. C., 1994. Development of a k-ε Model for Bubbly Two-Phase Flow, *J. Fluids Eng.* 116, 128-134.

[50] Drew, D.A., Lahey Jr., R.T., 1979. Application of general constitutive principles to the derivation of multidimensional two-phase flow equation. *International Journal of Multiphase Flow* 5, 243–264.

[51] Wosnik, M., Fontecha, L.G., Arndt, R.E., 2005. Measurements in High Void-fraction Bubbly Wakes Created by Ventilated Supercavitation, in *Proceedings of FEDSM*.

[52] Schauer, T.J., 2003. An Experimental Study of a Ventilated Supercavitating Vehicle, MS Thesis in Aerospace Engineering, University of Minnesota.

[53] Johanssona, P.B.V. and George, W.K., 2003. Equilibrium similarity, effects of initial conditions and local Reynoldsnumber on the axisymmetric wake. *PHYSICS OF FLUIDS* 15, 603-617.

[54] Deutsch, S. and Casano, J., 1986. Microbubble skin friction reduction on an axisymmetric body. *Phys. Fluids* 29, 3590-3597.

In: Computational Fluid Dynamics
Editor: Alyssa D. Murphy

ISBN 978-1-61209-276-8
© 2011 Nova Science Publishers, Inc.

Chapter 7

COMPUTATIONAL MODELLING IN TISSUE ENGINEERING: PROVIDING LIGHT IN TIMES OF DARKNESS

Ryan J. McCoy*, Fergal J. O'Brien and Daniel J. Kelly

Department of Anatomy, Royal College of Surgeons in Ireland
123 St. Stephen's Green, Dublin 2, Ireland
Trinity Centre for Bioengineering, School of Engineering,
Trinity College Dublin, Dublin, Ireland

ABSTRACT

The human body can be considered an extremely complex machine. The demands placed on this machine over the lifetime of an individual or their natural genetic predisposition to particular conditions can result in the failure of native internal structures; for example, hips, knees, hearts or kidneys. Computational modelling has played a key role in the realisation of mechanically engineered replacements that can either be implanted internally (e.g ventricular assist devices and orthopaedic implants) or used to replace organ functionality externally (e.g kidney dialysis machines). This has enabled temporary ablation of the body's decline in some instances, but the provision of a biologically and anatomically relevant substitute capable of integrating into the host and continuing the original function unaided for the remaining lifetime of the individual, has yet to reach fruition. Interdisciplinary collaboration over the last 40 years between the engineering and scientific communities, has pursued and continues to pursue, the development of physiologically representative substitutes through the cultivation of tissue ex-vivo or through the induction of tissue growth in-vivo (tissue engineering and regenerative medicine). The intrinsic complexity of these biological systems means that experimentally prizing apart the individual mechanisms at play may not always be viable. Computational modelling tools thus have the potential to prove a very powerful addition once again; providing light where there is darkness in this emerging field and promoting furthered understanding.

---

* Tel.: +353 1 402 8508; E-mail address: ryanmccoy@rcsi.ie

This chapter endeavours to introduce the reader to computational modelling concepts undertaken within the tissue engineering and regenerative medicine fields through a series of specifically selected case studies based within the bone tissue engineering arena.

INTRODUCTION

Hermann M. Biggs (1859 – 1923), an American physician, once wrote;

"The human body is the only machine for which there are no spare parts."

Almost nine decades later and medical science is only now on the verge of providing a truly viable solution to the problem that affects almost every human being at some point in their lifetime – the wearing out of the human body; the prevalence of which is expected to grow as life expectancies continue to rise and thus the body's workload is further amplified.

This is not to say that advances have not been made, increased life expectancies are due in part to our ability to "patch up" the human body. Our understanding of its structure, development, internal signalling mechanisms and homeostatic regulation of macro- or micro-environments are more complete than ever before. Coupled with technological advances in imaging, medical device design and surgical techniques, computational modelling has played a key role in the realisation and continued improvement of mechanically engineered replacements that upon failure of native internal structures such as hips[1][2][3][4], knees[5][6][7], hearts[8][9][10], arteries[11][12][13][14][15][16][17] and kidneys[18], can either be implanted internally (e.g ventricular assist devices and stents) or used to replace organ functionality externally (e.g kidney dialysis machines). Such devices have enabled temporary ablation of the body's decline in some instances, but the provision of a biologically and anatomically relevant substitute capable of integrating into the host and continuing the original function unaided for the remaining lifetime of the individual, has yet to reach fruition. Interdisciplinary collaboration between the engineering and scientific communities over the last 40 years, aiming to build upon understanding of the founding principles in each field, has pursued and continues to pursue, the development of physiologically representative "spare parts" through the cultivation of tissue ex-vivo or induction of tissue growth and/or regeneration in-vivo (tissue engineering and regenerative medicine). The in-vitro growth of cells to form biological substitutes that restore, maintain or improve tissue function when implanted into the patient is built upon a triad of constituents employed either individually or in combination (Figure 1). The first component is a three-dimensional scaffold that serves as an analogue of in-vivo extracellular matrix (the proteins that bind the cells together), it acts as a physical support structure and an insoluble regulator of biological activity that affects cell processes such as migration, contraction, and division; it can be considered a template for tissue formation. Scaffolds have been fabricated using a range of procedures (freeze-drying, salt-leaching, freeze-immersion-precipitation, foaming albumin, gel casting, compression molding) and composite materials (polyurethane, titanium, hydroxyapatite and calcium phosphate, collagen-GAG) for a broad range of pore sizes (100-400µm) and porosities (71-99%). Scaffolds generated through free-form fabrication techniques can be designed to have a regular porous geometry, but the majority of fabrication techniques result in an irregular, albeit homogenous, highly porous structure.

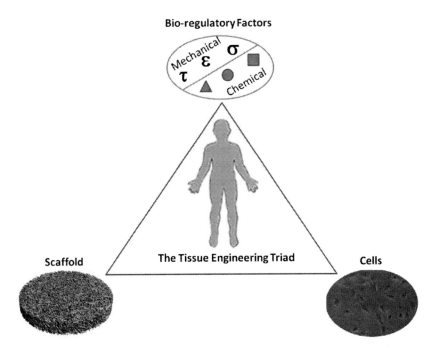

Figure 1. The tissue engineering triad. Physiologically representative "spare parts" are sought to replace failing original parts through the cultivation of tissue ex-vivo or induction of tissue growth in-vivo (tissue engineering and regenerative medicines). The philosophy is built upon a triad of constituents employed either individually or in combination; mechanically apt biocompatible composites with desired architectures (scaffolds), bio-regulatory factors (chemical and mechanical cues) and cells.

In our laboratory, we employ scaffolds composed of collagen and glycoaminoglycans (GAGs). Collagen is a significant constituent of natural extracellular matrix (ECM) and scaffolds made from collagen have been used in a variety of applications [19][20][21][22] as they possess requisite hemostatic properties, low antigenicity and mechanical characteristics. Furthermore, collagen has been observed to promote cell attachment and tissue growth. GAGs are negatively charged carbohydrates that are also common constituents of natural extracellular matrices and have been shown to be involved in cell adhesion, migration, proliferation and differentiation[23][24][25]. The presence of GAGs provide numerous side chains for cross-linking that can improve the mechanical properties of the scaffolds and help preserve the porous structure. The second component in the triad is bio-regulatory factors; these are usually either chemical/biological growth factors secreted by cells that control cell growth and differentiation (commitment towards a particular cell type – bone, vascular, fat, skin etc) and/or mechanical cues such as physical deformation, which stimulate the cell, activating internal pathways that subsequently direct the cell towards a particular cell type. The third component is the cells; they are either derived from the patient (autologous) or taken from other individuals of the same species (allogeneic). Implementing such a strategy requires an in depth prior understanding of the natural in-vivo development and homeostatic regulation at various hierarchal levels (organ, tissue, cellular and intra-cellular). Researchers have historically acquired this knowledge through the application of rigorous experimental methodologies. However, the intrinsic complexity of biological systems, in combination with current technological limitations, means that experimentally prizing apart the individual mechanisms at play may not always be feasible.

Bone is one such example of a complex tissue whose in-vivo regulation is still not fully understood. It is second only to blood on the list of transplanted materials (American Association of Orthopaedic Surgeons Report 2002) with approximately 50% of the 4 million bone replacement operations performed annually requiring a bone graft substitute. These bone graft substitutes are traditionally acquired through autograft or allograft routes. However, increasing limitations such as the restricted amount of bone removal, additional invasive surgery, risk of infectious disease transmission and lack of available donors is escalating the demand for an appropriate tissue engineered substitute. The generation of a tissue engineered bone graft substitute exemplifies the challenges presented in the broader context of the tissue engineering and regenerative medicine fields and thus bone tissue engineering has been selected as a case study to describe how computational modelling tools are showing the potential to be a very powerful addition once again; providing light where there is darkness in this emerging field and promoting furthered understanding.

The Importance of Computing Levels of Biomechanical Stimuli within Bone and Tissue Engineering Scaffolds

Biomechanical stimuli have been recognised as an important part of in-vivo bone remodelling for almost a century [26] and necessary for stimulating cells in order to elicit correct cellular differentiation/functioning [27]. The substantial loss of bone during spaceflight [28][29] or after significant periods of bed rest [30], where reduced mechanical loading environments are experienced for long periods of time, support the argument for the importance of biomechanical stimuli in bone homeostasis. Normal loading of bone in-vivo combines bending and compressive forces that under normal physiological conditions induce internal stresses and strains. The predicted levels of these forces are expected to be less than 2000 $\mu\varepsilon$ (μstrain) [31] in humans (1 μstrain is equal to 1 μm of deformation per metre length of bone). Additionally, these bending forces create pressure gradients within the canaliculae that drive extracellular/interstitial fluid flow, providing cells with nutrients and generating fluid shear stresses (0.8 – 3 Pa), which are an alternative means of mechanically stimulating the cells.

The development of non-invasive in-vivo models (experiments that use whole living organisms in highly controlled environments and don't require surgical procedures) for studying load induced bone adaption has provided evidence supporting the hypothesis that both strain magnitudes and cycle numbers [32][33] regulate bone mass and the up-regulation of bone remodelling associated genes [34]. Accurately predicting the mechanical environment within bone is central to understanding such processes. Imaging techniques such as MRI (Magnetic Resonance Imaging) and μCT (micro Computed Tomography)[35] allow the structural morphology of bones to be captured and subsequently meshed for computational analysis. Finite element methods (FEM), which render partial differential equations (PDEs) to an approximating system of ordinary differential equations (ODEs) that can be integrated using standard techniques, are then subsequently applied to the governing equations for the meshed representation of the bone and solved. Such approaches have shown that strains predicted by the more recently developed FEM techniques [36] correlate well with beam bending theory. The major advantage of the FEM analysis though is that it can provide indications of localised peak stress and strain distributions that occur as a result of non-uniform bone morphology [37]. It is these areas that are of interest to researchers as they tend to have increased levels of in-vivo bone remodelling (as indicated by enhanced levels of bone density), a process that allows the redistribution of strain and stress within the bone ensuring the maintenance of a continuum-level strain gradient[38].

In the highly porous scaffolds utilised for 3D tissue engineering, calculating the strain distribution at the micro-scale as a function of the internal scaffold architecture and the applied load at the structure surface (determining the magnitude of strain that an individual cell experiences), is a complex undertaking, one that would not be possible without the availability of non-destructive imaging and computational modelling techniques (see computational modelling of highly porous scaffolds for bone tissue engineering: modelling mechanical stress and strain). Being able to quantify the magnitude of these localised forces and subsequently replicating these levels of biomechanical stimuli in-vitro has shown positive effects upon cell proliferation [39][40] and regulatory pathways (the internal biochemical reactions and molecules that govern cell decision making) and/or genes associated with osteogenesis (the formation of bone)[39][41][42][43][44][45][46][47][48][49].

Evidence exists to suggest that strain alone (in-vivo) is not sufficient to induce bone remodelling and that fluid shear stress is of greater importance [50]. Cells are exposed to fluid shear stress (forces generated by the movement of fluid) in-vitro through the use of bioreactors. Bioreactors have been traditionally used in the pharmaceutical industry for the mass cultivation of genetically engineered cells to produce medicines (e.g insulin). In tissue engineering numerous bioreactors types have been designed to aid the cultivation of 3D tissue constructs [51] with the purpose of improving cellular distribution and mass transfer constraints within the scaffolds (removing issues associated with concentration gradients at fluid-construct boundary interfaces [52][53] under static culture conditions); diffusion of oxygen and soluble nutrients to the construct core can become critically limited in longer term static cultures as tissue growth and extracellular matrix/mineralisation occurs [54][55], resulting in a necrotic core with an external "shell" of viable tissue. Perfusion rigs are the most common type utilised with several reactor styles having been devised to date [52][56][57][58][59]. Perfusion appears to offer the greatest benefits as fluid is forced through the entire construct creating a more homogeneous micro-environment (mimicking the in-vivo scenario), rather than just improving convection at the construct surface. Furthermore, such bioreactor configurations are proving useful in the investigation of mechanical factors, such as fluid shear stress (generated by the flowing fluid within the reactor), under controlled defined chemical conditions that potentially allow the decoupling of mechanical and chemical signals, permitting the contribution of each individual factor with respect to osteogenic differentiation to be evaluated. The application of fluid shear stress has been shown to positively increase expression levels of known bone associated genes in-vitro using highly porous scaffolds [52][60][61][62][63][64][65][66][67]. However, calculation of fluid shear stress within irregular highly porous scaffolds is non-trivial. Analytical methods can be used, such as those employed by Goldstein et al[53] and Jungreuthmayer et al[68] based on crude assumptions about the scaffold geometry, where the porous scaffold is modelled as a bundle of parallel circular pipes with the width of a pipe equal to the mean pore size. However, this simplification only offers an estimation of the flow conditions, is inherently inaccurate, and does not provide a shear stress distribution. Numerical models are therefore preferred due to their greater level of accuracy, but can be time consuming and computationally expensive.

Computational Modelling of Highly Porous Scaffolds for Bone Tissue Engineering

Geometrical Reconstruction and Mesh Generation

A simplified reconstruction of the irregular highly porous scaffolds employed in tissue engineering approaches can be achieved using a cellular solids model based on polyhedral unit cells, more specifically, tetra-kaidecahedral polyhedrons; fourteen sided polyhedrons that pack to fill space almost satisfying minimum surface energy requirements and approximating to the morphological features of low density foams (14 faces per cell, 5.1 edges per face) (Figure 2). O'Brien et al used such an approach to gain an understanding of the relationship between mean pore size and the specific surface area for scaffolds of a uniform material (biological) composition, thus providing an indication of the surface ligand density available for cell attachment and migration [69]. Additionally, they applied the use of a tetra-kiadecahedral based model to the evaluation of scaffold permeability as a function of pore size and compressive strain [70]; a key variable in controlling the diffusion of nutrients in and waste out of the scaffold as well as influencing the pressure fields within the construct. Harley et al [71] assessed the dynamics of cell motility and migration as a function of strut junction spacing and density for scaffolds of different pores sizes using a similar approach to geometric reconstruction. More computationally focused approaches using this reconstruction technique, with specific reference to the influence of biomechanical stimuli, are discussed in a later section of this chapter (see computational modelling of highly porous scaffolds for bone tissue engineering: modelling mechanical stress and strain).

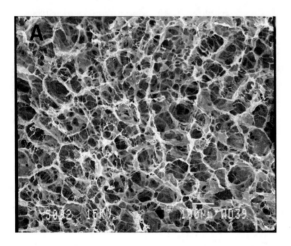
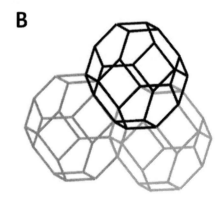

Figure 2. Scanning Electron Microscope image showing the structure of a highly porous collagen based scaffold (A). An example of tetra-kiadecahedral polyhedron packing as used by a number of authors to model scaffold structure (B).

Accurate reconstructions of these highly porous scaffold geometries is essential if computational modelling approaches are to precisely determine the levels of mechanical strain and fluid shear stress experienced at the micro-scale within these structures. Images of the structure are typically acquired through non-destructive imagining techniques such as μCT, computed tomography (CT) and MRI. In our laboratory, μCT scans (using a special filter technology to increase material contrast) of the collagen-GAG scaffolds are acquired and then reduced to smaller randomly chosen sub-volumes to lessen the associated computational costs. The grey-scale data relating to each sub-volume is Gaussian rank filtered prior to being subjected to a thresholding procedure that results in a black (scaffold) and white (void/interstice) image of the geometry. This image can then be imported into a range of commercial computational fluid dynamic (CFD) packages (Ansys, Fluent, Abaqus, Comsol etc) or open-source platforms (OpenFOAM) and meshed to create a numerical model of the scaffold (Figure 3); we used the mesh generation utility blockMesh, supplied with the opensource CFD tool box OpenFoam to decompose the domain geometry into a set of hexahedral elements with either line, arc or spline edges. Each element face is then assigned to a generic "patch" group, which is associated with a particular set of boundary or geometric conditions e.g inlet or symmetry plane. If structural analysis of the scaffold (quantification of mechanical stress and strain distributions) is to be undertaken, the black (scaffold) area of the image is meshed. If fluid dynamic simulations are to be conducted (quantification of fluid shear stresses) then the scaffold's interstice (white region of the image) is meshed (Figure 3).

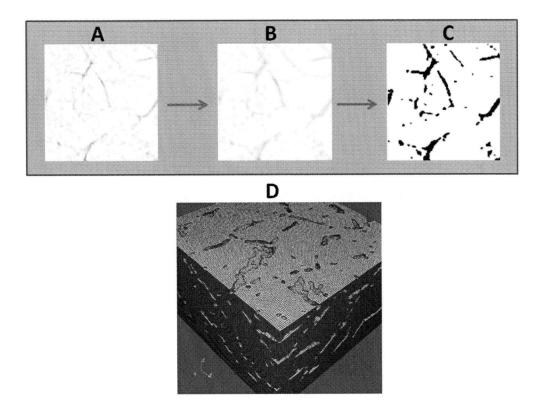

Figure 3. 2D representations of a collagen-GAG scaffold from (A) the initial data format as acquired from μCT imaging, (B) after Gaussian filtering to smooth the image and reduce background noise and (C) post thresholding to create a black and white image; scaffold (black) and void space (white). This data is then compiled for CFD analysis of the fluid flow through the porous structure by meshing the void space (D). Mesh was visualised using the open-source scientific visualisation software ParaView.

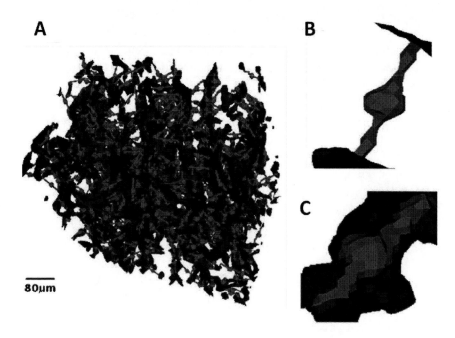

Figure 4. Representation of a collagen-GAG scaffold (blue) numerically seeded with cells (red) (A), using one of two cell morphologies, either bridged (B) of flat (C) reproduced from Jungreuthmayer et al[72] with permission from Elsiever.

Cell-scaffold interactions are often then inferred by analysing the nodal displacement of the numerical scaffold representation and assuming cells are attached between pairs of nodes. Alternatively, it is assumed the cell experiences the same level of fluid shear stress as that present in an empty scaffold simulation. A number of groups have tried to look at cell-scaffold interactions by numerically seeding cells onto the scaffold representation. In our group, the use of cell seeded scaffolds is preferred as they take into account different behaviours of cells and how they interact with scaffolds. In highly porous scaffolds, cells can attach to surfaces in one of two morphologies; flatly (binding to one strut only) or bridged (binding to more than one strut). The attachment type is important as it is hypothesised that the cells will behave differently based on their attachment morphology. Cells are numerically seeded onto the scaffold first (Figure 4) and then the scaffold/interstice is meshed. The cell seeding program used was developed in house [72] and can seed cells with both morphological attachments types – flat or bridged - with bridged cells being attached to two struts (Figure 4). The algorithms start by randomly picking a void-scaffold interface, they then try to propagate cell material through the scaffold void in a randomly chosen direction. If an entire cell can not be propagated without hitting an obstacle (scaffold material or seeded cell) the seeding attempt is aborted and a new propagation direction or starting void-scaffold interface is picked. The main difference between the two seeding algorithms is that for bridging cells a potential opposite attachment face is determined (the second strut) before the propagation attempt is started. The opposite attachment face is computed using the normal vectors of the attachment faces and the user-defined cell attachment lengths.

Filtering and thresholding procedures described previously are also usually conducted using scripts developed in-house. More recently however, commercial 3D image packages, with easy to navigate user interfaces, such as Mimics (Materialise, Leuven, Belgium) and

ScanIP (Simpleware, Exeter, UK) are proving powerful tools in aiding image visualisation, processing and segmentation for images derived from a multitude of platforms (μCT, MRI, confocal microscopy) and source data formats (DICOM, Interfile, Analyze, Meta-file, raw image data, stack of 2D images). Furthermore, in the case of Simpleware, the segmented 3D image can be directly translated to a high quality mesh and subsequently imported into a range of commercial CFD packages (Ansys, Fluent, Abaqus, Comsol etc) using the ScanFE module (Figure 5).

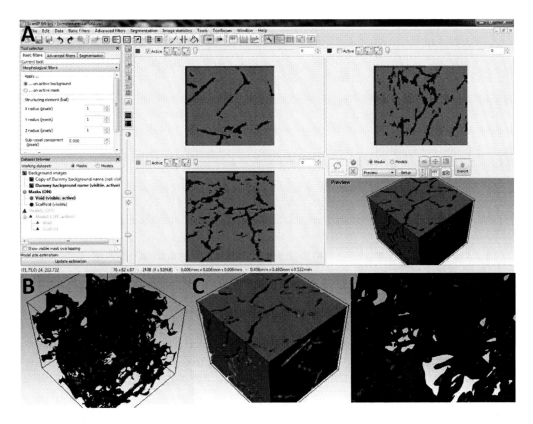

Figure 5. Illustration showing the Simpleware user interface displaying 2D μCT images in the x, y and z planes and a 3D reconstruction with segmentation of the images into scaffold (blue) and void space (green) (A). The segmentation masks can then be separated into their individual components (B and C) and subsequently meshed (D) for importation into commercial CFD packages.

If geometry reconstruction is not feasible, either because the composition of the scaffold material is not conducive to imaging, imaging capabilities are not sufficient to provide an accurate depiction of the scaffold architecture, or in the case of longer term studies incorporating cells, where tissue growth, extra-cellular matrix deposition and potentially scaffold degradation may have significantly altered the structure being modelled, then alternative approaches are implemented. Usually the scaffold/tissue construct is redefined as a homogenously porous material with material properties such as permeability that are defined through experimental evaluation.

MODELLING OF MECHANICAL STRESS AND STRAIN

Determining the maximum loads scaffolds can withstand and understanding why the macrostructure fails as a result of the microstructure is important as scaffold compaction (loss of porous structure) is undesirable; open pore structures enhance cell migration into the core of the construct and increase permeability. Additionally, greater mechanical strength is advantageous when considering surgical handling and the in-vivo load bearing capabilities once implanted. Alberich-Barrayi et al [73] used a computational-experimental approach to illustrate that FEM compressive stress-strain tests conducted using ANSYS (v.10) on 3D µCT reconstructions of scaffolds, meshed using a MATLAB algorithm that converted the voxel depiction directly into an isotropic hexahedral mesh, showed good correlation with experimental tests for different architectures; validating the use of computational models for studying the macro-scale dynamics of scaffolds and providing strong evidence for their application to evaluate dynamics at the micro-scale. Furthermore, distribution of principal stresses within the scaffold was shown to vary as a function of architecture. Pores acted as stress concentrators leading to localised high stresses; spherical pores provided a more homogeneous distribution with fewer areas of high stress compared to a more orthogonal architecture. Stops et al [74], using an in-house solver (FEEBE/linear) for the structural analysis, defined a critical stress magnitude for the buckling of a strut (1.78kPa) and analysed the percentage of elements within the structure that exceeded the limit, for different scaffold architectures, when a 1% compressive uni-axial strain was loaded in the z-axis. They showed pore architecture can lead to a 10 fold difference in the percentage of elements that exceeded the critical stress level, albeit only a very small amount of the total volume exceeded the limits under the compressive loads tested (0.02 and 0.2% respectively). Lacroix et al [75] investigated stress and strain distribution within scaffolds of different compositions and examined the effect of pore homogeneity. They pre-processed each scaffold using MIMICS into 7 layers in an effort to reduce the computational load. The layers were meshed using MARC (MSC software). They identified that the principal stresses were distributed differently with respect to material composition, but principal strains were similar, with regions of peak strain localised around areas of greater porosity.

Computational models have also been used to identify how scaffold architecture impacts upon the forces experienced by cells attached to the scaffold. Stops et al [76] investigated strain distribution using a tetrakiahedral representation of a collagen-GAG scaffold meshed using Timoshenko quadratic beam elements. They demonstrated that flat cells (represented as two nodal attachments along a single strut) had mean, minimum and maximum strains in the order of 10^{-6} % and bridged cells (three nodal attachments across multiple struts) had mean strains of 2%, minimums of -26% and maximums of 49%, with the majority experiencing less than 5% strain for an applied macroscopic (apparent) strain (uni-axial strain) of 10%. In an alternative model of a collagen-GAG scaffold, using a µCT geometric reconstruction, Stops et al [74] evaluated the inferred strain on cells of different morphologies for two scaffold architectures. The level of cellular deformation for bridged cells was assessed by comparing the displacement of pairs of nodes separated by approximately the length of a cell (60 µm) during scaffold deformation; percentage cell strain was calculated from the change in length between each pair of nodes. Flat cells were assessed by examining the loads (principle strains) imparted over the scaffold surface and were shown to experience very little strain, close to zero (a mean value of -10^{-6} and -10^{-7}% for the two different architectures), whilst the mean strain for bridging cells was -0.15 and -0.17%, and the maximum strains -35% and 28% respectively.

The cumulative findings of these studies show that non-uniform strain distributions exist throughout the scaffolds and are a function of composition and architecture. The strain experienced by cells attached within the porous scaffolds could potentially be much higher than the level of macroscopic strain applied, but equally could be much lower depending on how cells attach to the scaffold, which in turn is a function of the scaffold architecture. Quantitative analysis of cell attachment morphology and ratio of morphology types in 3D porous scaffolds has been limited to date. However, McMahon[77], through confocal microscopy, has suggested that 75% of cells are attached in a bridged morphology and 25% as a flat morphology (for a collagen-GAG scaffold of pore size 95 µm and porosity >98%). The complexity of such biophysical stimuli created within the scaffolds presents great difficulty for tissue engineers in ensuring population wide induction of the desired cellular responses in a uniform manner.

COMPUTATIONAL MODELLING OF FLUID SHEAR STRESS

The differences in geometrical configurations and operating conditions of perfusion bioreactor configurations [52][56][57][58][59](e.g input flow-rate) makes correlating observed biological outcomes to the application of fluid flow very difficult. Thus calculation of the forces (e.g fluid shear stresses) within the individual systems allows comparisons to be made across reactor configurations on a single scale. Furthermore being able to accurately calculate the level of fluid shear stress allows determination of the critical threshold limits of cell mechano-stimulation.

Finite difference methods for modelling fluid flow through porous scaffolds have been explored by a number of authors. Zhao et al [63] utilised the Lattice Boltzmann Method (LBM) on a two dimensional cubic lattice representing a simplified geometric reconstruction of their perfusion bioreactor and scaffold. They assumed flow in the traverse direction was negligible, allowing the system to be considered in two dimensions and defined the scaffold as a homogenously porous material with a permeability value determined at the end of the culture timeframe (i.e the most severe conditions when porosity is lowest and therefore shear stress is greatest). This approach was successful at reducing the computational costs but only provided a generalised macroscopic overview for the shear stress level in a gradient form from the surface of the scaffold to the centre with no indications of localised regions within the scaffold. Porter et al [78] also utilised LBM including the "reinterpreted bounce back condition" at solid-fluid boundaries for determining shear stress levels in porous scaffolds. In this instance a three dimensional cubic lattice was created from a µCT reconstruction for the void space within the scaffold micro-architecture. Validation of the computational approach was undertaken using a parallel plate example (where an analytical solution exists) and the method was found to underestimate the level of shear stress at the wall (because of the bounce back condition). The percentage error varied as a function of the lattice units (lu) per channel and distance from the wall. The estimation improved upon moving away from the channel wall and/or by increasing the number of lu per channel. For the actual simulations the resolution of the micro-CT scans (34 micron) were coarsened to 68 micron for a single lattice unit (~1/10 of the mean pore size), estimations of shear stress were not corrected for the porous scaffold based on findings in the parallel plate example, but if consistent should be accurate to within 5.7%. Finally, Voronov et al [79], using a similar computational set-up to Porter et al, showed that the magnitude of wall shear stress varied with pore size and porosity for a range of salt-leached scaffolds. More importantly however, they showed that small variation in manufacturing quality of porous scaffolds, which lead to a loss of homogeneity in the scaffold architecture, could consequently lead to zones of concentrated wall shear stress, which may explain heterogeneity of tissue growth observed within scaffolds experimentally; highlighting the importance of quality control in the manufacturing of scaffolds for tissue engineering applications.

Finite volume methods (FVM) for modelling fluid flow through porous scaffolds have also been explored [59][80][81][82][83][84][68]. Strategies commonly focus on modelling only a single scaffold or sub-volumes from a single scaffold with Fluent being the most popular computational fluid dynamics software utilised, although Marc-Mentat (MSC software) and OpenFOAM have also been utilised. In our group we employ the FVM approach to the modelling of fluid flow in highly porous structures. Using the laminar solver *icoFoam* (Transient solver for incompressible, laminar flow of Newtonian fluids) of the opensource CFD toolbox OpenFOAM we have determined the shear stress in collagen-GAG and calcium phosphate scaffolds. The model illustrated that a wide distribution of shear stresses were present in the scaffolds (Figure 6) and that analytical solutions (described previously) over-predict the shear stress levels by 12 – 21%[68]. Developing the model further to achieve a greater understanding of how the fluid forces interacted with cells seeded onto a scaffold, a collagen-GAG scaffold was numerically seeded with cells either adopting a bridging or flat morphology (Figure 4). Data pertaining to the levels of wall shear stress and wall hydrostatic pressure on the cells from the CFD simulation were then fed into a linear elastostatics model, where the cells were considered to have uniform material properties. The study showed that the displacement magnitude of the cells is primarily determined by the cell morphology; cells bridging two struts experienced deformations three orders of magnitude greater (micrometers) than cells only attached to one strut (nanometers)[72].

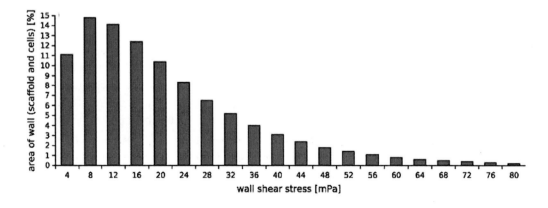

Figure 6. Shear stress distribution profile for a sub-volume of a collagen-GAG scaffold (pore size 120 μm, porosity 89-92%). The input flow-rate was 1mL/min. Reproduced from Jungreuthmayer et al[68] with permission from Elvesier.

The results from these studies have shown that a broad distribution of wall shear stresses (greater than one order of magnitude) can be present within a scaffold at a constant inlet velocity. Furthermore, as tissue growth occurs within the structure, porosity is reduced, and shear stress levels increase. If scaffold architecture results in a mixed population of attachment types, the large difference observed in displacement for bridging cells compared to flat cells means that if bioreactor operating conditions are optimized for stimulation of flatly attached cells this may have a negative impact on bridged cells, or if conditions are optimized for bridged cells, flat cells may lack the necessary stimulation to cause osteogenesis. This presents challenges to researchers trying to maintain a homogenous environment for the cultivation of tissue. However, through computational-experimental approaches the optimum conditions of bioreactor operation and scaffold architecture to create bone tissue engineered substitutes should be elucidated in due course.

COMPUTATIONAL MODELLING OF MECHANOTRANSDUCTION

Computational modelling of the mechanical stress/strain and fluid shear stress micro-environment in highly porous scaffolds has provided researchers with insight regarding the magnitude of biomechanical stimuli required to induce cellular biological responses that will lead to improved in-vitro bone tissue formation. However, how cells sense biomechanical stimuli and convert this to a biological response is an active area of research and have been reviewed in depth elsewhere [85][86]. An understanding of how this occurs may allow researchers to implement changes in the scaffold design to maximise the extent of cell stimulation or ensure greater levels of uniformity. Translating the physical force applied at the cell surface into a biological signal is termed mechanotransduction. The continuum model is one of a number of approaches being explored by various groups in an effort to understand mechanotransduction at the single cell level. The continuum model treats the cell cytoplasm as a continuum, representing cells as a two layer material, with unique properties for the cortical/membrane layer and for the cytoplasmic/cytoskeletal layer (a viscoelastic material) [87][88]. Karcher et al [89], developed a 3D FEM model for examining the intra-cellular strain and stress distributions as a function of micromanipulation by magnetocytometry (the deformation of cells by application of force through the movement of magnetic beads). The results suggested that intracellular stresses were localised in the vicinity of the applied force. A similar model was devised by Ferko et al [90] to examine shear stresses induced by both magnetic bead and steady state flow but included the addition of fixed focal adhesion sites and a nuclear compartment. Likewise, the model predicted that shear stress induces small heterogeneous deformations of the endothelial cell cytoplasm when single properties were applied to the whole cell and focal adhesion points were not considered. However, strain and stress were amplified 10–100-fold over apical values in and around the high-modulus nucleus and near focal adhesions (FAs) when they were incorporated into the model; suggesting a means by which sub-cellular localisation of mechanotransduction events may occur. The tensegrity model is an alternative approach that attributes the primary structural role to the cell cytoskeleton; a network of microfilaments and microtubules that distribute forces within the cell through a balance of compression and tension [91][92] (Figure 7).

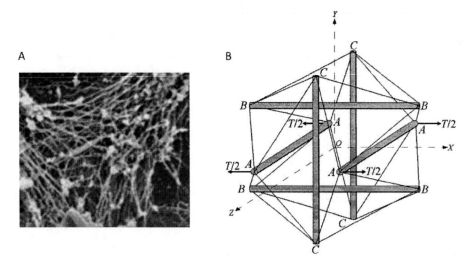

Figure 7. Field emission scanning electron microscopy image showing (A) a loose cytoplasmic meshwork of 10 nm and 25 nm filaments and (B) a schematic of a tensegrity based approach to the computational modelling of the cell cytoskeleton, where the cytoskeleton was considered as 6 struts interconnected with 24 elastic cables, representing the tubules and filaments respectively. Image (A) was reproduced from Tanaka-Kamioka et al [96] with permission from Wiley Publishers. Image (B) was reproduced from Stamenovic and Coughlin [93] with permission from ASME.

When uni-axially stretched they showed that the tubules carry the initial compression that exceeds the critical buckling force, but is much smaller than the yield force of the actin filaments. Shafrir and Forgacs [94], using a similar "rod and spring" approach for representing the cytoskeleton, placed the mesh in a viscous cytoplasm; an approach combining the tensegrity and continuum models. They showed that changing the ratio of linker molecules and associated elastic characteristics, the efficiency of mechanotransduction from perturbations at the cell surface, to the cell nucleus, could be altered. McGarry and Prendergast [95] developed a FEM model of an adherent cell using a combination of the models discussed previously. Numerous structural components were considered as individually significant and given separate material properties accordingly; the pre-stressed cytoskeleton, cytoplasm, nucleus and membrane components. The model was sufficiently complex to capture the non-linear behaviour of a cell in response to forces, highlighting the role of the cytoskeleton in stiffening a cell during the process of cell spreading when applying forces to five increasingly spread cell geometries, and revealing the influence of the cytoplasm in terms of its material properties (elasticity and compressibility) on cellular stiffness. Stamenovic and Coughlin [93] computationally modelled the cytoskeleton as 6 struts, interconnected with 24 elastic cables, representing the tubules and filaments respectively, to quantitatively assess the steady state elastic response of cells (Figure 7).

PREDICTIVE COMPUTATIONAL MODELLING OF DIFFERENTIATION BASED ON DETERMINISTIC FUNCTIONS

In this chapter we have discussed how computational modelling has been used in tissue engineering to calculate the magnitude of biomechanical forces present at the micro-scale and how these forces are potentially transmitted through the cell yielding a biological response. Such studies provide investigators utilising different experimental systems with a means to correlate results observed experimentally to defined system parameters, thereby aiding optimisation of experimental set-ups to achieve particular cellular responses. An understanding of this relationship between the mechanical stimuli (type and magnitude) and subsequent biological response has lead to the development of the next phase of computational models; models that use deterministic functions in association with force data to predict the temporal and spacial distribution of tissue formation. This approach was initially applied to fracture healing in-vivo with several mechanoregulation algorithms proposed to simulate tissue differentiation during normal fracture healing accurately, predicting temporal and spatial tissue distributions [97][98][99][100][101][102]. The underlining hypothesis for tissue-differentiation in the models was developed by comparing predicted levels of strain of fluid flow in regenerating tissue to histological patterns of tissue differentiation[103]. This approach has more recently been translated for the purpose of predicting population dynamics (how a group of cells will behave in response to a given set of conditions) or spatial and temporal tissue differentiation within highly porous scaffolds in-vitro [104][105][106][107][108][109][110] and in-vivo[105]. Stops et al [109] using a μCT reconstruction of a sub-volume of a collagen-GAG scaffold used a FEEBE/linear solver (an in-house Intels FORTRAN90 small-strain FE element-by-element solver) to simulate uni-axial compression of the Collagen-GAG scaffold and ANSYSs CFX to solve the governing equations relating to the fluid domain. Cells were modelled as a 2D virtual area of triangular line segments with lengths of 50 μm. From standard FE equations that govern 3-noded membrane deformation, in-plane octahedral shear strains were computed for each virtual cell from the scaffold nodal displacements calculated in the FE simulation. The fluid velocities experienced by the cell were determined from a volume (60 μm x 60 μm x 60 mm) around the cell. These values were then analysed by a purpose-written Intels FORTRAN90 code defining a cell mechanoregulatory algorithm, which predicted the fate of MSC cells and the specialised cells they divided into for a range of strains between 1-5% and fluid velocities between 1-10 μm s^{-1}. MSC differentiation patterns indicated that specific combinations of scaffold strains and inlet fluid flows cause phenotype assemblies dominated by single cell types with bone predominating under conditions of 1% scaffold strain and 1 μm s^{-1} inlet velocity. However, this model is yet to be validated experimentally and will no doubt require further refinement in light of experimental findings, but in the meantime shows the potential, once clear relationships are derived between environmental cues and biological outcomes, for translating efforts from an experimental platform to a computational platform, allowing the assessment of novel scaffolds with different architectures relatively quickly and cost effectively.

CONCLUSION

Whilst an understanding of the forces applied at the macro-scale is achievable experimentally for highly porous scaffolds that form the basis for tissue growth in-vitro, the micro-scale environment remains a black-box. Computational modelling has shown potential in granting researchers a means of exploring this black-box, allowing quantification of the heterogeneity existing at the micro-scale for key mechano-regulatory factors (mechanical strain and shear stress) in response to the uniform forces applied at the macro-scale. These findings have provided insight for the heterogeneous tissue formation observed experimentally as a function of changes in scaffold architecture and bioreactor operating conditions. Furthermore, computational modelling is helping to acquire an appreciation of the dynamics of cell-scaffold interactions and the transmission of intra-cellular forces; time will tell whether the use of bioreactors in will prove most beneficial as a means of creating tissue engineered substitutes in-vitro or as a model to investigate mechano-biology. As with any modelling approach the key is experimental validation, which will no doubt be forthcoming as technological advances allow experimental measurement of currently immeasurable variables, but raises the question, will the tools that are used to validate the model subsequently be capable of providing as much insight as the model itself? The key is to use the data being derived from these models to feedback into the design of scaffold architectures and bioreactor conditions to improve the quality and homogeneity of in-vitro engineered tissue substitutes. The increasing importance of computational modelling in the tissue engineering field can be recognised by the continued effort to supply tools to researchers in the field via collaborations between providers of imaging platforms, specialised image reconstruction software companies and providers of commercial computational modelling software. The future now relies on having collaborations between the fields of computer science and tissue engineering that will generate individuals with an in depth understanding of the cell biology and computational methods needed to execute the successful synergy between these two fields and ensure the success of tissue engineered substitutes.

REFERENCES

[1] Dopico-González C, New AM, Browne M. Probabilistic analysis of an uncemented total hip replacement. *Medical Engineering and Physics*. 2009 May;31(4):470-476.

[2] Pastrav L, Devos J, Van der Perre G, Jaecques S. A finite element analysis of the vibrational behaviour of the intra-operatively manufactured prosthesis-femur system. *Medical Engineering and Physics*. 2009 May;31(4):489-494.

[3] Rohlmann A, Mössner U, Bergmann G, Kölbel R. Finite-element-analysis and experimental investigation in a femur with hip endoprosthesis. *Journal of Biomechanics*. 1983;16(9):727-742.

[4] Zhao D, Moritz N, Laurila P, Mattila R, Lassila L, Strandberg N, et al. Development of a multi-component fiber-reinforced composite implant for load-sharing conditions. *Medical Engineering and Physics*. 2009 May;31(4):461-469.

[5] Fregly BJ, Sawyer WG, Harman MK, Banks SA. Computational wear prediction of a total knee replacement from in vivo kinematics. *Journal of Biomechanics*. 2005 Feb;38(2):305-314.

[6] Godest AC, Beaugonin M, Haug E, Taylor M, Gregson PJ. Simulation of a knee joint replacement during a gait cycle using explicit finite element analysis. *Journal of Biomechanics*. 2002 Feb;35(2):267-275.

[7] Halloran JP, Petrella AJ, Rullkoetter PJ. Explicit finite element modeling of total knee replacement mechanics. *Journal of Biomechanics*. 2005 Feb;38(2):323-331.

[8] Haut Donahue T, Dehlin W, Gillespie J, Weiss W, Rosenberg G. Finite element analysis of stresses developed in the blood sac of a left ventricular assist device. *Medical Engineering and Physics*. 2009 May;31(4):454-460.

[9] Hochareon P, Manning KB, Fontaine AA, Tarbell JM, Deutsch S. Fluid dynamic analysis of the 50 cc Penn State artificial heart under physiological operating conditions using particle image velocimetry. *J Biomech Eng*. 2004 Oct;126(5):585-593.

[10] Medvitz RB, Kreider JW, Manning KB, Fontaine AA, Deutsch S, Paterson EG. Development and validation of a computational fluid dynamics methodology for simulation of pulsatile left ventricular assist devices. *ASAIO J*. 2007 Apr;53(2):122-131.

[11] Bedoya J, Meyer CA, Timmins LH, Moreno MR, Moore JE. Effects of stent design parameters on normal artery wall mechanics. *J. Biomech. Eng*. 2006 Oct;128(5):757-765.

[12] De Beule M, Mortier P, Carlier SG, Verhegghe B, Van Impe R, Verdonck P. Realistic finite element-based stent design: The impact of balloon folding. *Journal of Biomechanics*. 2008;41(2):383-389.

[13] De Beule M, Van Cauter S, Mortier P, Van Loo D, Van Impe R, Verdonck P, et al. Virtual optimization of self-expandable braided wire stents. *Medical Engineering and Physics*. 2009 May;31(4):448-453.

[14] Migliavacca F, Petrini L, Colombo M, Auricchio F, Pietrabissa R. Mechanical behavior of coronary stents investigated through the finite element method. *Journal of Biomechanics*. 2002 Jun;35(6):803-811.

[15] Migliavacca F, Petrini L, Montanari V, Quagliana I, Auricchio F, Dubini G. A predictive study of the mechanical behaviour of coronary stents by computer modelling. *Medical Engineering and Physics*. 2005 Jan;27(1):13-18.

[16] Mortier P, De Beule M, Carlier SG, Van Impe R, Verhegghe B, Verdonck P. Numerical study of the uniformity of balloon-expandable stent deployment. *J. Biomech. Eng*. 2008 Apr;130(2):021018.

[17] Mortier P, De Beule M, Van Loo D, Verhegghe B, Verdonck P. Finite element analysis of side branch access during bifurcation stenting. *Medical Engineering and Physics*. 2009 May;31(4):434-440.

[18] Eloot S, Wachter DD, Tricht IV, Verdonck P. Computational Flow Modeling in Hollow-Fiber Dialyzers. *Artificial Organs*. 2002;26(7):590-599.

[19] Yannas I, Burke J, Orgill D, Skrabut E. Wound tissue can utilize a polymeric template to synthesize a functional extension of skin. *Science*. 1982 Jan 8;215(4529):174-176.

[20] Yannas IV, Lee E, Orgill DP, Skrabut EM, Murphy GF. Synthesis and characterization of a model extracellular matrix that induces partial regeneration of adult mammalian skin. *Proceedings of the National Academy of Sciences of the United States of America*. 1989 Feb;86(3):933-937.

[21] Yannas IV, Tzeranis DS, Harley BA, So PTC. Biologically active collagen-based scaffolds: advances in processing and characterization. *Philosophical Transactions of the Royal Society A: Mathematical, Physical and Engineering Sciences*. 2010 Apr 28;368(1917):2123-2139.

[22] Freyman TM, Yannas IV, Gibson LJ. Cellular materials as porous scaffolds for tissue engineering. *Progress in Materials Science*. 2001;46(3-4):273-282.

[23] Pieper JS, Oosterhof A, Dijkstra PJ, Veerkamp JH, van Kuppevelt TH. Preparation and characterization of porous crosslinked collagenous matrices containing bioavailable chondroitin sulphate. *Biomaterials.* 1999 May;20(9):847-858.

[24] Pieper JS, van Wachem PB, van Luyn MJA, Brouwer LA, Hafmans T, Veerkamp JH, et al. Attachment of glycosaminoglycans to collagenous matrices modulates the tissue response in rats. *Biomaterials.* 2000 Aug;21(16):1689-1699.

[25] Docherty R, Forrester JV, Lackie JM, Gregory DW. Glycosaminoglycans facilitate the movement of fibroblasts through three-dimensional collagen matrices. *J. Cell. Sci.* 1989 Feb;92 (Pt 2):263-270.

[26] Burr DB, Robling AG, Turner CH. Effects of biomechanical stress on bones in animals. *Bone.* 2002 May;30(5):781-786.

[27] Turner C, Forwood M, Otter M. Mechanotransduction in bone: do bone cells act as sensors of fluid flow? *FASEB J.* 1994 Aug 1;8(11):875-878.

[28] Bikle DD, Halloran BP. The response of bone to unloading. *Journal of Bone and Mineral Metabolism.* 1999 Nov 20;17(4):233-244.

[29] Morey E, Baylink D. Inhibition of bone formation during space flight. *Science.* 1978 Sep 22;201(4361):1138-1141.

[30] Zerwekh JE, Ruml LA, Gottschalk F, Pak CYC. The effects of twelve weeks of bed rest on bone histology, biochemical markers of bone turnover, and calcium homeostasis in eleven normal subjects. *Journal of Bone and Mineral Research.* 1998;13(10):1594-1601.

[31] Burr DB, Milgrom C, Fyhrie D, Forwood M, Nyska M, Finestone A, et al. In vivo measurement of human tibial strains during vigorous activity. *Bone.* 1996 May;18(5):405-410.

[32] Cullen DM, Smith RT, Akhter MP. Bone-loading response varies with strain magnitude and cycle number. *J. Appl. Physiol.* 2001 Nov 1;91(5):1971-1976.

[33] Turner CH, Akhter MP, Raab DM, Kimmel DB, Recker RR. A noninvasive, in vivo model for studying strain adaptive bone modeling. *Bone.* 1991;12(2):73-79.

[34] Miles RR, Turner CH, Santerre R, Tu Y, Mcclelland P, Argot J, et al. Analysis of differential gene expression in rat tibia after an osteogenic stimulus in vivo: Mechanical loading regulates osteopontin and myeloperoxidase. *Journal of Cellular Biochemistry.* 1998;68(3):355-365.

[35] van Rietbergen B, Majumdar S, Pistoia W, Newitt DC, Kothari M, Laib A, et al. Assessment of cancellous bone mechanical properties from micro-FE models based on micro-CT, pQCT and MR images. *Technol Health Care.* 1998 Dec;6(5-6):413-420.

[36] Akhter M, Raab D, Turner C, Kimmel D, Recker R. Characterization of in vivo strain in the rat tibia during external application of a four-point bending load. *Journal of Biomechanics.* 1992 Oct;25(10):1241-1246.

[37] Van Rietbergen B, Huiskes R, Eckstein F, Ruegsegger P. Trabecular Bone Tissue Strains in the Healthy and Osteoporotic Human Femur. *Journal of Bone and Mineral Research.* 2003 10;18(10):1781-1788.

[38] Turner CH, Anne V, Pidaparti RMV. A uniform strain criterion for trabecular bone adaptation: Do continuum-level strain gradients drive adaptation? *Journal of Biomechanics.* 1997 Jun;30(6):555-563.

[39] Koike M, Shimokawa H, Kanno Z, Ohya K, Soma K. Effects of mechanical strain on proliferation and differentiation of bone marrow stromal cell line ST2. *Journal of Bone and Mineral Metabolism.* 2005 May 1;23(3):219-225.

[40] Neidlinger-Wilke C, Wilke H, Claes L. Cyclic stretching of human osteoblasts affects proliferation and metabolism: A new experimental method and its application. *Journal of Orthopaedic Research.* 1994;12(1):70-78.

[41] You J, Yellowley CE, Donahue HJ, Zhang Y, Chen Q, Jacobs CR. Substrate deformation levels associated with routine physical activity are less stimulatory to bone cells relative to loading-induced oscillatory fluid flow. *J. Biomech. Eng.* 2000;122(4):387.

[42] Jagodzinski M, Drescher M, Zeichen J, Hankemeier S, Krettek C, Bosch U, et al. Effects of cyclic longitudinal mechanical strain and dexamethasone on osteogenic differentiation of human bone marrow stromal cells. *Eur. Cell Mater.* 2004 Apr 16;7:35-41; discussion 41.

[43] Brighton C, Strafford B, Gross S, Leatherwood D, Williams J, Pollack S. The proliferative and synthetic response of isolated calvarial bone cells of rats to cyclic biaxial mechanical strain. *J. Bone Joint Surg. Am.* 1991 Mar 1;73(3):320-331.

[44] Thomas G, El Haj A. Bone marrow stromal cells are load responsivein vitro. *Calcified Tissue International.* 1996 Feb 1;58(2):101-108.

[45] Sumanasinghe RD, Bernacki SH, Loboa EG. Osteogenic differentiation of human mesenchymal stem cells in collagen matrices: effect of uniaxial cyclic tensile strain on bone morphogenetic protein (BMP-2) mRNA expression. *Tissue Eng.* 2006 Dec;12(12):3459-3465.

[46] Klein-Nulend J, Roelofsen J, Semeins CM, Bronckers ALJJ, Burger EH. Mechanical stimulation of osteopontin mRNA expression and synthesis in bone cell cultures. *Journal of Cellular Physiology.* 1997;170(2):174-181.

[47] Murray D, Rushton N. The effect of strain on bone cell prostaglandin E2 release: A new experimental method. *Calcified Tissue International.* 1990 Jul 1;47(1):35-39.

[48] Jessop HL, Rawlinson SCF, Pitsillides AA, Lanyon LE. Mechanical strain and fluid movement both activate extracellular regulated kinase (ERK) in osteoblast-like cells but via different signaling pathways. *Bone.* 2002 Jul;31(1):186-194.

[49] Wozniak M, Fausto A, Carron CP, Meyer DM, Hruska KA. Mechanically strained cells of the osteoblast lineage organize their extracellular matrix through unique sites of alpha-1beta-1'-integrin expression. *Journal of Bone and Mineral Research.* 2000;15(9):1731-1745.

[50] Turner CH, Owan I, Takano Y. Mechanotransduction in bone: role of strain rate. *Am J Physiol Endocrinol Metab.* 1995 Sep 1;269(3):E438-442.

[51] Martin I, Wendt D, Heberer M. The role of bioreactors in tissue engineering. *Trends in Biotechnology.* 2004 Feb;22(2):80-86.

[52] Bancroft GN, Sikavitsas VI, van den Dolder J, Sheffield TL, Ambrose CG, Jansen JA, et al. Fluid flow increases mineralized matrix deposition in 3D perfusion culture of marrow stromal osteoblasts in a dose-dependent manner. *Proceedings of the National Academy of Sciences of the United States of America.* 2002 Oct 1;99(20):12600-12605.

[53] Goldstein AS, Juarez TM, Helmke CD, Gustin MC, Mikos AG. Effect of convection on osteoblastic cell growth and function in biodegradable polymer foam scaffolds. *Biomaterials.* 2001 Jun 1;22(11):1279-1288.

[54] Ishaug SL, Crane GM, Miller MJ, Yasko AW, Yaszemski MJ, Mikos AG. Bone formation by three-dimensional stromal osteoblast culture in biodegradable polymer scaffolds. *Journal of Biomedical Materials Research*. 1997;36(1):17-28.

[55] Martin I, Obradovic B, Freed LE, Vunjak-Novakovic G. Method for quantitative analysis of glycosaminoglycan distribution in cultured natural and engineered cartilage. *Annals of Biomedical Engineering*. 1999;27(5):656-662.

[56] Bonvin C, Overney J, Shieh AC, Dixon JB, Swartz MA. A multichamber fluidic device for 3D cultures under interstitial flow with live imaging: Development, characterization, and applications. *Biotechnology and Bioengineering*. 2010;105(5):982-991.

[57] Jaasma MJ, Plunkett NA, O'Brien FJ. Design and validation of a dynamic flow perfusion bioreactor for use with compliant tissue engineering scaffolds. *Journal of Biotechnology*. 2008 Feb 29;133(4):490-496.

[58] Zhao F, Ma T. Perfusion bioreactor system for human mesenchymal stem cell tissue engineering: Dynamic cell seeding and construct development. *Biotechnology and Bioengineering*. 2005;91(4):482-493.

[59] Wendt D, Jakob M, Martin I. Bioreactor-based engineering of osteochondral grafts: from model systems to tissue manufacturing. *Journal of Bioscience and Bioengineering*. 2005 Nov;100(5):489-494.

[60] Mueller SM, Mizuno S, Gerstenfeld LC, Glowacki J. Medium perfusion enhances osteogenesis by murine osteosarcoma cells in three-dimensional collagen sponges. *Journal of Bone and Mineral Research*. 1999;14(12):2118-2126.

[61] Sikavitsas VI, Bancroft GN, Holtorf HL, Jansen JA, Mikos AG. Mineralized matrix deposition by marrow stromal osteoblasts in 3D perfusion culture increases with increasing fluid shear forces. *Proceedings of the National Academy of Sciences of the United States of America*. 2003 Dec 9;100(25):14683-14688.

[62] Cartmell, S., Porter, B.D, García, A.J., Guldberg, R.E. Effects of medium perfusion rate on cell-seeded three-dimensional bone constructs in vitro. *Tissue Engineering*. 2003 Dec;9(6):1197.

[63] Zhao F, Chella R, Ma T. Effects of shear stress on 3-D human mesenchymal stem cell construct development in a perfusion bioreactor system: Experiments and hydrodynamic modeling. *Biotechnology and Bioengineering*. 2007;96(3):584-595.

[64] Jaasma MJ, O'Brien FJ. Mechanical stimulation of osteoblasts using steady and dynamic fluid flow. *Tissue Eng Part A*. 2008 Jul;14(7):1213-1223.

[65] Plunkett NA, Partap S, O'Brien FJ. Osteoblast response to rest periods during bioreactor culture of collagen-glycosaminoglycan scaffolds. *Tissue Eng Part A*. 2010 Mar;16(3):943-951.

[66] Vance J, Galley S, Liu DF, Donahue SW. Mechanical stimulation of MC3T3 osteoblastic cells in a bone tissue-engineering bioreactor enhances prostaglandin E2 release. *Tissue Eng*. 2005 Dec;11(11-12):1832-1839.

[67] Partap S, Plunkett N, Kelly D, O'Brien F. Stimulation of osteoblasts using rest periods during bioreactor culture on collagen-glycosaminoglycan scaffolds. *J. Mater Sci: Mater Med*. 2009 12;21(8):2325-2330.

[68] Jungreuthmayer C, Donahue SW, Jaasma MJ, Al-Munajjed AA, Zanghellini J, Kelly DJ, et al. A comparative study of shear stresses in collagen-glycosaminoglycan and

calcium phosphate scaffolds in bone tissue-engineering bioreactors. *Tissue Eng. Part A.* 2009 May;15(5):1141-1149.

[69] O'Brien FJ, Harley BA, Yannas IV, Gibson LJ. The effect of pore size on cell adhesion in collagen-GAG scaffolds. *Biomaterials.* 2005 Feb;26(4):433-441.

[70] O'Brien FJ, Harley BA, Waller MA, Yannas IV, Gibson LJ, Prendergast PJ. The effect of pore size on permeability and cell attachment in collagen scaffolds for tissue engineering. Technology and Health *Care.* 2007 Jan 1;15(1):3-17.

[71] Harley BA, Kim H, Zaman MH, Yannas IV, Lauffenburger DA, Gibson LJ. Microarchitecture of three-dimensional scaffolds influences cell migration behavior via junction interactions. *Biophysical Journal.* 2008 Oct 15;95(8):4013-4024.

[72] Jungreuthmayer C, Jaasma M, Al-Munajjed A, Zanghellini J, Kelly D, O'Brien F. Deformation simulation of cells seeded on a collagen-GAG scaffold in a flow perfusion bioreactor using a sequential 3D CFD-elastostatics model. *Medical Engineering and Physics.* 2009 May;31(4):420-427.

[73] Alberich-Bayarri A, Moratal D, Ivirico JLE, Hernández JCR, Vallés-Lluch A, Martí-Bonmatí L, et al. Microcomputed tomography and microfinite element modeling for evaluating polymer scaffolds architecture and their mechanical properties. *Journal of Biomedical Materials Research Part B: Applied Biomaterials.* 2009;91B(1):191-202.

[74] Stops A, Harrison N, Haugh M, O'Brien F, McHugh P. Local and regional mechanical characterisation of a collagen-glycosaminoglycan scaffold using high-resolution finite element analysis. *Journal of the Mechanical Behavior of Biomedical Materials.* 2010 May;3(4):292-302.

[75] Lacroix D, Chateau A, Ginebra M, Planell JA. Micro-finite element models of bone tissue-engineering scaffolds. *Biomaterials.* 2006 Oct;27(30):5326-5334.

[76] Stops AJF, McMahon LA, O'Mahoney D, Prendergast PJ, McHugh PE. A finite element prediction of strain on cells in a highly porous collagen-glycosaminoglycan scaffold. *J. Biomech. Eng.* 2008 Dec;130(6):061001.

[77] McMahon L. The effect of cylic tensile loading and growth factors on the chondrogenic differentiation of bone-marrow derived mesenchymal stem cells in a collagen-glycosaminoglycan scaffold. 2007.

[78] Porter B, Zauel R, Stockman H, Guldberg R, Fyhrie D. 3-D computational modeling of media flow through scaffolds in a perfusion bioreactor. *Journal of Biomechanics.* 2005 Mar;38(3):543-549.

[79] Voronov R, VanGordon S, Sikavitsas VI, Papavassiliou DV. Computational modeling of flow-induced shear stresses within 3D salt-leached porous scaffolds imaged via micro-CT. *Journal of Biomechanics.* 2010 May 7;43(7):1279-1286.

[80] Cioffi M, Boschetti F, Raimondi MT, Dubini G. Modeling evaluation of the fluid-dynamic microenvironment in tissue-engineered constructs: A micro-CT based model. *Biotechnology and Bioengineering.* 2006;93(3):500-510.

[81] Cioffi M, Küffer J, Ströbel S, Dubini G, Martin I, Wendt D. Computational evaluation of oxygen and shear stress distributions in 3D perfusion culture systems: Macro-scale and micro-structured models. *Journal of Biomechanics.* 2008 Oct 20;41(14):2918-2925.

[82] Maes F, Ransbeeck PV, Oosterwyck HV, Verdonck P. Modeling fluid flow through irregular scaffolds for perfusion bioreactors. *Biotechnology and Bioengineering.* 2009;103(3):621-630.

[83] Raimondi MT, Moretti M, Cioffi M, Giordano C, Boschetti F, Laganà K, et al. The effect of hydrodynamic shear on 3D engineered chondrocyte systems subject to direct perfusion. *Biorheology.* 2006 Jan 1;43(3):215-222.

[84] Sandino C, Planell J, Lacroix D. A finite element study of mechanical stimuli in scaffolds for bone tissue engineering. *Journal of Biomechanics.* 2008;41(5):1005-1014.

[85] Hoffman BD, Crocker JC. Cell mechanics: Dissecting the physical responses of cells to force. *Annu. Rev. Biomed. Eng.* 2009 8;11(1):259-288.

[86] Janmey PA, McCulloch CA. Cell mechanics: Integrating cell responses to mechanical stimuli. *Annu. Rev. Biomed. Eng.* 2007 8;9(1):1-34.

[87] Dong C, Skalak R, Sung KL. Cytoplasmic rheology of passive neutrophils. *Biorheology.* 1991;28(6):557-567.

[88] Tsai MA, Frank RS, Waugh RE. Passive mechanical behavior of human neutrophils: power-law fluid. *Biophys J.* 1993;65(5):2078-2088.

[89] Karcher H, Lammerding J, Huang H, Lee RT, Kamm RD, Kaazempur-Mofrad MR. A Three-Dimensional Viscoelastic Model for Cell Deformation with Experimental Verification. *Biophysical Journal.* 2003 Nov;85(5):3336-3349.

[90] Ferko M, Bhatnagar A, Garcia M, Butler P. Finite-Element Stress Analysis of a Multicomponent Model of Sheared and Focally-Adhered Endothelial Cells. *Annals of Biomedical Engineering.* 2007 Feb 1;35(2):208-223.

[91] Ingber DE. Tensegrity: the architectural basis of cellular mechanotransduction. *Annu. Rev. Physiol.* 1997;59:575-599.

[92] Wang N, Naruse K, Stamenović D, Fredberg JJ, Mijailovich SM, Tolić-Nørrelykke IM, et al. Mechanical behavior in living cells consistent with the tensegrity model. *Proceedings of the National Academy of Sciences of the United States of America.* 2001 Jul 3;98(14):7765-7770.

[93] Stamenović D, Coughlin MF. A quantitative model of cellular elasticity based on tensegrity. *J. Biomech. Eng.* 2000 Feb;122(1):39-43.

[94] Shafrir Y, Forgacs G. Mechanotransduction through the cytoskeleton. *Am. J. Physiol. Cell Physiol.* 2002 Mar 1;282(3):C479-486.

[95] McGarry JG, Prendergast PJ. A three-dimensional finite element model of an adherent eukaryotic cell. *Eur. Cell Mater.* 2004 Apr 16;7:27-33; discussion 33-34.

[96] Tanaka-Kamioka K, Kamioka H, Ris H, Lim SS. Osteocyte shape is dependent on actin filaments and osteocyte processes are unique actin-rich projections. *J. Bone Miner. Res.* 1998 Oct;13(10):1555-1568.

[97] Carter DR, Blenman PR, Beaupré GS. Correlations between mechanical stress history and tissue differentiation in initial fracture healing. *J. Orthop. Res.* 1988;6(5):736-748.

[98] Claes L, Heigele C. Magnitudes of local stress and strain along bony surfaces predict the course and type of fracture healing. *Journal of Biomechanics.* 1999 Mar;32(3):255-266.

[99] Lacroix D, Prendergast PJ. A mechano-regulation model for tissue differentiation during fracture healing: analysis of gap size and loading. *Journal of Biomechanics.* 2002 Sep;35(9):1163-1171.

[100] Lacroix D, Prendergast P, Li G, Marsh D. Biomechanical model to simulate tissue differentiation and bone regeneration: Application to fracture healing. *Medical and Biological Engineering and Computing.* 2002 Jan 12;40(1):14-21.

[101] Isaksson H, Donkelaar CCV, Huiskes R, Ito K. Corroboration of mechanoregulatory algorithms for tissue differentiation during fracture healing: comparison with in vivo results. *Journal of Orthopaedic Research*. 2006;24(5):898-907.

[102] Nagel T, Kelly DJ. Mechano-regulation of mesenchymal stem cell differentiation and collagen organisation during skeletal tissue repair. *Biomech Model Mechanobiol*. 2009 12;9(3):359-372.

[103] Prendergast PJ, Huiskes R, Søballe K. Biophysical stimuli on cells during tissue differentiation at implant interfaces. *Journal of Biomechanics*. 1997 Jun;30(6):539-548.

[104] Byrne DP, Lacroix D, Planell JA, Kelly DJ, Prendergast PJ. Simulation of tissue differentiation in a scaffold as a function of porosity, Young's modulus and dissolution rate: Application of mechanobiological models in tissue engineering. *Biomaterials*. 2007 Dec;28(36):5544-5554.

[105] Kelly DJ, Prendergast PJ. Prediction of the optimal mechanical properties for a scaffold used in osteochondral defect repair. *Tissue Eng*. 2006 Sep;12(9):2509-2519.

[106] Galbusera F, Cioffi M, Raimondi MT, Pietrabissa R. Computational modeling of combined cell population dynamics and oxygen transport in engineered tissue subject to interstitial perfusion. *Computer Methods in Biomechanics and Biomedical Engineering*. 2007;10(4):279.

[107] Tang L, Youssef B. A 3-D Computational Model for Multicellular Tissue Growth [Internet]. In: Biomedical Simulation. 2006 [cited 2010 Jul 2]. p. 29-39.Available from: http://dx.doi.org.elib.tcd.ie/10.1007/11790273_4.

[108] Flaibani M, Magrofuoco E, Elvassore N. Computational Modeling of Cell Growth Heterogeneity in a Perfused 3D Scaffold. *Industrial and Engineering Chemistry Research*. 2010 Jan 20;49(2):859-869.

[109] Stops A, Heraty K, Browne M, O'Brien F, McHugh P. A prediction of cell differentiation and proliferation within a collagen-glycosaminoglycan scaffold subjected to mechanical strain and perfusive fluid flow. *Journal of Biomechanics*. 2010 Mar 3;43(4):618-626.

[110] Olivares AL, Marsal È, Planell JA, Lacroix D. Finite element study of scaffold architecture design and culture conditions for tissue engineering. *Biomaterials*. 2009 Oct;30(30):6142-6149.

In: Computational Fluid Dynamics
Editor: Alyssa D. Murphy

ISBN 978-1-61209-276-8
© 2011 Nova Science Publishers, Inc.

Chapter 8

MODELLING AND SIMULATION OF AN ELECTROSTATIC PRECIPITATOR

Shah M. E. Haque[1], M. G. Rasul[2] and M. M. K. Khan[2]*

[1]Process Engineering and Light Metals (PELM) Centre, Faculty of Sciences, Engineering and Health, CQUniversity Gladstone, Queensland 4680, Australia
[2]School of Engineering and Built Environment, Faculty of Sciences, Engineering and Health, CQUniversity, Rockhampton, Queensland 4702, Australia

INTRODUCTION

Electrostatic precipitators (ESPs) are the most common, effective and reliable particulate control devices that are mainly used in power plants and other process industries. The ESP works as a cleaning device and uses electrical forces to separate the dust particles from the flue gas. A typical ESP consists of an inlet diffuser, known as inlet evase, a rectangular collection chamber and an outlet convergent duct, known as outlet evase. Perforated plates are placed inside both the inlet and outlet evase for the purpose of flow distribution. Inside the collection chamber are placed a number of discharge electrodes (DEs) and collection electrodes (CEs). Discharge electrodes are suspended vertically between two collection electrodes. While the flue gas flows through the collection area, electrostatic precipitators accomplish particle separation by the use of an electric field in three steps. At first it imparts positive or negative charges to the particles by discharge electrodes. In the second step it attracts the charged particles to oppositely charged or grounded collection electrodes. And finally it removes the collected particles by vibrating or rapping the collection electrodes or by spraying with liquid.

This chapter presents a Computational Fluid Dynamics (CFD) model for an ESP. An ESP system consists of flow field, electrostatic field and particle dynamics. But they cannot be simultaneously applied to the modelling of an industrial ESP. The main reason for this is the lack of computational resources. A one-step algorithm incorporating all the above systems would require excessive computational memory and unacceptably long calculation time.

---

* Corresponding author: Tel: + 61 7 49309676; Fax: + 61 7 49309382, E-mail address: m.rasul@cqu.edu.au

Therefore, the ESP was modelled in two steps. Firstly, the 3D fluid (air) flow was modelled considering the detailed geometrical configuration inside the ESP. The model was then validated with the measured data obtained from the geometrically similar laboratory experiments. Numerical calculations for the gas flow were carried out by solving the Reynolds-averaged Navier-Stokes equations, and turbulence was modelled using the realizable k-ε turbulence model. In the second step, as the complete ESP system consists of an electric field and a particle phase in addition to the fluid flow field, a two dimensional ESP model was developed. An additional source term was added to the gas flow equation to capture the effect of the electric field. This additional source term was obtained by solving a coupled system of the electric field and charge transport equations. The electrostatic force was applied to the flow equations by using User Defined Functions (UDFs). A discrete phase model (DPM) was incorporated with this 2D model to study the effects of particle size, electric field and flue gas flow on the collection efficiency of particles inside the ESP.

The CFD model thus developed was successfully applied to a prototype ESP at the power plant and used to recommend options for improving the efficiency of the ESP. The aerodynamic behaviour of the flow was improved by geometrical modifications in the existing 3D numerical model. In particular, the simulation was performed to improve and optimize the flow in order to achieve uniform flow and to increase particle collection inside the ESP. The particles injected in the improved flow condition were collected with higher efficiency after increasing the electrostatic force inside the 2D model. The approach adopted in this chapter to optimize flow and electrostatic field properties is a novel approach for improving the performance of an electrostatic precipitator.

FLOW FIELD MODELLING

A detailed numerical simulation method and an approach adopted to predict the flow pattern inside a laboratory scale ESP are presented in this section. The FLUENT Inc. geometry and mesh generation software "GAMBIT 2.3.16" was used as a pre-processor to create the geometry, discretise the fluid domain into small cells to form a volume mesh or grid and set up the appropriate boundary conditions. The flow properties were then specified, the equations were solved and the results were analysed by "FLUENT 6.2.16" solver. A limited number of research works could be found in the open literature for the prediction of turbulent flow behaviour inside the ESP. Most were focused on 2D models based on simplified geometrical arrangements and ignored the effect of sudden expansion in geometrical configuration of an ESP. Most of the research (Zhao *et al.*, 2006; Skodras *et al.*, 2006; Schwab and Johnson, 1994; Nikas *et al.*, 2005; Varonos *et al.*, 2002) assumed an ideal (uniform) velocity distribution at the inlet boundary condition. The accurate aerodynamic characteristics of the flow inside an ESP during operation may not be obtained without considering all its major physical details and the real velocity distribution at the inlet boundary. Limited research (Yang *et al.*, 2006; Ojha *et al.*, 2001; Xia *et al.*, 1997; Ikeda *et al.*, 1992) exists in the literature that considered the influence of inlet conditions on the simulation results of the flow in the duct, pipe or combustor. However, no research on ESPs was reported where the experimentally measured inlet velocity profile was applied for an accurate prediction of flow behaviour inside a real ESP.

The novel nature of this study is to develop a new 3D fluid flow model of a full scale ESP which considers all its major physical features. In this study all the collecting electrodes (CEs), baffles, gas deflectors etc. are taken into account in this 3D model, and have not been replaced by any equivalent porous region as other researchers have done in their studies. A detailed numerical approach and simulation procedure is presented to predict the flow behaviour inside the ESP after adopting a novel technique that overcomes the previous assumptions on the velocity inlet boundary condition of an ESP model. The predicted results are compared with the measured data to validate the numerical model. The flow model developed has the potential to better predict the effect of possible modifications and improvements in ESP design.

Governing Equations

Numerical computation of fluid transport includes conservation of mass, momentum and turbulence model equations. The air inside the laboratory ESP was treated as an incompressible Newtonian fluid due to the small pressure drop across it. The flow was assumed to be steady and can be described by the conservation of mass equation:

$$\frac{\partial \rho}{\partial t} + \vec{\nabla} \cdot (\rho \vec{U}) = 0 \qquad (1)$$

and the momentum equation:

$$\frac{\partial \vec{U}}{\partial t} + \vec{U} \cdot \vec{\nabla} \vec{U} = -\frac{\vec{\nabla} p}{\rho} + \nu \vec{\nabla}^2 \vec{U} + \vec{g} \qquad (2)$$

For the turbulent flow inside the ESP, the key to the success of CFD lies with the accurate description of the turbulent behaviour of the flow. To model the turbulent flow in an ESP, there are a number of k-ε turbulence models available in FLUENT. The realizable k-ε model is used in this study because this model contains a new formulation for the turbulent viscosity and a new transport equation for the dissipation rate, ε. The governing equations of the realizable k-ε model can be written as follows (FLUENT 6.2 User's Guide, 2005):

$$\frac{\partial}{\partial t}(\rho k) + \frac{\partial}{\partial x_j}(\rho k u_j) = \frac{\partial}{\partial x_j}[(\mu + \frac{\mu_t}{\sigma_k})\frac{\partial k}{\partial x_j}] + G_k + G_b - \rho\varepsilon - Y_M + S_k \qquad (3)$$

$$\frac{\partial}{\partial t}(\rho\varepsilon) + \frac{\partial}{\partial x_j}(\rho\varepsilon u_j) =$$
$$\frac{\partial}{\partial x_j}[(\mu + \frac{\mu_t}{\sigma_\varepsilon})\frac{\partial \varepsilon}{\partial x_j}] + \rho C_1 S\varepsilon - \rho C_0 \frac{\varepsilon^2}{k + \sqrt{v\varepsilon}} + \qquad (4)$$
$$C_{1\varepsilon}\frac{\varepsilon}{k}C_{3\varepsilon}G_b + S_\varepsilon$$

where,

$$C_1 = \max[0.43, \frac{\eta}{\eta + 5}], \quad \eta = S\frac{k}{\varepsilon}, \quad S = \sqrt{2S_{ij}S_{ij}}$$

The turbulence intensity, which is defined as the ratio of the root-mean-square of the velocity fluctuations to the mean flow velocity, can be estimated from the following formula derived from an empirical correlation for pipe flows (FLUENT 6.2 User's Guide, 2005) as,

$$I = \frac{u'}{u_{avg}} = 0.16(\text{Re}_{D_H})^{-1/8} \qquad (5)$$

The turbulent kinetic energy k and turbulent dissipation rate ε were calculated according to the following relations,

$$k = \frac{3}{2}(u_{avg}I)^2 \qquad (6)$$

$$\varepsilon = C_\mu^{3/4}\frac{k^{3/2}}{l} \qquad (7)$$

where C_μ is an empirical constant specified in the turbulence model (approximately 0.09) and l is the turbulence length scale obtained from the relation $l = 0.07 D_H$.

A source term was added to the momentum equation to estimate the pressure drop across the perforated plates. In the CFD simulation, the perforated plates are modelled as thin porous media of finite thickness with directional permeability over which the pressure change is defined as a combination of a viscous loss term and an inertial loss term, and is given by (FLUENT 6.2 User's Guide, 2005),

$$\Delta p = -(\frac{\mu}{\alpha}U + C_2 \frac{1}{2}\rho U^2)\Delta m \qquad (8)$$

Haque et al., (2006) found from their study, which has been included later in this section, that the pressure drop across the perforated plate is mainly due to the inertial loss at turbulent flow conditions. The viscous loss term was considered negligible and hence was eliminated from equation (x.8). Appropriate values of C_2 for each plate were then calculated for the corresponding Reynolds number and plate thickness from the literature presented by Idelchik (1994).

Boundary Conditions

Due to the symmetry in geometry, only one-half of the physical model was considered for the simulation. The finite volume method was used to discretise the partial differential equations of the model using the SIMPLEC method for pressure–velocity coupling and the second order upwind scheme to interpolate the variables on the surface of the control volume. The segregated solution algorithm was selected to solve the governing equations sequentially (i.e., segregated from one another). Non-equilibrium wall functions, which are a collection of semi-empirical formulas and functions, were applied to bridge the viscosity-affected region between the wall and the fully-turbulent region. Two simulations were performed with two different inlet boundary conditions. One was done with a velocity profile consisting of experimentally measured data and the other with uniform velocity distribution after calculating the mean value of the measured velocities. The measured inlet velocity profile consists of a set of velocities measured at 45 points inside the duct upstream of the ESP at the inlet boundary (see Figure 1) of the ESP.

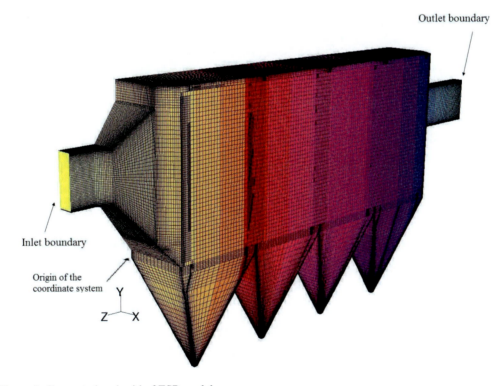

Figure 1. Computational grid of ESP model.

The direction of the velocity was normal to the inlet boundary. The turbulence intensity at the inlet boundary was set as 3%, based on the Reynolds number of the flow (Re = 3.1 x 10^5) and calculated using equation x.5. An atmospheric pressure boundary located downstream of the outlet duct was specified as pressure-outlet. The pressure-outlet boundary was placed far away from the outlet evase so as not to affect the flow inside the ESP. The no-slip boundary condition was used for all the walls. A porous jump boundary condition was used for the perforated plates.

Modelling Approach

Figure 1 shows the computational domain with grids of ESP model. The figure also shows the coordinate axes used to present the results in this chapter. It is to be noted that the direction of flow was normal to the inlet boundary and towards the negative-z axis. The plane at z = 0.3055 m (plane 1) corresponds to the inlet into the computational domain and the plane at z = -2.1805 m corresponds to the outlet from the computational domain.

It is necessary to refine the grid near the wall region for the boundary layer treatment. As the geometry of the ESP is quite complex, each part of the ESP (inlet evase, collection chamber, hoppers, outlet duct etc.) was created individually using GAMBIT and some faces of the computational domain were set as walls to represent collection walls, baffles and gas deflectors. Figure 2 shows a typical computational domain where a bell shaped grading was chosen for the baffle surfaces to create fine grids near the wall region. The individual mesh files thus created were then merged using the 3D version of 'tmerge' utility of FLUENT software.

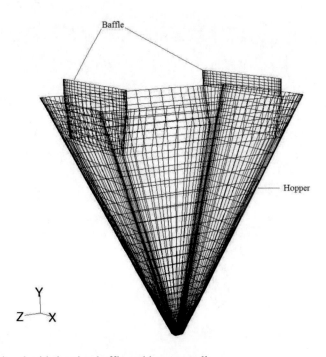

Figure 2. Computational grid showing baffle and hopper walls.

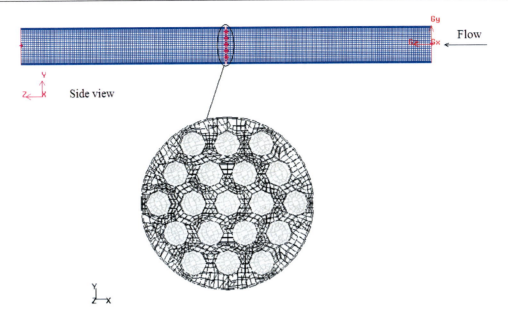

Figure 3. Computational grid of perforated plate model.

The number of nodes of the complete model thus developed was about 1.2 million. This was the maximum number of nodes the available computer could handle for numerical calculation and simulation. For standard or non-equilibrium wall functions, each wall-adjacent cell's centroid should be located within the log-law layer: y+ ≈ 30 – 300 (FLUENT user services center, 2005). As the y+ value of this model, which was 279, was in the recommended range, no further refinement was carried out.

A separate CFD study was conducted to determine the effect of the viscous loss term in equation x.8 at turbulent flow conditions. A small piece of an original 8 mm thick perforated plate, which was placed inside a round duct, was modelled for this purpose. The grid of this model is presented in Figure 3.

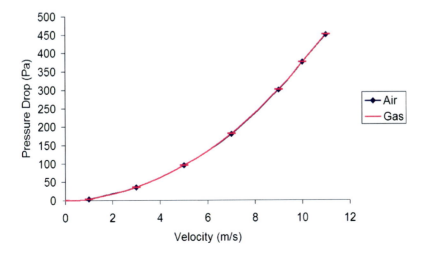

Figure 4. Predicted pressure drop across the perforated plate at different velocities for air and gas.

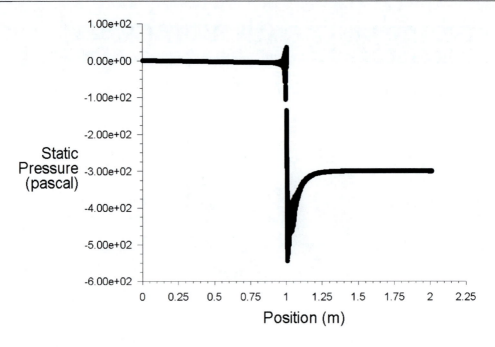

Figure 5. Predicted pressure drop across the perforated plate (inlet velocity 9 m/s).

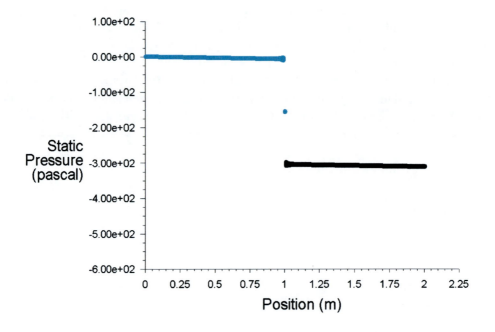

Figure 6. Predicted pressure drop across the porous jump boundary condition (inlet velocity 9 m/s).

Figure 4 represents the predicted pressure drop across the perforated plate at different velocities for air and for a fictitious gas with one hundred times lesser viscosity than air. Since the change of shear viscosity by the factor of one hundred did not result in any measurable difference in the predicted pressure drop, it was concluded that the pressure drop across the perforated plate is mainly due to the inertial loss at turbulent flow conditions. Hence the viscous loss term can be eliminated from equation 8. Appropriate values for C_2 for the same Reynolds number and plate thickness are then calculated from the literature (Idelchik, 1994).

Figure 5 represents the predicted pressure drop across the perforated plate, placed inside the round duct with an average inlet velocity of 9 m/s. Another CFD model was developed to justify the simplification of perforated plates as porous surfaces in the ESP model. The pressure drop across the porous surface incorporated into this model using the literature (Idelchik, 1994) value of C_2 for the same inlet and outlet boundary condition is shown in Figure 6. The method of calculating C_2 was found to be justifiable as the predicted pressure drop across the porous surface varied from that across the perforated plate by only 2%.

Simulation Results

Simulations were performed with two different velocity profiles at the inlet boundary (one with a uniform (ideal) velocity profile and the other with a non-uniform (real) velocity profile) to demonstrate the effect of velocity inlet boundary condition on the flow simulation results inside an ESP. The real velocity profile was obtained from the velocity measured at different points of the inlet boundary whereas the ideal velocity profile was obtained by calculating the mean value of the measured data.

The mean speed of the air flow was 30.62 m/s, which corresponds to a Reynolds number of $Re = 3.1 \times 10^5$ based on the hydraulic diameter of the inlet duct. The centreline velocity distribution measured at the inlet boundary was nonlinear and is shown with the average duct velocity at the same plane in Figure 7. The real velocity profile measured at plane 1 and a mean inlet velocity at the same plane were used as the two different inlet boundary conditions for the two CFD models.

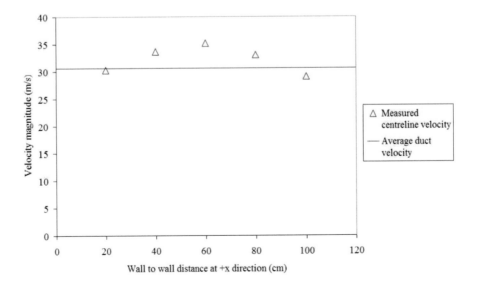

Figure 7. Velocity distribution at the inlet boundary (measurement plane 1).

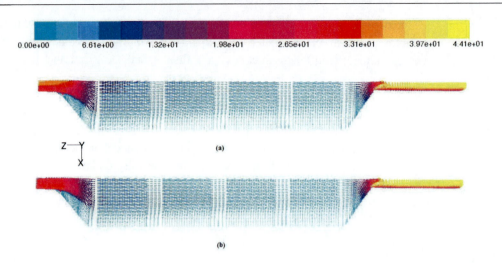

Figure 8. Velocity vectors at y = 0.36m (a) Using measured velocity profile at the inlet boundary, (b) Using uniform velocity profile at the inlet boundary.

The velocity vectors and velocity contours at height y = 0.36 m of the computational domain using both uniform (ideal) and measured (real) velocity profile at the inlet boundary are presented in Figures 8 and 9 respectively. The simulation with the real velocity profiles at the inlet boundary (Figures 8a and 9a) shows non uniform velocity distribution inside the ESP with high core velocities at the centre region, whereas the simulation with the ideal velocity profile at the inlet boundary (Figures 8b and 9b) shows almost uniform velocity distribution inside the ESP. The results obtained from this simulation confirmed that the properties of the inlet velocity profile extends throughout the ESP chamber and influence the flow distribution inside the ESP accordingly.

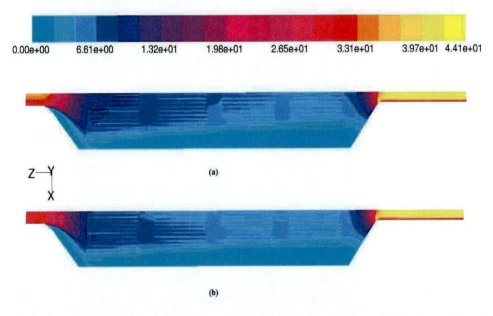

Figure 9. Velocity contours at y = 0.36m (a) Using real velocity profile at the inlet boundary, (b) Using uniform velocity profile at the inlet boundary.

Flow visualization with fog/smoke revealed vortices and flow separation in some areas particularly near the wall of the ESP. No measurements were carried out near the wall at those planes because correct measurements could not be achieved by using the Cobra probe in the presence of vortices (TFI, 2005).

Model Validation

The predicted velocities at height y = 0.36 m in the planes z = - 0.04575 m (plane 2), z = 0.759 m (plane 3) and z = - 1.47225 m (plane 4) are compared with the measured velocities and are presented in Figures 10, 11 and 12 respectively. It can be seen from these figures that the prediction lines are not as smooth as they usually appear in flows through ducts or pipes. This is due to the influence of the collection plates and other components which are fitted inside the ESP. It can be seen from Figure 10 that simulated velocity profiles did not match with the measured data at the third and fourth points from the centre of the inlet boundary. This can be attributed to the reverse flow phenomena which were observed at those points through flow visualization and were beyond the measurement range of the Cobra probe (Haque *et al.*, 2009).

The velocity distribution is found to be affected by the swirling flow phenomena near the wall region arising as a result of flow separation. Concave velocity distributions are noticed near the wall region (Figures 11 and 12). Kito and Kato (1984) observed a similar type of concave flow pattern in the swirling flow region in a pipe.

Negligible difference is observed between the two prediction lines at the location from x = 0.1 m to the wall in Figures 10, 11 and 12. But 21% deviation is observed between them near the centre (at x < 0.1m) in Figure 11, and in all cases the prediction lines with uniform inlet velocity profile underestimate the velocity distribution inside the ESP. Both prediction lines are compared with the experimentally measured data.

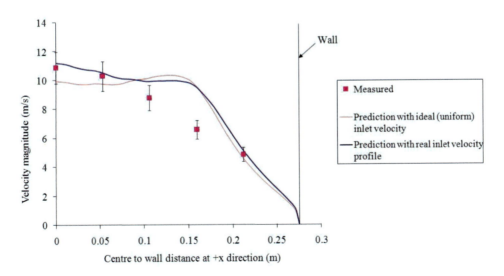

Figure 10. Velocity distribution from centre to wall at y = 0.36 m (plane 2) – Comparison of the measured data with CFD predictions.

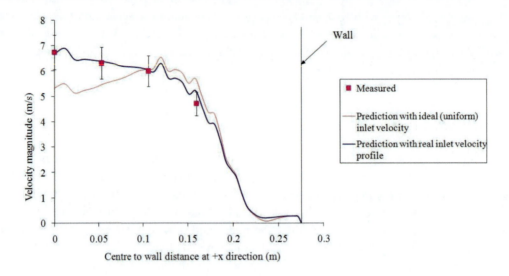

Figure 11. Velocity distribution from centre to wall at y = 0.36 m (plane 3) – Comparison of the measured data with CFD predictions.

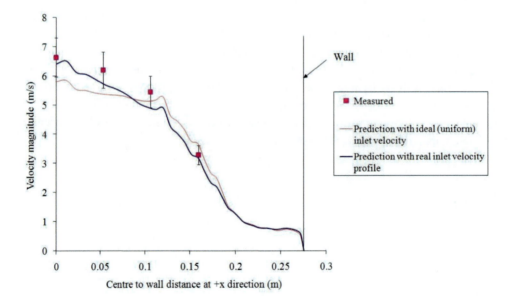

Figure 12. Velocity distribution from centre to wall at y = 0.36 m (plane 4) – Comparison of the measured data with CFD predictions.

The prediction line obtained from the simulation with a real velocity profile at the inlet boundary produces a better match with the experimental data. The reason for this better agreement can be attributed to the characteristics of the real velocity profile at the inlet boundary, which is found to be a nearly developed flow and hence produced high velocity at the centre and low velocity near the wall. This attribute of the flow continues through the total length of the ESP and causes high centreline velocity inside the ESP.

Predictions were found to be in very good agreement with the experimental data when the real velocity profile was used at the inlet boundary of the CFD model. It is therefore suggested, that for the industrial application, the experimentally measured velocity distribution should be used at the inlet boundary for an accurate and realistic flow simulation. As the complete ESP system consists of an electrostatic field and a particle phase in addition to the fluid flow field, a 2D ESP has been modelled in the next section where particle collection has been calculated after adding electrostatic force effects in the momentum equation.

NUMERICAL MODELLING OF ELECTROSTATIC FIELD AND PARTICLE DYNAMICS

The precipitation process involves charging particles of the flue gas by applying high voltage to the DE and driving the charged particles towards the grounded CE by the electric field produced. The collected particles are then removed from the collection electrodes by the rapping process. There is limited research found in the literature on ESP simulation. Most researchers simplified their models by creating a single DE ESP configuration (e.g., Lami *et al.*, 1995; Zhao *et al.*, 2006; Park and Kim, 2003; Anagostopoulos and Bergeles, 2002). An ESP model developed by Suda *et al.* (2006) consists of seven DEs to analyse the gas flow field, but they considered only a single DE segment for their ionic wind analysis. The single DE model may not be capable of capturing the wake of the wires properly, and it can not be assumed that the particle motions will be repeated in the gas flow direction (Haque *et al.*, 2007). Hence three wires were taken into consideration for the development of a representative numerical model to predict the motion of gas, ions and particles inside an ESP channel. The two-dimensional Navier–Stokes equations were used to model the gas flow where source terms were included in the transport equation to introduce the electric field and a Lagrangian discrete phase model was developed to calculate and monitor particle trajectories. The simulated result was verified with the available literature data. The study revealed that the particle collection and its movement depend not only on the size of the particle but also on the velocity of the gas flow and the electric potential applied at the discharge electrodes. The following considerations were adopted for modelling the electrostatic field and particle dynamics.

Modelling Approach

The full scale ESP consists of a number of identical flow passages made up of parallel collection electrodes having 400 mm spacing between them. A series of DEs are placed between every two CEs. The following considerations were taken into account for the modelling a 2D ESP:

- The height of the collecting plates in relation to the distance between them justifies a two-dimensional approach, where the change of the electric field in the vertical direction was considered negligible.

- The charge of the particles does not influence significantly the electric-field properties. This assumption is justified by the fact that critical precipitator sections, which control the final efficiency, operate under very small particle loads.
- Due to the symmetry between the plates, only half of the distance between the plates was required to model the electric field.
- The UDFs were used to provide a link between the electric field and the flow field. The UDF allows the user to input the electrical body force calculated into every single volume cell of the discretised FLUENT model.

A small section of the typical CE-DE arrangement of the local power station's ESP was considered for CFD modelling. The configuration of the model geometry is discussed in the next section.

Geometry Configuration

Two CEs and three DEs were considered for creating the geometry of an ESP channel. Due to the symmetry of the geometry, only half a channel was modelled in this study. The configuration and the dimensions of the ESP are shown in Figure 13. The computational mesh consists of 15000 cells and is shown in Figure 14. It also shows the boundary conditions that were used in this study for solving the problem. Unstructured quadrilateral/hexahedral elements have been chosen for meshing the 2D ESP model as they permit a much larger aspect ratio than triangular/tetrahedral cells for a relatively simple geometry (FLUENT 6.2 User's Guide, 2005).

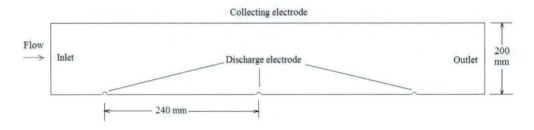

Figure 13. 2D geometry configuration.

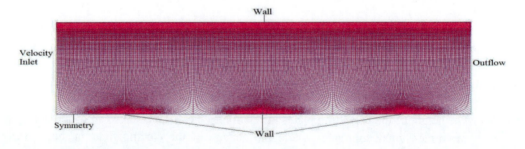

Figure 14. Computational mesh.

Governing Equations

Gas Dynamics

Numerical calculations for the gas flow are carried out by solving the Reynolds-averaged Navier-Stokes equations. An additional source term is added to the gas flow equation to capture the effect of the electric field.

$$\rho \vec{V} \cdot \vec{\nabla} \vec{V} = -\vec{\nabla} p + \mu \vec{\nabla}^2 \vec{V} + S \qquad (9)$$

where ρ is the fluid density (kg/m^3), μ is the dynamic viscosity (kg/m/s) of the fluid, p is the fluid pressure (Pa) and $\vec{V}$ is the fluid velocity (m/s). S is the source term, which expresses the momentum force (N/m^2) on the gas flow due to the electric field and can be expressed as (Choi and Fletcher, 1998),

$$S = \rho_{ion} \vec{E} \qquad (10)$$

where ρ_{ion} is the ion charge density (C/m^3) and $\vec{E}$ is the electric field intensity (V/m).

Electrostatic Field

The electric field intensity $\vec{E}$ inside the ESP can be described by the Gauss's law equation (Kallio and Stock, 1986),

$$\nabla \cdot \vec{E} = \frac{\rho_{ion}}{\varepsilon_0} \qquad (11)$$

where

$$\vec{E} = -\nabla \phi \qquad (12)$$

Combining Equations (x.11) and (x.12) gives the well known Poisson equation which is defined as

$$\nabla^2 \phi = -\frac{\rho_{ion}}{\varepsilon_0} \qquad (13)$$

where $\vec{E}$ is the electric field intensity (V/m), ϕ is the electric potential (Volt) and ε_0 is the permittivity of the free space. Under stationary conditions, the electrical flux density is divergence-free and can be written as (Poppner et al., 2005),

$$\nabla \cdot \vec{J} = 0 \qquad (14)$$

where $\vec{J}$ is the density of ionic current.

Assuming ion diffusion is of negligible importance compared to conduction (Lami et al., 1995; Poppner et al., 2005), $\vec{J}$ can be expressed as,

$$\vec{J} = \rho_{ion} b_{ion} \vec{E} \tag{15}$$

where b_{ion} is the ion mobility.

Combining Equation (x.14) and (x.15) gives the following expression,

$$\nabla \cdot (\rho_{ion} b_{ion} \nabla \phi) = 0 \tag{16}$$

Equation (x.16) can be transformed to the following form of transport equation

$$\nabla(\rho \vec{V} \rho_{ion} - D_{\rho_{ion}} \nabla \rho_{ion}) = S_{\rho_{ion}} \tag{17}$$

where $D_{\rho_{ion}}$ is the diffusion coefficient = ρD_e where D_e is the effective ion diffusivity = 10 m^2/s (Skodras et al., 2006) and source term $S_{\rho_{ion}} = -\nabla(\rho \rho_{ion} b_{ion} \vec{E})$. Two transport variables ϕ and ρ_{ion} were numerically calculated with the appropriate boundary conditions and solution methods. The details are provided later in this section.

Particle Dynamics

FLUENT predicts the trajectory of a discrete phase particle by integrating the force balance on the particle, which is written in a Lagrangian reference frame. This force balance equates the particle inertia with the forces acting on the particle, and can be written as (FLUENT 6.2 User's Guide, 2005),

$$\frac{du_{p,i}}{dt} = F_D(u_i - u_{p,i}) + \frac{g_i(\rho_p - \rho)}{\rho_p} + F_i \tag{18}$$

$$\frac{dz_i}{dt} = u \tag{19}$$
$$i = x, y, z$$

where ρ_p and $u_{p,i}$ denote particle density and velocity respectively. $F_D(\vec{u}_i - \vec{u}_{p,i})$ is the drag force per unit particle mass, given by

$$F_D = \frac{3 \mu C_D \text{Re}}{4 \rho_p d_p^2} \tag{20}$$

here, u_i is the fluid phase velocity, μ is the molecular viscosity of the fluid and ρ is the fluid density. $u_{p,i}$, d_p, m_p and ρ_p denote the velocity, diameter, mass and density of the particle respectively. Re is the relative Reynolds number, which is defined as

$$\mathrm{Re} \equiv \frac{\rho d_p |u_{p,i} - u_i|}{\mu} \tag{21}$$

The drag coefficient C_D can be calculated from the following equation (Haider and Levenspiel, 1989):

$$C_D = \frac{24}{\mathrm{Re}_{sph}}(1 + b_1 \mathrm{Re}_{sph}^{b_2}) + \frac{b_3 \mathrm{Re}_{sph}}{b_4 + \mathrm{Re}_{sph}} \tag{22}$$

where

$$b_1 = \exp(2.3288 - 6.4581\phi + 2.4486\phi^2)$$

$$b_2 = 0.0964 + 0.5565\phi \tag{23}$$

$$b_3 = \exp(4.905 - 13.8944\phi + 18.4222\phi^2 - 10.2599\phi^3)$$

$$b_3 = \exp(1.4681 + 12.2584\phi - 20.7322\phi^2 + 15.8855\phi^3)$$

The shape factor, ϕ, is defined as

$$\phi = \frac{s}{S} \tag{24}$$

where s is the surface area of a sphere having the same volume as the particle, and S is the actual surface area of the particle. The Reynolds number Re_{sph} is computed with the diameter of a sphere having the same volume (FLUENT 6.2 User,s Guide, 2005).

F_i corresponds to external forces exerted on the particle that, in the present study, are the electrostatic forces (Varonos et al., 2002) given by,

$$F_i = \frac{E_i q_p}{m_p} \tag{25}$$

where, E_i is the electric field component [V/m] and m_p is the particle mass [kg]. The particle charge, q_p, determines the Coulomb force which is exerted on the particle. It is calculated in the Lagrangian framework, as the particle moves through the gas by the following expression,

$$\frac{dq_p}{dt} = \frac{1}{\tau q_{max}}(q_{max} - q_p) \qquad (26)$$

here, τ is the time taken for the particle to reach half the saturation charge [s] and is defined as,

$$\tau = \frac{4\varepsilon_0}{\rho_{ion} b_{ion}} \qquad (27)$$

where b_{ion} is the ion mobility = 0.00022 (for air ions) [m²C/J/s] = [m²/V/s].

q_{max} is the saturation charge by the local electric field at the particle location and is expressed as,

$$q_{max} = q_{sat} = \frac{3\pi\varepsilon_o \varepsilon_r E d_p^2}{\varepsilon_r + 2} \qquad (28)$$

where ε_o is the permittivity of free space and is equal to 8.854×10^{-12} [C²/N/m²] and ε_r is the relative permittivity of the gas (or dielectric constant of the gas) which for air is 1.000590.

Boundary Conditions

The geometry developed by GAMBIT is exported to the FLUENT solver to specify the flow properties, solve the problems and analyse the results. The finite volume method was used to discretise the partial differential equations of the model using the SIMPLEC method for pressure–velocity coupling and the second order upwind scheme to interpolate the variables on the surface of the control volume. The segregated solution algorithm is selected to solve the governing equations sequentially. Non-equilibrium wall functions were applied to bridge the viscosity-affected region between the wall and the fully-turbulent region.

The electric potential on the CEs is zero whereas at the DEs surface the value is 70 (kV). The charge density ρ_0 at the discharge electrode can be approximately calculated by (Ye and Domnick, 2003) as,

$$\rho_0 = -\frac{\varepsilon(E - E_0)}{ds} \qquad (29)$$

where E is the field strength in the cell adjacent to the emitting electrode, ds is the distance between the cell and the electrode surface. The corona onset field E_0 along the corona-emitting surface is assumed constant, and it can be obtained according to Peek's law (Peek, 1929),

$$E_o = E_{Peek} = 3.1 \times 10^6 \delta (C_1 + C_2 / \sqrt{\delta r}) \tag{30}$$

where E_{Peek} is the ion current threshold value for an electrode of radius r and $C_1 = 1$ V/m and $C_2 = .031$ V/√m in air of relative density δ with respect to the normal temperature and pressure conditions.

A number of subroutines were written and compiled by using UDFs compatible with FLUENT. The UDFs are then linked to the standard fluid model to solve the Poisson's equation and the charge density equation.

The operating gas was ambient air while the particles were assumed to be ash with density equal to 600 kg/m³. Turbulent intensity at the inlet was 5%. Boundary conditions used to solve this problem are summarised in Table 1.

Table 1. Boundary conditions applied to the ESP model

	Gas dynamics	Electric potential (kV)	Ion charge density C/m³	Particle dynamics
Velocity Inlet	$u_x = 1.0 m/s$ $u_y = 0.0 m/s$	$\nabla \phi = 0$	$\nabla \rho_{ion} = 0$	$u_{px} = 1.0 m/s$ $u_{py} = 0.0 m/s$ Escape
Outflow	Mass conservation	$\nabla \phi = 0$	$\nabla \rho_{ion} = 0$	Escape
Wall-Collecting electrode	No slip	$\Phi = 0$	$\nabla \rho_{ion} = 0$	Trap
Wall-Discharge electrode	No slip	$\Phi = 60, 70, 80, 90$	Peek's law	Reflect

Numerical Simulation and Data Comparison

Velocity contours of the gas field are shown in Figure 15. As expected, flow separation occurred behind the DEs which are shown in Figure 16. The boundary layer, which starts along the CE, is found to suppress the flow separation that occurs after the discharge electrode. Figures 17 and 18 show the contours of equipotential spaced at 3.5 kV intervals and ion charge density spaced at 4 μC/m³ in the ESP channel. Both the electric potential and ion charge density distributions have periodic patterns as expected.

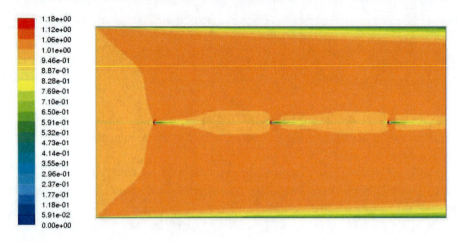

Figure 15. Contours of gas velocity magnitude (m/s).

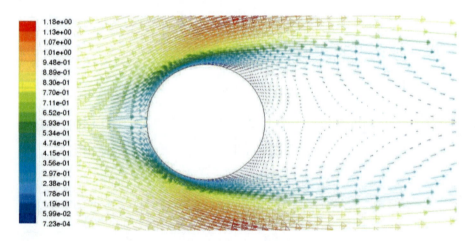

Figure 16. Velocity vectors around the discharge electrode (m/s).

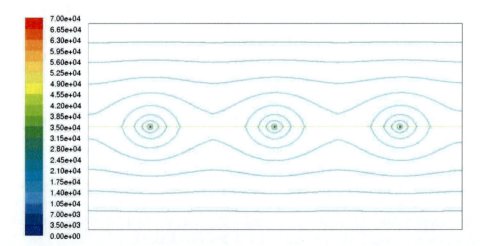

Figure 17. Contours of electric potential (Volt).

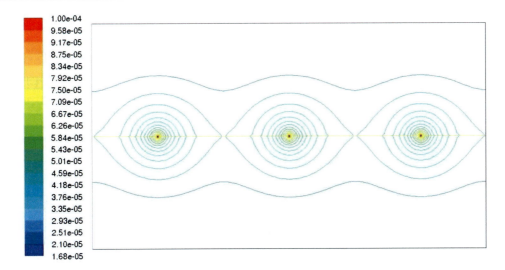

Figure 18. Contours of ion charge density (C/m^3).

The numerical prediction of the model was validated with the literature data as there was no facility available for experimental measurements. Choi and Fletcher (1998) developed a model of an ESP channel which also consisted of three discharge electrodes as does the model of this study. The predicted electric potential distribution in this study yields a good agreement with their data and is shown in Figure 19. This validated model could be used for studying the effect of various process parameters on the collection of particles inside the ESP channel.

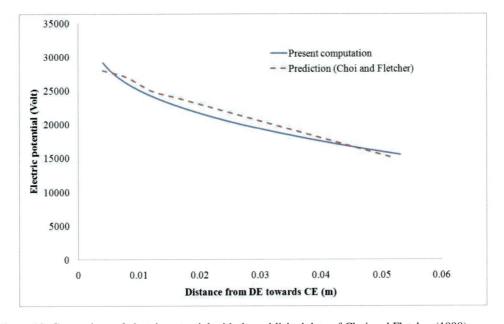

Figure 19. Comparison of electric potential with the published data of Choi and Fletcher (1998).

Effect of Particle Diameter on the Collection Efficiency of an ESP

The effect of particle diameters on particle collection is shown in Figure 20 where 500 particles with a size range from 1μm to 141 μm were injected from the inlet surface with the velocity of 1 m/s. The Rosin-Rammler distribution was adopted with the mean particle diameter of 18.11 μm. The electric potential at the boundaries of the DEs is set at 70 kV.

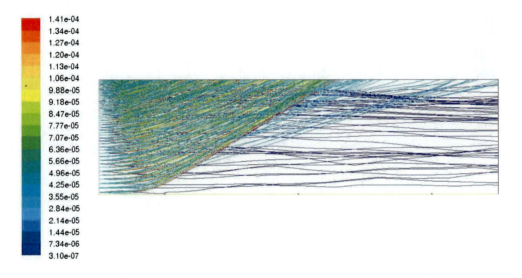

Figure 20. Particle trajectories coloured by particle diameter (Gas velocity 1m/s, Electric potential 70 kV).

The simulation shows that the current operating condition will not be able to capture PM2.5 (particles with size equal to or less than 2.5 μm) effectively as most of these particles have escaped from the collection area. The collection efficiency was calculated by dividing the number of particles collected inside the ESP with the number of particles injected. The result of the simulation shows that the collection efficiency of the ESP channel decreases with the decrease in particle size. The collection efficiencies are 30% for particles with a size of 2.5 μm and 18% for 1 μm sized particles. Figure 21 presents the trajectories of the particles with a size of 2.5 μm where 15 particles were collected after injecting 50 particles inside the ESP.

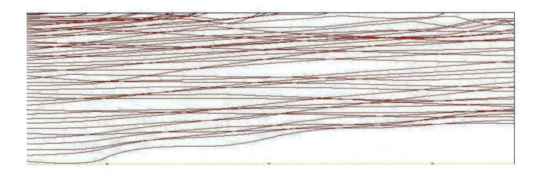

Figure 21. Particle trajectories (Size 2.5 μm, Gas velocity 1 m/s and Electric potential 70 kV).

Influence of Gas Velocity on the Collection Efficiency of ESP

The influence of gas velocity on particle collection was studied. The 3D ESP model of this study revealed that the velocity is not uniformly distributed across the collection chamber. The average gas velocity inside the collection chamber of an ESP varies from 0.5m/s to 2 m/s. However, it is crucial to optimize the flow for improving the collection of particles. A flow stream with high velocity leaves the collection chamber with poor particle collection. The particles need sufficient treatment time to get charged and collected inside an ESP. The collection efficiencies for all sizes of particles have significantly dropped while the velocity of the flow at the inlet surface is increased to 1.5 m/s. The efficiency dropped to 18% for particles with a size of 2.5 μm and 10% for 1 μm sized particles. Figure 22 shows the trajectories of the particles with a size of 2.5 μm where only 9 particles were collected after injecting 50 particles inside the ESP.

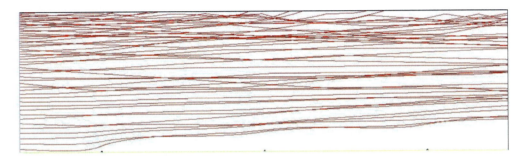

Figure 22. Particle trajectories (Size 2.5 μm, Gas velocity 1.5 m/s and Electric potential 70 kV).

The simulated results show significant improvement in the collection of fine particles while the velocity of the flow at the inlet surface is reduced to 0.5 m/s. The collection efficiencies for the 2.5 μm and 1 μm sized particles have increased to 42% and 24%. Figure 23 shows the trajectories of the particles of the size of 2.5 μm where 21 particles where captured after injecting 50 particles inside the ESP.

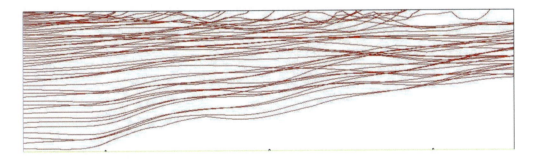

Figure 23. Particle trajectories (Size 2.5 μm, Gas velocity 0.5 m/s and Electric potential 70 kV).

Influence of Electric Potential on Collection Efficiency of ESP

Particle collection inside the ESP greatly depends on the electrostatic force of the ESP system. The electrostatic force, which is generated after applying electric potential on the DEs are strongly related to the movement of the particles. A range from 60 kV to 90 kV electric potential is applied to the discharge electrode to study its influence on the particle collection inside the ESP. The results of the simulation revealed that the particle collection efficiency is reduced with decreased electric potential at the surface of the DEs. The collection efficiencies are 20% and 12% for particles with a size of 2.5 μm and 1 μm respectively. Figure 24 presents the trajectories of the particles of the size of 2.5 μm where only 10 particles were capture after injecting 50 particles inside the ESP.

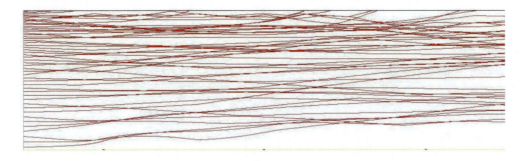

Figure 24. Particle tajectories (Size 2.5 μm, Gas velocity 1m/s and Electric potential 60 kV).

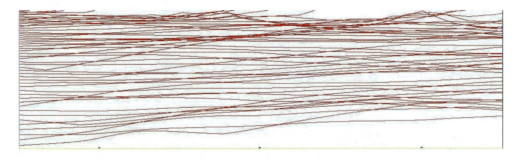

Figure 25. Particle trajectories (Size 2.5 μm, Gas velocity 1m/s and Electric potential 80 kV).

The collection of particles could be improved by increasing the electric potential at the DEs. The collection efficiencies are 34% and 20% for the particles with a size of 2.5 μm and 1 μm respectively while the applied electric potential at the discharge electrodes is 80 kV. Figure 25 shows the trajectory of the particles with the size of 2.5 μm where 17 out of 50 particles were collected. No further improvements on particulate collection were observed after increasing the electric potential to 90 kV.

The result of the simulation revealed that PM2.5 are found to be difficult to capture as most of them escaped from the collection chamber, whereas 100% collection could be achieved for the large particles (>18 μm). This is because the higher charge of the large particles forces them towards the CE for deposition. The large particle acquires more charge than a small one due to its larger surface area; the larger the diameter of the particle, the higher is the charge and the more efficient is the particle collection. However, it is evident from the simulated results that the increased electric potential cannot improve the collection of fine particles unless the flow velocity is adjusted and optimized. The optimized flow velocity results in fine particles to spend more time inside the collection chamber and so collect a sufficient charge, which is crucial for their movement towards the collection electrodes. The results obtained from the simulations are applied to identify options for improving the performance of the industrial electrostatic precipitator and are presented in the next section of this chapter.

APPLICATION OF CFD MODELS IN AN ESP OF A POWER PLANT

The geometry of the full scale ESP unit of a local power plant is similar to that of the laboratory scale ESP model. The numerical model presented in this chapter is the scaled-up version of the validated model discussed in the first section of this chapter. A novel approach is introduced in this section for modelling the perforated plate as a set of porous surface with variable porosity. This gives the ESP model flexibility to optimize the flow by adjusting the resistance of an individual surface. A novel guide vane is placed inside the inlet evase of the ESP model. The simulation results show improved flow distribution. The optimized velocity obtained from the three dimensional model is then applied to the two dimensional model. It is revealed from the CFD models that the particle collection efficiency could be improved by optimizing flow the field and increasing the electric force.

Optimisation of Flow Field – Using Variable Porosity of Perforated Plate

Three perforated plates are located within the inlet evase and one outlet screen is located within the outlet evase. The perforated plates are modelled as thin porous surfaces of finite thickness with directional permeability. In the simulations presented in this section, a unique approach was adopted to model the third perforated plate. The third perforated plate is the closest to the ESP chamber. This plate is modelled as a set of porous surfaces with variable porosity. This novel approach gives the ESP model versatility to optimize the flow by controlling the resistance of these surfaces. Due to the symmetry in geometry, the numerical model was constructed to represent only half a casing. Figure 26 shows the geometrical representation of the third perforated plate inside the inlet evase.

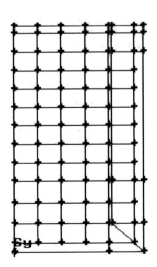

Figure 26. Perforated plate with variable porosity.

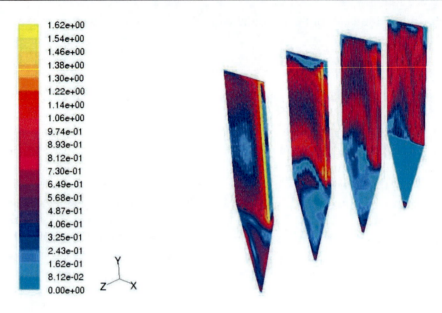

Figure 27. Velocity contours inside the ESP using a variable porosity of perforated plate.

The velocity contours with a variable porosity of the perforated plate are presented in Figure 27. The nonuniformity of the velocity distribution inside the ESP was removed after introducing a variable porosity of perforated plate inside the inlet evase.

Optimisation of Flow Field – Insertion of Guide Vanes Inside the Inlet Evase

It is found that when the flow medium spreads over the front of the grid, the streamlines become distorted.

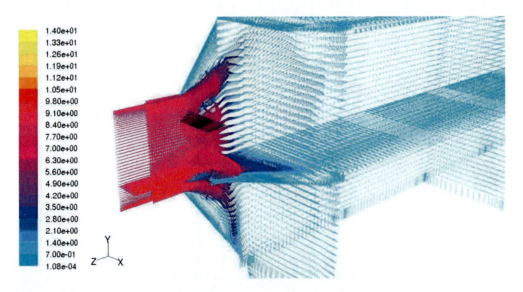

Figure 28. Velocity vectors with guide vane.

The higher the resistance coefficient of the grid, the sharper is the distortion of streamlines, and consequently the greater is the departure of the jets from the orifices to the periphery of the grid. With an increase of the resistance coefficient of the grid up to a certain value, the velocity profile becomes reversed. To avoid this situation guide vanes could be inserted over the cross section of the inlet evase.

Figure 28 presents the velocity vectors once a guide vane is introduced to distribute the flow all over the third perforated plate. The velocity contours inside the ESP after introducing the guide vane inside the inlet evase are presented in Figure 29.

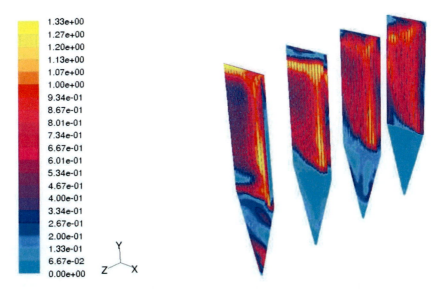

Figure 29. Velocity contours inside the ESP with inlet evase guide vane.

It is found from the simulated results that this new geometrical feature guides the flow uniformly over the third perforated plate. The third perforated plate then gives the flow an additional uniformity which continues throughout the total collection area of the ESP.

ESP Efficiency Improvement – Effect of Gas Velocity and Electric Potential

It can be seen from the 3D ESP model that the sudden expansion in geometrical configuration that is present just before the collecting section region inlet, along with the blockade of the vertical plates' structure inside the collection section, generates large flow velocities between the collecting plates. This poor aerodynamic quality of the flow was improved after introducing variable porosity of the perforated plate and horizontal guide vanes inside the inlet evase of the ESP model. The particle collection could be further improved with increased electric potential at the discharge electrodes while maintaining optimised flow velocity between the collection plates.

The collection efficiency of PM2.5 was found to be increased to 48% in Figure 29 after setting the optimized gas velocity (0.5 m/s) at the inlet boundary and an increase of electric potential at the discharge electrode from 70 kV to 80 kV.

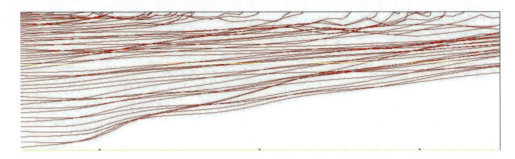

Figure 29. Particle trajectories (Size 2.5 µm, Injected-50, Collected- 24).

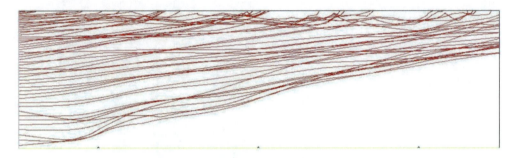

Figure 30. Particle trajectories (Size 2.5 µm, Injected-50, Collected- 29).

Figure 30 shows the trajectories of the particles while the flow velocity is maintained 0.5 m/s at the inlet boundary but the electric potential is increased to 90 kV. A further increase in electric potential is found to create improved collection efficiency of the ESP. The collection efficiency is increased to 58% for the particles with a size of 2.5 µm.

CONCLUSION

This chapter presented a detailed method for developing a numerical model of an ESP. The model thus developed was used to achieve an aerodynamic optimisation of the velocity profile inside the collecting section of the 3D CFD model. The novelty of this numerical model was the inclusion of the major physical features of a full scale ESP. All the CEs, baffles, gas deflectors etc were taken into account in this 3D flow model, and were not replaced by any equivalent porous region as previously done by others. The predicted results were found to be in reasonable agreement with the measured data. The CFD model was then scaled up to simulate the fluid flow through an ESP of a thermal power plant. A novel approach was adopted for modelling the perforated plate as a set of porous surfaces with variable porosity. This gave the ESP model flexibility to optimize the flow by adjusting the resistance of an individual surface. A guide vane was placed inside the inlet evase of the ESP model. The simulation results showed improved flow distribution inside the ESP. The evenly distributed velocity across the collection area obtained from the 3D model was then applied to a 2D ESP model which consisted of 3 DEs. An additional source term was added to the gas flow equation to capture the effect of the electric field on particle collection. The model was validated by the published data. A DPM model was used for predicting particle trajectories. It is revealed from the 2D ESP model that the particulate emission control was successfully improved by coupling the increased electric potential with decreased gas velocity.

REFERENCES

Anaglostopoulos, J. and Bergeles, G. 2002. Corona discharge simulation in wire-duct electrostatic precipitator. *Journal of Electrostatics*, 54, 129 – 147.

Choi, B. S. and Fletcher, C. A. J. 1998. Turbulent particle dispersion in an electrostatic precipitator. *Applied Mathematical Modeling*, 22, 1009 – 1021.

FLUENT 6.2 User's Guide 2005, FLUENT Inc.

FLUENT user services center. 2005. Turbulence Wallfunction, Lecture 6, Introductory FLUENT Notes, FLUENT V6.2.

Haider, A. and Levenspiel, O. 1989. Drag Coefficient and Terminal Velocity of Spherical and Nonspherical Particles. *Powder Technology*, 58, 63-70.

Haque, S. M. E., Rasul, M. G., Deev, A., Khan, M. M. K. and Zhou, J. 2006. The influence of flow distribution on the performance improvement of electrostatic precipitator. Conference proceedings, *The 10th International Conference on Air Pollution Abatement Technologies – Future Challenges?*, 25 – 29 June 2006, Cairns, Australia.

Haque, S. M. E., Rasul, M. G., Khan, M. M. K., Deev, A. V. and Subashchandar, N. 2009. Influence of the Inlet Velocity Profiles on the Prediction of Velocity Distribution inside an Electrostatic Precipitator, *Experimental Thermal and Fluid Science, 33, 322-328*.

Haque, S. M. E., Rasul, M. G., Khan, M. M. K., Deev, A. V. and Subaschandar, N. 2007, A Numerical Model of Particle Motion in an Electrostatic Precipitator, Conference proceedings, *The 16th Australasian Fluid Mechanics Conference*, CD-ROM, 1050-1054, 3 – 7 December 2007, Crowne Plaza, Gold Coast, Australia.

Idelchik, I. E. 1994. Handbook of hydraulic resistance. 3rd edition, CRC press, Inc., Florida, USA.

Kallio, G. A. and Stock, D. E. 1986. Computation of electrical conditions inside wire-duct electrostatic precipitators using a combined finite-element, finite-difference technique. *Journal of Applied Physics*, 59 (6), 1799 – 1806.

Kito, O. and Kato, T. 1984. Near wall velocity distribution of turbulent swirling flow in circular pipe. *Bulletin of JSME*, 27 (230), 1659 – 1666.

Lami, E., Mattachini, F., Gallimberti, I., Turri, R. and Tromboni, U. 1995. A numerical procedure for computing the voltage-current characteristics in electrostatic precipitator configurations. *Journal of Electrostatics,* 34, 385 – 399.

Nikas, K.S.P., Varnos, A.A. and Bergeles, G.C., 2005. Numerical simulation of the flow and the collection mechanisms inside a laboratory scale electrostatic precipitator. *Journal of Electrostatics*, 63, 423-443.

Park, S. J. and Kim, S. S. 2003. Effects of Electrohydrodynamic Flow and Turbulent Diffusion on Collection Efficiency of an Electrostatic Precipitator with Cavity Walls. *Aerosol Science and Technology*, 37, 574 – 586.

Peek, F. W. 1929. Dielectric Phenomena in High Voltage Engineering, 3rd edition, McGraw-Hill, New York.

Poppner, M., Sonnenschein, R. and Meyer, J. 2005. Electric field coupled with ion space charge. Part 2: computation, *Journal of Electrostatics*, 63, 781 – 787.

Schwab, M. J. and Johnson R. W. 1994. Numerical design method for improving gas distribution within electrostatic precipitators, *Proceedings of the American Power Conference*, 56, 882-888.

Skodras, G., Kalidas, S. P., Sofialidis, D., Faltsi, O., Grammelis, P. and Sakellaropoulos, G. P. 2006. Particulate removal via electrostatic precipitators – CFD simulation, *Fuel Processing Technology*, 87, 623 – 631.

Suda, J. M., Ivancsy, T., Kiss, I. and Berta, I. 2006. Complex Analysis of Ionic Wind in ESP Modeling, *The 10th International Conference on Electrostatic Precipitator*, Australia.

Turbulent Flow Instrumentation (TFI) Pty Ltd. http://www.turbulentflow.com.au.

Varonos, A.A., Anagnostopoulos, J.S. and Bergeles, G.C. 2002. Prediction of the Cleaning Efficiency of an Electrostatic Precipitator, *Journal of Electrostatics*, 55, 111 – 133.

Ye, Q. and Domnick, J. 2003. On the simulation of space charge in electrostatic powder coating with a corona spray gun, *Powder Technology*, 135-136, 250 – 260.

Zhao, L., Dela Cruz, E., Adamik, K., Berezin, A. A. and Chang, J. S. 2006. A numerical model of a wire-plate electrostatic precipitator under electrohydrodynamic flow conditions, Conference proceedings, *The 10th International Conference on Air Pollution Abatement Technologies – Future Challenges?*, Cairns, Australia.

In: Computational Fluid Dynamics
Editor: Alyssa D. Murphy

ISBN 978-1-61209-276-8
© 2011 Nova Science Publishers, Inc.

Chapter 9

APPLICATIONS OF COMPUTATIONAL FLUID DYNAMICS IN FOOD PROCESSING OPERATIONS

C. Anandharamakrishnan[*]

Human Resource Development, Central Food Technological Research Institute,
Council of Scientific and Industrial Research (CSIR), Mysore – 570 020, India

ABSTRACT

Computational fluid dynamics (CFD) has been used extensively by the scientific community worldwide for optimal design of industrial processes. CFD is a simulation tool, which uses powerful computers in combination with applied mathematics to model fluid motion. CFD is increasingly used in the design, scale-up and trouble-shooting of different unit operations in various processing industries. In recent years, a rapid development in the application of CFD in food processing operations has been witnessed. This chapter reviews the application of CFD in selected food processing operation such as spray drying, bread baking and pasteurization/sterilization processes. It also discusses the different modeling approaches and reference frames used for the CFD simulations, particles histories (temperature, velocity residence time and impact positions) during spray drying and spray-freezing, application of different radiation models (discrete transfer radiation, surface to surface and discrete ordinates) for electrical heating baking oven simulations and modeling of inactivation of enzymes during pasteurization of canned liquid food and eggs. In addition, the challenges involved in the CFD modelling of food processing, recent developments in this area and future applications are highlighted.

INTRODUCTION

The food processing industries are still facing some technological challenges including moving boundary value problems, simultaneous heat and mass transfer phenomena, health and safety issue and appearance. The conventional experimental design optimization

---

[*] E-Mail: anandharamakrishnan@cftri.res.in, Tel: 91-821-2514310; Fax: 91-821-2517233

procedure involves time consuming expensive trial and error experimental methods. Computational Fluid Dynamics (CFD) simulation method can be an effective alternative to the conventional experimental methods. CFD is a simulation tool, which uses powerful computers in combination with applied mathematics to model fluid flow situations and aid in the optimal design of industrial processes (Anderson, 1984; Versteeg and Malalasekera, 1995; Kuriakose and Anandharamakrishnan, 2010). The role of CFD in engineering has become so strong that today it may be viewed as a new third dimension of fluid dynamics (the other two dimensions being classical cases of pure experiment and pure theory). The method comprises of solving equations for the conservation of mass, momentum and energy, using numerical methods to give predictions of velocity, temperature and pressure profiles inside the system. Its powerful graphics can be used to then show the flow behaviour of fluid with 3D images (Scott and Richardson, 1997).

CFD gained prominence with availability of computers in the early 1960s and it is applied to all aspects of fluid dynamics. Since the early 1970s, commercial software packages became available, making CFD an important component of engineering practice in industrial, defense and environmental organizations (Parviz and John, 1997). By the 1990s, advances in computing power produced a similar boom in software development and solutions. Since then CFD has been used extensively by the scientific community worldwide. For more detailed historical perspective the books by Roache (1976) and Tannehill et al. (1997) are highly recommended. Today, CFD finds extensive usage in basic and applied research, in design of engineering equipment and in calculation of environmental and geophysical phenomena (Kuriakose and Anandharamakrishnan, 2010).

Although the origin of CFD can be found in the automotive, aerospace, and nuclear industries and also many other processing industries, it is only in the recent years that CFD has been applied to the food processing area (Scott, 1994). The ability of CFD to predict the performance of new designs or processes before they are ever manufactured or implemented make them a integral part of engineering design and analysis (Schaldach et al., 2000). The general application of CFD to the food industry was reviewed by many researchers (Scott and Richardson, 1997; Xia and Sun, 2002; Anandharamakrishnan, 2003; Norton and Sun, 2006; Sun, 2007). All the above reviews concluded that CFD is a powerful and pervasive tool for process and product improvement in food processing sector. This chapter mainly focuses on the recent advances in the application of CFD in selected food processing unit operations.

CFD ANALYSIS

CFD analysis involves following three main steps. The first step is *Pre- processing*, which includes problem definition, geometry, meshing (this can usually be done with the help of a standard CAD programme) and generation of a computational model. The second step is *Processing,* which uses a computer to solve the mathematical equations of fluid flow. The final step of *Post-processing* is used to evaluate and visualize the data generated by the CFD analysis (Xia and Sun, 2002) and validate the simulation results with experimental data.

APPLICATION OF CFD TO THE FOOD INDUSTRY

The past few years have witnessed a rapid development in the application of CFD in many areas of food processing such as spray drying, bread baking, heat exchangers, spray cooling, biosensors, high pressure processing, jet impingement oven, retail cabinet design, refrigerated truck, extrusion, mixing, pasteurization and sterilization. The list given above is non-exhaustive and for detailed review of CFD applications to food processing, reader may refer elsewhere (Anandharamakrishnan, 2003; Sun, 2007). The following sub-sections will discuss the recent advances of CFD applications in spray drying of food ingredients, bread baking process, pasteurization of eggs and sterilization of canned liquid foods.

SPRAY DRYING OF FOOD INGREDIENTS

Spray drying is a well established method for converting liquid feed materials into a dry powder form. Spray drying is widely used to produce powdered food products such as whey, instant coffee, milk, tea and soups, as well as healthcare and pharmaceutical products, such as vitamins, enzymes and bacteria. Normally, spray dryer comes at the end-point of the processing line, as it is an important step to control the final product quality. It has some advantages such as, rapid drying rates, a wide range of operating temperatures and short residence times (Anandharamakrishnan et al., 2007). In spray-drying operations, CFD simulation tools are now often used because measurements of airflow, temperature, particle size and humidity within the drying chamber are very difficult and expensive to obtain in large-scale dryer (Anandharamakrishnan et al., 2010). Langrish and Fletcher (2001) and Fletcher et al. (2003) earlier reviewed the applications of computational fluid dynamics in spray drying operations, to predict the flow patterns and temperature distributions of gas and droplets. Recently, Kuriakose and Anandharamakrishnan (2010) comprehensively reviewed the CFD application in spray drying operations including particle histories. The understanding of particle histories such as, velocity, temperature, residence time and the particle impact position are important to design and operating spray drying. Moreover, final product quality is depending on these particle histories. These particle histories can be tracked with the help of CFD simulations.

Crowe et al. (1977) first proposed the particle source in cell (PSI-Cell) model. This is the basis for the discrete phase model (DPM). The Eulerian-Lagrangian frame (DPM) provides better residence times of individual particles with a large range of particle sizes. In the DPM, the flow field is divided into a grid defining computational cells around each grid point. Each computational cell is treated as a control volume for the continuous phase (gas phase). The droplets are treated as sources of mass, momentum and energy inside the each control volume. The gas phase is regarded as a continuum (Eulerian approach) and is described by first solving the gas flow field assuming no droplets are present. Using this continuous phase flow field, droplet trajectories together with size and temperature histories along the trajectories are calculated. The mass, momentum and energy source terms for each cell throughout the flow field is then determined. The source terms are evaluated from the droplet equation and are integrated over the time required to cross the length of the trajectory inside each control volume. The results are multiplied (scaled up) by the number flow rate of drops associated with this trajectory (Crowe et al., 1977; Papadakis and King, 1988; Kuriakose and Anandharamakrishnan, 2010). The gas flow field is solved again, incorporating these source terms and then new droplet trajectories and temperature histories are calculated. This approach provides the influence of the droplets on the gas velocity and temperature fields. The method proceeds iteratively calculating gas and particle velocity fields.

In the CFD simulation, combined Eulerian and Lagrangian model was used to obtain particle trajectories by solving the force balance equation:

$$\frac{d\underline{u}_p}{dt} = \frac{18\mu}{\rho_p d_p^2} \frac{C_D \, Re}{24} (\underline{v} - \underline{u}_p) + \underline{g}\left[\frac{\rho_p - \rho_g}{\rho_p}\right] \qquad \text{Eq (1)}$$

where, v is the fluid phase velocity, u_p is the particle velocity, ρ_g is the density of the fluid and ρ_p is the density of the particle.

The particle force balance (equation of motion) includes discrete phase inertia, aerodynamic drag and gravity. The slip Reynolds number (Re) and drag coefficient (C_D) are given in the following equations.

$$Re = \frac{\rho_g d_p |\underline{u}_p - \underline{v}|}{\mu} \qquad \text{Eq (2)}$$

$$C_D = a_1 + \frac{a_2}{Re} + \frac{a_3}{Re^2} \qquad \text{Eq (3)}$$

where, d_p is the particle diameter, and a_1, a_2 and a_3 are constants that apply to smooth spherical particles over several ranges of Reynolds number (Re) given by Morsi and Alexander (1972).

The velocity of particles relative to air velocity was used in the trajectory calculations (equation 1). Turbulent particle dispersion was included in this model as discrete eddy concept (Langrish and Zbicinski, 1994). In this approach, the turbulent air flow pattern is assumed to be made up of a collection of randomly directed eddies, each with its own lifetime and size. Particles are injected into the flow domain at the nozzle point and envisaged to pass through these random eddies until they impact the wall or leave the flow domain through the product outlet.

The heat and mass transfer between the particles and the hot gas is derived following the motion of the particles.

$$m_p c_p \frac{dT_p}{dt} = hA_p (T_g - T_p) + \frac{dm_p}{dt} h_{fg} \qquad \text{Eq (4)}$$

where, m_p is the mass of the particle, c_p is the particle heat capacity, T_p is the particle temperature, h_{fg} is the latent heat, A_p is the surface area of the particle and h is the heat transfer co-efficient.

The heat transfer coefficient (h) is obtained from the Ranz-Marshall equation.

$$Nu = \frac{hd_p}{k_{ta}} = 2 + 0.6(Re_d)^{1/2}(Pr)^{1/3} \qquad \text{Eq (5)}$$

where, Prandtl number (Pr) is defined as follows

$$Pr = \frac{c_p \mu}{k_{ta}} \qquad \text{Eq (6)}$$

where, d_p is the particle diameter, k_{ta} is the thermal conductivity of the fluid, μ is the molecular viscosity of the fluid.

The mass transfer rate (for evaporation) between the gas and the particles is calculated from the following equation.

$$\frac{dm_p}{dt} = -k_c A_p (Y_s^* - Y_g) \qquad \text{Eq (7)}$$

where, Y_s^* is the saturation humidity, Y_g is the gas humidity and k_c is the mass transfer coefficient and it can be obtained from Sherwood number

$$Sh = \frac{k_c d_p}{D_{i,m}} = 2 + 0.6(Re_d)^{1/2}(Sc)^{1/3} \qquad \text{Eq (8)}$$

where, $D_{i,m}$ is the diffusion coefficient of water vapour in the gas phase and Sc is the Schmidt number, defined as follows

$$Sc = \frac{\mu}{\rho_g D_{i,m}} \qquad \text{Eq (9)}$$

The values of vapour pressure, density, specific heat and diffusion coefficients can be obtained from various sources like Perry (1984), Incropera (2006) etc.

When the temperatures of the droplet has reached the boiling point and while the mass of the droplet exceeds the non-volatile fraction. The boiling rate model was applied (Kuo, 1986):

$$\frac{d(d_p)}{dt} = \frac{4k_{ta}}{\rho_p c_g d_p}(1 + 0.23\sqrt{Re})\ln\left[1 + \frac{c_g(T_g - T_p)}{h_{fg}}\right] \qquad \text{Eq (10)}$$

where, k_{ta} is the thermal conductivity of the gas and c_g is the heat capacity of the gas.

Anandharamakrishnan et al. (2010 a) developed CFD models for short-form and tall-form spray dryers, assuming constant rate drying and including particle tracking using the source-in-cell method. The predictions from these models have been validated against published experimental data and other simulations. This study differed from previous work in that particle time-histories for velocity, temperature and residence time and their impact positions on walls during spray drying have been extracted from the simulations. Figure 1 shows the particles trajectories in a short-form and tall-form spray driers. These short-form and tall-form spray dryer comparative studies suggested that an increase in the chamber diameter: (1) may reduce the particle deposition rates on the cylindrical wall (e.g. in short-form) and (2) can accommodate a wider atomiser spray angle, which improves heat and mass transfer rates. Moreover, this study indicated that a bent outlet pipe inside the chamber increases the gas and particle recirculation; consequently, cold gas was mixed with down-flowing hot inlet gas and dried particles will be exposed to the high inlet gas temperatures.

In the recent years, the application of Reaction Engineering Approach (REA), drying kinetics model, droplet-droplet interactions, unsteady state modelling and population balance model for the simulation of spray dryers is increasing. In the Reaction Engineering Approach, it assumes that evaporation is an activation process to overcome an energy barrier, while it is not the case for condensation or adsorption. REA is a two-way model as compared to the one-way characteristic drying rate curve (CDRC) for predicting the single droplet drying. The basic of REA was described by Chen and Xie (1997) and Chen et al. (2001). The REA model was used by Chen and Xie (1997) for the simulation of drying of thin layer food materials such as kiwifruit, silica gel, potato and apple slices.

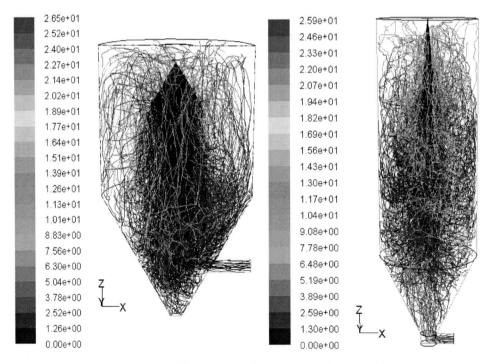

Figure 1. Particle trajectories coloured by residence time (s) (Anandharamakrishnan, et al., 2010 a).

Moreover, Huang et al. (2004) found that this approach (REA) fits in well with the fluent commercial CFD code for spray drying. A comparison of two drying kinetics models namely, characteristic drying curve (CDC) and REA was performed by Woo et al. (2008) and found that the different responses of the REA and CDC to different initial feed moisture conditions due to the respective formulation of the driving forces in the two models. The modelling of spray dryers using population balance method is gaining importance nowadays because this model accounts for the droplet growth, coalescence and break up during the spray drying process. Recently, Handscomb et al. (2009) included source term for droplet coalescence and break up. This approach gives a better understanding of complexities of two phase flows in spray dryers. The model includes nucleation and growth of suspended solids from an ideal binary solution.

SPRAY-FREEZING OPERATIONS

Freeze-drying is a popular method of producing shelf stable particulate products and is of particular value for drying thermally sensitive materials (usually biologically based), which can be heat damaged by higher temperature methods, such as spray-drying. Porous structures are formed from the creation of ice crystals during the freezing stage, which subsequently sublime during the drying stage and this often leads to good rehydration behaviour of the product. It is possible to produce freeze dried produce in powdered form using a technique known as spray freeze drying (Anandharamakrishnan et al., 2010 b), in which a liquid stream containing a dissolved solid is atomised in a manner similar to spray drying, then contacted with a cold fluid to freeze the droplets. These are finally freeze dried, either conventionally or in a fluidised bed. One method of spray-freezing is by contacting with a cold gas. This is a complex process which involves a number of mechanisms: (i) the formation and the motion of individual drops with respect to each other and the gas is determined by the fluid mechanics of the spray, (ii) heat transfer between the gas and the droplets depends on the local conditions, e.g. gas temperature, droplet temperature and droplet-gas slip velocity and (iii) the freezing and ice crystallization within the drops (Anandharamakrishnan et al., 2010 c).

A CFD model for spray-freezing has been developed by Anandharamakrishnan et al. (2010 c) for solid and hollow cone spray operations including with latent heat effects during phase change. The solid cone spray predictions of gas temperature and droplet velocities agreed fairly well with the experimental results. A comparative study with a hollow cone spray suggested that a hollow cone yields lower particle temperatures and a greater extent of freezing. A knowledge of particle impact positions is important for designing and operating spray-freezing equipment.

Figure 2 depicts the particle impacts on the chamber walls and this simulation indicated that a large fraction of the particles (65%) strike the conical part of the spray-freezing chamber; 11% of particles hit the cylindrical part of the wall, and only a small proportion (22%) of the particles come directly out of the chamber. These results were in reasonably good agreement with the experimental observations.

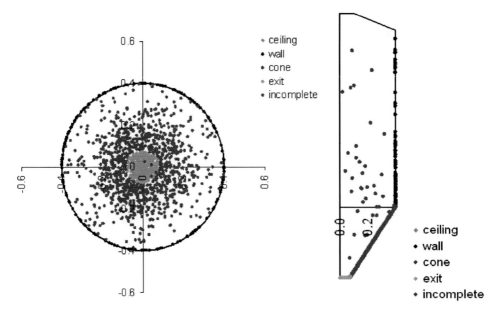

Figure 2. CFD simulation of particle impact position on the cone (left) and walls (right) during spray freezing operation (Anandharamakrishnan et al., 2010c).

BREAD BAKING PROCESS

Bread baking is a complex process where various physiochemical and biological transformations take place simultaneously such as evaporation of water, gelatinization of starch, volume expansion, crust formation, denaturation of protein and browning reactions etc. (Purlis and Salvadori 2009). Producing high quality bread is a great challenge in the baking process; because unsatisfactory product has to be discarded as baking is a irreversible process and this could be economically unfavorable. Water content and temperature are responsible for physiochemical and biological processes. Variation in color and moisture content occurs due to the non-uniform temperature distribution inside the oven. These variations can be minimized by the proper design of the oven as well as maintaining proper processing conditions such as air temperature, heating power, baking time and bread size (Therdthai and Zhou 2003; Zhou and Therdthai 2007; Wong and others 2007; Chhanwal et al., 2010). The placement of the bread is also important, as it changes flow pattern of heat and thus affects various transformation processes of bread baking. In continuous process, bread is exposed to different heating cabinets due to U-turn movement etc; hence it gets uniform heating, where as in electrical heating ovens are operated in batch mode and bread is almost stationary throughout the baking process. Therefore, placement of bread plays a vital role in deciding final quality of bread (Anishaparvin et al., 2010). However, studies on electrical heating batch baking process are very few in the literature.

During baking process, heat transfer occurs by radiation, conduction and convection. Radiation is the most dominant heat transfer mode in an electrical heating oven. Radiative heat transfer occurs from red hot heating coils and hot metal surfaces in the form of electromagnetic waves to the surroundings. Radiative heat transfer occurs by photons which were emitted by the respective surface and travel in straight lines without attenuation. Heating by radiation depends on emissivity of surface and higher the emissivity higher is the heating rate (Abraham and Sparrow, 2002). Air comes in contact with the heat source and hot metal part of walls and heat is transferred through convection mode. Heat transfer from air to the product surface occurs by convection mode, which is transferred from metal container to the bread by conduction (Sablani et al., 1998). Abraham and Sparrow (2002; 2004); Sparrow and Abraham (2003) studied extensively the heat transfer in an electrical heating oven with a variety of geometrical, radiative source and operating conditions of oven and also placement of thermal load.

In recent years, CFD has been widely used to design and development of baking ovens and also to study the baking process. Three dimensional CFD model for industrial continuous bread baking process was developed to study temperature profile of the bread in the moving trays (Therdthai et al., 2004b). Later, Wong et al. (2007) simulated industrial continuous bread baking process involving U-turn movement using discrete ordinates (DO) radiation model. They also analyzed the physical properties of dough/bread using combined experimental and CFD.

Recently, Chhanwal et al. (2010) developed a CFD model for the electrical heating domestic baking oven using DTRM, S2S and DO radiation models and validated with the experimental results. This simulation study showed that the electric oven attained uniformity of heating in twenty minutes (pre-heating cycle) and the slowest heating zone was present near the oven wall due to the lower air flow pattern. The DO model was selected for the product simulation based on the applicability. The simulation of the oven with bread indicated that both heating pattern and time were most important in influencing the temperature profile inside the bread. This developed model was validated with the experimental measurement of bread temperature at different locations. The time, temperature variation in the bread was obtained from the simulation. This 3D model predicted better approximation of crust and crumb temperature and baking time during the bread baking at stationary position as compared to 1D mathematical model.

A combined CFD and kinetic model was developed by Chhanwal et al. (2010) for starch gelatinization at crumb and crust. Starch gelatinization is the disaggregation or irreversible swelling of starch granules within an aqueous medium at suitable temperature. It is a progressive swelling due to hydration of starch molecules in a temperature range of 50-85°C (Lund, 1984; Zanoni et al.,1995a). Starch gelatinization process involves swelling, melting, disruption of starch molecule, and exudation of amylose (Zanoni et al., 1995a). Therdthai et al. (2004a) integrated the starch gelatinization kinetic model developed by Zanoni et al. (1995a) to obtain the baking index of the product.

Gelatinization properties depend on the kind and origin of the starch. Gelatinization of starch causes disruption of the intermolecular hydrogen bonds which maintain structural integrity of the granule and an exudation of amylose. During gelatinization, granules hydrate and swell and there is an increase in viscosity of the medium (Lund, 1984). Extent of starch gelatinization can be used as minimum baking index in industrial baking process which decides the baking time of bread.

Zanoni et al. (1995 a,b) studied starch gelatinization using differential scanning calorimetric (DSC) method and calculated the model parameters such as k_0 and E_a.

$$1 - \alpha = \exp(-kt) \qquad \text{Eq (11)}$$

where α is degree of starch gelatinization, k the reaction rate constant and t the time in seconds. The reaction rate constant (k) can be calculated using Arrhenius equation,

$$k = k_o \exp\left(-\frac{E_a}{RT}\right) \qquad \text{Eq (12)}$$

where, $k_0 = 2.8 \times 10^{18}$ s^{-1} and $E_a = 138$ kJ/mol (Zanoni et al., 1995b).

These equations were used by Chhanwal et al. (2010) to develop a CFD simulation model for the prediction of starch gelatinization as shown in Figure 3. Starch gelatinization completes within 900s of baking process as temperature at center of bread crosses 80°C. Formation of different starch gelatinization layers were observed identical to temperature layers inside bread slice.

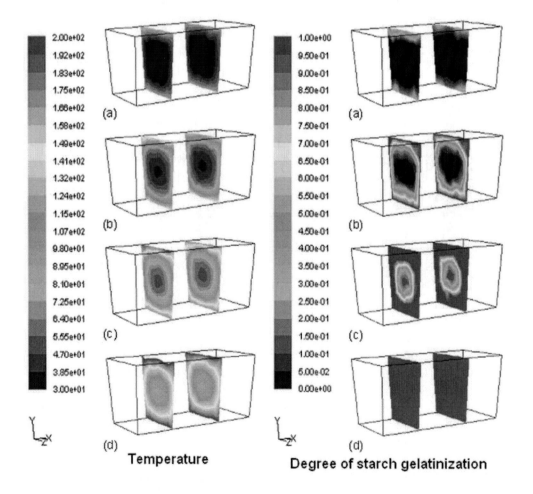

Figure 3. Comparison of temperature and starch gelatinization profile in crumb during bread baking process at (a) 60 s (b) 300 s and (c) 600 s (d) 900 s at 4 cm from center of the bread in radial direction (Chhanwal et al., 2010).

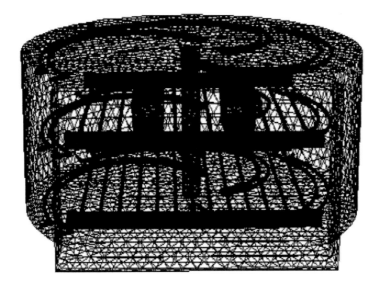

(a)

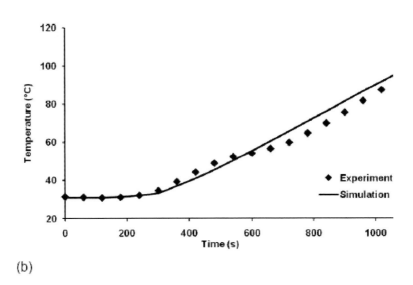

(b)

Figure 4. (a) Geometry of oven meshed with bread; (b) Experimental validation (Anishaparvin et al., 2010).

Baking process involves complex heat and mass transfer and air flow. Factors affecting the heat distribution in the oven chambers are location of heat source, air flow, product load and placement and baking time. Anishaparvin et al. (2010) developed a three dimensional CFD model for pilot-scale electric heating baking oven and validated with the experimental measurement of bread temperature. Figure 4 shows the geometry of the oven and experimental validation of the simulation results. This study was extended also to investigate the effect of placement of bread on temperature and baking time. This study indicated that vent position and placement of bread are the most important factors in influencing the air temperature profile inside the oven cavity and in turn bread quality eventually. The baking time, temperature variations in the bread were obtained from the simulation. Due to air flow pattern, breads placed in top tray bakes quickly as compared to breads placed in bottom tray. This model can be used for predicting crust and crumb temperatures during the bread baking in an electrical heating batch oven. Anishaparvin et al. (2010) studies indicated that apart from temperature and time, the placement of bread also influences the final bread quality during batch baking process in an electrical heating oven.

PASTEURIZATION OF EGG

Eggs are consumed all over the world and they have been recognized as a highly nutritive food. Intact, fresh, clean shell eggs and derived egg products (health food using raw eggs, salads, sauces, ice-cream and mayonnaise) can get spoiled easily and may also contain *Salmonella enteritidis* bacteria, which can be a major cause of food-borne illnesses. Previous studies have demonstrated that *Salmonella enteritidis* in raw eggs can be adequately killed by thermal pasteurization (Hou, et al., 1996; Schuman et al., 1997). Pasteurization of liquid egg products has been in practice since 1960s. Thermal energy is normally used for pasteurization since number of pathogenic microorganisms is drastically reduced by applying heat either using hot water or steam. The pasteurization of individual components of egg such as egg white and yolk is carried out routinely using heat exchanger devices. Although, egg products in the form of frozen, liquid or dried products are available in plenty, but consumers still prefer eggs in the intact form rather than the liquid egg products because of the flexibility of consumption in terms of handling and storage. Therefore, the possibility of pasteurizing intact eggs was explored in the 1990's (Denys et al., 2004).

The main need for CFD analysis of pasteurization is to determine the uniform and effective heat distribution in the egg and to examine the position of slowest heating zone (SHZ). SHZ is defined as the location, which receives minimum heating. During pasteurization of egg, heat transfer occurs through conduction as well as convection. Relatively, few works have been published related to applications of CFD in thermal processing of egg. Denys et al. (2003) proposed the CFD and experimental approach for determining the surface heat transfer co-efficient during thermal processing. A combined conductive and convective heat transfer model was developed for egg by Denys et al. (2004) and they found good agreement between experimental and simulation results. Further, the same group (Denys et al., 2005) adopted a first order kinetic model for studying the inactivation of *Salmonella enteritidis* with respect to position of the yolk.

Recently, Radhika et al. (2010) investigated the effects of pasteurization of egg at 55.6°C on pasteurization time (i.e. time to attain the maximum temperature of 55.6°C in whole egg) in stationary as well as two different rpms (2.5 and 5 rpm) in rotation modes. The temperature profiles of a stationary and rotating egg during the thermal pasteurization process at various process times of 30 s, 150 s, and 300 s were shown in Figure 5 (a-f). It shows that the egg white reaches the set pasteurization temperature much faster in a rotating egg due to heat penetration throughout the egg. However, yolk portion took more time to heat up compared to egg white in rotating position (Figure 5 e-f), but lesser than the time taken in a stationary egg (Figure 5 c). The SHZ was concentrated at the center of the yolk throughout the rotating process and maintained a circular shape (Figure 5 e) unlike the SHZ in a stationary egg (Figure 5 b). Thus, a rotating egg achieves almost uniform temperature in approximately 300 s (Figure 5 f), while a stationary egg still possesses a SHZ at a temperature of 37°C at the end of 300 s processing time (Figure 5 c). Thus, rotation improves the efficiency of a heating process by ensuring more uniform heating as compared to static heating. The studies by Radhika et al. (2010) indicated that thermal pasteurization under rotated egg reduced the time of pasteurization by a large extent. Moreover, based on temperature predictions locally calculated F-value indicated that the time required for inactivated the *Salmonella* was drastically reduced by rotation of egg compared to static position. Hence, it is advantageous to adopt rotation as an additional unit operation during thermal pasteurization of eggs for producing high quality intact eggs without affecting its functional properties (Radhika et al., 2010).

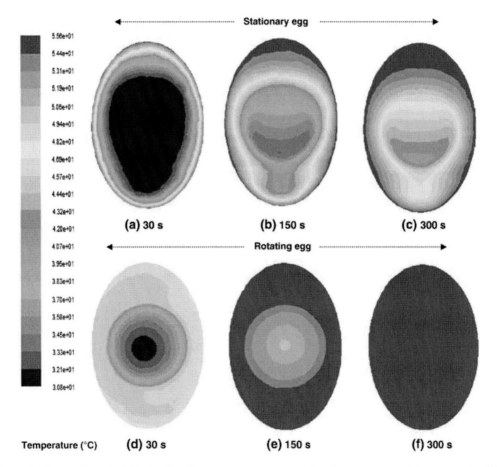

Figure 5. Comparison of CFD simulated temperature contours of stationary (a-c) and 5 rpm rotating (d-e) egg heated at 55.6°C (Radhika et al., 2010).

THERMAL PROCESSING OF CANNED LIQUID FOODS

During thermal processing of canned liquid foods, the processing time and temperature should be maintained at shorter intervals due to the heat sensitivity of the various nutrients. In this principle, the heat treatment should be designed in such a way that all the components with a positive effect on the product quality should not be damaged (e.g. vitamins, proteins). On the other hand, components with a negative impact (e.g. spoilage bacteria and enzymes) should be destroyed. Moreover, uniform and effective heat distribution in the canned liquid foods during pasteurization process is very important to improve the thermal processing.

Table 1. CFD simulations studies on sterilization of canned liquid foods

Product	Model description	Findings	Reference
Carrot orange Soup	Analysis of the thermal destruction of Vitamin C in food pouches	Migration of the slowest heating zone (SHZ) towards the bottom of the pouch as heating proceeds. They found that vitamin profile depend fluid velocity profiles in the pouch apart from the temperature distributions.	Ghani et al. (2002a)
Carrot orange Soup	Sterilization of canned liquid food (carrot orange soup) in a metal can horizontally heated	The faster heating was observed in vertical heated can than the horizontally heated.	Ghani et al. (2002b)
Beef-Vegetable soup	Analysis of thermal inactivation of *Bacillus stearothermophilus* in food pouches	The study indicated that concentration of live bacteria depends on both the temperature distribution and flow patterns as sterilization proceeds.	Ghani et al. (2002c)
Carrot orange Soup	Effect of can rotation on liquid food sterilization	Significant effect of can rotation on the shape, size and location of the SHZ. The combined effect of natural and forced convection splits the SHZ into two distinct regions	Ghani et al. (2003)
Carrot Orange Soup	Technique of estimating the sterilization time	A new correlation has been developed for the prediction of the sterilization time of liquid food in vertical and horizontal cans.	Farid and Ghani (2004)
Carboxy-methyl cellulose	Effect of sterilization through wall inclination and can geometry modifications	Considerable effect of wall inclination and orientation angle in can geometry on the performance of the sterilization of CMC	Varma and Kannan (2005)
Carboxy-methyl cellulose	Natural convective heating of canned food in conical and cylindrical containers	Enhancement of sterilization process without agitation through simple modifications of conventional cylindrical can.	Varma and Kannan (2006)
Pineapple slices	Analysis of thermal sterilization of solid–liquid food mixture in cans	Natural convection is strongly controls the rate of heating, liquid flow pattern and on the shape and movement of the SHZ.	Ghani and Farid (2006)
Tomato soup	Reusable pouch development for long term space missions	Presence of significant hot and cold zones, Requires further optimized pouch design for more uniform heating.	Jun and Sastry (2007)
Carboxy-methyl cellulose	Heat transfer analysis of canned food in a still retort	Comparison of heat transfer coefficients on absolute mass flow averaged and volume averaged temperatures. Influence of temperature difference on particular Nusselt numbers is discussed.	Kannan and Gourisankar (2008)
Canned peas	To determine temperature changes inside a can containing solid–liquid food mixtures (peas in water)	The CFD simulation predictions were validated with the experimental measurements. This 2D CFD simulation study provides some insights for modeling of canned solid–liquid mixtures.	Kiziltas et al. (2010)

Thermal processing time will depend upon the time required to achieve uniform temperature inside the can. It is very difficult to find out temperature profile inside the canned liquid foods in industrial scale continuous operation. Moreover, it is hard to experimentally measure temperature profile at different location inside the can as insertion of thermocouples will disturb the natural convection heating. CFD can predict the temperature distribution inside the can along with position of SHZ at various time stages. The detail descriptions of CFD models and findings related to the sterilization of canned liquid foods are given in the Table 1.

Relatively, few works have been published related to CFD simulation of canned foods: natural convective heating of canned foods (Ghani et al., 1999); effect of temperature on bacterial deactivation and vitamin destruction in 3D pouches (Ghani et al., 2002 a); technique to determine the sterilization time (Farid and Ghani, 2004). Varma and Kannan (2005) studied the effect of sterilization through wall inclinations and can geometry modifications, besides they also simulated natural convective heating of canned food in conical and cylindrical pouches (Varma and Kannan, 2006). Ghani and Farid (2006) developed a CFD model for the thermal sterilization of solid-liquid mixture food in cans. Kannan and Gourisankar (2008) simulated heat transfer analysis of canned food in a still retort. Kiziltas (2010) modeled the sterilization of canned solid-liquid (peas-water) using CFD. This model prediction was validated with the experimental measurements. Moreover, this 2D assumption of canned solid-liquid approach may help in reducing the computational requirements for further studies. Recently, Anandpaul et al. (2010) studied the pasteurisation of canned milk using CFD simulations. Their model predictions were validated with the experimental measurements. was extended from stationary can position heating to the rotational processing. In their study, a uniform heating (due to absence of SHZ) of milk was achieved in the rotated canned milk processing with reduced processing time.

CONCLUSIONS

In this chapter, recent advances in the application of CFD in spray drying operations, spray-freezing, bread baking and sterilization of canned liquid foods were illustrated. However, more works needs to be performed with population balance modeling in spray drying simulations for the account of droplet growth and breakage. In a bread baking modeling, the volume expansion of bread needs to be included for better predictions more close to the reality. Still, modeling of pasutrization/sterilization of canned solid food materials with liquid is challenging for CFD modelers. Nevertheless, it is clear from the above reports, CFD can be used for analyzing the flow problems in food processing operations. The present commercial CFD codes have some limitations to model simultaneous heat and mass transport during drying and baking process. Now, with the increasing computing power, the CFD will be a valuable tool for the food industry in solving complex fluid flow, heat and mass transfer problems and aid in better design and process control of unit operation in food processing. Moreover, these CFD modelling approach may leads to higher productivity and better quality food products.

ACKNOWLEDGMENTS

Authors wish to thank Dr. V. Prakash, Director, CFTRI, for the encouragement and shown keen interest in CFD simulation studies for food processing. We wish to acknowledge the Department of Science and Technology (DST), Government of India for the financial support to this work. Authors also acknowledge the CSIR for the financial support through Network project (NWP-02) for fluent 6.3 software licensing.

REFERENCES

Abraham, J.P., and Sparrow, E.M. (2002). Heat transfer characteristics of vented/unvented enclosures for various radiation surface characteristics of the thermal load, enclosure temperature sensor, and enclosure walls. *International Journal of Heat and Mass Transfer, 45*, 2255–2263.

Abraham, J.P., and Sparrow, E.M. (2004). A simple model and validating experiments for predicting the heat transfer to a load situated in an electrically heated oven. *Journal of Food Engineering, 62*, 409–415.

Anandharamakrishnan, C. (2003). Computational fluid dynamics (CFD) – applications for the food industry. *Indian Food Industry, 22* (6), 62-68.

Anandharamakrishnan, C., Rielly, C. D., and Stapley, A. G. F. (2007) Effects of process variables on the denaturation of whey proteins during spray-drying. *Drying Technology, 25,* 799-807.

Anandharamakrishnan, C., Gimbun, J., Stapley, A. G. F., and Rielly, C. D. (2010 a). A study on particle histories during spray drying using computational fluid dynamic simulations (CFD), *Drying Technology, 28*, 566-576.

Anandharamakrishnan, C., Stapley, A. G. F., and Rielly, C. D. (2010 b). Spray-freeze-drying of whey proteins at sub-atmospheric pressures, *Dairy Science and Technology, 90*, 321-334.

Anandharamakrishnan, C., Gimbun, J., Stapley, A. G. F., and Rielly, C. D. (2010 c). Application of computational fluid dynamic (CFD) simulations to spray-freezing operations. *Drying Technology, 28*, 94-102.

Anandpaul, D., Anishaparvin, A., and Anandharamakrishnan, C. (2010) "Computational Fluid Dynamics Studies on Pasteurization of Canned Milk" International Journal of Dairy Technology (In-Press) doi: 10.1111/.j.1471-0307.2010.00663.x.

Anderson JD. (1984). Computational fluid dynamics - The basics with applications. New York: McGraw-Hill Inc.

Anishaparvin, A., Chhanwal, N., Indrani, D., Raghavarao, K.S.M.S., and Anandharamakrishnan, C. (2010). An investigation of bread baking process in a pilot-scale electrical heating oven using computational fluid dynamics. *Journal of Food Science (In- press)* doi:10.1111/j.1750-3841.2010.01846.x .

Chen, X. D., Pirini, W., and Ozilgen, M. (2001). The reaction engineering approach to modeling drying of thin layer of pulped kiwi fruit flesh under conditions of small biot numbers. *Chemical Engineering and Progress, 40*, 165–181.

Chen, X. D., and Xie, G. Z. (1997). Fingerprints of the drying behavior of particulate or thin layer food materials established using a reaction engineering model. *Transactions of the Institution of Chemical Engineers, Part C, 75,* 213–222.

Chhanwal, N., Anishaparvin, A., Indrani, D., Raghavarao, K.S.M.S., and Anandharamakrishnan, C. (2010). Computational fluid dynamics (CFD) modeling of an electrical heating oven for bread baking process. *Journal of Food Engineering, 100,* 452-60.

Crowe, C. T., Sharam, M. P., and Stock, D. E. (1977). The particle source in cell (PSI-Cell) model for gas-droplet flows. *Journal of fluid Engineering, 9,* 325-332.

Denys, S., Dewettinck, K., and Pieters, J. G. (2005). CFD analysis for process impact assessment during thermal pasteurization of intact eggs. *Journal of Food Protection, 68,* 366-374.

Denys, S., Pieters, J. G., and Dewettinck, K. (2003). Combined CFD and experimental approach for determination of the surface heat transfer coefficient during thermal processing of eggs. *Journal of Food Science, 68,* 943-951.

Denys, S., Pieters, J. G., and Dewettinck, K. (2004). Computational fluid dynamics analysis of combined conductive and convective heat transfer in model eggs. *Journal of Food Engineering, 63,* 281-290.

Farid, M. M., and Ghani, A. G. (2004). A new computational technique for the estimation of sterilization time in canned food. *Chemical Engineering and Processing, 43,* 523-531.

Fletcher, D., Guo, B., Harvie, D., Langrish, T., Nijdam, J., and Williams, J. (2003). What is important in the simulation of spray dryer performance and how do current CFD models perform. *3rd International Conference on CFD in the Minerals and Process Industries,* CSIRO, Melbourne, Australia, 10-12 December 2004.

Ghani, A. G., and Farid, M. M. (2006). Using the computational fluid dynamics to analyze the thermal sterilization of solid-liquid food mixture in cans. *Innovative Food Science and Emerging Technologies, 7,* 55-61.

Ghani, A. G., Farid, M. M., Chen, X.D., and Richards P. (1999). Numerical simulation of natural convection heating of canned food by computational fluid dynamics. *Journal of Food Engineering, 41,* 55-64.

Ghani, A. G., Farid, M. M., and Chen, X. D. (2002a). Theoretical and experimental investigation of the thermal destruction of Vitamin C in food pouches. *Computers and Electronics in Agriculture, 34,* 129–143.

Ghani, A. G., Farid, M. M., and Chen, X. D. (2002b). Numerical simulation of transient temperature and velocity profiles in a horizontal can during sterilization using CFD. *Journal of Food Engineering, 51,* 77-83.

Ghani, A. G., Farid, M. M., and Chen, X. D. (2002c). Theoretical and Experimental Investigation of Thermal Inactivation of *Bacillus stearothermophilus* in Food Pouches. *Journal of Food Engineering, 51,* 221–228.

Ghani, A. G., Farid M. M., and Zarrouk, S. J. (2003). The effect of can rotation on sterilization of liquid food using computational fluid dynamics. *Journal of Food Engineering, 57,* 9–16.

Handscomb, C. S., Kraft, M., and Bayly, A. E. (2009). A new model for the drying of droplets containing suspended solids. *Chemical Engineering Science, 64,* 628 – 637.

Hou, H., Singh, R. K., Muriana, P. M., and Stadelman, W. J. (1996). Pasteurization of intact shell eggs. *Food Microbiology, 13,* 93–101.

Huang, L. X., Kumar, K., and Mujumdar, A. S. (2004). Simulation of a spray dryer fitted with a rotary disk atomizer using a three-dimensional computational fluid dynamic model. *Drying Technology, 22 (6)*, 1489 -1515.

Incropera (2006). Fundamentals of heat and mass transfer. 6th ed. John Wiley and Sons Inc, New York.

Jun, S. and Sastry S. (2007). Reusable pouch development for long term space missions: A 3D ohmic model for verification of sterilization efficacy. *Journal of Food Engineering, 80,* 1199–1205.

Kannan, A., and Gourisankar, S.P.Ch. (2008). Heat transfer analysis of canned food sterilization in still retort. *Journal of Food Engineering, 88,* 213–228.

Kiziltas, S., Erdogdu, F., and Palazoglu, T. K (2010). Simulation of heat transfer for solid–liquid food mixtures in cans and model validation under pasteurization conditions. *Journal of Food Engineering, 97,* 449–456.

Kuo, K. K. Y. (1986). Principles of combustion. *John Wiley and Sons,* New York.

Kuriakose, R., and Anandharamakrishnan, C. (2010). Computational fluid dynamics (CFD) application in spray drying of food products, *Trends in Food Science and Technology 21,* 383-398.

Langrish, T. A. G., and Zbicinski, I. (1994). The effects of air inlet geometry and spray cone angle on the wall deposition rates in spray dryer. *Transactions of the Institution of Chemical Engineers, Part A, 72,* 420-430.

Langrish, T. A. G., and Fletcher, D. F. (2001). Spray drying of food ingredients and applications of CFD in spray drying. *Chemical Engineering and Processing, 40,* 345-354.

Lund D. B. (1984). Influence of time, moisture, ingredients, and processing conditions on starch gelatinization. *Critical Reviews in Food Science and Nutrition 20,* 249-73.

Morsi, S. A., and Alexander, A. J. (1972). An investigation of particle trajectories in two-phase flow systems. *Journal of Fluid Mechanics, 55(2),* 193-208.

Norton, T., and Sun, D. W. (2006). Computational fluid dynamics (CFD) - an effective and efficient design and analysis tool for the food industry: A review. *Trends in Food Science and Technology, 17,* 600-620.

Papadakis, S. E., and King, C. J. (1988). Air temperature and humidity profiles in spray drying. 1. Features predicted by the particle source in cell model. *Industrial Engineering Chemistry Research, 27,* 2111-2116.

Parviz, M., and John, K. (1997). Tackling turbulence with supercomputers. *Scientific American, 1,* 276.

Perry, R. H., and Chilton, C. H. (1984). Perry's chemical engineers handbook. *McGraw-Hill,* London.

Purlis, E., and Salvadori, V. O. (2009). Bread baking as a moving boundary problem. Part 1: Mathematical modeling. *Journal of Food Engineering, 91,* 428 – 433.

Radhika R., Deepti, M., Anishaparvin, A., and Anandharamakrishnan, C. (2010). Computational fluid dynamics simulations studies on pasteurization of egg in stationary and rotation modes. *Innovative Food Science and Emerging Technology*, In-Press, doi:10.1016/j.ifset.2010.11.008.

Roache P. J. (1976). Computational fluid dynamics. *Albuquerque*, NM: Hermosa Publishers.

Sablani, S. S., Marcotte, M., Baik, O. D., and Castaigne, F. (1998). Modeling of simultaneous heat and water transport in the baking process. *LWT-Food Science and Technology, 31,* 201 – 209.

Schaldach, G., Berger, L., Razilov, I., and Berndt, H. (2000). Computer simulation for fundamental studies and optimization of ICP spray chambers. *ISAS (Institute of Spectrochemistry and Applied Spectroscopy) Current Research Reports*, Berlin, Germany.

Schuman, J. D., Sheldon, B. W., Vandepopuliere, J. M., and Ball, H. R. (1997). Immersion heat treatments for inactivation of *Salmonella Enteritidis* with intact eggs. *Journal of Applied Microbiology*, *83*, 438–444.

Scott, G. M. (1994). Computational fluid dynamics for the food industry. *Food Technology International*, Europe, 49–51.

Scott, G. M., and Richardson, P. (1997). The application of computational fluid dynamics in the food industry. *Trends in Food Science and Technology*, 8 (4), 119–124.

Sparrow, E. M., and Abraham, J. P. (2003). A computational analysis of the radiative and convective process that take place in preheated and non-preheated ovens. *Heat Transfer Engineering, 24 (5),* 25 – 37.

Sun, D. W. (2007). Computational fluid dynamics in food processing. *CRC Press,* Taylor and Francis group, Boca Raton.

Tannehill J. C., Anderson D. A., and Pletcher R. H. (1997). Computational fluid mechanics and heat transfer, *Taylor and Francis*, 2nd ed. Philadelphia.

Therdthai, N., and Zhou, W. (2003). Recent advances in the studies of bread baking process and their impact on the bread baking technology. *Food Science and Technology Research, 9 (3),* 219 – 226.

Therdthai, N., Zhou, W., and Adamczak, T. (2004a). Simulation of starch gelatinization during baking in a traveling – tray oven by integrating a three dimensional CFD model with a kinetic model. *Journal of Food Engineering, 65,* 543 – 550.

Therdthai, N., Zhou, W., and Adamezak, T. (2004b). Three-dimensional CFD modeling and simulation of the temperature profiles and airflow patterns during a continuous industrial baking process. *Journal of Food Engineering, 65,* 599 – 608.

Varma, M. N., and Kannan, A. (2005). Enhanced food sterilization through inclination of the container walls and geometry modifications. *International Journal of Heat and Mass Transfer, 48,* 3753–3762.

Varma, M. N., and Kannan, A. (2006). CFD studies on natural convective heating of canned food in conical and cylindrical containers. *Journal of Food Engineering, 77,* 1024–1036.

Versteeg, H. K., and Malalasekera, W. (1995). An Introduction to computational fluid dynamics, Pearson Education Ltd, Essex, England.

Wong, S.Y., Zhou, W., and Hua, J. (2007). CFD modeling of an industrial continuous bread-baking process involving U-movement. Journal of Food Engineering, 78, 888–96.

Woo, M. W., Daud, W. R. W., Mujumdar, A. S., Talib, M. Z. M., Wu, Z. H., and Tasirin, S. M. (2008). Comparative study of droplet drying models for CFD modelling. *Chemical Engineering Research and Design, 86(9),* 1038-1048.

Xia, B., and Sun, D. W. (2002). The application of computational fluid dynamics (CFD) in the food industry: A review, *Computers and Electronics in Agriculture, 34,* 5-24.

Zanoni, B., Schiraldi, A. and Simonetta, R. (1995a). A naive model of starch gelatinization kinetics. *Journal of Food Engineering, 24,* 25 – 33.

Zanoni, B., Peri, C., Bruno, D. (1995b). Modeling of starch gelatinization kinetics of bread crumb during baking. *LWT-Food Science and Technology, 28,* 314 – 318.

Zhou, W., and Therdthai, N. (2007). Three-dimensional modeling of a continuous industrial baking process. In D.W. Sun, *Computational Fluid Dynamics in Food Processing.* Boca Raton: CRC Press: p.287-312.

In: Computational Fluid Dynamics
Editor: Alyssa D. Murphy

ISBN 978-1-61209-276-8
© 2011 Nova Science Publishers, Inc.

Chapter 10

NUMERICAL STUDY ON STEAM-JET VACUUM PUMP BY COMPUTATIONAL FLUID DYNAMICS APPROACH

*Xiaodong Wang[*1], Jiangliang Dong[1,2], Hongjian Lei and Jiyuan Tu[2]*

[1]Vacuum and Fluid Engineering Research Centre,
School of Mechanical Engineering and Automation,
Northeastern University, Shenyang 110004, P.R. China
[2]School of Aerospace, Mechanical and Manufacturing Engineering,
RMIT University, Victoria 3083, Australia

ABSTRACT

Steam-jet vacuum pump is one of the important equipments widely used in industry to obtain a vacuum environment for various special techniques. The primary fluid (steam) with high pressure is accelerated through a nozzle to obtain supersonic speed. The supersonic motive steam and secondary fluid mix in mixing chamber with energy and momentum exchanging. A normal shock wave is induced in throat and the flow speed suddenly drops to subsonic value. Further compression is achieved when the mixed stream passes through diffuser. The flow is complicated in the pump due to the transonic flow and difficult to be described by traditional methods. Computational fluid dynamics (CFD) can be used to investigate and predict the complicated flow problems in steam-jet pump.

A mathematic model for transonic flow was proposed to investigate the mixing flow behaviors of primary and secondary fluids in steam-jet vacuum pump. The simulation was carried out to predict the state pressure distribution among mixing chamber wall. Close agreements between the predicted results and experimental data validates the theoretical model. The velocity vectors and Mach number profiles in mixing chamber at different back pressures and the secondary fluid pathlines and mass flux profiles at different suction pressures were predicted. It is found that there are swirls separated from secondary fluid near the wall and the velocity of secondary fluid was fallen down obviously when the back pressure was bigger than critical back pressure. The above two

[*] E-mail: xdwang@mail.neu.edu.cn

factors lead to the entrainment ratio reduced rapidly. The flow structure in mixing chamber would be broken down and secondary flow would reverse to upstream if back pressure is in excess of the break down pressure, and the steam-jet pump lost pumping ability completely. It is also found that the mass flux increased with the increasing of suction pressures which made the entrainment ratios increased. The prediction results show that the pressure ratio is a dominant position in affecting the pump's performances.

It is found that there are spontaneously condensing phenomena in the nozzle supersonic flow process at different primary fluid initial parameters, as the primary fluid is not assumed as perfect gas. The outlet pressure of nozzle predicted is higher than that as the primary fluid assumed as perfect gas, and the outlet velocity is lower than that of general simulations, and then the efficiency of nozzle and steam-jet pump would be reduced.

Keywords: Steam-jet pump; Transonic flow; Spontaneous condensation; CFD modeling; Numerical simulations.

NOMENCLATURE

ρ	density
u_i, u_j	velocity
u'	fluctuation velocity
P	pressure
τ_{ij}	stress tensor
E	total energy
α_{eff}	effective thermal conductivity
μ_{eff}	effective dynamic viscosity
k	turbulent kinetic energy
T	thermal temperature
μ_t	eddy viscosity
	turbulence kinetic energy dissipation
S_{ij}	strain rate
$C_2, C_{1\varepsilon}, C_{3\varepsilon}, \sigma_k, \sigma_\varepsilon$	model coefficients
v	kinematic viscosity
S_k, S_ε	source terms
β	mass fraction
ρ, ρ_l, ρ_v	mixture density, liquid density, vapour density
Γ	mass generation rate
$\bar{r}$	droplet average radius
r^*	critical droplet radius
I	nucleation rate
η	droplet number density
q_c	evaporation coefficient
θ	non-isothermal correction factor
σ	droplet surface tension
K	Boltzmann constant

M	molecular mass
γ	specific heat capacities ratio
h_{lv}	specific enthalpy of evaporation
R	gas-law constant
S	super saturation ratio
C_p	isobaric heat capacity
T_0	droplet temperature
V_d	average droplet volume
B, C	virial coefficients
μ	dynamic viscosity

1 INTRODUCTION

Steam-jet vacuum pump is one of the important equipments widely used in chemistry, petroleum, metallurgy, refrigeration and food industry to obtain a vacuum environment for various special techniques. A steam-jet pump consists of a nozzle, a mixing chamber, throat and a diffuser as shown in Figure 1 [1], and pressure and velocity profiles along the pump axis are described in the same diagram. The primary fluid (steam) with high pressure (0.3-1.6MPa) is accelerated through nozzle and obtained supersonic speed and results in a low pressure (vacuum) at the outlet of nozzle. The supersonic primary steam entrains and draws the secondary fluid into the mixing chamber, where the secondary steam is accelerated. The mixing process with energy and momentum exchanging between primary and secondary flow happened through mixing chamber. A normal shock wave is induced in the throat and the speed of the mixing fluid suddenly drops to subsonic value. Further compression is achieved when the mixed fluid passes through the diffuser. To attain lower suction pressure (higher vacuum), multistage steam-jet system can be established by series ejectors.

The main pumping performance can be described by entrainment ratio (E_m, the ratio of mass flow rate of secondary fluid to that of primary fluid) affected by pressure ratio (K, the ratio of back pressure to the suction pressure) and expanding ratio (B, the ratio of primary fluid pressure to the secondary fluid pressure) separately [2].

The typical performance curve for special primary fluid parameters and secondary fluid pressure is shown in Figure 2 [3]. There are three regions: choked flow in the mixing chamber (critical mode), unchoked flow in the mixing chamber (subcritical mode) and reversed flow in the mixing chamber (malfunction). At the back pressure (P_B, outlet pressure of pump) below the critical value (P_B^*), the entrainment ratio remains constant. When the P_B is increased higher than the P_B^*, the secondary flow varies and the entrainment ratio begins to fall down rapidly.

The mixing flow structure is complicated in the pump due to the transonic flow and it is difficult to be described by traditional methods [2]. Computational fluid dynamics (CFD) is a good choice as a research tool to investigate and predict the complicated flow problems in steam-jet pump [4-6].

The following focuses on the relationship between operating conditions and the flow features in vacuum pump to understand the inner flow mechanism and its affection to the pumping performances.

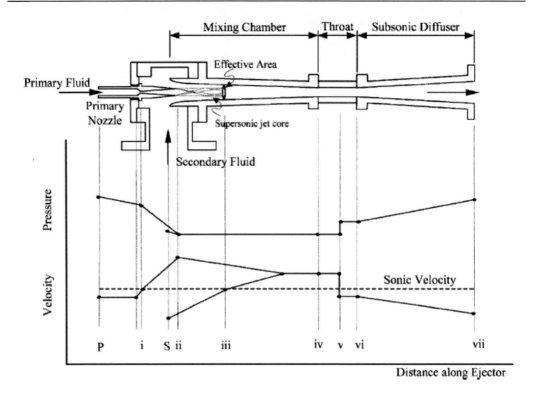

Figure 1. Steam-jet vacuum pump and flow characteristics.

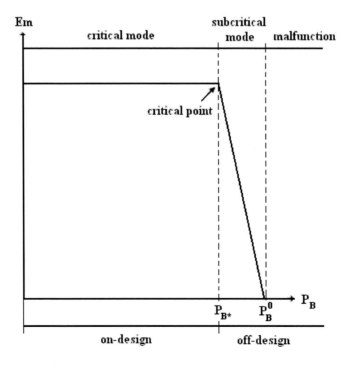

Figure 2. Typical performances of steam-jet pump.

2. MATHEMATICAL MODELS

2.1. Governing Equations

The flow in steam-jet pump is governed by the compressible steady-state axisymmetric form of the fluid flow conservation equations. For variable density flows, the Favre averaged Navier–Stokes equations are more suitable and will be used in this chapter. The total energy equation including viscous dissipation is also included and coupled to the set with the perfect gas law. The governing equations can be written in their compact Cartesian form:

The continuity equation:

$$\partial \rho / \partial t + \partial (\rho u_i) / \partial x_i = 0 \tag{1}$$

The momentum equation:

$$\partial (\rho u_i) / \partial t + \partial (\rho u_i u_j) / \partial x_j = -\partial P / \partial x_i + \partial \tau_{ij} / \partial x_j \tag{2}$$

The energy equation:

$$\partial (\rho E) / \partial t + \partial (u_i (\rho E + P)) / \partial x_i = \vec{\nabla}\left(\alpha_{eff} \partial T / \partial x_i\right) + \vec{\nabla}\left(u_j (\tau_{ij})\right) \tag{3}$$

where

$$\tau_{ij} = \mu_{eff}\left(\partial u_i / \partial x_j + \partial u_j / \partial x_i\right) - 2/3 \mu_{eff} \partial u_k / \partial x_k \delta_{ij} \tag{4}$$

with

$$\rho = P/(RT) \tag{5}$$

2.2. Turbulence Modeling

The *realizable k-ε* turbulence model used in present study relies on the Boussinesq hypothesis [7]. It means that they are based on an eddy viscosity assumption, which makes the Reynolds stress tensor coming from equation averaging, to be proportional to the mean deformation rate tensor:

$$-\rho \overline{u_i' u_j'} = \mu_t \left(\partial u_i / \partial x_j + \partial u_j / \partial x_i\right) - 2/3(\rho k + \mu_t \partial u_i / \partial x_i) \delta_{ij} \tag{6}$$

The advantage of this approach is the relatively low computational cost associated with the determination of the turbulent viscosity, and suitable for industrial applications. The *realizable k-ε* model can be described as below:

$$\partial(\rho k)/\partial t + \partial(\rho k u_j)/\partial x_j = \partial((\mu + \mu_t/\sigma_k)\partial k/\partial x_j)/\partial x_j + G_k + G_b - \rho\varepsilon - Y_M + S_k \quad (7)$$

where G_k, G_b and Y_M are the generation of turbulence kinetic energy and the contribution of the fluctuating dilatation respectively.

$$\partial(\rho\varepsilon)/\partial t + \partial(\rho\varepsilon u_j)/\partial x_j \\ = \partial((\mu + \mu_t/\sigma_\varepsilon)\partial\varepsilon/\partial x_j)/\partial x_j + \rho C_1 S_\varepsilon - \rho C_2 \varepsilon^2/(k+\sqrt{\nu\varepsilon}) + C_{1\varepsilon}\varepsilon/k C_{3\varepsilon} G_b + S_\varepsilon \quad (8)$$

where $C_1 = \max(0.43, \mu/(\eta_1 + 5))$, $\eta_1 = S_1 \cdot k/\varepsilon$, $S_1 = \sqrt{2S_{ij}S_{ij}}$.

3. EXPERIMENTAL SETUP

3.1. Experimental Equipment

The schematic diagram of an experimental steam ejector system established by Sriveerakul T, Aphormratana S and Chunnanond K is shown in Figure 3 [8]. This steam ejector system used to simulate a steam-jet refrigeration system consisted of 7 major components: a steam boiler, an evaporator, a condenser, a receiver tank, a pumping system, a pressure manifold and an ejector.

Two electric immersion heaters were used as simulated heat source and cooling load at the boiler and the evaporator, respectively. The condenser was a water-cooled shell and coil type. A liquid refrigerant in the receiver tank was returned back to the boiler and the evaporator via an air-driven diaphragm pump. The boiler was covered by glass fiber wool with aluminum foil backing to prevent the thermal loss. The evaporator shell was well insulated, by neoprene foam rubber, from an unexpected heat gain from the environment.

3.2. Measurement System

Along the wall of the ejector, a static pressure at each operating condition was tapped and measured using a pressure transducer attached at the pressure manifold in the experimental system of reference [8] (as shown in Figure 4). This information was used to create the plots of the static pressure distribution along the ejector and they became the significant information for the validation of the simulated CFD ejector models.

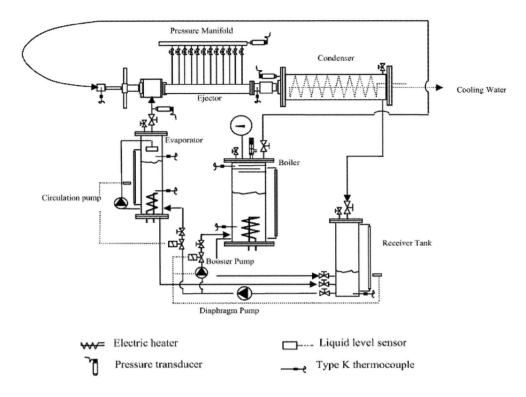

Figure 3. Experimental equipment.

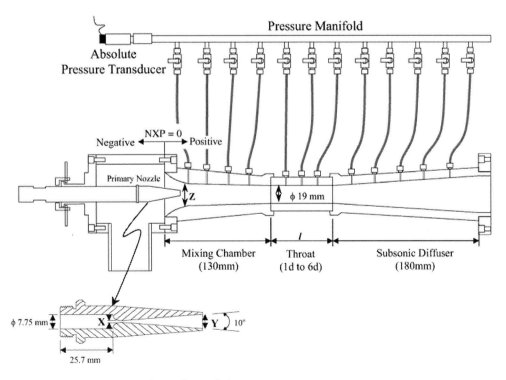

Figure 4. Schematic diagram of experimental ejector.

4. Numerical Simulation Methods

The geometric parameters of the steam-jet pump experiment model set-up by T. Sriveerakul are summarized in Table1 [8]. The primary and secondary fluids are steam and the parameters summarized in Table 2. The study by K. pianthong, W. Seehanam was shown that the simulation results for flow properties by the 2D axisymmetric model (ASXM) and three dimensional model (3D) were very close [9], and means that the ASXM is good enough to get accurate results. We use ASXM in this chapter to simulate and analyse the flow in steam-jet pump which would save lots of computing cost.

The commercial CFD code FLUENT6.3 was employed as a platform for CFD simulation. Quadrilateral structure meshes were used in 2D axis symmetric model, and the dense meshes are preset at the mixing zone as shown in Figure 5 (a) [10]. Figure 5 (b) shows the adapted meshes where it is possible to guess the location of incident and reflected waves in this chapter [10]. Boundary conditions are two pressure inlet boundaries and one pressure outlet boundary.

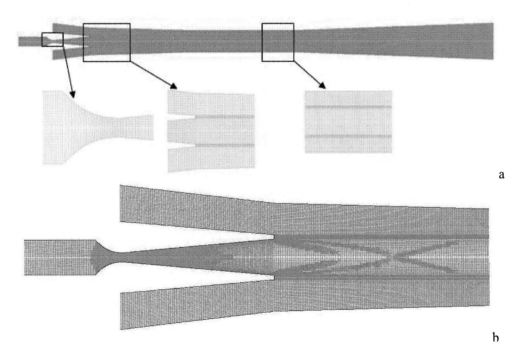

Figure 5. The structure of steam-jet pump and grid of CFD calculating domain (a) dense meshes (b) adapted meshes.

The governing equations were solved by a finite-volume approach; the convection terms were discretized with second-order upwind scheme and a central difference discretization was used in the diffusive terms. The discretized system was solved by a Gauss Seidel method. The couple-implicit solver was chosen to solve the governing equations, enhanced wall functions were used to describe the near-wall flow. Computations are stopped when residues fall below 10^{-6} and the mass imbalance (difference between mass flow on inlet and outlet boundaries) falls below 10^{-7}.

5 NUMERICAL STUDIES ON THE PERFORMANCES AT DIFFERENT OPERATING CONDITIONS

5.1. Validation of CFD Simulations

The static pressure of simulation results and the experimental data from T. Sriveerakul [8] along steam-jet pump at specific based on standard $k-\varepsilon$ turbulence model and realizable $k-\varepsilon$ turbulence model are shown in Figure 6. Close agreement between the simulations based on realizable $k-\varepsilon$ turbulence model and measurements validates the mathematic models, and the follow simulation results are obtained based on the present model and numerical strategy.

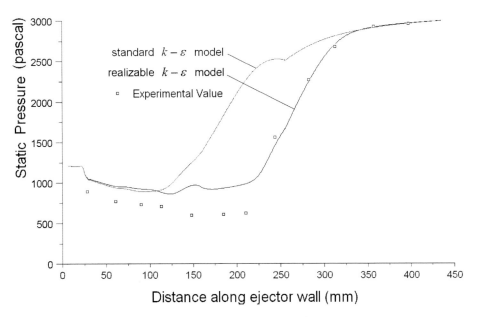

Figure 6. Static pressure profile along the steam-jet pump at primary fluid pressure 200000Pa, secondary fluid pressure 1200Pa and back pressure 3000Pa.

5.2. Flow Characteristics at Different Back Pressure

5.2.1. Velocity Vectors in Mixing Chamber

The velocity vectors in mixing chamber at different back pressures are shown in Figure 7 [10]. It is observed that the flow characteristics in mixing chamber are not affected when the value of back pressure is below 4500Pa (as shown in Figure 7 (a) and (b)), and the pump operates at critical mode in which operating condition the shock waves occur in the downstream of mixing chamber and do not affect the entrainment ratio of ejector. There are swirls separated from secondary fluid near the wall (as shown in Figure 7 (c)) which would reduce the effective area [11] of the pump when the back pressure is increase to 5000 Pa (bigger than the critical pressure, P_B^*=4500Pa), and the pump operates at subcritical mode which would lead to the entrainment ratio (*Em*) reduced rapidly (as shown in Figure 2).

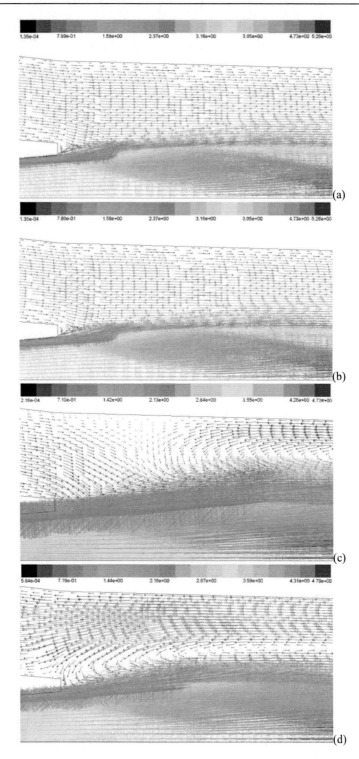

Figure 7. Velocity vectors in mixing chamber (a portion near the outlet of nozzle) at different back pressure (a) P_B=3500Pa (b) P_B=4500Pa (c) P_B=5000Pa (d) P_B=5500Pa.

The flow structure in mixing chamber would be break down and secondary flow would reverse to upstream when back pressure increases to 5500Pa (in excess of the break down pressure) (as shown in Figure 7 (d)), and the steam-jet pump lost pumping ability completely.

5.2.2. Mach Number Profile in the Mixing Chamber

The Mach number profiles of primary fluid along axis and secondary fluid along a line 4mm from pump wall in mixing chamber at different back pressures are shown in Figure 8 [10].

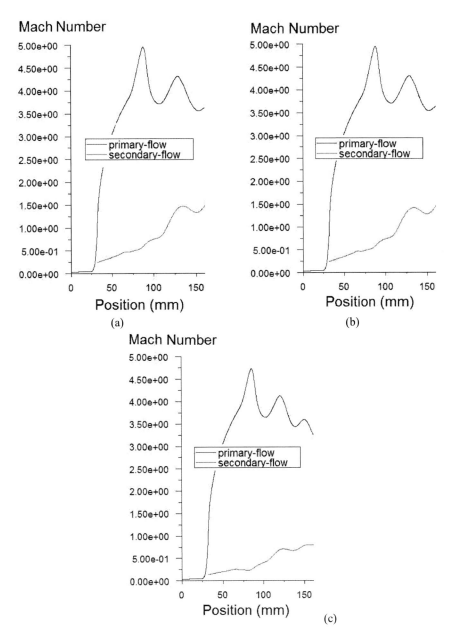

Figure 8. Mach number profiles in mixing chamber at different back pressure (P_B) (a) P_B=3500Pa (b) P_B=4500Pa (c) P_B=5000Pa.

It can be found that the Mach number profile of secondary fluid are not affected when the back pressure below 4500Pa (operating on design condition) and the secondary fluid is dragged by the primary fluid and gets supersonic flow (as shown in Figure 8 (a) and (b)). The Mach number is fallen down below 1 when the back pressure increased to 5000Pa (as shown in Figure 8 (c)), the pump operates at subcritical mode operating condition. The velocity of subsonic flow is much lower than that of supersonic flow, and leads to the E_m reduced rapidly.

5.3. Flow Characteristics at Different Suction Pressure

5.3.1. Secondary Fluid Pathlines in Steam-Jet Pump

The pathlines of secondary fluid in the steam-jet pump at different suction pressure (P_S) is shown in Figure 9 [10]. It can be found that the flow domain of secondary fluid in the pump keep constant generally when the suction pressure increased from 1200Pa to 1600Pa. It means that the primary fluid is the dominant flow in the pump which is almost not affected by the change of suction pressure.

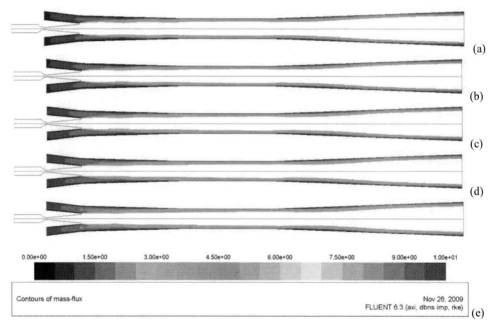

Figure 9. Secondary fluid pathlines at different suction pressures (P_S) (a) P_S=1200Pa (b) P_S=1300Pa (c) P_S=1400Pa (d) P_S=1500Pa (e) P_S=1600Pa.

5.3.2. Mass Flux Profiles in Steam-Jet Pump

To evaluate the pumping efficiency of steam-jet pump, we introduce a variable, *mass flux* (M_f), defined as the secondary fluid mass flow rate at per unit section area and described as: $M_f = \rho v$. The *mass flux* implemented through FLUENT User-Defined Function (UDF). The mass flux profiles near the interface of primary and secondary fluid at different suction pressures are shown in Figure 10 [10].

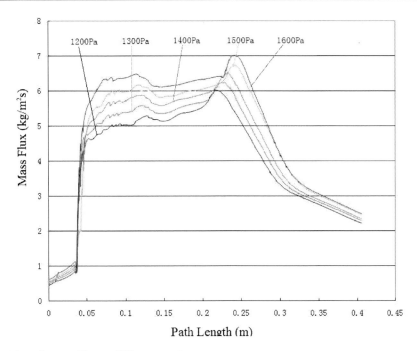

Figure 10. Mass flux profiles at different suction pressures.

It can be found that the mass fluxes increase with the increasing of suction pressures. Therefore the main pumping performance of steam-jet pump, entrainment ratio, should increase with the increasing of suction pressure, and this result can be found in Figure 11 [10].

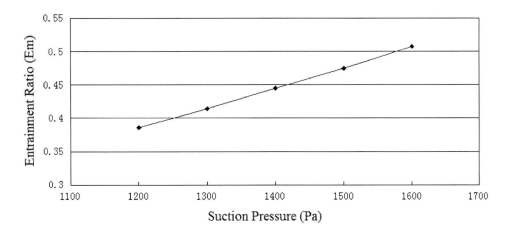

Figure 11. Relationship curve between suction pressure and entrainment ratio.

5.4. Flow Characteristics at Different Primary Steam Temperature

The primary fluid temperature (T_P) adopt in a steam-jet pump is an important factor which would affect the pumping performance of steam-jet pump, entrainment ratio (Em), mainly. The contours of Mach number of primary fluid in the steam-jet pump at different T_P are shown in Figure 12 (P_S=1228Pa, P_B=3000Pa). It can be found that the flow region of primary fluid increased and the shocking position moved to downstream as the primary fluid temperature increased. The relationship between primary fluid temperature and entrainment ratio is shown in Figure 13.

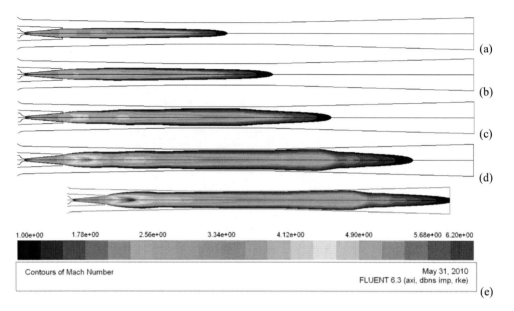

Figure 12. Contours of Mach numbers at different primary fluid temperature (T_P) (P_S=1228Pa, P_B=3000Pa) (a) T_P=110 °C (b) T_P=115°C (c) T_P=120°C (d) T_P=130°C (e) T_p=140°C.

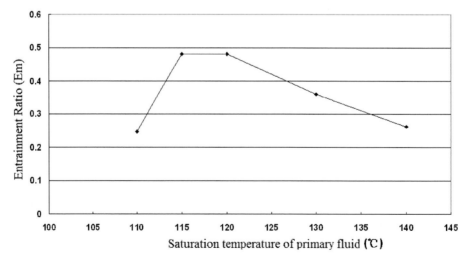

Figure 13. Relationship curve between primary fluid temperature and entrainment ratio (P_S=1228Pa, P_B=3000Pa).

The normal shock would be occurred at upstream when the primary steam temperature is lower than 115°C, in which the momentum of primary fluid is low and the steam-jet pump operated at un-chock state with lower *Em*; the normal shock would be occurred in throat as the primary steam temperature between115-120°C, the *Em* is keeping constant, in which the pump operated at chock state; the shocking position would be moved into diffuser when the temperature of primary steam is higher than 120°C, the increasing of momentum makes the primary fluid leaving nozzle with bigger under-expand degree which causes jet core enlarging, therefore, the annulus effective area is reduced, less secondary fluid can be entrained and accelerated through converging duct, the *Em* would be reduce.

6. SPONTANEOUSLY CONDENSING PHENOMENA IN NOZZLE

Nozzle in a steam-jet pump is the key part in which steam is accelerated from stagnation state to supersonic flow to obtained vacuum at nozzle outlet and the temperature of steam passing through a nozzle would be much lower than the stagnation temperature due to internal energy partly transformed into kinetic energy, which would induce condensing phenomena when the temperature of steam is lower than that corresponding to the saturation pressure. The condensation process can be understood by the paths of expansion on the enthalpy-entropy diagram shown in Figure 14 [12].

It can be found from Figure 14 that primary fluid expanding in nozzle from point "O" to point "A" (isentropic expansion), "B" or "C" (consideration of nozzle efficiency) would transfer from steam-gas into supersaturation flow (dryness is lower than 1) which is in a metastable state and reverts to an equilibrium state via formation and growth of vast numbers of extremely small liquid droplets (a homogeneous nucleation process). A latent heat would be released during the spontaneously condensing process which will affect the flow behaviour of wet steam.

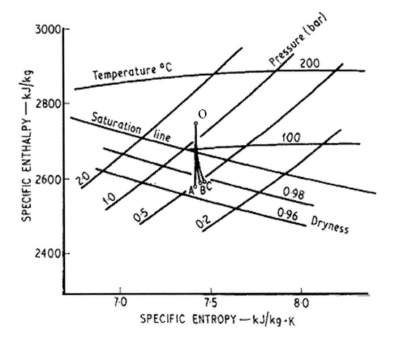

Figure 14. Paths of expansions on enthalpy-entropy diagram.

Almost all the researches on steam-jet pump are based on the assumption of primary fluid steam as perfect gas, and paid attention to the flow characteristics in mixing chamber [13, 14]. The wet steam flow in nozzle will be discussed in this section to investigate the effect of condensation to the primary fluid flow behaviour, and the study results can be used to transvaluation the flow behaviour in mixing chamber.

6.1. Mathematical Models

The mass and heat transfer between gas and liquid phases during wet steam flow should be described, and two transport equations for the liquid-phase mass-fraction, and the number of liquid-droplets per unit volume added to close the continuity, momentum and energy conservation equations. The phase change model, which involves the formation of liquid-droplets in a homogeneous nonequilibrium condensation process, is based on the classical nonisothermal nucleation theory.

Assumptions are made as following:

1. The velocity slip between the droplets and gaseous-phase is negligible.
2. The interactions between droplets are neglected.
3. The mass fraction of the condensed phase is small, the volume of the condensed liquid phase is negligible.
4. The heat capacity of the fine droplet is negligible compared with the latent heat released in condensation.
5. The droplet is assumed to be spherical, and its growth is based on average representative mean radii.

6.1.1. Transport Equation of Mass Fraction of the Condensed Liquid Phase

The mass fraction of the condensed liquid phase transport equation can be expressed [15]:

$$\frac{\partial(\beta\rho)}{\partial t} + \nabla \cdot (\rho\beta\bar{u}) = \Gamma \tag{9}$$

The mass generation rate Γ is given by the sum of mass increase due to nucleation (the formation of critically sized droplets) and also due to growth of these droplets, and can be written as [15]:

$$\Gamma = \frac{4}{3}\pi\rho_l \mathrm{I} r^{*3} + 4\pi\rho_l \eta \bar{r}^2 \frac{\partial \bar{r}}{\partial t} \tag{10}$$

The nucleation rate I is given by reference [16]:

$$I = \frac{q_c}{(1+\theta)}\left(\frac{\rho_v^2}{\rho_l}\right)\sqrt{\frac{2\sigma}{M^3\pi}}\exp(-\frac{4\pi \cdot r^{*2}\sigma}{3K\,T}) \tag{11}$$

Nonisothermal correction factor θ can be written [16]:

$$\theta = \frac{2(\gamma-1)}{(\gamma+1)}\left(\frac{h_{lv}}{RT}\right)\left(\frac{h_{lv}}{RT} - 0.5\right) \tag{12}$$

The mixture density ρ can be related to the vapor density ρ_v and the mass fraction of the condensed phase β by the following equation:

$$\rho = \frac{\rho_v}{(1-\beta)} \tag{13}$$

The critical droplet radius r^* (above which the droplet will grow and below which the droplet will evaporate) is given by reference [16]:

$$r^* = \frac{2\sigma}{\rho_l RT \ln S} \tag{14}$$

where S is the super saturation ratio (defined as the ratio of vapor pressure to the equilibrium saturation pressure).

The average radius of the droplet $\bar{r}$ is related to the transfer of mass from the vapor to the droplets and the transfer of heat from the droplets to the vapor in the form of latent heat, and can be described as [15]:

$$\frac{\partial \bar{r}}{\partial t} = \frac{P}{h_{lv}\rho_l\sqrt{2\pi RT}} \cdot \frac{\gamma+1}{2\gamma} C_p(T_0 - T) \tag{15}$$

6.1.2. Transport Equation of the Number Density of Droplets

Combining Equation 13 and the average droplet volume V_d (defined as: $V_d = \frac{4}{3}\pi \cdot \bar{r}^3$), the number of droplets per unit volume η can be determined in the following expression:

$$\eta = \frac{\beta}{(1-\beta)V_d(\rho_l/\rho_v)} \tag{16}$$

The transport equation of the number density is given by reference [15]:

$$\frac{\partial(\rho\eta)}{\partial t} + \nabla \cdot (\rho\eta\bar{u}) = \rho I \tag{17}$$

6.1.3. Wet Steam State Equation

The steam state equation which relates the pressure to the vapor density and the temperature is given by reference [17]:

$$P = \rho_v RT(1 + B\rho_v + C\rho_v^2) \tag{18}$$

where B and C are the second and the third virial coefficients given by the following empirical functions:

$$B = a_1(1+\frac{\tau}{\alpha})^{-1} + a_2 e^{-\tau}(1-e^{-\tau})^{5/2} + a_3\tau \tag{19}$$

$$C = a(\tau_1 - \tau_0)e^{-\alpha_1\tau_1} + b \tag{20}$$

where $\tau=1500/T$, $\alpha=1000$, $a_1=0.0015$, $a_2=-0.000942$, $a_3=-0.0004882$, $\tau_1=T/647.286$, $\tau_0=0.8978$, $\alpha=11.16$, $a=1.772$ and $b=1.5E-06$.

The vapor isobaric specific heat capacity which is a function of temperature is given below:

$$Cp_v = Cp_v(T) + R\left(\left((1-\alpha_v T)(B-B_1)-B_2\right)\rho_v + \left((1-2\alpha_v T)C + \alpha_v TC_1 - C_2/2\right)\rho_v^2\right) \tag{21}$$

The vapor specific enthalpy h_v and specific entropy s_v which are functions of temperature and density of vapor are given by [17]:

$$h_v = h_0(T) + RT\left[(B-B_1)\rho_v + (C-C_1)/2\rho_v^2\right] \tag{22}$$

$$S_v = S_0(T) + R\left[\ln\rho_v(B+B_1)\rho_v + (C+C_1)/2\rho_v^2\right] \tag{23}$$

The vapour dynamic viscosity and thermal conductivity are functions of temperature and can be obtained from reference [15].

6.2. Numerical Simulations

The nozzle geometric parameters of a steam-jet pump experiment model set-up by F Bakhtan and M T Mohammadi Tochai [18]. The primary fluid is superheated steam, which total pressure P_0 and temperature T_0 at the nozzle inlet were P_0=87KPa and T_0=390.15K.

The commercial CFD code FLUENT6.3 was employed as a platform for CFD simulation. Quadrilateral structure meshes were used in 2D axis symmetric model, boundary conditions are pressure inlet boundary and pressure outlet boundary. The governing equations were solved by a finite-volume approach, the density-solver was chosen to solve the governing equations, computations are stopped when residues fall below 10^{-6} and the mass imbalance (difference between mass flow on inlet and outlet boundary) falls below 10^{-7}.

6.3. Results and Discussion

6.3.1. Validation of CFD Simulations

The static pressure distributions of simulation results and the experimental data from reference [18] along nozzle axis are shown in Figure 15. It is found that the CFD simulation captured the pressure enhancing (Wilson point at 20mm from nozzle throat section) as condensation occurred. Close agreement between the simulations and measurements validates the mathematic models, and the follow simulation results were obtained based on the present model and numerical strategy.

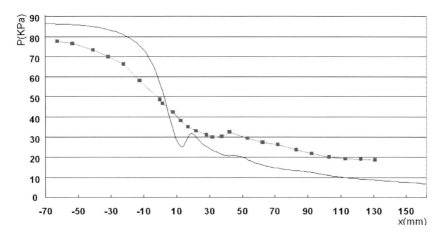

Figure 15. Static pressure along nozzle axis.

6.3.2. Simulation Results Based on Perfect Gas and Wet Steam Model

6.3.2.1. Pressure Distributions along Nozzle Axis

The pressure distributions along nozzle axis are shown in Figure 16 based on wet steam model and perfect gas assumption separately. It is observed that there was a pressure sudden enhancing of wet steam to higher than that of perfect gas as condensation occurred, and this differentia was keep to the outlet of nozzle.

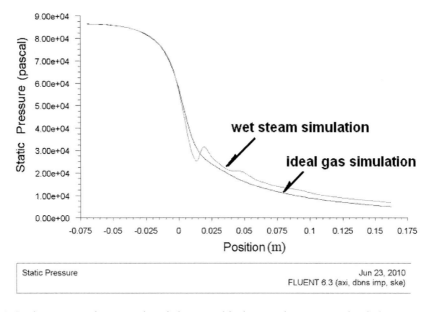

Figure 16. Static pressure along nozzle axis between ideal gas and wet steam simulation.

6.3.2.2. Mach Number Profiles along Nozzle Axis

The Mach number profiles along nozzle axis are shown in Figure 17 based on wet steam model and perfect gas assumption separately. It is found that there was a Mach number sudden dropping of wet steam to lower than that of perfect gas as condensation occurred and this differentia escalated from Wilson point to the outlet of nozzle, and the maximum of differentia was about 10%. The droplets grown up with thermal-positive process from Wilson point to outlet hindered the increasing of velocity along nozzle axis.

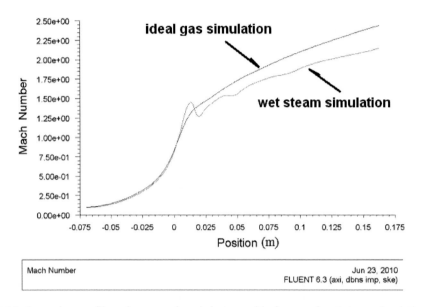

Figure 17. Mach number profiles along nozzle axis between ideal gas and wet steam simulation.

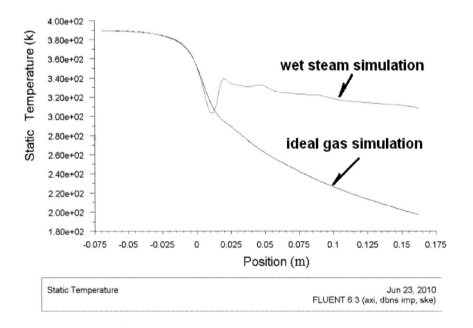

Figure 18. Temperature along nozzle axis between ideal gas and wet steam simulation.

6.3.2.3. Temperature Profiles along Nozzle Axis

The temperature profiles along nozzle axis are shown in Figure 18 based on wet steam model and perfect gas assumption separately. It is found that there was a temperature jumping of wet steam due to large thermal-positive induced by spontaneous condensation, the condensation heat produced from droplets growing up keep the wet steam temperature much higher than that of perfect gas and the maximum of differentia was above 35%. The pressure increasing and velocity reducing of steam supersonic flow as spontaneous condensation occurred in nozzle would reduce the efficiency of nozzle and the pumping performance of steam-jet vacuum pump would be evaluated deeply.

6.3.3. Influences to Steam Condensation by Operating Conditions

6.3.3.1. Influences to Steam Condensation by Steam Initial Pressure

The influences to steam condensation by steam initial pressure are shown in Figure 19-21.

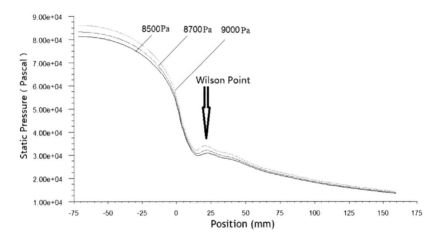

Figure 19. Static pressure along nozzle axis at different initial pressure.

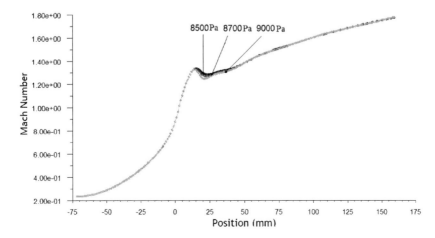

Figure 20. Mach number along nozzle axis at different initial pressure.

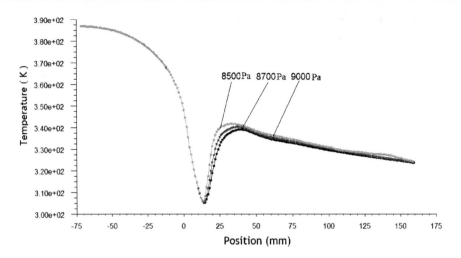

Figure 21. Temperature along nozzle axis at different initial pressure.

The static pressure along nozzle axis at different initial pressure is shown Figure 19, and can be found that the static pressure increased and the Wilson point moves to throat as steam initial pressure increasing. The Mach number along nozzle axis at different initial pressure is shown Figure 20. It can be found that the Mach number near the Wilson point reduced as steam initial pressure increasing. The temperature along nozzle axis at different initial pressure is shown Figure 21. It can be found that there are same temperature profiles at different initial pressure before steam condensation occurred, the temperature near Wilson point increased and the Wilson point moves to throat as steam initial pressure increasing.

6.3.3.2. Influences to Steam Condensation by Steam Inlet Temperature

The influences to steam condensation by steam inlet temperature are shown in Figure 22-24. It can be found that the Wilson point moves from nozzle throat to outlet with the increasing of steam inlet temperature, which means that the high inlet temperature is useful to postpone the steam condensation occurred in nozzle.

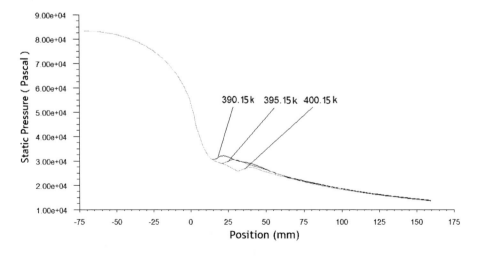

Figure 22. Static pressure along nozzle axis at different inlet temperature.

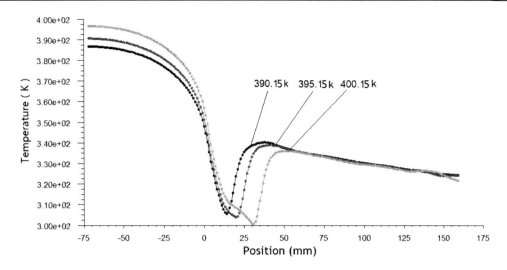

Figure 23. Temperature along nozzle axis at different inlet temperature.

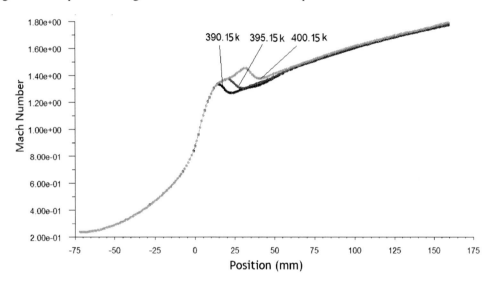

Figure 24. Mach number along nozzle axis at different inlet temperature.

CONCLUSION

(1) CFD approach has been used successfully to simulate and capture the transonic flow features in ejector, allowing analyzing and understanding the pumping performances of steam-jet pump.
(2) The pumping performance of steam-jet pump is affected by back pressure:
   a) The entrainment ratio (Em) keeps constant when the value of back pressure is less than that of critical pressure, and steam-jet pump should be operated at critical mode conditions to obtain stable pumping performances.
   b) Swirls were found in mixing chamber separated from secondary fluid near the wall which would reduce the effective area of the ejector. The velocity of

secondary fluid was fallen down obviously when the back pressure was bigger than the critical pressure. These two factors lead to worse and unstable pumping performances of steam-jet pump.

c) The flow structure in mixing chamber would be broken down and secondary flow would reverse to upstream if back pressure is in excess of the break down pressure, and the steam-jet pump lost pumping ability completely.

(3) The pumping performance of steam-jet pump is affected by suction pressure. The increasing of suction pressure leads to the reducing of expanding ratio and pressure ratio synchronously, and the entrainment ratio is mainly affected by the latter.

(4) The mass flux introduced in present study is an available variable and can be used to evaluate the pumping efficiency.

(5) The pumping performance of steam-jet pump is affected by primary fluid temperature. An appropriate temperature would make pump operating with high entrainment ratio and the Em would be affected disadvantageously by lower or higher temperature of primary fluid.

(6) The mass fraction transport equation and droplet number density transport equation were introduced into Eulerian-Eulerian two-fluid model and successfully simulated and captured the spontaneously condensing phenomena in nozzle. The simulation results from this chapter are significant to analyzing and understanding the flow characteristics of steam-jet pump in next study.

(7) The influences to steam transonic flow and spontaneously condensing phenomena in nozzle at different steam inlet pressure and temperature were studied by numerical simulation methods. It can be found that the flow parameters and spontaneous condensation in nozzle are affected obviously by operating conditions, and it must be considered the influences of nozzle inlet pressure and temperature to the flow features of nozzle during the design process of nozzle.

(8) The condensation heat produced during wet steam condensation in nozzle made pressure increased and velocity dropped at outlet of nozzle which would reduce pumping performance of steam-jet pump.

(9) The supersonic steam condensation in nozzle maybe avoided or weakened by enhancing steam superheat degree, the appropriate superheat degree of steam is chosen according to geometric parameters of nozzle and operating conditions of steam-jet pump which needs to be further study.

(10) It is important for deep research to capture the exact position of spontaneous condensation, the distribution and its evolvement of droplets in nozzle, which need to investigate in next study.

Acknowledgment

The financial support provided by the Liaoning natural science foundation (ID20102073), the foundation of international cooperation project (ID7440015), the doctoral foundation for returning-back scholar of Northeastern University (ID18504032) and Australian Research Council (ARC project ID DP0877734) is gratefully acknowledged.

REFERENCES

[1] Kanjanapon Chunnanond; Satha Aphornratana. *App. Therm. Eng.* 2004, *24*, 311-322.
[2] Xiaodong Wang; Dechun Ba; Shiwei Zhang; Yichen Zhang. *Vacuum Technology*; ISBN 7-5024-3571-9; Publishing House of Metallurgy Industry: Beijing, CH, 2006; Vol.1, pp 167-177.
[3] Huang BJ; Chang JM; Wang CP; Petrenko VA. *Int. J. Refrig* 1999, *22*: 354-364.
[4] Bartosiewicz Y; Aidoun Z; Desevaux P; Mercadier Y. *Int. J. Heat Fluid Flow* 2005, *26*: 56–70.
[5] Hemidi A; Henry F; Leclaire S; Seynhaeve J; Bartosiewicz Y. *App. Therm. Eng.* 2009, *29*: 1523–1531.
[6] Ouzzane M; Aidoun Z. *App. Therm. Eng.* 2003, *23*: 2337–2351.
[7] Launder BE; Spalding DB. *Comp. Meth Appl. Mech. Eng* 1974, *3*: 269-289.
[8] Sriveerakul T; Aphormratana S; Chunnanond K. *Int. J. Therm. Sci* 2007, *46*: 812–822.
[9] Pianthong K; Seehanam W; Behnia M; Sriveerakul T; Aphornratana S. *Energy Convers Manage* 2007, *48*: 2556–2564.
[10] Wang Xiaodong, Dong Jingliang. *Vacuum* 2010, *84*, 1341-1346.
[11] Munday JT; Bagster DF. *Ind. Eng. Chem Proc Des* 1997, *16*: 442–449.
[12] F. Bakhtar; BA Campbell. *Proc. Lnstn. Mech. Engrs.* 1970; *185*: 395-405.
[13] Dong Jingliang;Wang Xiaodong. *Chinese J Vac Sci Tech* 2010, *30*: 455-458.
[14] S Varga; AC Oliveira; B Diaconu. *Int refrige* 2009, *32*:1203–1211.
[15] K Ishazaki; T Ikohagi; H Daiguji. *6th Int Sym CFD* 1995, 1: 479–484.
[16] JB Young. *J. Turbo mach* 1992, *114*: 569–579.
[17] JB Young. *J. Eng Gas Turb Power* 1988, *110*: 1–7.
[18] Bakhtan F; Mohammadi Tochai MT. *Int. J .Heat Fluid Flow* 1980, *2*: 5-18.

Chapter 11

COMPUTATIONAL AND RAMAN SPECTRAL STUDIES ON PROBING AZAINDOLE ADSORPTION ON Ag NANOMETAL DOPED SOL - GEL SUBSTRATES

B. Karthikeyan, S. K. Thabasum Sheerin and M. Murugavelu

Department of Chemistry, Annamalai University,
Annamalai Nagar, India

ABSTRACT

Computational and Surface enhanced Raman spectral studies are used to probe the adsorption of 7-azaindole on the Ag doped sol-gel substrates. The adsorption mode and the vibrational spectral futures are discussed[1]. This chapter gives a further account of application of theoretical calculations and modeling of the adsorption of azaindole on the Ag nanometal surface. The abinito and DFT level calculated Ag complexed and experimentally obtained SERS results are well agreed. It is proposed that the similar methods can be implemented in the study of surface vibrational properties of adsorbates adsorbed on nanometal embedded sol-gel substrates.

INTRODUCTION

To tailor the new generation of nanodevices and "smart" materials to organize the nanoparticles into controlled architecture must be found. This conundrum has been addressed to some extent by a large body of recent work describing the construction of colloidal dyads and triads,[2,3] "string,"[4] clusters,[5] and other architectures[5-8] in solution and at liquid-liquid interfaces. These constructions will not be described in any detail here, since this chapter is dedicated to surface-bound nanostructures. Likewise, we will not discuss about the highly active research area of sol-gel preparative method in great depth, as excellent reviews already exist. Here we will focus on efforts to construct a model of adsorption on organized nanoparticle arrays. Although work on surface vibrational spectral studies are reported for solution phase but in sol-gel substrates is quite less. This chapter deals one such study of adsorptoin of 7- azaindole on Ag nanometal doped sol-gel substrates.

SOL-GEL PROCESSING

Sol-Gel Processing is used extensively in creating polymer- ceramic composites. In this method, a combination of chemical reaction turns a homogeneous solution of reactants into an infinite molecular weight oxide polymer as seen in the following scheme,

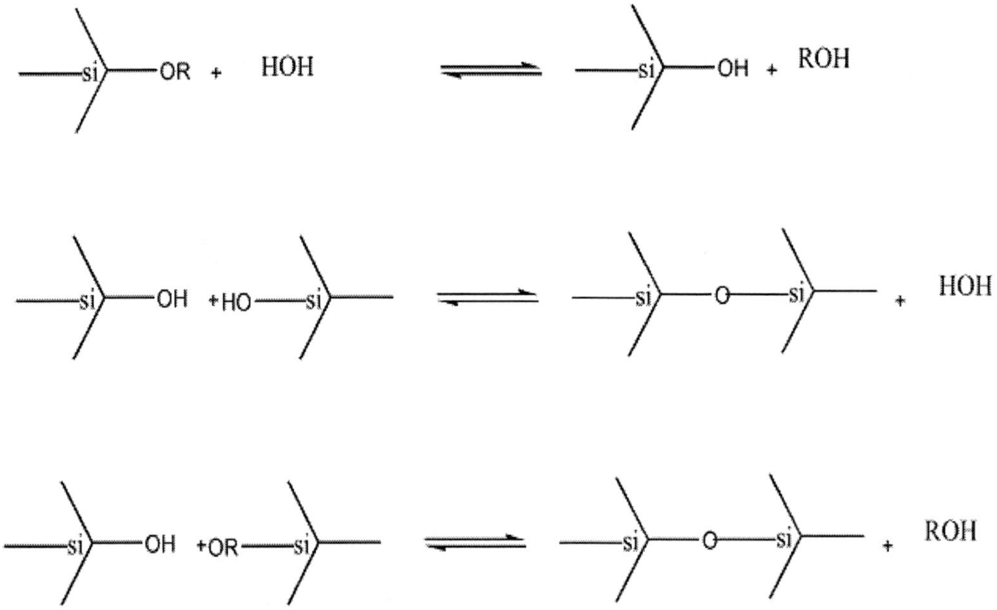

Scheme 1. Sol-gel route.

Today's sol-gel processing is a form of nanostructure processing. It begins with a nanometer-sized unit and undergoes reaction on the nanometer scale, resulting in a material with a nanometer features. Sol-gel processing results in highly pure, uniform nanostructure. The general steps in the sol-gel process regardless of the starting materials are indicated below sequentially.

i. (i). Mixing
ii. (ii). Gelling-defined empirically as the time when the solution shows no flows
iii. (iii). Shape forming- three different shapes; thin film, fiber, bulk
iv. (iv). Drying
v. (v). Densification

Simplicity and flexibility of the process, small capital investment as well the possibilities of carrying out the process at low temperature are the unique advantage offered by the sol-gel process.

Self-Assembly has only recently emerged as a viable new strategy for synthesis of nanoscale materials. These processes are commonly found in nature. Biological organisms have mastered the art of assembling complex nanostructures from relatively simple building blocks. These structures range from strong and tough composites to highly selective sensing membranes. These are two major strategies of self-assembly. They are intercalation of single polymer chains as well as the dispersion of ceramic layers in a continuous polymer matrix.

The properties of a material in its nanoscopic phase are found to be very different from those of a bulk sample of the same material. These nanometals exhibit various electronic, optical and catalytic functions. Hence the preparation of this kind of materials has become one of the important and active areas of today's research [9]. Interestingly, the molecule adsorbed on the surface of the nanometal give rise to a phenomenon called Surface Enhanced Raman Scattering (SERS) in which the Raman signals are found to be highly enhanced with an intensity being 10^6 times that of the normal Raman ones[10]. This effect has gained the attention from both basic and practical viewpoints. It takes Raman spectroscopy to the rank of single molecule detection technique [11]. The resulting SERS spectra contain the structural information of vibrational spectroscopy with the ability of detecting molecules up to their atto molar level. Since from the discovery of SERS, lot of SERS active substrates have been reported which include roughened electrodes and metal foils, colloids, metal island films, etc. and each of these categories has its own advantages. Among these substrates silver sol has been widely used because of its easy of preparation and convenience to use [12,13]. 7-Azaindole (7-AI) is an important bicyclic aromatic molecule and has a close relationship with the nucleic bases adenine and guanine. We have reported the SERS results of 7-AI adsorbed on the surface of silver nanometal present in the silver sol [1]. In continuation with this here we report the computational modeling of the obtained SERS which will further confirm the chemisorptions of the 7-AI on the silver nanometal surface.

EXPERIMENTAL

Sodium chloride, methanol, silver nitrate and 7-AI (99% Aldrich) were of analytical grade and used as received. Silver sol was prepared by the procedure reported by Lee and Miesel[14] and activated by the addition of sodium chloride (10^{-3} M) before taking the spectral measurements. For the SERS measurements, methanolic solution of 7-AI of concentration of 0.1 M was used (6 mM of 7-AI in the resulting solution). Surface Plasmon of the silver sol was identified by recording the absorption spectra using Hitachi U-3000 UV-visible spectrophotometer. Raman spectra were recorded using Bruker RFS 100/s FT-Raman spectrometer, which provides an excitation wavelength of 1064 nm. The typical power at the sample was 100 mW and the samples were scanned for duration of 300 seconds. Sol-gel substrates were prepared by a known method [15].

DFT Methods

The density functional calculations presented here were performed with the Gaussian-03W [16] programme on an Pentium (IV) computer system. The molecular geometry of 7-AI was optimized using the method B3LYP with the basis set 6-31G**. A complete geometry optimization was carried out employing Berny's optimization algorithm, which resulted in C_S symmetry. For the Ag-AI Lanl2Dz basis set is used to optimize the structure and for the corresponding vibrational frequency calculations. The vibrational frequencies and corresponding normal modes were then evaluated at the optimized geometry using analytical differentiation algorithms contained within the program.

RESULTS AND DISCUSSION

7-AI whose optimized molecular structure is shown in Figure.1, which closely resembles the molecule indole, has one pyridine and one pyrrole ring fused together. In the present study one of the hydrogen atom is replaced by silver atom i.e. silver complexed model (Figure.2) is proposed. The vibrational frequencies are calculated of this model and compared with the experimentally obtained SERS frequencies. SERS enhancement factors for isolated small spherical silver colloidal particles have been estimated to be the order of 10^6. The normal Raman spectrum of 7-AI and the SERS of 7-AI adsorbed on silver nanometal is given in Figure 3. Some of the bands which appear very weak in the normal Raman spectrum are found to be enhanced in the observed SERS spectrum. For the better assignment of 7-AI, DFT calculation has been done at B3LYP/6-31G** level. The obtained important results are given in the Table 1.

The intense band at 778 cm^{-1} in the SERS spectrum can be assigned to the 7-AI breathing vibration (C-H stretching) which appears at 768 cm^{-1} in the normal Raman spectrum. The bands assigned to the out of plane symmetric vibration that are very weak in the normal Raman spectra, appear with medium intensities in the SERS spectrum (428, 436, 463, 571, 625 and 925 cm^{-1}). In plane symmetric vibrations are also quite intense in the range from 1000 to 1600 cm^{-1} while the band at 1506 cm^{-1} which is quite intense in the normal Raman spectra shows a decrease in its intensity.

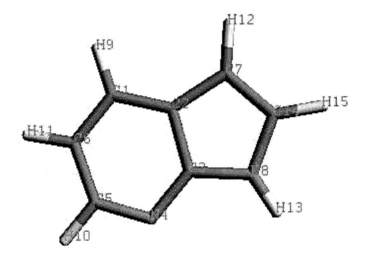

Figure 1. Structure of 7-Al and numbering of atoms.

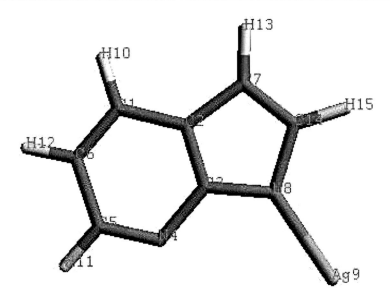

Figure 2. Structure of silver complexed 7-AI and numbering of atoms.

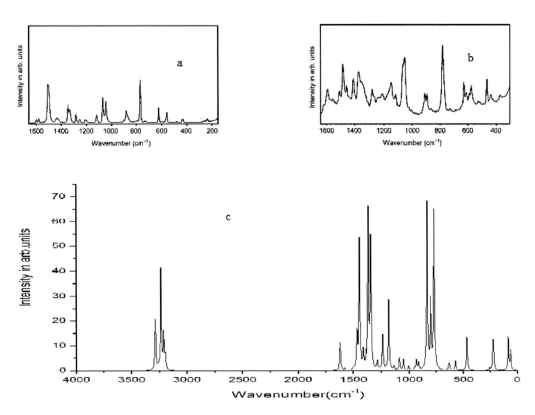

Figure 3. (a) Normal Raman of 7-AI, (b) SERS of 7-AI adsorbed on silver, (c) computed Raman of silver 7-AI complex.

This band has been assigned to the in plane symmetric C-N stretching vibration[17].The observed enhancement of both totally symmetric and non totally symmetric bands and the decrease in the intensity of the 1506 cm^{-1} band suggest that the orientation of 7-AI on the metal surface may neither be parallel nor be perpendicular but tilted towards the surface.

Table 1 Observed and calculated Raman frequencies of 7-AI and Ag complexed 7-AI

2	22 m	625 m	622	627	a' C-C bending	C-C (45)
3	768 s	778 vs	775	772	a' breathing vibration	CH (76)
4	882 m	888 m	900	909	a' CH bending	CH (65)
5	1044 s	1047 vs	1063	1051	a' C-C stretching	CC (41)
6	1070 s	1066 vs	1092	1089	a' C-H stretching	CH (40)
7	1333 m	1346 s	1324	1345	a' C-H bending	CH (61)
8	1349 m	1352 s	1362	1368	a' C-N stretching	–
9	1506 s	1506 w	1537	1497	a' C-N stretching	CN (31)

m: medium, w: weak , s: strong, vs: very strong.
a': in-plane, a'': out-of-plane.

B3LYP level calculated vibrational frequencies of free 7-AI are also compared with the silver complexed 7-AI in the table 1. The obtained vibrational frequencies of the latter one matches with the observed SERS frequencies .The computed Raman spectrum of the complex is given in the Figure. 3. From that it is concluded that nitrogen atom of azaindole forms a chemical bond with the silver surface. Formation of a bond with the metal surface shifts the Raman frequency which clearly reveals from the Table 1. The calculated frequencies of Ag complexed 7-AI well matches with the observed frequencies that can be taken as a model of the substrate adsorption on the metal surface. The enhancements of observed SERS bands are mainly contributed by the chemical enhancement which requires the formation of a chemical bond between the metal atom and the substrate. Similar calculation are reported in the literature for benzene and benzonitrile [18].

CONCLUSION

A systematic SERS study of 7-AI adsorbed of silver sol has been reported by us is further confirmed by doing the theoretical spectral computation of the silver complexed model of 7-AI. This model supports the role of chemical enhancement in the 7-AI SERS.

ACKNOWLEDGMENTS

The authors are highly thankful to Prof. S.Umapathy, S.Sampath of IPC and IISC for helpful discussions and one of the authors (BK) is thankful to UGC, New Delhi, for granting a major research project, F.No. 37-33/2009 (SR).

REFERENCES

[1] Karthikeyan, B. *J. Sol. Gel. Sci. Tech.* 2008,45(1),79-82.
[2] Alivisatos, A.P.; Johnsson, K.P.; Peng, X. ; Wilson, T.E.; Loweth, C.J.; Bruchez, M.P. ; Schultz, Jr.,P.G. *Nature* . 1996, 382, 609-611.
[3] a) Peng, X.; Wilson, T.E.; Alivisatos, A.P.; Schultz, P.G. ; Angew. Chem. 1997, 109 ,113. *Angew. Chem . Int. Ed. Engl*, 1997, 36 ,145-147. b) Brousseau lll, L.C.; Novak, J.P.; Marinakos ,S.M.; Feldheim, D.L. *Adv. Mater.* 1999, 11, 447-449.
[4] a) Marinakos, S.M.; Broussseau lll, L.C.; Jones, A .; Feldheim, D.C. *Chem Mater* 1998, 10, 1214-1219. b) Hornyak, G.; Kroll, M.; Pugin, R.; Sawitowski, T.; Schmid, G.; Bovin, J.-O. ; Korsson, G.; Hofmeister, H.; Hopfe, S. *Chem Eur J.* 1997, 3, 1951-1956. c) Chung, S.-W.; Markovich, G.; Heath, J.R. *J. Phys. Chem B.* 1999, 102, 6685. Sano, M.; Kamino, A. ; Shinkai, S. *Langmuir* .1999, 15, 13-15.
[5] Antonietti, M.; Goltner, C. *Angew. Chem.* 1997, 109, 944-964. ; *Angew. Chem . Int . Ed. Engl* . 1997,36, 910-928.
[6] Feldheim, D. L. ; Keating, C. D. *Chem. Soc .Rev.* 1998, 27, 1-12.
[7] a) Westcott, S. L.; Oldenburg, S. J.; Lee, T. R;. Halas, N. J. Langmuir. 1998, 14, 5396-5401.; b) Sato, T;. Ahmed, H.; *Appl. Phys. Lett.* 1997, 70, 2759-2761. ; c) Cusack, L.; Rizza, R. ; Gorelov, A. ; Fitzmaurice, D. *Angew. Chem .* 1997,109, 887-890. ; *Angew .Chem. Int. Ed. Engl.* 1997, 36, 848-851. ; d) Whetten, R. L. ; Khoury, J. T. ; Alvarez, M.M.; Murthy, S. ;Vezmar, I.; Wang, Z. L.; Stephens, P. W. ; Cleveland, C. L.; Luedke, W.D. ; Landman, U. *Adv. Mater.* 1996,8, 428-433. ; e) Chemseddine, A.; Jungbult, H. ; Boulmazz, S. *J. Phys. Chem.* 1996, 100 ,12546-12551. ; f) Lin, X.M.; Sorenson, C.M. ; Klabunde, K. J. Chem. Mater. 1999, 11, 198-202. ; g) Shenton, W. ; Davis, S.A. ; Mann, S. *Adv. Mater.* 1999, 11, 449-452. ; h) Wang, Z. L. *Adv .Mater.* 1998, 10, 13-30. ; i) Li, M.; Schnablegger, H.; Mann, S. *Nature.* 1999 ,402, 393-395.
[8] Mucic, R.C.; Storhoff, J. L.; Mirkin, C. A. ; Letsinger, R. L. *J. Am. Chem .Soc.*1998,120, 12674-12675.
[9] Karthikeyan, B. *Ind.J.Chem. 2007,*46A, 929-931
[10] Shipway, A. N.; Katz, E.; Willner, I. *Chem. Phys. Chem.* 1, 2000, 18-52.
[11] Fleishmann, M.; Hendra, P. J.; McQuillan, A. J. *Chem. Phys. Lett.* 1974, 26, 163-166.
[12] Nie, S.; Emory, S. R. *Science.* 1997, 275, 1102-1106.
[13] Tanaka, T.; Nakajima, A.; Watanabe ,A.; Ohno, T.; Ozaki, Y. *J. Mo.l Struct.* 2003, 661, 437-449.
[14] Vickova, B.; Solecka-Cermakova, K.; Matejka, P.;Baumruk, V. *J. Mol. Struct.* 1997, 408, 149-154.
[15] Lee, P. C.; Meisel, D. *J. Phys. Chem.* 1982, 86, 3391-3395.
[16] Frisch, M. J.; Trucks, G .W.; Schlegal, H.B.; Gill, P. M.W.; Johnson, B. G.; Robb, M. A.; Cheeseman, J. R.; Keith, T.; Peterson, G. A.; Montgomery, J. A.; Raghavachari, K.; Al-Laham, M. A.; Zakrzewski, V. G.; Ortiz, J. V.; Foresman, J. B.; Cioslowski, Stefanov, B. B. ;Nanayakkara, A.; Challacombe, M. ;Peng, C. Y.; Ayala, P. Y.; Chen, W.; Wong, M .W. ;Andres, J. L.; Replogle, E. S.; Gonperts, R.; Martin, R .L.; Fox, D. J.; Binkley, J. S. ;Defrees, D.J.; Baker, J.; Stewart, J. J. P.; Head-Gordon, M.; Gonzalez, C.; Pople, J.A. *Gaussian 94*, Revision C.2; Gaussian Inc. Pittsburgh, PA, 1995.

[17] Cane, E.; Palmieri, P. ;Tarroni ,R.; Trombetti, A. *J. Chem. Soc. Faraday Trans.*1994, 90 , 3213-3219.
[18] Guillermo, D. F.; Italo, G.; Andres. A .; Freddy, C.; Leticia, V.; Rainer, K.; Marcelo, C. V. *Spectro.Chim. Acta.* 2008 , 71(3) ,1074-1079.

In: Computational Fluid Dynamics
Editor: Alyssa D. Murphy

ISBN: 978-1-61209-276-8
© 2011 Nova Science Publishers, Inc.

Chapter 12

NUMERICAL APPROACHES FOR SOLVING FREE SURFACE FLUID FLOWS

L. Battaglia[1,2,*], *J. D'Elía*[1,†] *and M. Storti*[1,‡]

[1] Centro Internacional de Métodos Computacionales en Ingeniería (CIMEC)
Instituto de Desarrollo Tecnológico para la Industria Química (INTEC)
Universidad Nacional del Litoral - CONICET
Güemes 3450, 3000-Santa Fe, Argentina
web page: http://www.cimec.org.ar

[2] Grupo de Investigación en Métodos Numéricos en Ingeniería (GIMNI)
Facultad Regional Santa Fe - Universidad Tecnológica Nacional
Lavaise 610, 3000-Santa Fe, Argentina.

Abstract

A free surface is defined as an interface between two fluids, where the lighter phase, which is usually a gas, has negligible effect over the other due to its very low values of density and viscosity.

Free surface flows are common issues among several engineering disciplines, such as civil, mechanical or naval. Some typical problems are open channel flows, sloshing in tanks for storing or transporting liquids, and mold filling, among others.

Different numerical methods have been developed for solving these kind of flows, being the most popular classification of techniques the one that refers to "interface tracking" and "interface capturing" methods. On one side, the interface tracking approaches are based on considering the free surface as a boundary of the domain, and defining over that boundaries some entities such as nodes or element edges of a finite element method mesh, in such a way that the fluid flow problem is solved over a single liquid phase. On the other side, interface capturing approximations are based on marking functions that indicate which part of the domain is occupied by the liquid, and which other is occupied by the gaseous phase, such that the interface position is "captured" over certain values of the marking function. In the present chapter, two finite element methods for solving free surface flows are described: an interface tracking

---

[*]E-mail address: lbattaglia@santafe-conicet.gob.ar
[†]E-mail address: jdelia@intec.unl.edu.ar
[‡]E-mail address: mstorti@intec.unl.edu.ar

technique developed over an arbitrary Lagrangian-Eulerian framework, and a level set interface capturing proposal. Each approach has been considered for solving different free surface flow problems, regarding the capabilities of the methods.

Key Words: fluid mechanics, free surface, level set method, arbitrary Lagrangian-Eulerian method, computational fluid dynamics, finite elements.

1. Introduction

In a multiphase fluid flow context, free surface (FS) flows constitute a particular case where a liquid phase coexists with a gaseous one. This kind of flows are found in several cases, such as in open channel flows [1], naval hydrodynamics [2] or sloshing in liquid storage tanks [3, 4]. Such variety led to the development of several numerical methods to predict the FS behavior. In this chapter, two numerical methods for solving FS incompressible flows of Newtonian viscous fluids are summarized, without considering surface tension effects in any case.

Numerical approaches for solving free surface fluid or two-fluid flows can be grouped on either *interface tracking* or *interface capturing* methods [5]. On the one hand, in interface tracking methods, the interface is explicitly followed because it is defined over certain entities of the discretization. Some typical interface tracking methods are Marker-and-Cell (MAC) [6, 7], particle methods [8], and Arbitrary Lagrangian-Eulerian (ALE) methods [9–13]. In the last case, in a Finite Element Method (FEM) context, the FS is defined over nodes or faces of a mesh, and the computational domain includes only the fluid phase in such a way that the interface displacements induce the domain deformation. These displacements should be considered through a remeshing or a mesh-update procedure which, in turn, constitutes a limitation regarding the merging or the breaking up of the interface when FS suffers relatively large movements. On the other hand, interface capturing methods typically solve two-fluid flows over fixed meshes, such that the FS crosses a set of elements and its position is "captured" by an additional scalar field, and naturally allows the interface breaking or folding. A classical interface capturing method is the so-called Volume-Of-Fluid (VOF) [14, 15], where a fluid fraction F defined over every element corresponds to one fluid or the other with either $F = 0$ or $F = 1$, while $0 < F < 1$ identifies the transition elements; finally, the FS precise position is recovered by specialized algorithms [16]. In the case of most interface capturing Level Set (LS) methods [17, 18], a continuous LS function ϕ identifies each phase with a positive value or a negative value, $\phi > 0$ or $\phi < 0$ respectively, while $\phi = 0$ corresponds to the interface position. Furthermore, these methodologies for solving FS problems are part of some hybrid methods, as the combined LS and VOF (CLSVOF) method [19], the Particle Level Set [20] or mixed front-tracking and front-capturing approaches [21].

In the following sections, two FE methods for solving FS incompressible viscous fluid flows are described: an ALE methodology and a LS technique. The ALE approach is developed for solving small and moderate FS displacements, while the LS one is proposed for problems with folding and breaking up of the interface. In both cases, numerical results are compared to either analytical, semi-analytical or experimental results.

The implementation of the methodologies has been developed as part of the PETSc-

FEM code [22], which are based on the Message Passing Interface [23] and the Portable Extensible Toolkit for Scientific Computation (PETSc) libraries [24], allowing the use of parallel computing for the solution of large systems of equations.

2. Governing Equations

2.1. Domains and Kinematic Descriptions

As usual in continuum mechanics, there are three main descriptions of motion that can be used to describe the fluid flow: Lagrangian, Eulerian and ALE, which are represented in Fig. 1.

The material or Lagrangian description $\Omega_\mathbf{X}$ is constituted by the system of material particles associated to the coordinates $\mathbf{X}$. The spatial or Eulerian description $\Omega_\mathbf{x}$ is associated to the spatial coordinates $\mathbf{x}$. Finally, the ALE reference configuration Ω_χ is the one associated to the χ coordinates: this reference can coincide with one of the others, but in general it is used as an intermediate (arbitrary) configuration, where the numerical entities are allowed to move independently from the material particles motion [25].

The linear transformations that relate the three configurations, represented in Fig. 1, are the following:

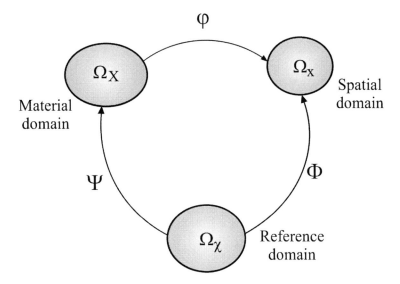

Figure 1. Reference configurations and transformations for describing the flow field.

- the linear transformation $\varphi : \Omega_\mathbf{X} \times [t_0, T) \to \Omega_\mathbf{x} \times [t_0, T)$, that transforms the material coordinates $\mathbf{X}$ in spatial ones $\mathbf{x}$,

$$(\mathbf{X}, t) \to \varphi(\mathbf{X}, t) = (\mathbf{x}, t); \tag{1}$$

where the existence of the inverse $(\mathbf{X}, t) = \varphi^{-1}(\mathbf{x}, t)$ requires a positive value of the Jacobian of the transformation $J = \det\left(\frac{\partial \mathbf{x}}{\partial \mathbf{X}}\right) > 0$. Otherwise, $J < 0$ leads to a change in the orientation for the reference axes, and $J = 0$ represents a singular transformation.

- the linear transformation $\Phi : \Omega_\chi \times [t_0, T] \to \Omega_\mathbf{x} \times [t_0, T]$ from the arbitrary domain of reference Ω_χ to the spatial one $\Omega_\mathbf{x}$,

$$(\chi, t) \to \Phi(\chi, t) = (\mathbf{x}, t); \qquad (2)$$

which, as the first one, requires $\tilde{J} = \det\left(\frac{\partial \mathbf{x}}{\partial \chi}\right) > 0$;

- the linear transformation $\Psi : \Omega_\chi \times [t_0, T] \to \Omega_\mathbf{X} \times [t_0, T]$ from the arbitrary coordinates χ to the material domain $\mathbf{X}$,

$$(\chi, t) \to \Psi(\chi, t) = (\mathbf{X}, t); \qquad (3)$$

with $\bar{J} = \det\left(\frac{\partial \mathbf{X}}{\partial \chi}\right) > 0$, as before.

The absolute velocity of a particle with respect to the Eulerian framework, or material velocity, is defined as:

$$\mathbf{v}(\mathbf{X}, t) = \left.\frac{\partial \mathbf{x}}{\partial t}\right|_\mathbf{X}; \qquad (4)$$

and coincides with the velocity of a material particle $\mathbf{X}$ which is placed on the spatial coordinate $\mathbf{x}$ in time t, where Eq. (4) is evaluated. This partial derivative evaluated on fixed $\mathbf{X}$ produces the convective effect when the fluid equations are proposed over an Eulerian referential, and it indicates the displacement velocity between the fluid and that frame.

2.2. Navier–Stokes Equations

The Navier–Stokes (NS) equations represent the flow of a viscous and incompressible fluid, and for an Eulerian framework these are written as follows,

$$\rho\left(\partial_t \mathbf{v} + \mathbf{v} \cdot \nabla \mathbf{v} - \mathbf{f}\right) - \nabla \cdot \boldsymbol{\sigma} = 0 ; \qquad (5)$$

$$\nabla \cdot \mathbf{v} = 0 ; \qquad (6)$$

over the flow domain Ω in time $t \in [0, T]$, where $\mathbf{v}$ is the flow velocity, $\mathbf{f}$ is the body force, ρ the fluid density and T is a given final time. Furthermore, $\nabla = \nabla_\mathbf{x}$ is the gradient operator with respect to the spatial coordinates $\mathbf{x}$, and ∂_t indicates a partial derivative with respect to the time. The fluid stress tensor $\boldsymbol{\sigma} = \boldsymbol{\sigma}(\mathbf{v}, p)$ can be decomposed into an isotropic and a deviatoric part, $-p\mathbf{I}$ and $\mathbf{T}$, respectively,

$$\boldsymbol{\sigma} = -p\mathbf{I} + \mathbf{T} ; \qquad (7)$$

with the pressure indicated with p and the identity tensor with $\mathbf{I}$. Since the fluids are considered Newtonian, the deviatoric part $\mathbf{T}$ is linearly related to the strain rate $\boldsymbol{\epsilon} = \boldsymbol{\epsilon}(\mathbf{v})$ through

$$\mathbf{T} = 2\mu\boldsymbol{\epsilon} ; \qquad (8)$$

being μ the dynamic viscosity of the fluid, and

$$\epsilon = \frac{1}{2}\left[\nabla \mathbf{v} + (\nabla \mathbf{v})^T\right] ; \qquad (9)$$

the strain rate, where $(...)^T$ indicates transposition.

The boundary conditions over the contours Γ of the domain Ω are given by

$$\mathbf{v} = \mathbf{v}_D \qquad \text{on } \Gamma_D; \qquad (10)$$
$$\sigma \cdot \mathbf{n} = \mathbf{t} \qquad \text{on } \Gamma_t; \qquad (11)$$

such that the Dirichlet condition is the known velocity $\mathbf{v}_D$ of the solid walls included in Γ_D, and the traction forces $\mathbf{t}$ are given over the surface Γ_t, which is generally an interface between fluids. Furthermore, it is verified that $\Gamma = \Gamma_D \cup \Gamma_t$ and $\Gamma_D \cap \Gamma_t = \emptyset$.

The most common boundary conditions for the velocity are the following.

- *Fixed walls,*

$$\mathbf{v}_D = \mathbf{0}, \qquad (12)$$

which responds to the viscous fluid hypothesis for the governing equations.

- *Low viscosity fluids,* when the kinematic fluid viscosity is very low, and the perfect slip condition $\mathbf{v}_D \cdot \mathbf{n} = 0$ is suitable [10, 12, 26] due to the reduction of the computational effort required by Eq. (12).

- *Symmetry planes and other artificial boundaries,* which can be imposed when the fluid does not circulate through the (fictitious) contours; again, the condition to be imposed is the perfect slip one, i.e. $\mathbf{v}_D \cdot \mathbf{n} = 0$.

The initial conditions for solving the transient NS equations, Eqs. (5-6), are given on the velocity,

$$\mathbf{v}(\mathbf{x}, 0) = \mathbf{v}_0(\mathbf{x}) ; \qquad (13)$$

being $\mathbf{v}_0$ the initial velocity field defined over the whole domain at time $t = 0$.

3. Arbitrary Lagrangian-Eulerian Method

The ALE methodology exposed in this section is a technique that allows to consider small or moderate FS displacements, such that no breaking up or folding of the interface can be modeled.

This ALE procedure involves three consecutive instances in each time step: (i) the fluid flow state, solving the NS equations to obtain the velocity and the pressure fields; (ii) the FS displacement, determined by the velocities from the previous stage; (iii) the coordinates update of the nodes that belong to the interior of the domain, keeping the topology of the mesh. The consecutive execution of these instances constitutes a weak-coupling procedure, and they are solved using the FE method.

The numerical examples solved allow the validation of the procedure, and show their ability to solve the FS displacement, mostly in transient sloshing problems.

3.1. Fluid Flow Equations

When an ALE formulation is employed to solve the fluid flow in a reference framework which is different from the spatial one, the relative velocity between the mesh and the Eulerian frameworks is given by [25, 27]

$$\hat{\mathbf{v}}(\chi,t) = \left.\frac{\partial \mathbf{x}}{\partial t}\right|_{\chi}. \tag{14}$$

Then, the convective velocity $\mathbf{c}$ is defined as follows,

$$\mathbf{c} := \mathbf{v} - \hat{\mathbf{v}}; \tag{15}$$

and constitutes the relative velocity between the material and the reference configurations.

The NS equations given in Eqs. (5-6) are referred to spatial coordinates $\mathbf{x}$, and they are not suitable to consider the deformation of the domain Ω due to the FS displacement and shape changes. Then, the ALE form of the NS equations considers Eqs. (14-15) for relating the Eulerian framework $\Omega_{\mathbf{x}}$ and the arbitrary referential Ω_{χ}, such that the momentum equation and the incompressibility condition in NS equations are, respectively [25],

$$\rho\left(\partial_t \mathbf{v} + \mathbf{c} \cdot \nabla \mathbf{v} - \mathbf{f}\right) - \nabla \cdot \boldsymbol{\sigma} = 0 ; \tag{16}$$

$$\nabla \cdot \mathbf{v} = 0 . \tag{17}$$

Note that the only affected equation is the momentum one, where the advective term velocity is $\mathbf{c}$, regarding that the reference domain velocity $\hat{\mathbf{v}}$ is included through the Eq. (15).

Furthermore, in an ALE representation of a FS fluid problem, such interface, named Γ_{FS}, constitutes part of the boundary, particularly as a subset of a traction surface, $\Gamma_{FS} \subseteq \Gamma_t$. Then, for this methodology, the condition to be applied over Γ_{FS} is given by Eq. (11), which consists of the equilibrium between the stress tensors on the fluid and on the gas, σ_l and σ_g, respectively, projected onto the normal to the interface,

$$\sigma_l \cdot \mathbf{n} = \sigma_g \cdot \mathbf{n} \qquad \text{on } \Gamma_{FS}. \tag{18}$$

Since the viscosity and the density of the gas are negligible, then

$$\mathbf{T} \cdot \mathbf{n} = 0 \qquad \text{on } \Gamma_{FS}; \tag{19}$$

$$p = P_{atm} \qquad \text{on } \Gamma_{FS}; \tag{20}$$

where P_{atm} is the pressure exerted by the gaseous phase over the liquid, usually the atmospheric, and the tensor $\mathbf{T}$ represents the tractions produced by the gas over the interface. It follows that the traction forces over the FS are $\mathbf{t} = -P_{atm}\, \mathbf{n}$, i.e.,

$$\sigma_g \cdot \mathbf{n} = -P_{atm}\, \mathbf{n} \quad \text{over } \Gamma_{FS}. \tag{21}$$

If $P_{atm} = 0$ is adopted, the boundary term over Γ_{FS} is null. Note that surface tension effects are not considered in this model, and that the pressure on the fluid side is different from the pressure on the gas side, due to the condition of equal tractions given in Eq. (18).

Regarding the condition given in Eq. (10), it is worth mentioning that FS problems solved with an ALE strategy can present drawbacks during the mesh update process when walls are considered fixed. Then, the contact line is fixed, while the FS in its neighborhood moves according to the kinematics of the flow, and large elevation gradients arise. In order to prevent these large gradients, the non-slip condition close to the contact line is relaxed, e.g. by using the so-called "Navier slip condition" [1],

$$(\mathbf{I} - \mathbf{nn}) \cdot (\mathbf{n} \cdot \boldsymbol{\sigma}) = -\frac{1}{\beta}(\mathbf{I} - \mathbf{nn}) \cdot (\mathbf{v}_S - \mathbf{v}_D) ; \tag{22}$$

in which $\mathbf{v}_S$ is the fluid velocity on the contact line and $\mathbf{I} - \mathbf{nn}$ projects the velocity over the tangent plane. The arbitrary coefficient β takes values from $\beta = 0$ to $\beta \to \infty$, recovering the non-slip and the perfect slip conditions, respectively.

The Eqs. (16-17) are solved by FE with the NS module of the PETSc-FEM libraries [28, 29]. In this case, stabilization with streamline upwind/Petrov-Galerkin (SUPG) [30] and pressure stabilizing/Petrov-Galerkin (PSPG) [31] are applied, given the presence of the advective term and the use of linear elements with equal velocity and pressure interpolations.

3.2. Free Surface Displacement

The displacements of the FS nodes are restricted to a fixed direction represented by a unit vector $\hat{\mathbf{s}}_j$, in such a way that the updated position of the j node over Γ_{FS} in time t is given as [13]

$$\mathbf{x}_j(t) = \mathbf{x}_{0,j} + \eta_j(t)\,\hat{\mathbf{s}}_j ; \tag{23}$$

in which the fixed direction is given by $\hat{\mathbf{s}}_j$, also called *spine*, $\mathbf{x}_{0,j}$ is the initial position of the node j, and η_j is the scalar coordinate over the spine, as represented in Fig. 2. The reference direction $\hat{\mathbf{s}}_j$ is usually adopted as normal to the FS at rest.

Note that the reference directions are used exclusively for the nodal FS displacements, while in the rest of the domain the nodal positions are calculated independently of such directions. This procedure is different from the typical methods of spines, like the one proposed by Saito and Scriven [32]. This independence in the kinematics of the FS nodes and the internal nodes leads to lower distortions of the mesh elements in comparison to the spines methods.

The FS displacement is ruled by the kinematic condition, that physically indicates that there is no material exchange through the interface, and it is expressed by [33, 34]

$$\mathbf{v} \cdot \mathbf{n} = \partial_t \boldsymbol{\eta} \cdot \mathbf{n} \quad \text{over } \Gamma_{FS}; \tag{24}$$

where $\mathbf{v}$ is the velocity and $\mathbf{n}$ is the direction normal to the FS, with $\boldsymbol{\eta} = \eta\,\hat{\mathbf{s}}$ as the FS elevation along the $\hat{\mathbf{s}}$ direction, being η the scalar displacement, as represented in Fig. 2. By replacement in Eq. (24), it is

$$\partial_t \eta = \frac{\mathbf{v} \cdot \mathbf{n}}{\hat{\mathbf{s}} \cdot \mathbf{n}} . \tag{25}$$

Note that only the displacements over the direction of $\mathbf{n}$ are considered, meanwhile the tangential ones are considered irrelevant.

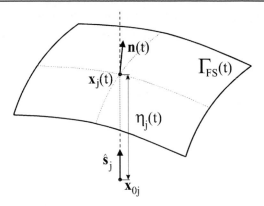

Figure 2. Nodal displacements over the FS.

From Eq. (23), and considering that in three spatial dimensions the scalar field can be expressed as $\eta = \eta(x_1, x_2, t)$, the FS can be implicitly represented by

$$F(\mathbf{x}, t) = x_3 - \eta = 0; \qquad (26)$$

which implies that $\hat{\mathbf{s}} \parallel x_3$, i.e. the spines direction is adopted parallel to the vertical axis x_3. Then, the normal to the FS can be calculated as

$$\mathbf{n} = \nabla F(\mathbf{x}) = \left[-\frac{\partial \eta}{\partial x_1}; \; -\frac{\partial \eta}{\partial x_2}; \; +1 \right]^T. \qquad (27)$$

By introducing Eq. (27) in Eq. (25), and defining the projection parameter $H = \hat{\mathbf{s}} \cdot \mathbf{n}$, it results in

$$\partial_t \eta = \frac{1}{H} \mathbf{v} \cdot \mathbf{n} = \frac{1}{H} \left[-v_1 \frac{\partial \eta}{\partial x_1} - v_2 \frac{\partial \eta}{\partial x_2} + v_3 \right]; \qquad (28)$$

which is the representation of η as an advection equation.

Re-arranging Eq. (28), the scalar field η can be determined by solving the following advective system of equations,

$$\partial_t \eta + \mathbf{v}_\parallel \cdot \hat{\nabla} \eta = s \qquad \text{in } \Omega'_{FS}; \qquad (29)$$
$$\eta = \eta_D \qquad \text{over } \Gamma'_D; \qquad (30)$$

where the velocity in the plane normal to the spines is

$$\mathbf{v}_\parallel = \frac{1}{H} [v_1; \; v_2]^T; \qquad (31)$$

the gradient of η is

$$\hat{\nabla} \eta = \left[\frac{\partial \eta}{\partial x_1}; \; \frac{\partial \eta}{\partial x_2} \right]^T; \qquad (32)$$

and the source term $s = v_3/H$ represents the fluid velocity in the spine direction for each point of the FS. Furthermore, $\Omega'_{FS} = \Gamma_{FS}$ in Eq. (21) is a domain of $n_{\dim} - 1$ spatial dimensions, where $n_{\dim}$ indicates the number of spatial dimensions considered in the complete

FS flow problem. Due to the hyperbolic character of Eq. (29), the corresponding Dirichlet boundary conditions are defined over the inflow section, i.e., $\Gamma'_D = \{\Gamma'_{FS} \mid \mathbf{v}_\| \cdot \hat{\mathbf{n}} < 0\}$.

The FS formulation given by Eqs. (29,25) is numerically unstable for high-frequency surface gravity waves, given their advective character [13]. Then, the numerical solution of problems with dominant advective terms requires an additional step, that can take either the form of a typical stabilized advective system or a smoothing operation. In the first case, some typical stabilization procedures can be applied for the numerical solution, such as SUPG [30], as made by Soulamani *et al.* [26], Gler *et al.* [1] and Battaglia [35], or the *Galerkin/Least-Squares* (GLS) strategy, as in Behr and Abraham [36]. As an alternative, a smoothing operator based on the heat equation has been also used for overcoming this issue [13]; however, it depends on arbitrary parameters, and is not suitable for highly advective problems.

3.3. Mesh Updating

Once the FS nodal displacements are determined as explained in Sec. 3.2., and the contact line displacements are determined with Eq. (22), the full set of data for updating the mesh is available, independently of the mesh update method adopted.

As the FS displacements produce a deformation of the shape of the domain, the tessellation of the reference domain requires to be adapted to the new shape each time step. A general classification of alternatives for mesh updating is the following, given by Behr and Abraham [36],

- **Algebraic**: the spatial coordinates of the internal nodes are determined by explicit algebraic expressions as a function of the FS displacements, keeping the topology of the (structured) mesh unchanged; this is the case of the spines methods [32].

- **Relocalization of internal nodes**: the internal nodes are relocated inside the domain by keeping the topology, through the solution of pseudo-elastic problems [1, 12, 13, 36, 37], or by minimizing the distortion of the mesh elements [38, 39].

- **Remeshing**: every time the domain is deformed, a new mesh is generated, requiring the interpolation or extrapolation of the fields from the initial discretization to the new one.

From these alternatives, the second one was adopted for solving FS fluid flows within an ALE framework, particularly the one that consists in solving a pseudo-elastic mesh update. The auxiliary problem is solved over the reference domain Ω, generally the initial one, where the only boundary conditions to be applied are Dirichlet kind. The standard formulation is

$$\sigma_{ij,j} = 0 \, ;$$
$$\sigma_{ij} = 2\tilde{\mu}\varepsilon_{ij} + \tilde{\lambda}\delta_{ij}\varepsilon_{kk} \, ; \tag{33}$$
$$\varepsilon_{ij} = \frac{1}{2}(u_{i,j} + u_{j,i}) \, ;$$

being $\tilde{\mu}$ and $\tilde{\lambda}$ the artificial Lamé elastic properties of the material, the Kronecker tensor is δ_{ij} and the known nodal displacements are

$$\mathbf{u}_j = \mathbf{x}_j^{n+1} - \mathbf{x}_j^0 \, ; \tag{34}$$

given by the FS displacements. The remaining contours, which generally correspond to solid walls, are proposed either as $\mathbf{u} = 0$ for the non-slip parts or $\mathbf{u} \cdot \mathbf{n} = 0$ for the slip ones. Note that the mechanical properties of the material can be given also as the Poisson coefficient $\tilde{v}$ and the longitudinal elasticity module $\tilde{E}$. On the one hand, the numerical value of $\tilde{E}$ is irrelevant, given the purpose of determining the deformed shape of the domain, independently of the stresses on the artificial problem; on the other hand, a common value is $\tilde{v} = 0.3$, considering that $\tilde{v} \to 0.5$ would lead to an ill-conditioned elastic problem.

The pseudo-elastic mesh update is capable of solving problems with small to moderate FS deformations. Nevertheless, when deformations increase and the mesh quality is extremely compromised, another approach has to be used; for example, the method proposed by López et al. [38, 39], which is suited for computational mesh dynamics in general, and applicable to FS fluid flow problems within an ALE framework, see Battaglia [35].

3.4. Numerical Examples

3.4.1. Two-Dimensional Viscous Sloshing

The performance of the ALE method is shown through a two-dimensional (2D) viscous sloshing of known analytical solution. This test consists of solving the small amplitude motion of a FS of an incompressible viscous fluid in a rectangular tank submitted to a gravity acceleration, see Fig. 3. The displacements of one node of the FS are compared to the analytical solution given by Prosperetti [40], which predicts the frequency and the viscous damping of the system, as referred by other authors [12, 41, 42].

The problem consists of releasing the FS from an initial sinusoidal position with amplitude a_0, shown in Fig. 3, given by

$$h(x) = 1.5 + a_0 \sin[\pi(1/2 - x)]. \tag{35}$$

Then, due to the viscous forces, the amplitude of the displacements of the FS is damped. The boundary conditions over the lateral walls and the bottom of the tank are slip, i.e. $\mathbf{v}_D \cdot \mathbf{n} = 0$, meanwhile over the FS the conditions are given as $p = P_{\text{atm}} = 0$ and $\mathbf{T} \cdot \mathbf{n} = \mathbf{0}$.

The analytical expression for the displacement of one point of the FS is as follows [40],

$$a(t) = \frac{4v^2 k^4}{8v^2 k^4 + \omega_0^2} a_0 \, \text{erfc}(vk^2 t)^{1/2} + \sum_{i=1}^{4} \frac{z_i}{Z_i} \left(\frac{\omega_0^2 a_0}{z_i^2 - vk^2} \right) \exp[(z_i^2 - vk^2)t] \, \text{erfc}(z_i t^{1/2}) \, ; \tag{36}$$

with v the kinematic viscosity of the fluid, k the wave number, and $\omega_0^2 = gk$ the inviscid natural frequency, where g is the gravity acceleration. Moreover, each z_i is a root of the equation

$$z^4 + k^2 v z^2 + 4(k^2 v)^{3/2} z + v^2 k^4 + \omega_0^2 = 0 \, ; \tag{37}$$

in which $Z_1 = (z_2 - z_1)(z_3 - z_1)(z_4 - z_1)$, obtaining Z_2, Z_3, Z_4 by circular permutation of the indices, and $\text{erfc}(\ldots)$ is the error function for complex variable. This expression corresponds to the linearized case, and considering small amplitude flat waves in a domain of infinite depth.

The numerical solution of the problem was performed in a tank of $h = 1.5$ m height and $d = 1.0$ m wide, such that $k = 2d$. with an initial amplitude $a_0 = 0.01$ m, unit gravity acceleration $g = 1.0$ m/s^2, and a fluid kinematic viscosity $\nu = 0.001$ m^2/s. The finite element mesh consists of 40×60 quadrangular elements in width and height, respectively, and the time step was chosen as $\Delta t = 2.22 \; 10^{-2}$ s.

Once the problem was solved, the vertical displacement of the top left node as a function of the time is plotted in Fig. 4 together with the curve analytically determined with Eq. (36). Note that the frequency and the rate of viscous damping are well captured by the numerical method.

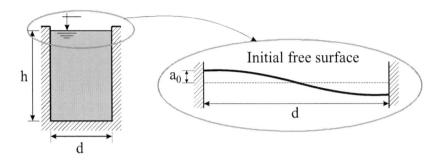

Figure 3. Geometry and initial free surface position for the small-amplitude sloshing problem.

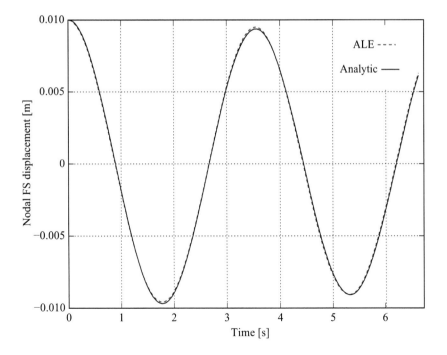

Figure 4. Analytic solution curve and numerical results calculated with ALE for the small-amplitude sloshing problem.

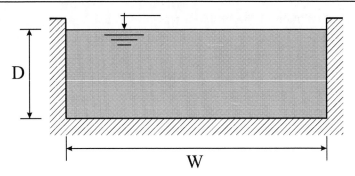

Figure 5. Geometry of the large-amplitude sloshing problem.

3.4.2. Two-Dimensional Large-Amplitude Sloshing

The numerical example consists of solving the large-amplitude sloshing in a tank filled with water submitted to a periodic horizontal acceleration. The container is modeled as a two-dimensional domain of $W = 0.80$ m width and the fluid depth is $D = 0.30$ m, as shown in Fig. 5.

Boundary conditions are $p = P_{atm} = 0$ and $\mathbf{T} \cdot \mathbf{n} = \mathbf{0}$ over Γ_{FS}, and over the solid walls are perfect slip, i.e. $\mathbf{v} \cdot \mathbf{n} = 0$, due to the low kinematic viscosity of the fluid, $\nu = 10^{-6}$ m^2/s, as chosen from the references [10, 26, 43]. Furthermore, the density is $\rho = 1000$ kg/m^3.

Following Huerta and Liu [10], the acceleration acting over the fluid is $G = [g_1; g_2]^T$, being the vertical component of the gravity acceleration $g_2 = -g = -9.81$ m/s^2. The horizontal component of the gravity acceleration, g_1, is given with a sinusoidal variation in time

$$g_1 = A\, g\, \sin \omega t\,; \tag{38}$$

in which the amplitude coefficient is $A = 0.01$, t is the time and ω is the circular frequency, such that the frequency of the acceleration coincides with the first sloshing mode of the system, Then, considering a wave length $\lambda = 2W$, is

$$f = \left(\frac{g}{4\pi W} \tanh \frac{\pi D}{W} \right)^{0.5} \approx 0.89825\ 1/\text{s}\,; \tag{39}$$

and the circular frequency results in $\omega = 2\pi f \approx 5.64$ rad/s.

The problem was solved on a mesh of 39×104 quadrangles in depth and width, respectively, with a time step of $\Delta t = 0.009$ s and time integration with a Crank-Nicolson scheme, up to a final time of $T = 15$ s. The vertical displacement of the top left node of the domain relative to the water depth D is plotted as a function of time in Fig. 6. Taking into account that the tank is induced with the first sloshing mode from the rest position, the amplitude of the movement is increased up to either the viscous dissipation avoids further increment, or the FS breaks up.

The results obtained show the ability of the method for solving large amplitude of the FS displacements. Taking into account the relatively high horizontal velocities over the FS, it was necessary to use the FS stabilization procedure mentioned in Sec. 3.2.; otherwise, numerical instabilities would produce the failure of the method.

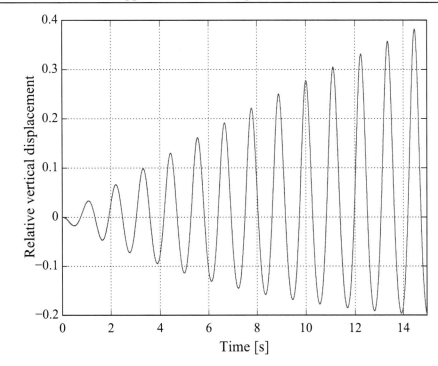

Figure 6. Vertical displacement of the top left node, over the FS, referred to the depth D of the tank.

A mesh convergence study was performed with this example for the ALE method, considering the typical element size of each mesh, h, and time step values Δt summarized in Table 1, as well as the absolute value of the error E taking as a reference the results obtained with the most refined mesh. Furthermore, the absolute error E is plotted as a function of the number of nodes in the mesh, see Fig. 7.

3.4.3. Three-Dimensional Viscous Sloshing

This three-dimensional case is a straight cylindrical tank with an annular base, represented in Fig. 8. The problem consists of releasing the FS from an initial position different from the rest height, producing the fluid circulation from one side of the tank to the other due to a gravity acceleration effect. A similar case was presented in Battaglia *et al.* [13], and was

Table 1. Mesh convergence for the tank submitted to horizontal periodic acceleration.

Mesh	Discretization		Nodes	Absolute Error
	h [m]	Δt [s]	[]	[m]
1 (reference)	0.0038	0.0045	16511	-
2 (results)	0.0077	0.0090	4200	0.0050
3	0.0103	0.0135	2370	0.0106
4	0.0154	0.0180	1113	0.0163

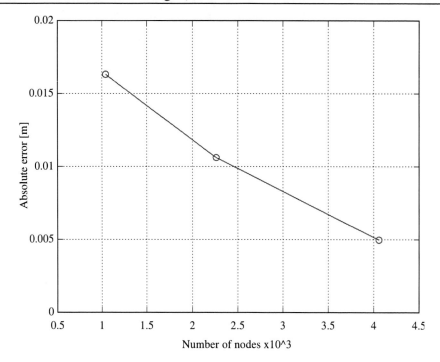

Figure 7. Mesh convergence for the horizontally accelerated tank.

contrasted with semi-analytical results.

The internal and external radius of the cylinder are $R_i = 1$ m and $R_e = 2$ m, respectively, with a fluid height at rest of $H = 1.0$ m, see Fig. 8. For the numerical solution, boundary conditions adopted are $p = P_{atm} = 0$ and $\mathbf{T} \cdot \mathbf{n} = \mathbf{0}$ over Γ_{FS}. Over the bottom and the lateral walls, the condition adopted is perfect slip, i.e. $\mathbf{v} \cdot \mathbf{n} = 0$. Regarding the mesh update instance, the nodal displacements over the rigid walls are free on the vertical direction, meanwhile the bottom nodes are fixed. The FS initial position is proposed as a plane deformation of maximum amplitude $a_0 = 0.05$ m over the height of the interface at rest.

The finite element model was solved with a mesh composed by 32000 hexahedral elements of 8 nodes, and time stepping of $\Delta t = 0.0375$ s. The vertical gravity acceleration was fixed on $g = -1.0$ m/s^2, with a kinematic viscosity $\nu = 10^{-3}$ m^2/s, and a fluid density $\rho = 1$ kg/m^3.

The vertical nodal displacements of some representative points are plotted in Fig. 9. The curves of displacement show the superposition of different eigenmodes of sloshing, being two of them the most notorious; other higher frequency modes are damped due to the high kinematic viscosity adopted. The amplitude of the displacements also diminishes in time as a consequence of viscous effects.

4. Level Set Method

The interface capturing LS method presented in this section was proposed with the aim of solving two-fluid incompressible flows with large interface deformations, including folding

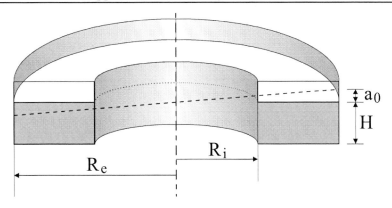

Figure 8. Geometry of a straight vertical cylinder with an annular base for the three-dimensional sloshing test. Initial free surface (trace line) and reference axis.

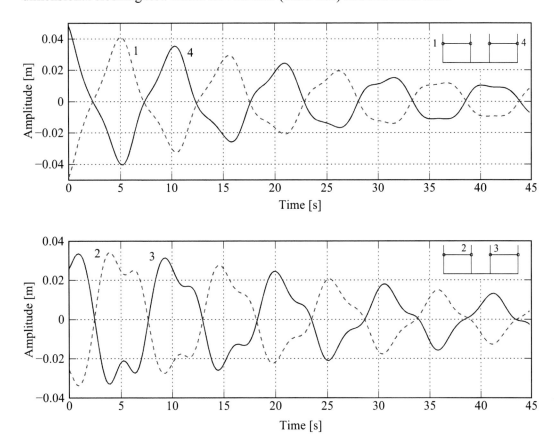

Figure 9. Time evolution of representative nodal displacements for the cylindrical tank with an annular base obtained with ALE.

and breaking up, which allows the solution of problems that cannot be solved by the ALE methodology described in Sec. 3. The procedure is proposed over an Eulerian framework, and the reference configuration coincides with such referential. Furthermore, in the general case, the method solves both the liquid and the gaseous phase, in case FS flows were considered.

This kind of approximation has numerous variants which are generated in order to avoid some problems associated to the advection of the LS function ϕ, such as volume variation in one of the phases involved and the loss of the smoothness of the transition between fluids [44].

The computational procedure consists of three parts [35,45], being each of them: (i) the NS solver for non-homogeneous fluid flows; (ii) the advective (ADV) solver which solves the transport of the LS function field ϕ; and (iii) the reinitialization procedure over the field of ϕ, named Bounded Renormalization with Continuous Penalization (BRCP), to keep the regularity of the LS function [46].

4.1. Level Set Function

The LS function field $\phi = \phi(\mathbf{x},t)$ is defined as continuous over the domain of analysis Ω for any time t and for every coordinate $\mathbf{x}$. This function allows the identification of each phase involved in the simulation, identified here as liquid and gaseous phases. In the most common case, the LS function is proposed as

$$\phi(\mathbf{x},t) \begin{cases} > 0 & \text{if } \mathbf{x} \in \Omega_l \\ = 0 & \text{if } \mathbf{x} \in \Gamma_I \\ < 0 & \text{if } \mathbf{x} \in \Omega_g \end{cases} \tag{40}$$

being Ω_l the subspace corresponding to the liquid phase and Ω_g the one for the gaseous domain region. These domains verify both the conditions $\Omega = \Omega_l \cup \Omega_g$ and $\Omega_l \cap \Omega_g = \varnothing$, meanwhile the interface is identified as [17]

$$\Gamma_I = \{\mathbf{x} \mid \phi(\mathbf{x},t) = 0\}. \tag{41}$$

As mentioned before, the LS function ϕ is continuous over the whole domain Ω, and is employed as reference for identifying the fluid properties at the time when the flow stage is computationally solved.

In the particular case of the reinitialization stage employed in this section, see Sec. 4.4., the LS function is bounded to ± 1, with a smooth transition between bounds.

4.2. Non-homogeneous Fluid Flow Equations

In the LS approach, the domain where the NS Eqs. (5-6) are solved involves two different immiscible, incompressible and viscous fluids. Then, the properties of the fluids involved are taken into account by employing the LS function ϕ and time t as reference, such that

$$\rho(\phi(\mathbf{x},t))\left(\partial_t \mathbf{v} + \mathbf{v} \cdot \nabla \mathbf{v} - \mathbf{f}\right) - \nabla \cdot \boldsymbol{\sigma} = 0 \,; \tag{42}$$

$$\nabla \cdot \mathbf{v} = 0 \,; \tag{43}$$

in which the position vector is $\mathbf{x} \in \Omega$, the fluid density is $\rho(\phi(\mathbf{x},t))$ and ϕ is the LS function, while the remaining parameters are those given in Sec. 2.2. The fluid properties are involved also in the viscous forces tensor,

$$\mathbf{T} = 2\,\mu(\phi(\mathbf{x},t))\,\boldsymbol{\epsilon} \,; \tag{44}$$

where the dynamic viscosity $\mu = \mu(\phi(\mathbf{x},t))$ depends also on the LS function and the time of evaluation, and the strain rate tensor ϵ is calculated as in Eq. (9).

Even when the transition between $\phi > 0$ and $\phi < 0$ is smooth, the interpolation of the fluid properties requires a thinner transition width in order to reach a more precise fluid flow state. Hence, the fluid properties for Eqs. (42) and (44) are interpolated as follows,

$$\rho(\phi) = \frac{1}{2}\left[(1+\tilde{H}(\phi))\rho_l + (1-\tilde{H}(\phi))\rho_g\right];$$
$$\mu(\phi) = \frac{1}{2}\left[(1+\tilde{H}(\phi))\mu_l + (1-\tilde{H}(\phi))\mu_g\right]; \quad (45)$$

where the key is a smeared Heaviside function $\tilde{H}(\phi)$ given by

$$\tilde{H}(\phi) = \tanh\left(\frac{\pi\phi}{\tilde{\varepsilon}}\right); \quad (46)$$

in which the reference parameter $\tilde{\varepsilon}$ controls the transition width. From Eq. (46), if $|\phi| \to \tilde{\varepsilon}$ then $\tilde{H}(\phi) \to 1$. This reference value is adopted here as $\tilde{\varepsilon} = 0.5$, giving a NS transition 70% thinner than the one measured over the ϕ field.

There are other smeared Heaviside functions used in LS approaches [44,47,48], generally defined as piecewise functions. However, the expression from Eq. (46) is preferable in the present method because it automatically fits the bounds proposed for the LS function, $-1 \leq \phi \leq +1$, without any additional definition.

Equations (42-43) are solved by FE with the NS module of the PETSc-FEM libraries [28, 29], where the SUPG [30] and the PSPG [31] strategies are used in order to avoid the typical instabilities produced by the advective term and by the use of linear elements with equal velocity and pressure interpolations.

4.3. Advection of the LS Field

The evolution of the LS function in time and space is ruled by the nature of the physic problem to be solved [49]. In the case of two-fluid or FS incompressible flows, the velocity of the fluid $\mathbf{v}$ from Eqs.(42-43) produces the transport of ϕ inside the domain Ω as follows,

$$\partial_t \phi + \mathbf{v} \cdot \nabla \phi = 0; \quad (47)$$

with boundary conditions over $\Gamma_{LS} = \partial\Omega$ such that

$$\phi = \bar{\phi} \quad \text{over} \quad \Gamma_{in}; \quad (48)$$

with inflow sections given as $\Gamma_{in} = \{\Gamma_{LS} \mid \mathbf{v} \cdot \mathbf{n} < 0\}$, being $\mathbf{n}$ the outgoing unit normal of the boundary. As the transport is solved over the whole domain, the interface Γ_{FS} is displaced automatically, and there is no need to define an additional condition over it, regarding that the surface tension is neglected.

The computation of the state of the LS function ϕ from Eq. (47) is made through the FE advection-diffusion module from the PETSc-FEM code [22, 29]. As in the previous step, the instabilities which arise from the use of a Galerkin formulation are avoided by using an SUPG strategy [30, 50]. Nevertheless, in cases where the BRCP procedure is applied, the SUPG results unnecessary.

4.4. Reinitialization of the Level Set Function Field

Generally speaking, LS methods have a stage additional to the fluid state and the advection instances. This additional step, a reinitialization of ϕ, focuses on keeping the smoothness and regularity of the LS function field, which is deteriorated as a consequence of the numerical errors arising from the advection. Furthermore, the reinitialization sometimes helps to compensate volume loss [44, 51]. Some strategies are based on solving the distance function from a point **x** to the interface ϕ = 0, in time t, or some kind of variant there of [44, 51, 52]. Other proposals, such as the BRCP, are oriented to the use of an auxiliary advection-diffusion-like strategy [47, 53].

The ϕ reinitialization employed in this chapter is based on solving a Partial Differential Equation (PDE) system, particularly by the FE method, taking the LS function ϕ as the variable [35, 46].

Once Eq. (47) is solved, there is an updated value of the LS field, which in this section is named ϕ^0. Then, the BRCP operator is applied over ϕ^0 as follows,

$$\phi\,(\phi^2 - \phi_{\text{ref}}^2) - \kappa\Delta\phi + M\left[\check{H}(\phi) - \check{H}(\phi^0)\right] = 0\,; \tag{49}$$

with an artificial diffusivity κ, or diffusive parameter, a penalty coefficient M, and a reference value for the variable, ϕ_{ref}, that is adopted as $\phi_{\text{ref}} = 1$. The diffusivity κ is given in the order of a representative element size of the mesh h, in length^2 units. Appropriate values for κ are h^2 to $(3h)^2$, considering that the lower the value of κ, the thinner the transition width. A similar approach is given by Olsson et al. [47, 53], who solve a similar PDE system; however, their formulation lacks a penalizing term.

The values ϕ^0, which are the data from the advection stage, are considered as a part of the penalizing term, affected by the penalty M, see Eq. (49). The objective of introducing this term is avoiding the interface ϕ = 0 displacement during the reinitialization of the field by weighting $\check{H}(\phi) - \check{H}(\phi^0)$. The penalization function is given by the continuous expression

$$\check{H}(\phi) = \tanh(2\pi\phi); \tag{50}$$

where, the positive penalty coefficient M is non-dimensional. Such penalization function was chosen due to the advantages of this smooth, bounded, and infinitely differentiable expression in contrast with the discontinuous ones, e.g. the Heaviside function, which is harder to manipulate from the numerical point of view. Moreover, the argument proposed for the penalization in Eq. (50) allows concentrating the effect of such term nearby ϕ = 0. Numerical experiments show that appropriate values for the penalty parameter M are $O(10^{n_d+2})$, where n_d represents the spatial dimensions for the whole problem to be solved. On the one hand, if M values are lower than such order, high errors on the position of the interface arise, while the algorithm fails when $M = 0$ due to the lack of reference from ϕ^0. On the other hand, when $M > O(10^{n_d+2})$, the reinitialized LS function field tends to the same value given by the advection stage, $\phi \to \phi^0$, and the reinitialization effect is lost. Other penalty term used in interface position preservation is found in the Edge-Tracked Interface Locator Technique (ETILT) method [54, 55].

The effect of the first two terms of Eq. (49) on the LS function field can be analyzed by considering the following expression, applied over a one-dimensional domain $0 \leq x \leq L$,

$$C_r\,\phi\,(\phi^2 - \phi_{\text{ref}}^2) - \kappa\Delta\phi = 0. \tag{51}$$

where ϕ_{ref} is known, as well as the non-dimensional coefficient C_r. If $\kappa = 0$, the Laplacian term is null, while the roots of the remaining equation are $\phi = \pm\phi_{ref} = \pm 1$, the stable roots, and $\phi = 0$. Then, there are infinite piecewise-constant solutions with these roots values. By reintroducing the diffusivity, the consecutive segments of $\phi = \pm\phi_{ref} = \pm 1$ of the previous solutions are smoothly connected, as shown in Fig. 10. Further references about this subject are given in [45, 46].

Regarding the transition width γ from $-\phi_{ref}$ to ϕ_{ref}, in Eq. (51) it is

$$\gamma = O\left(\sqrt{\frac{\kappa}{C_r}}\right) ; \qquad (52)$$

i.e. it depends on both the diffusivity κ and the coefficient C_r. Then, by adopting $C_r = 1$, as in Eq. (49), γ is proportional to the square root of κ, as can be seen in Fig. 10.

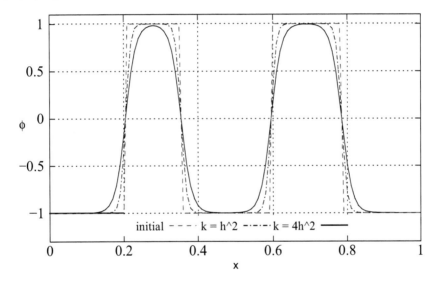

Figure 10. Artificial reaction-diffusion problem renormalized with BRCP for different values of diffusivity κ.

The choice of the diffusive parameter κ and the penalty coefficient M has already been addressed. However, it should be noted that in the cases where the interface breaks up or folds, it is convenient to adopt a low κ and a high M, so that the transition width remains thin and the $\phi \neq 0$ regions are conserved, which are typically drops or bubbles.

The domain regions placed far from the FS are not sensitively modified by this reinitialization proposal. This is a consequence of the operator of Eq. (49), because all the terms tend to zero when $\phi \to \phi_{ref}$. On the contrary, the BRCP affects mostly the neighborhood of $\phi = 0$, which is the region where the mass and precision loss are observed in the numerical methods [51, 55].

The numerical solution of the Eq. (49) is performed as a steady problem through a standard FE method developed as part of the PETSc-FEM code [22, 29], with the initial condition given by the results of the previous advective stage, ϕ^0. This stage is proposed in such a way that it can be executed after each global time step or every n_{reno} global time steps.

4.5. Numerical Examples: Level Set Advection and Reinitialization

4.5.1. Two-Dimensional Single Vortex Flow

The LS function transport and reinitialization are evaluated over a typical two-dimensional problem, considering the cases of pure advection (ADV), advection plus numerical stabilization (ADV+SUPG), and advection with reinitialization of ϕ (ADV+BRCP). Several authors show the capabilities of their interface-capturing methods by performing one of the different variants of this example [20, 21, 47, 56–58].

In this section, the test consists of deforming a disk identified with $\phi \geq 0$ by applying certain velocity field on a unit-side square domain, where the x and y-values vary from 0 to 1 m. The disk of radius $r = 0.15$ m is initially centered in $(x_c, y_c) = (0.50, 0.75)$ m. and the velocity field components are the following,

$$v_x = -\sin(2\pi y)\, \sin^2(\pi x) \tag{53}$$

$$v_y = \sin(2\pi x)\, \sin^2(\pi y). \tag{54}$$

The deformation of the disk is evaluated at a final time $t_f = 5$ s, when the final form of the area involving $\phi > 0$ is a thin strip which should be properly represented by the method with minimum area loss.

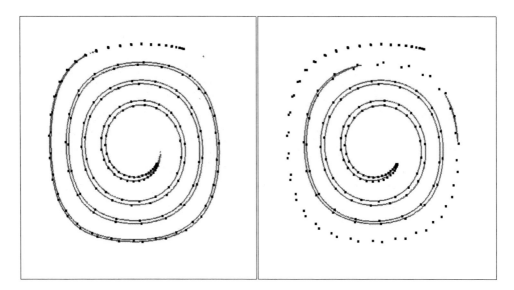

Figure 11. Curves of $\phi = 0$ in $t_f = 5$ s for the 2D vortex solved with ADV (left), and ADV+SUPG (right), solved with meshes of 256 elements by side. Reference particles are represented with dots.

The different strategies employed for solving this problem were applied over a structured tessellation of 131 000 linear triangular elements, corresponding to 256 elements over each side of the domain. The Courant number was fixed in Co = 0.5, with time stepping of $\Delta t = 1.95 \times 10^{-3}$ s, and performing an implicit time integration with $\theta = 0.5$ for the trapezoidal rule. The initial condition for the LS function ϕ was given as a linear variation

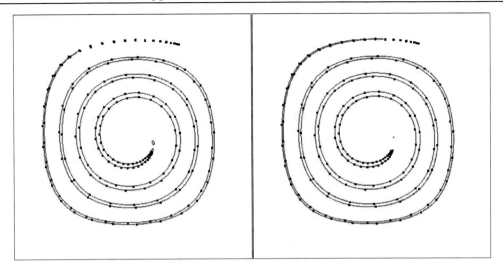

Figure 12. Curves of $\phi = 0$ in $t_f = 5$ s for the 2D vortex solved with ADV+BRCP over the mesh of 256 elements by side (left), and over the refined mesh with 512 elements by side (right). Reference particles are represented with dots.

from $\phi = -1$ to $\phi = 1$ on the circumference of the circle, so that the interface semi-width is 0.05 m.

The problem was solved by using the ADV stage without any stabilization, leading to the results shown in Fig. 11 (left) with dots representing the reference position. Then, it was solved considering an ADV stage with a SUPG stabilization technique, in order to reduce numerical instabilities registered in the former numerical experiment, see Fig. 11 (right). Finally, an ADV + BRCP strategy was used, see Fig. 12 (left), where the adopted BRCP parameters were diffusivity $\kappa = 2h^2 = 3.06 \times 10^{-5}$ m^2 and penalization $M = 10000$, performed every $n_{reno} = 10$ time steps.

The mesh convergence of the BRCP method is verified by solving the problem with $N = 512$ elements by side, with diffusivity $\kappa = 2(2h)^2 = 3.06 \times 10^{-5}$ m^2, the same as in the previous example, penalization $M = 10000$ and $\Delta t = 9.77 \times 10^{-4}$ s, for a fixed Co = 0.5. The results are plotted in Fig. 12 (right) together with dots representing the expected final shape.

The final shapes in $t_f = 5$ s obtained with pure advection and with ADV+BRCP plotted in Figs. 11 (left) and 12 (left), respectively, show little difference between them, while results obtained over the refined mesh are the most precise, as depicted in Fig. 12 (right). On the one hand, the shape obtained with ADV+SUPG shows 27% of area loss, see Fig. 11 (right), while the area of $\phi > 0$ in the ADV case diminishes 3%. On the other hand, the same problem solved with ADV+BRCP show area variations lower than 0.5%.

In addition to the shape of the interface $\phi = 0$ and area preservation, the LS function transitions obtained with each strategy present other differences. In Fig. 13, the LS function variation along a section at $y = 0.5$ m show numerical instabilities in the solution with ADV, which are not present in the ADV+SUPG solution. Finally, the ADV+BRCP gives a smooth transition on the LS function field, which is a desirable property at the time of solving a FS fluid flow with a LS-like method.

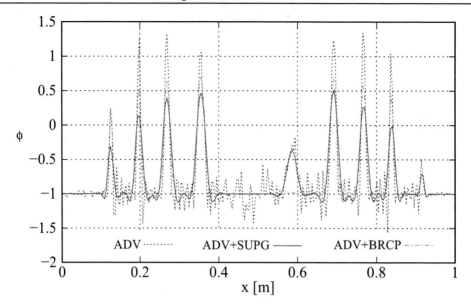

Figure 13. Section at $y = 0.5$ m for the two-dimensional vortex problem in $t_f = 5$ s for ADV, ADV+SUPG and ADV+BRCP.

4.5.2. Three-Dimensional Sphere Deformation

A typical three-dimensional test for advection and renormalization of interfaces is the vortex example introduced by LeVeque [59], that is reproduced by several authors [20, 21, 57, 58].

The test consists of deforming a sphere by the application of a time-dependent three-dimensional velocity field, given as follows,

$$\begin{cases} v_x &= 2\sin^2(\pi x)\,\sin(2\pi y)\,\sin(2\pi z)\,\cos\left(\frac{\pi t}{T_e}\right) \\ v_y &= -\sin(2\pi x)\,\sin^2(\pi y)\,\sin(2\pi z)\,\cos\left(\frac{\pi t}{T_e}\right) \\ v_z &= -\sin(2\pi x)\,\sin(2\pi y)\,\sin^2(\pi z)\,\cos\left(\frac{\pi t}{T_e}\right) \end{cases} \quad (55)$$

with a period $T_e = 3$ s for the time dependency. This velocity field produces an increasing deformation of the sphere, given by the surface of $\phi = 0$, from $t = 0$ to $t = 1.5$ s, resulting in a thin shape. Furthermore, from $t = 1.5$ s to $t = 3$ s, the spherical surface should be recovered.

The initial conditions are given by a sphere with radius $r_e = 0.15$ m whose center is placed in coordinates $(0.35, 0.35, 0.35)$ m of a unit cube, where $0 \leq (x_1, x_2, x_3) \leq 1$ m. The spherical surface corresponds to the zero LS, i.e. $\phi = 0$. The transition from outside the surface, with $\phi < 0$, to the interior of the sphere, with $\phi > 0$, is given by a linear variation.

The problem was solved over 128^3 hexahedral elements, which compose a regular mesh, giving an element size of $h = 7.8 \times 10^{-3}$ m. The BRCP instance is performed every $n_{\text{reno}} = 10$ time steps, being the time step $\Delta t = 0.00195$ s. The parameters adopted for the renormalization stage are the diffusivity $\kappa = 6h^2 = 3.66 \times 10^{-4}$ m^2 and the penalty coefficient $M = 100000$, following the recommendations of Sec. 4.4..

Figure 14 shows the evolution of the interface $\phi = 0$ in different time instants. Due to

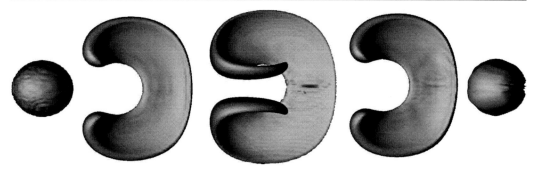

Figure 14. Deformed shape of the sphere of the three-dimensional advection-renormalization test. Zero level set ($\phi = 0$) at times $t = 0.02$ ss, $t = 0.76$ s, $t = 1.50$ s, $t = 2.25$ s, and $t = 3$ s.

the periodic character of the velocity field, the shapes registered in $t = 3$ s and $t \approx 0$ should be equal, as well as $t = 0.76$ s and $t = 2.25$ s.

4.6. Numerical Examples: Free Surface Flows

4.6.1. Two-Dimensional Dam-Break Problem

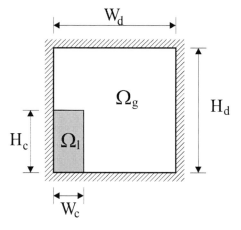

Figure 15. Geometry for the two-dimensional dam-break problem.

The solution of the dam-break problem constitutes a classic test for interface-capturing methods [55, 58, 60]. The test consists of releasing a water column and following the fluid front displacement along the bottom of the domain, or the fluid descent over one of the walls. There are experimental results available for different column sizes and fluids [55, 61] which allow the validation of the numerical approaches.

In the present case, the domain Ω is square, with width and height $W_d = 0.228$ m and $H_d = 0.228$ m, respectively, see Fig. 15. The initial position of the water is given in the column of $W_c = 0.057$ m width and $H_c = 0.114$ m height, Ω_l, which is sketched in the same figure, with an aspect ratio of $r_a = H_c/W_c = 2$.

The region Ω_l corresponds to water, with density and dynamic viscosity $\rho_l = 1000$ kg/m^3 and $\mu_l = 1.0 \times 10^{-3}$ kg/(m s), respectively, meanwhile the rest of the domain, Ω_g, represents the air phase, with density $\rho_g = 1$ kg/m^3 and dynamic viscosity $\mu_g = 1.0 \times 10^{-5}$ kg/(m s).

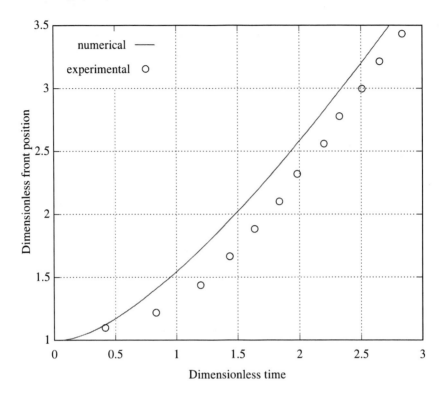

Figure 16. Experimental measurements and simulation results for the two-dimensional dam-break problem. Dimensionless front position $x_f(t^*)/W_c$ as a function of the dimensionless time $t^* = t\sqrt{2g/W_c}$.

The boundary conditions over the contours are considered as perfect slip, i.e. $\mathbf{v} \cdot \mathbf{n} = 0$, and pressure $p = 0$ on top, considering the NS stage. For the ADV stage, since there are no inflow sections on Ω, boundary conditions are unnecessary.

The initial conditions for the NS and the ADV stages consider a null velocity field, $\mathbf{v}_0 = \mathbf{0}$. The LS function field marks the presence of water with $0 < \phi \leq 1$, and the region Ω_g, filled with air, with $-1 \leq \phi < 0$, being $\phi = 0$ over the FS.

For the simulation, the gate is instantaneously released and the water column collapses due to the effect of the gravity acceleration, given as $g = 9.81$ m/s^2.

The FE mesh employed for the three stages of the numerical solution is composed by quadrilateral elements of mean size $h \approx 0.0023$ m. The simulation was performed during 1000 time steps of $\Delta t = 0.002$ s, considering implicit temporal integration for the NS and ADV stages, with an integration parameter $\theta = 0.7$ for the trapezoidal rule.

The renormalization of the LS function ϕ is performed every $n_{\text{reno}} = 2$ time steps, considering as parameters for the BRCP process the diffusivity $\kappa = 2h^2 = 1.04 \times 10^{-5}$ m^2 and the penalization $M = 10000$.

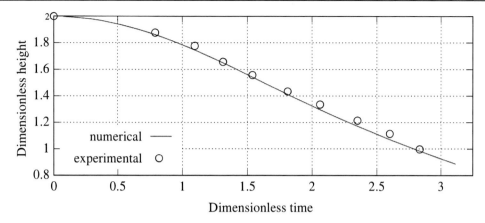

Figure 17. Experimental measurements and simulation results for the two-dimensional dam-break problem. Dimensionless column height $h_c(t^*)/W_c$ versus dimensionless time $t^* = t\sqrt{2g/W_c}$.

On one side, the numerical results are compared to the experimental measurements taken from Martin and Moyce [61], considering the dimensionless front position $x_f(t^*)/W_c$ as a function of the dimensionless time $t^* = t\sqrt{2g/W_c}$, as depicted in Fig. 16, where the slope of the curve represents the advancing front velocity.

On the other side, the dimensionless water height $h_c(t^*)/W_c$ over the left side of the domain is plotted as a function of the dimensionless time $t^* = t\sqrt{2g/W_c}$ in Fig. 17, where experimental measurements are also included [61]. Again, the mean descending velocity given by the slope of the curve is well captured.

The instantaneous position of the interface is shown in Fig. 18 for some instants of the numerical simulation: early stages without interface breaking, and stages with FS folding and bubbles.

4.6.2. Three-Dimensional Dam-Break Problem

A three-dimensional version of the dam-break problem, also studied by Martin and Moyce [61], consists of the collapse of a cylindrical column of water. Typical numerical simulations of this example are performed considering one fourth of the column in a cubic domain [62, 63].

The domain Ω is a cube of coordinates $0 \leq (x_1, x_2, x_3) \leq b$ with edge length $b = 0.2284$ m, see Fig. 19, with a water subdomain Ω_l as one fourth of a cylinder of radius and height $r_0 = 0.0571$ m and $h_0 = 0.1142$ m, respectively, whose longitudinal axis is placed at coordinates $(x_1, x_2) = (0.2284, 0.2284)$ m. The aspect ratio of the column is $r_a = h_0/r_0 = 2$.

The column is released at time $t = 0$ for the simulation, and the gravity acceleration $g = 9.81 \text{m/s}^2$ in $-x_3$ direction produces the collapse. As in the former case, the properties correspond to water for Ω_l and gas for Ω_g, i.e. density $\rho_l = 1000$ kg/m^3 and dynamic viscosity $\mu_l = 1.0 \times 10^{-3}$ kg/(m s) for the water and $\rho_g = 1$ kg/m^3 and $\mu_g = 1.0 \times 10^{-5}$ kg/(m s) for the air.

The numerical simulation is performed over a regular mesh with 50^3 hexahedral elements and a mean element size of $h \approx 4.5 \times 10^{-3}$ m. The time stepping is adopted as

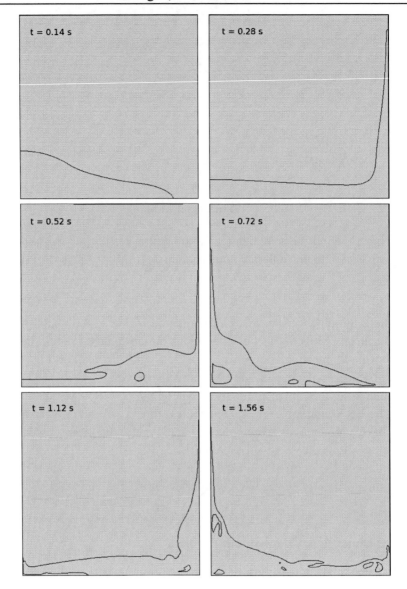

Figure 18. Free surface positions in the two-dimensional dam-break problem.

$\Delta t = 0.001$ s, and 1000 time steps were solved. Numerical integration is implicit for the NS and the ADV stages. The BRCP is executed each time step, i.e. $n_{\text{reno}} = 1$, and the parameters adopted are a penalization $M = 500000$ and a diffusivity $\kappa = 2h^2 = 4.17 \times 10^{-5}$ m^2.

Boundary conditions are given as perfect slip over the whole contour for the NS instance, while no condition has to be imposed for ADV or BRCP.

The dimensionless front displacement of the water column $r_f(t^*)/r_0$ is represented as a function of the dimensionless time $t^* = t\sqrt{2g/r_0}$ in Fig. 20, together with the experimental results [61]. Furthermore, the dimensionless height of the column $h_c(t^*)/r_0$ is represented as a function of t^* in Fig. 21. However, there are not experimental data available to compare these last results. The early stages of the collapse of the column are represented in Fig. 22.

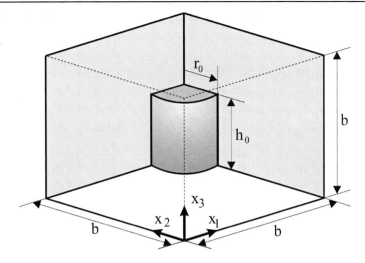

Figure 19. Geometry for the dam-break problem with the cylindrical column.

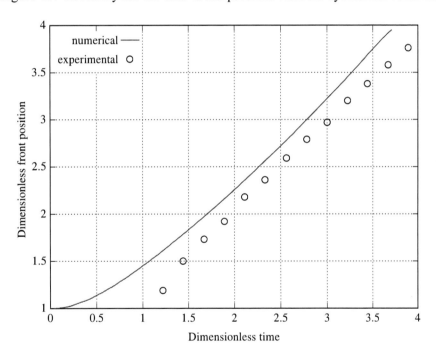

Figure 20. Numerical results and experimental data in the three-dimensional dam-break problem. Dimensionless front position $r_f(t^*)/r_0$ versus dimensionless time $t^* = t\sqrt{2g/r_0}$.

5. Conclusion

In this chapter, two methodologies for solving transient free surface fluid flows were described, considering Newtonian fluids in the incompressible case for two and three spatial dimensions.

The arbitrary Lagrangian-Eulerian approach depicted in Sec. 3. is well suited for solving small-displacements one-phase fluid flows, giving high-precision free surface displace-

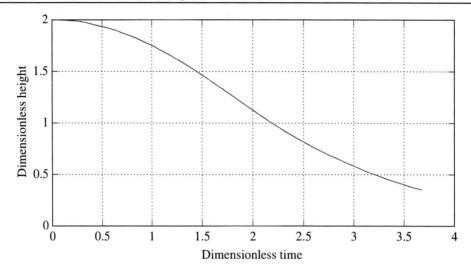

Figure 21. Numerical results in the three-dimensional dam-break problem. Dimensionless column height $h_c(t^*)/r_0$ as a function of the dimensionless time $t^* = t\sqrt{2g/r_0}$.

ments values. The examples presented covered issues on both two and three dimensional simulations, as in the sloshing examples presented in Secs. 3.4.1. and 3.4.3.. Furthermore, for higher free surface displacements, as in the horizontally excited tank of Sec. 3.4.2., the interface instabilities due to the advective character of the interface behavior are overcome through a SUPG stabilization technique.

The level set proposal described in Sec. 4. is appropriate for solving large free surface deformations, being capable to reproduce surface breaking and folding. This approach solves three numerical stages: (i) the fluid state; (ii) the level set function advection; and (iii) the renormalization of the level set field. The stages (ii) and (iii) were evaluated through typical two- and three-dimensional problems in order to establish their performance, see Secs. 4.5.1. and 4.5.2.. Then, free surface flow simulations were made over two classical dam-break problems in Secs. 4.6.1. and 4.6.2., respectively, giving good agreement with experimental results taken from the bibliography.

Acknowledgments

This chapter has received financial support from Consejo Nacional de Investigaciones Científicas y Técnicas (CONICET, Argentina, grant PIP 5271/05), Universidad Nacional del Litoral (UNL, Argentina, grants CAI+D 2009–III-4–2 and CAI+D 2009 65/33) and Agencia Nacional de Promoción Científica y Tecnológica (ANPCyT, Argentina, grants PICT 1141/2007, PICT 1506/2006), and was partially performed with the *Free Software Foundation GNU-Project* resources such as GNU/Linux OS and GNU/Octave, as well as other Open Source resources such as PETSc, MPICH, OpenDX, ParaView, LATEX and JabRef.

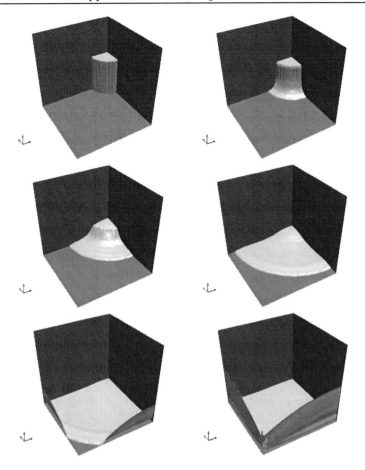

Figure 22. Numerical results in the three-dimensional dam-break problem. Free surface at times $t = 0.002$ s. $t = 0.063$ s, $t = 0.125$ s, $t = 0.188$ s, $t = 0.250$ s, and $t = 0.313$ s, from left to right and from top to bottom.

References

[1] Güler, I., Behr, M., and Tezduyar, T. E., 1999. "Parallel finite element computation of free-surface flows", pp. 117–123.

[2] D'Elía, J., Storti, M. A., and Idelsohn, S. R., 2000. "A panel-Fourier method for free surface methods", pp. 309–317.

[3] Papaspyrou, S., Valougeorgis, D., and Karamanos, S. A., 2004. "Sloshing effects in half-full horizontal cylindrical vessels under longitudinal excitation", pp. 255–265.

[4] Hernández-Barrios, H., Heredia-Zavoni, E., and Aldama-Rodríguez, A. A., 2007. "Nonlinear sloshing response of cylindrical tanks subjected to earthquake ground motion", pp. 3364 – 3376.

[5] Shyy, W., Udaykumar, H. S., Rao, M. M., and Smith, R. W., 1996. *Computational Fluid Dynamics with Moving Boundaries*. Taylor and Francis.

[6] Harlow, F. H., and Welch, J. E., 1965. "Numerical calculation of time-dependent viscous incompressible flowof fluid with free surface", pp. 2182–2189.

[7] Oishi, C. M., Tome, M. F., Cuminato, J. A., and McKee, S., 2008. "An implicit technique for solving 3D low Reynolds number moving free surface flows", pp. 7446–7468.

[8] Idelsohn, S. R., Oñate, E., and Del Pin, F., 2004. "The Particle Finite Element Method: a powerful tool to solve incompressibleflows with free-surfaces and breaking waves", pp. 964–984.

[9] Hughes, T. J. R., Liu, W. K., and Zimmermann, T. K., 1981. "Lagrangian-Eulerian finite element formulation for incompressible viscous flows", pp. 329–349.

[10] Huerta, A., and Liu, W. K., 1988. "Viscous flow with large free surface motion", pp. 277–324.

[11] Chippada, S., Jue, T. C., Joo, S. W., Wheeler, M. F., and Ramaswamy, B., 1996. "Numerical simulation of free-boundary problems", pp. 91–118.

[12] Rabier, S., and Medale, M., 2003. "Computation of free surface flows with a projection FEM in a moving mesh framework", pp. 4703–4721.

[13] Battaglia, L., D' Elía, J., Storti, M. A., and Nigro, N. M., 2006. "Numerical simulation of transient free surface flows using a moving mesh technique", pp. 1017–1025.

[14] Hirt, C. W., and Nichols, B. D., 1981. "Volume of fluid (VOF) method for the dynamics of free boundaries", pp. 201–225.

[15] Scardovelli, R., and Zaleski, S., 1999. "Direct numerical simulation of free-surface and interfacial flow", pp. 567–603.

[16] Pilliod, J. E., and Puckett, E. G., 2004. "Second-order accurate volume-of-fluid algorithms for tracking material interfaces", pp. 465–502.

[17] Osher, S., and Sethian, J. A., 1988. "Fronts propagating with curvature-dependent speed: Algorithms based on Hamilton-Jacobi formulations", pp. 12–49.

[18] Sethian, J. A., 1995. *A Fast Marching Level Set Method for Monotonically Advancing Fronts*. National Academy of Sciences.

[19] Sussman, M., and Puckett, E. G., 2000. "A coupled level set and volume-of-fluid method for computing 3D and axisymmetric incompressible two-phase flows", pp. 301–337.

[20] Enright, D., Fedkiw, R., Ferziger, J., and Mitchell, I., 2002. "A hybrid particle level set method for improved interface capturing", pp. 83–116.

[21] Shin, S., and Juric, D., 2009. "A hybrid interface method for three-dimensional multiphase flows based on front tracking and level set techniques", pp. 753–778.

[22] PETSc-FEM, 2010. A general purpose, parallel, multi-physics FEM program. GNU General Public License (GPL).

[23] MPI, 2008. Message Passing Interface. http://www.mpi-forum.org/docs/docs.html.

[24] Balay, S., Buschelman, K., Eijkhout, V., Gropp, W., Kaushik, D., Knepley, M., McInnes, L., Smith, B., and Zhang, H., 2008. PETSc Users Manual. ANL 95/11 - Revision 3.0.0, Argonne National Laboratory.

[25] Donea, J., and Huerta, A., 2003. *Finite Element Methods for Flow Problems*. John Wiley & Sons.

[26] Soulaïmani, A., Fortin, M., Dhatt, G., and Ouellet, Y., 1991. "Finite element simulation of two- and three-dimensional free surface flows", pp. 265–296.

[27] Folch Duran, A., 2000. "A numerical formulation to solve the ALE Navier–Stokes equations applied to the withdrawal of magma chambers". PhD thesis, Universitat Politécnica de Catalunya.

[28] Sonzogni, V. E., Yommi, A. M., Nigro, N. M., and Storti, M. A., 2002. "A parallel finite element program on a Beowulf Cluster", pp. 427–443.

[29] Storti, M. A., Nigro, N. M., Paz, R. R., and Dalcín, L. D., 2008. "Dynamic boundary conditions in computational fluid dynamics", pp. 1219–1232.

[30] Brooks, A., and Hughes, T. J. R., 1982. "Streamline upwind/Petrov-Galerkin formulations for convection dominatedflows with particular emphasis on the incompressible Navier-Stokes equations", pp. 199–259.

[31] Tezduyar, T. E., Mittal, S., Ray, S. E., and Shih, R., 1992. "Incompressible flow computations with stabilized bilinear and linear-equal-order interpolation velocity-pressure elements", pp. 221–242.

[32] Saito, H., and Scriven, L. E., 1981. "Study of coating flow by the finite element method", pp. 53–76.

[33] Pedlosky, J., 2003. *Waves in the ocean and atmosphere. Introduction to wave dynamics*. Springer.

[34] Stoker, J. J., 1957. *Water waves. The mathematical theory with applications*, Vol. IV. Interscience.

[35] Battaglia, L., 2009. "Stabilized finite elements for free surface flows: Tracking and capturing of interface". PhD thesis, Facultad de Ingeniería y Ciencias Hídricas, Universidad Nacional del Litoral.

[36] Behr, M., and Abraham, F., 2002. "Free surface flow simulations in the presence of inclined walls", pp. 5467–5483.

[37] Johnson, A. A., and Tezduyar, T. E., 1994. "Mesh update strategies in parallel finite element computations of flow problems with moving boundaries and interfaces", pp. 73–94.

[38] López, E. J., Nigro, N. M., Storti, M. A., and Toth, J. A., 2007. "A minimal element distortion strategy for computational mesh dynamics", pp. 1898–1929.

[39] López, E. J., Nigro, N. M., and Storti, M. A., 2008. "Simultaneous untangling and smoothing of moving grids", pp. 994–1019.

[40] Prosperetti, A., 1981. "Motion of two superposed viscous fluids", pp. 1217–1223.

[41] Braess, H., and Wriggers, P., 2000. "Arbitrary Lagrangian Eulerian finite element analysis of free surface flow", pp. 95 – 109.

[42] Ramaswamy, B., 1990. "Numerical simulation of unsteady viscous free surface flow", pp. 396–430.

[43] Souli, M., and Zolesio, J. P., 2001. "Arbitrary Lagrangian-Eulerian and free surface methods in fluid mechanics", pp. 451–466.

[44] Sussman, M., and Smereka, P., 1997. "Axisymmetric free boundary problems", pp. 269–294.

[45] Battaglia, L., Storti, M. A., and D'Elía, J., 2010. "Simulation of free-surface flows by a finite element interface capturing technique", pp. 121–133.

[46] Battaglia, L., Storti, M. A., and D'Elía, J., 2010. "Bounded renormalization with continuous penalization for level set interface capturing methods".

[47] Olsson, E., and Kreiss, G., 2005. "A conservative level set method for two phase flow", pp. 225–246.

[48] Kurioka, S., and Dowling, D. R., 2009. "Numerical simulation of free surface flows with the level set method using an extremely high-order accuracy weno advection scheme", pp. 233–243.

[49] Osher, S., and Fedkiw, R. P., 2001. "Level set methods: An overview and some recent results", pp. 463–502.

[50] Tezduyar, T. E., and Osawa, Y., 2000. "Finite element stabilization parameters computed from element matrices and vectors", pp. 411–430.

[51] Mut, F., Buscaglia, G. C., and Dari, E. A., 2006. "New mass-conserving algorithm for level set redistancing on unstructured meshes", pp. 1011–1016.

[52] Elias, R. N., Martins, M. A. D., and Coutinho, A. L. G. A., 2007. "Simple finite element-based computation of distance functions in unstructured grids", pp. 1095–1110.

[53] Olsson, E., Kreiss, G., and Zahedi, S., 2007. "A conservative level set method for two phase flow II", pp. 785–807.

[54] Tezduyar, T. E., 2006. "Interface-tracking and interface-capturing techniques for finite element computation of moving boundaries and interfaces", pp. 2983–3000.

[55] Cruchaga, M. A., Celentano, D. J., and Tezduyar, T. E., 2007. "Collapse of a liquid column: numerical simulation and experimental validation", pp. 453–476.

[56] Di Pietro, D. A., Lo Forte, S., and Parolini, N., 2006. "Mass preserving finite element implementations of the level set method", pp. 1179–1195.

[57] Gois, J. P., Nakano, A., Nonato, L. G., and Buscaglia, G. C., 2008. "Front tracking with moving-least-squares surfaces", pp. 9643–9669.

[58] Elias, R., and Coutinho, A. L. G. A., 2007. "Stabilized edge-based finite element simulation of free-surface flows", pp. 965–993.

[59] LeVeque, R. J., 1996. "High-resolution conservative algorithms for advection in incompressible flow", pp. 627–665.

[60] Marchandise, E., and Remacle, J.-F., 2006. "A stabilized finite element method using a discontinuous level set approach for solving two phase incompressible flows", pp. 780–800.

[61] Martin, J. C., and Moyce, W. J., 1952. "An experimental study of the collapse of liquid columns on a rigid horizontal plane", pp. 312–324.

[62] Akin, J. E., Tezduyar, T. E., and Ungor, M., 2007. "Computation of flow problems with the mixed interface-tracking/interface-capturing technique (MITICT)", pp. 2–11.

[63] Cruchaga, M., Celentano, D., Breitkopf, P., Villon, P., and Rassineux, A., 2010. "A surface remeshing technique for a Lagrangian description of 3D two-fluid flow problems", pp. 415–430.

Chapter 13

NUMERICAL SIMULATION OF MULTIPHASE FLOW IN CHEMICAL REACTORS

Chao Yang, Yumei Yong and Zai-Sha Mao

Institute of Process Engineering, Chinese Academy of Sciences, Beijing, China

1 INTRODUCTION

More and more attentions have been paid to better understanding of the mechanisms of multiphase flow and interphase mass and heat transfer on the mesoscale and macroscale systems. The development of computational fluid dynamics (CFD) and related sciences make it possible to quantitatively describe complicated multiphase flow and solve the problems of scale-up effect and macro-control in the chemical industry processes. The current state of the art in CFD is the Reynolds-averaged Navier-Stokes method (RANS), but the marked disadvantage with this method is the modeling limited generality and unsatisfactory accuracy. Direction numerical simulation (DNS) methods completely simulate turbulent flow with high precision by a large number of grids, but up to now it is still impossible to apply it into industrial processes. The large eddy simulation (LES) method instead can be used for very general applications (geometry as well as flow fields) and the accuracy is dramatically better compared to RANS turbulence modeling. The undesirable drawback of LES is the higher computational time required for the solution.

The goal of DNS of multiphase flows is both to generate insight and understanding of the basic behavior of multiphase flow—such as the forces on a single bubble or a drop, how bubbles and drops affect the flow, and how many bubbles and drops interact in dense dispersed flows—as well as to provide data for the generation of closure models for engineering simulations of the averaged flow field. Unlike the turbulent flow of a single-phase fluid, multiphase flows generally possess a large range of scales, ranging from the sub-millimeter size of a bubble or an eddy to the size of the system under investigation. Multiphase flows, like single-phase turbulent flows, exhibit a great deal of universality and it is almost certain that re-computing small-scale behavior that is already understood is not necessary. DNS should be able to provide both the insight and the data for the modeling at the smallest scales.

In the last 20 years or so, there has been rapid progress in developing the method of lattice Boltzmann method (LBM) as a possible alternative to these methods for DNS in complex geometries to solve a variety of fluid dynamic problems. The multi-block method can greatly improve the numerical efficiency. To further improve the computational efficiency and the flexibility of the method, the adaptive mesh refinement is more desirable. The adaptive mesh refinement method has been successfully applied in solving Euler and Navier-Stokes equation using a Cartesian grid or unstructured grid. Based on the authorsulation of jet noise, at the multi-block LBM, it is anticipated that the successful application of adaptive mesh refinement in the LBM will require careful treatments at the block interface in order to reduce the numerical inaccuracy. Second-order accurate boundary conditions have been developed for curved solid body moving with a given velocity. However, when boundary moves and crosses the nodes, there is a continuous conversion between solid nodes and their neighboring fluid nodes and the total number of the boundary nodes is not fixed. This poses a challenge to the accurate evaluation of the fluid dynamic force on the body. Furthermore, it brings in extra computational noise near the body. This issue has not been adequately addressed at high Reynolds number. Multi-relaxation-time (MRT) models have been demonstrated to be beneficial in many respects for single-phase flows. It is strongly desirable to extend the MRT model to multi-component flows. For high Reynolds number flows, while LBM combining with LES have been carried out, engineering turbulence models in the context of the LBM are lacking and more research efforts are required.

In recent years, LES has been an increase in usage for flows, for which previously RANS methods were typically applied, for the following reasons : (1) rapid increases in computing power, memory and storage; (2) realization that RANS methods inherently cannot handle certain classes of complex turbulent flow problems such as massively separated flows; (3) the availability of LES and hybrid RANS/LES methods in applications. LES was already used successfully: simple benchmark flows, complex geometries in aeronautical and marine applications, active flow control problems, calculation of jet noise, wall bounded flows at Reynolds numbers of engineering interest, shock waves, prediction of reacting flows and so on. LES has been successfully applied to model single phase flows in stirred vessels and the comparison with RANS simulation and data clearly demonstrates the superiority of LES in terms of predicting the turbulent quantities in the reactor. However, it should be noted that the improvement in the predictions obtained in a large eddy simulation is at the expense of the computational cost associated with it. As a result these simulations are still limited to smaller reactor sizes and relatively lower volume fractions of solids.

We have to explore the multiphase flow, transport and reaction mechanics, build up the multi-scale mathematical models and develop new numerical simulation methods with high efficiency in order to strengthen the processes and promote the technology by gaining and retrieving the macroscopic performance of process equipments, the aggregation and breakup mechanics of the mesoscopic bubbles, droplets and particles and the micro processes of thermal transfer, mass diffusion and multiphase flow in microchannels. Many progresses have been achieved in this line.

2. NUMERICAL SIMULATION OF FLOW AND TRANSPORT ON MICROSCALE AND PARTICLE SCALE

Some micro or meso processes and effects will be amplified suddenly when the sizes and structures of equipments are changed, so more and more attentions have been paid to better understanding of the mechanisms of multiphase flow and interphase mass transfer on the particle scale (solid particles, bubbles and droplets).

We simulate the process that oxygen molecular diffuses while the natural convection develops in a simplified container, coupling with the effects of thermal and mass diffusion. Immiscible two-phase flows in microchannels are numerically studied by a LBM based on field mediators. The quantitative and qualitative agreement between the simulations and the experimental data show the effectiveness of the numerical method. This work demonstrates that the developed LBM simulator is a viable tool to study thermal transfer and mass diffusion and immiscible two-phase flows in microchannels.

Stream function-vorticity formulation/boundary-fitted orthogonal reference frame may be used for slightly deformed bubbles and droplets, but the level set approach is more suitable for fluid particles with large deformation even topological change. The level set approach is extended to interphase mass transfer, but a variable transform and higher order numerical scheme are required. We simulate the motion and mass transfer of a single liquid drop and the drop behavior are interpreted with reasonable precision.

The numerical approach such as the level set method is now available to simulate the surface tension gradient induced Marangoni convection at the interface and the thereby increased interphase mass transfer. The influence of surfactant transport on Marangoni effect is also investigated by numerical simulation. The resultant information can shed light to the mechanisms based on which the interphase interaction is to be either promoted or suppressed.

2.1. Lattice Boltzmann Method

2.1.1. Numerical Method

2.1.1.1. Lattice-Gas Cellular Automata and Lattice-Boltzmann Model

An alternative way to computational fluid dynamics is the LBM. Lattice-gas cellular automata (LGCA) and LBMs are methods for the simulation of fluid flows, which are quite distinctive from not only molecular dynamics (MD) but also methods based on the discretization of partial differential equations (finite differences, finite volumes, finite elements, or spectral methods).

Historically, the method is derived from the lattice-gas automata, but it is an obvious relationship to the Boltzmann equation always used as an inspiration. The fact that different microscopic interactions can lead to the same form of macroscopic equations is the starting point for the development of LGCA. In addition to real gases or real liquids one may consider artificial micro-worlds of particles 'living' on lattices with interactions that conserve mass and momentum. The microdynamics of such artificial micro-worlds should be very simple in order to run it efficiently on a computer. LGCA can be characterized as follows:

[1] LGCA are regular arrangements of single cells of the same kind;
[2] Each cell holds a finite number of discrete states;
[3] The states are updated simultaneously at discrete time levels;
[4] The update rules are deterministic and uniform in space and time;
[5] The rules for the evolution of a cell depend only on a local neighborhood of cells around it.

Lattice-gas cellular automata for Navier-Stokes equations are plagued by several diseases. Only for some of them therapies and cures could be found (as compared in Table 2-1).

Table 2- 1. Diseases of lattice-gas cellular automata

Disease	Cause	Therapy/cure
Non-isotropic advection term	Lattice tensor of 4th rank is non-isotropic	Higher symmetry of lattice add inner degree of freedom multi-speed models
Violation of Galilei invariance	Fermi-Dirac distribution	Rescaling (symptomatic treatment)
Noise	Boolean variables	Averaging (coarse graining)
Pressure depends explicitly on velocity		Multi-speed models

LB models vastly simplify Boltzmann's original conceptual view by reducing the number of possible particle spatial positions and microscopic momentum from a continuum to just a handful and similarly discretizing time into distinct steps. Particle positions are confined to the nodes of the lattice. Variations in momentum that could have been due to a continuum of velocity directions and magnitudes and varying particle mass are reduced to 8 directions (19 directions for 3D), three magnitudes and a single particle mass. Historically the following stages in the development of LB models can be distinguished:

[1] LB equations have been used already at the cradle of lattice-gas cellular automata to calculate the viscosity of LGCA.
[2] he main motivation for the transition from LGCA to LB method was the desire to get rid of the noise. The Boolean fields were replaced by continuous distributions over the FHP and FCHC lattices. Fermi-Dirac distributions were used as equilibrium functions.
[3] Linearized collision operator.
[4] Boltzmann instead of Fermi-Dirac distributions.
[5] The collision operator, which is based on the collisions of a certain LGCA, has been replaced by the BGK (also called single time relaxation) approximation. These lattice BGK models mark a new level of abstraction: collisions are not anymore defined explicitly.

2.1.1.2. BGK Lattice Boltzmann Mothed

The BGK (Bhatnagar-Gross-Krook) approximation is used in the simplest LB method. Succi [1] provided excellent discussions of more complex models and the path to BGK. Streaming and collision (i.e., relaxation towards local equilibrium) look like

$$f_i(\mathbf{x}+\mathbf{e}_i\Delta t, t+\Delta t) = \underbrace{f_i(\mathbf{x},t)}_{\text{Streaming}} - \underbrace{\frac{[f_i(\mathbf{x},t) - f_i^{eq}(\mathbf{x},t)]}{\tau}}_{\text{Collision}} \qquad (2\text{-}1)$$

where $f_i(\mathbf{x}+\mathbf{e}_i\Delta t, t+\Delta t) = f_i(\mathbf{x},t)$ is the streaming part, and in streaming we move the direction-specific densities to the nearest neighbor lattice nodes. And $[f_i(\mathbf{x},t) - f_i^{eq}(\mathbf{x},t)]/\tau$ is the collision term, collision of fluid particles is considered as a relaxation towards a local equilibrium. Although they can be combined into a single statement as they are in Eq. (2- 1), collision and streaming steps must be separated if solid boundaries are present because the bounce back boundary condition is a separate collision.

The most popular lattice BGK model is DnQb model proposed by Qian [2], and the equilibrium distribution function of DnQb model is given by the following expression:

$$f_i^{eq} = \rho\omega_i\left[1 + \frac{c\mathbf{e}_i \cdot \mathbf{u}}{c_s^2} + \frac{(c\mathbf{e}_i \cdot \mathbf{u})^2}{2c_s^4} - \frac{|\mathbf{u}|^2}{2c_s^2}\right] \qquad (2\text{-}2)$$

There are D1Q3 and D1Q5 models for one dimensional model, D2Q7 and D2Q9 models for two dimensional model, and D3Q15 and D3Q19 models for three dimensional model. Here we applied the D2Q9 model at first.

The 9-velocity square lattice and the corresponding lattice velocity vectors are shown in Figure 2- 1.

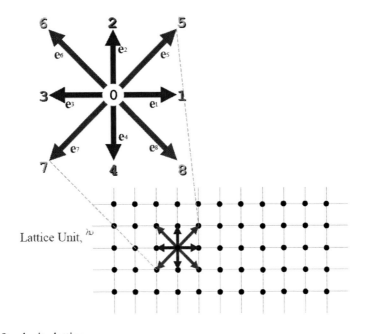

Figure 2- 1. 2D 9-velocity lattice.

There are 8 sites around a site, so there are 9 velocities for a particle, including itself, which constitute a group, $\mathbf{e}_i$. Weighting factors ω_i are chosen so that the moments of the equilibrium distribution function will satisfy Eq. (2-1). ω_i have the following values for the 2D 9-velocity model respectively: $\omega_0 = 4/9$, $\omega_{1-4} = 1/9$ and $\omega_{5-8} = 1/36$. The equilibrium distribution function of D2Q9 LB model is

$$f_i^{eq} = \frac{4}{9}\rho\left[1 - \frac{3}{2}\mathbf{u}^2\right] \quad i=0$$

$$f_i^{eq} = \frac{1}{9}\rho\left[1 + 3(\mathbf{e}_i \cdot \mathbf{u}) + \frac{9}{2}(\mathbf{e}_i \cdot \mathbf{u})^2 - \frac{3}{2}\mathbf{u}^2\right] \quad i=1,2,3,4 \qquad (2\text{-}3)$$

$$f_i^{eq} = \frac{1}{36}\rho\left[1 + 3(\mathbf{e}_i \cdot \mathbf{u}) + \frac{9}{2}(\mathbf{e}_i \cdot \mathbf{u})^2 - \frac{3}{2}\mathbf{u}^2\right] \quad i=5,6,7,8$$

At a given lattice site $(\mathbf{x}, t)$, the mass density ρ and fluid velocity u may be computed as follows:

$$\rho = \sum_i f_i \qquad \rho\mathbf{u} = \sum_i f_i \mathbf{e}_i \qquad (2\text{-}4)$$

2.1.1.3. Boundary Conditions

The processing of boundary conditions is very important for LBM because it has great effect on the accuracy and stability of LBM. The boundary conditions for macroscopic equations of fluid mechanics, such as Navier-Stokes equations, can be conveniently and directly specified, but we have to specify the microcosmic distribution functions which are unknown on the boundary for LBM. Therefore we construct certain scheme and deduce the distribution functions while the macroscopic properties of boundary conditions calculated from LBM must meet with real macroscopic flow properties of boundary conditions.

Guo et al.[3] put forward to the basic idea of non-equilibrium extrapolation boundaries schemes, in which distribution functions are divided into equilibrium and non-equilibrium parts, i.e., new equilibrium distribution functions are decided by detailed definition of boundary conditions and non-equilibrium distribution functions are extrapolated. Therefore the final distribution functions are second order accuracy approximately. Here we use the second order form of non-equilibrium extrapolation boundaries schemes from Du [4], in which the boundary nodes are related with the fluid nodes near boundaries so the numerical stability is improved. The basic idea of non-equilibrium extrapolation boundaries schemes is shown in Figure 2-2.

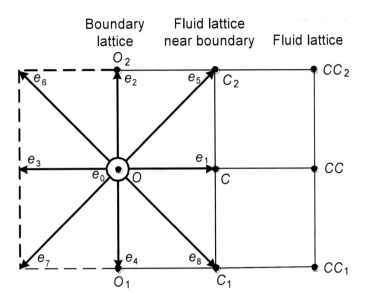

Figure 2-2. Non-equilibrium extrapolation boundary.

The west boundary is taken as an example here to explain the non-equilibrium extrapolation schemes. O_2, O and O_1 are on the west boundary, C, C_1 and C_2 are fluid nodes nearest the boundary, CC, CC_1 and CC_2 are fluid nodes second nearest the boundary. We make boundary node O as an example, divide the distribution functions on O node into two parts:

$$f_i(O, t) = f_i^{eq}(O, t) + f_i^{neq}(O, t) \tag{2-5}$$

In Eq.(2-5), $f_i^{eq}(O,t)$ is the equilibrium part, and $f_i^{neq}(O,t)$ is the non-equilibrium part.

If the velocities $\mathbf{u}(O,t)$ on the west boundary are known and the pressure $p(O,t)$ is unknown, the equilibrium part is approximately by

$$f_i^{eq}(O,t) = f_i^{eq}[p(C,t), \mathbf{u}(O,t)] \tag{2-6}$$

The non-equilibrium part is approximately formulated according to the known distribution functions, macroscopic velocities and pressures on C and CC nodes:

$$\begin{aligned} f_i^{neq}(O,t) &= 0.5 f_i^{neq}(C,t) + 0.5 f_i^{neq}(CC,t) \\ &= 0.5\left(f_i(C,t) - f_i^{eq}(C,t)\right) + 0.5\left(f_i(CC,t) - f_i^{eq}(CC,t)\right) \end{aligned} \tag{2-7}$$

We can get all other variables on the different boundaries in same way.

2.1.2. Heat Transfer and Mass Diffusion

2.1.2.1. Thermal Convection in Triangular Enclosure

The natural convection in triangular enclosures occurs widely in the fields of electronic device cooling, solar energy collector and so on. The two-dimensional impressible lattice Boltzmann model is built in this work for its simulation. On the basis of D2G9 model, coupling with the two-dimensional thermal lattice Boltzmann model TD2Q5 and non-equilibrium extrapolation boundary scheme, the temperature field in Couette flow is simulated.

There are two thermal LB models: multi-speed (MS) and multi-distribution function (MDF) thermal LB models. MS thermal LB model is limited to the thermal flow with constant Prandtl number primitively, and later the LB model was improved [5] but the numerical stability is not good so the application of it has been limited. MDF thermal LB model is a kind of LB model coupling with the temperature, which uses a new distribution function independent on the density distribution function to simulate the temperature [6], and then it mainly simulates the flow of thermal fluids. The MDF thermal LB model has a main advantage of good stability and disadvantage of compressible effect [7].

Guo et al. [8] put forward to a new way which inherit the advantages of MDF model and completely overcome the compressible effect of LB models. Guo et al. used two independent lattice BGK equations to respectively simulate the velocity and temperature fields, and coupled the two equations with Boussinesq approximation. Here we use the thinking to solve the heat and mass transfer problem of fluids.

The impressible LB model from Guo et al. [2] is widely used. Then on the basis of D2G9 model, coupling with two-dimensional thermal lattice Boltzmann model TD2Q5 and non-equilibrium extrapolation boundary scheme on the wall boundary condition, the impressible thermal flow with high Raleigh number in triangle enclosure is simulated.

The difference of D2G9 model from other impressible LB models lies in its constant equilibrium distribution function, where the continuity equation deduced by the D2G9 model satisfies the impressible conditions, so the D2G9 model eliminates the compressible effect. The D2G9 model introduces the new distribution function, g_i, its equilibrium distribution function g_i^{eq} is defined as

$$g_i^{eq} = \begin{cases} \rho_0 - 4\sigma(p/c^2) + s_i(u) & (i=0) \\ \lambda(p/c^2) + s_i(u) & (i=1\sim 4) \\ \gamma(p/c^2) + s_i(u) & (i=5\sim 8) \end{cases} \quad (2\text{-}8)$$

where ρ_0 is the lattice density and a constant, c is the lattice velocity and equal one, p is the local pressure, and the constants σ, λ and γ need to meet

$$\begin{cases} \lambda + \gamma = \sigma \\ \lambda + 2\gamma = 0.5 \end{cases} \quad (2\text{-}9)$$

$$s_i(u) = \omega_i \left[3(e_i u)/c + 4.5(e_i u)^2/c^2 - 1.5|u|^2/c^2 \right] \tag{2-10}$$

The weight factors ω_i are

$$\omega_i = \begin{cases} 4/9 & (i=0) \\ 1/9 & (i=1\sim 4) \\ 1/36 & (i=5\sim 8) \end{cases} \tag{2-11}$$

The evolution equation of the D2G9 model is

$$g_i[x+ce_i\delta_t, t+\delta_t - g_i(x,t)] = -\tau_u^{-1}[g_i(x,t) - g_i^{eq}(x,t)] \tag{2-12}$$

where $i=0\sim 8$.

Macro velocity and pressure are

$$u = \sum_{i=1}^{8} ce_i g_i \tag{2-13}$$

$$p = [c^2/(4\sigma)]\left[\sum_{i=1}^{8} g_i + s_0(u)\right] \tag{2-14}$$

Because the energy equation can be written as

$$\frac{\partial T}{\partial t} + \nabla(uT) = \alpha \nabla^2 T + \phi \tag{2-15}$$

then the density distribution function g_i and the energy distribution function f_k respectively meet

$$g_i(x+ce_i\delta_t, t+\delta_t) - g_i(x,t) = -\tau_u^{-1}[g_i(x,t) - g_i^{eq}(x,t)] + \Delta t F_i(x,t) \tag{2-16}$$

$$f_k(x+ce_k\delta_t, t+\delta_t) - f_k(x,t) = -\tau_T^{-1}[f_k(x,t) - f_k^{eq}(x,t)] + \Delta t R_k(x,t) \tag{2-17}$$

where τ_u and τ_T are the relaxation factors of velocity and temperature equations, and $F_i(x, t)$ and $R_k(x, t)$ are external force and source term, respectively.

The viscosity ν and D of fluids can be gotten by the two relaxation factors:

$$v = c^2(\tau_u - 0.5)\delta_t \qquad (2\text{-}18)$$

$$D = c^2(\tau_T - 0.5)\delta_t \qquad (2\text{-}19)$$

We used the TD2Q5 thermal LB model with the economic computational complexity and proper accuracy, and the energy distribution function of the TD2Q5 thermal LB model is

$$h_k = \frac{T}{5}\left[1 + 2.5\frac{e_k u}{c^2}\right] \qquad (2\text{-}20)$$

where $k=0\sim4$, and the macro temperature T is calculated by

$$T = \sum_{k=0}^{4} f_k^{eq}. \qquad (2\text{-}21)$$

In order to verify the reliability of the thermal LB model, first the Coutte flow with temperature gradient is simulated. We assume that the inferior wall is still and the superior wall moves with a velocity, so the analytical solution is

$$\frac{T - T_0}{T_1 - T_0} = \frac{y}{H} + \frac{Pr\, Ec}{2}\frac{y}{H}\left(1 - \frac{y}{H}\right) \qquad (2\text{-}22)$$

where T_0 and T_1 are respectively the temperatures on the superior and inferior wall, H is the distance between two walls, y is the distance along the y axis, Pr is Prandtl number defined as $Pr=v/D$, and Ec is Eckert number defined as $Ec=u^2/[c_p(T_1-T_0)]$. The inlet and outlet are set as the period boundary condition. The schematic diagram is shown in Figure 2-3.

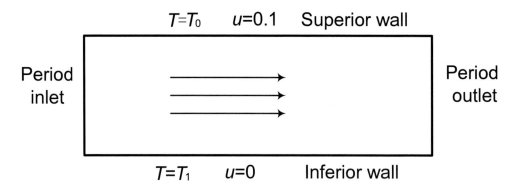

Figure 2-3. Boundary conditions of Couette flow.

In order to verify the stability of the LB model, we simulate the temperature distribution of Couette flow with the range of Eckert number $5\sim100$, at $Pr=1$, and the results are shown in Figure 2-4. At $Pr=1$ and $Ec=5\sim100$, numerical results agree with the analytical results well, which also shows that the thermal LB model owns the good stability and accuracy.

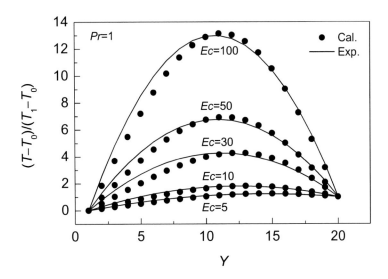

Figure 2- 4. Numerical stability in simulating temperature distribution in Couette flow.

The calculation domain is shown in Figure 2- 5, the height H is 1.0 m, and the base-side d is 0.5m. All walls are still, in another word, the velocity of each wall is zero. The temperature of the base-side is high, $T_h=0.5$ K, and the temperature of the two hypotenuses are low, $T_c=-0.5$ K. The fluid is air, so $Pr=0.71$. The initial temperature and velocity are zero. The mesh is 200×200, and the final convergent rules are that the relative errors of the whole velocity, pressure and temperature fields are less than 10^{-7}.

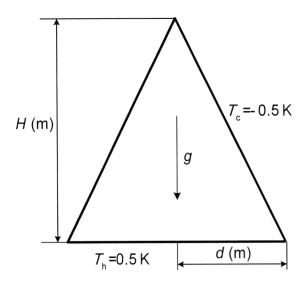

Figure 2- 5. Calculated domain and boundary conditions in a triangular cavity.

We simulate the thermal convection flow with Ra number of 10^3, 10^4, 10^5, 10^6, 10^7 and 10^8. The isotherm and stream lines are very smooth and close to the practical phenomena and the results are well agreeable with the published data [3] in the Table 2-2.

Table 2- 2. *Nu* **number and maximum velocity at different** *Ra* **numbers**

Ra	Source	τ_U^{-1}	Nu_{min}	Nu_{ave}	Y_{max} (M)
10^3	Calculated	0.16	2.773	15.84	0.268
	Ref.[3]	0.16	2.541	15.05	0.258
10^4	Calculated	0.4	2.29	17.64	0.265
	Ref.[3]	0.4	2.069	15.18	0.26
10^5	Calculated	1.1	2.58	15.60	0.285
	Ref.[3]	1.1	2.591	15.58	0.285
10^6	Calculated	1.5	3.58	20.16	0.333
	Ref.[3]	1.5	3.608	19.86	0.319
10^7	Calculated	1.7	4.36	28.38	0.358
	Ref.[3]	1.7	4.27	28.22	0.339
10^8	Calculated	1.95	5.84	46.23	0.306
	Ref.[3]	1.95	5.502	46.13	0.306

2.1.2.2. Mass Diffusion of Gas in an Incompressible Thermal Fluid

It is of great significance to study the mass transfer and mixing of gas in natural convection flow caused by density gradient resulting from thermal conditions, because chemical reactions mainly depend on the gas molecular concentration that resolves in fluids and natural convection flow can strengthen the transport of gas molecular.

According to the flow, heat transfer and gas mass diffusion processes in a closed container, the impressible D2G9 LB model, D2Q9 thermal and mass diffusion LB models are built and coupled with each other in order for velocity and convection fields of an impressible fluid and the gas mass fraction distribution in the thermal fluid. Three thermal boundary conditions are given to explore the influence of the convection way on gas diffusion with LB method, and comparing with the published data.

Here we add the mass diffusion LB evolution equation:

$$Cm_i(x+ce_i\Delta t, t+\Delta t) - Cm_i(x,t) = -\tau_C^{-1}[Cm_i(x,t) - Cm_i^{eq}(x,t)] \quad (2\text{-}23)$$

where $i=0\sim8$, τ_C is the relaxation time when the velocity distribution function meets with the equilibrium state. The mass equilibrium distribution function Cm_i^{eq} is

$$Cm_i^{eq} = \omega_i C \cdot (1+3e_i \cdot u) \quad (2\text{-}24)$$

where C is the local mass concentration of gas. The macro gas mass concentration is

$$C = \sum_{i=0}^{8} Cm_i \quad (2\text{-}25)$$

The diffusion coefficient factor of mass concentration D can be calculated by

$$D = c^2(\tau_C - 0.5)\Delta t \quad (2\text{-}26)$$

The calculated domain is a two-dimensional container, as shown in Figure 2-6. The top wall moves at a rightward speed, $u_0=0.001$; the gas concentration on the top wall C_0 is 1.0;

bottom wall is still and non-filtration; the left and right walls both are non-filtration; the temperature boundary conditions are designed as three kinds cases, shown in Table 2- 3. . .

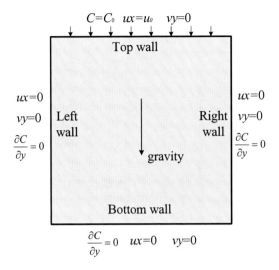

Figure 2-6. Schematic diagram of calculated domain.

Table 2- 3. gives three kinds of temperature boundary conditions on the four walls. For case A, the left wall is with high temperature, the right wall is with low temperature, and top and bottom walls both are adiabatic. For case B, the left and right walls both are with high temperature, the top wall is with low temperature, and the bottom wall is adiabatic. For case C, the superior 1/4 of the left wall is with low temperature, the inferior 1/4 of the left wall is with high temperature, the rest middle of the left wall is adiabatic, the right wall is same as the left, and the top and bottom walls both are adiabatic.

Table 2- 3. Three kinds of thermal boundary conditions

	case A	case B	case C	
Left wall	$T=T_{high}$	$T=T_{high}$	$\begin{cases} T = T_{high} \\ \partial T/\partial y = 0 \\ T = T_{low} \end{cases}$	$0 < y < L/4$ $L/4 < y < 3L/4$ $3L/4 < y < L$
Right wall	$T=T_{low}$	$T=T_{high}$	$\begin{cases} T = T_{high} \\ \partial T/\partial y = 0 \\ T = T_{low} \end{cases}$	$0 < y < L/4$ $L/4 < y < 3L/4$ $3L/4 < y < L$
Top wall	Adiabatic	$T=T_{low}$	Adiabatic	
Bottom wall	Adiabatic	Adiabatic	Adiabatic	

Here we calculate three mesh schemes as 50×50, 100×100 and 200×200 and the temperature distributions under three mesh schemes for case A are shown in Figure 2-7. The heat transfer in superior area is stronger than that in inferior area and the temperature along the y direction increases gradually because the top wall moves along the x direction. From

Figure 2-7, the results with 100×100 and 200×200 are almost same so the mesh of 100×100 is selected for saving the cost of computation.

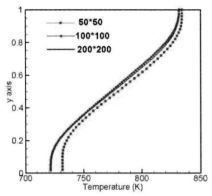

Figure 2-7. Temperature distributions of Case A in three mesh schemes.

Comparing the numerical results with the publish data [9], the same physical parameters are set, $Ra=10^4$ and $Sc=6.8$. The calculated results are shown and compared with the published data in Figure 2-8, and our calculated results agree with the published data qualitatively.

From Figure 2-8, a clockwise vortex is formed because the thermal boundary condition for case A drives the convection flow. The convective heat transfer is dominated mainly because of the big Ra number, and then the isotherm lines gradually become horizontal in the center of container but the isotherm lines near the hot and cold walls boundaries still keep vertical. The isotherm lines near the top and bottom walls should be vertical with the both walls because the top and bottom walls are adiabatic. The published isotherm lines [8] are not vertical strictly with the two walls so our calculated results are better than the data [8].

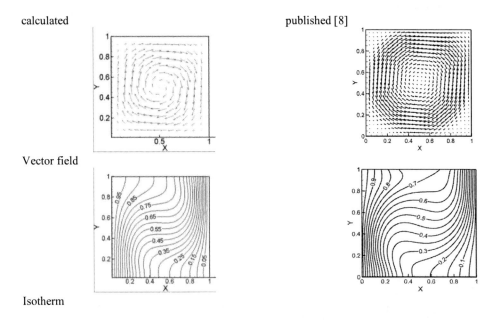

Figure 2-8. Comparison between computational and published results for case A (Ra=104).

The gas concentration distributions for three cases are shown in Figure 2-9. We can make a conclusion that the mass diffusion process depends on the convection process and the gas concentration distribution can be controlled by the fluid flow from the three figures.

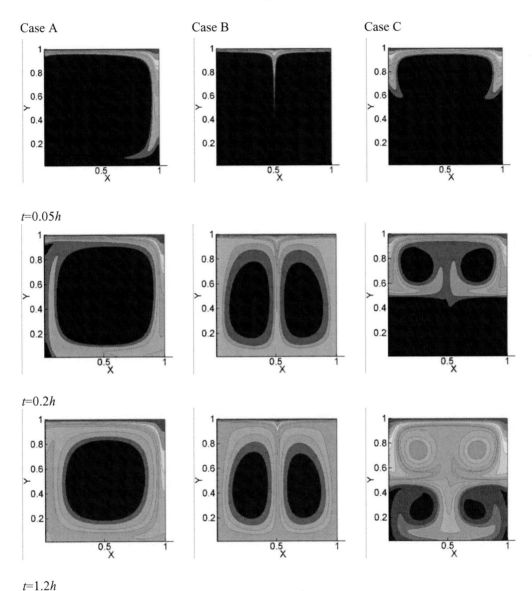

Figure 2-9. Concentration distribution of oxygen for three cases.

The gas average concentration distributions for three cases are compared under three thermal boundary conditions, as seen in Figure 2-10. When the diffusion time is longer, the average concentration is higher for three cases because the gas continuously diffuses into the container. At the beginning stage, the convection flow influences the mass transfer greatly so the average concentration increases greatly and then the average concentration increases slowly with the decrease of the convection effect. For three cases in our calculated time, the changes of gas average concentration for case A are the smallest, and those of cases B and C are bigger. We think case B is the best thermal boundary condition among three cases.

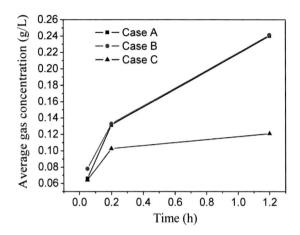

Figure 2-10. Average concentrations of oxygen at different time.

2.1.3. Liquid-Liquid Flow in Microchannels

2.1.3.1. Introduction

Microfluidic systems that involve gas-liquid or liquid-liquid two-phase flows have been extensively used as emulsification devices, bioassay and micro-reactors due to their advantages in achieving accurate flow control at the microscale, providing uniform mass and temperature distributions and small demand of the volume of working fluids [10-14]. The dynamics of gas-liquid or liquid-liquid mixtures near the junctions of microchannels – the most common component in a microfluidic system – is influenced by interfacial tension, viscosity, inertia, and channel walls and has been a significant subject of research [15, 16]. Among two-phase microfluidic systems, liquid-liquid and gas-liquid systems behave differently, and the results for gas-liquid flows cannot be generalized to liquid-liquid flows [17]. Compared with the vast amount of work on gas-liquid flows in microchannels [18-22], there are fewer experimental and numerical studies on the liquid-liquid flow in microchannels [23]. Several studies have indicated the flow patterns of liquid-liquid systems in microchannels are controlled by the channel geometry, the interfacial properties of channel surface and fluids, as well as the flow rate [15, 20, 24]. These variables are considered simultaneously to understand multiphase hydrodynamics in microchannels.

For such a system, Zhao et al. [26-28] have conducted experiments to study the flow, transport, and reaction processes. These studies have laid a good foundation for the validation of our numerical method. In this study, we use numerical simulation to study the flow patterns and volumetric distribution of the dispersed phase in order to obtain the hydrodynamics of a kerosene-water system because of its relevance to a large class of immiscible organics-aqueous flow problems.

The first class of immiscible multiphase LB models (color LB model) could be traced back to Gunstensen et al. [29] and Gunstensen and Rothman [30, 31]. Another class of immiscible multiphase LB models started with Shan and Chen [32], who proposed a multi-component LB model based on the microscopic interactions between fluid particles. Swift et al. [33] developed another class of immiscible multiphase LB models using the free-energy approach. One of the challenges in simulating immiscible multiphase flows is to determine the location of the interface. The interface changes position constantly, and when two interfaces meet, breakup and merging need to be modeled [9, 34]. The above LB models have great advantages over conventional methods because they do not track interfaces explicitly. Instead, interfaces arise naturally due to phase separation between fluid components. The drawbacks of these LB models are that the interfaces are diffuse and several-lattice-unit wide, and the models usually become unstable when density and viscosity ratios are too high. Some recent developments, however, have alleviated these constraints [35, 36]. The LB method used in this work was developed by Santos et al. [37], who extended the field mediator concept [38] for lattice-gas models of immiscible fluids to LB models. The conception of Santos' lattice-gas models of immiscible fluids to LB models. The conception of Santos does not track interfaces explicitly. Instead, interfaces arise naturally relaxation times related to the species diffusivity and to the viscosity of each fluid, so the LB model keeps the numerical stability at very high viscosity ratio of fluids. The interference between mediators and particles is modeled by considering a deviation in particle velocity, which is proportional to the mediators' distributions at the site. Interfacial tension is attained by modifying the collision term between two fluids, introducing long-range forces in the transition layer through field mediators. Mediators' action is restricted in the transition layer and ideal gas state equation is recovered for each fluid far from the interface. Santos et al. [37] applied the LB model to investigate the liquid junction potential at the interface between two electrolyte layers and the LB solutions were validated against the results of analytical and finite difference method for the evolution of concentration, net charge density and electrostatic potential. Santos et al. have applied the LB model to more fields [39-47].

The objective of this work is to apply the two-phase lattice-Boltzmann model based on the fluid mediator to study the immiscible kerosene-water flow in a T-shaped microchannel to validate the two-phase LB model. Typical flow patterns are simulated and compared with the available experimental results qualitatively and quantitatively after the two-phase LB model is built. The results are analyzed to explain the mechanisms for the formation of the various flow patterns for immiscible kerosene-water systems in the T-shaped microchannel. We predict how the static contact angle and wettability influence the flow patterns and volume of the dispersed phase. The numerical method and two-phase LB model based on the field mediators could be readily applied to simulate a class of microfluidic liquid-liquid two-phase flow problems.

2.1.3.2. Numerical Model and Method

The LB model based on field mediators for immiscible fluids from Santos et al. [37] is adopted in this simulation. We will repeat the significant aspects of the method. More detailed description could be found in the report of Santos et al. [37].

Considering $f_i^r(\mathbf{x},t)$ and $f_i^b(\mathbf{x},t)$ as the particle distributions of R and B fluid particles in site $\mathbf{x}$ at time t, and similarly $M_i(\mathbf{x},t)$ as the particle distribution of mediators that are created just before the propagation step, the evolution of the third distribution function besides the two fluids' evolution equations, $M_i(\mathbf{x},t)$, which models the long range interaction, is as follows:

$$M_i(\mathbf{x}+\mathbf{e}_i,t+1) = c_1 M_i(\mathbf{x},t) + c_2 \frac{\sum f_i^r(\mathbf{x},t)}{\sum f_i^r(\mathbf{x},t) + \sum f_i^b(\mathbf{x},t)} \tag{2-27}$$

where c_1 and c_2 are weights used for setting the interaction length, $c_1+c_2=1$. In Eq. (2-27), the first term on the right-hand side is a recurrence relation, because $M_i(\mathbf{x},t)$ depends on $M_i(\mathbf{x}-\mathbf{e}_i,t-1)$, $f_i^r(\mathbf{x}-\mathbf{e}_i,t-1)$ and $f_i^b(\mathbf{x}-\mathbf{e}_i,t-1)$ for all neighbor sites around site $\mathbf{x}-\mathbf{e}_i$. When $c_1=0$ (or $c_2=1$), mediators are created at site ($\mathbf{x}-\mathbf{e}_i$, $t+1$) with the sole information of the concentration of R particles on the next neighbor site. In this case, the interaction length corresponds to one lattice unit. By increasing c_1 with respect to c_2, the interaction length could be arbitrarily increased.

The LB evolution equations for R and B fluid particles are written as

$$f_i^r(\mathbf{x}+\mathbf{e}_i,t+1) - f_i^r(\mathbf{x},t) = \Omega_i^r(\mathbf{x},t) \tag{2-28}$$

$$f_i^b(\mathbf{x}+\mathbf{e}_i,t+1) - f_i^b(\mathbf{x},t) = \Omega_i^b(\mathbf{x},t) \tag{2-29}$$

Collision operators for R and B are Ω_i^r and Ω_i^b, which could model the particles' interaction, are required to satisfy mass and momentum conservation. Splitting the BGK collision operators, as proposed by Santos et al.[38], a three-parameter BGK collision term satisfies the above restrictions:

$$\Omega_i^r = m^r \frac{f_i^{r\,eq}(\rho^r,\mathbf{u}^r) - f_i^r(\mathbf{x},t)}{\tau^r} + m^b \frac{f_i^{r\,eq}(\rho^r,\boldsymbol{\theta}^b) - f_i^r(\mathbf{x},t)}{\tau^m} \tag{2-30}$$

$$\Omega_i^b = m^b \frac{f_i^{b\,eq}(\rho^b,\mathbf{u}^b) - f_i^b(\mathbf{x},t)}{\tau^b} + m^r \frac{f_i^{b\,eq}(\rho^b,\boldsymbol{\theta}^r) - f_i^b(\mathbf{x},t)}{\tau^m} \tag{2-31}$$

ρ^r, ρ^b, $\mathbf{u}^r$ and $\mathbf{u}^b$ are respectively the macroscopic densities and velocities of fluid R and B. τ_r, τ_b and τ_m are respectively the relaxation factors of the fluid R, B and field mediators, when the density distributions of them reach equilibrium state. Equilibrium distributions are derived based on small velocity expansions, and are consistent with the description of incompressible flows:

$$f_i^{r\,eq}(\rho^r,\mathbf{u}^r) = \rho^r \omega_i \left[1 + \frac{c\mathbf{e}_i \cdot \mathbf{u}^r}{c_s^2} + \frac{(c\mathbf{e}_i \cdot \mathbf{u}^r)^2}{2c_s^4} - \frac{|\mathbf{u}^r|^2}{2c_s^2} \right] \quad i=1,\cdots,18 \tag{2-32}$$

In Eq. (2-32), $\mathbf{e}_i$ is the discrete velocity of the LB model, particle velocity c is equal to 1.0 here, and c_s is the lattice sound velocity, $c_s = c/\sqrt{3} = 1/\sqrt{3}$. ω_i is the weight factors: $\omega_0=1/3$, $\omega_1...\omega_6=1/18$, $\omega_7...\omega_{18}=1/36$, and the equilibrium distributions of fluid B are similar to Eq. (2-32).

The first term on the right hand side of Eq. (2-30) is related to the relaxation of R particle distribution to an equilibrium state given by the R component density and velocity (R-R collisions). The second term considers R-B collisions and is related to the relaxation of R particles to the equilibrium state given by the density ρ^r and B velocity ($\boldsymbol{\theta}^b$) modified by the action of mediators in the same site.

$$\boldsymbol{\theta}^b = \mathbf{u}^b - A\hat{\mathbf{u}}^m \qquad (2\text{-}33)$$

$$\boldsymbol{\theta}^r = \mathbf{u}^r + A\hat{\mathbf{u}}^m \qquad (2\text{-}34)$$

where $\boldsymbol{\theta}^r$ and $\boldsymbol{\theta}^b$ are the local velocities modified by the action of mediators and constant A is related to interfacial tension. We could obtain the interfacial tension between two fluids by adjusting the constant A. For the ideal miscible fluids, $A=0$ and this collision term describes the relaxation of R particle distribution to an equilibrium state given by ρ^r and u^b, as a consequence of R-B cross collisions. In immiscible fluids, Eq. (2-33) means that R particles will be separated from B particles by long-range attractive forces from R phase, and Eq. (2-34) is similar to Eq.(2-33). The mediator velocity at site x in Eqs. (2-33) and (2-34) is given by

$$\hat{\mathbf{u}}^m = \sum_{i=1}^{18} M_i(\mathbf{x},t)\mathbf{e}_i \bigg/ \left|\sum_{i=1}^{18} M_i(\mathbf{x},t)\mathbf{e}_i\right| \qquad (2\text{-}35)$$

The mixture kinematics viscosity coefficient υ in lattice units is given by

$$\upsilon = \frac{c^2}{6}\left[\frac{1}{2} - \left(\frac{m^r}{\frac{m^r}{\tau^r}+\frac{m^b}{\tau^m}} + \frac{m^b}{\frac{m^b}{\tau^b}+\frac{m^r}{\tau^m}}\right)\right] \qquad (2\text{-}36)$$

Mixture viscosity appears as a function of three collision parameters of τ^r, τ^b and τ^m, and m^r and m^b, which are the mass fractions of fluids R and B, where $\rho = \rho^r + \rho^b$, $m^r = \rho^r/\rho$ and $m^b = \rho^b/\rho$.

The interfacial thickness is

$$\Lambda = \frac{c_s^2 (\tau^m - 1/2)}{A} \int_0^1 \frac{\delta m^r}{m^r (1 - m^r)} \tag{2-37}$$

Restricting the analysis to a plane interface and taking coordinate normal to the interface, the interfacial tension σ_{rb} is calculated by

$$\sigma_{rb} = \int_{-\infty}^{\infty} \left[\Pi_{yy}^*(y) - \Pi_{xx}^*(y) \right] \delta y \tag{2-38}$$

where $\Pi_{yy}^*(y)$ and $\Pi_{xx}^*(y)$ are the momentum fluxes along the y and x axes.

$$\Pi_{yy}^*(y) = c_s^2 \rho + \rho A^2 a_r a_{r1} + \rho A^2 a_b a_{b1} \tag{2-39}$$

$$\Pi_{xx}^*(y) = c_s^2 \rho \tag{2-40}$$

A closed form expression for the interfacial tension is obtained, in terms of lattice parameters and relaxation times by changing variables, considering that at each interface point x the diffusive flux of R or B particles promoted by concentration gradients is required to be concealed by long-range force:

$$\sigma_{rb} = c_s^2 (\tau^m - 0.5) A \rho^{out} \int_0^1 (a_r a_{r1} + a_b a_{b1}) \frac{\delta m^r}{m^r (1 - m^r)} \tag{2-41}$$

where ρ^{out} is the particle density outside the transition layer.

The boundary conditions for the mediator distribution are used for modulating the interaction process between the solid and the fluid. The wettability, or static contact angle, could be obtained by imposing, at each fluid site adjacent to a solid site, a constant value M_i^{solid} for the mediator distribution along the directions i leading to the fluid phase in the propagation step. This quantity determines the static contact angle, since it determines the interaction between the solid and the fluid. As the mediator distribution carries the values of mass fraction, $M_i^{solid} \in [-1, 1]$. When $M_i^{solid} = 0$, the static contact angle will be $\alpha = 90°$. When $M_i^{solid} > 0$, fluid R will behave as the wetting fluid, and when $M_i^{solid} < 0$, fluid B will behave as the wetting fluid. Nevertheless, the precise value of the dynamic contact angle also depends on the relaxation parameters, τ_r, τ_b and τ_m.

2.1.3.3. Results and Discussion

In order to study immiscible fluid flows and the flow patterns in a T-shaped mirochannel, we simulate cases from the experiments conducted by Zhao et al. [25]. Water and kerosene are used as work fluids whose physical properties are shown in Table 2-4.

Table 2-4 Physical properties of water and kerosene

Working medium	Density (kg/m³)	Viscosity (Pa·s)	Interfacial tension (N/m)
Water	998.2	0.001	0.045
Kerosene	780	0.00115	

Figure 2-11 shows a schematic of the microchannel. The width B is 600 μm, the height H is 300 μm, the length L_1 is 3 mm and L_2 is 6 mm.

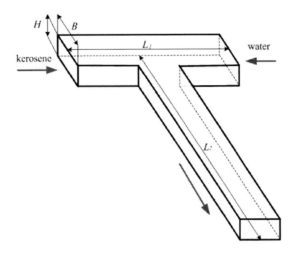

Figure 2-11. Simplified schematic and mesh of a 300μm× 600μm T-shaped microchannel.

We set the same Weber number as shown in Table 2-5 according to Zhao et al. [25] to compare the numerical results with the experimental data. The densities of fluids R and B are 1.0 and 0.7816, respectively. The characteristic scale is 20 lattices.

Table 2-5 Parameters of calculated cases

Case	We_{ws}	We_{ks}	Re_{ws}	Re_{ks}	q	τ_{ws}	τ_{ks}
a (a')	3.04×10^{-3}	5.94×10^{-5}	7.39	0.25	20	0.73	0.756
b (b')	3.04×10^{-1}	5.94×10^{-4}	2.96×10^{1}	1.0	20	0.68	0.701
c (c')	2.74	1.34×10^{-1}	3.70×10^{2}	2.51×10^{1}	40	0.63	0.645
d (d')	6.85×10^{-1}	5.35×10^{-1}	5.55×10^{2}	3.77×10^{2}	1	0.52	0.522
e (e')	7.61	5.94	7.39×10^{2}	1.0×10^{2}	1	0.51	0.511

The predicted five distinct flow patterns are shown in Figure 2-12. We reproduce the slug flow (ST) as shown in Figure 2-12(a) and Figure 2-12(f), monodispersed droplet (MDT) in Figure 2-12(b) and Figure 2-12(g), droplet population (DPM) in Figure 2-12(c) and Figure 2-12(h), and parallel flow with smooth interface (PFST) in Figure 2-12(d) and Figure 2-12(i) at different Weber numbers, and flows with wavy interface (PFWT) in Figure 2-12(e) and Figure 2-12(j).

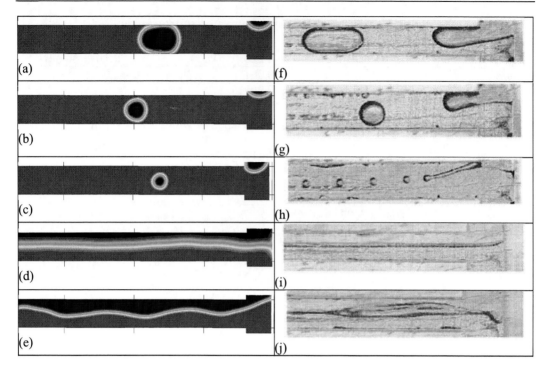

Figure 2-12. Typical flow patterns in a T-shaped microchannel.

In Figure 2-12(a) and Figure 2-12(f), the kerosene phase is observed to form slugs when the Weber number is low and water/kerosene superficial flow rate ratio is high. The formation of the slug is controlled by the dynamics at the microchannel and not so much by the microchannel downstream. In Figure 2-12(b) and Figure 2-12(g), the kerosene phase is observed to form discrete drops. The distance between two consecutive monodispersed droplets is larger than the microchannel width. Because the height (H) is smaller than the width (B), monodispersed droplets are squeezed between the top and bottom wall and the geometry of the droplets is like a disc. The formation of these large drops is an "entrance phenomenon" that is controlled by the hydrodynamics near the junction. In Figure 2-12(c) and Figure 2-12(h), a series of well-defined spherical droplets are formed. The distance between two consecutive droplets is smaller than the microchannel width in the experiments. In simulations, the distance between consecutive droplets is larger due to the limited resolution of the grid and the finite thickness of the interface. The formation of the drop is no longer an "entrance phenomenon", because the separation of the drop from the kerosene "tongue" extends far downstream in the microchannel. In Figure 2-12(d) and Figure 2-12(i), we observe that two fluids flow continuously side by side in the rectangular microchannel.

We observe slight differences between simulations and experiments for PFWT in Figure 2-12(e) and Figure 2-12(j), which is likely because of the limited grid resolution that is unable to resolve the Kolmogorov length scale which equals $L/Re^{3/4}$, where L is the integral length scale and could be taken as the width/height of the microchannel.

Figure 2-13 shows the formation process of a slug near the T-junction: the numerical results agree well with the experimental data. The kerosene slug is separated from the kerosene incoming stream by interfacial instability. The pressure drop along the axis of the microchannel forces the tip of the kerosene "tongue" to extend into the microchannel downstream. While the kerosene advances in the microchannel, it blocks the flow of water, increasing the pressure behind the "tongue" and eventually leading to separation of the slug from the incoming stream.

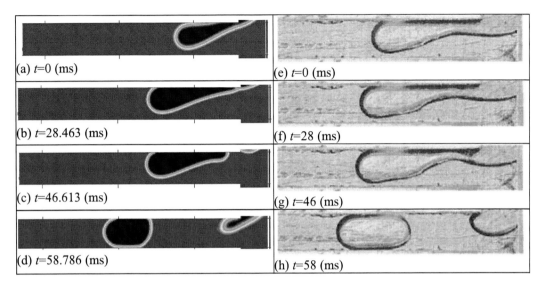

Figure 2-13. Kerosene slug formation mechanism in a T- shape microchannel $q = 10$, $We_{ks} = 2.383 \times 10^{-5}$, $We_{ws} = 3.043 \times 10^{-3}$.

For a fixed kerosene flow rate Q_{ks}, the increase in water flow rate results in the formation of monodispersed droplets. Figure 2-14 shows the process of the formation of monodispersed droplets, and the numerical results also agree well with experimental data. In this flow pattern, the low value of We_{ks} ensures that the interfacial tension overwhelms the inertial force for the kerosene phase. The value of We_{ws}, however, is 0.12, showing that the water phase inertial effects begin to play a role in kerosene droplet formation.

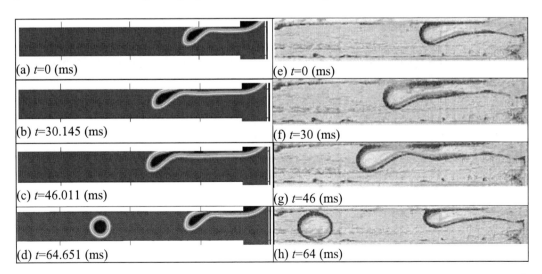

Figure 2-14. Monodispersed droplet formation mechanism in a microchannel $q = 25$, $We_{ks} = 1.483 \times 10^{-4}$, $We_{ws} = 0.12$.

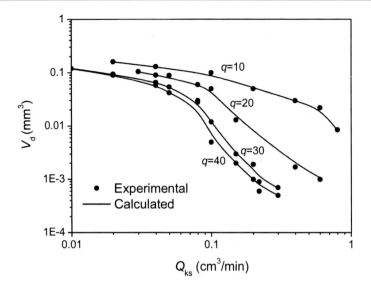

Figure 2-15. Evolution of the dispersed phase volumes (V_d) as a function of the kerosene flow rate in a microchannel.

In many emulsification and two-phase reaction applications, the size of dispersed droplets has the greatest effect on heat and mass transfer rates and the design of downstream separation processes. In Figure 2-15, we compare the volume of the dispersed droplets (in MDF, DPF and SF regimes) as a function of kerosene flow rate characterized from our simulations to that from the experiments of Zhao et al. [25]. For Figure 2-15, in order to get the accurate volume of the dispersed phase, we extend the mesh as $10(H) \times 20(B) \times 100(L_1) \times 1000(L_2)$ in y axis to get more slugs or droplets. Then we calculate the numbers of slugs and bubbles and averaged volumes. The numerically predicted volumes of kerosene droplets are in very good agreement with the experimental data. The volumes of the dispersed droplets decrease in all cases with the increase in kerosene flow rate, and for a given kerosene flow rate, the volumes of the dispersed droplets decrease with increasing volumetric flux ratio q. This demonstrates the feasibility of controlling the volume of dispersed droplets by adjusting the volumetric flux ratio q. The volumes of dispersed droplets depend on interfacial tension, viscosity, inertial force and the volumetric fraction of the dispersed phase. The average relative deviation of the volumes of dispersed droplets is 5.67%.

In order to investigate the influence of the contact angle and interfacial tension on the flow patterns, we change the wettability M_i^{solid} and constant A to get different contact angles and interfacial tensions while other parameters are not changed. As shown in Figure 2-16, we set five cases according to the values of wettability M_i^{solid}. When the wettability is more than 0 and less than 1.0, the flow pattern of the immiscible kerosene-water system in a T-shaped microchannel is slug flow. α is defined as the angle between the water phase and the wall. With the increase of wettability, the contact angle α becomes larger. When the wettability is less than 0 and more than -1.0, the flow pattern of the immiscible kerosene-water system in the T-shaped microchannel is stratified flow. With the decrease of wettability, the contact angle measured from the calculated pictures becomes larger. When the wettability nature is not changed (infiltration and non-infiltration), the flow pattern will not be changed.

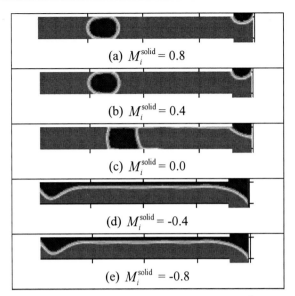

Figure 2-16. Contact angles versus flow patterns $q = 10$, $We_{ks} = 5.94 \times 10^{-2}$, $We_{ws} = 7.61$, $Re_{ks} = 19.45$, $Re_{ws} = 286.4$.

In Figure 2-17, we set the constant A as 0.28, 0.2, 0.12, 0.06 and 0.0, because we are not able to obtain a convergent solution when A is greater than 0.3. When the constant A decreases, the immiscible kerosene-water system in T-shaped microchannels keeps the dispersed flow but the size of the dispersed phase reduces and smaller droplets appear, the contact angle measured according to the calculated results increases slightly, and the interfacial layer becomes thicker. When the size of the dispersed phase becomes very small, the two-phase flow becomes unsteady. When the constant A is 0, it means the two phases become one phase so we cannot get the interface.

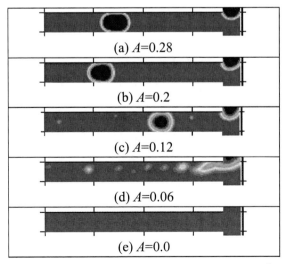

Figure 2-17. Interfacial tensions versus flow patterns $q = 10$, $We_{ks} = 5.94 \times 10^{-2}$, $We_{ws} = 7.61$, $Re_{ks} = 19.45$, $Re_{ws} = 286.4$.

2.2. Buoyancy-Driven Motion and Mass Transfer of Single Particles

Level set method has many advantages and allows for large surface tension and jump in density and viscosity across the interface without reconstructing the numerical grid. Unfortunately, the algorithm for the level set function is in a non-conservative form. The volume loss (or gain) of bubble/drop would occur as the computational time elapses to result in the deviation from the correct interface position and the distortion of the flow field. Up to now, three kinds of correcting methods have been reported for improving the mass conservation [48]. First of all, a variety of modified reinitialization procedures [49-52] have been adopted extensively. Secondly, several "hybrid" methods have been developed, by combining the level set approach with the volume-tracking [53-55] or front-tracking [56, 57] methods. Lastly, the mass conservation in the level set method can be significantly improved by reducing spatial discretization errors. The use of AMR (adaptive mesh refinement) for better resolution of interface structures has been attempted to demonstrate improvements in mass conservation [58, 59]. A volume-amending is proposed by Li et al. [60] to preserve mass (volume) conservation in the level set method, which can be accomplished without significant complication in coding or increase of computational burden.

2.2.1. Mathematical Model and Numerical Method

2.2.1.1. Conservation Equations

In a two-dimensional coordinate system, mass and momentum conservation with the level set approach incorporated [18] are written in terms of dimensionless variables as

$$\frac{\partial u'}{\partial X} + \frac{1}{r}\frac{\partial}{\partial Y}(rv') = 0 \qquad (2\text{-}42)$$

$$\frac{\partial}{\partial \theta}(\rho u') + \frac{\partial}{\partial X}\left(\rho u'u' - \frac{\mu}{Re}\frac{\partial u'}{\partial X}\right) + \frac{1}{r}\frac{\partial}{\partial Y}\left(r\rho v'u' - r\frac{\mu}{Re}\frac{\partial u'}{\partial Y}\right) = -\frac{\partial p'}{\partial X} + \frac{1}{Fr}\rho g_X - \frac{1}{We}\kappa(\phi)\delta_\varepsilon(\phi)\frac{\partial \phi}{\partial X} + \frac{1}{Re}\frac{\partial}{\partial X}\left(\mu\frac{\partial u'}{\partial X}\right) + \frac{1}{Re}\frac{1}{r}\frac{\partial}{\partial Y}\left(r\mu\frac{\partial v'}{\partial X}\right) \qquad (2\text{-}43)$$

$$\frac{\partial}{\partial \theta}(\rho v') + \frac{\partial}{\partial X}\left(\rho u'v' - \frac{\mu}{Re}\frac{\partial v'}{\partial X}\right) + \frac{1}{r}\frac{\partial}{\partial Y}\left(r\rho v'v' - r\frac{\mu}{Re}\frac{\partial v'}{\partial Y}\right) = -\frac{\partial p'}{\partial Y} + \frac{1}{Fr}\rho g_Y - \frac{1}{We}\kappa(\phi)\delta_\varepsilon(\phi)\frac{\partial \phi}{\partial Y} + \frac{1}{Re}\frac{\partial}{\partial X}\left(\mu\frac{\partial u'}{\partial Y}\right) + \frac{1}{Re}\frac{1}{r}\frac{\partial}{\partial Y}\left(r\mu\frac{\partial v'}{\partial Y}\right) - \left\{\frac{2}{Re}\frac{\mu v'}{r^2}\right\} \qquad (2\text{-}44)$$

where $r \equiv 1$ for Cartesian coordinates, $r \equiv Y$ for cylindrical coordinates and curly brackets indicate the term presents only in cylindrical coordinates. The dimensionless groups Re, Fr and We are Reynolds, Froude and Weber numbers, respectively,

$$Re \equiv \frac{\rho L V}{\mu}, \quad Fr \equiv \frac{V^2}{Lg}, \quad We \equiv \frac{L\rho_1 V^2}{\sigma} \qquad (2\text{-}45)$$

where the characteristic length L is chosen as the initial diameter of a spherical particle (d) and the corresponding reference velocity as $V = \sqrt{dg}$.

2.2.1.2. Level Set Method

The level set function ϕ is introduced into the formulation of multiphase flow and mass transfer systems to define and capture the interface between two fluids. The interface is defined as the zero set of the level set function, ϕ, which is defined as the signed algebraic distance of a node to the interface, being positive in the continuous fluid phase and negative in the drop.

The following equation is used to advance the level set function exactly as the drop moves:

$$\frac{\partial \phi}{\partial \theta} + \frac{\partial}{\partial X}(u'\phi) + \frac{1}{r}\frac{\partial}{\partial Y}(rv'\phi) = 0 \qquad (2\text{-}46)$$

$\kappa(\phi)$ is the curvature of drop surface defined as

$$\kappa(\phi) = -\nabla \cdot \mathbf{n} = -\nabla \cdot \left(\frac{\nabla \phi}{|\nabla \phi|}\right) \qquad (2\text{-}47)$$

where $\mathbf{n}$ is the unit vector normal to the interface pointing towards the continuous phase and $\delta_\varepsilon(\phi)$ is the regularized delta function defined as

$$\delta_\varepsilon(\phi) = \begin{cases} \frac{1}{2\varepsilon}\left(1 + \cos\left(\frac{\pi\phi}{\varepsilon}\right)\right) & \text{if } |\phi| < \varepsilon \\ 0 & \text{otherwise} \end{cases} \qquad (2\text{-}48)$$

here ε prescribes the finite 'half thickness' of the interface. Usually we take $\varepsilon = 1.5\Delta x$, where Δx is the dimensionless uniform mesh size near the interface. $H_\varepsilon(\phi)$ is the regularized Heaviside function expressed as

$$H_\varepsilon(\phi) = \begin{cases} 0 & \text{if } \phi < -\varepsilon \\ \frac{1}{2}\left(1 + \frac{\phi}{\varepsilon} + \frac{\sin\left(\frac{\pi\phi}{\varepsilon}\right)}{\pi}\right) & \text{if } |\phi| \leq \varepsilon \\ 1 & \text{if } \phi > \varepsilon \end{cases} \qquad (2\text{-}49)$$

or defining the corresponding regularized (smoothed) density function ρ and the regularized viscosity μ as

$$\rho_\varepsilon(\phi) = \rho_d/\rho_c + \left(1 - \rho_d/\rho_c\right)H_\varepsilon(\phi) \qquad (2\text{-}50)$$

$$\mu_\varepsilon(\phi) = \mu_d/\mu_c + \left(1 - \mu_d/\mu_c\right)H_\varepsilon(\phi) \qquad (2\text{-}51)$$

Generally, ϕ will no longer be a distance function (i.e., $|\nabla \phi| \neq 1$) after some iterations, even if Eq. (2-46) moves the interface ($\phi = 0$) at the correct velocity. Maintaining ϕ as a distance function is essential for providing the interface with an invariant width and a sound basis for estimating the surface curvature. This problem can be solved by adopting a reinitialization method to solve the following initial problem to steady state in a virtual time domain:

$$\frac{\partial \phi}{\partial \tau} = \text{sgn}(\phi_0)(1 - |\nabla \phi|) \qquad (2\text{-}52)$$

$$\phi(x, 0) = \phi_0(x) \qquad (2\text{-}53)$$

where τ is the virtual time for reinitialization, $\phi_0(x)$ the level set function at any computational instant, and $\text{sgn}(\phi_0)$ the sign function needed for enforcing $\nabla\phi = 1$. Eq. (2-52) and Eq. (2-53) have the property that ϕ remains unchanged at the interface, therefore, the zero level set of ϕ_0 and ϕ is the same. Away from the interface ϕ will converge to $\nabla\phi = 1$, i.e., the actual distance function.

The following perturbed Hamilton-Jacobi equation was introduced to guarantee mass conservation:

$$\frac{\partial \phi}{\partial \tau} + (A_0 - A(\tau))(-Q + \kappa(\phi))|\nabla\phi| = 0 \tag{2-54}$$

where A_0 denotes the initial total mass of both fluids at $\tau = 0$ and $A(\tau)$ the total mass corresponding to the level set function $\phi(\tau)$. The parameter Q is a positive constant and is usually set to be 1.

Yang and Mao [60] found that a considerable amount of mass in the fluid particle vanished gradually although an excellent conservation of total mass with time was observed with the reinitialization by Eq. (2-54). Therefore, the definition of $A(\tau)$ was modified to

$$A(\tau) = \sum_{\phi_{ij} \leq \varepsilon} \rho_\varepsilon(\phi_{ij}) R_j \Delta x \Delta y \tag{2-55}$$

where $\phi_{ij} \leq \varepsilon$ denotes the nodes in the fluid particle and the interface has a virtual thickness of 2ε, i.e., $A(\tau)$ in Eq. (2-55) is taken as the mass of drop/bubble instead of the total mass as in Eq. (2-54). The improved reinitialization procedure can maintain the level set function as a distance function and guarantee the drop mass conservation.

Since the fluids are incompressible, mass conservation is equivalent to volume conservation. Furthermore, a volume-amending method [61] was proposed, in which we reinitialize the level set function ϕ to an exact signed distance function by solving Eq. (2-52) firstly and then the position of the interface is corrected according to the loss or gain of the particle (bubble/drop) volume ($V = \sum_{\phi \leq 0} H(\phi) R \Delta x \Delta y$):

$$\Delta V = \frac{V(t) - V_0}{V_0} \tag{2-56}$$

Assuming the bubble/drop is spherical, the increment of the particle radius may be expressed as

$$\Delta R = R - R_0 = ((1 + \Delta V)^{1/\alpha} - 1)R_0 \tag{2-57}$$

where α is the dimension of the system simulated: $\alpha = 2$ for an axisymmetric fluid particle and $\alpha = 3$ for a three-dimensional case. Since the volume increase of the particle corresponds to that of the distance from the interface to the particle center, the correction to the level set function ϕ may be taken in proportion to ΔR. Thus, a relationship is established between the increment of the level set function ϕ and the loss or gain of the particle (bubble/drop) volume, so that the correction to the level set function ϕ may be expressed as

$$\delta\phi = \beta((1 + \Delta V)^{1/\alpha} - 1) \tag{2-58}$$

where β is an empirical coefficient, large β may cause the reinitialization divergent, and a small one leads to low computational efficiency. β is usually chosen between 0.01-0.1 by experience. Thus, the volume-amending equation can be written as

$$\phi = \phi_0 + \delta\phi = \phi_0 + \beta((1 + \Delta V)^{1/\alpha} - 1) \tag{2-59}$$

In the following, the "one-fluid" formulation coupled with the level set function for interphase mass transfer is derived. When the distribution coefficient of a solute, m, is unity, the solute concentration across the interface is continuous. The only difficulty is the discontinuity of molecular diffusion coefficients, which can easily be handled by the level set function with the same smoothed Heaviside function.

The transient mass transfer to/from a drop is governed by the convective diffusion equation in the vector form:

$$\frac{\partial c}{\partial t} + \mathbf{u} \cdot \nabla c = D\nabla^2 c \tag{2-60}$$

Eq. (2-60) for two phases may be expediently solved in a single domain by the level set method similar to the solution of multiphase flow, subject to two interfacial conditions:

$$D_c \frac{\partial c_c}{\partial n_c} = D_d \frac{\partial c_d}{\partial n_d} \quad \text{(flux continuity at the interface)} \tag{2-61}$$

$$c_d = mc_c \quad \text{(interfacial dissolution equilibtium)} \tag{2-62}$$

However, more general case is that m is not equal to unity. In this case some measures must be taken to make the concentration field become continuous across the interface, just as the continuity of fluid velocity at the interface. For this purpose, the concentration transformation method, such as $\hat{c}_1 = c_1\sqrt{m}$, and $\hat{c}_2 = c_2/\sqrt{m}$. The transformed Eq. (2-60) remains in the following form:

$$\frac{\partial \hat{c}_i}{\partial t} + \mathbf{u} \cdot \nabla \hat{c}_i = D_i \nabla^2 \hat{c}_i, \quad i = c, d \tag{2-63}$$

The difficulty that remains is only Eq. (2-61). When the transformations are utilized, Eq. (2-61) becomes

$$\frac{D_c}{\sqrt{m}} \frac{\partial \hat{c}_c}{\partial n_c} = \sqrt{m} D_d \frac{\partial \hat{c}_d}{\partial n_d} \tag{2-64}$$

Therefore, at the interface, which is of finite thickness, the diffusion coefficients in both phases should be locally replaced by $D_c/\sqrt{m}$ and $\sqrt{m}D_d$ to satisfy the original condition of mass flux continuity. This would make the diffusivity in the interface region different from that in the bulk domain and may result in unacceptable errors. In our studies, a simple transformation is applied to make the diffusivity being equal throughout a fluid phase. For the continuous and dispersed phases, Eq. (2-63) can be rewritten separately as

$$\frac{\partial \hat{c}_c}{\partial (\sqrt{m}t)} + \frac{1}{\sqrt{m}}\mathbf{u}\cdot\nabla\hat{c}_c = \frac{1}{\sqrt{m}}D_c\nabla^2\hat{c}_c \tag{2-65}$$

$$\frac{\partial \hat{c}_d}{\partial (\frac{1}{\sqrt{m}}t)} + \sqrt{m}\mathbf{u}\cdot\nabla d = \sqrt{m}\nabla^2\hat{c}_d \tag{2-66}$$

And then can be put into a common equation over the whole domain in the form analogous to the momentum equation:

$$\frac{\partial \hat{c}}{\partial \hat{t}} + \hat{\mathbf{u}}\cdot\nabla\hat{c} = \nabla\cdot(\widehat{D}\nabla\hat{c}) \tag{2-67}$$

through different definitions of $\hat{t}$, $\widehat{D}$ and $\hat{\mathbf{u}}$ by the regularized Heaviside function $H_\varepsilon(\phi)$:

$$\hat{t}(\phi) = \begin{cases} \sqrt{m}t, & \text{if } \phi \geq 0 \\ \frac{1}{\sqrt{m}}t, & \text{if } \phi < 0 \end{cases} \tag{2-68}$$

$$\widehat{D}(\phi) = \sqrt{m}D_2 + (\frac{1}{\sqrt{m}}D_1 - \sqrt{m}D_2)H_\varepsilon(\phi) \tag{2-69}$$

$$\hat{\mathbf{u}}_\varepsilon(\phi) = \sqrt{m}\mathbf{u} + (\frac{1}{\sqrt{m}}\mathbf{u} - \sqrt{m}\mathbf{u})H_\varepsilon(\phi) \tag{2-70}$$

where $\hat{\mathbf{u}}$ should be the velocity field in a frame of reference moving with the drop.

The governing equations for interphase mass transfer are nondimensionalized and the expanded expressions in a two-dimensional axisymmetric coordinate system are

$$\frac{\partial C}{\partial \theta} + u'\frac{\partial C}{\partial X} + v'\frac{\partial C}{\partial Y} = \frac{1}{Pe_1}\left[\frac{\partial}{\partial X}\left(D\frac{\partial C}{\partial X}\right) + \frac{1}{Y}\frac{\partial}{\partial Y}\left(rD\frac{\partial C}{\partial Y}\right)\right] \tag{2-71}$$

where C is the dimensionless concentration based on the reference concentration, using c_c^∞ for solute transportation from the continuous phase to the drop and the dimensionless group Pe_c is the Peclet number denoted as

$$Pe_c \equiv \frac{LV}{D_c} \tag{2-72}$$

where D_c is chosen as the characteristic molecular diffusivity.

Besides, the other transformation forms of concentration, such as $\hat{c}_c = mc_c$ and $\hat{c}_d = c_d$, or $\hat{c}_c = c_c$ and $\hat{c}_d = c_d/m$ have also been tested. The overall interphase mass transfer can be solved in the similar way, and completely identical results were obtained. Therefore, the transformation form has no effect on the numerical solutions of interphase mass transfer.

2.2.2. Drop And Bubble Motion

2.2.2.1. Motion of Single Bubble in Liquid

In the past decades, many experimental studies have been focused on the motion of single bubbles and drops [62-66]. The present test is focused on the determination of the fluid particle shape and terminal velocity at steady state motion.

However, it is a greater challenge to be satisfied in the fully three-dimensional problems [67]. Wang et al. [69] have successfully simulated the interactional motion of two three-dimensional buoyancy-driven deformable bubbles. In order to verify the present method to be effective in fully three-dimensional complex free-surface problems, the interactive motion of a pair of bubbles that are not vertically in line is tested. It involves complex topological deformation of the interfaces as shown by the experimental results [69]. In the simulation by Oka and Ishii [65], mass conservation was well kept but it was very luxurious to use three reinitialization procedures simultaneously.

Two spherical bubbles with the radius ratio $R_1/R_2 = 1.5$ are initially at rest in the vertical center plane of the computational box. Most of the conditions are set just as those in Manga and Stone's experiments [67] other than the distance between two bubbles, such as $We \approx 50$ and $Re \approx 5 \times 10^{-3}$. Some of the important stages of the predicted interactive behavior are demonstrated in Figure 2- 18. The two bubbles deform considerably while rising, but the deformations of two bubbles are quite different, which resemble the experimental result. The upper bubble deforms towards the steady-state shape of a single bubble in free rise. The trailing bubble deforms a little. The lower part of upper bubble becomes concave while that of the trailing one becomes protuberant. In the following time, the distance between two bubbles turns shorter and shorter. The trailing bubble is entrained into the wake of the preceding one and then they integrate together. The numerical result is in qualitatively agreement with the experimental data [67].

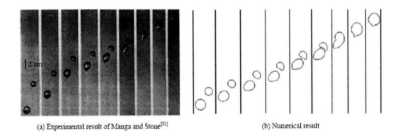

(a) Experimental result of Manga and Stone[31] (b) Numerical result

Figure 2- 18. Experimental and numerical results for interactive behavior of two bubbles rising in corn syrup. Photograghs and simulated results are shown at 10s intervals.

2.2.2.2. Parasitic Flow

Parasitic flows may occur in the numerical simulation of incompressible multiphase flow with level set method due to errors in the calculation of surface tension terms, specifically for the curvature and unit normal vector. A simple weighted integration method based on the level set method is proposed by Wang *et al.*[70], in which the contribution of not only the center node but also the rest area of a control volume to the calculation of surface tension is considered in a balanced manner. The weighted integration method (WIM) is more consistent with the concept of a banded interface in the level set method. It is applied to the temporal evolution of a two-dimensional neutrally buoyant liquid drop and a buoyancy driven deformable bubble in an immiscible fluid for the validation of WIM. Figure 2-18 shows that the parasitic flows are evidently suppressed by the weighted integration method.

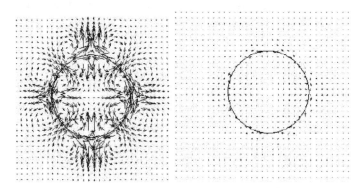

Figure 2-18. Spurious currents (with the same velocity scales) at the 1000th time steps for a stationary, neutrally buoyant drop ($(\varepsilon = \Delta x$, $Oh = 2.88 \times 10^{-2}$, $\rho_b/\rho_c = 1$, $\mu_b/\mu_c = 1$, $\gamma = 0.02$ N·m^{-1}, $R = 0.01$ mm).

2.2.2.3. A Bubble in a Non-Newtonian Fluid

It is highly important to obtain the knowledge of bubble behavior not only in a Newtonian continuous phase but also in a non-Newtonian one, because the investigation on bubble motion provides useful and essential information for realizing suitable process design and operation. It is well known that a large number of investigations concerning various aspects of the bubble or drop motion in non-Newtonian fluids have been reported theoretically and experimentally in the past. Many of them have been well summarized and reviewed by Chhabra [71, 72] and Kulkarni and Joshi [73].

The motion of a single bubble rising freely through shear-thinning fluids represented by the Carreau model is computed numerically using a level set method for tracking the bubble interface. The changes in the local viscosity field around a bubble rising in shear-thinning non-Newtonian fluids are examined. Real density and viscosity ratios of gas and liquid are used to describe the shear-thinning non-Newtonian two-phase flow systems.

The terminal rising velocity of air bubbles in five non-Newtonian liquids (Table 2- 6 and Table 2- 7) with shear-thinning property are determined and these bubbles are simulated. Table 2- 7 shows the numerical and experimental results for a single bubble rising freely in pure shear-thinning non-Newtonian solutions. The calculated terminal velocities are in close agreement with the present experimental values with the relative deviation below 10%. The satisfactory agreement indicates the validity of numerical procedure for the numerical simulation of a single bubble rising in a shear-thinning non-Newtonian continuous phase. In these shear-thinning solutions the apparent viscosity around the bubble decreases due to the shear, so the bubble can rise more and more easily. It is interesting to note that the zero-shear viscosity of CMC2 and XG is roughly equal (Table 2- 6), but U_T of the bubble in XG is much larger than that in CMC2, since XG has stronger shear-thinning property and the liquid around the bubble thus becomes less viscous. Hence, understanding the viscosity distribution over the whole domain is important to elucidate the bubble behaviors in shear-thinning non-Newtonian liquids.

Table 2-6 Parameters for Carreau model and physical properties of liquids

Liquid	η_0 (Pa s)	λ (s)	s	ρ_l (kg m^{-3})	σ (mN m^{-1})
CMC1	0.0377	0.512	0.89	1000.0	72
CMC2	0.122	0.207	0.90	1000.0	72
HEC1	0.047	0.0624	0.79	1000.0	72
HEC2	0.315	0.260	0.68	1000.0	72
XG	0.115	0.984	0.49	1000.0	72

Table 2-7 Comparisons between experimental and numerical results for a single bubble rising in Carreau shear-thinning liquids

Liquid	d_e (mm)	$U_{T,exp}$ (cm s^{-1})	$U_{T,cal}$ (cm s^{-1})	$Re_{M,exp}$	$Re_{M,cal}$	D_{ev}^* %
CMC1	4.19	21.5	20.87	121.27	116.29	2.9
CMC2	3.96	9.88	9.39	4.05	3.832	4.4
HEC1	2.84	11.88	11.6	10.19	9.91	2.4
HEC2	3.11	5.25	4.74	1.041	0.91	9.7
XG	2.29	26.5	24.6	84.07	75.15	7.16

*$D_{ev} = |U_{cal} - U_{exp}|/U_{exp} \times 100$.

Table 2-8 Carreau model parameters and numerical results of bubbles with $d_e = 1$ cm

System	η_0 (Pa s)	λ (s)	s	U_T (cm s^{-1})	$2\lambda U_T/d_e$	Re_M
1	0.5	-	1.0	11.2	-	2.24
2	0.5	1.00	0.8	16.5	33.0	6.65
3	0.5	1.00	0.5	22.7	45.4	30.69
4	0.5	1.00	0.4	23.9	47.8	48.8
5	0.1	1.00	0.8	25.3	50.6	55.5
6	0.5	1.00	0.3	25.6	51.2	80.85

The shear-thinning liquids whose properties are shown in are tested in order to elucidate the shear-thinning effect on bubble rising. The numerical results are also listed in Table 2-8. The gas-related physical properties of the systems are $\rho_g = 1.2$ kg m^{-3}, $\eta_g = 1.8 \times 10^{-5}$ Pa s, and $\sigma = 7.2 \times 10^{-2}$ N m^{-1}.

Figure 2-20. shows the viscosity distribution (non-dimensionalized by η_0) around a bubble for systems 1 through 6. The colorbar shown at the right of the figures depicts the range of reduced apparent viscosity around the bubble. The real shape of the bubble is also displayed in Figure 2-19. . System 1 corresponds to the bubble rising in a Newtonian fluid with the same zero shear-rate viscosity and the viscosity around the bubble is constant. As for systems 2 through 6, the viscosity around the bubble varies in correspondence to the bubble shape and the shear-thinning property of the liquid, and the degree of the decrease in viscosity is the largest in close vicinity to the bubble. The change in viscosity becomes gradually drastic from system 3 to system 6, showing that a confined region with high viscosity exists in the wake of these bubbles and it becomes finally detached from the bubble rear surface. As seen in

Table 2- 8, it is obvious that U_T (or Re_M) increases gradually due to the stronger shear-thinning effect. The bubble can rise faster due to the large extent of decrease in viscosity around the bubble. Obviously, the bubble shape in the shear-thinning non-Newtonian fluid differs much from that in the Newtonian case. As the shear-thinning effect becomes intensive gradually, the bubble takes the more oblate shape, with the front surface flatter than the rear part.

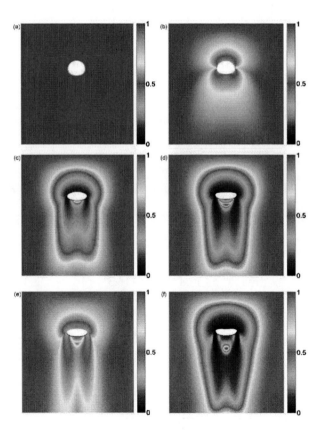

Figure 2-19. Viscosity distribution and shape of a bubble rising in a solution. (a) System 1: Newtonian fluid, $Re_M = 2.24$, $\eta_0 = 0.5$ Pa s. (b) System 2: Carreau shear-thinning fluid, $Re_M = 6.65$ $\eta_0 = 0.5$ Pa s. (c) System 3: Carreau shear-thinning fluid, $Re_M = 30.69$, $\eta_0 = 0.5$ Pa s. (d) System 4: Carreau shear-thinning fluid, $Re_M = 48.8$, $\eta_0 = 0.5$ Pa s. (e) System 5: Carreau shear-thinning fluid, $R_M = 55.5$, $\eta_0 = 0.5$ Pa s. (f) System 6: Carreau shear-thinning fluid, $Re_M = 80.85$, $\eta_0 = 0.5$ Pa s.

2.2.2. Mass Transfer to a Drop

The convective diffusion equation for a single particle can be numerically solved when the fluid dynamic is resolved. Numerical simulation of interphase mass transfer in a boundary-fitted orthogonal reference frame has been successfully adopted for mass transfer to/from a steady or accelerating drop in solvent extraction systems. The boundary-fitted orthogonal coordinate method [74] is used for the numerical simulation of interphase mass transfer at low Reynolds numbers. The level set approach [75] is extended to interphase mass transfer, but a variable transform and higher order numerical scheme are required.

The main difficulty in numerical simulation is that the motion of a deformed drop with simultaneous mass transfer must be solved with the unknown shape of free surface [76]. Numerical method must ensure accurate estimation of the interfacial position and the concentration so that the mass transfer calculation remains accurate. In recent years, several different methods for solving multiphase flow have been developed. Among them, the level set method is stable, surface geometric parameters such as curvature become easy to be calculated and three-dimensional calculation problems become easy to be implemented. The major shortcoming of the level set approach is less accurate treatment of interface compared with the boundary-fitted coordinate method.

2.2.2.1. Mass Transfer Dominated by External Resistance

In a boundary-fitted orthogonal coordinate system, the general governing equations for steady and transient axisymmetrical cases of mass transfer in the external region around single buoyancy-driven drops in steady motion under low or intermediate Reynolds numbers are expanded and numerically solved by the control volume formulation.

For the situation of mass transfer dominated by the external resistance, the partition coefficient of solute between the drop and the continuous phase is rather large, so that the solute concentration in the drop and at the surface can be reasonably assumed to be zero during the extraction. Considering that the concentration of solute at the remote boundary far away from the drop may be maintained at a constant c^{∞}, the mass transfer to/from the drop will eventually approach a steady state as the mass transfer proceeds. Thus, the following initial and boundary condition may be assumed for mass transfer from the continuous phase to drops (mass transfer direction c → d):

$$C(\xi, \eta, 0) = 0 \quad \text{at } \theta = 0 \tag{2-73}$$

$$C(1, \eta, \theta) = 0 \quad \text{at the drop surface, } \xi = 1 \tag{2-74}$$

$$C(0, \eta, \theta) = 1 \quad \text{at the remote boundary, } \xi = 0 \tag{2-75}$$

$$\frac{\partial C}{\partial \eta} = 0 \quad \text{on the axis of symmetry, } \eta = 0, \eta = 1 \tag{2-76}$$

As compared with the empirical correlation of external mass transfer for liquid drops with intermediate Reynolds numbers developed for the case of $Re > 70$ and $\lambda < 2$ by Harper and Moore[76],

$$Sh = \frac{2}{\sqrt{\pi}}\left[1 - \frac{1}{Re^{0.5}}(2.89 + 2.15\lambda^{0.64})\right]^{0.5} Pe^{0.5} \qquad (2\text{-}77)$$

good agreement in the range of $0.1 < Pe < 2000$ is evidenced in Figure 2-20. When Pe is very large, the transport is dominated by convection, this may lead to very large concentration gradient at the surface. Since the mesh density is limited at the interface, the large gradient of concentration as $D \to 0$ is not accounted for with sufficient accuracy, and hence, the discrepancy becomes obvious at high Pe region.

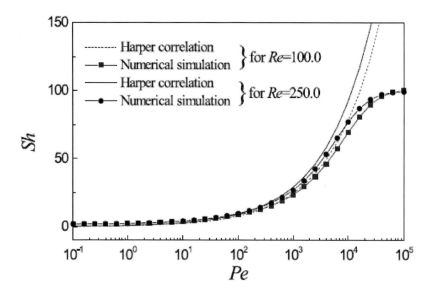

Figure 2-20. Comparison of numerical prediction and empirical correlation by Harper and Moore [74].

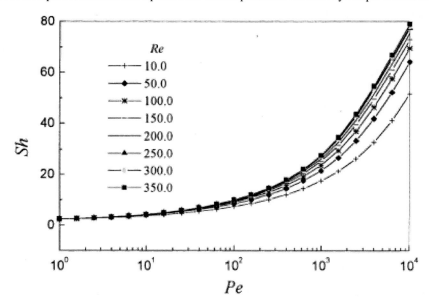

Figure 2-21. Sh for single drops in laminar flow regime as a function of Pe with different Reynolds number ($We = 4.0, \lambda = 5.0, \zeta = 2.0$).

It is easy to investigate the effect of relevant physical parameters on the average Sh by means of numerical experimentation. Figure 2-21 shows that the average Sh increases along with Pe at different Reynolds numbers. When Pe is small, the mass transfer is dominated by molecular diffusion, Sh is hence very close to the lower limit of $Sh = 2$. However, Re plays a significant role in promoting the rate of mass transfer when Pe is large, since the convective transport controls the rate process in this case. It should be kept in mind that when interpreting mass transfer for a given extraction system, Sc is predescribed as a constant, where Re and Pe $(= ReSc)$ work in a synchronous way, and higher Re indicates stronger convection and higher mass transfer rate.

2.2.2.2. Mass Transfer with Resistance in Both Phases

A level set approach is applied for simulating the interphase mass transfer of single drops in an immiscible liquid with resistance in both phases. In Figure 2-22, based on the comparison of different spatial discretization schemes such as the power-law, second-order ENO (essentially nonoscillatory), fifth-order WENO (weighted essentially nonoscillatory) and WENO coupled with ACM technique for computing the governing equations of mass transfer, the fifth-order WENO scheme performs better and is applied for the present simulation including the solution of the evolution and reinitialization equations of level set function.

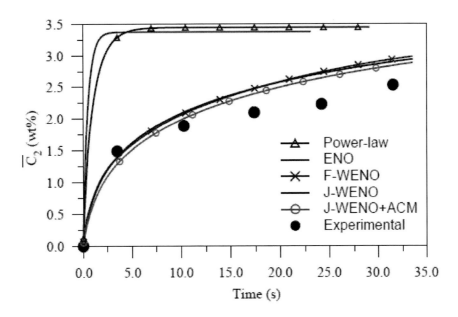

Figure 2-22. Comparison of predicted average drop concentrations for different spatial discretization schemes with experimental data.

The Reynolds number and time history of concentration and mass transfer coefficient are well predicted by this algorithm as verified against the reported experimental data of the *n*-butanol–succinic acid–water system recommended by EFCE. As shown in Figure 2-23, for example, the predicted drop average concentrations and overall mass transfer coefficients for mass transfer direction from the drop to the continuous phase is in reasonably good agreement with the corresponding experimental data. In Figure 2-23, although better agreement between computational and experimental Reynolds numbers is observed, most predicted overall mass transfer coefficients, including the result in a boundary-fitted orthogonal coordinate system, are slightly higher than the experiments. Li et al. [77] conjectured the experimental system might not be very pure, though serious care was taken in conducting experiments. Despite the fact that the boundary-fitted orthogonal coordinate method offers more accurate numerical results, it is very difficult to construct orthogonal curvilinear coordinates for highly deformed drops and be extended to the simulation of three-dimensional problems.

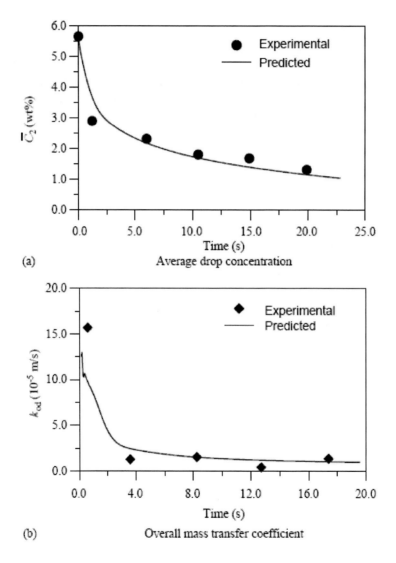

Figure 2-23. Predicted and experimental concentrations and overall mass transfer coefficients as a function of mass transfer time for serial No.BD16-1: (a) average drop concentration, (b) overall mass transfer coefficient ($d = 1.69$ mm, $\rho_1 = 988.2$ kg·m^{-3}, $\rho_2 = 841.1$ kg·m^{-3}, $\mu_1 = 1.44 \times 10^{-3}$ Pa·s, $\mu_2 = 3.34 \times 10^{-3}$ Pa·s, $\sigma = 1.00 \times 10^{-3}$ N·m^{-1}, $D_1 = 5.5 \times 10^{-10}$ m^2·s^{-1}, $D_2 = 2.1 \times 10^{-10}$ m^2·s^{-1}, $m = 1.16$, $C_1^0 = 0$, $C_2^0 = 5.66\%$; solute transfer direction: d → c).

Table 2-9. Comparison of predict and experimental Reynolds numbers and average overall mass transfer coefficients for *n*-butanol-succinic acid-water system (the same conditions as in Figure 2-22)

No.	Re_{exp}	Re_{pre}	$\bar{k}_{od,exp}$ $\times 10^5$ (m/s)	$\bar{k}_{od,pre}$ $\times 10^5$ (m/s) [10]	$\bar{k}_{od,pre}$ $\times 10^5$ (m/s) [75]
BC4-1	18.9	19.0	0.449	0.709	0.370
BC9-1	37.7	38.4	0.589	1.126	0.807
BC12-1	49.5	48.5	0.897	1.400	1.278
BC16-1	61.9	59.6	2.566	1.900	2.130
BD4-1	21.2	18.8	0.374	0.728	0.408
BD9-1	39.0	37.3	0.752	1.275	0.907
BD12-1	48.1	50.6	0.661	1.543	1.137
BD16-1	55.2	56.9	1.108	1.739	2.340

For two limiting cases of mass transfer into or out of single drops with resistance dominated only in the stagnant continuous phase or in the stagnant spherical drop, the numerical results are almost identical to the corresponding analytical solutions. The good agreements between the results of numerical simulation and those from experiments and theoretical analysis for the transient interphase mass transfer of a drop also confirm the correct jump conditions at the interface are satisfied by the necessary transformation making the concentration continuous. These numerical tests indicate the present level set approach is simple and effective in simulating the interphase mass transfer of single drops.

2.2.2.3. Interphase Mass Transfer to/from an Unsteady Drop

Figure 2-24 shows the predicted solute distribution in both phases and the velocity vector field relative to the drop motion at selected dimensionless times for Case A (in Table 2-10.). The results show the shape of the drop is gradually deformed into a spheroid from its initial spherical shape, and a toroidal recirculation is formed within the drop. In the center part of the drop the concentration of the solute progressively increases as it is swept to the downstream direction, suggesting that convective mass transfer does play an important role. The convection at the drop nose promotes the transport of solute to the drop.

Figure 2-25 shows the predicted solute distribution in both phases and the velocity vector field relative to the drop motion at different dimensionless time for Case B (Table 2-10.) (the direction of mass transfer: d → c). The results show that in the center part of the drop the solute concentration gradually decreases as it is swept by the recirculation towards the drop surface and then towards the rear of the drop here it enters the continuous phase as a plume. The plume is ore distinct at early time when the concentration of the drop is the highest.

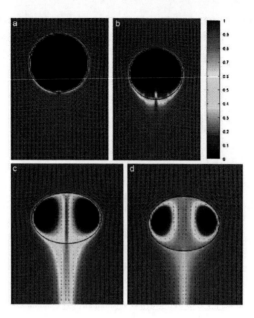

Figure 2-24. Predicted fractional solute concentration distribution and velocity vector field relative to the motion of the drop in Case A (Table 2-10.) at various dimensionless time: (a) $t = 0.02$ s ($v_d = 1.18$ cm s^{-1}), (b) $t = 0.04$ s, ($v_d = 1.65$ cm s^{-1}) concentration, (c) $t = 0.13$ s, ($v_d = 2.33$ cm s^{-1}) and (d) $t = 0.49$ s, ($v_d = 2.47$ cm s^{-1}).

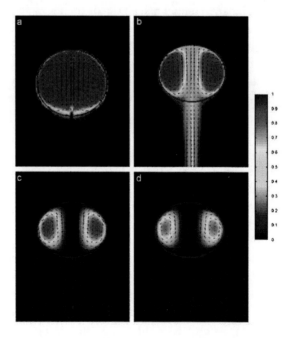

Figure 2-25. Predicted fractional solute concentration distribution and velocity vector field relative to the motion of the drop in Case B at various dimensionless times: (a) $t = 0.04$ s($v_d = 0.76$ cm s^{-1}), (b) $t = 0.14$ s, ($v_d = 1.75$ cm s^{-1}) concentration, (c) $t = 1.39$ s, ($v_d = 2.76$ cm s^{-1}) and (d) $t = 2.10$ s, ($v_d = 2.78$ cm s^{-1}).

Table 2-10. Relative deviations of predicted and experimental average overall mass transfer coefficients of the present study

Case	Experimental average $k_{od} \times 10^5$ (m/s)	Predicted average $k_{od} \times 10^5$ (m/s)	Deviation (%)
A	4.57	2.16	16.41
	3.23	2.75	11.99
	2.90	3.67	5.17
	1.99	3.82	7.87
B	3.23	4.31	25.53
	2.81	3.51	19.94
	2.59	2.80	7.5
	2.54	2.53	0.39
	2.36	2.31	2.12

3. NUMERICAL SIMULATION OF MULTIPHASE FLOW IN CHEMICAL REACTORS

Multiphase flows (including two-and three-phase flows) are discussed in detail based on numerical methods using the Eulerian multi-fluid approach, k-ε turbulence model and large eddy simulation. Good agreement between predictions and experiments indicates that CFD has become a powerful tool in the research of chemical reactors such as stirred vessels. To account for the anisotropy of turbulence in the flow field, the algebraic stress model and large eddy simulation are adopted in the numerical study of hydrodynamics in multiphase stirred tanks as a result of their capacity on retrieving turbulent properties more accurately.

3.1. Mathematical Model and Numerical Method

3.1.1. Mathematical Model

In 1982, Harvey et al. [78] attempted two-dimensional numerical simulation of a stirred tank. Since then, numerical simulation has developed very quickly. A plenty of literature reports demonstrate that the numerical simulation of single phase flow is generally successful. When a second phase is introduced, the mathematical treatment becomes more complicated. Generally, there are two major approaches for modeling two-phase flow, differing from each other by treatment of the dispersed phase, the Eulerian-Eulerian and the Eulerian-Lagrangian approaches. In the Eulerian-Lagrangian approach the trajectory of each discrete particle is calculated by its motion equation subject to respective initial conditions and exerted force, while the motion of the continuous phase is followed in a fixed Eulerian grid. On the other hand, in the Eulerian-Eulerian approach the dispersed phase is also treated as a continuum.

In most works, the "two-fluid" approach based on a Eulerian-Eulerian approach [79] and the two-phase k-ε turbulence model are usually employed to simulate two-phase flow. In this approach, there are some assumptions: (1) the continuous phase and dispersed phase are considered to be interpenetrating continua with interaction; (2) the pressure field of the system is assumed to be the same for two phases. With such considerations, the time averaged governing equations of momentum balance for phase k are

$$\frac{\partial(\rho_k \alpha_k)}{\partial t} + \frac{\partial\left(\rho_k \alpha_k u_{kj} + \rho_k \overline{\alpha'_k u'_{kj}}\right)}{\partial x_j} = 0 \tag{3-1}$$

$$\rho_k \frac{\partial(\alpha_k u_{ki})}{\partial t} + \rho_k \frac{\partial(\alpha_k u_{ki} u_{kj})}{\partial x_j} = -\alpha_k \frac{\partial P}{\partial x_i} + \frac{\partial(\alpha_k \overline{\tau_{kij}})}{\partial x_j} + F_{ki} + \rho_k \alpha_k g_i -$$
$$\rho k \frac{\partial}{\partial x_j}(\alpha_k \overline{u'_{kj} u'_{ki}} + u_{ki} \overline{\alpha'_k u'_{kj}} + u_{kj} \overline{\alpha'_k u'_{ki}} + \overline{\alpha'_k u'_{kj} u'_{ki}}) \tag{3-2}$$

$$\sum \alpha_k = 1.0 \tag{3-3}$$

where α_k is the phase volume fraction (or holdup) and F_{ki} is the interphase momentum exchange term.

For the closure of momentum equations, the turbulent fluctuation correlation terms should be related to the known or calculable quantities via either algebraic or differential equations. The triple and higher order correlations are usually omitted and the viscous shear terms are also neglected compared with the turbulent shear terms. With the Boussinesq gradient transport hypothesis, the velocity correlations, named the Reynolds stresses, are modeled following the practice of single-phase flow as

$$\overline{u'_{ki} u'_{kj}} = \frac{2}{3} k \delta_{ij} - \nu_{k,t} (\frac{\partial u_{kj}}{\partial x_i} + \frac{\partial u_{ki}}{\partial x_j}) \tag{3-4}$$

The correlation between velocity fluctuations and phase holdup fluctuation $\overline{u'_{ki} \alpha'_k}$, which appears in both continuity and momentum equations due to the presence of a second phase, represents the transport of both mass and momentum respectively by dispersion. However, it has not been included in the mathematical model [90,91]. The simplest way to model $\overline{u'_{ki} \alpha'_k}$ is to assume the same gradient transport, which gives

$$\overline{u'_{ki} \alpha'_k} = -\frac{\nu_{k,t}}{\sigma_t} \frac{\partial \alpha_k}{\partial x_i} \tag{3-5}$$

where σ_t is the turbulent Schmidt number for phase dispersion, whose value is related with the scale of turbulent flow, but few reports are available regarding the systematical experiments and theoretical models on σ_t. It was found that the simulation results were sensitive to σ_t in solid-liquid simulation and the value between 1.0 and 2.0 was suggested [61]. In gas-liquid systems, the value of 1.0 was recommended [81, 82], but Wang and Mao [83] suggested the value of 1.6.

The final closure momentum equation reads

$$\rho_k \left[\frac{\partial(\alpha_k u_{ki})}{\partial t} + \frac{\partial(\alpha_k u_{ki} u_{kj})}{\partial x_j} \right] = -\alpha_k \frac{\partial P}{\partial x_i} + F_{ki} + \rho_k \alpha_k g_i + \frac{\partial}{\partial x_j} \left[\alpha_k \mu_{k,\text{eff}} \left(\frac{\partial u_{ki}}{\partial x_j} + \frac{\partial u_{kj}}{\partial x_i} \right) \right] + \frac{\partial}{\partial x_j} \left[\frac{\mu_{k,t}}{\sigma_t} \left(u_{ki} \frac{\partial \alpha_k}{\partial x_j} + u_{kj} \frac{\partial \alpha_k}{\partial x_i} \right) \right] \qquad (3\text{-}6)$$

For the LES, the space-filtered equations for the conservation of mass and momentum of an incompressible Newtonian fluid can be written as, after the terms with $\overline{u_{mi} u_{mj}}$ are formulated by a LES subgrid model,

$$\frac{\partial(\rho_k \bar{\alpha}_k)}{\partial t} + \frac{\partial}{\partial x_j}(\rho_k \overline{\alpha_k u_{ki}}) = 0 \qquad (3\text{-}7)$$

$$\frac{\partial}{\partial t}(\rho_k \bar{\alpha}_k \bar{u}_{ki}) + \frac{\partial}{\partial x_j}(\rho_k \bar{\alpha}_k \bar{u}_{ki} \bar{u}_{kj}) = -\frac{\partial(\bar{p})}{\partial x_i} + \rho_k \bar{\alpha}_k g_i + \bar{F}_{ki} + \frac{\partial}{\partial x_j}\left(\mu_k \left(\frac{\partial(\bar{\alpha}_k \bar{u}_{kj})}{\partial x_i} + \right.\right.$$
$$\partial \alpha mumi \partial xj - \partial \rho k \alpha k \tau kij \partial xj \qquad (3\text{-}8)$$

$$\tau_{kij} = \overline{u_{ki} u_{kj}} - \bar{u}_{ki} \bar{u}_{kj} \qquad (3\text{-}9)$$

where τ_{kij} is the subgrid scale stress tensor, which reflects the effect of the unresolved scales on the resolved scales. The effect of the sub-grid scales on the large scales of the liquid phase is accounted for based on the standard Smagorinsky model:

$$\tau_{lij} - \frac{1}{3\tau_{lkk}} = -(c_s \Delta)^2 |\bar{S}_l| \bar{S}_{lij} \qquad (3\text{-}10)$$

The effective gas viscosity is calculated based on the following formula:

$$\mu_{eff,g} = \frac{\rho_g}{\rho_l} \mu_{eff,l} \qquad (3\text{-}11)$$

The interphase coupling terms make two-phase flow fundamentally different from single-phase flow. The interphase interaction term, F_{ki}, satisfies the following relationship:

$$F_{c,i} = -F_{d,i} \tag{3-12}$$

where subscripts c and d denote continuous and dispersed phases, respectively. For two-phase flow, F_{ki} is modeled as a linear combination of some terms, i.e., interphase drag force, virtual mass force, Basset force and lift force. In most cases, the magnitude of the Basset force and lift force is much smaller than that of the interphase drag force. A report [84] indicated that the effect of virtual mass force is not significant in the bulk region of a stirred tank. Therefore, only the interphase drag force is always considered in two-phase flows. In addition, in a non-inertial reference frame rotating with the impeller axis, the centrifugal force, $\mathbf{F}_{r,k}$, and the Coriolis force, $\mathbf{F}_{c,k}$, must be included:

$$\mathbf{F}_{r,k} = \alpha_k \rho_k (\boldsymbol{\omega} \times \mathbf{r}) \times \boldsymbol{\omega} \tag{3-13}$$

$$\mathbf{F}_{c,k} = 2\alpha_k \rho_k (\boldsymbol{\omega} \times \mathbf{u}_k) \tag{3-14}$$

In solid-liquid systems, the expression of drag force is usually as follows:

$$F_{c,\text{drag}} = \frac{3}{4} \alpha_c \alpha_d C_D \frac{|\mathbf{u}_d - \mathbf{u}_c|(u_{d,i} - u_{c,i})}{d_d} \tag{3-15}$$

$$F_{d,\text{drag}} = -F_{c,\text{drag}} \tag{3-16}$$

where C_D, the interphase drag coefficient, is a complex function dependent on the dispersed phase holdup and the turbulence:

$$C_D = \frac{24(1 + 0.15 Re_d^{0.687})}{Re_d} \quad \text{(for } Re_d < 1000\text{)} \tag{3-17}$$

$$C_D = 0.44 \quad \text{(for } Re_d \geq 1000\text{)} \tag{3-18}$$

in which Re_d is the Reynolds number of the particles:

$$Re_d = \frac{d_d |\mathbf{u}_d - \mathbf{u}_c| \rho_c}{\mu_{c,\text{lam}}} \tag{3-19}$$

In gas-liquid flow, the interphase drag force on a bubble in the control volume is

$$F_n = \frac{1}{2}\rho_c C_D A_n |\mathbf{u}_c - \mathbf{u}_d|(u_c - u_d) \tag{3-20}$$

where A_n is the projected area of the bubble and C_D is the drag coefficient. The Ishii's expression [85] for C_D, which takes account of bubble-bubble interaction and bubble distortion, can be used here. Assuming that bubbles have the same u_c-u_d in an infinitesimal unit volume, then

$$C_D = \frac{4}{3}r_b\sqrt{\frac{g\Delta\rho}{\sigma}}(1-\alpha_d)^{-0.5} \tag{3-21}$$

The bubble shape is assumed to be spherical, and Eq.(3-20) becomes

$$F_n = \frac{3}{4}\rho_c C_D \frac{V_n}{d_{b,n}}|\mathbf{u}_c - \mathbf{u}_d|(u_c - u_d) \tag{3-22}$$

Substituting C_D into the above equation, then it reads

$$F_n = \frac{1}{2}\rho_c\sqrt{\frac{g\Delta\rho}{\sigma}}(1-\alpha_d)^{-0.5}|\mathbf{u}_c - \mathbf{u}_d|(u_c - u_d)V_n \tag{3-23}$$

The total interphase force in the control volume becomes

$$F_d = \sum_{n=1}^{N_b} F_n = \frac{1}{2}\rho_c\sqrt{\frac{g\Delta\rho}{\sigma}}(1-\alpha_d)^{-0.5}|\mathbf{u}_c - \mathbf{u}_d|(u_c - u_d)\sum_{n=1}^{N_b} V_n \tag{3-24}$$

where N_b is the bubble number in the control volume. Finally, the force component in the ith coordinate direction per unit volume reads

$$F_{d,i} = \frac{1}{2}\alpha_d\rho_c\sqrt{\frac{g\Delta\rho}{\sigma}}(1-\alpha_d)^{-0.5}|\mathbf{u}_c - \mathbf{u}_d|(u_{c,i} - u_{d,i}) \tag{3-25}$$

It should be noted that the interphase drag force becomes independent of bubble size.

To treat the turbulent two-phase flow rigorously, the turbulent model adopted should include interphase turbulence transfer terms accounting for the turbulence promotion or damping due to the presence of the dispersed phase. However, there is no reliable information on such terms, and the proper turbulence model for turbulent multiphase systems is absent. In two-phase stirred tanks, the turbulence is mainly attributed to velocity fluctuation of liquid phase because the holdup of the dispersed phase is often quite low in most parts of the tank. The dispersed phase can affect the turbulence of the system through interphase momentum exchange. The liquid phase turbulence effect is modeled using a suitable extension of the standard k-ε turbulence model, written in a general form as [79].

$$\frac{\partial}{\partial t}(\rho_c \alpha_c k) + \frac{\partial}{\partial x_i}(\rho_c \alpha_c u_{ci} k) = \frac{\partial}{\partial x_i}\left(\alpha_c \frac{\mu_{ct}}{\sigma_k}\frac{\partial k}{\partial x_i}\right) + \frac{\partial}{\partial x_i}\left(k\frac{\mu_{ct}}{\sigma_k}\frac{\partial \alpha_c}{\partial x_i}\right) + S_k \quad (3\text{-}26)$$

$$\frac{\partial}{\partial t}(\rho_c \alpha_c \varepsilon) + \frac{\partial}{\partial x_i}(\rho_c \alpha_c u_{ci} \varepsilon) = \frac{\partial}{\partial x_i}\left(\alpha_c \frac{\mu_{ct}}{\sigma_\varepsilon}\frac{\partial \varepsilon}{\partial x_i}\right) + \frac{\partial}{\partial x_i}\left(\varepsilon\frac{\mu_{ct}}{\sigma_\varepsilon}\frac{\partial \alpha_c}{\partial x_i}\right) + S_\varepsilon \quad (3\text{-}27)$$

in which the values of the Schmidt number are given as $\sigma_k=1.3$ and $\sigma_\varepsilon=1.0$. The source terms in the above equations are

$$S_k = \alpha_c\left[(G+G_e) - \rho_c \varepsilon\right] \quad (3\text{-}28)$$

$$S_\varepsilon = \alpha_c \frac{\varepsilon}{k}\left[C_1(G+G_e) - C_2 \rho_c \varepsilon\right] \quad (3\text{-}29)$$

where G is the turbulent generation and G_e is the extra dissipation term due to the dispersion phase. Based on the analysis of Kataoka et al.[86], G_e is mainly dependent on the drag force between the continuous phase and the dispersed phase:

$$G = -\rho_c \alpha_c \overline{u'_{ci} u'_{cj}} \frac{\partial u_{ci}}{\partial x_j} \quad (3\text{-}30)$$

$$G_e = \sum_d C_b |\mathbf{F}| \left(\sum (u_{di} - u_{ci})^2\right)^{0.5} \quad (3\text{-}31)$$

where C_b is an empirical coefficient. When $C_b=0$, the energy induced by the dispersed phase dissipates at the interface and has no influence on the turbulent kinetic energy of the continuous phase. According to the analysis in the literature, the value of C_b has been always set at 0.02 or 0.03.

The reference values of the model constants are the well-accepted ones: $C_\mu=0.09$, $C_1=1.44$ and $C_2=1.92$. Another value of $C_2=1.60$ [87] is also suggested. In consideration of the strong vortex in the discharge zone, C_1 is always modified as follows:

$$C_1 = 1.44 + \frac{0.8 R_f \rho_c \varepsilon}{(G+G_e)} \quad (3\text{-}32)$$

$$R_f = \begin{cases} \dfrac{1}{\varepsilon}\left[\overline{u'_{c,r} u'_{c,\theta}} r \dfrac{\partial}{\partial r}\left(\dfrac{u_{c,\theta}}{r}\right)\right], & C-1.5w < z < C+1.5w \\ 0, & z < C-1.5w,\, z > C+1.5w \end{cases} \quad (3\text{-}33)$$

where C is the clearance between impeller and tank bottom and w represents the axial width of blades.

It is known that the flow in stirred tanks is unsteady because of the interaction of the rotating impeller blades with the stationary baffles. However, the flow pattern will become axisymmetrically repeating once it has fully developed. A snapshot of this flow can describe the flow within the impeller blades at this particular instant. Ranade and van den Akker [81] suggested that the time derivative terms in the governing equations can be ignored without much error in most regions of the tank except for the impeller swept volume. The flow field in the impeller swept volume was simulated in a non-inertial reference frame rotating with the impeller. Therefore, the temporal terms in the equations can be omitted. In this way, the resulted model formulation of the mass and momentum conservation equations for phase k in a general form in the cylindrical coordinate system reads:

$$\frac{1}{r}\frac{\partial}{\partial r}\left(\rho_k r \alpha_k u_{kr} \varphi\right) + \frac{1}{r}\frac{\partial}{\partial \theta}\left(\rho_k \alpha_k u_{kq} \varphi\right) + \frac{\partial}{\partial z}\left(\rho_k \alpha_k u_{kz} \varphi\right)$$
$$= \frac{1}{r}\frac{\partial}{\partial r}\left(\alpha_k \Gamma_{\varphi,\text{eff}} r \frac{\partial \varphi}{\partial r}\right) + \frac{1}{r}\frac{\partial}{\partial \theta}\left(\frac{\alpha_k \Gamma_{\varphi,\text{eff}}}{r}\frac{\partial \varphi}{\partial \theta}\right) + \frac{\partial}{\partial z}\left(\alpha_k \Gamma_{\varphi,\text{eff}}\frac{\partial \varphi}{\partial z}\right) + S_\varphi \tag{3-34}$$

The governing equations can be summarized in Table 3-1, with

$$G = 2\mu_{ct}\left[\left(\frac{\partial u_{cr}}{\partial r}\right)^2 + \left(\frac{1}{r}\frac{\partial u_{c\theta}}{\partial \theta} + \frac{u_{cr}}{r}\right)^2 + \left(\frac{\partial u_{cz}}{\partial z}\right)^2\right]$$
$$+ \mu_{ct}\left[r\frac{\partial}{\partial r}\left(\frac{u_{c\theta}}{r}\right) + \frac{1}{r}\frac{\partial u_{cr}}{\partial \theta}\right]^2 + \mu_{ct}\left[\frac{1}{r}\frac{\partial u_{cz}}{\partial \theta} + \frac{\partial u_{c\theta}}{\partial z}\right]^2 \tag{3-35}$$
$$+ \mu_{ct}\left[\frac{\partial u_{cr}}{\partial z} + \frac{\partial u_{cz}}{\partial r}\right]^2$$

$$\mu_{c,\text{eff}} = \mu_{c,t} + \mu_{c,\text{lam}} \tag{3-36}$$

$$\mu_{c,t} = \frac{C_\mu k^2 \rho_c}{\varepsilon} \tag{3-37}$$

$$\mu_{d,\text{eff}} = \mu_{d,t} + \mu_{d,\text{lam}} \tag{3-38}$$

Table 3-1. Source terms and diffusion coefficients in the equation

Equation	ϕ	$\Gamma_{\varphi,\text{eff}}$	S_ϕ
Continuity	1	0	$\dfrac{1}{r}\dfrac{\partial}{\partial r}\left(r\dfrac{\mu_{k,t}}{\sigma_t}\dfrac{\partial \alpha_k}{\partial r}\right)+\dfrac{1}{r}\dfrac{\partial}{\partial \theta}\left(\dfrac{\mu_{k,t}}{\sigma_t}\dfrac{\partial \alpha_k}{r\partial \theta}\right)+\dfrac{\partial}{\partial z}\left(\dfrac{\mu_{k,t}}{\sigma_t}\dfrac{\partial \alpha_k}{\partial z}\right)$
Radial momentum	$u_{k,r}$	$\mu_{k,\text{eff}}$	$\dfrac{1}{r}\dfrac{\partial}{\partial r}\left(\alpha_k\mu_{k,\text{eff}}r\dfrac{\partial u_{k,r}}{\partial r}\right)+\dfrac{1}{r}\dfrac{\partial}{\partial \theta}\left(\alpha_k\mu_{k,\text{eff}}r\dfrac{\partial}{\partial r}\left(\dfrac{u_{k,\theta}}{r}\right)\right)+\dfrac{\partial}{\partial z}\left(\alpha_k\mu_{k,\text{eff}}\dfrac{\partial u_{k,z}}{\partial r}\right)$ $+\dfrac{\partial}{\partial z}\left(u_{k,r}\dfrac{\mu_{k,t}}{\sigma_t}\dfrac{\partial \alpha_k}{\partial z}\right)+\dfrac{1}{r}\dfrac{\partial}{\partial r}\left(ru_{k,r}\dfrac{\mu_{k,t}}{\sigma_t}\dfrac{\partial \alpha_k}{\partial r}\right)+\dfrac{1}{r^2}\dfrac{\partial}{\partial \theta}\left(u_{k,r}\dfrac{\mu_{k,t}}{\sigma_t}\dfrac{\partial \alpha_k}{\partial \theta}\right)$ $-\dfrac{2\alpha_k\mu_{k,\text{eff}}}{r^2}\dfrac{\partial u_{k,\theta}}{\partial \theta}-\dfrac{2\alpha_k\mu_{k,\text{eff}}u_{k,r}}{r^2}+\dfrac{\rho_k\alpha_k u_{k,\theta}^2}{r}$ $+\dfrac{\partial}{\partial z}\left(u_{k,r}\dfrac{\mu_{k,t}}{\sigma_t}\dfrac{\partial \alpha_k}{\partial z}\right)+\dfrac{1}{r}\dfrac{\partial}{\partial r}\left(ru_{k,r}\dfrac{\mu_{k,t}}{\sigma_t}\dfrac{\partial \alpha_k}{\partial r}\right)+\dfrac{1}{r^2}\dfrac{\partial}{\partial \theta}\left(u_{k,r}\dfrac{\mu_{k,t}}{\sigma_t}\dfrac{\partial \alpha_k}{\partial \theta}\right)$ $+\dfrac{\partial}{\partial z}\left(u_{k,z}\dfrac{\mu_{k,t}}{\sigma_t}\dfrac{\partial \alpha_k}{\partial r}\right)+\dfrac{1}{r}\dfrac{\partial}{\partial r}\left(ru_{k,r}\dfrac{\mu_{k,t}}{\sigma_t}\dfrac{\partial \alpha_k}{\partial r}\right)+\dfrac{1}{r}\dfrac{\partial}{\partial \theta}\left(u_{k,\theta}\dfrac{\mu_{k,t}}{\sigma_t}\dfrac{\partial \alpha_k}{\partial r}\right)$ $-\dfrac{2}{r^2}\left(u_{k,\theta}\dfrac{\mu_{k,t}}{\sigma_t}\dfrac{\partial \alpha_k}{\partial \theta}\right)$ $-\alpha_k\dfrac{\partial p}{\partial r}+F_{k,r}\{+\rho_k\alpha_k(\omega^2 r+2\omega u_\theta)\}-\rho_k\dfrac{2}{3}\dfrac{\partial(\alpha_k k)}{\partial r}$
Azimuthal momentum	$u_{k,\theta}$	$\mu_{k,\text{eff}}$	$\dfrac{1}{r}\dfrac{\partial}{\partial r}\left(\alpha_k\mu_{k,\text{eff}}\dfrac{\partial u_{k,r}}{\partial \theta}\right)+\dfrac{1}{r}\dfrac{\partial}{\partial \theta}\left(\dfrac{\alpha_k\mu_{k,\text{eff}}}{r}\dfrac{\partial u_{k,\theta}}{\partial \theta}\right)+\dfrac{1}{r}\dfrac{\partial}{\partial \theta}\left(2\alpha_k\mu_{k,\text{eff}}\dfrac{u_{k,r}}{r}\right)+\dfrac{\partial}{\partial z}\left(\dfrac{\alpha_k\mu_{k,\text{eff}}}{r}\dfrac{\partial u_{k,z}}{\partial \theta}\right)$ $+\alpha_k\mu_{k,\text{eff}}\dfrac{\partial}{\partial r}\left(\dfrac{u_{k,\theta}}{r}\right)-\dfrac{1}{r}\dfrac{\partial}{\partial r}(\alpha_k\mu_{k,\text{eff}}u_{k,\theta})-\dfrac{\rho_k\alpha_k u_{k,r}u_{k,\theta}}{r}+\dfrac{\alpha_k\mu_{k,\text{eff}}}{r^2}\dfrac{\partial u_{k,r}}{\partial \theta}$ $+\dfrac{\partial}{\partial z}\left(u_{k,\theta}\dfrac{\mu_{k,t}}{\sigma_t}\dfrac{\partial \alpha_k}{\partial z}\right)+\dfrac{1}{r}\dfrac{\partial}{\partial r}\left(ru_{k,\theta}\dfrac{\mu_{k,t}}{\sigma_t}\dfrac{\partial \alpha_k}{\partial r}\right)+\dfrac{1}{r^2}\dfrac{\partial}{\partial \theta}\left(u_{k,\theta}\dfrac{\mu_{k,t}}{\sigma_t}\dfrac{\partial \alpha_k}{\partial \theta}\right)$ $+\dfrac{\partial}{\partial z}\left(u_{k,z}\dfrac{\mu_{k,t}}{\sigma_t}\dfrac{\partial \alpha_k}{r\partial \theta}\right)+\dfrac{1}{r}\dfrac{\partial}{\partial r}\left(ru_{k,r}\dfrac{\mu_{k,t}}{\sigma_t}\dfrac{\partial \alpha_k}{r\partial \theta}\right)+\dfrac{1}{r^2}\dfrac{\partial}{\partial \theta}(u_{k,\theta}\dfrac{\mu_{k,t}}{\sigma_t}\dfrac{\partial \alpha_k}{\partial \theta})$ $+\dfrac{u_{k,\theta}}{r}\dfrac{\mu_{k,t}}{\sigma_t}\dfrac{\partial \alpha_k}{\partial r}+\dfrac{u_{k,r}}{r^2}\dfrac{\mu_{k,t}}{\sigma_t}\dfrac{\partial \alpha_k}{\partial \theta}$ $-\dfrac{\alpha_k}{r}\dfrac{\partial p}{\partial \theta}+F_{k,\theta}\{+\rho\alpha_k(-2\omega u_{k,r})\}-\rho_k\dfrac{2}{3}\dfrac{1}{r}\dfrac{\partial(\alpha_k k)}{\partial \theta}$
Axial momentum	$u_{k,z}$	$\mu_{k,\text{eff}}$	$\dfrac{1}{r}\dfrac{\partial}{\partial r}\left(\alpha_k\mu_{k,\text{eff}}r\dfrac{\partial u_{k,r}}{\partial z}\right)+\dfrac{1}{r}\dfrac{\partial}{\partial \theta}\left(\alpha_k\mu_{k,\text{eff}}\dfrac{\partial u_{k,\theta}}{\partial z}\right)+\dfrac{\partial}{\partial z}\left(\alpha_k\mu_{k,\text{eff}}\dfrac{\partial u_{k,z}}{\partial z}\right)$ $+\dfrac{\partial}{\partial z}\left(u_{k,z}\dfrac{\mu_{k,t}}{\sigma_t}\dfrac{\partial \alpha_k}{\partial z}\right)+\dfrac{1}{r}\dfrac{\partial}{\partial r}\left(ru_{k,z}\dfrac{\mu_{k,t}}{\sigma_t}\dfrac{\partial \alpha_k}{\partial r}\right)+\dfrac{1}{r^2}\dfrac{\partial}{\partial \theta}\left(u_{k,z}\dfrac{\mu_{k,t}}{\sigma_t}\dfrac{\partial \alpha_k}{\partial \theta}\right)$ $+\dfrac{\partial}{\partial z}\left(u_{k,z}\dfrac{\mu_{k,t}}{\sigma_t}\dfrac{\partial \alpha_k}{\partial z}\right)+\dfrac{1}{r}\dfrac{\partial}{\partial r}\left(ru_{k,r}\dfrac{\mu_{k,t}}{\sigma_t}\dfrac{\partial \alpha_k}{\partial z}\right)+\dfrac{1}{r}\dfrac{\partial}{\partial \theta}\left(u_{k,\theta}\dfrac{\mu_{k,t}}{\sigma_t}\dfrac{\partial \alpha_k}{\partial z}\right)$ $-\alpha_k\dfrac{\partial p}{\partial z}+F_{k,z}-\rho_k\alpha_k g-\rho_k\dfrac{2}{3}\dfrac{\partial(\alpha_k k)}{\partial z}$
Turbulent kinetic energy	k	$\dfrac{\mu_{ct}}{\sigma_k}$	$\alpha_c[(G+G_e)-\rho_c\varepsilon]$
Turbulent energy dissipation	ε	$\dfrac{\mu_{ct}}{\sigma_\varepsilon}$	$\alpha_c\dfrac{\varepsilon}{k}[C_1(G+G_e)-C_2\rho_c\varepsilon]$

Note: the terms in the curly brackets are present only when a non-inertial reference frame is used

The formulations for $\mu_{d,t}$ take the following forms in different systems:

[1] In gas-liquid systems, $\mu_{d,t}$ is

$$\mu_{d,t} = \rho_d \nu_{d,t} \tag{3-39}$$

where $\nu_{d,t}$ is given by the Hinze-Tchen formulation:

$$\frac{\nu_{d,t}}{\nu_{c,t}} = \frac{1}{1+\dfrac{\tau_{r,k}}{\tau_t}} \tag{3-40}$$

with τ_t being the turbulence fluctuation time:

$$\tau_t = 1.5^{0.5} \frac{k}{\varepsilon} C_\mu^{0.75} \tag{3-41}$$

and $\tau_{r,k}$ the bubble relaxation time:

$$\tau_{r,k} = \frac{\rho_d d_b^2}{18\mu_c} \tag{3-42}$$

The numerical value of the term $\tau_{r,k}/\tau_t$ is about order 10^{-2} in most parts of a stirred tank, so d_b can be approximately replaced by the Sauter mean diameter d_{32} without introducing much error. Some investigators estimated the ratio between d_{32} and d_{max} for bubbles. Hesketh et al. [88] proposed the ratio between d_{32} and d_{max} to be 0.62 for the bubble breakup in a horizontal pipe flow. Parthasarthy and Ahmed [89] did experiments on different positions in a gas-liquid stirred tank, and found the value of d_{32}/d_{max} for bubbles to be 0.785. Various d_{32}/d_{max} values have also been reported for droplets in liquid-liquid dispersions. Brown and Pitt [90] found a linear relationship between d_{32} and d_{max} in the kerosene-water system in a stirred tank and found the value of d_{32}/d_{max} to be 0.7. Calabrese et al. [91] reported a value of 0.6 for the dispersion of moderate viscosity oils in a stirred vessel. It ought to be noted that despite the diversity of systems, the reported d_{32}/d_{max} values do not show considerable scatter. An average value of 0.68 is resulted for the above mentioned values and is often used. The d_{max} can be obtained by the next equation:

$$d_{max} = 0.725 \left(\frac{\sigma}{\rho_c}\right)^{0.6} \varepsilon^{-0.4} \tag{3-43}$$

[2] In solid-liquid systems, the turbulent viscosity of the dispersed phase, $\mu_{d,t}$ is expressed in terms of its counterpart of the continuous phase as

$$\mu_{d,t} = K\mu_{c,t} \qquad (3\text{-}44)$$

$$K = \frac{\rho_d \overline{u'_{d,i} u'_{d,i}}}{\rho_c \overline{u'_{c,i} u'_{c,i}}} \qquad (3\text{-}45)$$

Gosman et al. [92] proposed a correlation of u'_d to u'_c derived from a Lagrangian analysis of particle response to eddies which are much larger than the particle diameter:

$$u'_{d,i} = u'_{c,i}\left[1 - \exp(-\frac{t_1}{t_p})\right] \qquad (3\text{-}46)$$

with $t_1 = 0.41 k/\varepsilon$ being the mean eddy lifetime, and t_p, the particle response time, obtained by the Lagrangian integration of the equation of motion of a swarm of particles moving through a fluid eddy of given velocity distribution with the expression:

$$t_p = \frac{4\rho_d d_d}{3\rho_c C_D \alpha_d |\mathbf{u}_d - \mathbf{u}_c|} \qquad (3\text{-}47)$$

The laminar viscosity of the solid phase is not directly available from the literature, and the determination of the laminar viscosity coefficient $\mu_{d,\,lam}$ of the dispersion phase is a hot topic, when studying two-phase flow, especially in gas-liquid fluidized beds. In solid-liquid flow simulations, $\mu_{d,\,lam}$ is set as a parameter, varying between 10^{-4} Pa·s and 10^{-2} Pa·s [78]. The influence of $\mu_{d,\,lam}$ on the predicted local solid concentration is very small. Therefore, $\mu_{d,\,lam}$ is always set to be that of the continuous phase laminar viscosity, namely $\mu_{d,\,lam} = \mu_{c,\,lam} = 10^{-3}$ Pa·s.

3.1.2. Numerical Method

There are some difficulties in resolving the complex 3-D turbulent flow in a baffled stirred tank as a result of the rotating impeller, which makes the treatment of the impeller region critical. Gosman et al. [90] calculated the flow in solid-liquid stirred tanks treating the Rushton impeller region as a 'black box', hence the experimental data had to be imposed on the surface swept by the impeller blades as boundary conditions. The obvious shortcoming of such an approach is that the experimental data in this region is crucially needed for initiating the simulation and it is not applicable to novel operation conditions without experimental backup. Micale et al. [93] and Montante et al. [94] applied the inner-outer (IO) iterative procedure developed by Brauto et al. [95] to simulate the flow in solid-liquid stirred vessels. In this approach, the whole vessel volume is subdivided into two partly overlapping zones: an 'inner' domain, containing the impeller, and an 'outer' one, extending from the inner region border to the wall with baffles. The 'inner' and 'outer' simulations are conducted separately under the steady-state assumption in their own reference frames. Since the two frames are different, the information (velocity, turbulence energy and dissipation) is iteratively exchanged on the overlapping region, and is corrected for the relative motion and averaged over the azimuthal direction so as to provide the boundary conditions for the solution of the other domain.

The inner-outer-iterative method does not need experimental data as the impeller region boundary conditions. However, the information on the surfaces of the 'inner' and 'outer' domains is averaged over the azimuthal direction, thus some important features for the flow in the stirred vessel generated by the periodical rotation of the impeller are ignored. Wang and Mao [83] improved the inner-outer iterative procedure by combining it with a snapshot approach proposed by Ranade and van den Akker [81] to keep the unsteady turbulent properties not averaged and applied it to simulate the flow in single-phase and gas-liquid stirred tanks. In this improved approach, keeping the unsteady turbulent properties enables the simulation of the flow parameters with better accuracy.

Since the reference frames in the 'inner' and 'outer' domains are different, the values of flow parameters have to be converted when exchanging between two domains. During the simulation, the parameters are the same as those in the inertial frame, except the tangential velocity in the non-inertial frame. The latter is transformed back onto the inertial frame by adding the product of the angular velocity of the frame and the radial position:

$$u_{k,\theta,\text{in}} = u_{k,\theta,\text{out}} + \omega r \tag{3-48}$$

where $u_{k,\theta,\text{in}}$ is the tangential velocity in the inertial frame and $u_{k,\theta,\text{out}}$ is the tangential velocity in the non-inertial frame.

A crucial feature of the IO approach is the existence of an overlapping region, which requires the iterative match of the two solutions. The width of this region and the exact location of its boundaries are largely arbitrary. On the contrary, in the 'multi frame of reference' (MFR) method by Luo et al. [96], the 'inner' and 'outer' steady-state solutions are implicitly matched along a single boundary surface. The choice of this surface is not arbitrary, since it has to be assumed a priori as a surface where flow variables do not change appreciably either with θ or with time.

In order to solve the governing equations numerically, some approaches have been proposed for discretizing the partial differential equations. The control volume approach using a staggered arrangement of variables with a power-law scheme is popularly adopted in simulation of two-phase flow in stirred tanks. The first step in the control volume method is to divide the domain into discrete control volumes using the staggered grid. The idea is to evaluate scalar variables, such as pressure, density, temperature etc., at ordinary nodal points but to calculate velocity components on staggered grids centered around the cell faces. The staggering of the velocity avoids the unrealistic behavior of the spatially oscillating pressure field. A further advantage of the staggered grid arrangement is that it generates velocities at exactly the locations where they are required for the scalar transport-convection-diffusion-computations. Hence, no interpolation is needed to calculate velocities at the cell faces [97].

The second step is the integration of the governing equations over a control volume to yield a discretized equation at its nodal point P:

$$a_\text{P} \phi_\text{P} = \sum_{nb} a_{nb} \phi_{nb} + b \tag{3-49}$$

where b is the source term. Besides, the central, upwind and hybrid differencing schemes are used for discretization. The central differencing method is not suitable for general purpose convection-diffusion problems because it generates large error for solution at large values of the cell Peclet number. The upwind, hybrid or power-law differencing scheme is highly stable, but suffers from false diffusion in multi-dimensional flows if the velocity vector is not parallel to one of the coordinate directions. Higher order schemes, such as QUICK, can minimize false diffusion errors but are computationally less stable. However, if used with care and judgment, the QUICK scheme can give very accurate solutions of convection-diffusion problems.

The final step is the solution to the algebraic equations. The SIMPLE algorithm is often adopted to resolve the coupling between pressure and velocity. The acronym SIMPLE stands for Semi-Implicit Method for Pressure-Linked Equations. The algorithm was originally put forward by Patankar and Spalding [98] and is essentially a guess-and-correct procedure for the calculation of pressure on the staggered grid arrangement. To initiate the SIMPLE calculation process, a pressure field p^* is guessed. Discretized momentum equations are solved using the guessed pressure field to yield the velocity field. Then, the correction p' is defined, as the difference between the correct pressure field p and the guessed pressure field p^*, so that

$$p = p^* + p' \tag{3-50}$$

Similarly we define velocity corrections u' and v' to relate the correct velocities u and v to the guessed velocities u^* and v^*

$$u = u^* + u' \tag{3-51}$$

$$v = v^* + v' \tag{3-52}$$

Substituting the correct pressure field p into the momentum equations yields an equation for deciding the field of correction p'. The pressure correction equation is susceptible to divergence unless some under-relaxation is used during the iterative process. The new, improved, pressure p^{new} is obtained with

$$p^{new} = p^* + \alpha_p p' \tag{3-53}$$

where α_p is the pressure under-relaxation factor. The velocities are also under-relaxed. The iteratively improved velocity components u^{new} are obtained from

$$u^{new} = \alpha_u u + (1 - \alpha_u) u^{(n-1)} \tag{3-54}$$

where α_u is the velocity under-relaxation factors with values between 0 and 1.0, u is the corrected velocity component without relaxation) and $u^{(n-1)}$ represents its value in the previous iteration. A proper value of the under-relaxation factor α is essential for cost-effective simulation. Too large a value of α may lead to oscillatory or even divergent iterative solutions, while too small a value will result in extremely slow convergence. Unfortunately, the optimal value of under-relaxation factor is flow dependent and must be sought on a case-by-case basis [95].

For two phase flow sharing the same pressure field, two continuity equations should be satisfied at the same time. They are combined together so as to obtain the pressure-correction formula [99]:

$$\left[\left(a_p\right)_c + \left(a_p\right)_s\right] p'_P = \sum_{nb}\left[\left(a_{nb}\right)_c + \left(a_{nb}\right)_s\right] p'_{nb} + f_c + f_s + D_c + D_s \qquad (3\text{-}55)$$

where f_k is the turbulent diffusive term due to the asymmetry of the mass diffusive term, and D_k is the mass imbalance term reflecting the phase conservation over a cell due to the pressure field being not compatible with the velocity field. The volume fraction of the continuous phase is obtained by solving its continuity equation, and the volume fraction of the dispersed phase can then be obtained from Eq. (3-3). The non-linearity in the phase momentum and turbulence equations is handled by standard under-relaxation techniques. The solution is considered converged when the residual in the equations solved become smaller than a prescribed tolerance.

3.1.3. Boundary Conditions

[1] Symmetry axis: $u_{c,r}=u_{c,\theta}=u_{d,r}=u_{d,\theta}=0$, and otherwise $\partial \phi / \partial r = 0$.
[2] Free surface: the surface is assumed to be flat, then $u_{c,z}=u_{d,z}=0$, and otherwise $\partial \phi / \partial z = 0$.
[3] The solid surface: no-slip condition is the appropriate condition for the velocity components at solid walls including the wall, bottom, baffles, shaft and impeller.
[4] The wall function is adopted to resolve the flow velocity and turbulent properties at the nodes adjacent to all solid walls when the turbulent flow in stirred tanks is simulated.

3.2. Liquid-Liquid Flow in Stirred Tanks

Dispersion of two immiscible liquids in mechanically stirred tanks is commonly encountered in chemical and biochemical processes, for instance, in liquid-liquid extraction, in suspension polymerization, etc. The purpose of such an operation is to mix two phases and increase the interfacial area by intensifying the dispersion of one liquid into another and to enhance consequently the interphase heat/mass transfer and the chemical reaction.

Investigations previously conducted on liquid-liquid dispersion in stirred tanks were mainly focused on the measurements of the drop size distribution [89, 100], which is a crucial factor for heat and mass transfer. However, the rate of transfer, heavily dependent on the breakage and coalescence of droplets [101], is controlled by the phase dispersion and distribution in the macroscopic flow field [102]. In the past studies, LDA technique was often used to measure the velocity field. On the other hand, more attempts were made to measure the holdup profiles of the dispersed phase by using, for example, the sample withdrawal technique [103-105], a conductivity probe and the ultrasonic method.

CFD method was also employed to predict the hydrodynamic characteristics of liquid-liquid flow in stirred tanks. Compared with other two-phase flows, the difficulties are mainly due to the additional complexity generated by the drop deformation, breakage, coalescence, and occurrence of circulation inside the drops. In the numerical simulations, the Eulerian-Lagrangian approach has been adopted. Zhu and Vigil [106] employed an algebraic slip mixture model to simulate the banded liquid-liquid Taylor-Couette-Poiseiulle flow. Wang and Mao [107] simulated the three- dimensional flow field of a stirred tank (Rushton impeller, $T=H=0.154$ m) using the two-fluid approach with incorporation of the phase holdup fluctuation correlations appearing in the Reynolds time-averaged governing equations. The k-ε turbulence model is employed to describe the turbulence in the system.

Eulerian-Lagrangian and Eulerian-Eulerian approaches are both used in liquid-liquid two-phase flow simulation. However, the latter shows many computational advantages over the first one in the case of high dispersed-phase concentration. In the Eulerian-Eulerian approach, the droplets are considered as rigid spheres without deformation as far as the interphase friction is concerned, and the fluid flow inside the droplet as well as the processes of breakage and coalescence is neglected.

The analysis presented by Joshi [108] suggested that the lift force and the added mass force were important in numerical simulation of multiphase flow in a bubble-column reactor. The drag force and the added mass force proposed by Gosman *et al.* [90], had much influence on gas-liquid and solid-liquid two-phase flow in stirred tanks; Ljungqvist and Rasmuson [109] reported that the added mass force and the lift force had no remarkable influence on the slip velocities in solid-liquid flows in stirred tanks except in the impeller region.

The above three forces were included in the simulation by Wang and Mao [107]. The drag force between the continuous and dispersed phases can be expressed as [79]

$$F_{ci,\mathrm{drag}} = -F_{di,\mathrm{drag}} = \frac{3\rho_c \alpha_c \alpha_d C_\mathrm{D} |\mathbf{u}_d - \mathbf{u}_c|(u_{di} - u_{ci})}{4 d_d} \qquad (3\text{-}56)$$

where C_D is the nondimensional drag coefficient depending on the droplet Reynolds number:

$$Re_d = \frac{\rho_c d_d |\mathbf{u}_d - \mathbf{u}_c|}{\mu_{c,\mathrm{lam}}} \qquad (3\text{-}57)$$

and the local value of droplet diameter d_d is suggested by Nagata[110]:

$$d_d = 10^{-2.316+0.672\alpha_d} v_{c,\mathrm{lam}}^{0.0722} \varepsilon^{-0.914} (\sigma g/\rho_c)^{0.196} \qquad (3\text{-}58)$$

The correlation is adopted for the advantage that the deformation of droplets was taken into account [83]:

$$C_D = \frac{24}{Re_m}\left(1+0.1Re_m^{0.75}\right) \tag{3-59}$$

$$Re_m = \frac{\rho_c d_d |\mathbf{u}_d - \mathbf{u}_c|}{\mu_m} \tag{3-60}$$

$$\mu_m = \mu_{c,lam}\left(1-\frac{\alpha_d}{\alpha_m}\right)\exp\left(-2.5\alpha_m \frac{\mu_{d,lam}+0.4\mu_{c,lam}}{\mu_{d,lam}+\mu_{c,lam}}\right) \tag{3-61}$$

The added mass force originated from the acceleration of one phase relative to another can be calculated from [111]

$$F_{ci,am} = -F_{di,am} = -C_{am}\alpha_d\rho_c\left[\left(\frac{\partial u_{ci}}{\partial t}+u_{cj}\frac{\partial u_{ci}}{\partial x_j}\right)-\left(\frac{\partial u_{di}}{\partial t}+u_{dj}\frac{\partial u_{di}}{\partial x_j}\right)\right] \tag{3-62}$$

where C_{am} is expressed as [112]

$$C_{am} = 0.5(1+2.78\alpha_d) \tag{3-63}$$

For a moving rigid spherical particle, a lift force perpendicular to the average flow direction occurs. The lift force is given by [113]

$$F_{ci,lift} = -F_{di,lift} = C_{lift}\alpha_d\rho_c\varepsilon_{ijk}\varepsilon_{klm}(u_{dj}-u_{cj})\frac{\partial u_{cm}}{\partial x_l} \tag{3-64}$$

where C_{lift} is the lift force coefficient with a value of 0.5.

The influence of the dispersed phase on turbulence structure was experimentally investigated and reported in terms of the root-mean-square (rms) of the fluctuating velocities [100]. The rms of the continuous phase with various values of the average dispersed-phase holdup is calculated through Eq. (3-20) given $i=j$. It has been observed that the rms values of the continuous phase decrease with the increase of the average holdup of the dispersed phase, especially in the impeller region, implying that the presence of a larger amount of the dispersed phase produces a stronger suppression on the turbulence of the continuous phase.

The predicted contour plots of the normalized dispersed phase holdup are shown in [105]. With a relatively lower impeller speed, the dispersed phase seems to accumulate around the impeller shaft at the top of the tank, in accordance with experimental findings. With an increased impeller speed, the distribution of the dispersed phase becomes gradually homogeneous.

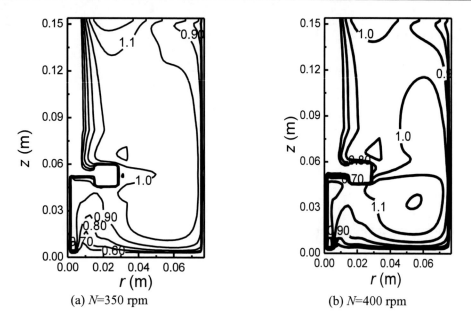

Figure 3-1. Predicted contour plots of normalized holdup of the dispersed phase.

3.3. Gas-Liquid Flow in Stirred Tanks

Rushton disk turbine is often employed in gas-liquid stirred tanks, because it can provide powerful shear force, which can be used to break up bubbles into smaller ones. Many studies on the numerical simulation of the flow in gas-liquid stirred tanks with a Rushton disk impeller, and the gas holdup distribution in the impeller stream and bulk flow has been reported. Flooding of the tank means that the gas is not well dispersed and rises up in a limited region around the shaft. To show the different gas-liquid flow patterns including flooding, the spatial distribution of gas holdup has been measured in a wide range of stirring speed and gas holdup level in some works. Although the standard k-ε model is the most popular one in use, it often leads to poor results for the flow subjected to complicated distribution of strain rate, e.g., in swirling flow and curved streamline flow.

Improvement of numerical methods and turbulence models is very crucial for obtaining reasonable simulation of stirred tanks. In a stirred tank, the flow around the rotating impeller blades interacts with the stationary baffles and generates a complex, three-dimensional, recirculating turbulent flow. A new swirl number, R_s, is proposed for the gas-liquid flow in a stirred tank, and the k-ε model is modified accordingly by introducing R_s into the energy dissipation equation [114]. The LES model has shown great potential in understanding the fluid flow behaviors in recent years. A Eulerian-Eulerian two-fluid model has also been proposed for gas-liquid flow using LES for both gas and liquid phases in a three-dimensional frame [115].

The typical maps for gas and liquid velocity vectors in stirred tanks are shown in Figure 3-2 (Rushton impeller, T=450 mm, ω=27.8 rad/s, Q=1.67×10^{-3} m$^3\cdot$s^{-1}). Two large eddies are formed close to the outer verge of the impeller blades, similar to these in the single phase stirred tank. In the right upper corner, another smaller eddy is formed due to the buoyancy of rising gas. It is also observed that the impeller discharged stream is a little inclined upward by the buoyant action of gas in Figure 3- 2a. Gas and liquid behave differently in their motion in stirred tanks. Gas bubbles out of the sparger rise up to the impeller and are dispersed by the impeller to other regions. Compared with the eddies of liquid phase, they are smaller, and located at different positions as in Figure 3- 2b.

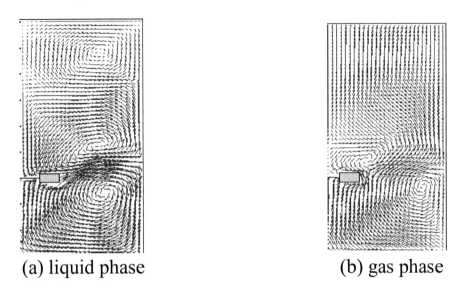

(a) liquid phase (b) gas phase

Figure 3- 2. Velocity vector maps of gas-liquid flow in a stirred tank [83].

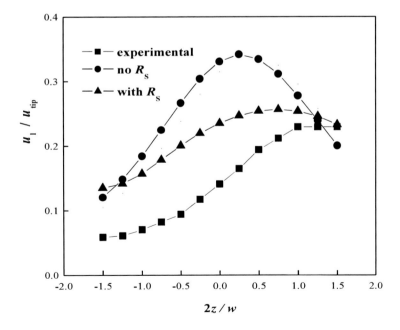

Figure 3- 3. Predictions of mean liquid velocity with swirl modifications compared with the experimental data [112].

It seems that the swirl modification of the mathematical model is necessary to match better the simulation results with the experimental data. Figure 3-3 shows the comparison between the predicted results and experimental data of the resultant velocity u along a vertical line at $r=73$ mm in a stirred tank with the diameter of 288 mm driven by a single Rushton impeller. It is seen that the swirl modification model provides more reasonable results.

The instantaneous velocity vector maps of the gas and liquid phases in the r-z plane located midway between two blades are presented in Figure 3-4 for a tank with the diameter of 288 mm. It is obvious that the flow pattern in the tank is dynamic and very complex, and such fine structures could not be well predicted by the k-ε model. There are many small vortices in both the gas and the liquid flow fields. Furthermore, the flow in the tank is not symmetrical as most literature presumed it to be.

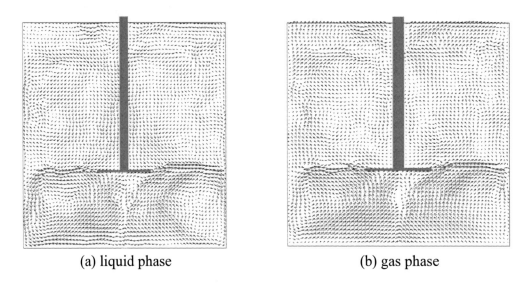

(a) liquid phase (b) gas phase

Figure 3-4. Instantaneous velocity fields in r-z plane by the LES [115].

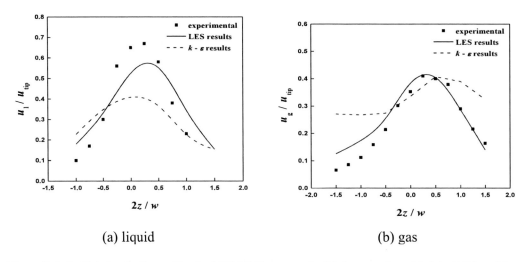

(a) liquid (b) gas

Figure 3-5. Predicted velocity profiles by LES [115] compared with the experimental data [116] and the standard k-ε model prediction at different vertical positions ($\omega=62.8$ rad·s^{-1}).

Figure 3-5 shows the profiles of the gas resultant velocities at different radial positions compared with the experimental data. As the gas-liquid turbulent flow in a stirred tank is quite difficult to be measured, especially at high gas holdup, few experiments have been reported on gas or liquid velocity field and phase holdup. In comparison to the predictions by the k-ε model, the predictions from the LES are closer to the experimental data, especially in the positions near the impeller tips. This is because the LES simulation is more powerful in capturing the anisotropic nature of turbulence in the impeller region and the impeller stream, much superior to the k-ε model based on the assumption of isotropic turbulent flow.

3.4. Solid-Liquid Flow in Stirred Tanks

The main purpose of the widely used solid-liquid agitated tanks in process industry is to enhance the heat and mass transfer between two phases. A number of investigations were focused on achieving empirical correlations, mostly on the distribution of solid holdup in stirred tanks and the criteria for off-bottom solid suspension. In recent years, reports about the above topics are available through experimental methods such as LDV, PDA and CT instruments. However, experimental measurements are insufficient to provide the insight of solid-liquid suspension in stirred tanks or for designing and scaling-up purposes. Numerical methods are now more and more popularly adopted to simulate solid-liquid flow in stirred tanks to provide significant details of hydrodynamic information complimentary to experimental methods and empirical correlations. However, there are still some difficulties in simulating multi-phase flows in stirred tanks, for example, in providing an accurate representation of the impeller action and an appropriate model of interphase interaction and flow turbulence.

3.4.1. Radial Flow Impeller

In Figure 3- 6, it is generally observed that the flow pattern of solid phase is similar to that of liquid, revealing that fine solid particles follow the liquid closely. It also shows that the impeller discharged stream is somewhat inclined downward for the settling of solid particles. The circulation patterns are similar for both phases: there are two primary circulation loops in the upper and lower parts of the impeller in the *r-z* plane. In the velocity vector plot of the solid phase, a small recirculation loop appears near the vertical axis of the tank below the impeller which may be the result of the accumulation of the solid particles in this region. This observation is also true for stirred tanks operated under laminar or mild flow condition.

The distribution of solid particles in the stirred tank with a radial impeller in Figure 3- 7 suggests that the maximum solid concentration occurs on the center of the tank bottom, and the level gradually decreases from the bottom to the free liquid surface. The solid concentration contour maps show a small circular region with low concentration below the impeller plane. In the region above the impeller plane, there is also a circulation flow but no region with low solid concentration.

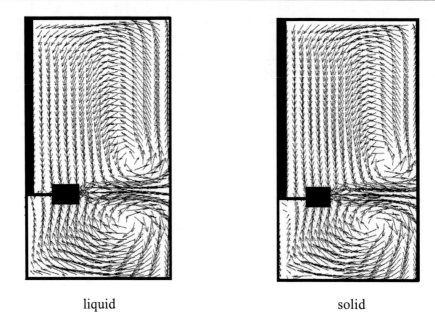

Figure 3-6. Velocity vector plots of continuous and dispersed phases ($T=H=0.294$ m, $D=C=T/3$, $d_s=232.5$ μm, $N=300$ rpm) [117].

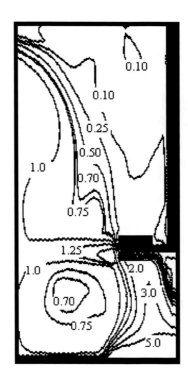

Figure 3-7. Contour plots of normalized concentration of solid phase ($T=H=0.294$ m, $D=C=T/3$, $d_s=232.5$ μm, $N=300$ rpm)[117].

An important criterion in designing a solid-liquid stirred reactor is the complete suspension of solid particles off the vessel bottom, at which point all of the particles are in motion, and all the solid surface are available for mass and heat transfer and chemical reaction. The impeller speed in accordance with this status is identified as the critical impeller speed N_{js}. Upon further increasing of the impeller speed, the transport rate between two phases will increase only slightly.

The work conducted by Zwietering [118] in the past decades on the critical impeller speed is considered the most systematic work performed even viewed nowadays. By visual observation of the solid particle motion on the stirred tank floor, the complete suspension state is defined as this in which all particles are in motion and no particle remained on the floor for more than 1 to 2 s. An empirical correlation of N_{js} is proposed as

$$N_{js} = S \frac{v_{c,lam}^{0.1} d_s^{0.2} (g\Delta\rho/\rho_c)^{0.45} X^{0.13}}{D^{0.85}} \qquad (3\text{-}65)$$

where S is a dimensionless parameter defined as a function of the tank configuration and the impeller type, i.e.,

$$S = f\left(\frac{T}{D}, \frac{T}{C}\right) \qquad (3\text{-}66)$$

The work of Armenante *et al.* [119, 120] focused on the effect of impeller clearance off the tank bottom on N_{js}. The correlation of Rushton impeller for $C/D > 1.5$ is

$$N_{js} = 3.06 \left(\frac{T}{D}\right)^{1.33} \exp\left(0.44\frac{C_b}{T}\right) D^{-0.63} X^{0.13} d_s^{0.20} v_{c,lam}^{0.1} \left(\frac{g\Delta\rho}{\rho_c}\right)^{0.45} \qquad (3\text{-}67)$$

Another approach to determine the critical impeller speed is to examine the variation of solid concentration profile with impeller speed. It was found that more solid particles will be suspended and the solid concentration will increase as the impeller speed is increased gradually. When the critical impeller speed is reached, few solid particles can be added to the bulk liquid flow. Further increase of the impeller speed will result in the homogenization of the suspension and the solid concentration will decrease in some locations.

Baldi *et al.* [121] introduced a theoretical model to determine N_{js} based on the assumption that the energy needed to suspend solid particles was proportional to that of turbulent vortices. Another theoretical model was proposed based on the balance between the upward flow velocity and the particle setting velocity [122] Wang *et al.* [117] employed CFD to analyze the solid-liquid flow in stirred tanks and adopted the following three criteria for predicting N_{js}:

Criterion 1: By observing the solid concentration profiles against the varying impeller speeds at a specific monitor point, N_{js} is defined as the impeller speed corresponding to the peak concentration or that where a sudden change in the slope of the profile occurs.

Criterion 2: N_{js} is determined by inspecting the axial velocity of solid phase in cells closet to the tank bottom at different impeller speeds. When the speed is lower, the velocity in a fairly large region in the vicinity of the tank floor center is negative, meaning that the solid particles settle down. With the increased impeller speed, the velocity at the center of the tank bottom becomes positive, indicating the solid particles are suspended.

Criterion 3: Because the solid particles tend to deposit at the center of tank floor, it is possible to determine N_{js} by inspecting the variation of solid concentration against the impeller speed at such position, experimentally or numerically. The impeller speed to decrease the solid concentration at tank bottom below that of compact sediment is defined as N_{js}.

The predicted values of N_{js} by the above criteria are compared with empirical correlations in Table 3- 2. The results by Criteria 2 and 3 are both close to the correlations but give somewhat lower predictions. It seems that Criteria 2 and 3 are more consistent and reasonable.

Table 3- 2. Comparison of N_{js} calculated with results from correlations

	Case a	Case b	Case c
Zwietering's Eq. (3- 65)	425	585	717
Armenante's Eq.(3- 67)	472	661	808
Criterion 1	350	485	550
Criterion 2	412	562	637
Criterion 3	412	562	637

3.4.2. Axial Flow Impeller

Compared with a Rushton turbine, it is more difficult to build the suitable computational grids for an axial flow impeller when dealing with the edge surface of the impeller. It is known that the accuracy of CFD predictions critically relies on the manner in which complex impeller and tank configurations are represented by regular Eulerian computational grids and numerical discretization schemes. In the early 1990s, fully predictive modeling methodologies that were known as the sliding-mesh method [123], the inner-outer approach [124], and the multiple-frames of reference method [123] were developed for the explicit odeling of impeller geometry.

Although, baffled stirred tanks are widely used for better mixing of solid particles and liquid, there are cases in which the use of unbaffled tanks may be desirable. Baffles are usually unnecessary in the case of very viscous fluids ($Re < 20$), because they result in dead zones where vortex formation is inactive by the low stirring speeds and the high friction on the cylindrical wall. Consequently, the mixing performance may actually be worsened. Unbaffled tanks are also used as crystallizers, in which the presence of baffles may promote particle attrition. Besides, higher fluid-particle mass-transfer rates may be obtained in unbaffled tanks for a given power consumption, which may be desirable in many processes. Shan et al. [80] conducted both experiments and numerical simulation of an unbaffled stirred tank of 300 mm diameter agitated with a pitched-blade turbine downflow.

There are many experimental studies focused on solid-liquid flow in stirred tanks. Also, a lot of methods, for example, optics using an endoscope technique [125] and pressure gauge technique (PGT) based on the measurement for pressure variation on the bottom as a result of the presence of suspended solids [126, 127], have been developed to investigate the local solid hold-up and the critical speed for just-off-bottom suspension. One of the optical methods for local solid concentration was introduced and performed as follows [80]. Firstly, the impeller off-bottom clearance and agitation speed were set and the overall solid holdup was set at a desired level. The measurement of the local solids concentrations, using a PC-6A fiber optic probe (manufactured by Institute of Process Engineering, Chinese Academy of Sciences), was conducted in an obscured environment to prevent daylight from interfering with the optical measurements. The details of this method may be referred to other literature [80, 128, 129].

Both the Rushton turbine and the pitched-blade impeller are widely used in stirred tanks. However, the simulation of the latter is obviously more difficult, because there may be some difficulties in dealing with the moving blade surface with irregular geometry. The so-called 'vector distance' is introduced to determine whether a node under consideration is in the liquid domain or not [80]. For example, if point A is outside the impeller, as shown in Figure 3-8a, and its distances to the two surfaces of the impeller blade are expressed by vectors $\vec{a}_1$ and $\vec{a}_2$, respectively, their dot product $\vec{a}_1 \cdot \vec{a}_2$ is positive; if point A is inside a blade, as shown in Figure 3-8b, the dot product of two distance vectors $\vec{a}_3 \cdot \vec{a}_4$ would be negative; if A is just on the surface, the dot product is equal to zero. With such a simple geometric rule, all the nodes in the liquid domain can be identified, given that all surfaces of the impeller are already specified. Thus, the smooth blade surface is now approximated by a rough one. This would produce some numerical errors to the simulation results, but the errors are expected to decrease as the grid is refined further.

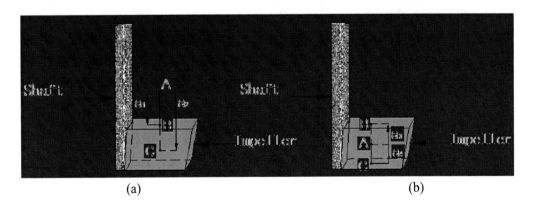

Figure 3-8. Geometric rule for identifying an active node for velocity components and pressure: (a) node outside impeller; (b) node inside impeller.

Shan et al. [80] provided the two-dimensional vector plots of the continuous phase (water) and dispersion solids (α_{av}=0.005 (volume fraction), ρ_d=1970 kg/m^3, d_s=80 μm). The flow patterns for the two phases seem to be quite similar, naturally because the solid particles have a terminal velocity close to liquid flow. Secondary circulation loops in the flow field are revealed both above and below the impeller, in addition to the main circulation near the blade tips. These circulation regions are the primary cause of the segregation of solid particles in the system, and their positions are dependent on the impeller speed. In the r-θ plane, a vortex is formed behind the impeller in both continuous and dispersion phases, because of relatively low pressure in this region.

From the contour profiles of the solids concentration (Figure 3-9), a relatively high concentration region exists below the impeller, which is in accordance with the zone of low pressure in the flow field. The high concentration near the wall in the upper tank region can be attributed to the circumferential flow and centrifugal force. A similar observation was made with a Lagrangian simulation approach [130. 131]. The solids collided with the wall, thus losing momentum, resulting in inability of the liquid to carry them through. Consequently, the particles have a tendency to settle instead of moving with their initial trajectory. The concentration near the shaft is very low, caused by the central vortex. With the increased impeller speed, the concentration below the impeller is also increased, and the concentration of the regions near the free surface and shaft shrank. Furthermore, more and more particles are transformed to the upper tank region. However, further increasing of the impeller speed above the critical suspension speed will not result in higher homogeneity in the tank, which implies that the efficiency that is involved in achieving higher homogeneity simply by increasing the impeller speed above N_{js} is not obvious. Besides, a vortex exists in the lower impeller zone near the bottom of the tank, which can be the result of the high shear stress of the continuous phase.

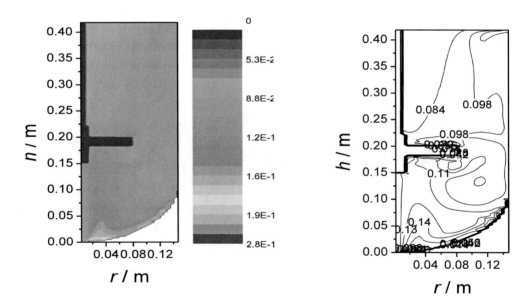

Figure 3-9. Solid particle concentration distribution (α_{av}=0.1, N=500 rpm) [80].

3.5. Liquid-Liquid-Solid Flow in Stirred Tanks

Liquid-liquid-solid three-phase stirred tanks are popular in process industries. Typical applications include reactive flocculation and solid catalyzed liquid-liquid reaction, etc. The understanding of the hydrodynamic characteristics, such as suspension of solid particles, dispersion of dispersed liquid phases and their spatial distribution in a stirred tank is critical for the determination of the rates of heat/mass transfer, and therefore of great importance for the reliable design and scale-up of such chemical reactors[132].

For liquid-liquid-solid three-phase flow in stirred tanks, it is desirable to obtain the information on the state of the dispersion of the dispersed phases. Wang *et al.* [132]provided experimental measurements of axial and radial variations of phase holdups of the two dispersed phases in a lab-scale stirred tank under different operating conditions by the sample withdrawal method. Tap water, *n*-hexane and glass beads were chosen for liquid, liquid and solid phases, respectively.

The measurement results normalized with the respective phase volume fraction are presented in Figure 3-10 . It can be seen that the lower local holdup of solids appears close to the tank bottom, and the higher local holdup of oil at the top surface, indicating that both solid and oil phases are not sufficiently dispersed at such impeller speed. Increasing the stirring speed can obviously promote the dispersion of oil phase in continuous phase. It is also observed that the local holdup of the oil phase below the impeller is larger than that at the upper sections, which can be explained by the fact that some droplets adhere to the surface of solid particles as a result of better wettability to oil. This suggests that more ways of phase interaction may appear when a new phase is introduced.

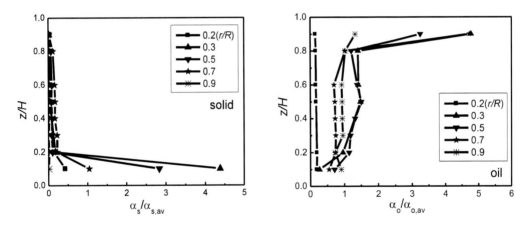

Figure 3-10 Axial profiles of normalized holdup of solid and oil phases ($\alpha_{s,av}$=0.10 and $\alpha_{o,av}$=0.10). Left: solid; right: oil. (*N*=300 rpm).

For the numerical simulation of liquid-liquid-solid three-phase flow, the Eulerian multi-fluid model is preferred. Wang *et al.* [132]treated water, oil and solid phases as different continua, interpenetrating and interacting with each other everywhere in the computation domain under consideration. The oil and solid phases are in the form of spherical dispersed droplets and particles, respectively. The effect of breakup and coalescence of droplets is neglected. The pressure field is shared by all three phases, which are exerted, respectively, by the pressure gradient multiplied by respective volume fraction. Motion of each phase is governed by respective mass and momentum conservation equations.

The RANS version of governing equations for three-phase flow is similar to that of two-phase flows. Reynolds stresses are always modeled by introducing the Boussinesq hypothesis. Although different from two-phase flows, there are still interactions between dispersed droplets and solid particles in liquid-liquid-solid three-phase flow, which have to be modeled.

However, this factor was not always included in the reported literature. In the model developed by Padial et al. [133] for a gas-liquid-solid three-phase bubble column, the drag between solid particles and gas bubbles was modeled identically to drag between liquid and gas bubbles based on the notion that particles in the vicinity of gas bubble tend to follow the liquid. In the simulation by Michele and Hempel [134], the momentum exchange terms between the dispersed gas and solid phases were expressed as

$$F_{\text{g-s},i} = -F_{\text{s-g},i} = \frac{3\rho_g \alpha_s c_{\text{g-s}} |\mathbf{u}_s - \mathbf{u}_g|(u_{si} - u_{gi})}{4d_p} \tag{3-68}$$

The combination of $c_{\text{g-s}}|\mathbf{u}_s - \mathbf{u}_g|$ was defined to be a fitting parameter determined by fitting model predictions to measured local solid holdups.

Since the two dispersed phases are presumed to be continua, as mentioned above, it is reasonable to expect the drag between solid particles and droplets behaving in the similar way as the drag force between the continuous and the dispersed phase:

$$F_{\text{o-s, drag},i} = -F_{\text{s-o, drag},i} = \frac{3\rho_o \alpha_o \alpha_s C_{\text{D, o-s}} |\mathbf{u}_s - \mathbf{u}_o|(u_{si} - u_{oi})}{4d_s} \tag{3-69}$$

Numerical simulation of such a liquid-liquid-solid system seems to be successful. In the flow fields of both continuous and dispersed phases, the well-documented flow pattern generated by a disc turbine in a stirred tank is clearly illustrated. Two large ring eddies exist, above and below the impeller plane respectively, and a high-velocity radial impeller stream is also predicted. Overall, the flow fields of the three phases are very similar to each other in most parts of the domain. The velocity field of the dispersed oil phase shows a trend of drifting upwards at lower impeller speeds. This might be due to the fact that the oil phase with lower density accumulates easily in the top section of the tank. The time-averaged flow field of the solid phase reveals a small recirculation zone above the center of the tank bottom, implying that the solid particles tend to settle down in this zone.

The numerically computed axial profiles of the dispersed oil and solid phases are compared in Figure 3-11 with the present measurements. The simulation results are generally in reasonable agreement with the experimental data, especially for higher impeller speeds. The comparison indicates that the computational approach adopted here is suitable for predicting the dispersed phase distribution in the liquid–liquid–solid three-phase stirred tank. The model prediction for the solid phase, however, is notably above the experimental data especially at low impeller speeds. With the increase of N, the agreement improves slightly. The discrepancy between the prediction and the experimental results is probably because the model of inter-phase momentum exchange employed in the present work is too simple to describe the real complex inter-phase interaction coupling in liquid–liquid–solid three-phase flows. Additionally, the isotropic k-ε two equation turbulence model is deficient in describing the well-recognized anisotropic nature of turbulent flow in stirred tanks.

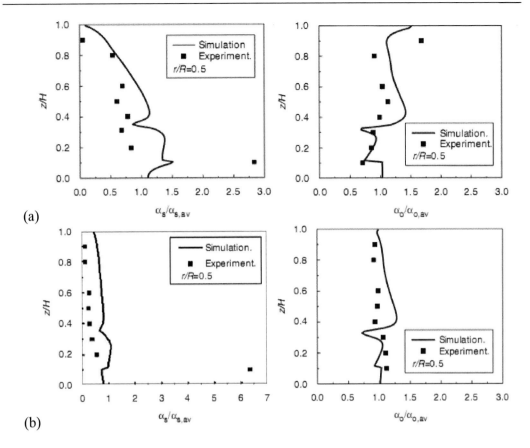

Figure 3-11. Comparison of simulated normalized holdup profiles of oil and solid phases with the experimental data for different impeller speeds. ($\alpha_{s,av}$=0.1, $\alpha_{o,av}$=0.1). Left: solid; right: oil. (a) N=300rpm, (b) N=500rpm.

The calculated normalized local holdups of the two dispersed phases are presented as contour plots in [132]. It is easy to observe that the distributions of oil and solid phases are both less homogeneous at low impeller speeds. The maximum oil phase holdup occurs in the center of the free surface because of the ring vortices in the upper bulk zone, in qualitative agreement with experimental investigations. The distributions of both dispersed phases become more uniform at higher agitation speeds. The maximum solid concentration is located at the center of the tank bottom due to the density difference and the ring eddy at the bottom, as it is confirmed above.

4. NUMERICAL SIMULATION OF MULTIPHASE FLOW IN LOOP REACTORS

Airlift loop reactors (ALRs) are also widely used in process industries in recent years because of the following advantages [135, 136] simple construction without moving parts, good mass and heat transfer characteristics, efficient mixing with low energy consumption, simple operation with low cost and so on. For instant, as an intrinsic element of the new technology of coal liquefaction in China, internal airlift loop reactor (IALR) was tested for cutting the operating cost and reducing the mechanical breakdown of expensive recirculating pumps by replacing bubble columns with IALR. Although great achievements have been made on IALRs, the design and scale-up of these reactors still remain difficult due to the nature of complex multiphase flow. How to design and optimize the IALRs is a key issue for the application of these reactors in new fields. As we know, the optimization and the upgrade of these industrial processes need a better knowledge of the physical mechanisms at the local scale [137].

4.1. Mathematical Model and Numerical Method

The two-phase k-ε turbulence model is usually employed to simulate two-phase flow so the conservation equations for mass and momentum are same as Eqs. (3-1) and (3-1). When the gas volume fraction is solved for gas-liquid two-phase flow, the liquid voidage can be obtained by the volumetric conservation in Eq. (3-3). The drawback of this procedure is that the volume-fraction field of one phase will not be influenced by the other phase. This can affect the overall convergence rate negatively. The improved procedure is to solve the gas and liquid phase continuity equations separately and then enforce the geometric conservation constraint on the solved volume-fraction values using the following correction:

$$\alpha_g = \frac{\alpha_g}{\alpha_g + \alpha_l} \qquad \alpha_l = \frac{\alpha_l}{\alpha_g + \alpha_l} \tag{4-1}$$

Thus, the volumetric conservation is always satisfied and the mutual influence of volume fractions is also taken into account.

The discretization of the governing equations is accomplished by volume integration over each cell with a staggered arrangement of the primary variables. The power-law scheme [138] is adopted to discretize all the equations except the continuity equations that are by the upwind scheme.

In the case of a relatively high density ratio for two phases (e.g., $\rho_l/\rho_g \approx 1000$ for air-water bubbly flow) and the pressure-correction equation being derived from the global mass conservation equation, the pressure correction will tend to drive the high-density fluid to conservation and the lighter phase may be less conserved. Darwish et al. [139] argued that this problem can be considerably alleviated by normalizing the individual continuity equations, and hence the global mass conservation equation becomes a global geometric conservation equation by means of a weighting factor of the reference density (which is fluid dependent). The mass conservation-based SIMPLE algorithm (MCBA-SIMPLE) [140] is adopted in this work.

The boundary conditions are very important to getting correct simulation results of real flow. The axis of the pipe is treated as the one of symmetry of the cylindrical coordinate system. A no-slip boundary is used and the standard wall functions are adopted for all phases at the solid wall [141]. The boundary conditions at the air inlet are set by prescribing a fixed inlet velocity related to the superficial velocity with the gas fraction set to 1.0. The inlet conditions for the turbulent kinetic energy and the dissipation rate are estimated by

$$k_{l,in} = 0.004 u_{g,in}^2 \qquad (4\text{-}2)$$

$$\varepsilon_{l,in} = C_\mu^{\frac{3}{4}} k_{in}^{\frac{3}{2}} / (0.07D) \qquad (4\text{-}3)$$

where D is the hydraulic diameter of the air inlet.

The outflow boundary of the reactor is a free surface, and its treatment is essential to get realistic results in the vicinity of the free surface. Mudde et al. [142] regarded the free surface as shear free for the liquid phase with zero normal velocity, while for the gas phase it is presumed as an outlet with a fixed velocity of 0.20 m/s, which is approximately the terminal slip velocity of a single bubble in water. Huang et al. [143] argued that it is more reasonable for taking the relative velocity between gas and liquid equal to the terminal velocity in accordance to the bubble size in the steady simulation of ALRs. Here we take the same boundary for the vertical velocity component, and all other variables are subject to

$$\frac{\partial \phi}{\partial z} = 0 \qquad (4\text{-}4)$$

The average diameter of bubbles, d_b, is calculated approximately by

$$d_b = 2.9 \left(\frac{\sigma d_0}{g \rho_l} \right)^{1/3} \qquad (4\text{-}5)$$

where d_0 is the diameter of the orifice.

However, different people obtained different models to estimate the mass transfer coefficient. For air-water dispersions and for suspensions in which the suspending fluid is water-like, Acién Fernández et al. [144] calculated the mass transfer coefficient from

$$k_L a = 3.378 \times 10^{-4} \left(\frac{g D_L \rho_l^2 \sigma}{\mu_l^3} \right)^{0.5} \alpha_g e^{-0.131 C_S^2} \qquad (4\text{-}6)$$

where C_S is the concentration of solids in suspension (wt/vol%), which is zero in the present gas-liquid systems. D_L is the molecular diffusivity of gas in liquid.

Cockx et al. [145, 146] and Talvy et al. [146] proposed a time model based on the penetration theory of Higbie [147] to estimate the local mass transfer as follows:

$$k_L a = \frac{12 \alpha_g}{d_b} \sqrt{\frac{D_L U_{slip}}{\pi d_b}} \qquad (4\text{-}7)$$

Xue and Yin [148] derived an expression for the mass transfer coefficient according to Boussinesq hypothesis and the classical penetration theory of Higbie [148]. The final form of the time model is

$$k_L a = \sqrt{\frac{D_L}{\pi}} \left(\frac{\rho_l \varepsilon}{\mu_l}\right)^{\frac{1}{4}} \frac{12\alpha_g}{d_b} \qquad (4\text{-}8)$$

For a mobile interface, the contact time with the liquid is short, the Higbie's theory is valid, according to which Vasconcelos et al.[149] achieved the following expression for the mass transfer coefficient:

$$k_L a = 6.78 \frac{\alpha_g}{d_b} \sqrt{\frac{D_L U_{slip}}{d_b}} \qquad (4\text{-}9)$$

This formula can also be used for deforming bubbles. For a rigid interface, the bubble behaves like a solid sphere. The mass transfer coefficient is then obtained theoretically from laminar boundary layer theory and a spatial model is achieved:

$$k_L a = c \frac{6\alpha_g}{d_b} \sqrt{\frac{U_{slip}}{d_b}} D_L^{2/3} v_l^{-1/6} \qquad (4\text{-}10)$$

4.2. Numerical Simulation of Hydrodynamics in Internal Airlift Loop Reactor

The configuration of an internal airlift loop reactor shown in Figure 4- 1, which is described by van Baten et al. [150] and Huang et al.[151], is chosen to test the sensitivity of different models. The IALR consists of a polyacrylate column with an inner diameter of 0.15 m and a height of 2 m. A draft tube of 0.10 inner and 0.11 m outer diameter with a height of 2.02 m is mounted into the column 0.10 m above the gas distributor. A gas-liquid separator is mounted at the top of the column with 1 m in height and 0.38 m in diameter. The column is filled with water to a height of 2.5 m and the experiments were carried out at ambient conditions. .

The sensitivity of simulation results with respect to the bubble size is studied with a superficial gas velocity of 0.06 m/s and the simulation results are shown in Figure 4- 2. The bubble size has important impact on the simulation results. The smaller the bubbles in bubbly flow, the stronger the influence of the specified bubble size on the simulation results. It is evident that the gas holdup and the liquid velocity decrease with the increase of bubble size in airlift loop reactors. This is due to large bubbles have a higher slip velocities comparing to small ones and cannot be entrapped into the downcomer. It is demonstrated that the correct designation of bubble size is an important factor to bubbly flow. For bubbly flow with a wide range of bubble size, modeling the bubble size variable with respect to the flow and location correctly, or modeling the flow with a multiple size group model, sounds a prerequisite.

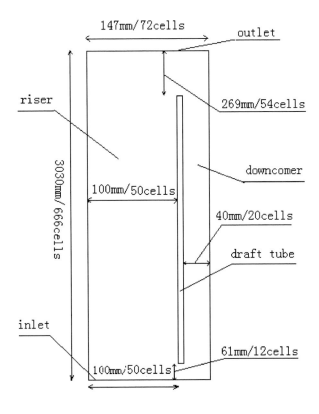

Figure 4-1. Configuration of the reactor and details of the grid distribution in the radial and axial directions.

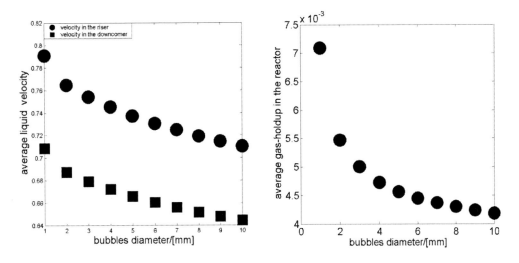

Figure 4-2. Impact of bubbles diameter on the gas-liquid flow in 2D internal airlift loop reactor.

Figure 4-3 shows that the cross-sectional area averaged liquid velocities in the riser and downcomer are all well predicted. The maps for liquid velocity vectors in the reactor are shown in Figure 4-4 with the superficial gas velocity of 0.06 m/s. It is seen from Figure 4-4 that there are two large eddies in the reactor. At the lower part of the reactor, the liquid gets into the riser from the downcomer by a path just beneath the draft tube. In the upper, the liquid reaches the free surface and turns to the downcomer.

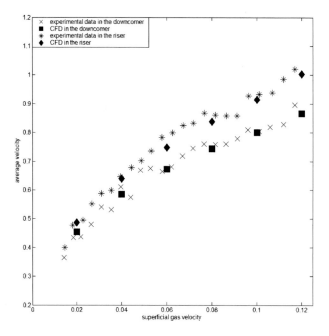

Figure 4-3. Comparison of predicted average liquid velocity with experimental data by van Baten et al. [150].

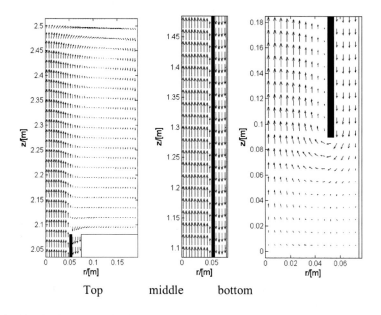

Figure 4-4. Liquid velocity vectors in the reactor.

4.3. Mass Transfer of Gas-Liquid Internal Airlift Loop Reactor

More comprehensive understanding of the two-phase hydrodynamics in IALRs is needed in order to know the effects of flow and geometric parameters on the mass transfer and ultimately on the reactor design, optimization, and performance. Great progress has been made in modeling the hydrodynamics of bubbly flow in ALRs. However, the prediction of mass transfer and mixing of chemical species in bubbly flow is still a great challenge, mainly because the interfacial transfer rates in a bubble swarm do not follow the laws of interfacial transfer valid for isolated bubbles in general [136]. In chemical engineering, the global mass transfer efficiency of an airlift reactor is usually expressed by a volumetric mass transfer coefficient $k_L a$. However, different people obtained different models to estimate the mass transfer coefficient.

The comparisons of the predicted average gas holdup in the reactor under different superficial velocities with experimental data by Juraščík et al. [152] are shown in Figure 4-5. It is seen that the predicted average fractions of gas in the reactor agree well with experimental data for a wide range of superficial gas velocities.

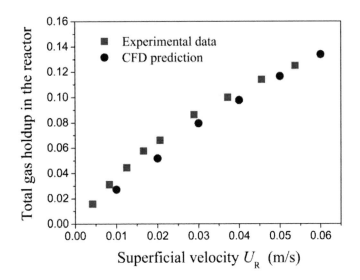

Figure 4- 5. Comparison of predicted average gas holdup with experimental data [153].

As mentioned above, the mass transfer coefficient of oxygen in air dissolved in water can be computed by Eqs. (4-6) ~ (4-10) for each cell and then the average $k_L a$ over the whole reactor volume is taken, which depends highly on the well predicted gas holdup. The comparison of the predicted mass transfer coefficients using different mass transfer models with experimental data by Jurar th et al. is shown in Figure 4-6. All the predicted volumetric mass transfer coefficients using different models increase with the increase of superficial gas velocity almost linearly in all tested cases. It is surprising that though Eq. (4-6) is a phenomenological correlation but it predicts the mass transfer coefficient well in all the test cases. Among these models, Eq. (4-7) has no adjustable regression coefficient and is deduced with a mechanism-based model, so it is recommended for modeling the mass transfer of bubbly flow in airlift loop reactors.

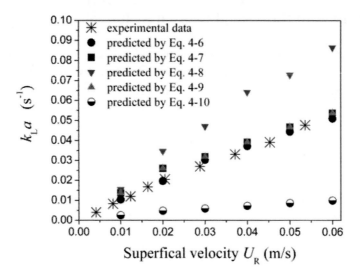

Figure 4-6. Comparison of predicted average mass transfer coefficients by different mass transfer models with experimental data [153].

The distributions of local gas holdup, local volumetric mass transfer coefficient and liquid-side mass transfer coefficient under the condition of the superficial gas velocity equaling to 0.03 m/s are shown in Figure 4-7. The spatial distribution pattern of mass transfer coefficient is very similar to that of gas holdup, indicating that the value of k_L varies only slightly throughout the reactor. It is seen that the mass transfer coefficient is higher in the riser than that in the downcomer, and the more the gas fraction, the bigger the mass transfer coefficient in the reactor. Because large bubbles cannot go to the bottom of the downcomer and gather in the top of it, large mass transfer coefficients in the top and small values in the bottom are resulted. The bottom of downcomer with negligible value of mass transfer coefficient seems useless for mass transfer.

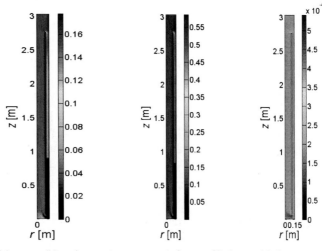

(a) gas holdup   (b) volumetric mass transfer coefficient   (c) k_L

Figure 4-7. Distribution of gas holdup, mass transfer coefficient and k_L in the reactor [153].

5. PERSPECTIVE

The computational fluid dynamics (CFD) method has showed great prospect and attracted extensive attention in recent years owing to its powerful capacity for understanding the physical reality of multiphase flows in macro-system and micro-system.

As a new numerical method to simulate the fluid flow, lattice Boltzmann method (LBM) is already used to reconstruct many flow phenomena, such as Newtonian and non-Newtonian fluids, multi-phase flow [154] and multi-component flow[155], compressible [156] and incompressible fluids, and make many practical applications, such as aerodynamics, aeroacoustics [157], thermodynamics [158], chemical reactions [159] and hydraulic problems[160]. LBM is so diverse and interdisciplinary that it is not possible to include all interesting topics but it can model the problems wherein both macroscopic hydrodynamics and microscopic statistics are important because the LBM is a mesoscopic and dynamic description of the physics of fluids. It is a powerful tool for modeling new physical phenomena that are not yet easily described by macroscopic equations. From a computational point of view, the lattice Boltzmann equation is hyperbolic and can be solved locally, explicitly and efficiently on parallel computers. Not only is the scheme computationally comparable with traditional numerical methods, but it is also easy to program and include new physics because of the simplicity of the form of the LBM equations. At present, the LBM is still undergoing development and the popular issues of LBM are: how to solve the forces in the lattice Boltzmann equation; how to use uneven mesh to solve the flow near the wall and far from the wall in different domains; second order boundary conditions near the curved walls; analyze the LB models with single and multi relaxation time and so on.

The level set method has been used to capture rather than track interfaces and the advantages of this capturing approach are well known by now [161]. Level set method is a numerical technique which offers remarkably powerful tools for understanding, analyzing, and computing interface motion in a host of settings. At their core, they rely on a fundamental shift in how one views moving boundaries; rethinking the Lagrangian geometric perspective and exchanging it for an Eulerian, initial value partial differential equation perspective [162] On the theoretical side, considerable analysis of level set methods has been performed in recent years[163-166]. This work has concentrated on many aspects, including questions of existence and uniqueness, pathological cases, extensions of these ideas to fronts of co-dimension greater than one (such as evolving curves in three dimensions), coupling with diffusion equations, links between the level set technique and Brakke's groundbreaking original varifold approach. On the theoretical/numerical analysis side, level set techniques exploit the considerable technology developed in the area of viscous solutions to Hamilton-Jacobi equations[167, 168]. A wide range of applications relate to level set methods, including problems in the geometric evolution of curves and surfaces, combustion problems that the interface is a flame and both exothermic expansion along the front and flame-induced vorticity drive the underlying fluid mechanics[169], crystal growth and dendritic formation that the interface is the solid/liquid boundary and is driven by a jump condition related to heat release along the interface [170], two-fluid flow problem that the interface represents the boundary between two immiscible fluids of significantly different densities and viscosities and the surface tension along this interface plays a significant role in the motion of the fluids[172], work on minimal arrival time by Falcone[172], flame propagation work by Zhu and Ronney[173], a wide collection of applications from computer vision by Kimmel[174], gradient flows applied to geometric active contour models[175], work on offine invariant scale space[176], and some related work on the scalar wave equation[177].

To solve the time-averaged equations of the turbulent flow field in reactors, a turbulent model is needed for the calculation of Reynolds stress terms. The standard k-ε turbulence model, based on the Boussinesq hypothesis of eddy viscosity, is the most widely employed isotropic turbulence model. The Reynolds stress model (RSM) is established when the Reynolds stress terms are solved from Reynolds stress transport equations. It requires too much computing resources to solve the Reynolds stress transport equations in engineering applications. In addition, the RSM simulation is heavily dependent on suitable initial conditions. The approximate and compromising form of RSM is the algebraic stress model (ASM), in which the Reynolds stress terms are expressed by a set of implicit algebraic equations. ASM has been applied successfully in many approximately parallel flows, where Reynolds stresses are almost explicitly related to the mean velocity field. In the case of general and complex three-dimensional flow in a stirred vessel, it is very difficult to keep the numerical solution stable, because there is no diffusion or damping in the ASM equation set. Consequently, it is important to employ very small time step or very small under-relaxation factors for achieving a convergent solution. This may increase computational load so that the advantage of simplicity of the ASM model is lost. For a remedy, an explicit form of the stress-strain relation is required. Pope [178] firstly proposed a methodology to obtain an explicit relation for the Reynolds stress tensor from the implicit algebraic stress model that Rodi [179] developed. The methodology producing explicit algebraic stress model is based on the principle of linear algebra. Because of the complexity of the algebra, only a solution of two-dimensional turbulent flows was obtained by Pope. Since then, some research [180-184] extended the results of Pope to three-dimensional turbulence flows.

It is known that a direct numerical simulation (DNS) of turbulence has to take into account explicitly all scales of motion, from the smallest (Kolmogorov dissipation scale for instance) to the largest (imposed by the existence of boundaries). For the flow with a high Reynolds number in a stirred tank, there are large and intermediate scale eddies, which are dependent on the tank and impeller geometries. There are also smaller vortices of Kolmogorov scale, driven by the turbulence cascade and are therefore likely to be independent of the species of the stirred tank. It is difficult in the near future to simulate explicitly all the scales of motion of fluids. However, it is the large scales that are more important and contain the desired information about turbulent transfer of momentum and mass. The motion on this scale is desired to be simulated on the computer. The LES was born in 1970, seeking to predict the motion of eddies which are larger than the user-chosen length scale (mesh size). In LES, not only the mean flow field, but also the dynamics of a range of energetic large scales of motion is directly computed. The effect of the unresolved small-scale turbulence needs to be accounted for through subgrid models[185, 186]. The large eddy simulation was successfully applied in the single phase simulation in stirred tanks[187-192]. Some groups [193, 194, 195] have attempted to employ the LES in two-phase flow simulation and in most of the reported literature the Lagrangian approach was used to model the dispersed phase. Derksen [129] simulated the liquid-solid flow describing the solid phase by the Lagrangian frame. The Eulerian formulation is preferred in the engineering application. Smith and Milelli [196] and Deen *et al.* [197]used a combination of the large eddy simulation and the Eulerian-Eulerian two phase flow approach to simulate the gas-liquid bubble column. Recently, Zhang *et al.* [115] firstly attempted to apply the large eddy simulation to a gas-sparged stirred tank with two different models for turbulent viscosity (LES model and standard k-ε model). In fact, it is well known that the plain Smagorinsky model is too dissipative close to a wall. This can be shown through expansions of the velocity components in powers of the wall distance. So the model does not work for transition in a boundary layer on a flat plate, starting with a laminar profile to which a small perturbation is added: the flow remains laminar, due to an excessive eddy viscosity coming from the mean shear [164]. A nice alternative to Smagorinsky model is its dynamic version. The dynamic model relies on a LES using a "base" subgrid-scale model such as Smagorinsky's model.

ACKNOWLEDGMENTS

The authors acknowledge the financial supports from the National Science Fund for Distinguished Young Scholars (No. 21025627), the National Natural Science Foundation of China (Nos. 20976177, 20990224, 20906090), 973 Program (2010CB630904, 2007CB613507, 2009CB623406, 2007CB714305) and the Jiangsu Province Projects (BY2009133, BE2008086). Authors wish to thank Prof. Jiayong Chen in our Institute, for his valuable advice and continuous encouragement. We are grateful to many other former Ph.D. and M.S. students of Prof. Mao and Prof. Yang for their contributions to the experimental and numerical studies. We also thank Prof. Santos from University of Federal de Santa Catariana (UFSC) for valuable discussions.

NOMENCLATURE

	a	acceleration, m/s^2
	Ac	acceleration number, $d_e a/u_b^2$
	Ar	Archimedes number, dimensionless
	c	concentration, mol/m^3
	C	dimensionless concentration
	Ca	Capillary number
	C_D	drag coefficient
	C_{DA}	total drag coefficient
C_1, C_2		coefficients in k-ε equation
$C.V.$		coefficient of variation in crystallization
d		diameter of particle
	d_e	volume-equivalent particle diameter, m
D		diameter of impeller
	De	Deborah number, $\lambda U_T/d_e$
D_k		mass imbalance term
d_{32}		mean particle size
	f	distortion function
f_k		turbulent diffusive term
F		interphase force
	F_A	added mass force, N
	F_B	buoyancy force, N

		F_D	drag force, N
$F_{r,k}$			centrifugal force
$F_{c,k}$			Coriolis force
Fr			Froude number
Fl			flow number
g			acceleration of gravity
G			dissipation function
G_e			extra dissipation function
H			liquid depth
		h_ξ, h_η	scaling factor, m
		H_ξ, H_η	dimensionless scaling factor
$k_L a$			mass transfer coefficient
N			impeller speed
		K	power-law consistency coefficient, $Pa \cdot s^n$
		L	characteristic length, m
		m	distribution coefficient
		Mo	Morton number,
		$\mathbf{n}$	unit vector normal to a surface
		N	diffusive flux
N_{js}			impeller speed for critical suspension
$N\theta_m$			dimensionless mixing time
		Oh	Ohnesorg number, $[\mu_c^4/(\gamma\rho_b R)]^{1/2}$
P			pressure
		Pe	Peclet number for mass transfer, $2RU_T/D$
P_V			power consumption per volume
Q			gas flow rate
		R	volume-equivalent particle radius, m
rpm			revolutions per minute
rad			radian
Re			Reynolds number
		s	Carreau model parameter
		$\mathbf{S}$	source term
		Sc	Schmidt number, $\mu/\rho D$
		Sh	Sherwood number, $2Rk/D$
t			time
t_l			mean eddy lifetime
t_p			particle response time
T			tank meter
u			velocity
		U_T	terminal velocity, m/s
		v	radial or transverse velocity component, m/s
		V	reference velocity, $\sqrt{dg}$
		V_p	particle volume, m^3
		We	Weber number, $2R\rho_1 V^2/\sigma$
		x, y	coordinate in physical plane, dimensionless
		$\Delta x, \Delta y$	grid spacing in x and y directions, m
Y			degree of mixing

GREEK SYMBOLS

ρ	density
α	phase volume fraction
τ	shear force
τ_t	turbulence fluctuation time
$\tau_{r,k}$	bubble relaxation time
δ	Kronecker delta
ν	kinetics viscosity, $m^2 \cdot s^{-1}$
ν_t	turbulent kinetics viscosity, $m^2 \cdot s^{-1}$
σ	surface tension
σ_t	turbulent Schmidt number
μ	dynamic viscosity, $Pa \cdot s$
ω	angle velocity, $rad \cdot s^{-1}$
ε	turbulent energy dissipation, $m^2 \cdot s^{-3}$
Γ	turbulent diffusivity, $Pa \cdot s$

SUBSCRIPTS

am	added mass force
av	averaged
b	bubble
c	continuous phase
d	dispersed phase
drag	drag force
eff	effective
F	flooding
g	gas phase
i, j	principal direction
k	phase
l	liquid phase
lam	laminar
lift	lift force
o	oil phase
r	radial coordinate
s	solid phase
t	turbulent
z	axial coordinate
θ	azimuthal coordinate

REFERENCES

[1] Succi, S. *The lattice Boltzmann equation for fluid dynamics and beyond.* Clarendon Press, Oxford, 2001

[2] QianY.; d' Humires D.; Lallemand P. Lattice BGK models for Navier-Stockes equation. *Europhys. Lett.* 1992, 17, 497-484

[3] Guo, Z. L.; Zheng, C. G.; Shi, B. C. Non-equilibrium Extrapolation Method for Velocity and Pressure Boundary Conditions in the Lattice Boltzmann Method. *Chinese Phys.* 2002, 11(4), 366–374.

[4] Du, R. *An implicit scheme for incompressible LBGK model and analysis on complex boundary processing.* Wuhan: Huazhong University of Science and Technology. 2004, 33.

[5] Chen, Y.; Ohashi, H.; Akiyama, M. Thermal Lattice Bhatnagar– Gross–Kook Model without Nonlinear Deviations in Macro-dynamic Equations. *Phys. Rev. E.* 1994, 50(4), 2776–2783.

[6] He, X.; Chen, S.; Doolen, G. D. A Novel Thermal Model for the Lattice Boltzmann Method in Incompressible Limit. *J. Comput. Phys.* 1998, 146(1), 282–300.

[7] Shi, B. C.; Guo, Z. L. Thermal Lattice BGK Simulation of Turbulent Natural-convection due to Internal Heat. *Int. J. Mod. Phys. B.* 2003, 17(1/2), 173–177.

[8] Guo, Z. L.; Shi, B. C.; Zheng, C. G. A Coupled Lattice BGK Model for the Boussinesq Equation. *Int. J. Numer. Method Fluids.* 2002, 39(4), 325–342.

[9] Chen, H. J.; Chen, Y. T.; Hsieh, H. T.; Zhang J. S. A lattice Boltzmann modeling of corrosion behavior and oxygen transport in the natural convection lead-alloy flow, *Nucl. Eng. Des.* 2007, 237(18), 1987-1998.

[10] Cristini, V.; Tan, Y. C. Theory and numerical simulation of droplet dynamics in complex flows–a review. *Lab Chip.* 2004, 4(4), 257-264.

[11] Gunther, A.; Jensen, K. F. Multiphase microfluidics: from flow characteristics to chemical and materials synthesis. *Lab Chip.* 2006, 6(12), 1487-503.

[12] Lomel, S.; Falk, L.; Commenge, J. M.; Houzelot, J. L.; Ramdani, K. The microreactor: A systematic and efficient tool for the transition from batch to continuous process. *Chem. Eng. Res. Des.* 2006, 84(A5), 1-7.

[13] Kiwi Minsker, L.; Renken A. Micro-structured reactors for catalytic reactions. *Catal. Today.* 2005, 110(1-2), 2-14

[14] Christopher, G. F.; Anna, S. L. Microfluidic methods for generating continuous droplet streams. J. Phys. D. *Appl. Phys.* 2007, 40(19), 319–336.

[15] Nisisako, T.; Torii, T.; Higuchi, T. Droplet formation in a microchannel network. *Lab Chip.* 2002, 2, 24-26.

[16] Wang, K.; Lu, Y. C.; Xu, J. H.; Luo, G. S. Determination of dynamic interfacial tension and its effect on droplet formation in the T-shaped microdispersion process. *Langmuir,* 2009, 25(4), 2153-2158.

[17] Dessimoz, A. L.; Cavin, L.; Renken, A.; Kiwi Minsker, L. Liquid–liquid two-phase flow patterns and mass transfer characteristics in rectangular glass micro-reactors. *Chem. Eng. Sci.* 2008, 63(16), 4035-4044.

[18] Triplett, K. A.; Ghiaasiaan, S.; M.; Abdel Khalik, S. I.; Sadowski, D. L. Gas–liquid two-phase flow in microchannels, Part I: two-phase flow pattern. *Int. J. Multiphase Flow.* 1999, 25(3), 377–394.

[19] Xu, J. L.; Cheng, P.; Zhao, T. S. Gas–liquid two-phase flow regimes in rectangular channels with mini/microgaps. *Int. J. Multiphase Flow.* 1999, 25, 411–432.

[20] Yu, Z.; Hemminger, O.; Fan, L. S. Experiment and lattice Boltzmann simulation of two-phase gas-liquid flows in microchannels. *Chem. Eng. Sci.* 2007, 62(24), 7172-7183.

[21] Takamasa, T.; Hazuku, T.; Hibiki, T. Experimental study of gas-liquid two-phase flow affected by wall surface wettability. *Int. J. Heat Fluid Flow*. 2008, 29(6), 1593-1602.

[22] Donata, M. F.; Franz, T.; Philipp, R. V. R. Segmented gas-liquid flow characterization in rectangular microchannels. *Int. J. Multiphase Flow*. 2008, 34(12), 1108-1118.

[23] Pandey, S.; Gupta, A.; Chakrabarti, D. P.; Das, G.; Ray, S. Liquid-liquid two phase flow through a horizontal T-junction. *Chem. Eng. Res. Des*. 2006, 84(10), 895-904.

[24] Kashid, M. N.; Agar, D. W. Hydrodynamics of liquid–liquid slug flow capillary microreactor: flow regimes, slug size and pressure drop. *Chem. Eng. J*. 2007, 131(1-3), 1-13.

[25] Xu, J. H.; Luo, G. S.; Li, S. W.; Chen, G. G. Shear force induced monodisperse droplet formation in a microfluidic device by controlling wetting properties. *Lab Chip*. 2006, 6(1),131-136.

[26] Zhao, Y. C.; Chen, G. W.; Yuan, Q. Liquid-liquid two-phase flow patterns in a rectangular microchannel. *AIChE J*. 2006, 52(12), 4052-4060.

[27] Zhao, Y. C.; Ying, Y.; Chen, G. W.; Yuan, Q. Characterization of micro-mixing in T-shaped micromixer. *J. Chem. Ind. Eng*. (China). 2006, 57(8), 1184-1190.

[28] Zhao, Y. C.; Chen, G. W.; Yuan, Q. Liquid-liquid two-phase mass transfer in the T-junction microchannels. *AIChE J*. 2007, 53(12), 3042-3053

[29] Gunstensen, A. K.; Rothman, D. H.; Zaleski, S.; Zanetti, G. Lattice Boltzmann model of immiscible fluids. *Phys. Rev. A*. 1991, 43(18), 4320-4327.

[30] Gunstensen, A. K.; Rothman, D. H. A lattice-gas model for three immiscible fluids. *Phys. D: Nonlinear Phenomena*. 1991, 47(1-2), 47-52.

[31] Gunstensen, A. K.; Rothman, D. H. A Galilean-invariant immiscible lattice gas. *Phys. D: Nonlinear Phenomena*. 1991, 47(1-2), 53-63.

[32] Shan, X.; Chen, H. Lattice Boltzmann model for simulating flows with multiple phases and components. *Phys. Rev. E*. 1993, 47(3), 1815–1819.

[33] Swift, M. R.; Osborn, W. R.; Yeomans, J. M. Lattice Boltzmann simulation of nonideal fluids. *Phys. Rev. Lett*. 1995, 75(5), 830–833.

[34] Junseok, K. A diffuse-interface model for axisymmetric immiscible two-phase flow. *Appl. Math. Comput*. 2005, 160(2), 589–606.

[35] Sankaranarayanan, K.; Kevrekidis, I. G.; Sundaresan, S.; Lu J.; Tryggvason, G. A comparative study of lattice Boltzmann and front-tracking finite-difference methods for bubble simulations. *Int. J. Multiphase Flow*. 2003, 29(1),109-116.

[36] Sankaranarayanan, K.; Shan, X.; Kevrekidis, I. G.; Sundaresan, S. Analysis of drag and virtual mass forces in bubbly suspensions using an implicit formulation of the lattice Boltzmann method. *J. Fluid Mech*. 2002, 452, 61–96.

[37] Santos, L. O. E.; Facin, P. C.; Philippi, P. C. Lattice-Boltzmann model based on field mediators for immiscible fluids. *Phys. Rev. E*. 2003, 68(5), 056302.

[38] Santos, L. O. E.; Wolf, F. G.; Philippi, P. C. Dynamics of interface displacement in capillary flow. *J. Stat. Phys*. 2005, 121(1-2), 197-207.

[39] Surmas, R.; Santos, L. O. E.; Philippi, P. C. Lattice Boltzmann simulation of the flow interference in bluff body wakes. *Future Gener. Comp. Sy*. 2004, 20 (6), 951-958.

[40] Facin, P. C.; Philippi, P. C.; Santos, L. O. E. A non-linear lattice-Boltzmann model for ideal miscible fluids. *Future Gener. Comp. Sy*. 2004, 20 (6), 945-949.

[41] Philippi, P. C.; Hegele, L. A. Jr.; Santos, L. O. E.; Surmas, R. From the continuous to the lattice Boltzmann equation: The discretization problem and thermal models. *Phys. Rev. E*. 2006, 73 (5), 056702

[42] Surmas, R.; Pico, C. E.; Santos, L. O. E.; Philippi, P. C. Volume exclusion for reducing compressibility effects in lattice Boltzmann models. *Int. J. Mod. Phys. C.* 2007, 18 (4), 576-584.
[43] Pico, C. E; Santos, L. O. E.; Philippi, P. C. A two-fluid BGK lattice Boltzmann model for ideal mixtures. *Int. J. Mod. Phys. C.* 2007, 18 (4), 566-575.
[44] Philippi, P. C.; Hegele, L. A. Jr.; Surmas, R.; Siebert, D. N.; Santos, L. O. E. From the Boltzmann to the lattice-Boltzmann equation: Beyond BGK collision models. *Int. J. Mod. Phys. C.* 2007, 18 (4), 556-565.
[45] Siebert, D. N.; Hegele, L. A. Jr.; Surmas, R.; Santos, L. O. E. Philippi, P. C. Thermal lattice Boltzmann in two dimensions. *Int. J. Mod. Phys. C.* 2007, 18 (4), 546-555.
[46] Wolf, F. G.; Santos, L. O. E.; Philippi, P. C. Micro-hydrodynamics of immiscible displacement inside two-dimensional porous media. *Microfluid. Nanofluid.* 2008, 4(4), 307-319.
[47] Wolf, F. G.; Santos, L. O. E.; Philippi, P. C. Modeling and simulation of the fluid-solid interaction in wetting. *J. Stat. Mech-Theory E.* 2009, Article Number: P06008
[48] Sussman, M.; Fatemi, E. An efficient, interface-preserving level set redistancing algorithm and its application to interfacial incompressible fluid flow. *SIAM J. Sci. Comput.* 1999, 20(4), 1165-1191.
[49] Chang, Y. C.; Hou, T. Y.; Merriman, B.; Osher, S. A level set formulation of Eulerian interface capturing methods for incompressible fluid flows. *J. Comput. Phys.* 1996, 124(2), 449-464.
[50] Sussman, M.; Smereka, P. Axi-symmetric free boundary problems. *J. Comput. Phys.* 1997, 341(1), 269-294.
[51] Sussman, M.; Almgren, A. S.; Bell, J. B.; Colella, P.; Howell, L. H.; Welcome, M. L. An adaptive level-set approach for incompressible two-phase flows. *J. Comput. Phys.* 1999, 148(1), 81-124.
[52] Wang, Z. L.; Zhou, Z. W. An improved level-set re-initialization solver. *Appl. Math. Mech.* 2004, 25(10), 1083-1088.
[53] Sussman, M. A second order coupled level set and volume-of-fluid method for computing growth and collapse of vapor bubbles. *J. Comput. Phys.* 2003, 187(1), 110-136.
[54] Sussman, M.; Puckett, E. G. A coupled level set and volume-of-fluid method for computing 3D and axisymmetric incompressible two-phase flows. *J. Comput. Phys.* 2000, 162(2), 301-337.
[55] Sussman, M.; Smith, K. M.; Hussaini, M. Y.; Ohta, M.; Zhi-Wei, R. A sharp interface method for incompressible two-phase flows. *J. Comput. Phys.* 2007, 221(2), 469-505.
[56] Enright, D.; Fedkiw, R.; Ferziger, J.; Mitchell, I. A hybrid particle level set method for interface capturing. *J. Comput. Phys.* 2002, 183(1), 83-116.
[57] Enright, D.; Losasso, F.; Fedkiw, R. A fast and accurate semi-Lagrangian particle level set method. *Comput. Struct.* 2005, 83, 479-490.
[58] Strain, J. Tree methods for moving interfaces. *J. Comput. Phys.* 1999, 151(2), 616-648.
[59] Nourgaliev, R. R.; Wiri, S.; Dinh, N. T.; Heofanous, T. G. On improving mass conservation of level set by reducing spatial discretization errors. *Int. J. Multiphase Flow.* 2005, 31(12), 1329-1336.
[60]

[61] Li, X. Y.; Wang, Y. F.; Yu, G. Z.; Yang, C.; Mao, Z. S. A volume-amending method to improve mass conservation of level approach for incompressible two-phase flows. *Sci. China Ser. B.* 2008, 51(11), 1132-1140.

[62] Osher, S.; Sethian, J. A. Fronts propagating with curvature-dependent speed: algorithms based on Hamilton-Jacobi formulations. *J. Comput. Phys.* 1988, 143(2), 495-518.

[63] Yang, C.; Mao, Z. S. An improved level set approach to the simulation of drop and bubble motion. *Chinese J. Chem. Eng.* 2002, 10(3), 263-272.

[64] Bhage, D.; Webber, M. E. Bubbles in viscous liquids: Shapes, wakes and velocities. *J. Fluid Mech.* 1981, 105(1), 61-85.

[65] Ryskin, G.; Leal, L. G. Numerical solution of free boundary problems in fluid mechanics, Part 2. Buoyancy-driven motion of a gas bubble through a quiescent liquid. *J. Fluid Mech.* 1984, 148(1), 19-35.

[66] Tomiyama, A.; Zun, I.; Sou, A. Numerical analysis of bubble motion with the VOF method. *Nucl. Eng. Des.* 1993, 141(1-2), 69-82.

[67] Sankaranarayanan, K.; Shan, X.; Kevrekidis, I. G.; Sundaresan, S. Bubble flow simulation with the lattice Boltzmann method. *Chem. Eng. Sci.* 1999, 54(21), 4817-4823.

[68] Oka, H.; Ishii, K. Numerical analysis on the motion of gas bubbles using level set method. *J. Phys. Soc. Japan.* 1999, 68(3), 823-832.

[69] Wang, Y. F., Yang, C., Mao, Z.-S., Chen, J. Y., Application of the level set approach for numerical simulations of two-phase flow (in Chinese), *Prog. Nat. Sci.,* 2004, 14(2): 220-222.

[70] Manga, M.; Stone, H. A. Buoyancy-driven interactions between two deformable viscous drops. *J. Fluid Mech.* 1993, 256(3), 647-683.

[71] Wang, J. F.; Yang, C.; Mao, Z. S. A simple weighted integration method for calculating surface tension force to suppress parasitic flow in the level set approach. *Chinses J. Chem. Eng.* 2006, 14(6), 740-746.

[72] Chhabra, R. P. *Bubbles, Drops, and Particles in Non-Newtonian Fluids*, Boca Raton, FL, CRC Press, 1993.

[73] Chhabra, R. P., *Bubbles, Drops, and Particles in Non-Newtonian Fluids* (second ed.), Boca Raton, FL, CRC Press, 2006.

[74] Kulkarni, A. A.; Joshi, J. B. Bubble formation and bubble rise velocity in gas-liquid system: A review. *Ind. Eng. Chem. Res.* 2005, 44(16), 5873-5931.

[75] Mao, Z. S.; Li, T. W.; Chen, J. Y. Numerical simulation of steady and transient mass transfer to a single drop dominated by external resistance. *Int. J. Heat Mass Transfer.* 2001, 44(6),1235-1247.

[76] Yang, C.; Mao, Z. S. Numerical simulation of interphase mass transfer with the level set approach. *Chem. Eng. Sci.* 2005, 60(10), 2643-2660.

[77] Harper, J. F.; Moore, D. W. The motion of a spherical liquid drop at high Reynolds number. *J. Fluid Mech.* 1968, 32, 367-391.

[78] Li, T. W.; Mao, Z. S.; Chen, J. Y.; Fei, W. Y. Experimental and numerical investigation of single drop mass transfer in solvent extraction systems with resistance in both phases. *Chinese J. Chem. Eng.* 2002,10 (1), 1–14.

[79] Harvey P.; Greaves M. Turbulent flow in an agitated vessel-Part II: Numerical solution and model predictions. *Chem. Eng. Res. Des.* 1982, 60, 201-210.

[80] Ishii M. *Thermo-Fluid Dynamic Theory of Two-Phase Flow.* Paris: Eyrolles, 1975

[81] Shan X.; Yu G.; Yang C.; Mao Z. S.; Zhang W. Numerical simulation of liquid solid flow in an unbaffled stirred tank with a pitched-blade turbine downflow. *Ind. Eng. Chem. Res*. 2008, 47, 2926-2940.

[82] Ranade V.; Van den Akker H. A computational snapshot of gas-liquid flow in baffled stirred reactors. *Chem. Eng. Sci*. 1994, 49, 5175-5192.

[83] Lin W.; Mao Z. S.; Chen J. Hydrodynamic studies on loop reactors (II) Airlift loop reactors. *Chinese J. Chem. Eng*. 1997, 5, 11-22.

[84] Wang W.; Mao Z. S. Numerical simulation of gas-liquid flow in a stirred tank with a Rushton impeller. *Chinese J. Chem. Eng*. 2002, 10, 385-395.

[85] Khopkar A.; Rammohan A.; Ranade V.; Dudukovic, M. Gas–liquid flow generated by a Rushton turbine in stirred vessel: CARPT/CT measurements and CFD simulations. *Chem. Eng. Sci*. 2005, 60, 2215-2229.

[86] Ishii M.; Zuber, N. Drag coefficient and relative velocity in bubbly, droplet or particulate flows. *AIChE J*. 1979, 25, 843-855.

[87] Kataoka, I.; Besnard, D.; Serizawa, A. Basic equation of turbulence and modeling of interfacial transfer terms in gas-liquid two-phase flow. *Chem. Eng. Commun*. 1992, 118, 221-236.

[88] Ranade, V.; Joshi, J. Flow generated by a disc turbine: Part II Mathematical modelling and comparison with experimental data. *Chem. Eng. Res. Des*. 1990, 68, 34-50.

[89] Hesketh, R.; Russell, T.; Etchells, A. Bubble size in horizontal pipelines. *AIChE J*. 1987, 33, 663–667.

[90] Parthasarathy R.; Ahmed, N. Sauter mean and maximum bubble diameters in aerated stirred vessels. *Chem. Eng. Res. Des*. 1994, 72, 565-572.

[91] Brown D.; Pitt, K. Drop size distribution of stirred non-coalescing liquid-liquid system. *Chem. Eng. Sci*. 1972, 27, 577-583.

[92] Calabrese, R.; Chang, T.; Dang, P. Drop breakup in turbulent stirred-tank contactors. I: Effect of dispersed-phase viscosity. *AIChE J*. 1986, 32, 657-666.

[93] Gosman, A.; Lekakou, C.; Politis, S.; Issa, R.; Looney, M. Multidimensional modeling of turbulent two-phase flows in stirred vessels. *AIChE J*. 1992, 38, 1946-1956.

[94] Micale, G.; Montante, G.; Grisafi, F.; Brucato, A.; Godfrey, J. CFD simulation of particle distribution in stirred vessels. *Chem. Eng. Res. Des*. 2000, 78, 435-444.

[95] Montante, G.; Micale, G.; Magelli, F.; Brucato, A. Experiments and CFD predictions of solid particle distribution in a vessel agitated with four pitched blade turbines. *Chem. Eng. Res. Des*. 2001, 79, 1005-1010.

[96] Brucato, A.; Ciofalo, M.; Grisafi, F.; Micale, G. Numerical prediction of flow fields in baffled stirred vessels: A comparison of alternative modelling approaches. *Chem. Eng. Sci*. 1998, 53, 3653-3684.

[97] Luo, J.; Issa, R.; Gosman, A. Prediction of impeller induced flows in mixing vessels using multiple frames of reference. *IChemE Symp. Ser*. 1994, 136, 549-556.

[98] Versteeg H.; Malalasekera, W. *An introduction to computational fluid dynamics: the finite volume method:* Prentice Hall, 1995.

[99] Patankar S.; Spalding, D. A calculation procedure for heat, mass and momentum transfer in three-dimensional parabolic flows. *Int. J. Heat Mass Transfer*. 1972, 15, 1787–1806.

[100] Carver M.; Salcudean, M. Three dimensional numerical modeling of phase distribution of two-fluid flow in elbaus and return bends. *Numer. Heat Transfer*. 1986, 10, 229-251.

[101] Zhou G.; Kresta, S. Correlation of mean drop size and minimum drop size with the turbulence energy dissipation and the flow in an agitated tank. *Chem. Eng. Sci.* 1998, 53, 2063-2079.

[102] Wichterle, K. Drop breakup by impellers. *Chem. Eng. Sci.* 1995, 50, 3581-3586.

[103] Svensson F.; Rasmuson, A. LDA-measurements in a stirred tank with a liquid-liquid system at high volume percentage dispersed phase. *Chem. Eng. Technol.* 2004, 27, 335-339.

[104] Skelland A.; Lee, J. Agitator speeds in baffled vessels for uniform liquid-liquid dispersions. *Ind. Eng. Chem. Process Des. Develop.* 1978, 17, 473-478.

[105] Okufi, S.; de Ortiz, E.; Sawistowski, H. Scale-up of liquid-liquid dispersions in stirred tanks. *Can. J. Chem. Eng.* 1990, 68, 400-406.

[106] Armenante P.; Huang, Y. Experimental determination of the minimum agitation speed for complete liquid-liquid dispersion in mechanically agitated vessels. *Ind. Eng. Chem. Res.* 1992, 31, 1398-1406.

[107] Zhu X.; Vigil, R. Banded liquid–liquid Taylor-Couette-Poiseuille flow. *AIChE J.* 2001, 47, 1932-1940.

[108] Wang F.; Mao, Z.-S. Numerical and Experimental Investigation of Liquid Liquid Two-Phase Flow in Stirred Tanks. *Ind. Eng. Chem. Res.* 2005, 44, 5776-5787.

[109] Joshi, J. Computational flow modelling and design of bubble column reactors. *Chem. Eng. Sci.* 2001, 56, 5893–5933.

[110] Ljungqvist M.; Rasmuson, A. Numerical simulation of the two-phase flow in an axially stirred vessel. *Chem. Eng. Res. Des.* 2001, 79, 533-546.

[111] Nagata, S. *Mixing: Principles and Applications*: Halsted Press, 1975.

[112] Anderson T.; Jackson, R. A fluid mechanical description of fluidized beds. *Ind. Eng. Chem. Fund.* 1967, 6, 527–538.

[113] [Van Wijngaarden, L. Hydrodynamic interaction between gas bubbles in liquid. *J. Fluid Mech.* 1976, 77, 27-44.

[114] Auton, T. The lift force on a spherical rotational flow. *J. Fluid Mech.* 1987, 183, 199-218.

[115] Zhang, Y.; Yong, Y.; Mao, Z.-S.; Yang, C.; Sun, H.; Wang, H. Numerical Simulation of Gas-Liquid Flow in a Stirred Tank with Swirl Modification. *Chem. Eng. Technol.* 2009, 32, 1266-1273.

[116] Zhang, Y.; Yang, C.; Mao, Z.-S. Large eddy simulation of the gas-liquid flow in a stirred tank. *AIChE J.* 2008, 54, 1963-1974.

[117] Lu W.; Ju, S. Local gas holdrup, mean liquid velocity and turbulence in an aerated stirred tank using hot-film anemometry. *Chem. Eng. J.* 1987, 35, 9-17.

[118] Wang, F.; Wang, W.; Mao, Z.-S. Numerical study of solid-liquid two-phase flow in stirred tanks with Rushton impeller (I) Formulation and simulation of flow field. *Chinese J. Chem. Eng.* 2004, 12, 599-609.

[119] Zwietering, T. Suspending of solid particles in liquid by agitators. *Chem. Eng. Sci.* 1958, 8, 244-253.

[120] Armenante, P.; Nagamine, E.; Susanto, J. Determination of correlations to predict the minimum agitation speed for complete solid suspension in agitated vessels. *Can. J. Chem. Eng.* 1998, 76, 413-419.

[121] Armenante P.; Nagamine, E. Effect of low off-bottom impeller clearance on the minimum agitation speed for complete suspension of solids in stirred tanks. *Chem. Eng. Sci.* 1998, 53, 1757-1775.

[122] Baldi, G.; Conti, R.; Alaria, E. Complete suspension of particles in mechanically agitated vessels. *Chem. Eng. Sci.* 1978, 33, 1-25.

[123] Wichterle, K. Conditions for suspension of solids in agitated vessels. *Chem. Eng. Sci.* 1988, 43, 467-471.

[124] Luo, J.; Gosman, A.; Issa, R.; Middleton, J.; Fitzgerald, M. Full flow field computation of mixing in baffled stirred vessels. *Chem. Eng. Res. Des.* 1993, 71, 342-344.

[125] Brucato, A.; Ciofalo, M.; Grisafi, F.; Micale, G. Complete numerical simulation of flow fields in baffled stirred vessels: the inner-outer approach, *Can. J. Chem. Eng.* 1994, 71, 269-289.

[126] Angst R.; Kraume, M. Experimental investigations of stirred solid/liquid systems in three different scales: Particle distribution and power consumption. *Chem. Eng. Sci.* 2006, 61, 2864-2870.

[127] Micale, G.; Carrara, V.; Grisafi, F.; Brucato, A. Solids suspension in three-phase stirred tanks. *Chem. Eng. Res. Des.* 2000, 78, 319-326.

[128] Micale, G.; Grisafi, F.; Brucato, A. Assessment of particle suspension conditions in stirred vessels by means of pressure gauge technique. *Chem. Eng. Res. Des.* 2002, 80, 893-902.

[129] Zhang, H.; Johnston, P.; Zhu, J.; De Lasa, H.; Bergougnou, M. A novel calibration procedure for a fiber optic solids concentration probe. *Powder Technol.* 1999, 100, 260-272.

[130] Zheng, Y.; Zhu, J.; Marwaha, N.; Bassi, A. Radial solids flow structure in a liquid–solids circulating fluidized bed. *Chem. Eng. J.* 2002, 88, 141-150.

[131] Ochieng A.; Lewis, A. Nickel solids concentration distribution in a stirred tank. *Mineral Eng.* 2006, 19, 180-189.

[132] Derksen, J. Numerical simulation of solids suspension in a stirred tank. *AIChE J.* 2003, 49, 2700–2714.

[133] Wang, F.; Mao, Z.-S.; Wang, Y.; Yang, C. Measurement of phase holdups in liquid-liquid-solid three-phase stirred tanks and CFD simulation. *Chem. Eng. Sci.* 2006, 61, 7535-7550.

[134] Padial, N.; VanderHeyden, W.; Rauenzahn, R.; Yarbro, S. Three-dimensional simulation of a three-phase draft-tube bubble column. *Chem. Eng. Sci.* 2000, 55, 3261-3273.

[135] Michele V.; Hempel, D. Liquid flow and phase holdup-measurement and CFD modeling for two-and three-phase bubble columns. *Chem. Eng. Sci.* 2002, 57, 1899-1908.

[136] Bendjaballah, N.; Dhaouadi, H.; Poncin, S.; Midoux, N.; Hornut, J. M.; Wild, G. Hydrodynamics and flow regimes in external loop airlift reactors. *Chem. Eng. Sci.* 1999, 54, 5211-5221

[137] Freitas, C.; Fialová, M.; Zahradnik, J.; Teixeira, J. A. Hydrodynamics of a three-phase external-loop airlift bioreactor. *Chem. Eng. Sci.* 2000, 55, 4961-4972.

[138] Ayed, H.; Chahed, J.; Roig, V. Hydrodynamics and mass transfer in a turbulent buoyant bubbly shear layer. *AIChE J.* 2007, 53, 2742-2753.

[139] Patankar, S. V. *Numerical Heat Transfer and Fluid Flow*. 1980, McGrwa-Hill, New York.
[140] Darwish, M.; Moukalled, F. Sekar, B. A unified formulation of the segregated class of algorithms for multifluid flow at all speeds. *Numer. Heat Transfer* B 2001, 40, 99-137.
[141]]Moukalled, F.; Darwish, M.; A comparative assessment of the performance of mass conservation-based algorithms for incompressible multiphase flows. *Numer. Heat Transfer B*. 2002, 42, 259-283.
[142] Versteeg, H. K.; Malalasekera, W. *An introduction to computational fluid dynamics: the finite volume method.* 1995, Wiley, New York.
[143] Mudde, R. F.; Van Den Akker, H. E. A. 2D and 3D simulations of an internal airlift loop reactor on the basis of a two-fluid model. *Chem. Eng. Sci.* 2001, 56, 6351-6358.
[144] Huang, Q. S.; Yang, C.; Yu, G. Z.; Mao, Z.-S. Sensitivity study on modeling an internal airlift loop reactor using a steady 2D two-fluid model. *Chem. Eng. Technol.* 2008, 31, 1790-1798
[145] Acién Fernández, F. G.; Fernández Sevilla, J. M.; Sánchez Pérez, J. A.; Molina Grima, E.; Chisti, Y. Airlift-driven external-loop tubular photobioreactors for outdoor production of microalgae: assessment of design and performance. *Chem. Eng. Sci.* 2001, 56, 2721-2732
[146] Cockx, A.; Do-Quang, Z.; Line, A.; Roustan, M. Use of computational fluid dynamics for simulating hydrodynamics and mass transfer in industrial ozonation towers. *Chem. Eng. Sci.* 1999, 54, 5085-5090.
[147] Talvy, S.; Cockx, A.; liné, A. Modeling of oxygen mass transfer in a gas-liquid airlift reactor. *AIChE. J.* 2007, 53, 316-326.
[148] Higbie, R. The rate of absorption of a pure gas into a still liquid during short periods of exposure. *Trans. Am. Inst. Chem. Eng.* 1935, 35, 365-389.
[149] Xue, S. W.; Yin, X. Numerical simulation of flow behavior and mass transfer in internal airlift-loop reactor. *Chem. Eng.* (China) 2006, 34, 23-27.
[150] Vasconcelos, J. M. T.; Rodrigues, J. M. L.; Orvalho, S. C. P.; Alves, S. S.; Mendes R. L.; Reis, A. Effect of contaminants on mass transfer coefficients in bubble column and airlift contactors. *Chem. Eng. Sci*. 2003, 58, 1431-1440.
[151]]van Baten, J. M.; Ellenberger, J.; Krishna, R. Hydrodynamics of internal air-lift reactors: experiments vs. CFD simulations. *Chem. Eng. Process.* 2003, 42, 733-742.
[152] Huang Q. S.; Yang C.; Yu G. Z.; Mao Z.-S. 3-D simulations of an internal airlift loop reactor using a steady two-fluid model. *Chem. Eng. Technol.* 2007, 30(7), 870-879
[153] Juraščík, M.; Blažej, M.; Annus, J.; Markoš, J. Experimental measurements of volumetric mass transfer coefficient by the dynamic pressure-step method in internal loop airlift reactors of different scale. *Chem. Eng. J.* 2006, 125, 81-87.
[154] Huang Q. S.; Yang C.; Yu G. Z.; Mao Z.-S. CFD simulation of hydrodynamics and mass transfer in an internal airlift loop reactor using a steady two-fluid model. *Chem. Eng. Sci.* 2010, 65, 5527-5536
[155] Zheng H. W.; Shu C.; Chew Y. T. A lattice Boltzman model for multiphase flows with large density ratio. *J. Comput. Phys.* 2006, 218, 353-371.
[156] Onishi J.; Chen Y.; Ohashi H. Dynamic simulation of multi-component viscoelastic fluids using the lattice *Boltzmann method*. *Physica A*. 2006, 362, 84-92.
[157] Mason R. J. A. Multi-speed compressible lattice-Boltzmann model. *J. Stati. Phys.* 2002, 107, 385-400.

[158] Lallemand P.; Luo L. S. Theory of the lattice Boltzmann method-acoustic and thermal properties in two and three dimensions. *Phys. Rev. E.* 2003, 68, 1-25.

[159] Gupa N.; Chaitanya G. R.; Mishra S. C. Lattice Boltzmann method applied to variable thermal conductivity conduction and radiation problems. *Int. J. Heat Mass Transfer.* 2006, 20(4), 895-902.

[160] Sullivan S. P.; Gladden L. F. 3D chemical reactor LB simulations. *Math. Comput. Simulat.* 2006, 72, 206-211.

[161] Chen, Y. G.; Zhang, S. H; Chen, J. Z. Water hammer simulation by the lattice Boltzmann method. *J. Hydraul. Eng.* 1998, 29(6):25-31

[162] Peng, D. P.; Merriman, B.; Osher, S.; Zhao H. K.; Kang, M. A PDE-based fast local level set method. *J. Comput. Phys.* 1999, 155, 410-438

[163] Sethian, J. A. Theory, algorithms and applications of level set methods of propagating interfaces. Technical Report PAM-651, *Center for Pure and Applied Mathematics*, University of California, Berkeley, August 1995

[164] Giga, Y., and Goto, S., Motion of hypersurfaces and geometric equations, *J. Math. Soc. Jpn.* 1992, 44, 99.

[165] Evans, L.C., and Spruck, J., Motion of level sets by mean curvature IV, *J. Geom. Anal.* 1995, 5(1), 77-114.

[166] Evans, L.C., and Spruck, J., Motion of level sets by mean curvature I, *J. Diff. Geom.* 1991, 33, 635.

[167] Ambrosio, L.; Sonar, H.M., Level set approach to mean curvature flow in arbitrary codimension. *J. Diff. Geom.* 1996, 43(4):693-737

[168] Crandall, M.G.; Ishii, H.; Lions, P L. User's guide to viscosity solutions of second order partial differential equations, *Bull. AMS*, 1992, 27/1, 1-67.

[169] Barles, G. Discontinuous viscosity solutions of first order Hamilton-Jacobi equations: a guided visit. *Nonlinear Anal.* 1993, 20(9), 1123-1134.

[170] Sethian, J.A. *A brief overview of vortex methods, in vortex methods and vortex motion*, Eds. K. Gustafson and J.A. Sethian, SIAM Publications, Philadelphia, 1991.

[171] Sethian, J.A.; Strain, J.D. Crystal growth and dendritic solidification. *J. Comp. Phys.* 1992, 98, 231-253.

[172] Sussman, M.; Smereka, P.; Osher, S.J. A level set method for computing solutions to incompressible two-phase flow. *J. Comp. Phys.* 1994, 114, 146-159.

[173] Falcone, M. The minimum time problem and its applications to front propagation, in "Motion by mean curvature and related topics". *Proceedings of the International Conference at Trento*, 1992

[174] Zhu, J.; Ronney, P. D. Simulation of front propagation at large non-dimensional flow disturbance intensities. *Comb. Sci. Tech.* 1995, 100:183–201

[175] Kimmel, R. *Curve Evolution on Surfaces*, Ph.D. Thesis, Dept. of Electrical Engineering, Technion, Israel, 1995.

[176] Kichenassamy, S.; Kumar, A.; Olver, P.; Tannenbaum, A.; Yezzi, A. Gradient flows and geometric active contours. Fifth International Conference on Computer Vision, *IEEE Computer Soc. Press*, Cambridge, Mass., 1995, 810-815.

[177] Sapiro, G.; Tannenbaum, A. Affine Invariant Scale-Space. *Int. Jour. Comp. Vision.* 1993, 11(1), 25-44.

[178] Fatemi, E.; Engquist, B.; Osher, S. J. Numerical solution of the high frequency asymptotic wave equation for the scalar wave equation. *J. Comp. Phys.* 1995, 120, 145-155.
[179] Pope, S. A more general effective-viscosity hypothesis. *J. Fluid Mech.* 1975, 72, 331-340.
[180] Rodi, W. *A new algebraic relation for calculating the Reynolds stresses.* Zeitschrift fuer angewandte Mathematik und Mechanik. 1976, 56, 219-221.
[181] Yoshizawa, A. Statistical analysis of the deviation of the Reynolds stress from its eddy-viscosity representation. *Phys. Fluids.* 1984, 27, 1377-1387.
[182] Speziale, C. On nonlinear k-l and k-ε models of turbulence. *J. Fluid Mech.* 1987, 178, 459-475.
[183] Gatski, T.; Speziale, C. On explicit algebraic stress models for complex turbulent flows. *J. Fluid Mech.* 1993, 254, 59-78.
[184] Rubinstein, R.; Barton, J. Nonlinear Reynolds stress models and the renormalization group. *Phys. Fluids A: Fluid Dynamics.* 1990, 2, 1472-1476.
[185] Taulbee, D. An improved algebraic Reynolds stress model and corresponding nonlinear stress model. *Phys. Fluids A: Fluid Dynamics.*1992, 4, 2555-2561.
[186] Lesieur, M. *Turbulence in Fluids.* Dordrecht, The Netherlands: Springer. 2008, 84.
[187] Yoon, H.; Balachandar, S.; Ha, M.; Kar, K. Large eddy simulation of flow in a stirred tank. *J. Fluids Eng.* 2003, 125, 486-499.
[188] Eggels, J. Direct and large-eddy simulation of turbulent fluid flow using the lattice-Boltzmann scheme. *Int. J. Heat Fluid Flow.* 1996, 17, 307-323.
[189] Derksen J.; Van den Akker, H. Large eddy simulations on the flow driven by a Rushton turbine. *AIChE J.* 1999, 45, 209-221.
[190] Revstedt J.; Fuchs, L. Large eddy simulation of flow in stirred tanks. *Chem. Eng. Technol.* 2002, 25, 443-446.
[191] Yeoh, S.; Papadakis, G.; Lee, K.; Yianneskis, M. Large eddy simulation of turbulent flow in a Rushton impeller stirred reactor with sliding-deforming mesh methodology. *Chem. Eng. Technol.* 2004, 27, 257-263.
[192] Alcamo, R.; Micale, G.; Grisafi, F.; Brucato, A.; Ciofalo, M. Large-eddy simulation of turbulent flow in an unbaffled stirred tank driven by a Rushton turbine. *Chem. Eng. Sci.* 2005, 60, 2303-2316.
[193] Zhang, Y.; Yang, C.;. Mao, Z. S. Large eddy simulation of liquid flow in a stirred tank with improved inner-outer iterative algorithm. *Chin. J. Chem. Eng.* 2006, 14, 321-329.
[194] Wang, Q.; Squires, K.; Simonin, O. Large eddy simulation of turbulent gas-solid flows in a vertical channel and evaluation of second-order models. *Int. J. Heat Fluid Flow.*1998, 19, 505-511.
[195] Apte, S.; Gorokhovski, M.; Moin, P. LES of atomizing spray with stochastic modeling of secondary breakup. *Int. J. Multiphase Flow.* 2003, 29, 1503-1522.
[196] Shotorban B.; Mashayek, F. Modeling subgrid-scale effects on particles by approximate deconvolution. *Phys. Fluids.* 2005, 17, 1-4.
[197] Smith B.; Milelli, M. An investigation of confined bubble plumes. in Proc. 3rd *International Conference on Multiphase Flow*. Lyon, France, 1998, 8-12.
[198] Deen, N.; Solberg, T.; Hjertager, B. Large eddy simulation of the gas–liquid flow in a square cross-sectioned bubble column. *Chem. Eng. Sci.* 2001, 56, 6341-6349.

Chapter 14

MODELING AND OPTIMISATION OF MICROFLUIDIC PASSIVE MIXER USING CFD ANALYSIS

Karol Malecha[1*] *and Ziemowit M. Malecha*[2†]
[1]Faculty of Microsystem Electronics and Photonics,
Wrocław University of Technology,
Janiszewskiego 11/17, 50-372 Wrocław, Poland.
[2]Faculty of Mechanical and Power Engineering,
Wrocław University of Technology,
Wybrzeże Wyspiańskiego 27, 50-370 Wrocław, Poland.

Abstract

This chapter includes basic information on the scaling effect in flui mechanics. Computational flui dynamics (CFD) is used to simulate and study the fl w and mixing process of two incompressible and miscible fluid in microscale. The mathematical model of the considered fl w, along with the computational procedure used to solve it, is also described. The microscale geometry of interest is a sequence of the different microconduits with a rectangular cross-section. Moreover, the influenc of fl w conditions and various arrangements of microconduits on mixing efficien y is also presented and discussed.

PACS 47.11.-j, 47.11.Df, 47.15.Rq.

Keywords: microfluidic mixing, diffusion, chaotic advection, numerical modeling.

1. Introduction

Analytical procedures applied in modern chemistry, biology and medicine consist of several steps: sample collection, carrying out the appropriate (bio)chemical reaction, product separation, and detection of the analyte. Classical laboratory equipment is characterized by relatively long time analysis, considerable reagent consumption and large

---
[*]E-mail address: karol.malecha@pwr.wroc.pl
[†]E-mail address: ziemowit.malecha@pwr.wroc.pl

amounts of wastes. All mentioned disadvantages can be eliminated through the use of microfluidi systems. These miniature devices have dimensions ranging from millimeters to micrometers and are capable of handling fluid in the micro- and nanoliter scale (10^{-6} - 10^{-9} litre)[1]. Microfluidi systems fin practical application in analytical diagnosis, DNA (Deoxyribonucleic acid) sequencing, cell separation, and environmental monitoring because of their potential and advantages: small amounts of sample, reagents, and wastes, low time consumption, low cost, and high throughput. One of the essential parts of the modern microfluidi system is an efficien micromixer. For example, enzymatic reactions require the intermix of all reagents for initiation. In general, micromixers can be classifie as either active or passive[2]. An active micromixer requires external energy (pressure, temperature, electro-hydrodynamics, acoustics, etc.) to generate the disturbances necessary for achieving the efficien mixing process[3, 4]. In contradistinction to active micromixers, passive mixing structures do not require external forces[5, 6]. Their mixing efficien y relies entirely on molecular diffusion and chaotic advection. Passive micromixers are generally easier to fabricate and integrate with other microfluidi devices. A good mixing structure should homogenize two (or more) fluid in micro- or nanoliter volume scale without taking up too much space in acceptable time-scales. However, due to very small dimensions of the microfluidi systems, the fl w is mostly in laminar regime[7].

In this chapter, practical application of the CFD (computational flui dynamics) analysis for the design process of the microfluidi passive mixer is described. The CFD modeling is carried out for a transient, laminar fl w of miscible and incompressible fluid through a sequence of microconduits. The numerical simulations are used for optimization of the microfluidi system's geometry and consideration of the influenc of operating condition (e.g. fl w rate) on mass transfer in microscale. Results of the CFD analyses are used to determine an optimal micromixer design.

2. Basic Terms and Equations in Fluid Dynamics

Chemical and process engineering divide flui fl w patterns into two categories:

1. laminar fl w,

2. and turbulent fl w.

In laminar fl w, forces connected with flui viscosity dominate. As a result of this, flui particles move calmly and orderly according to the streamlines. In the case of turbulent fl w, inertia forces exceed viscous effects. The flui particles move chaotically in all directions without any traceable pattern. The transition from laminar to turbulent fl w is given by the Reynolds number:

$$Re = \frac{\rho U L}{\mu} \quad (1)$$

where ρ is the flui density [kg/m^3], U is the flui mean velocity [m/s], L is the characteristic length [m], and μ is the flui dynamic viscosity [$Pa \cdot s$]. The laminar flui fl w for fl w in circular pipes occurs for $Re < 2400$ [16, 11]. In this case, viscous effects dominate over inertia forces and flui re-circulations and vortices decay rapidly. One phenomenon which arises from the equation (1) is that the Reynolds number is inversely proportional to the second power of the conduit characteristic dimension:

$$Re \sim L^2 (U \sim L) \qquad (2)$$

this means that turbulence disappears in microsystems in which flui fl ws[7].
In practice, the value of the Re at which occurs the transition from laminar to turbulent fl w, depends on conduit geometry. Therefore, in chemical and process engineering, the parameter L from equation (1) is replaced by hydraulic diameter D_h. It is a commonly used term when handling fl w in non-circular conduits. It can be calculated using equation:

$$D_h = \frac{4A}{O} \qquad (3)$$

where A is the cross sectional area of the channel [m^2] and O is the „wet perimeter" [m], meaning the perimeter that is in contact with the fluid One can notice that parameter D_h in the above expression for a circular tube is equal to its diameter. The hydraulic diameter for a rectangular channel with width w and height h is equal to:

$$D_h = \frac{4 \cdot w \cdot h}{2(w+h)} = \frac{2wh}{w+h} \qquad (4)$$

The second parameter from equation (1) which should be define more precisely is the mean flui velocity. It is define as:

$$U = \frac{Q}{A} \qquad (5)$$

where Q is a flui fl w rate [m^3/s].
On the basis of equations (1), (4), and (5), it can be noticed that in most cases of microfluidics a low Reynolds number is expected. In consequence, the mass transfer is based mainly on molecular diffusion. The diffusive mixing process is relatively slow in comparison with the rate at which flui is convected along the conduit. This makes effective mixing of initially segregated fluid difficult The Péclet number represents the ratio between the mass transport due to convection and that of diffusion, and is define as:

$$Pe = \frac{UD_h}{D} \qquad (6)$$

where D is a coefficien of molecular diffusion [m^2/s]. It varies from approximately 10^{-9} for small molecules (e.g. water) to 10^{-11} for larger ones (e.g. glucose). Convection is dominant at higher Péclet number. Typical values of the Pe for microfluidi systems are high and vary from 10^1 to 10^6[2, 7]. The Péclet number can be interpreted as a ratio of diffusive mixing time:

$$\tau_{diff} \sim \frac{D_h^2}{D} \qquad (7)$$

using equations (5) and (7) we are able to observe that fluid which move with a mean velocity U have to pass through a distance of $U(D_h^2/D)$ to be completely mixed. For example, the mixing time and distance required to homogenize the concentrations for a mean fl w velocity of 1 mm/s and a channel hydraulic diameter of 200 μm may vary from

40 s and 0.4 m for a diffusion coefficien of 10^{-9} m^2/s to 4000 s and 40 m for diffusivity of 10^{-11} m^2/s. The magnitude of the mass flu (J) of the flui particles due to molecular diffusion is equal to:

$$J = -D\nabla c \quad (8)$$

where c is the concentration $[m^{-3}]$. ∇ (nabla) symbol is a vector differential operator. For the three-dimensional Cartesian coordinate system R^3 with (x, y, z) coordinates, ∇ is define as:

$$\nabla = (\frac{\partial}{\partial x}, \frac{\partial}{\partial y}, \frac{\partial}{\partial z}) \quad (9)$$

From equation (8), it is clearly seen that the key to efficien mixing in microscale relies mainly on two points:

1. the amount of interface area between two initially segregated fluids

2. and the ability to create high concentration gradients between two mixing fluids

On one hand, a larger interface area results in larger area for mass transfer. On the other hand, very high concentration gradients accelerate the diffusion process. Elongation of the inter-material area and the increase of concentration gradients can be achieved with the stretch and fold phenomena, characteristic of chaotic advection. These effects can be obtained in microfluidic through the appropriate design of the micromixer's channel geometry. Spatial changes along the fluidi channel axes result in frequent changes of fl w direction. According to Ottino and Wiggins[8, 9] it may lead to chaotic mixing.

The mixing process in microfluidi system can be modeled and visualized by solving the convection-diffusion equation:

$$\frac{\partial c}{\partial t} + \mathbf{u} \cdot \nabla c = D\nabla^2 c \quad (10)$$

where $\mathbf{u} = (u, v, w)$ is the flui velocity vector with $[m/s]$ unit. The velocity fiel can be calculated from the Navier-Stokes equations which describe the conservation of momentum [16, 15]:

$$\frac{\partial(\rho \mathbf{u})}{\partial t} + \nabla \cdot (\rho \mathbf{u}\mathbf{u}) = -\nabla p + \nabla \cdot \mu \nabla \mathbf{u} + F \quad (11)$$

where p and F denote respectively the pressure $[Pa]$ and the external driving force $[N]$. In equations (11) both velocity $\mathbf{u}$ and pressure p are unknown and that is why one more equation is needed to close the system. In the case of the incompressible fl ws, it is a continuity equation:

$$\nabla \cdot (\rho \mathbf{u}) = 0 \quad (12)$$

which describes the mass conservation.

Although equations (11) and (12) describe the incompressible fl w, they are also valid for fl ws with spatially varying density $\rho = \rho(x, y, z)$ and viscosity $\mu = \mu(x, y, z)$. For fl ws with constant density and viscosity, the Navier-Stokes system can be simplifie to:

$$\begin{array}{c} \frac{\partial(\mathbf{u})}{\partial t} + \nabla \cdot (\mathbf{u}\mathbf{u}) = -\nabla p + \nu \nabla^2 \mathbf{u} + F \\ \nabla \cdot (\mathbf{u}) = 0 \end{array} \quad (13)$$

where $\nu = \mu/\rho$ denotes kinematic viscosity $[m^2/s]$. In some cases of the microfluidic systems, it happens that the Reynolds number is very low $Re \ll 1$. In consequence, the convection term $\nabla \cdot (\rho \mathbf{u}\mathbf{u})$ in equation (11) can be neglected, and the equations simplify to Stokes flow:

$$\frac{\partial(\rho \mathbf{u})}{\partial t} = -\nabla p + \nabla \cdot \mu \nabla \mathbf{u} + F. \tag{14}$$

The presented Navier-Stokes equations are partial differential equations which have analytical solutions only for a few very simple cases. Therefore, more sophisticated cases have to be solved by using numerical methods such as the finite difference method (FDM), finite element method (FEM), and finite volume method (FVM), or others [13, 14, 15]. To model the microfluidic mixer performance, it is necessary to apply the mathematical model for two mixing fluids with differing density and viscosity. This is possible by combining the incompressible flow equations (11), (12) and transport equation (10).

3. Two Mixing Fluids Model

From the macroscopic perspective, incompressible fluid can be characterized by its density and viscosity. Two different fluids can be distinguished through these physical properties. For the first fluid we can write: ρ_1 and μ_1 and for the second: ρ_2 and μ_2. If we denote the concentration of the fluid in the mixture by c (see equation 10), then $c = 1$ means pure phase 1 (only first fluid) and $c = 0$ means pure phase 2 (only second fluid). Any value between 0 and 1 relates to the mixture of a certain concentration c of the first fluid within the second fluid. The density and viscosity of the mixture at any point in the domain are calculated as a weighted average of the concentration of the two fluids

$$\rho = c\rho_1 + (1-c)\rho_2 \tag{15}$$

and

$$\mu = c\mu_1 + (1-c)\mu_2 \tag{16}$$

Regardless of whether the density or viscosity of any mixing phases are constant, the resulting mixture density ρ and viscosity μ are functions of the space (they are only constant if $\rho_1 = \rho_2 = const$ and $\mu_1 = \mu_2 = const$). That is why in order to model incompressible fluid mixture dynamics, we have to solve Navier-Stokes equations in the form (11).

In general form, the computational algorithm goes as follows:

1. set the initial and boundary conditions for $\mathbf{u}, p, c$

2. calculate the mixture density ρ and viscosity μ using eq. (15), (16)

3. solve next time step of the Navier-Stokes equation for mixture (eq. 11, 12)

4. solve next time step of the transport equation for concentration c (eq. 10)

5. increase simulation time by one time step and go to step 2.

As mentioned before, the unknown in the momentum and continuity equations are velocity and pressure, but any of these equations can be used directly to calculate the pressure field That is why the pressure fiel can only be calculated by applying some additional procedure. One of the possibilities are projection methods, such as SIMPLE or PISO [12, 15, 13].

4. Optimization of the Microfluidic Passive Mixer

To model and optimize the microfluidi passive mixer, we used OpenFOAM (Open Source Field Operation and Manipulation) [10], a FVM-based, open-source C++ library for solving partial differential equations (PDEs) in arbitrary 3-D geometries (*http://www.openfoam.com/*). OpenFOAM is distributed along with many ready-to-use solvers. One of them, which has been used in present simulations, is called *twoMixingFluidsFoam* and is designed to model the mixing process of two different fluid and is implemented along with different turbulence models. In the laminar version, it is based on the model described in the previous section. The momentum equation in *twoMixingFluidsFoam* is solved using the PISO method.

The main part of CFD modeling of the fl w and mixing process for the two fluid was carried out via four varied geometries of the micromixer. Each mixer consisted of T-junction and a sequence of microchannels. Every micromixing structure was 10 mm long, 300 μm wide, and 150 μm high. The CFD modeling was made for laminar and incompressible fl w through a sequence of micro-conduits with rectangular cross-sections. The Navier-Stokes (16) and diffusion-convection (10) equations were solved using the FVM method which is implemented in the OpenFOAM software package. The numerical computations were made to show the impact of the micromixer's geometry and flui fl w condition (Reynolds number) upon mixing process in microscale. The parameters of the modeled fluid are given in table1.

Table 1. modeled fluid parameters

Fluid parameter	Value
ρ, density (kg/m^3)	10^3
μ, dynamic viscosity ($Pa \cdot s$)	10^{-3}
D, diffusion coefficien (m^2/s)	10^{-9}

A numerical mesh grid was composed of approximately 35,000 hexahedral elements. A single element was equal to $1.5 \cdot 10^{-6}$ m. The results for the mixing model of the T-mixer with a straight microfluidi channel for $Re = 1, 5, 10$, and 100 are presented in Figure (1). As can be noted, there are no significan differences in the concentration distribution for all investigated values of the Reynolds number. For this micromixer structure, the mixing process relied entirely on molecular diffusion. Therefore, mixing was very poor.

Improvement of the mixing can be obtained by increasing the contributions of the convection part in equation (10). That can be achieved through the proper design of the mixing structure. Yi and Bau [6] have shown that the appropriate number of bends can enhance the mass transfer in microscale. The results of CFD modeling for a micromixer which consists of five bends (meander) is presented in Figure (2). The shape of the inter-material layer for meander structure of the micromixer for $Re = 1, 5, 10$, and 100 after the flow has passed through n bends is illustrated in Figures (3)-(6), respectively.

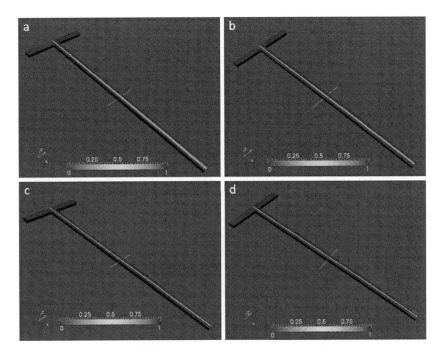

Figure 1. Concentration distribution in micromixer with a straight channel for (a) $Re = 1$, (b) $Re = 5$, (c) $Re = 10$, and (d) $Re = 100$. Results of CFD computations.

At a low Reynolds number (up to 100), the presence of the bends do not contribute significantly to improve the mixing process. The amelioration of mass transfer, due to the development of the inter-material layer, can only be noticed for a relatively high Reynolds number ($Re = 100$). However, flows in microfluidic systems are typically at very low Reynolds numbers (up to 10) and the mixing induced by a single bend is insufficient to ensure efficient mixing. Thus, a larger number of bends is needed. The results for the mixing model of the meander micromixer, which is composed of a larger number of bends, are presented in Figure (7). For this structure, viscous forces still dominate the inertial forces and bend-induced fluid re-circulations and vortices decay rapidly. Consequently, the mixing between both streams is still almost purely diffusive. The shape of the inter-material layer for the meander micro-mixer with a larger number of bends for $Re = 1, 5, 10$, and 100 is presented in Figures (8)-(12).

In the case of all previously considered micromixer structures, the bends were arranged in the plane, and the symmetry with respect to the „x" axis was preserved. Fluid elements which flowed on the side of the symmetry line remained on the same side of the symmetry line indefinitely. Therefore, it was desirable to break this symmetry to obtain a more efficient mixing process. This could be achieved by introducing so-called „out-of-plane" bends [6]. Figure (12) presents results of the CFD modeling for a micromixer structure which is composed of a sequence of the „out-of-plane" bends.

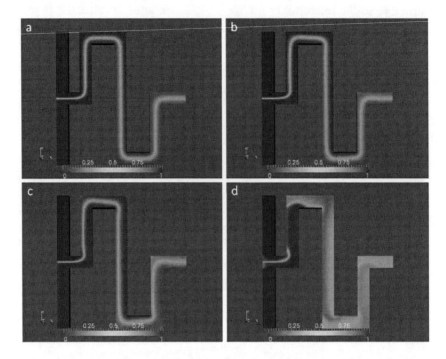

Figure 2. Concentration distribution in meander micromixer for (a) $Re = 1$, (b) $Re = 5$, (c) $Re = 10$, and (d) $Re = 100$. Results of CFD computations.

The bends are arranged in an I-shape serpentine configuratio [5]. The serpentine micromixer consists of a T-junction and nine segments. Each segment is $900\ \mu m$ long, $300\ \mu m$ wide, and $150\ \mu m$ high. For $Re \leq 5$ (Figure 12a, b, 13, 14) both fluid remain separate even after going through all bends. However, mixing efficien y is much better in comparison to previously presented meander structures. When $Re = 10$ and 100 (Figure 12c, d) fluid remain separate only in the vicinity of the T-junction, but appear to be well homogenized after passing through 4 or 5 bends, respectively. No separation between fluid can be noticed. For those values of the Reynolds number (fl w rate), the inter-material layer enlarges due to the stretch and fold phenomena. As the length of the inter-material layer increases, the two materials are brought into closer contact creating more space for flui particles for diffusion. The shape of the inter-material layer after the fl w passes through n bends for $Re = 10$ and 100 is presented in Figures (15) and (16), respectively.

5. Conclusions

The application of the CFD analysis for modeling of the mixing process in microchannels was presented.
The influenc of the microfluidi system geometry and its operation condition upon flui fl w and mixing efficien y in microscale was described.
The CFD analysis of flui fl w and micromixing allowed us to estimate the best conditions

Modeling and Optimization of Microfluidi Passive Mixer Using CFD Analysis   483

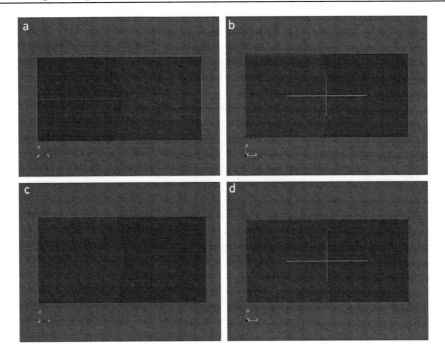

Figure 3. Structure of the inter-material layer for the meander micromixer as the fl w passes through n bends: (a) $n = 1$; (b) $n = 2$; (c) $n = 3$; and (d) $n = 4$, $Re = 1$. Results of CFD computations.

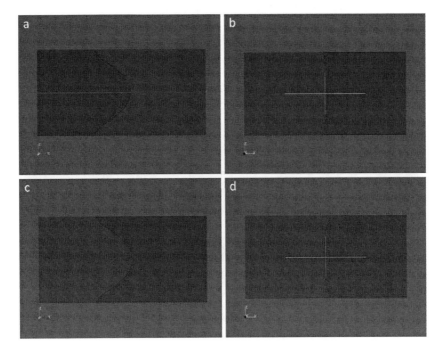

Figure 4. Structure of the inter-material layer for the meander micromixer as the fl w passes through n bends: (a) $n = 1$; (b) $n = 2$; (c) $n = 3$; and (d) $n = 4$, $Re = 5$. Results of CFD computations.

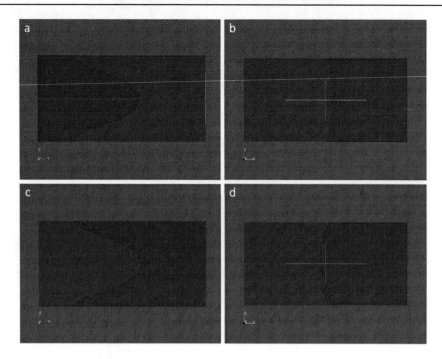

Figure 5. Structure of the inter-material layer for the meander micromixer as the fl w passes through n bends: (a) $n = 1$; (b) $n = 2$; (c) $n = 3$; and (d) $n = 4$, $Re = 10$. Results of CFD computations.

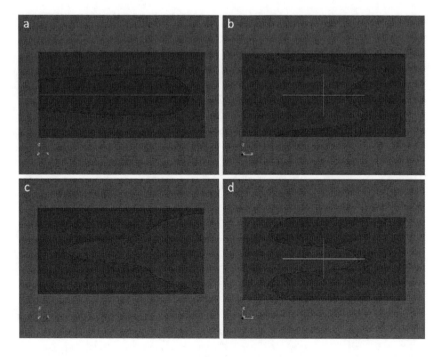

Figure 6. Structure of the inter-material layer for the meander micromixer as the fl w passes through n bends: (a) $n = 1$; (b) $n = 2$; (c) $n = 3$; and (d) $n = 4$, $Re = 100$. Results of CFD computations.

Modeling and Optimization of Microfluidi Passive Mixer Using CFD Analysis 485

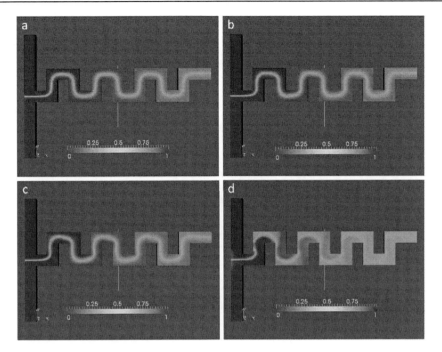

Figure 7. Concentration distribution in meander micromixer with larger number of bends for (a) $Re = 1$, (b) $Re = 5$, (c) $Re = 10$, and (d) $Re = 100$. Results of CFD computations.

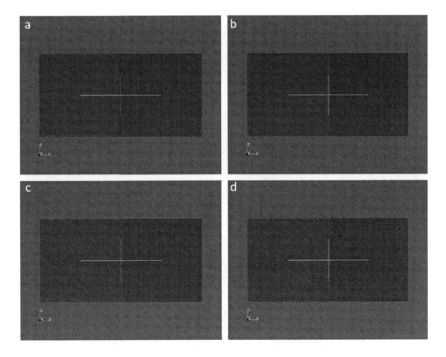

Figure 8. Structure of the inter-material layer for the meander micromixer with larger number of bends as the fl w passes through n bends: (a) $n = 1$; (b) $n = 2$; (c) $n = 3$; and (d) $n = 4$, $Re = 1$. Results of CFD computations.

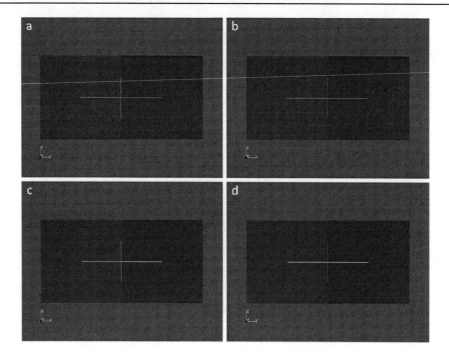

Figure 9. Structure of the inter-material layer for the meander micromixer with larger number of bends as the fl w passes through n bends: (a) $n = 1$; (b) $n = 2$; (c) $n = 3$; and (d) $n = 4$, $Re = 5$. Results of CFD computations.

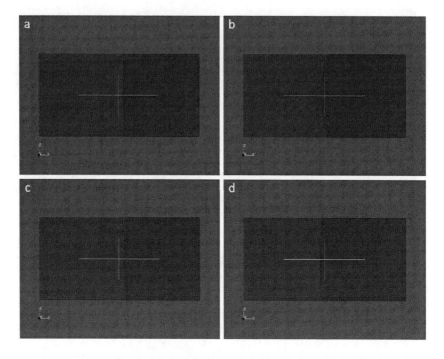

Figure 10. Structure of the inter-material layer for the meander micromixer with larger number of bends as the fl w passes through n bends: (a) $n = 1$; (b) $n = 2$; (c) $n = 3$; and (d) $n = 4$, $Re = 10$. Results of CFD computations.

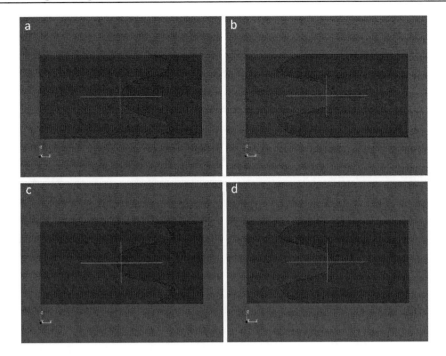

Figure 11. Structure of the inter-material layer for the meander micromixer with larger number of bends as the fl w passes through n bends: (a) $n = 1$; (b) $n = 2$; (c) $n = 3$; and (d) $n = 4$, $Re = 100$. Results of CFD computations.

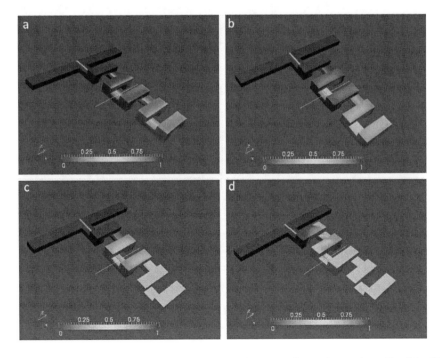

Figure 12. Concentration distribution in I-shaped serpentine micromixer for (a) $Re = 1$, (b) $Re = 5$, (c) $Re = 10$, and (d) $Re = 100$. Results of CFD computations.

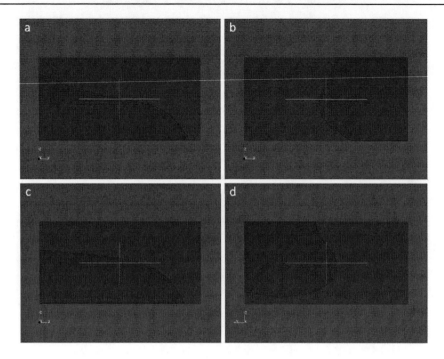

Figure 13. Structure of the inter-material layer for the I-shaped serpentine micromixer as the flow passes through n bends: (a) $n = 1$; (b) $n = 2$; (c) $n = 3$; and (d) $n = 4$, $Re = 1$. Results of CFD computations.

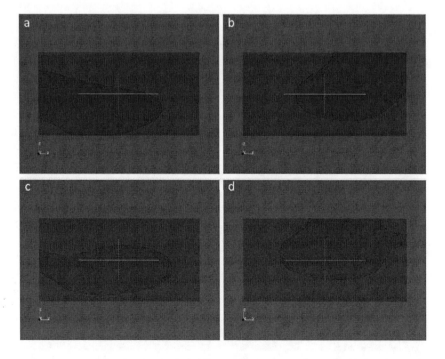

Figure 14. Structure of the inter-material layer for I-shaped serpentine micromixer as the flow passes through n bends: (a) $n = 1$; (b) $n = 2$; (c) $n = 3$; and (d) $n = 4$, $Re = 5$. Results of CFD computations.

Modeling and Optimization of Microfluidi Passive Mixer Using CFD Analysis 489

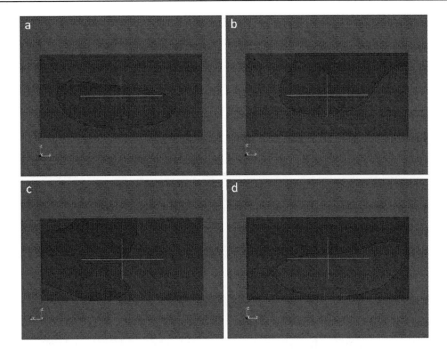

Figure 15. Structure of the inter-material layer for I-shaped serpentine micromixer as the fl w passes through n bends: (a) $n = 1$; (b) $n = 2$; (c) $n = 3$; and (d) $n = 4$, $Re = 10$. Results of CFD computations.

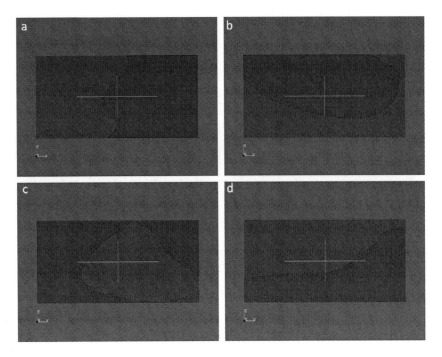

Figure 16. Structure of the inter-material layer for the I-shaped serpentine micromixer as the fl w passes through n bends: (a) $n = 1$; (b) $n = 2$; (c) $n = 3$; and (d) $n = 4$, $Re = 100$. Results of CFD computations.

needed to significantl improve mixing efficien y.
According to the CFD results presented in this chapter, the I-shaped serpentine micromixer needs to operate at $Re \geq 10$.
CFD modeling can be applied to arrive at various optimal microfluidi device designs.

Acknowledgment

The authors wish to thank the Polish Ministry of Science and Higher Education (grants No. N N515 607 639) for the financia support.
The Foundation for Polish Science (FNP) is also acknowledged for financia support for Karol Malecha.

References

[1] Dongqing L (ed.). *Encyclopedia of microfluidics and nanofluidics,* New York, Springer, 2008.

[2] Nguyen N-T; Wu Z. Micromixers - a review. *Journal of Micromechanics and Microengineering,* 2005, 15, R1-R16.

[3] Oberti S; Neild A; Ng TW. Microfluidi mixing under low frequency vibration. *Lab on a Chip,* 2009, 9, 1435-1438.

[4] Zhong J; Yi M; Bau HH. Magneto hydrodynamic (MHD) pump fabricated with ceramic tapes. *Sensors and Actuators A,* 2002, 96, 59-66.

[5] Malecha K; Golonka LJ; Bałdyga J; Jasińska M; Sobieszuk P. Serpentine microflu idic mixer made in LTCC. *Sensors and Actuators B,* 2009, 143, 400-413.

[6] Yi M; Bau HH. The kinematics of bend-induced mixing in micro-conduits. *International Journal of Heat and Fluid Flow,* 2003, 24, 645-656.

[7] Wautelet M. Scaling laws in the macro-, micro- and nanoworlds. *European Journal of Physics,* 2001, 22, 601-611.

[8] Ottino JM; Wiggins S. Introduction: mixing in microfluidics *Royal Society of London Transactions Series A,* 2004, 362, 923-935.

[9] Wiggins S; Ottino JM. Foundations of chaotic mixing. *Royal Society of London Transactions Series A,* 2004, 362, 937-970.

[10] OpenFOAM. *The Open Source CFD Toolbox User Guide*. Free Software Foundation, Inc., 2009.

[11] Oertel H (ed.). *Prandtl's Essentials of Fluid Mechanic* s, Springer-Verlag, 2000.

[12] Issa R. Solution of implicitly discretized flui fl w equations by operator-splitting. *J. Comput. Phys.,* 1986, 62, 40-65.

[13] Chung T. *Computational fluid dynamics*, Cambridge University Press, 2002.

[14] Versteeg H; Malalsekera W. *An introduction to computational fluid dynamics,* Longman Scientifi & Technical, 1995.

[15] Ferziger J; Peric M. *Computational methods for fluid dynamics,* Springer Berlin, 1999.

[16] Batchelor GK. *Introduction to Fluid Dynamics*, Cambridge University Press, 1967.

Reviewed by NOVA Publishers.

Chapter 15

COMPUTER SIMULATION AND OPTIMIZATION OF STIRRER HYDRODYNAMICS AT HIGH REYNOLDS NUMBERS

Ö. Uğur[a], K. Yapıcı[b], Y. Uludağ[c] and B. Karasözen[a,d]
[a] Institute of Applied Mathematics, Middle East Technical University,
06531 Ankara, Turkey
[b] Department of Chemical Engineering, Cumhuriyet University, Turkey
[c] Department of Chemical Engineering, Middle East Technical University,
06531 Ankara, Turkey
[d] Department of Mathematics, Middle East Technical University,
06531 Ankara, Turkey

Abstract

Computational fluid dynamics (CFD) is a powerful tool for solving problems associated with flow, mixing, heat and mass transfer and chemical reaction. Hence they are useful alternatives to experimental techniques that are expensive and time consuming in the simulation of mixing processes. In this chapter, computational analysis of turbulent flow field in a mixing tank and a numerical approach for the optimization of stirrer configuration parameters are presented. Velocity field and power requirement are obtained using the FASTEST, a CFD package, which employs a fully conservative finite volume method for the solution of the Navier-Stokes equations. The inheritably time-dependent geometry of stirred vessel is simulated by clicking mesh method.

In the simulation of the turbulent flow field in a mixing tank with six bladed Rushton turbine, the effects of impeller clearance and disc thickness on the power number are determined and it is found that the power number decreases with decreasing clearance and increasing disc thickness. The results are comparable with those of well established measurement techniques in terms of time-averaged velocity field, turbulent kinetic energy, dissipation rate, and power number.

The methodology for the optimization of stirrer configuration is based on a flow solver FASTEST, and a mathematical optimization tool, which are integrated into an automated procedure. Two trust region based derivative free optimization algorithms, the DFO and CONDOR are considered. Both are designed to minimize smooth functions whose evaluations are considered to be expensive and whose derivatives are not available or not desirable to approximate. An exemplary application for a standard stirrer configuration illustrates the functionality and the properties of the proposed methods.

Keywords: Navier-Stokes equations, computational fluid dynamics, stirrer, agitation, hydrodynamics, simulation, multivariate interpolation, derivative free optimization

AMS Subject Classification: 76D05, 76D55, 76B75, 35Q93, 49Q10, 65D05, 90C56

1. Introduction

Mixing of liquids in stirred tanks is one of the most important unit operations for many industries such as chemical, biotechnological, pharmaceutical, and food processing to homogenization, to increase heat and mass transfer rates, to prevent particle settling, to obtain emulsions and to even out all physical property gradients. Hence from the product quality and process economics point of views, determining the level of mixing and overall behavior and performance of the stirred tanks are crucial. Depending on purpose of the operation carried out in an agitated tank, the best choice for the geometry of the tank and impeller type can vary widely. Different materials require different types of impellers and tank geometries in order to achieve the desired product quality and reasonable operational costs. One of the most fundamental needs for analysis of these processes from both theoretical and industrial perspectives is the knowledge of the agitation hydrodynamics.

Agitation is carried out using stirred tanks. Basic parts of a stirred tank are a cylindrical tank and an impeller driven by a motor. To enhance agitation efficiency baffles are also installed on the interior wall of the tank. During the agitation process, fluid around the rotating impeller blades interacts with the stationary baffles and generates three-dimensional turbulent flow structure. Along with fluid physical properties, the parameters including tank dimensions, impeller angular velocity, type of the impeller, impeller clearance from the tank bottom, impeller diameter, proximity of the vessel walls, baffle length affect the hydrodynamics of the flow. Therefore, presence of such a large number of design parameters makes the flow hydrodynamics in stirred tank very complicated. Understanding the fluid mechanics and developing rational design procedures associated with the agitated tanks have been subject of many studies. In these studies the main employed techniques are experimental fluid dynamics (EFD) and computational fluid dynamics (CFD). Experimental studies have proven to be successful in measuring the flow field accurately [9, 37, 44]. However, they are neither economical nor practical when the large numbers of design and process variables are considered. With the advances in computer technology and numerical techniques, CFD has been increasingly used to solve hydrodynamics in agitated tanks since it offers quick, practical and economical simulation of the process.

Various approaches have been introduced to simulate the flow pattern induced by a revolving impeller in stirred tank. Most of the previous studies have treated rotating impeller as a 'black box'. This approach requires impeller boundary conditions as input which needs to be determined experimentally. Though this approach is successful in predicting the flow characteristics in the bulk of the vessel, its convenience is inherently limited due to the dependency on the availability of the experimentally obtained velocity and turbulence quantities around the impeller [31]. Detailed review of this method has been done by Brucato et al. [3] and Ranade [30]. In order to overcome these drawbacks, five different explicit methods have been introduced in the literature to predict the flow fields around the impeller blades. The chronological order of these explicit methods are sliding mesh method [16], multiple frames of reference approach [17], snapshot flow model [31], momentum source method [41] and clicking mesh method [2]. Moreover, three-dimensional, highly turbulent flow structure in mixing tanks has been simulated by turbulence models, e.g., direct numerical simulation (DNS), Reynolds averaged Navier-Stokes (RANS) and large eddy simulation (LES). Hartman et al. [12] assessed LES and RANS turbulence models on the flow in a baffled stirred tank driven by a Rushton turbine at Reynolds number being 7200 (Re = 7200). They reported that the agreement between experiment and LES predictions was better than those obtained by RANS for both velocity components and the turbulent kinetic energy near the impeller region. Yeoh et al. [43] employed LES approach to predict the turbulent flow in mixing tank generated by Rushton type turbine. They concluded that LES approach is amenable to predict three-dimensional turbulent and anisotropic flow in stirred tank.

Montante et al. [19] simulated the effects of different clearances on the circulation pattern and power number for the Rushton turbine and compared the results with measurements. They observed double to single loop flow structure and decreasing power number when the impeller clearance from the vessel is reduced. Recently, Yapıcı et al. [42] have investigated effect of the impeller clearances and disc thickness of bladed Rushton turbine impeller on the power number as well as flow structure. They used clicking mesh method to capture relative motion between rotating impeller and the stationary baffle. They found that power number decreases with decreasing clearance and increasing with disc thickness.

All these studies have shown that even quite complex flow fields, such as that of an agitated tank, can be handled using CFD, and this should be an appropriate numerical tool employed.

2. Model Equations and Numerical Methodology

2.1. Governing Equations

The equations describing Newtonian fluid flow are derived from the conservation of mass and momentum which are also known as the Navier-Stokes equations for fluids of constant physical properties:

$$\frac{\partial u_i}{\partial x_i} = 0, \tag{1}$$

$$\frac{\partial \rho u_i}{\partial t} + \frac{\partial \rho u_j u_i}{\partial x_j} - \frac{\partial \tau_{ij}}{\partial x_j} + \frac{\partial p}{\partial x_i} = \rho f_i, \tag{2}$$

where u_i is the velocity vector component with respect to the Cartesian coordinates x_i, p is the pressure, ρ is the fluid density, and τ_{ij} is the deviatoric (i.e., deformation dependent) part of the total stress tensor. Finally, f_i is the ith component of the external body force per unit mass.

For closure an additional relation is also needed to determine the stress term. It is provided through a number of constitutive equations depending on the fluid type. For simple (Newtonian) fluids, Newton's law of viscosity provides the needed relation between the stress, rate of deformation, and material property.

Owing to the very nature of turbulence, flow quantities, such as velocity, pressure and stress become time dependent. In many applications, such as agitation, time smoothed forms of these quantities are useful as opposed to their instantaneous values. To determine the forms, the governing equations are time averaged over a meaningful time scale. As a result of the averaging additional terms appear that are referred to as Reynolds averaged Navier-Stokes (RANS). Similar to the deviatoric stresses, a turbulence model is also needed to simulate typical agitation hydrodynamics. Large eddy simulation is well suited for three dimensional, unsteady, complex flow fields. LES is based on decomposition of the dependent variables into large or grid scale components and small or sub-grid scale components which represent the unresolved fraction of the turbulence. A standard Smagorinsky model (see Smagorinsky [35]) is represented by eddy viscosity ν_t as

$$\nu_t = (C_s \Delta)^2 |S|, \qquad (3)$$

where C_s is the Smagorinsky constant that is commonly chosen to be 0.1 or 0.12. Here, Δ is the lattice spacing and S^2 is the resolved deformation rate tensor:

$$S^2 = \frac{1}{2}\left(\frac{\partial u_i}{\partial x_j} + \frac{\partial u_j}{\partial x_i}\right). \qquad (4)$$

2.2. Numerical Methodology for the Flow Problem

In CFD the first step is to divide the flow geometry into smaller domains, grids, through a number of available discretization methods. Then, differential equations governing the fluid flow are converted into a coupled algebraic set of equations which are solved by using a numerical method along with the appropriate boundary conditions.

In what follows, the agitated tank is discretized using a multi-block boundary fitted grid which allows finer grids in regions where higher spatial resolutions (e.g., at the vicinity of the impeller) are required. Moreover, a multi-block grid construction allows the accurate representation of the complex geometries associated with stirrer configurations [13]. Another advantage is that it is well-suited for parallelization of the computation by means of grid partitioning techniques. Here, the stirred tank driven by six bladed Rushton turbine impeller is divided into two symmetric parts: each encompasses two baffles and three impeller blades. A sketch of the corresponding surface mesh of stirred tank with bladed Rushton turbine is depicted in Figure 1 for a three-dimensional visualization. Solution domain consists of 22 blocks, 17 of which are used to construct the impeller and its close surroundings while reaming 5 blocks are defined to be stationary. The blocks are further divided into finer grids so that the total number of computational grids is 238,996.

In order to solve the Navier-Stokes equations numerically, the finite volume method is applied on non-orthogonal block-structured grids [11]. A special interpolation is employed for approximating cell face values by nodal values within the finite volume discretization [15]. The pressure-velocity coupling is established by using a pressure correction technique, called SIMPLE [25]. ILU decomposition method (see Stone [36]) is used for solving various sparse linear systems within the pressure correction scheme; and for accelerating the convergence, a nonlinear multi-grid scheme is implemented.

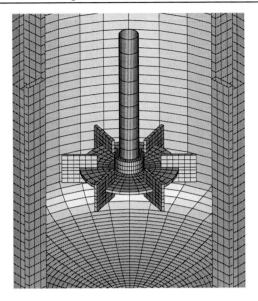

Figure 1. Sketch of the surface mesh for standard stirrer configuration driven by six bladed Rushton turbine impeller.

Accordingly, to handle the rotation of the impeller two reference frame were employed: one of the frames was rotating with the impeller and compassed the impeller and its surroundings. The other, being stationary, was used for the rest of the regions in the tank. Interface between these frames was handled by allowing flow of momentum and mass between the interface grids.

Discretization in time is also required as well as spatial one, to solve an unsteady problem. Time may be regarded as an additional coordinate, therefore a spatial problem can be considered as sequence of levels at several times, so-called time levels. In contrast to spatial discretization the variable values have to be determined before moving to the next time level. New time levels are always extrapolated from the older ones.

It is worth mentioning that we use strongly implicit (SIP) method [36] to solve the linear system of algebraic equations, which is obtained from the spatial and temporal discretization process. Main reason for using SIP method is that it provides a better convergence result than other methods such as Gauss-Seidel, conjugate gradient methods, or the tridiagonal-matrix algorithm.

2.3. Derivative Free Optimization Methods

Growing sophistication of computer hardware as well as mathematical algorithms together with the software developments opens new possibilities for optimization. Derivative free optimization methods are designed for solving nonlinear problems where the derivatives of the objective function are not available nor desired to compute. We consider formally the problem

$$\underset{x \in \mathbb{R}^n}{\text{minimize}} \ f(x),$$

where f is a smooth nonlinear objective function from $\mathbb{R}^n$ into $\mathbb{R}$ and is bounded below. We thus assume that the gradient $\nabla f(x)$ and the Hessian $\nabla^2 f(x)$ cannot be computed for any x.

There is a high demand from practitioners for such algorithms because this kind of problems occur relatively frequently in the industry. In our context of stirrer configuration, the objective function f may correspond to the Newton number, mixing time, stresses, etc., which are derived from the flow field. The design variables, the components of the x-vector, describe the geometry of the stirrer and/or operating parameters, such as the rotational speed. Constraints to the formal optimization problem can also be defined via the natural restrictions of the stirrer geometry, of operational parameters, or of other quantities that might depend on the flow characteristics.

In many applications, the derivatives are either unavailable or the objective function is non-differentiable. Even more, accurate approximation of derivatives by finite differences is prohibitive or function evaluations are too costly and/or noisy. Automatic differentiation might be impossible to apply or source code resulting from a large and expensive computer simulation may not be available.

Recent years have seen the emergence of model-based trust region algorithms for optimization where the derivatives of the objective function are unavailable. These methods were introduced by Winfield [40] and by Powell [26, 28, 29] and developed for constrained and unconstrained problems, by a number of authors [1, 4–6, 18, 22, 27]. Many of the developed versions are not only numerically efficient [21] but also used widely in practice [10, 23, 24, 34, 38]. Details of various derivative free methods can be found in the recent book of Conn et al. [8] as well as in the survey article [14].

Derivative free algorithms belong to the class of trust-region methods [7] where cheaper models are iteratively built around the current iterate. The model objective is then easier to minimize than the objective function. At the kth step the quadratic model within the trust-region $\mathcal{B}_k$,

$$\mathcal{B}_k = \{x_k + s : s \in \mathbb{R}^n \text{ and } \|s\| \leq \Delta_k\},$$

with the trust-region radius Δ_k may be given as

$$m_k(x_k + s) = f(x_k) + \langle g(x_k), s \rangle + \frac{1}{2} \langle s, H(x_k)s \rangle \tag{5}$$

for some $g \in \mathbb{R}^n$ and some symmetric $n \times n$ matrix H, where $\langle \cdot, \cdot \rangle$ denotes the inner product. The vector g and the matrix H do not necessarily correspond to the first and second derivatives of the objective function f. They are determined by requiring that the model (5) interpolates the function f at a set $Y = \{y_i\}$ of points containing the current iterate x_k:

$$f(y_i) = m_k(y_i) \quad \text{for all} \quad y_i \in Y. \tag{6}$$

Here, Y denotes the set of interpolation points, which is a subset of the set of points at which the values of f are known, including the current iterate. Building the full quadratic model (5) requires the determination of $f(x_k)$, the components of the vector g and the entries of the matrix H; thus the cardinality of Y must be equal to $\frac{1}{2}(n+1)(n+2)$. However if $n > 1$, the condition (6) is not sufficient for the existence and uniqueness of the interpolant and to guarantee the good quality of the model. Geometric conditions (known as poisedness) on the set Y are required to ensure the existence and uniqueness of the interpolant. In other words, the points of Y do not collapse into a lower dimensional subset or lie on a quadratic curve. If $\{\phi_i(\cdot)\}_{i=1}^{p}$ denotes a basis of the linear space of n-dimensional quadratics, the model (5) and the interpolation conditions (6) can be written as

$$m_k(x) = \sum_{i=1}^{p} \alpha_i \phi_i(x), \quad \sum_{i=1}^{p} \alpha_i \phi_i(y) = f(y) \quad \text{for all} \quad y \in Y$$

and some scalars α_i, $i = 1, \ldots, p$. Then the set $Y = \{y_1, \ldots, y_p\}$ is poised if the matrix

$$M = \begin{pmatrix} \phi_1(y_1) & \cdots & \phi_1(y_p) \\ \vdots & \ddots & \vdots \\ \phi_p(y_1) & \cdots & \phi_p(y_p) \end{pmatrix} \tag{7}$$

is non-singular, i.e., $\delta(Y) := \det(M)$ is non-zero [5, 6]. The quality of the model (5) around the current iterate x_k depends on the geometry of the interpolation set. The interpolation set is referred to as *well-poised* if $|\delta(Y)| \geq \varepsilon$ for some threshold $\varepsilon > 0$.

General form of the derivative free algorithm can be given as follows: let

$$0 < \eta_0 \leq \eta_1 < 1 \quad \text{and} \quad 0 \leq \gamma_1 < 1 \leq \gamma_2.$$

- *Setting an initial poised interpolation set*

    For a given x_s and a computed $f(x_s)$ choose an initial interpolation set Y. Determine $x_0 \in Y$ such that $f(x_0) = \min_{y_i \in Y} f(y_i)$. Choose an initial trust region radius $\Delta_0 > 0$, and set $k = 0$.

- *Building the model*

    Build a quadratic model $m_k(x_k+s)$ according to (5) for the objective function around the iterate x_k based on a well-poised interpolation set Y.

- *Minimization of the model within the trust-region*

    Calculate a step s_k by solving the constrained problem

    $$\underset{s \in \mathcal{B}_k(x_k;\Delta_k)}{\text{minimize}} \; m_k(x_k + s)$$

    within the trust region $\mathcal{B}_k(x_k; \Delta_k)$.

- *Comparison of the achieved and the predicted reductions*

    Evaluate $f(x_{k+1})$ and compute the ratio

    $$\rho_k = \frac{f(x_k) - f(x_{k+1})}{m_k(x_k) - m_k(x_{k+1})},$$

    where $x_{k+1} = x_k + s_k$.

- *Updating the trust-region radius*

- *for a successful step:*

    If $\rho_k \geq \eta_1$, then define $x_{k+1} = x_k + s_k$ and increase the trust region radius to

    $$\Delta_{k+1} \in [\Delta_k, \gamma_2 \Delta_k].$$

- *for an unsuccessful step:*

    If $\rho_k < \eta_1$, then let $x_{k+1} = x_k$ and decrease the trust region radius to

    $$\Delta_{k+1} \in [\gamma_0 \Delta_k, \gamma_1 \Delta_k].$$

- *Updating the interpolation set*

    Include the new point x_{k+1} in the interpolation set Y_{k+1} depending on its success, and exclude the worst point if needed or required.

- *Computing a new interpolation model*

    Using the interpolation set Y_{k+1}, if $Y_{k+1} \neq Y_k$ compute a new interpolation model m_{k+1} around x_{k+1}. Then, start for a new iterate.

For the model to be well defined, the interpolation points must be poised, meaning that they must be compatible with the interpolation conditions imposed on them. In practice, it is common for the interpolation points to lose poisedness as the iteration progresses, and this is reflected in an increasing ill-conditioning of the interpolation. To prevent such a difficulty, a geometry phase is included in contemporary model-based algorithms: it typically replaces one or more of the interpolation points by other points that are selected with the exclusive goal of improving the position of the interpolation set. As those geometry restoration steps are expensive, one may ask if they are really necessary. Morales et al. [22] observed that an algorithm which simply ignores the geometry considerations may in fact perform quite well in practice, that is, they have a self-correcting geometry mechanism, but may lose the property of provable global convergence to first-order critical points [33]. Based on this considerations, Scheinberg and Toint [33] developed a new derivative free algorithm including the geometry restoration. Design and convergence properties of the new algorithm depend on a self-correcting mechanism combining trust-region mechanism with polynomial interpolation setting.

There are several implementations of interpolation based on derivative free methods which differ in detail. The oldest one is the DFO (Derivative Free Optimization) package developed by Conn and co-workers [5, 6]. The newly developed CONDOR (Constrained, Non-linear, Direct, parallel Optimization using trust Region method for high-computing load function) package by Berghen (see [1] and references therein) is based on the UOBYQA of Powell [29].

Differences between the DFO and the CONDOR packages that are used for the stirrer optimization, in this chapter, are summarized as follows.

- The DFO package uses Newton polynomials while CONDOR uses Lagrange polynomials as the basis for the space of quadratic polynomials. The choice of Newton polynomials (in DFO) causes the matrix in (7) to be lower triangular with a special block diagonal structure, while the use of Lagrange polynomials (in CONDOR) results in an identity matrix.

- The minimization of the possibly quadratic model in the DFO package is done by applying a standard optimization procedure, for instance, sequential quadratic programming (SQP) [32], interior-point optimizer (IPOPT) [39]. The CONDOR package uses the Moré and Sorenson algorithm [20] for the computation of the trust region radius and the minimization of the quadratic model, which is very stable.

- Another essential difference between DFO and CONDOR algorithms is that the former uses the smallest Frobenius norm of the Hessian matrix to minimize the local model, which may cause numerical instability; whereas CONDOR uses the Euclidean norm which is considered to be more robust.

- The DFO package uses interpolating polynomials that alternates between linear and quadratic specifically at the initialization steps. However, the CONDOR algorithm uses full quadratic polynomials even initially.

3. Automated Control Procedure

The components, grid generator, fluid solver and optimizer, described in the preceding sections are combined within an integrated control script in a modular form so that each component can be replaced by an alternative scheme. The automated control procedure is schematically illustrated in Figure 2.

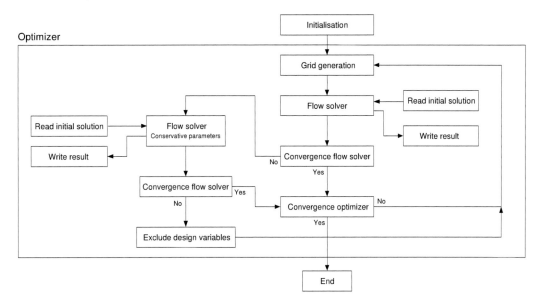

Figure 2. Flow chart of the automated control script.

Having initialized the necessary inputs, the control procedure involves the following major steps:

1. *Optimizer*: At each iteration the optimizer computes a new set of design variables unless it converges.

2. *Grid Generation*: When new design variables are available, the grid generation algorithm creates the corresponding numerical grid, hence the new geometry.

3. *Flow Solver*: Flow solver computes the flow field and the corresponding value of the objective function for the new geometry. If the flow solver converges, then the optimizer produces new geometrical design variables unless the objective value is found to be optimum. Otherwise, FASTEST3D runs with more conservative parameters to achieve convergence in its inner calculations. Even if this is not possible, the optimizer may reject the use of those design variables.

4. *Convergence of Optimizer*: The optimizer decides, by a given criterion, whether the current value of the objective function is accepted as optimum. The new design variables are calculated in case the optimum is not achieved, and the automated procedure is continued to the next iterate. The procedure is finished when the convergence criteria are satisfied.

Note that the flow-chart in Figure 2 represents a systematic way for optimization problems in industrial applications: instead of experimenting with several control variables, the presented automated procedure can easily be altered to be used, for instance, in shape-optimizations.

4. Numerical Results and Conclusion

In the following we consider a Rushton turbine (as described in Section 2.2.) for a practical stirrer configuration to illustrate the functionality of the proposed approach [34, 38]. Prior to the optimization task driven by the automated control procedure of the preceding section, at first, computations for validation are carried out. Such numerical experiments versus validations surely help describe the *conservative parameters* for which the flow solver most probably converges. See Figure 2 as well as the explanation for the *Flow Solver* (item 3 on page 502).

Computations, in this chapter, are carried out on an eight processor Redstone cluster machine. Communication between processing units is controlled by the MPI (Message-Passing Interface). FASTEST3D (Flow Analysis by Solving Transport Equation Simulating Turbulence) which was developed at the Institute for Fluid Mechanics, University of Erlangen and in the Department of Numerical Methods in Mechanical Engineering at Darmstadt University of Technology, was used in the numerical simulations. Convergence criterion used in all flow simulation is 10^{-4}, a value generally sufficient to assume that the solution is satisfactory.

4.1. Optimized Stirred Tank Configuration

A schematic sketch of the considered standard stirred tank configuration with impeller is shown in Figure 3. All geometrical quantities are expresses in terms of the tank diameter T. Whenever any of the geometrical parameters is changed the spatial discretization of the new configuration is updated through the grid generation procedure.

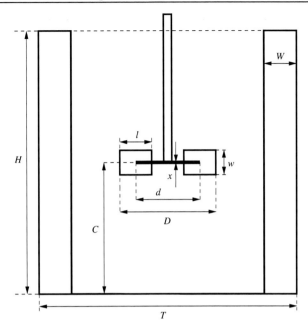

Figure 3. A schematic representation of a standard baffled stirred tank.

The system investigated consists of a standard flat bottomed cylindrical vessel, the diameter ($T = 0.15$ m) of which equals to the height of the solution ($H = T$). Four baffles having width W of $T/10$ are spaced equally around the vessel. The shaft of the impeller is concentric with the axis of the vessel. The impeller is the six bladed Rushton turbine with diameter D which is equivalent to $T/3$. Blade height w, blade length l, and the disc thickness x are set to $D/5$, $D/4$ and $7D/200$, respectively. Distance between tank bottom and impeller position C is set to $T/2$. The actual geometrical parameters which we consider (initially) are summarized in Table 1.

Table 1. Geometrical parameters of standard stirrer configuration

Parameter	Value
Tank diameter	$T = 0.15$ m
Impeller diameter	$D = T/3 = 0.05$ m
Bottom clearance	$C = T/2 = 0.075$ m
Height of the liquid	$H = T = 0.15$ m
Length of the baffles	$W = T/10 = 0.015$ m
Height of the blade	$w = D/5 = 0.01$ m
Length of the blade	$l = D/4 = 0.0125$ m
Disc thickness	$x = 7D/200 = 0.00175$ m
Diameter of the disk	$d = 3D/4 = 0.0375$ m

The working Newtonian fluid inside the stirrer tank under consideration is a glucose solution with density $\rho = 1330\,\text{kg/m}^3$ and viscosity $\mu = 0.105\,\text{Pas}$. The dimensionless Newton number, Ne, which relates the resistance force to the inertia force,

$$Ne = \frac{P}{\rho N^3 D^5}, \qquad P = -\int_S (pu_j + \tau_{ij}u_i)\,n_j\,\mathrm{d}S,$$

is considered to be the characteristic reference quantity to be minimized. Here N denotes the rotational speed of the impeller and P is the power computed from the flow quantities over the surface S of the impeller. Although the other variables were defined before, for ease of access, u_i is the component of the velocity vector corresponding to the Cartesian coordinate x_i, p is the pressure, τ_{ij} is the viscous part of the stress tensor for incompressible Newtonian fluids, and n_j is the component of the outer unit normal vector.

For derivative free optimization purposes we consider the design variables to be the disk thickness x, the bottom clearance C, and the baffle length W, for which the inequality constraints $0.001 \leq x \leq 0.005, 0.02 \leq C \leq 0.075$, and $0.005 \leq W \leq 0.03$ are prescribed. All other parameters are kept fixed according to the standard configuration presented in Table 1.

Because the gradient information is not available, for both algorithms, DFO and CONDOR, the minimum trust region radius is used as a stopping criterion. However, their default stopping criteria were selected in both algorithms for the sub-minimization problem. We investigate the case when the Reynolds number, $Re = \rho N D^2/\mu$ is 1000. Figure 4 shows the Newton number versus the number of cycles for the two optimization algorithms. For both algorithms the Newton number attains the same optimum. CONDOR terminates earlier than the DFO due to the fact that the latter oscillates around the optimum unnecessarily. Figure 5 depicts the corresponding changes of the considered design variables during the function evaluations. The three design parameters assign almost the same optimal values.

We remark that the two optimization tools differ in building the starting model: DFO starts to build the model objective function approximation (unique or not) from the very beginning. That is, its starting objective polynomial approximation is not fully quadratic. On the other hand, CONDOR starts with a fully quadratic model for the objective function. Despite its long initialization it turns out that CONDOR needs less function evaluations than DFO to reach an optimum point after completing the initial quadratic approximation. DFO, as time passes, oscillates around the minimum although it approaches the minimum very sharply at the beginning. CONDOR waits for a full quadratic polynomial to build its model, and then gradually approaches the minimum, using a complete interpolating bases. In fact, this situation might be disadvantageous for large scale optimization problems. In any case, however, both algorithms reach the same optimized Newton number which is significantly (about 37%) lower than that of the non-optimized standard tank configuration in Table 1.

Furthermore, it should be remarked that; since no rigorous convergence properties for globally optimal solutions are available, the optimal solutions obtained with both optimization tools must be considered as local ones. This aspect can be investigated by variations of the starting value and/or the trust region radius.

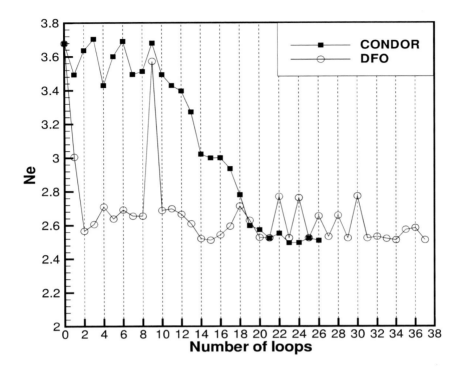

Figure 4. Newton number versus number of function evaluations

4.2. Conclusion

In this chapter, we have presented a numerical approach for optimizing practical stirrer configurations. The automated integrated procedure consists of a combination of a parametrized grid generator, a parallel flow solver, and a derivative free optimization procedure. For the latter, two different methods (DFO and CONDOR) have been investigated.

The numerical experiments have shown the principle applicability of the considered approach. For the Rushton turbine, it has been possible to achieve a significant reduction of the Newton number with relatively low computational effort. Of course, in a mixing process the power consumption is important but not the sole quantity in obtaining an optimum stirrer configuration. In particular, a satisfactory mixing should be achieved: due to the generality and modularity of the considered approach, other objectives and/or other design variables can be handled straightforwardly in a similar way as illustrated in Figure 2.

As the number of design variables increases, due to the large number of function evaluations, the computational cost of the algorithms becomes very high. For structured problems this cost can be reduced by using the separability of the objective function (see for example [4]) and implemented in the package PSDFO. Indeed, the FASTEST3D flow solver, with finite volume discretization, may give us the opportunity to compute the desired function values within blocks and their neighbors allowing us to use the concept of partially

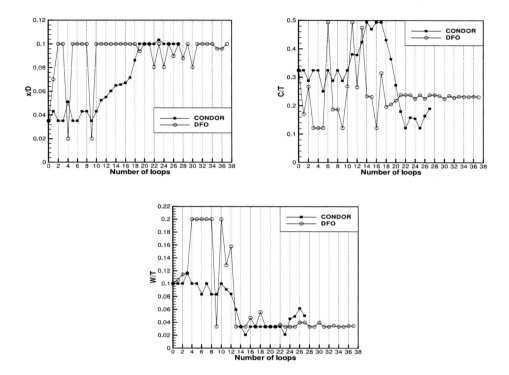

Figure 5. Design variables (disc thickness, bottom clearance, baffle length) versus number of function evaluations.

separability. A future application would be to apply PSDFO using the structure of the discretized Navier-Stokes equations by the finite volume method within the concept of domain decomposition.

References

[1] F. V. Berghen and H. Bersini. CONDOR a new parallel, constrained extension of powell's UOBYQA algorithm: Experimental results and comparison with the dfo algorithm. *Journal of Computational and Applied Mathematics*, **181**:157–175, 2005.

[2] M. Böhm, K. Wechsler, and M. Schäfer. A parallel moving grid multigrid method for flow simulation in rotor-stator configuration. *Int. J. Numer. Meth. Eng.*, **42**:175–189, 1998.

[3] A. Brucato, M. Ciofalo, F. Grisafi, and G. Micale. Numerical prediction of flow fields in baffled stirred vessels: a comparison of alternative modeling approaches. *Chemical Engineering Science*, **53**:3653–3684, 1998.

[4] B. Colson and Ph. L. Toint. Optimizing partially separable functions without derivatives. *Optimization Methods and Software*, **20**:493–508, 2005.

[5] A. R. Conn, K. Scheinberg, and Ph. L. Toint. On the convergence of derivative-free methods for unconstrained optimization. In A. Iserles and M. Buhmann, editors, *Approximation Theory and Optimization: Tribute to M. J. D. Powell*, pages 83–108. Cambridge University Press, 1997.

[6] A. R. Conn, K. Scheinberg, and Ph. L. Toint. Recent progress in unconstrained nonlinear optimization without derivatives. *Mathematical Programming*, **79**:397–414, 1997.

[7] A. R. Conn, N. I. M. Gould, and Ph. L. Toint. *Trust-Region Methods*, volume 1 of *MPS-SIAM Series on Optimization*. SIAM, Philadelphia, USA, 2000.

[8] A. R. Conn, K. Sheinberg, and L. N. Vicente. *Introduction to Derivative-Free Optimization*. SIAM Series on Optimization, 2009.

[9] J. Costes and J. P. Couderc. Study by Laser Doppler Anemometry of the turbulent flow induced by a Rushton turbine in a stirred tank: Influence of the size of the units I. mean flow and turbulence. *Chemical Engineering Science*, **43**:2751–2764, 1988.

[10] L. Driessen. *Simulation-based Optimization for Product and Process Design*. Phd thesis, University of Tilburg, 2006.

[11] J. Ferziger and M. Peric. *Computational Methods for Fluid Dynamics*. Springer, Berlin, 1996.

[12] H. Hartman, J. J. Derksen, C. Montavon, J. Pearson, I. S. Hamill, and H. E. A. Akker. Assessment of large eddy and RANS stirred tank simulations by means of LDA. *Chemical Engineering Science*, **59**:2419–2432, 2004.

[13] A. D. Harvey and S. E. Rogers. Steady and unsteady computation of impeller stirred reactor. *AIChE journal. American Institute of Chemical Engineers*, **42**:2701–2712, 1996.

[14] B. Karasözen. Survey of trust-region derivative free optimization methods. *Journal of Industrial and Management Optimization*, **3**:321–334, 2007.

[15] T. Lehnhäuser and M. Schäfer. Improve linear interpolation practice for finite volume scheme on complex grids. *International Journal of Numerical Methods in Fluids*, **38**: 625–645, 2002.

[16] J. Y. Luo, A. D. Gosman, R. I. Issa, J. C. Middleton, and M. K. Fitzgerald. Full flow field computation of mixing in baffled stirred tank vessels. *Inst. Chem. Eng. Res. Des.*, **71**:342–344, 1993.

[17] J. Y. Luo, R. I. Issa, and A. D. Gosman. Prediction of impeller-induced flows in mixing vessels using multiple frame of reference. *Institution of Chemical Engineers Symposium Series*, **136**:549–556, 1994.

[18] M. Marazzi and J. Nocedal. Wedge trust region methods for derivative free optimization. *Mathematical Programming Ser. A*, **91**:289–305, 2002.

[19] G. Montante, K. C. Lee, A. Brucato, and M. Yianneskis. Numerical simulations of the dependency of flow pattern an impeller clearance in stirred vessel. *Chemical Engineering Science*, **56**:3751–3770, 2001.

[20] J. J. Moré and D. C. Sorenson. Computing a trust region step. *SIAM J. Sci. Statist. Comput.*, **4**:553–572, 1983.

[21] J. J. Moré and S. M. Wild. Benchmarking derivative-free optimization algorithm. Preprint ANL/MCS-P1471-1207, Argonne National Laboratory, Mathematics and Computer Science Division, Argonne, Illinois, December 2007.

[22] J. L. Morales, G. Fasano, and J. Nocedal. On the geometry phase in model-based algorithms for derivative-free optimization. *Optimization Methods and Software*, **24**: 145–154, 2009.

[23] P. Mugunthan, C. A. Shoemaker, and R. G. Regis. Comparison of function approximation, heuristic and derivative-based methods for automatic calibration of computationally expensive groundwater bioremediation models. *Water Resources Research*, **41**(W11427):1–17, 2005.

[24] R. Oeuvray. *Trust-Region Methods Based on Radial Basis Functions with Application to Biomedical Imaging*. Phd thesis, Institut de Mathmatiqués, École Polytechnique Fédéerale de Lausanne, Switzerland, 2005.

[25] S. Patankar and D. Spalding. A calculation procedure for heat, mass and momentum transfer in three dimensional parabolic flows. *Int. J. Heat Mass Transfer*, **15**:1787–1806, 1972.

[26] M. J. D. Powell. UOBYQA: unconstrained optimization by quadratic approximation. *Mathematical Programming*, **92**:555–582, 2002.

[27] M. J. D. Powell. Developments of NEWUOA for minimization without derivatives. *IMA Journal of Numerical Analysis*, **28**(4):649–664, 2008.

[28] M. J. D. Powell. An efficient method for finding the minimum of a function of several variables without calculating derivatives. *Computer Journal*, **17**:175–162, 1964.

[29] M. J. D. Powell. A direct search optimization method that models the objective and constraint functions by linear interpolation. In S. Gomez and J. P. Hennart, editors, *Advances in Optimization and Numerical Analysis*. Kluwer Academic Publishers, 1994.

[30] V. V. Ranade. Computational fluid dynamics for reactor engineering. *Rev. Chem. Eng.*, **11**:229–289, 1995.

[31] V. V. Ranade and S. M. S. Dommetti. Computational snapshot of flow generated by axial impellers in baffled stirred vessels. *Trans. Inst. Chem. Eng.*, **74**:476–484, 1996.

[32] S. S. Rao. *Engineering Optimization*. Wiley, Chichester, 1996.

[33] K. Scheinberg and Ph. L. Toint. Self-correcting geometry in model-based algorithms for derivative-free unconstrained optimization. Technical Report 09/06, Department of Mathematics, University of Namur, Namur, Belgium, 2009.

[34] M. Schäfer, B. Karasözen, Y. Uludağ, K. Yapıcı, and Ö. Uğur. Numerical method for optimizing stirrer configurations. *Computers & Chemical Engineering*, **30**(2):183–190, 2005.

[35] J. Smagorinsky. General circulation experiments with the primitive equations: I. the basic experiment. *Monthly Weather Review*, **91**:99–164, 1969.

[36] H. L. Stone. Iterative solution of implicit approximation of multi-dimensional partial differential equations. *SIAM Journal on Numerical Analysis*, **5**:530–558, 1968.

[37] C. M. Stoots and R. V. Calabrese. Mean velocity field relative to a Rushton turbine blade. *AIChE journal. American Institute of Chemical Engineers*, **41**:1–11, 1995.

[38] Ö. Uğur, B. Karasözen, M. Schäfer, and K. Yapıcı. Derivative free optimization methods for optimizing stirrer configurations. *European Journal of Operational Research*, **191**(3):855–863, 2008.

[39] A. Waechter. *An Interior Point Algorithm for Large-Scale Nonlinear Optimization with Applications in Process Engineering*. Phd thesis, Department of Chemical Engineering, Carnegie Mellon University, 2002.

[40] D. Winfield. Function minimization by interpolation in a data table. *Journal of the Institute of Mathematics and its Applications*, **12**:339–347, 1973.

[41] Y. Xu and G. McGrath. CFD predictions of stirred tank flows. *Trans. Inst. Chem. Eng.*, **74**:471–475, 1996.

[42] K. Yapıcı, B. Karasözen, M. Schäfer, and Y. Uludağ. Numerical investigation of the effect of the Rushton type turbine design factors on agitated tank flow characteristics. *Chemical Engineering and Processing*, **47**:1346–1355, 2008.

[43] S. L. Yeoh, G. Papadakis, and M. Yianneskis. Numerical simulation of turbulent flow characteristic in a stirred vessel using the LES and RANS approaches with the sliding/deforming mesh methodology. *Trans. IChemE. Part A: Chem. Eng. Res. Des.*, **82**:834–848, 2004.

[44] G. Zhou and S. M. Kresta. Impact of tank geometry on the maximum turbulence energy dissipation rate for impellers. *AIChE journal. American Institute of Chemical Engineers*, **42**:2476–2490, 1996.

In: Computational Fluid Dynamics
Editor: Alyssa D. Murphy

ISBN: 978-1-61209-276-8
© 2011 Nova Science Publishers, Inc.

Chapter 16

FLUID STRUCTURE INTERACTION AND GALILEAN INVARIANCE

Luciano Garelli[1], *Rodrigo R. Paz*[1], *Hugo G. Castro*[1,2]
Mario A. Storti[1] *and Lisandro D. Dalcin*[1]

[1]Centro Internacional de Métodos
Computacionales en Ingeniería (CIMEC)
INTEC - CONICET - UNL
Güemes 3450, Santa Fe (3000), Argentina

[2]Grupo de Investigación en Mecánica de Fluidos,
Universidad Tecnológica Nacional,
Facultad Regional Resistencia, Chaco, Argentina

Abstract

Multidisciplinary and Multiphysics coupled problems represent nowadays a challenging field when studying or analyzing even more complex phenomena that appear in nature and in new technologies (e.g. Magneto-Hydrodynamics, Micro-Electro-Mechanics, Thermo-Mechanics, Fluid-Structure Interaction, etc.). Particularly, when dealing with Fluid- Structure Interaction problems several questions arise, namely the coupling algorithm, the mesh moving strategy, the Galilean Invariance of the scheme, the compliance with the Discrete Geometric Conservation Law (DGCL), etc. Therefore, the aim of this chapter is to give an overview of the issues involved in the numerical solution of Fluid-Structure Interaction (FSI) problems.

Regarding the coupling techniques, some results on the convergence of the strong coupling Gauss-Seidel iteration are presented. Also, the precision of different predictor schemes for the structural system and the influence of the partitioned coupling on stability are discussed.

Another key point when solving FSI problems is the use of the "Arbitrary Lagrangian Eulerian formulation" (ALE), which allows the use of moving meshes. As the ALE contributions affect the advective terms, some modifications on the stabilizing and the shock-capturing terms, are needed. Also Dirichlet constraints at slip (or non-slip) walls must be modified when the ALE scheme is used. In this chapter the presented ALE formulation is invariant under Galilean transformations.

1. Introduction

Multidisciplinary and multiphysics coupled problems represent nowadays a paradigm when studying or analyzing even more complex phenomena that appear in nature and in new technologies. There exists a great number of problems where different physical processes (or models) converge, interacting in a strong or weak fashion (e.g. acoustics/noise disturbances in flexible structures, magneto-hydrodynamics devices, micro-electro-mechanical devices, thermo-mechanical problems like continuous casting process, fluid-structure interaction (FSI) like wing flutter or flow-induced pipe vibrations).

In the FSI area, the dynamic interaction between an elastic structure and a compressible or incompressible fluid has been the subject of intensive investigations in the last years. In civil engineering, wind flow may lead to aeroelastic instabilities due to the structural motion on long-span bridges, high-rise buildings and light-weight roof structures and flexible silo [1, 2]. In biomechanical applications [3, 4, 5] the blood flow interacts with thin vascular walls, so it is important to know the hemodynamics and wall shear stress to understand the mechanisms leading to various complications in cardiovascular function, like aneurysms. In aeronautical engineering [6, 7] it is essential to understand the interaction between a high Mach number fluid with an elastic structure, like the aircraft wing, to predict the flutter velocity and avoid this flight condition. Other process like the start up of a rocket engine [8, 9, 10, 11] has been widely study considering the interaction between the fluid and the nozzle.

In general, when solving FSI problems, the characteristic scales in time and space vary widely, so the simulations tends to be arduous in the coupling process and time-consuming in the solution process, because the whole coupled problem is characterized by the smaller time scale to simulate. However there is a great interest in solving coupled multiphysics problems, with the aim to obtain reliable predictions. This interest has been accompanied by the increase in computer power in terms of CPU and memory, allowing to obtain results in reasonable time.

Today, the methodologies and techniques behind each discipline involved in FSI problems (e.g. Computational Fluid Dynamics (CFD) and Computational Structural Dynamics (CSD)) have been developed enough to allow the reliable simulation of a great variety of problems. These advances in the numerical simulation have been implemented in specialized codes. Therefore, an alternative is to couple these specialized codes with the objective of solving FSI problems, but in this case new questions arise, such as the coupling techniques [12, 13, 14, 15, 16, 6] and the motion of the fluid mesh [17]. Another possibility is to combine in a single formulation the fluid and the structural governing equations [18, 19], but this monolithic scheme may be mathematically unmanageable or its implementation can be a laborious task. Furthermore, the monolithic coupled formulation would change significantly if different fluid and/or structure models were considered.

With the aim of using existing codes several procedures have been proposed about coupling fluid and structure solvers. In this chapter a staggered fluid-structure coupling algorithm is considered and for stability reasons, often a fully implicit formulation has to be used. In this approach, we have to solve a large system of non-linear equations with the use of the (iterative) solvers for the subsystems. Usually, this is done with Block-Jacobi, Block-Gauss-Seidel or related relaxation methods [20, 21]. An extended description of the "state of the art" in the computational FSI area can be also found in works [22, 13] and the references therein. When a partitioned coupling technique is used, it involves a three-field system: the structure, the fluid and the moving mesh solver. The governing equations of the fluid are written, in general, in an Eulerian framework so it must be rewritten to allow the motion of the mesh, for which one of the most popular methods is the so-called Arbitrary Lagrangian Eulerian (ALE). The idea behind the ALE formulation is to introduce a computational mesh which moves with a velocity independent of the speed of the material particles. The ALE method was first proposed in the context of finite differences [23, 24], then it was extended to finite elements [25, 26] and to finite volumes [27].

When an ALE formulation is used, additional terms related to the mesh velocity and position, are introduced. Those related to the mesh velocity affect the advective terms and some modifications on the stabilizing and the shock-capturing terms are needed. Also boundary conditions at walls (slip or non-slip) must be modified when an ALE formulation is used. Since all these features are well known in a non-ALE frame, one can simply compute the appropriate objects (stabilization operators) in the non-ALE frame and transform them to the ALE frame. If this is properly done, then a formulation that is invariant to ALE transformations is obtained. In this chapter an ALE formulation that is invariant under Galilean transformations, which are the simplest of all the possible ALE transformations, is presented. A different approach to Galilean invariance has been proposed in [28].

Also, the movements of the fluid mesh produces a volume change in time and in this context is where the Discrete Geometric Conservation Law (DGCL) arises. This law was introduced by Thomas and Lombard in [29] and it is a consistency criterion in which the numerical method must be able to reproduce exactly a constant solution on a moving domain [30, 31, 32]. The impact of the DGCL on the stability and precision of the numerical methods is still unclear, but there is a general consensus in the development of schemes that satisfy the DGCL, in particular for FSI problems [33, 34, 3, 35, 36].

Regarding the movement of the mesh, it can be performed with a general strategy using both, nodal relocation or re-meshing. This methodology maintains the topology unchanged, simplifying the algorithms and their implementation. To perform the nodal relocation of the mesh there exist several strategies, some of them use a tension or torsion spring system [37] to propagate the boundary motion into the volume mesh and others solve a linear elastic or pseudo-elastic problem [38, 39, 40] to deform the mesh. A more sophisticated strategy to define the mesh motion as an optimization problem. This strategy may be classified as a particular case of an elastic problem where the material constitutive law is defined in terms of the minimization of certain energy functional that takes into account the degree of element distortion [17, 41].

2. Strongly Coupled Partitioned Algorithm via Fixed Point Iteration

In this section the temporal algorithm that performs the coupling between the structure and the fluid codes is sketched. It is a fixed point iteration scheme over the variables of both fluid and structure systems. Inside of the time step loop the algorithm is equipped with an inner loop called *'stage'*, so if the *'stage loop'* converges, then a *'strongly coupled'* algorithm is obtained. Hereafter, this algorithm is called *'staggered algorithm'*.

The basic scheme considered in this chapter proceeds as follows:

i) Transferring the motion of the wet boundary of the solid to the fluid problem.

ii) Updating the position of the fluid boundary and the bulk fluid mesh accordingly.

iii) Advancing the fluid system and compute new pressures.

iv) Converting the new fluid pressure (and stress field) into a structural load.

v) Advancing the structural system under the flow loads.

In this algorithm the three codes CFD (Computational Fluid Dynamics), CSD (Computational Structure Dynamics) and CMD (Computational Mesh Dynamics) are running simultaneously. For simplicity, the basic algorithm can be thought as if there were no *'concurrence'* between the codes, i.e. at a given time only one of them is running. This can be controlled using *'semaphores'* and this is done using Message Passing Interface (MPI) *'synchronization messages'*. A schematic diagram is shown in Figure (1).

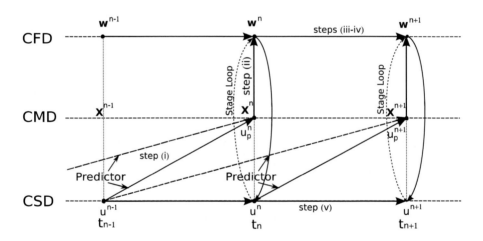

Figure 1. Synchronous FSI partitioned scheme.

At time t_n, we define $\mathbf{w}^n$ to be the fluid state vector $(\rho, \mathbf{v}, p)$, $\mathbf{u}^n$ to be the displacement vector (structure state vector), $\dot{\mathbf{u}}^n$ the structure velocities and $\mathbf{X}^n$ the fluid mesh node positions. In this case, both fluid and structure partitions are integrated with the trapezoidal algorithm (with trapezoidal parameter $0 < \alpha_{\text{trap}} \leq 1$) but another integration scheme could be used, such as linear multisteps methods, depending on the particular application [35].

In each time step the fluid is first advanced using the previously computed structure state $\mathbf{u}^n$ and the current estimated value $\mathbf{u}_p^{n+1}$. In this way, a new estimation for the fluid state $\mathbf{w}^{n+1}$ is computed. Next the structure is updated using the forces of the fluid from states $\mathbf{w}^n$ and $\mathbf{w}^{n+1}$. The estimated state $\mathbf{u}_p^{n+1}$ is predicted using a second or higher order approximation (1), were α_0 and α_1 are two real constants. The predictor (1) is trivial if $\alpha_0 = \alpha_1 = 0$, first-order time-accurate if $\alpha_0 = 1$ and second-order time-accurate if $\alpha_0 = 1$ and $\alpha_1 = 1/2$.

In reference [6] there is an extended description about the use of the predictor (1) on FSI problems and the energy transfer between the fluid and the structure. It was proved that monolithic schemes and strongly-coupled staggered schemes conserve energy-transfer at the fluid-structure interface boundary, whereas weak-coupled algorithms introduce after a certain amount of time t an artificial energy $E = \mathcal{O}(\Delta t^p)$, where p is the order of the prediction.

$$\mathbf{u}_p^{(n+1)} = \mathbf{u}^n + \alpha_0 \Delta t \dot{\mathbf{u}}^n + \alpha_1 \Delta t (\dot{\mathbf{u}}^n - \dot{\mathbf{u}}^{n-1}). \tag{1}$$

Once the coordinates of the structure are known, the coordinates of the fluid mesh nodes are computed by a *'Computational Mesh Dynamics'* code, which is symbolized as:

$$\mathbf{X}^n = \text{CMD}(\mathbf{u}^n). \tag{2}$$

At the beginning of each fluid stage there is a computation of skin normals and velocities. This is necessary due to the time dependent slip boundary condition for the inviscid case, and also when using a non-slip boundary condition, where the fluid interface has the velocity of the moving solid wall, i.e., $\mathbf{v}|_\Gamma = \dot{\mathbf{u}}|_\Gamma$.

3. Variational Formulation for Advective Diffusive System for Moving Meshes Using ALE

Let us start with the derivation of the ALE formulation for a general advective-diffusive system [25, 34]. The governing equation is

$$\frac{\partial U_j}{\partial t} + \left(\mathcal{F}_{jk}^c(\mathbf{U}) - \mathcal{F}_{jk}^d(\mathbf{U}, \nabla \mathbf{U}) \right)_{,k} = 0, \quad \text{in } \Omega^t \tag{3}$$

where $1 \leq k \leq n_d$, n_d is the number of spatial dimensions, $1 \leq j \leq m$, m is the dimension of the state vector (e.g. $m = n_d + 2$ for compressible flow), t is time, $(\ )_{,j}$ denotes derivative with respect to the j-th spatial dimension, $\mathbf{U} \in \mathbb{R}^n$ is the state vector, and $\mathcal{F}_{jk}^{c,d} \in \mathbb{R}^{n \times n_d}$ are the convective and diffusive fluxes, respectively. Appropriate Dirichlet and Neumann conditions are imposed at the boundary.

As the problem is posed in a time-dependent domain Ω^t, it can not be solved with standard fixed-domain methods, so that it is assumed that there is an inversible and continuously differentiable map $\mathbf{x} = \chi(\boldsymbol{\xi}, t)$ between the current domain Ω^t and a reference domain Ω^ξ, which can be for instance the initial domain $\Omega^\xi = \Omega^{t=0}$, and $\boldsymbol{\xi}$ is the coordinate in the reference domain. The Jacobian of the transformation is

$$J = \left| \frac{\partial x_j}{\partial \xi_k} \right|, \tag{4}$$

and satisfies the following volume balance equation

$$\left. \frac{\partial J}{\partial t} \right|_{\boldsymbol{\xi}} = J \frac{\partial v_k^*}{\partial x_k}, \tag{5}$$

where
$$v_k^* = \left.\frac{\partial x_k}{\partial t}\right|_{\boldsymbol{\xi}}, \qquad (6)$$
are the components of the mesh velocity.

The variational formulation of (3) is obtained multiplying with a weighting function $w(\mathbf{x}, t) = w(\boldsymbol{\chi}(\mathbf{x}, t))$ and integrating over the current domain Ω^t

$$\int_{\Omega^t} w \frac{\partial U_j}{\partial t} \, \mathrm{d}\Omega^t + \int_{\Omega^t} \left[\mathcal{F}_{jk}^c - \mathcal{F}_{jk}^d\right]_{,k} w \, \mathrm{d}\Omega^t = 0. \qquad (7)$$

The integrals are brought to the reference domain Ω^ξ

$$\int_{\Omega^\xi} w \frac{\partial U_j}{\partial t} J \, \mathrm{d}\Omega^\xi + \int_{\Omega^\xi} \left[\mathcal{F}_{jk}^c - \mathcal{F}_{jk}^d\right]_{,k} wJ \, \mathrm{d}\Omega^\xi = 0, \qquad (8)$$

and the temporal derivative term can be converted to the reference mesh by noting that the partial derivative of U_j is in fact a partial derivative at $\mathbf{x}$ = constant, and then can be converted to a partial derivative at $\boldsymbol{\xi}$ = constant with the relation

$$\left.\frac{\partial U_j}{\partial t}\right|_{\mathbf{x}} = \left.\frac{\partial U_j}{\partial t}\right|_{\boldsymbol{\xi}} - v_k^* \frac{\partial U_j}{\partial x_k}. \qquad (9)$$

So the temporal derivative term in (8) can be transformed, using (5), as follows

$$\begin{aligned} J \left.\frac{\partial U_j}{\partial t}\right|_{\mathbf{x}} &= J \left.\frac{\partial U_j}{\partial t}\right|_{\boldsymbol{\xi}} - J v_k^* \frac{\partial U_j}{\partial x_k}, \\ &= \left.\frac{\partial (JU_j)}{\partial t}\right|_{\boldsymbol{\xi}} - JU_j \frac{\partial v_k^*}{\partial x_k} - J v_k^* \frac{\partial U_j}{\partial x_k}, \\ &= \left.\frac{\partial (JU_j)}{\partial t}\right|_{\boldsymbol{\xi}} - J \frac{\partial (U_j v_k^*)}{\partial x_k}. \end{aligned} \qquad (10)$$

Replacing (10) in (8),

$$\int_{\Omega^\xi} w(\boldsymbol{\xi}) \left.\frac{\partial}{\partial t}(JU_j)\right|_{\boldsymbol{\xi}} \mathrm{d}\Omega^\xi + \int_{\Omega^\xi} \left(\mathcal{F}_{jk}^c - v_k^* U_j - \mathcal{F}_{jk}^d\right)_{,k} w(\boldsymbol{\xi}) J \, \mathrm{d}\Omega^\xi = 0. \qquad (11)$$

The temporal derivative can be commuted with the integral and the weighting function since both do not depend on time, so that

$$\frac{\mathrm{d}}{\mathrm{d}t}\left(\int_{\Omega^\xi} w \, JU_j \, \mathrm{d}\Omega^\xi\right) + \int_{\Omega^\xi} \left(\mathcal{F}_{jk}^c - v_k^* U_j - \mathcal{F}_{jk}^d\right)_{,k} wJ \, \mathrm{d}\Omega^\xi = 0, \qquad (12)$$

and the integrals can be brought back to the Ω^t domain

$$\frac{\mathrm{d}}{\mathrm{d}t}\left(\int_{\Omega^t} w \, U_j \, \mathrm{d}\Omega^t\right) + \int_{\Omega^t} \left(\mathcal{F}_{jk}^c - v_k^* U_j - \mathcal{F}_{jk}^d\right)_{,k} w \, \mathrm{d}\Omega^t = 0. \qquad (13)$$

The variational formulation can be obtained by integrating by parts, so that

$$\frac{\mathrm{d}}{\mathrm{d}t}\left(H(w, U)\right) + F(w, U) = 0, \qquad (14)$$

where

$$H(w, U) = \int_{\Omega^t} w\, U_j \, d\Omega^t,$$
$$F(w, U) = A(w, U) + B(w, U) + S(w, U),$$
$$A(w, U) = -\int_{\Omega^t} \left(\mathcal{F}_{jk}^c - v_k^* U_j - \mathcal{F}_{jk}^d \right) w_{,k} \, d\Omega^t, \quad (15)$$
$$B(w, U) = \int_{\Gamma^t} \left(\mathcal{F}_{jk}^c - v_k^* U_j - \mathcal{F}_{jk}^d \right) n_k w \, d\Gamma,$$

Γ^t is the boundary of Ω^t, and n_k is its unit normal vector pointing to the exterior of Ω. Also, a consistent stabilization term $S(w, U)$ is included in order to avoid numerical problems for advection dominated problems [42].

Finally (13) is discretized in time with the *trapezoidal rule*.

$$\begin{aligned} H(w, U^{n+1}) - H(w, U^n) &= -\int_{t=t^n}^{t^{n+1}} F(w, U^{t'}) \, dt', \\ &\approx -\Delta t\, F(w, U^{n+\theta}). \end{aligned} \quad (16)$$

with $0 \leq \theta \leq 1$. During the time step it is assumed that the nodal points move with constant velocity, i.e.

$$\left. \begin{aligned} v_k^*(\boldsymbol{\xi}) &= \frac{x_k(\boldsymbol{\xi}, t^{n+1}) - x_k(\boldsymbol{\xi}, t^n)}{\Delta t}, \\ x_k(\boldsymbol{\xi}, t) &= x_k(\boldsymbol{\xi}, t^n) + (t - t^n) v_k^*(\boldsymbol{\xi}), \end{aligned} \right\} \quad \text{for } t^n \leq t \leq t^{n+1}. \quad (17)$$

4. Frame Invariance of ALE Formulation

As was introduced in §1. the ALE contributions affect the advective terms, some modifications are needed to the standard stabilization terms in order to get the correct amount of stabilization. Also boundary conditions at walls (slip or non-slip) must be modified when ALE is used.

Since all these features are well known in a non-ALE frame, one can simply compute the appropriate objects (stabilization operators, absorbing boundary condition projections) in the non-ALE frame and transform them to the ALE frame. If this is properly done, then a formulation that is "invariant" to ALE transformations is obtained. In this section a ALE formulation that is invariant under Galilean transformations, which are the simplest of all the possible ALE transformations, is described.

4.1. Transformations Laws for Jacobians

Consider the governing equations for compressible flow of an ideal gas in a Galilean frame $S = (\mathbf{x}, t)$ derived from (3)

$$\frac{\partial \mathbf{U}_c}{\partial t} + \mathcal{F}_k^c(\mathbf{U})_{,k} = 0 \quad (18)$$

where

$$\begin{aligned}
\mathbf{U}_c &= [\rho, \rho u, \rho e]^T, &\text{(conservative variables)}\\
\mathcal{F}^c &= [\rho u, \rho u^2 + p, \rho h]^T, &\text{(convective flux)}\\
e &= p[\rho/(\gamma-1)]^{-1} + \tfrac{1}{2}\rho u^2, &\text{(specific total energy)}\\
h &= e + p/\rho, &\text{(specific total enthalpy)}\\
\mathbf{U} &= [\rho, \mathbf{u}, p]^T, &\text{(primitive variables)}
\end{aligned} \quad (19)$$

ρ is the density, u velocity, p pressure and $\gamma = C_p/C_v$ is the *"adiabatic index"* of the gas.

Now consider the corresponding system of equations in a second Galilean frame $S' = (\mathbf{x}', t)$ moving with a constant velocity $\mathbf{v}$ with respect to S, i.e. $\mathbf{x}' = \mathbf{x} - \mathbf{v}t$. The fluid states $\mathbf{U}'$, $\mathbf{U}_c'$ in a given point $(\mathbf{x}', t)$ in the S' frame, are related to the states in the original S frame through the relations

$$\begin{aligned}
\mathbf{U}_c' &= [\rho', \rho' u', \rho' e']^T = [\rho, \rho(\mathbf{u}-\mathbf{v}), \rho e - u'v + \tfrac{1}{2}v^2]^T, &\text{(conservative variables)},\\
\mathbf{U}' &= [\rho', u', p']^T = [\rho, u-v, p]^T, &\text{(primitive variables)},
\end{aligned} \quad (20)$$

so that the Jacobian of the transformations are

$$\begin{aligned}
\frac{\partial \mathbf{U}'}{\partial \mathbf{U}} &= \mathbf{I},\\
\frac{\partial \mathbf{U}_c'}{\partial \mathbf{U}_c} &= \begin{bmatrix} 1 & 0 & 0 \\ -\mathbf{v} & \mathbf{I} & 0 \\ \tfrac{1}{2}v^2 & -\mathbf{v}^T & 1 \end{bmatrix} = \mathbf{T}(-\mathbf{v}).
\end{aligned} \quad (21)$$

It can be easily verified that they are a *"group"* of transformations, i.e.

$$\begin{aligned}
\mathbf{T}(\mathbf{v}+\mathbf{w}) &= \mathbf{T}(\mathbf{v})\,\mathbf{T}(\mathbf{w}),\\
\mathbf{T}(-\mathbf{v}) &= \mathbf{T}(\mathbf{v})^{-1}
\end{aligned} \quad (22)$$

Now, if $\mathbf{U}(\mathbf{x}, t)$ is a solution of (18,19) then $\mathbf{U}'(\mathbf{x}', t)$ obtained with the transformations (20) must be a solution of the governing equations in frame S'. In order to simplify the algebra, consider the quasi-linear form of the governing equations

$$\mathbf{C}\frac{\partial \mathbf{U}}{\partial t} + \mathbf{A}_x \frac{\partial \mathbf{U}}{\partial x} = 0 \quad (23)$$

where

$$\begin{aligned}
\mathbf{C} &= \frac{\partial \mathbf{U}_c}{\partial \mathbf{U}}, &\text{(enthalpy Jacobian)},\\
\mathbf{A}_x &= \frac{\partial \mathcal{F}^c x}{\partial \mathbf{U}}. &\text{(advective Jacobian)}.
\end{aligned} \quad (24)$$

The temporal derivative transforms in the following way,

$$\begin{aligned}
\left.\frac{\partial \mathbf{U}}{\partial t}\right|_x &= \left.\frac{\partial \mathbf{U}}{\partial t}\right|_{x'} + \mathbf{v}\frac{\partial \mathbf{U}}{\partial x},\\
&= \frac{\partial \mathbf{U}}{\partial \mathbf{U}'}\left(\left.\frac{\partial \mathbf{U}'}{\partial t}\right|_{x'} + \mathbf{v}\frac{\partial \mathbf{U}'}{\partial x'}\right) = \left.\frac{\partial \mathbf{U}'}{\partial t}\right|_{x'} + \mathbf{v}\frac{\partial \mathbf{U}'}{\partial x'}.
\end{aligned} \quad (25)$$

Replacing in (23) the following expression is obtained,

$$\mathbf{C}\left(\left.\frac{\partial \mathbf{U}'}{\partial t}\right|_{x'} + \mathbf{v}\frac{\partial \mathbf{U}'}{\partial x'}\right) + \mathbf{A}_x \frac{\partial \mathbf{U}}{\partial \mathbf{U}'}\frac{\partial \mathbf{U}'}{\partial x} = 0, \qquad (26)$$

and multiplying by $(\partial \mathbf{U}_c'/\partial \mathbf{U}_c)$ at the left,

$$\left(\frac{\partial \mathbf{U}_c'}{\partial \mathbf{U}_c}\frac{\partial \mathbf{U}_c}{\partial \mathbf{U}}\frac{\partial \mathbf{U}}{\partial \mathbf{U}'}\right)\left(\left.\frac{\partial \mathbf{U}'}{\partial t}\right|_{x'} + \mathbf{v}\frac{\partial \mathbf{U}'}{\partial x'}\right) + \left(\frac{\partial \mathbf{U}_c'}{\partial \mathbf{U}_c}\mathbf{A}_x \frac{\partial \mathbf{U}}{\partial \mathbf{U}'}\right)\frac{\partial \mathbf{U}'}{\partial x} = 0,$$
$$\mathbf{C}'\frac{\partial \mathbf{U}'}{\partial t} + \mathbf{A}'\frac{\partial \mathbf{U}'}{\partial x} = 0, \qquad (27)$$

from which the following transformation laws for the Jacobians are deduced,

$$\begin{aligned}
\mathbf{C}' &= \frac{\partial \mathbf{U}_c'}{\partial \mathbf{U}_c}\mathbf{C}\frac{\partial \mathbf{U}}{\partial \mathbf{U}'}, \\
\mathbf{C} &= \frac{\partial \mathbf{U}_c}{\partial \mathbf{U}_c'}\mathbf{C}'\frac{\partial \mathbf{U}'}{\partial \mathbf{U}}, \\
\mathbf{A}_x' &= v_x \mathbf{C}' + \frac{\partial \mathbf{U}_c'}{\partial \mathbf{U}_c}\mathbf{A}_x \frac{\partial \mathbf{U}}{\partial \mathbf{U}'}, \\
\mathbf{A}_x &= v_x \mathbf{C} + \frac{\partial \mathbf{U}_c}{\partial \mathbf{U}_c'}\mathbf{A}_x'\frac{\partial \mathbf{U}'}{\partial \mathbf{U}}.
\end{aligned} \qquad (28)$$

It can be shown that source and diffusive terms also transform in the appropriate way. We say that $\mathbf{C}$ and $\mathbf{A}$ *transform as* $\mathbf{U}_c \times \mathbf{U}$, i.e. as $\mathbf{U}_c$ (conservative variables) at left and as $\mathbf{U}$ (primitive variables) at right.

4.2. Definition of ALE Invariance

Of course, the discrete equations are not invariant under an arbitrary Galilean transformation, mainly because the importance of the advective terms are relative to the frame of reference. For instance, a fluid which is at rest in frame S does not need stabilization, whereas in a frame S' with relative velocity $\mathbf{v}$ it may have a high Pèclet number and then it will need stabilization. However, when using ALE formulations with moving domains, stabilization is based on the velocity of the *fluid relative to the mesh*. With this additional degree of freedom introduced with moving meshes a physical problem can be posed in different Galilean frames and in such a way that the velocity of the fluid *relative to the mesh is the same*. Then the question can be posed of whether discrete stabilized equations give the same solution (after appropriate transformation laws) in these equivalent situations. If the scheme is not invariant then great chances exist that the scheme adds more diffusion in one frame than in other, and then to be unstable or too diffusive. If the discrete formulation pass the test we say that it is *"ALE invariant"*. In this article only invariance under Galilean (i.e. constant velocity) transformations between the systems is considered. As ALE can be applied to more general mesh movements involving, for instance, rotations and accelerations, the same question can be posed for those cases as well.

4.3. Transformation of the Stabilization Term

The key point in obtaining an ALE invariant discrete system is to make the stabilization numerical diffusion operator invariant. Recall that the *"Streamline Upwind Petrov Galerkin"* (SUPG, see [43]) stabilization term is obtained by weighting the governing equations with a weight function W that is the sum of the interpolation function and a perturbation function P defined (in the scalar case) as

$$W = N + P,$$
$$P \propto \nabla N \cdot \mathbf{a} \tag{29}$$

where $\mathbf{a}$ is the scalar version of $\mathbf{A}$ which is, for scalar problems, a vector of scalar matrices. Similarly, c will denote the scalar version of $\mathbf{C}$. The $\mathbf{a}$ factor in the definition of P introduces the bias of the weight function and guarantees that the formulation tends to its *Galerkin ("centered")* form when the advective term is negligible. As P should be nondimensional it can be shown that for scalar problems P should be of the form

$$P = (\tau/c)\nabla N \cdot \mathbf{a}, \tag{30}$$

where τ has dimensions of time (the so called *"intrinsic time"*). Typically

$$\tau = \frac{hc}{|a|} \tag{31}$$

where h is the mesh size. In the S frame, where the domain is fixed, the stabilized equations are

$$\mathbf{C}\frac{\partial \mathbf{U}_j}{\partial t} + \mathbf{A}\frac{\mathbf{U}_{j+1} - \mathbf{U}_j}{2h} = (\mathbf{K}_{\text{num}} + \mathbf{K})\frac{\mathbf{U}_{j+1} - 2\mathbf{U}_{j-1} + \mathbf{U}_{j-1}}{h^2} \tag{32}$$

where, in the scalar case

$$\mathbf{K}_{\text{num}} = (\tau/c)a^2. \tag{33}$$

The objective of this section is how to extend (31,33) to systems as in the case of the gas-dynamics equations with an ALE formulation. It will be shown that if P is chosen in the form

$$P = \nabla N \cdot \tilde{\mathbf{A}}\tau\mathbf{C}^{-1},$$
$$\tilde{\mathbf{A}} = \mathbf{A} - v_*\mathbf{C}, \tag{34}$$

then the discrete scheme is ALE invariant, provided that τ transforms as $\mathbf{U} \times \mathbf{U}$, i.e.

$$\tau' = \frac{\partial \mathbf{U}'}{\partial \mathbf{U}}\tau\frac{\partial \mathbf{U}}{\partial \mathbf{U}'}. \tag{35}$$

In (34) $\tilde{\mathbf{A}}$ is the ALE corrected advective Jacobian.

After some algebra it can be shown that the numerical diffusion operator produced by a stabilization term like (34) is

$$\mathbf{K}_{\text{num}} = \tilde{\mathbf{A}}\tau\mathbf{C}^{-1}\tilde{\mathbf{A}}. \tag{36}$$

Now, let us transform the discrete equations (32) to the S' frame. Note that, as in the S frame the mesh is fixed, the equation has not ALE term (i.e. $v_* = 0$). To transform the equation, multiply the equations at left by $(\partial \mathbf{U}_c'/\partial \mathbf{U}_c)$ and by $(\partial \mathbf{U}/\partial \mathbf{U}')$ at right, resulting in

$$\left(\frac{\partial \mathbf{U}_c'}{\partial \mathbf{U}_c}\mathbf{C}\frac{\partial \mathbf{U}}{\partial \mathbf{U}'}\right)\left(\frac{\partial \mathbf{U}'}{\partial \mathbf{U}}\frac{\partial \mathbf{U}_j}{\partial t}\right) + \left(\frac{\partial \mathbf{U}_c'}{\partial \mathbf{U}_c}\mathbf{A}\frac{\partial \mathbf{U}}{\partial \mathbf{U}'}\right)\left(\frac{\partial \mathbf{U}'}{\partial \mathbf{U}}\frac{\mathbf{U}_{j+1}-\mathbf{U}_{j-1}}{2h}\right) =$$
$$\left[\frac{\partial \mathbf{U}_c'}{\partial \mathbf{U}_c}(\mathbf{K}_{\text{num}}+\mathbf{K})\frac{\partial \mathbf{U}}{\partial \mathbf{U}'}\right]\left(\frac{\partial \mathbf{U}'}{\partial \mathbf{U}}\frac{\mathbf{U}_{j+1}-2\mathbf{U}_j+\mathbf{U}_{j-1}}{h^2}\right) \quad (37)$$

and using the transformation law for the Jacobians (28),

$$\mathbf{C}'\frac{\partial \mathbf{U}_j'}{\partial t} + (\mathbf{A}'-v\mathbf{C}')\frac{\mathbf{U}_{j+1}'-\mathbf{U}_{j-1}'}{2h} = (\mathbf{K}_{\text{num}}'+\mathbf{K}')\frac{\mathbf{U}_{j+1}'-2\mathbf{U}_j'+\mathbf{U}_{j-1}'}{h^2} \quad (38)$$

The term added to the advective Jacobian stands for the ALE formulation (since $v_* = v$ in frame S'). The expression for $\mathbf{K}_{\text{num}}'$ is

$$\begin{aligned}
\mathbf{K}_{\text{num}}' &= \frac{\partial \mathbf{U}_c'}{\partial \mathbf{U}_c}\mathbf{A}\tau\mathbf{C}\mathbf{A}\frac{\partial \mathbf{U}}{\partial \mathbf{U}'}, \\
&= \left(\frac{\partial \mathbf{U}_c'}{\partial \mathbf{U}_c}\mathbf{A}\frac{\partial \mathbf{U}}{\partial \mathbf{U}'}\right)\left(\frac{\partial \mathbf{U}'}{\partial \mathbf{U}}\tau\frac{\partial \mathbf{U}}{\partial \mathbf{U}'}\right)\left(\frac{\partial \mathbf{U}'}{\partial \mathbf{U}}\mathbf{C}^{-1}\frac{\partial \mathbf{U}_c}{\partial \mathbf{U}_c'}\right)\left(\frac{\partial \mathbf{U}_c'}{\partial \mathbf{U}_c}\mathbf{A}\frac{\partial \mathbf{U}}{\partial \mathbf{U}'}\right), \quad (39)\\
&= \tilde{\mathbf{A}}'\tau'\mathbf{C}'^{-1}\tilde{\mathbf{A}}',
\end{aligned}$$

showing the invariance of the numerical diffusion term. As $\mathbf{C}$, $\tilde{\mathbf{A}}$ and $\mathbf{K}$ all transform as $\mathbf{U}_c \times \mathbf{U}$, the combinations $\mathbf{C}^{-1}\tilde{\mathbf{A}}$ and $\mathbf{C}^{-1}\mathbf{K}$ transform as $\mathbf{U} \times \mathbf{U}$, so equation (35) is verified if it is computed as a matrix function of $\mathbf{C}^{-1}\tilde{\mathbf{A}}$ and $\mathbf{C}^{-1}\mathbf{K}$. For instance, typical extensions of (31) to systems of equations in the inviscid case are

$$\begin{aligned}
\tau &= \frac{h}{\max|\lambda_j|}\mathbf{I}, \quad \lambda_j = \text{eig}(\mathbf{C}^{-1}\tilde{\mathbf{A}}), \quad \text{(maximum eigenvalue, in magnitude)} \\
\tau &= h|\mathbf{C}^{-1}\tilde{\mathbf{A}}|^{-1}. \quad\quad\quad\quad\quad\quad\quad\quad\quad\quad (|\cdot| \text{ in matrix sense})
\end{aligned} \quad (40)$$

5. The Discrete Geometric Conservation Law Condition

A discrete formulation is said to satisfy the DGCL condition if it solves exactly a constant state regime, i.e. not depending on space or time, for a general mesh movement $\mathbf{x}(\boldsymbol{\xi},t)$.

As noted in [32] the effect of the DGCL on the stability of ALE schemes is still unclear and somewhat contradictory. In the work [44], it has been observed that the movement of the domain can degrade the accuracy and stability of the numerical scheme with respect to their counterpart on fixed domains. In this direction, many researchers have been working with the aim of linking the accuracy and the stability of numerical schemes on an ALE framework with the discrete version of the Geometric Conservation Law [44, 31, 45, 32]. In several works [44, 45] it is recommended to employ numerical schemes that satisfy the DGCL, in particular for FSI problems. This may help in improving the precision and the stability.

Now, by replacing $U_j = $ constant in (13) and after some manipulations it can be shown that the DGCL is satisfied if

$$\int_{\Omega^{n+1}} w \, d\Omega - \int_{\Omega^n} w \, d\Omega = \Delta t \int_{\Omega^{n+\theta}} v_k^* w_{,k} \, d\Omega. \tag{41}$$

A similar restriction holds for the boundary term. The stabilization term $S(w, U)$ normally satisfies automatically the DGCL since it involves gradients of the state, and then it is null for a constant state.

Note that this previous equation holds if the right hand side is evaluated as an integral instead of being evaluated at $t^{n+\theta}$, i.e. the DGCL error comes from the approximation that was made in (16), i.e. it is always true that

$$\int_{\Omega^{n+1}} w \, d\Omega - \int_{\Omega^n} w \, d\Omega = \int_{t=t^n}^{t^{n+1}} \left\{ \int_{\Omega^{n+\theta}} v_k^* w_{,k} \, d\Omega \right\} dt. \tag{42}$$

A straightforward way to satisfy the DGCL is to use a time integration rule with degree of precision $n_d \cdot s - 1$, where n_d is the spatial dimension and s is the order of the polynomial used to represent the time evolution of the nodal displacement within each time step. For example, in 3D problems with a linear in time reconstruction a rule with degree of precision 2 should be used. Alternatively, the methodology proposed in [46] to obtain an ALE extension for a given time-integrator in fixed meshes, could be used.

6. Numerical Tests

In this section a set of numerical tests are performed. First, the flutter of a flat solid plate is estudy. This test is relevant because the onset of physical instabilities is very sensitive to the precision in the transfer of forces and displacements. Then, is carried out the DGCL validation for 2D scalar diffusion problem with internal node movement. The problem is solved in an unit square using three differents time integration method are used.

With regard to the ALE invariance the problem of a sudden stop of gas container is solved using two different reference frames. Of course, in the continuum both systems of equations are completely equivalent, this can be verified by writing down the equations for one frame an transform to the other.

Finally, the simulation of the start-up of a rocket engine nozzle is performed. This problem involves all the issues discussed in this chapter.

6.1. Flutter of a Flat Solid Plate

The flutter of a flat solid plate aligned with a gas flow at supersonic Mach numbers (see Fig. (2)) is studied and the critical Mach number is computed.

It must be stressed that in this test case we deal with both physical and numerical instabilities. They are, in general, uncorrelated, i.e. a temporal integration scheme can be unstable for a configuration of physical parameters that is well outside the flutter region and *vice versa*.

This test is relevant because the onset of physical instabilities is very sensitive to the precision in the transfer of forces and displacements. Afterwards the stability of the numerical scheme will be assessed in a region far from the physically unstable region (flutter) in order to be sure that if any instabilities are detected, they come exclusively from the coupling process.

A uniform fluid at state $(\rho_\infty, U_\infty, p_\infty)$ flows over an horizontal rigid wall $y = 0$ parallel to it. This test case has been studied also in [6].

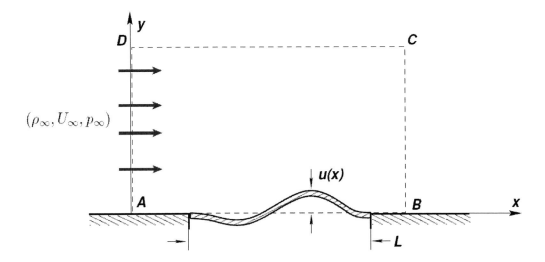

Figure 2. Description of test.

In a certain region of the wall ($0 \leq x \leq L$) the wall deforms elastically following thin plate theory, i.e.

$$m\ddot{u} + D\frac{\partial^4 u}{\partial x^4} = -(p - p_\infty) + f(x, t), \qquad (43)$$

where m is the mass of the plate per unit area in Kg/m^2, $D = Et^3/12(1-\nu^2)$ the *bending rigidity* of the plate module in Nm, E is the Young modulus in Pa, t the plate thickness in m, ν the Poisson modulus, u the normal deflection of the plate in m, defined on the region $0 \leq x \leq L$ and null outside this region, p the pressure exerted by the fluid on the plate in Pa, f is an external force in N and will be described later. The plate is clamped at both ends, i.e. $u = (\partial u/\partial x) = 0$ at $x = 0, L$. For the sake of simplicity the fluid occupying the region $y > 0$ is inviscid. The compressible Euler model with SUPG stabilization and 'anisotropic shock-capturing' method is considered (see Reference [47]). A slip condition is assumed

$$(\mathbf{v} - \mathbf{v}_{\text{str}}) \cdot \hat{\mathbf{n}} = 0 \qquad (44)$$

on the (curved) wall $y = u(x)$, where

$$\mathbf{v}_{\text{str}} = (0, \dot{u}),$$
$$\hat{\mathbf{n}} \propto (-\frac{\partial u}{\partial x}, 1) \qquad (45)$$

are the velocity of the plate and its unit normal. Finally, initial conditions for both the fluid and the plate are taken as

$$u(x, t = 0) = u_0(x),$$
$$\dot{u}(x, t = 0) = \dot{u}_0(x), \tag{46}$$
$$(\rho, \mathbf{v}, p)_{\mathbf{x}, t=0} = (\rho, \mathbf{v}, p)_0, \quad \text{for } y \geq u_0(x).$$

Note that for the fluid pressure load on the plate the free stream fluid pressure is subtracted so that in the absence of any external perturbation ($f \equiv 0$) the undisturbed flow $(\rho, \mathbf{v}, p)_{x,t} \equiv (\rho, \mathbf{v}, p)_\infty$ is a solution of the problem for the initial conditions

$$u \equiv 0,$$
$$\dot{u} \equiv 0, \tag{47}$$
$$(\rho, \mathbf{v}, p)_{\mathbf{x}, t=0} \equiv (\rho, \mathbf{v}, p)_\infty.$$

The study of the flutter instability is carried out by means of the modal analysis, assuming the *'Houbolt approximation'* for the fluid (see Reference [48]), where the pressure acting on the plate surface is a function of the plate deflection derivatives, i.e., $p - p_\infty = f(\frac{\partial u}{\partial x}, \frac{\partial u}{\partial t})$. Then a Galerkin method is used and the normal displacement is expanded in a global basis. These basis functions satisfy the essential boundary conditions for the plate equation $u = \frac{\partial u}{\partial x} = 0$ at $x = 0, L$. Replacing the Houbolt approximation in Equation (43), using Galerkin method and integrating by parts as needed, leads to the eigenvalue problem.

Flutter is detected whenever the real part of some eigenvalue λ changes its sign.
In order to determine the critical Mach number M_{cr}, the interval $1.8 \leq M \leq 3.0$ was swept with increments of 0.01.

For this problem all the eigenvalues have negative real part for $M_\infty < M_{cr} = 2.265$ which results in a stable system. For $M_\infty > M_{cr} = 2.265$ there are two complex conjugate roots with positive real parts. The computed value of M_{cr} is in agreement with the result given in [6] (i.e., $M_{cr} = 2.23$). These results will be used to validate the coupling code.

6.2. Dimensionless Parameters

As the fluid is inviscid, it is determined by the *'adiabatic index'* $\gamma = C_p/C_v = 1.4$ for air, and the Mach number $M_\infty = U_\infty/c_\infty$, where c_∞ is the speed of sound $c = \sqrt{\gamma p/\rho}$ for the undisturbed state.

Another dimensionless parameter can be built by taking the ratio between the characteristic time of the fluid which is $T_\text{fl} = L/U_\infty$ and the characteristic time of the structure $T_\text{str} = \sqrt{mL^4/D}$. For practical reasons we take the square of this ratio

$$N_T = \left(\frac{T_\text{fl}}{T_\text{str}}\right)^2 = \frac{D}{mL^2 U_\infty^2}. \tag{48}$$

Finally, a (dimensionless) number can be formed by taking the ratio between the mass of the fluid being displaced by the structure and the structure mass

$$N_M = \frac{\rho_\infty L^3}{mL^2} = \frac{\rho_\infty L}{m}. \tag{49}$$

The same parameters as reported in Reference [6] are considered. In this contribution, flutter was studied near the point $M_\infty = 2.27$, $N_T = 4.3438 \times 10^{-5}$ and $N_M = 0.054667$. The flutter region was studied by varying the M_∞ value while keeping ρ_∞ and the structure parameters (m, L, D) constant (so that N_M constant and $N_T \propto M_\infty^{-2}$), and the same approach is taken here. The dimensionless parameters are obtained by choosing the following dimensional values

$$\begin{aligned}
\rho_\infty &= 1 \text{ Kg/m}^3, \\
p_\infty &= 1/\gamma = 0.71429 \text{ Pa}, \\
U_\infty &= M_\infty, \quad (\text{since } c_\infty = \sqrt{\gamma p_\infty / \rho_\infty} = 1 \text{ m/sec}), \\
D &= 0.031611 \text{ Nm}, \\
m &= 36.585 \text{ Kg/m}^2, \\
L &= 2 \text{ m}.
\end{aligned} \quad (50)$$

6.2.1. FSI Code Results

The aeroelastic problem defined above was modeled with the strongly coupled partitioned algorithm described in section §2. with a mesh of 12800 quadrilateral elements for the fluid and 5120 for the plate. As the flow is supersonic only a small entry section of $\frac{1}{8}L$ upstream the plate and $\frac{1}{3}L$ downstream is considered. The vertical size of the computational domain was chosen as $0.8L$. It is assured that no reflection from the upper boundary affects the plate itself when considering these sizes for the fluid domain.

6.2.2. Determination of Flutter Region

This section presents some results obtained with PETSc-FEM code [49] using the weak coupling between fluid and structure, i.e. $n_{\text{stage}} = 1$. The physical characteristics of the plate are the same as in previous section. In order to find (numerically) the *critical Mach number* for this problem a sweep in the Mach number in the range of 1.8 to 3.2 was done. Results for some Mach numbers can be seen in Figs. (3) to (7). In these plots the time evolution of displacements of several points distributed along the skin plate are shown. The fluid density field and the structure displacement at Mach=3.2 (flutter region) for a given time step is shown in Figs. (8) and (9).

For Mach numbers below the M_{cr}, Figs. (3) and (4), the maximum plate displacement grows until the forces exerted by the fluid dump the plate displacements. The damping rate (measured for instance as the time needed to reduce the amplitude by a given fixed factor, say 30%) grows with the Mach number. For Mach numbers near the M_{cr}, Fig. (5), the maximum amplitude grows slightly. The flutter mode is triggered at this point. For Mach numbers above the M_{cr}, Figs. (6), (7), (8) and (9), the fluid forces cannot damp the structure response and displacements grow without limit in an unstable fashion according to the theory.

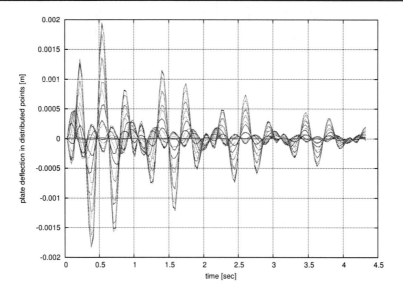

Figure 3. Plate deflection in distributed points along plate at M=1.8.

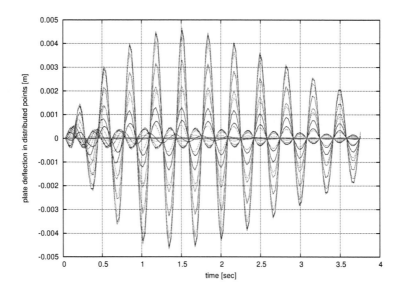

Figure 4. Plate deflection in distributed points along plate at M=2.25.

6.2.3. Time Accuracy

If the stage loop converges, i.e. $(\mathbf{u}, \mathbf{w})^{n+1,i} \to (\mathbf{u}, \mathbf{w})^{n+1,*}$, then it can be shown that the limit states $(\mathbf{u}, \mathbf{w})^{n+1,*}$ satisfy the fully implicit, strong coupled equations. The main effect of the staged algorithm is to have a strong coupling and then, enhanced stability, regardless the time accuracy.

To understand how to specify the parameters in Eq. (1) for predictors a simple two *dofs* wake oscillator model represented by two second order differential equations as follow

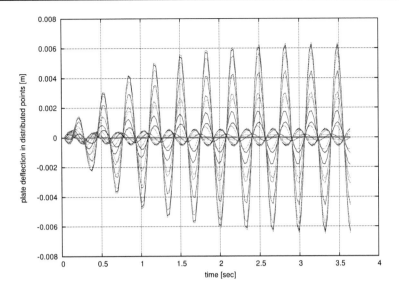

Figure 5. Plate deflection in distributed points along plate at M=2.275.

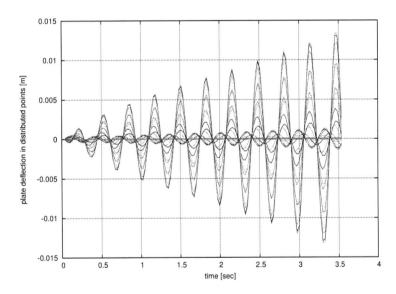

Figure 6. Plate deflection in distributed points along plate at M=2.3.

$$m_z \ddot{z} + c_z \dot{z} + k_z z = f_z(\ddot{y}, \dot{y}, y, t),$$
$$m_y \ddot{y} + c_y \dot{y} + k_y y = f_y(\ddot{z}, \dot{z}, z, t), \tag{51}$$

with (m, c, k) the mass, the damping and the stiffness parameters for each degree of freedom. In this simple model, y, z represent the structure and fluid states in the CFD and CSD codes in the algorithm. The forcing terms at the right hand side contain the coupling between the two blocks. This coupling may be formulated in terms of the main variables and their two first derivatives, generally velocities and accelerations. If the coupling contains only the main variables, i.e. $f_z(y, t)$ and $f_y(z, t)$, the predictor with $\alpha_0 = 1$ and $\alpha_1 = 0$

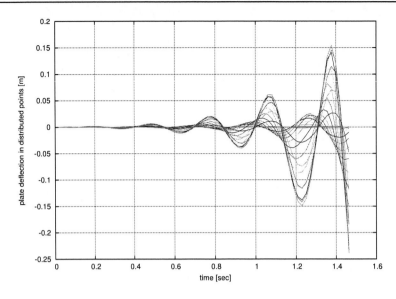

Figure 7. Plate deflection in distributed points along plate at M=3.2.

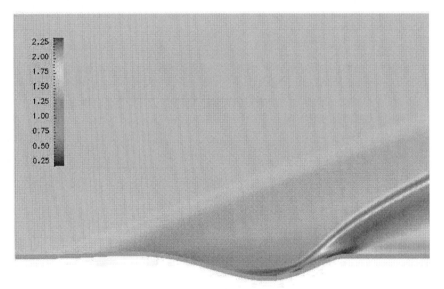

Figure 8. Fluid and structure fields at M=3.2.

achieves second order accurate in time solutions. If the coupling contains velocities it is necessary to use $\alpha_1 = 1/2$ to recover second order in time. In fluid structure interaction problems solved via ALE it is known that the mesh velocity dependent of the fluid-solid interface velocity is incorporated in the formulation, therefore to guarantee second order in time accuracy it is necessary to use $\alpha_0 = 1$ and $\alpha_1 = 1/2$ for the predictor.

Note that, if the Crank-Nicolson scheme is used for the time integration of both the

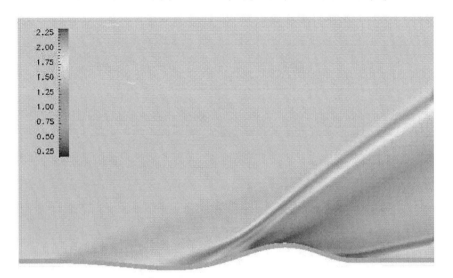

Step 327, time 1.68176 secs, Color=density

Figure 9. Fluid and structure fields at M=3.2.

structure and the fluid equations and the predictor is chosen with at least second order precision, then the whole algorithm is second order, *even if only one stage is performed* (see [50]).

In Fig. (10) the error obtained after the simulation of a certain fixed amount of time t_0 and increasing time refinement is shown. The exact solution is estimated through a Richardson extrapolation with the two more refined simulations for the more accurate scheme ($\alpha_{\text{trap}} = 0.5$). The error at t_0 is evaluated for a certain number of different Δt values. It is verified that for $\alpha_{\text{trap}} = 0.6$ the scheme is first order accurate, whereas for $\alpha_{\text{trap}} = 0.5$ precision is $O(\Delta t^2)$. When using $\alpha_{\text{trap}} = 0.5$ with no predictor (Eq. (1)), a second order convergence is still obtained.

6.2.4. Convergence of Stage Loop

The convergence of the stage loops has been assessed by running the test case over 20 time steps and performing 10 stages at each time step. In Fig. (11) the convergence of the fluid state (i.e. $\|\mathbf{u}^{n+1,i+1} - \mathbf{u}^{n+1,i}\|$) for all the time steps (convergence curves of the time steps are concatenated) is shown. Analogously, the convergence of the structure is plotted in Fig. (12). The average convergence is one order of magnitude per stage or higher, suggesting that for such a situation a small n_{stage} (2 or 3) would be enough.

6.2.5. Stability of the Staged Algorithm

The following numerical test allows to evaluate the stability of the staged algorithm presented in section §2.. The example is similar to the aeroelastic test case presented in sec-

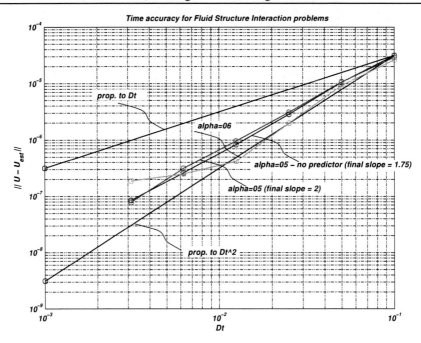

Figure 10. Experimentally determined order of convergence with Δt for the uncoupled algorithm with fourth order predictor.

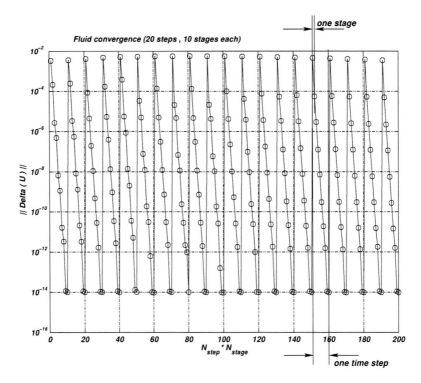

Figure 11. Convergence of fluid state in stage loop.

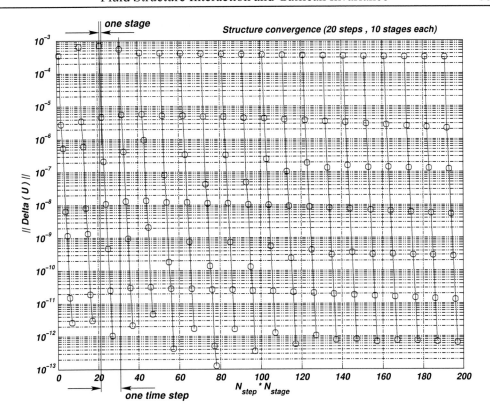

Figure 12. Convergence of structure state in stage loop.

tion §6.1. with some different parameters for the plate in order to produce larger plate deformations and stronger physical instabilities.

$$\begin{aligned} U_\infty &= \mathrm{M}_\infty = 2 \\ t &= 0.06 \\ \nu &= 0.33 \\ m &= 0.002 \\ E &= 39.6 \\ D &= 8.0 \, 10^{-4} \\ N_T &= \frac{D}{mL^2 U_\infty^2} = 0.025 \\ N_M &= \frac{\rho_\infty L}{m} = 1000.0 \end{aligned} \qquad (52)$$

The following figures show results obtained with both strategies, the staged and non-staged algorithms.

The vertical displacements on some points of the plate for the staged algorithm using $n_{\text{stage}} = 5$ after approximately 1300 time steps are shown in Fig. (13). The results for the non-staged algorithm diverge at 40 time steps and are shown in Fig. (14). The time step chosen for the non-staged algorithm was one fifth of the time step used with the staged one (i.e. $\Delta t_{\text{non-staged}} = \Delta t_{\text{staged}}/n_{\text{stage}}$). In this way the comparison of the stability of the two strategies is performed *at the same computational cost* (assuming that the computational cost of a *stage* in the *staged* algorithm is similar to the cost of a *time step* in the *non-staged* one).

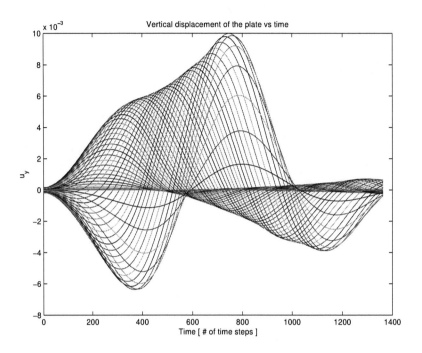

Figure 13. Stability analysis - Staged algorithm with $n_{\text{stage}} = 5$. Vertical displacements of the plate vs time.

Even though the staged algorithmic shows an extra stability compared with the non-staged one, the conclusions about this numerical experiment are not obvious because the flow regime is in a flutter condition. Further work needs to be done towards the understanding about how the staged algorithmic improves the stability of the whole coupled problem.

6.3. DGCL Validation for 2D Scalar Diffusion Problem with Internal Node Movement

As stated in section §5., a discrete formulation is said to satisfy the DGCL condition if it solves exactly a constant state regime, i.e. not depending on space or time, for a general mesh movement $\mathbf{x}(\boldsymbol{\xi}, t)$. In this section a DGCL compliance test is performed. For the sake of clarity, let us consider, the scalar diffusion version of the equation (3).

$$\begin{aligned} \frac{\partial u}{\partial t} - \mu \Delta u &= 0 \quad \text{for } \mathbf{x} \in \Omega^t,\ t \in (0, T] \\ u &= u_0 \quad \text{for } \mathbf{x} \in \Omega^0,\ t = 0 \\ u &= u_D \quad \text{for } \mathbf{x} \in \partial\Omega^t,\ t \in [0, T] \end{aligned} \quad (53)$$

where μ is the constant diffusivity and Δ is the Laplacian operator.

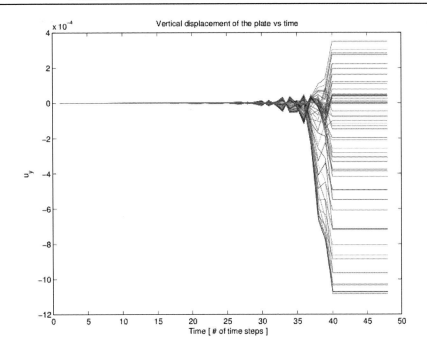

Figure 14. Stability analysis - Non-staged algorithm. Vertical displacements of the plate vs time.

To carry out the DGCL compliance test, the problem (53) is solved on an unit square domain with $\mu = 0.01$, so that

$$
\begin{aligned}
u_t - 0.01\Delta u &= 0 \quad \text{for } \mathbf{x} \in \Omega^t,\ t \in (0, T], \\
u_0 &= 1 \quad \text{for } \mathbf{x} \in \Omega^0,\ t = 0, \\
u &= 1 \quad \text{for } \mathbf{x} \in \partial\Omega^t,\ t \in [0, T],
\end{aligned}
\tag{54}
$$

being the mesh deformed according to the following rule

$$
\begin{aligned}
\chi(\xi, t) &= x = \xi + 0.125 \sin(\pi t) \sin(2\pi \xi). \\
\chi(\eta, t) &= y = \eta + 0.125 \sin(\pi t) \sin(2\pi \eta).
\end{aligned}
\tag{55}
$$

Figure (15) shows the reference domain and the deformed mesh for $t = 0.5\ [s]$ where the maximum deformation occurs. The problem is solved using piecewise linear triangles for the spatial discretization, a piecewise linear interpolation of the mesh movement and for the time integration the *Backward Euler* ($\theta = 1$), *Crank-Nicolson* ($\theta = 0.5$) and *Galerkin* ($\theta = 2/3$) schemes are considered with $\Delta t = \{0.15, 0.1, 0.05, 0.025\}$. Figure (16) reports the error $||u_h - u||_{L^2(\Omega^n)}$ for three periods of oscillation, which must be null to machine precision over time for a DGCL compliant scheme.

An error is introduced when using the *Backward Euler* or *Garlerkin* scheme due to lack in DGCL compliance. In Figure (17) the solution for times $t = \{0.1, 2.4, 5.4\}\ [s]$ is shown for the three different integration schemes. The error related to the constant solution is located in the zones of the domain where the element deformation is higher, as in the center and the corners.

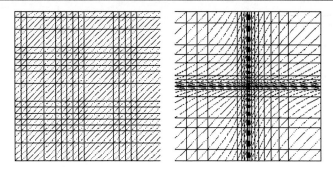

Figure 15. Reference and deformed mesh.

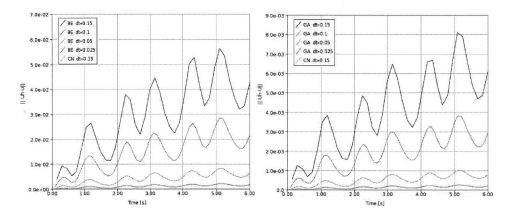

Figure 16. $||u_h - u||_{L^2(\Omega^n)}$ for *Garlerkin* (GA) and *Backward Euler* (BE) schemes compared with *Crank-Nicolson* (CN).

6.4. Sudden Stop of Gas Container. ALE Invariance Test Case

Consider the case of a 1D flow in a container, i.e. between two walls at $x = 0, L$. For $t < 0$ the container and the fluid are moving with constant velocity u_0 and at $t = 0$ the container is suddenly stopped. It is assumed that the fluid is in a homogeneous state for $t < 0$, so that the governing equations, along with the initial and boundary conditions are

$$\begin{aligned}\frac{\partial \mathbf{U}_c}{\partial t} + \frac{\partial \mathcal{F}_x^c}{\partial x} &= \frac{\partial \mathcal{F}_x^d}{\partial x}; \quad 0 \leq x \leq L, \\ u(x=0,t) &= u(x=L,t) = 0, \\ \mathbf{U}(x, t=0) &= [\rho, u, p]_0^T,\end{aligned} \quad (56)$$

where $\mathcal{F}_j^d$ are the *"diffusive fluxes"* and $u_0 > 0$. Due to the presence of the walls at $t = 0$ the fluid starts to compress at the $x = L$ wall starting eventually a shock (depending on the Mach number $\mathrm{M}_0 = u_0/c_0$, $c_0 = \sqrt{\gamma p_0/\rho_0}$). At the same time an expansion fan is formed at the $x = 0$ wall. In this frame S fixed to the container the domain $\Omega = [0, L]$ is fixed and no ALE terms are needed. But the same problem can be described in a frame S' moving with velocity $v = u_0$, so that the fluid in this frame is initially at rest. The walls move with

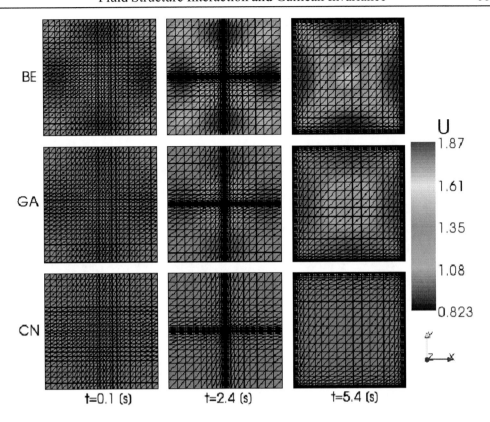

Figure 17. Solution for the *Backward Euler* (BE), *Galerkin* (GA) and *Crank-Nicolson* (CN) schemes.

velocity $-u_0$ with respect to this frame, so that the governing equations are

$$\frac{\partial \mathbf{U}_c'}{\partial t} + \frac{\partial \mathcal{F}_x^{c'}}{\partial x'} = \frac{\partial \mathcal{F}_x^{d'}}{\partial x}; \quad 0 \leq x' + u_0 t \leq L,$$
$$u(x' = -u_0 t, t) = u(x = L - u_0 t, t) = -u_0,$$
$$\mathbf{U}(x, t = 0) = [\rho, 0, p]_0^T.$$
(57)

Of course, in the continuum both systems of equations are completely equivalent, this can be verified by writing down the equations for one frame an transform to the other using the transformation rules described above. But the numerical stabilization terms can break the invariance of the discrete equations. For the case of inviscid flow with $\gamma = 1.4$ and $M_0 = 0.5$. The values are made nondimensional by selecting L, ρ_0 and c_0 as reference values for length, density and velocity, so that the nondimensional quantities are $\rho_0' = 1$, $p_0' = 1/\gamma$, $u_0' = 0.5$, and in the following the prime indicating nondimensional quantities is dropped.

The problem was simulated in both frames, with the results being equivalent to machine precision. In the following the energy balance in each frame is discussed in detail.

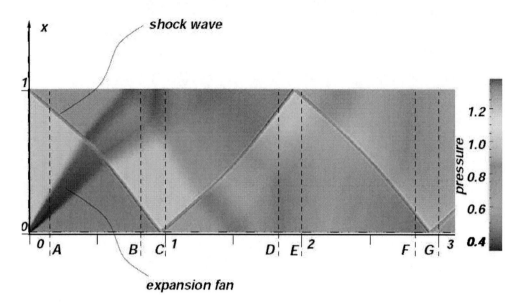

Figure 18. Suddenly stopped container test case. Color map of pressure in x, t plane.

In Fig. (18) the pressure in a $x - t$ axis in the frame fixed to the container walls is shown. A shock wave starts at the $x = 1$ wall and propagates backwards with a velocity of approximately 1. At the same time an expansion fan starts at the $x = 0$ wall and propagates forward. The shock wave reflects several times at the walls, with a background state that is not homogeneous due to the expansion fan. The intensity of the shock wave decays and eventually the system reaches a new homogeneous thermodynamic state with null velocity. In Fig. (19) the pressure profiles as a function of x at several stages can be seen. In Fig. (20) the energy balance in this frame is shown. As the container is at rest, the walls don't do work on the fluid and then the total energy is conserved.

$$E_{\text{kin}}(t) + E_{\text{int}}(t) = E_{\text{kin}}(0) + E_{\text{int}}(0), \quad (S \text{ frame})$$
$$E_{\text{kin}}(t) = \int_0^L (\tfrac{1}{2}\rho u^2)_{(x,t)} \, dx.$$
$$E_{\text{int}}(t) = \int_0^L \frac{1}{\gamma - 1} p(x, t) \, dx.$$
(58)

In the process the shock wave converts the initial kinetic energy in internal energy. The kinetic and the internal energy and the sum of the two are shown in the figure. This last one is constant up to a 1.5% of the mean total energy.

In frame S' the fluid is initially at rest, and at $t = 0$ the container starts moving to the left with velocity $-u_0$. As the container is moving, it does some work on the fluid

$$\dot{W} = (F_{\text{wall},x=L} - F_{\text{wall},x=0})u_0, \tag{59}$$

so that the energy balance is

$$E_{\text{kin}}(t) + E_{\text{int}}(t) = E_{\text{kin}}(0) + E_{\text{int}}(0) + \int_0^t \dot{W}(t')\,\mathrm{d}t'. \tag{60}$$

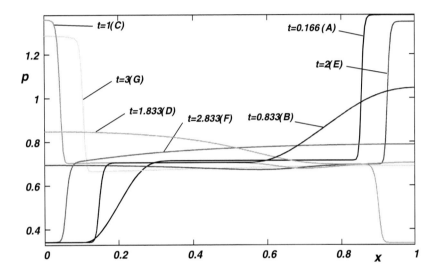

Figure 19. Suddenly stopped container test case. Pressure at several instants.

The energy balance is shown in Fig. (21). The internal energy has the same increment as in frame S, but now the kinetic energy also has a positive increment. The total energy of the fluid then increases, which is balanced with the work done by the container on the fluid. The error in the energy balance is the same as before, since the results are completely equivalent (up to machine precision).

6.5. Start-up of a Rocket Engine Nozzle

The algorithm described in section §2. will be used to obtain the deformation in the nozzle of a rocket engine during the ignition. This problem has been under study by many researchers over the years, carrying both numerical [51, 52, 53, 10] and experimental [54, 55] analysis.

Nozzles with high area ratio are used in the main space launchers (Space Shuttle Main Engine, Ariane 5). These engines must work in conditions ranging from sea level to orbital altitude but an efficient operation is reached only at high altitude. The nozzles contour is often designed according to the theory proposed by Rao [56] that results in TOP (Thrust Optimized Parabolic or Parabolic Bell Nozzle) nozzle, which has some advantages compared to the traditional conical shapes. These advantages are the smaller length, lower weight, as well as the reduction in energy losses in the expansion of gases [57, 58, 59, 60]. During the

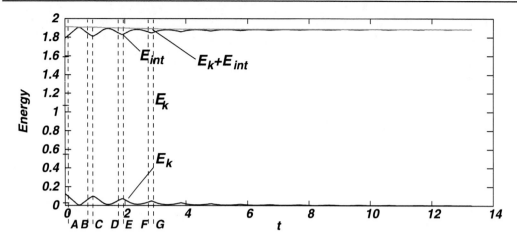

Figure 20. Energy balance at frame S. (Fixed with respect to the container).

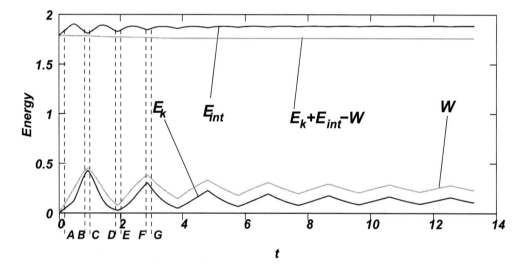

Figure 21. Energy balance at frame S'. (Moving with respect to the container).

start-up phase the structure is deformed due to the advance of a shock wave that is highly detrimental to the integrity and service life cycle of the rocket engine. Many problems have been encountered in the Space Shuttle Main Engine, European Vulcain (Ariane) and in the Japanese LE-7, all these were related to the ignition stage and side loads phenomena.

The nozzle under study has a bell-shape geometry which is generated by rotating a contour line around the x axis. In this way the 3D geometry is obtained (see Figure (22)).

The most relevant geometrical data are detailed below:

- Overall length: $l = 1810$ [mm].
- Throat diameter: $D_t = 304$ [mm].
- Exit diameter: $D_e = 1396$ [mm].
- Area ratio: $\epsilon = 21.1$.

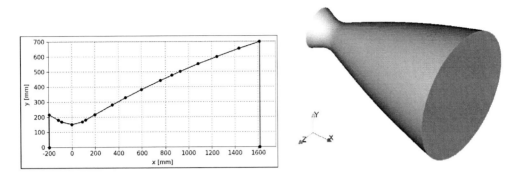

Figure 22. Contour line and 3D model.

6.5.1. Numerical Model

Starting from the three-dimensional model two independent meshes are generated, one for the fluid domain discretization and the other for the structure domain discretization. A mesh with 334700 tetrahedral elements is generated for the fluid with a linear interpolation of the variables. The structural mesh is composed of 59600 wedge (triangular base prismatic) elements. Detailed view of grid zones of both meshes is shown in Figure (23).

In FSI problems there is an information transfer in the fluid-structure interface. Using conforming meshes (node to node coincident) on the interface, the transmission is direct and does not need an algorithm to do a surface tracking, state interpolation and load projection, but the major drawback of this method is that refinement in the structure mesh will cause an increase in the fluid mesh and therefore in the overall problem size.

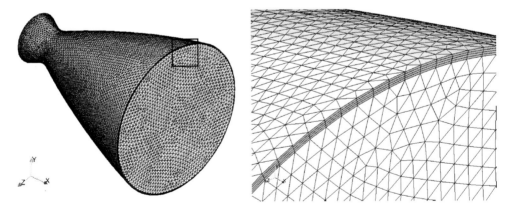

Figure 23. Spatial discretization for the fluid and for the structure.

The structural problem is solved using a PETSc-FEM module, which is based on the theory of constitutive linear elastic material, geometrically nonlinear and no material damping. The gasdynamics Euler equations are solved and SUPG stabilization is used together with the shock-capturing method which has been proposed by [47]. Using the Euler equations the CPU and memory costs can be significantly reduced comparatively to the viscous case. Furthermore, from previous works ([61, 62]) can be concluded that this equations

Table 1. Solid Properties

Young's modulus	Poisson's coeff.	Density	Thickness
$2.07 \cdot 10^{10}$ [N/m^2]	0.28	8400 [kg/m^3]	0.015 [m]

Table 2. Fluid Properties

R	γ	ρ_∞	T_∞	p_∞
287 [J/kg K]	1.40	1.225 [kg/m^3]	288 [K]	101253 [Pa]

correctly predict the main flows feature.

In order to solve the aeroelastic problem the material properties for the nozzle and fluid are summarized in Tables 1 and 2. In this work the nozzle is modeled with an homogeneous material, but more complex structural models can be similarly used.

6.5.2. Boundary and Initial Conditions for the Nozzle Ignition Problem

The FSI problem requires initial and boundary conditions for both, the structural and the fluid problem, separately. The nozzle is clamped (all displacements null) at the junction with the combustion chamber and the rest is left free. In the fluid flow problem a slip condition is applied to the wall of the nozzle, which is mathematically represented by the following equation.

$$(\mathbf{v} - \mathbf{v}_{str}) \cdot \hat{\mathbf{n}} = (\mathbf{v} - \dot{\mathbf{u}}) \cdot \hat{\mathbf{n}} = 0. \tag{61}$$

As mentioned above the slip condition must be applied dynamically because the normal to the wall and the structure velocity change during the simulation. For the fluid, (p_0, T_0) are imposed at the inlet. These conditions are taken from the stagnation condition of the combustion chamber (p_0, T_0), and then ρ_0 is computed from the state equation (see Table 3).

The modeling of the ignition of a rocket exhaust nozzle is challenging from several points of view. One of these points is the imposition of boundary conditions that at the outlet wall must be non-reflective. Moreover, in such case, the needed conditions at the outlet boundary change from rest (i.e., subsonic flow) to supersonic flow as a shock wave appears

Table 3. Stagnation values used for the combustion chamber

p_0	ρ_0	T_0
26 [MPa]	306.25 [kg/m^3]	299 [K]

Table 4. Characteristic Dimensions

Nozzle	Vulcain	S1	S3
Area ratio (ϵ)	45	20	18.2
Nozzle length (L) [mm]	2065.5	350	528.2
Throat diam. (D_t) [mm]	262.4	67.08	67.08
Nozzle exit diam.(D_e) [mm]	1760.2	300.0	286.5

at the throat and propagates toward the boundary. So, the condition must be capable of handling the dynamical change of the Jacobians matrix profile. During the flow computation inside the nozzle the number of incoming/outgoing characteristics, and therefore the number of Dirichlet conditions to be imposed, will change. Having a boundary condition that can automatically adapt itself to this change is essentially useful in such a problem. In addition, the computational domain can be limited to the nozzle interior up to the exit plane, with a significant reduction in CPU time and memory use. Imposing absorbent/dynamic boundary conditions is based on the analysis of the projection of the Jacobians of advective flux functions onto normal directions to fictitious surfaces. The advantage of the method is that it is very easy to implement and that it is based on imposing non-linear constraints via Lagrange Multipliers or Penalty Methods (see Reference [63] for a more detailed description).

Initial conditions must be established in both domains. The following are adopted for the fluid

$$\mathbf{v}(\mathbf{x}, t_0) = 0, \tag{62}$$
$$p(\mathbf{x}, t_0) = p_\infty, \tag{63}$$
$$\rho(\mathbf{x}, t_0) = \rho_\infty, \tag{64}$$

and the next for the structure

$$\mathbf{u}(\mathbf{x}, t_0) = 0, \tag{65}$$
$$\dot{\mathbf{u}}(\mathbf{x}, t_0) = 0. \tag{66}$$

6.5.3. Aeroelastic Behavior of the Nozzle

Before performing the aeroelastic analysis, the nozzle used in this work is compared to the Vulcain nozzle and to the sub-scale S1 and S3 nozzles (see Table 4) through a parametric study that was carried out in [64]. That is done because the fluid flow field is determined by the shape of the nozzle and this affects the pressure distribution on the wall from which the fluid loads are computed.

In the S1 sub-scale nozzle the characteristic length for the scaling was the nozzle exit radius (r_e) and in the S3 sub-scale nozzle was the throat radius (r_t), thus different contours are obtained. Therefore to perform an aeroelastic study of the proposed TOP (Thrust Optimized Parabolic or Parabolic Bell Nozzle) nozzle the radius and the wall pressure distribution (p_w) must be comparable to the Vulcain, S1 and S3 (see Figure (24)).

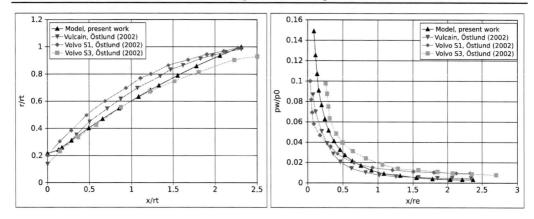

Figure 24. Radius and wall pressure distribution.

Figure (24) shows that the TOP nozzle under study has similar radius distribution than Vulcain, S1 and S3, which makes valid the comparison between the parietal pressures. Then, the computed wall pressure when the flow is completely developed is compared, showing a good agreement.

Having verified the pressure distribution when the flow is completely developed, the next step is to study qualitatively the evolution of the shock wave during the start-up. The behavior of the structure when a shock wave moves through the divergent zone of the nozzle is described and the process is outlined in Figure (25).

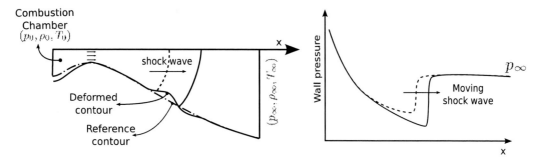

Figure 25. Schematic deformation of the structure.

During the start-up process the pressure increases linearly from p_∞ to p_0 in $1 \cdot 10^{-4}$ seconds. The formed shock moves rapidly (faster than speed of sound on the non-perturbed condition) trough the stagnant low pressure medium. Also a secondary left running (with respect to the fluid) shock wave appears and is carried to the right because of the supersonic carrier flow. This shock wave links the high Mach number, low pressure flow, with the lower velocity high pressure gas behind the primary shock. The results of the fluid structure interaction during this stage are shown in Figure (26), together with the pressure at the wall.

Note a large pressure jump across the secondary shock wave (see Figure (26)), which produces significant bending moments in the structure, changing the outflow pattern and the pressure downstream while the shock wave propagates towards outlet, making this process totally dynamic. First of all, a run is performed only considering the fluid problem (hereafter

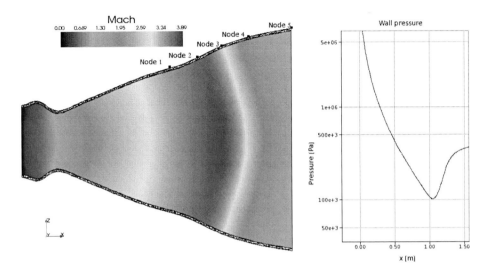

Figure 26. Structure deformation and pressure distribution for the moving shock wave.

case name **NO-FSI**) such that the parietal pressure is computed without the effect of the wall movement. Then, the coupling is performed (case name **FSI**) and the parietal pressures of both cases are compared. The temporal evolution of the pressure at the nodes (1-5) located at the positions shown in Figure (26) are plotted in Figure (27).

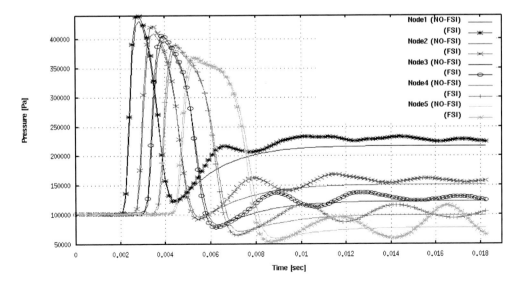

Figure 27. Wall pressures during the start-up. Comparison between **NO-FSI** and **FSI** cases.

As seen in the Figures, the wall displacements (shown in Figure (28)) produce oscillations in the fluid pressure which are not considered for the first case **(NO-FSI)**. As the plot shows, considering the wall displacement to compute the pressure acting in the nozzle is very important and this is one of the key points of this work.

In this case the shock wave is expelled from the nozzle but in certain operating condition, like overexpanded mode, the shock wave do not leave the nozzle. This kind of shock produce a strong pressure jump and with the structure deformation can cause an asymmetric pressure distribution as is mentioned in [64]. So, this is a first step in order to demonstrate the relation between the aeroelastic coupling and the acting lateral loads.

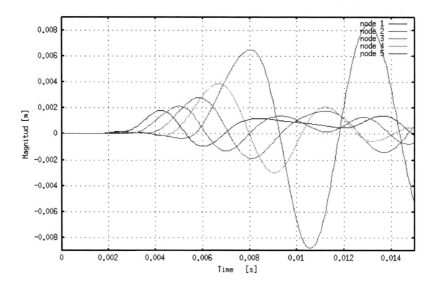

Figure 28. Displacement of nodes 1-5.

The sequence in Figure (29) shows the behavior of the structure as the shock wave moves through the divergent section of the nozzle. Also, the Mach number on the nozzle centerline is plotted in the right side.

7. Conclusions

In this chapter an overview of the issues involved in the numerical solution of Fluid-Structure Interaction (FSI) problems are given. Within these issues are the coupling techniques, the ALE formulation and the Galilean invariance of the scheme.

Regarding the coupling techniques, a strong coupled partitioned algorithm via fixed point iteration is presented, jointly with the variational formulation for advective diffusive system for moving meshes using an ALE scheme. As was mentioned, the ALE contributions affect the advective terms, so some modifications are needed to the standard stabilization terms in order to get the correct amount of stabilization. Also boundary conditions at walls (slip or non-slip) must be modified when ALE is used.

Finally, a set of numerical examples are performed using the presented techniques. In these test some results on the convergence when solving the flutter of a flat solid plate are

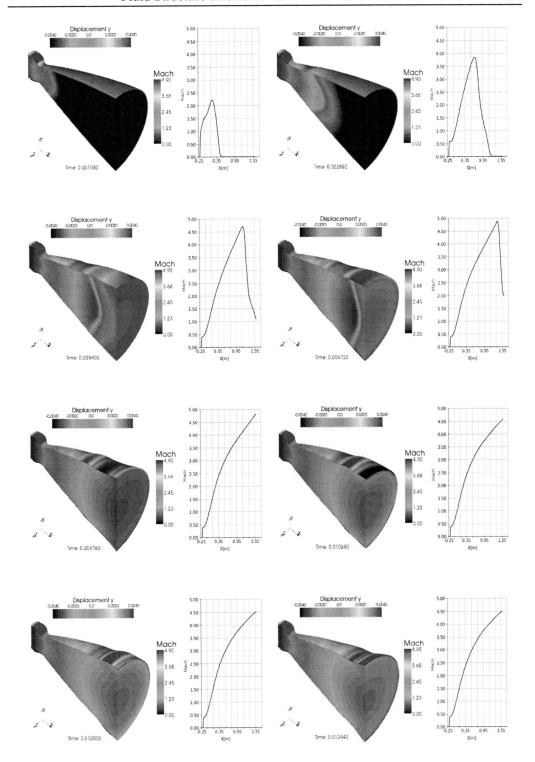

Figure 29. Ignition process of the rocket engine.

presented. In the start-up of a rocket engine nozzle test, the interaction between the fluid and the structure is critical in order to obtain the pressure on the wall and allows to know the deformation of the wall.

References

[1] Dooms, D., 2009. "Fluid-structure interaction applied to flexible silo constructions.". PhD thesis, Departement Burgerlijke Bouwkunde, Katholieke Universiteit Leuven.

[2] Koh, H., Kim, J., and Park, J., 1998. "Fluid-structure interaction analysis of 3-d rectangular tanks with a variationally coupled BEM-FEM and comparison with test results". *Earthquake Engng. and Structural Dynamics,* **27**, pp. 109–124.

[3] Nobile, F., 2001. "Numerical approximation of fluid-structure interaction problems with application to haemodynamics.". PhD thesis, Department of Mathematics, École Polytechnique Fédérale de Lausanne.

[4] Tezduyar, T., Schwaab, M., and Sathe, S., 2009. "Sequentially-coupled arterial fluid-structure interaction (SCAFSI) technique.". *Computers & Structures,* **198:45-46**, pp. 3524–3533.

[5] Hron, J., and Turek, S., 2006. "A monolithic FEM/Multigrid solver for an ALE formulation of fluid-structure interaction with applications in biomechanics.". *Lecture Notes in Computational Science and Engineering,* **53**, pp. 146–170.

[6] Piperno, R., and Farhat, C., 2001. "Partitioned procedures for the transient solution of coupled aeroelastic problems. Part II: energy transfer analysis and three-dimensional applications". *Computer Methods in Applied Mechanics and Engineering,* **190**, pp. 3147–3170.

[7] Gnesin, V., and Rzadkowski, R., 2005. "A coupled fluid structure analysis for 3-d inviscid flutter of IV standard configuration". *Jornal of Sound and Vibration.,* **49**, pp. 349–369.

[8] Garelli, L., Paz, R., and Storti, M., 2008. "Interacción fluido estructura en flujo compresible en régimen supersónico.". In XVII Congreso sobre Métodos Numéricos y sus Aplicaciones.

[9] Garelli, L., Paz, R. R., and Storti, M. A., 2010. "Fluid-structure interaction study of the start-up of a rocket engine nozzle". *Computers & Fluids,* **39:7**, pp. 1208–1218.

[10] Lefrancois, E., Dhatt, G., and Vandromme, D., 1999. "Fluid-structure interaction with application to rocket engines.". *International Journal for Numerical Methods in Fluids,* **30**, pp. 865–895.

[11] Ludeke, H., Calvo, J., and Filimon, A., 2006. "Fluid structure interaction at the ariane-5 nozzle section by advanced turbulence models". In European Conference on Computational Fluid Dynamics ECCOMAS CFD 2006, P. Wesseling, E. Onate, and J. Periaux, eds.

[12] Felippa, C.A. Park, K. F. C., 1999. "Partitioned analysis of coupled mechanical systems.". *Computer Methods in Applied Mechanics and Engineering.*, **190**, pp. 3247–3270.

[13] Park, K., and Felippa, C., 2000. "A variational principle for the formulation of partitioned structural systems". *International Journal for Numerical Methods in Engineering.*, **47**, pp. 395–418.

[14] Cebral, J., 1996. "Loose coupling algorithms for fluid structure interaction". PhD thesis, Institute for Computational Sciences and Informatics, George Mason University.

[15] Lohner, R., Yang, C., Cebral, J., Baum, J., Luo, H., Pelessone, D., and Charman, C., 1998. "Fluid-structure interaction using a loose coupling algorithm and adaptive unstructured grids". *AIAA paper AIAA-98-2419*.

[16] Storti, M., Nigro, N., Paz, R., and Dalcín, D., 2009. "Strong coupling strategy for fluid structure interaction problems in supersonic regime via fixed point iteration". *Journal of Sound and Vibration,* **30**, pp. 859–877.

[17] López, E., Nigro, N., Storti, M., and Toth, J., 2007. "A minimal element distortion strategy for computational mesh dynamics". *International Journal for Numerical Methods in Engineering,* **69**, pp. 1898–1929.

[18] Idelsohn, S. R., Oñate, E., del Pin, F., and Calvo, N., 2006. "Fluid-structure interaction using the particle finite element method". *Computer Methods in Applied Mechanics and Engineering,* **195:17-18**, pp. 2100–2123.

[19] Michler, C., Hulshoff, S. J., van Brummelen, E. H., and De Borst, R., 2004. "A monolithic approach to fluid-structure interaction.". *Computers & Fluids,* **33**, pp. 839–848.

[20] Codina, R., and Cervera, M., 1995. "Block-iterative algorithms for nonlinear coupled problems". *Advanced Computational Methods in Structural mechanics, CIMNE, Barcelona.*, pp. 115–134.

[21] Artlich, S., and Mackens, W., 1995. "Newton-coupling of fixed point iterations.". *Numerical Treatment of Coupled Systems*.

[22] Dettmer, W., and Peric, D., 2006. "A computational framework for fluid-rigid body interaction: Finite element formulation and applications". *Computer Methods in Applied Mechanics and Engineering,* **195**, pp. 1633–1666.

[23] Noh, W., 1964. "A time-dependent, two space dimensional, coupled Eulerian-Lagrange code.". *Methods in Computational Physics.,* **3**, pp. 117–179.

[24] Hirt, C., Amsden, A., and Cook, J., 1974. "An Arbitrary Lagrangian-Eulerian computing method for all flow speeds.". *Journal of Computational Physics,* **14:3**, pp. 227–253.

[25] Donea, J., 1983. *Arbitrary Lagrangian-Eulerian finite elements method.*, 1th ed. Belytschko, T. and Hughes, J.R., Amsterdam.

[26] Hughes, T., Liu, W., and Zimmermann, T., 1978. "Lagrangian-Eulerian finite elements formulations for incompressible viscous flows.". In US-Japan Interdisciplinary Finite Element Analysis.

[27] Trepanier, J., Reggio, M., Zhang, H., and Camarero, R., 1991. "A finite-volume method for the Euler equations on arbitrary Lagrangian-Eulerian grids". *Computers and fluids.*, **20:4**, pp. 399–409.

[28] Scovazzi, G., 2005. "A discourse on galilean invariance, SUPG stabilization, and the variational multiscale framework.". In Technical report, Sandia National Labs.

[29] Thomas, P., and Lombard, C., 1979. "Geometric conservation law and its applications to flow computations on moving grids.". *AIAA*, **17**, pp. 1030–1037.

[30] Garelli, L., Paz, R., and Storti, M., 2009. "Geometric conservation law in ALE formulations.". In XVIII Congreso sobre Métodos Numéricos y sus Aplicaciones.

[31] Boffi, D., and Gastaldi, L., 2004. "Stability and geometric conservation laws for ALE formulations". *Computer Methods in Applied Mechanics and Engineering*, **193**, pp. 4717–4739.

[32] Étienne, S., Garon, A., and Pelletier, D., 2009. "Perspective on the geometric conservation law and finite element methods for ALE simulations of incompressible flow.". *Journal of Computational Physics*, **228:7**, pp. 2313–2333.

[33] Ahn, H., and Kallinderis, Y., 2006. "Strongly coupled flow/structure interactions with a geometrically conservative ALE scheme on general hybrid meshes". *Journal of Computational Physics*, **219**, pp. 671–693.

[34] Lesoinne, M., and Farhat, C., 1996. "Geometric conservation laws for flow problems with moving boundaries and deformable meshes, and their impact on aeroelastic computations.". *Computer Methods in Applied Mechanics and Engineering*, **134**, pp. 71–90.

[35] Mavriplis, D. J., and Yang, Z., 2005. "Achieving higher-order time accuracy for dynamic unstructured mesh fluid flow simulations: Role of the GCL". *17th AIAA Computational Flow Dynamics Conference*, pp. 1–16.

[36] Farhat, C., Geuzaine, P., and Grandmont, C., 2001. "The discrete geometric conservation law and the nonlinear stability of ALE schemes for the solution of flow problems on moving grids". *Journal of Computational Physics*, **174**, pp. 669–694.

[37] Yang, Z., and Mavriplis, D. J., 2005. "Unstructured dynamic meshes with higher-order time integration schemes for the unsteady Navier-Stokes equations.". In AIAA 2005-1222.

[38] Xu, Z., and Accorsi, M., 2004. "Finite element mesh update methods for fluid-structure interaction simulations". *Finite Elements in Analysis and Design,* **40:9-10**, pp. 1259–1269.

[39] Johnson, A., and Tezduyar, T., 1994. "Mesh update strategies in parallel finite element computations of flow problems with moving boundaries and interfaces". *Computer Methods in Applied Mechanics and Engineering,* **119**, pp. 73–94.

[40] Stein, K., Tezduyar, T., and Benney, R., 2004. "Automatic mesh update with the solid-extension mesh moving technique". *Comput. Meth. Appl. Mech. Engrg.,* **192**, pp. 2019–2032.

[41] López, E., Nigro, N., and Storti, M., 2006. "Simultaneous untangling and smoothing of moving mesh". *International Journal for Numerical Methods in Engineering* .

[42] Franca, L., Frey, S., and Hughes, T., 1992. "Stabilized finite element methods: I. application to the advective-diffusive.". *Computer Methods in Applied Mechanics and Engineering,* **95**, pp. 253–276.

[43] Hughes, T., and Tezduyar, T., 1984. "Finite element methods for first-order hyperbolic systems with particular emphasis on the compressible Euler equations". *Comp. Meth. Applied Mechanics and Engineering,* **45**, pp. 217–284.

[44] Guillard, H., and Farhat, C., 2000. "On the significance of the geometric conservation law for flow computations on moving meshes". *Comput. Methods Appl. Mech. Engrg,* **190**, pp. 1467–1482.

[45] Formaggia, L., and Nobile, F., 2004. "Stability analysis of second-order time accurate schemes for ALE-FEM.". *Computer Methods in Applied Mechanics and Engineering.,* **193:39-41**, pp. 4097–4116.

[46] Farhat, C., and Geuzaine, P., 2004. "Design and analysis of robust ale time-integrators for the solution of unsteady flow problems on moving grids". *Computer Methods in Applied Mechanics and Engineering,* **193**, pp. 4073–4095.

[47] Tezduyar, T., and Senga, M., 2004. "Determination of the shock-capturing parameters in supg formulation of compressible flows.". In Computational Mechanics WCCM IV, Beijing, China 2004., T. U. P. . Springler-Verlag, ed.

[48] Houbolt, J., 1958. *A study of several aerothermoelastic problems of aircraft structures.* Mitteilung aus dem Institut fur Flugzeugstatik und Leichtbau 5, E.T.H., Zurich, Switzerland.

[49] Storti, M., Nigro, N., and Paz, R. Petsc-fem: A general purpose, parallel, multiphysics FEM program. http://www.cimec.org.ar/twiki/bin/view/cimec/petscfem.

[50] Farhat, C., Van der Zee, K. G., and Geuzaine, P., 2006. "Provably second-order time-accurate loosely-coupled solution algorithms for transient nonlinear computational aeroelasticity". *Computer Methods in Applied Mechanics and Engineering,* **195:17-18**, pp. 1973–2001.

[51] Lefrancois, E., 2005. "Numerical validation of a stability model for a flexible overexpanded rocket nozzle". *International Journal for Numerical Methods in Fluids.,* **49:4**, pp. 349–369.

[52] Taro Shimizu, M. K., and Tsuboi, N., 2008. "Internal and external flow of rocket nozzle". *Journal of the Earth Simulator,* **9**, pp. 19–26.

[53] Wang, T.-S., 2004. "Transient three dimensional analysis of side load in liquid rocket engine nozzle". *AIAA,* p. 3681.

[54] Shashi Bhushan Verma, Ralf Stark, a. O. H., 2006. "Relation between shock unsteadiness and the origin of side-loads inside a thrust optimized parabolic rocket nozzle". *Aerospace Science and Technology,* **10:6**, pp. 474–483.

[55] Moríñigo, J., and Salvá, J., 2008. "Numerical study of the start-up process in an optimized rocket nozzle.". *Aerospace Science and Technology,* **12:6**, pp. 485–486.

[56] Rao, S., 1996. *Engineering optimization*. Wiley and Sons.

[57] Sutton, G., and Biblarz, O., 2001. *Rocket propulsion elements*, 7th ed. John Wiley and Sons.

[58] Oates, G., 1997. *Aerothermodynamics of gas turbine and rocket propulsion*, 3rd ed. AIAA.

[59] Mattingly, J., and Ohain, H. V., 2006. *Elements of propulsion: gas turbines and rockets.*, 2nd ed. AIAA.

[60] Tuner, M., 2006. *Rockets and spacecraft propulsion*, 2nd ed. Springer.

[61] Prodromou, P., and Hillier, R., 1992. "Computation of unsteady nozzle flows". In Proceedings of the 18th. ISSW, Sendai, Japan, Vol. II.

[62] Igra, O., Wang, L., Falcovitz, J., and Amann, O., 1998. "Simulation of the starting flow in a wedge-like nozzle". *Int. J. Shock Waves,* **8**, pp. 235–242.

[63] Storti, M., Nigro, N., and Paz, R., 2008. "Dynamic boundary conditions in Computational Fluid Dynamics". *Computer Methods in Applied Mechanics and Engineering,* **197(13-16)**, pp. 1219–1232.

[64] Östlund, J., 2004. "Side-load phenomena in highly overexpanded rocket nozzles.". *Journal of Propulsion and Power,* **20:2**.

Chapter 17

LATTICE BOTLZMANN METHOD FOR FLUID DYNAMICS

Mojtaba Aghajani Delavar*, Mousa Farhadi‡ and Kurosh Sedighi‡
Faculty of Mechanical Engineering, Babol University of Technology,
Babol, Islamic Republic of Iran

1. INTRODUCTION

Lattice Boltzmann method is relatively new scheme that uses microscopic models to simulate macroscopic behavior of fluid flow and dependent phenomenons. LBM can be considered as discrete version of kinetics theory. The lattice-Boltzmann method, evolved out of ideas that have been extremely investigated since 1986, when it was discovered that very simple models of discrete particles restricted to a lattice can be used to yield Navier Stokes equation to solve complicated flow problems (Frisch et al., 1986 & 1987).

The lattice method can be concerned as one of the simplest microscopic approaches for simulation of macroscopic models. It is based on the Boltzmann transport equation, which concern about the time rate of change of the particle distribution function (probability to find particles with specific velocity range at the limited position at the given time) in a particular state.

This section presents basic concepts of Lattice Boltzmann Moethod (LBM) as the basis of simulation of fluid flows. It is assumed that the reader is somewhat familiar with the physics of fluid flow and related phenomena, so here the main concentration is on to describe basic concepts of lattice Boltzmann method, in such way that it can be understood and used by reader to solve different fluid flow problems. So authors avoided prolonged mathematical analysises.

In the lattice Boltzmann method the same calculations carry out at every lattice site and only nearby particles interact with each other so it is parallel and local, which make it very

---

* Corresponding author: Mojtaba Aghajani Delavar, Faculty of Mechanical Engineering, Babol University of Technology, Babol, Islamic Republic of Iran, P.O. Box: 484, E-mail: m.a.delavar@nit.ac.ir
‡ E-mail: mfarhadi@nit.ac.ir
‡ E-mail: ksedighi@nit.ac.ir

suitable for programming and efficient run in parallel processing. Complex boundary conditions are included in an uncomplicated approach. The method yields a good approximation to the Navier-Stokes equations.

It is hard to shift from microscopic lattice gas model to macroscopic fluid dynamics governing equations, Navier-Stokes (NS) equations, using statistical methods for gases. After while that first model for uncompressible NS equataions was proposed by Frisch, Hasslacher and Pomeau (FHP) in 1986 (Frisch et. al, 1986), Lattice Gas Automata (LGA) has been attended as promising method to solve partial differential equations and simulation of natural phenomenons (Frisch et. al, 1987; Wolfarm 1986; Doolen 1989; Doolen, 1991).

Recently lattice Boltzmann equation has been successfully used for simulation of fluid flow and transport phenomenon. Inspite of ordinary CFD methods, LBM is based on microscopic models and mesoscopic kinetics equations, which in them behavior of collection of particle will be used to simulate continuum mechanics of system. Due to kintecis nature of LBM, it has cleared that it is appropriate for usages including interfacial dynamics, complex flows such as multi phase and multi component flows and complex boundary condition (Chen and Doolen, 1998). In addition, it must considered that modeling of the complex phenomenons like as multi phase and multi component flows, porous mediums, fluid flow in electrical and magnetic fields can be simpler in molecular view and discrete mechanics. This chapter concerns about the techniques and main concepts of LBM.

2. DIFFERENT METHOS FOR FLOW SIMULATION

Macroscopis systems and transport problems have discussed systematically since 19th century. Two basic methods were used to study: first macroscopic continuum media, including fluid mechanics and thermodynamics; and the other was microscopic scheme or kinetics theory. For systems involving larg number of particles, both schemes result same macroscopic governing equations.

Fluid mechanics investigates fluid systems macroscopically. This means that although the system consists of discrete particles, the behavior of each single particle is not treated. The main consideration is to macroscopic variables such as density, pressure, temperature and velocity, which show the macroscopic state of system.

Based on assumption of continuum media Navier Stokes equations can be derived using conservation laws (continuity and momentum conservation). Solving NS equations with given boundary and initial conditions and physical constrains are one of the major tasks of fluid mechanics researches. Thermodynamics and heat transfer in thermal system analyze based on some fundamental roles such as thermodynamics first, second and third laws, Forier law for conduction and Newton's law for cooling, these laws were found based on experimental results.

Afterward studies spread to solve these governing equations. Due to that, NS equations are nonlinear partial differential that corresponds to Mach number, the problems that can solve analytically are distinct to simple flows and geometries.

About 20 to 30 years ago, investigation of most practical problems was based on experimental data, which usually were expensive. Because that experimental solution of some important problems is impossible, it is appropriate to solve fluid problems using partial differential equation (PDEs), such as NS equation.

Appearance of digital computers makes available the numerical solution of Navier Stokes equations. In three recent decades different more accurate computational fluid dynamics methods developed. Almost all commercial CFD methods are on the basis of integral or differential form of conservation laws, and yield approximate results due to discretization of equations. These methods also are called "top-bottom" methods, as illustrated in Figure 1. These methods begin with a continuum description of governing PDEs (for example Navier-Stokes equations). Then numerical techniques, such as finite-volume method, are used (usually in the mesh with proper grid size) to transform the continuum description into a discrete one. Finally, these discretized equations are solved numerically on a computer to achieve approximate solution (Peyret and Taylor, 1983; Luo, 2000).

On the other hand the "bottom-up" approach is based on the microscopic particle description provided by the equations of Molecular Dynamics (MD); here the position and velocity of each atom or molecule in the system are closely relative and followed by its position and velocity on the previous time step and the Newton's equations of motion. These models simulate macroscopic behavior of fluid systems in the microscopic scale. Here multi scale analysis is used to achieve macroscopic PDEs.

It is evident that the molecular dynamics simulations are limited to very small systems because the huge number of involved particles (2×10^9 per cubic centimeter for gas in standard conditions) and very short time of particles interactions (a few picosends). For these reasons molecular dynamics simulations are more suitable for studies in material science, biologic researches and especially for investigation of structure, dynamics and thermodynamics of biologic molecules (Evans and Morris, 1983; Goodfellow, 1991; Hardy et al., 1973).

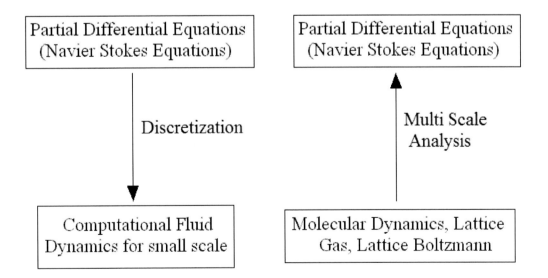

Figure 1. Two different methods, left) up-bottom, right) bottom-up.

Two practical modifications are used to decrease computational problems of MD methods:

1. First, instead of considering discrete molecules in microscopic scale, fluid particles in mesoscopic scale, which consist a group of molecules are treated in simulations.
2. Second, decreasing the degree of freedom of system by restriction of particle movements along finite number of directions.

Lattice gas and Lattice Boltzmann methods based on these two modifications and have been used successfully in fluid flow and transport phenomena.

3. LATTICE BOLTZMANN EQUATIONS

In lattice Boltzmann method the distribution function is treated as replacement of investigation of each particle in molecular dynamics. The distribution functions are defined as probability to find particles with specific velocity range at the limited position at the given time. Using the distribution functions will cause a lot of saving in computer sources. In this section, LBM equations for simulation of fluid flow, heat transfer, and porous media are discussed.

3.1. Boltzmann Transport Equation

As discussed above in contrast to the classical macroscopic Navier–Stokes (NS) approach, the lattice Boltzmann method uses a mesoscopic simulation model to simulate fluid flows. It models movement of fluid particles to capture macroscopic fluid quantities such as velocity and pressure. In LBM, the domain is discretized in uniform Cartesian cells, which each one holds a fixed number of distribution functions, which represent the number of fluid particles moving in these discrete directions. The distribution functions are calculated by solving the Lattice Boltzmann Equation (LBE), which is a special discretization of the kinetic Boltzmann equation.

The distribution function $f(\vec{r},\vec{c},t)$ is the number of molecules in time t, located in the position between $\vec{r}$ and $\vec{r}+d\vec{r}$, with velocity ranging from $\vec{c}$ to $\vec{c}+d\vec{c}$. If an external force acts on a particle (molecule) with unit mass, it's velocity and location chang from $\vec{c}$ to $\vec{c}+\vec{F}dt$, and $\vec{r}$ to $\vec{r}+d\vec{r}$, respectively. The change between initial and final statuses is called collision operator, Ω. So the equation of molecules number can be written as (Mohammad, 2007):

$$f\left(\vec{r}+\vec{c}dt,\vec{c}+\frac{\vec{F}}{m}dt,t+dt\right)d\vec{r}d\vec{c} - f(\vec{r},\vec{c},t)d\vec{r}d\vec{c} = \Omega(f)d\vec{r}d\vec{c}dt \qquad 1$$

Dividing both sides of above equation and limiting $dt \to 0$:

$$df/dt = \Omega(f) \qquad (2)$$

Above equation shows that the final rate of distribution function change is equal to collision rate.

Distribution function, f, is function of $\vec{c}$, $\vec{r}$ and t, so it's differential can be developed as:

$$df = \frac{\partial f}{\partial \vec{r}} d\vec{r} + \frac{\partial f}{\partial \vec{c}} d\vec{c} + \frac{\partial f}{\partial t} dt \qquad (3)$$

Dividing to dt yields:

$$\frac{df}{dt} = \frac{\partial f}{\partial \vec{r}} \frac{d\vec{r}}{dt} + \frac{\partial f}{\partial \vec{c}} \frac{d\vec{c}}{dt} + \frac{\partial f}{\partial t} \qquad (4)$$

Note that the $d\vec{r}/dt$ is velocity vector, $\vec{c}$, and acceleration, $\vec{a}$, is equal to $d\vec{c}/dt$, and in respect to Newton's second law $\vec{a} = \vec{F}/m$, so the above equation can be rewritten as:

$$\frac{df}{dt} = \frac{\partial f}{\partial \vec{r}} \vec{c} + \frac{\partial f}{\partial \vec{c}} \vec{a} + \frac{\partial f}{\partial t} \qquad (5)$$

Then the Boltzmann transport equation (Eq. 4) can be written as below:

$$\frac{\partial f}{\partial t} + \frac{\partial f}{\partial \vec{r}} \vec{c} + \frac{\vec{F}}{m} \frac{\partial f}{\partial c} = \Omega \qquad (6)$$

For a system without any external force, the Boltzmann equation is:

$$\frac{\partial f}{\partial t} + \vec{c}.\vec{\nabla}f = \Omega \qquad (7)$$

Note that $\vec{c}$ and $\vec{\nabla}f$ are vectors.

3.2. Bgk Approximation

The collision operator (Ω) is function of f and it must be given to solve Boltzmann equation. Bhatangar, Gross and Krook introduced a simple model for the collision operator. By using the BGK (Bhatnagar–Gross–Krook) approximation, the Boltzmann equation without external forces is as below (Bhatangar et al. 1954):

$$\frac{\partial f}{\partial t} + \vec{c}.\vec{\nabla} f = \frac{1}{\tau}\left(f^{eq} - f\right) \qquad 8$$

where $\omega = 1/\tau$, ω stands for collision frequency, and τ stands for relaxation factor, and f^{eq} stand for local equilibrium distribution function.

The discretized lattice Boltzmann equation without external forces is as below:

$$\frac{\partial f_k}{\partial t} + c_k.\nabla f_k = \frac{1}{\tau}\left(f_k^{eq} - f_k\right) \qquad 9$$

The above equation can be disretized as:

$$f_k(\vec{r} + \vec{c}_k \Delta t, t + \Delta t) = f_k(\vec{r}, t) + \frac{\Delta t}{\tau}\left[f_k^{eq}(\vec{r}, t) - f_k(\vec{r}, t)\right] \qquad 10$$

The general form of lattice Boltzmann equation with external force can be written as:

$$f_k(\vec{x} + \vec{c}_k \Delta t, t + \Delta t) = f_k(\vec{x}, t) + \frac{\Delta t}{\tau}\left[f_k^{eq}(\vec{x}, t) - f_k(\vec{x}, t)\right] + \Delta t F_k \qquad 11$$

where Δt denotes lattice time step, $\vec{c}_k$ is the discrete lattice velocity in direction k, F_k is the external force in direction of lattice velocity (c_k), τ denotes the lattice relaxation time, f_k^{eq} is the equilibrium distribution function which determines the type of problem that needs to be solved.

Equation (11) is usually solve in two steps:

$$\tilde{f}_k(\vec{x}, t + \Delta t) = f_k(\vec{x}, t) - \frac{\Delta t}{\tau}\left[f_k(\vec{x}, t) - f_k^{eq}(\vec{x}, t)\right] + \Delta t \vec{F}_k \qquad 12$$

$$f_k(\vec{x} + \vec{c}_k \Delta t, t + \Delta t) = f_k(\vec{x}, t + \Delta t) \qquad 13$$

Equations (12) and (13) are called the collision and streaming steps, respectively. The collision step models various fluid particle interactions like collisions and calculates new distribution functions according to the distribution functions of the previous time step. It also models the equilibrium distribution functions, which are calculated with Eq. (14):

$$f_k^{eq} = \omega_k \cdot \rho \left[1 + \frac{\vec{c}_k \cdot \vec{u}}{c_s^2} + \frac{1}{2}\frac{(\vec{c}_k \cdot \vec{u})^2}{c_s^4} - \frac{1}{2}\frac{\vec{u} \cdot \vec{u}}{c_s^2}\right] \qquad 14$$

where ω_k is a weighting factor depending on the used LB model, ρ is the lattice fluid density and u is the fluid macroscopic velocity.

According to the equations mentioned values of ρ and velocity component will determine by:

$$\rho = \sum_k f_k \qquad 15$$

$$\rho u_i = \sum_k f_k c_{ki} \qquad 16$$

Subscript i relates to the direction of the Cartesian axis coordinates that on it collection is summed.

4. Thermal LBM

To consider both fluid flow and temperature fields, the thermal LBM utilizes two distribution functions, f and g, for the flow and temperature fields respectively. The f distribution function is as same as discussed above; the g distribution is as below:

$$g_k(\vec{x} + \vec{c}_k \Delta t, t + \Delta t) = g_k(\vec{x},t) + \frac{\Delta t}{\tau_g}\left[g_k^{eq}(\vec{x},t) - g_k(\vec{x},t)\right] + \Delta t \, w_k S_T \qquad 17$$

where S_T is the heat source and will be determined by (Mohammad, 2007):

$$S_T = \frac{q_g}{\rho C} \qquad 18$$

where q_g is the macroscopic heat source.

The corresponding equilibrium distribution functions are defined as (Mohammad, 2007, Mezrhab et al. 2006, Delavar et al. 2009&2010):

$$g_k^{eq} = \omega_k . T . \left[1 + \frac{\vec{c}_k . \vec{u}}{c_s^2} \right] \qquad 19$$

Having computed the values of these local distribution functions, the temperature is defined as:

$$T = \sum_k g_k \qquad 20$$

where sub-index i denotes the component of the Cartesian coordinates which implied summation for repeated indices.

In order to incorporate buoyancy force in the model, the force term in the Eq. (12) needs to be calculated as below in vertical direction (y) (Mohammad, 2007):

$$F_k = 3\omega_k g_y \beta \rho \theta c_{ky} \qquad 21$$

where β is thermal expansion coefficient, c_{ky} is the y-component of c_k, ρ and θ are local density and dimensionless temperature respectively. The dimensionless temperature is calculated by:

$$\theta = \frac{T - T_c}{T_h - T_c} \qquad 22$$

5. MULTICOMPONENT LBM (SPECIES MODELING)

The species conservation equation is as below:

$$\rho \vec{V} . \nabla C_l = \rho D_l^{eff} \nabla^2 C_l + S_l \qquad 23$$

$$\sum_l C_l = 1 \qquad 24$$

where l represents chemical species, C_l the mole fraction of species l, S_l the mass generation rate for species l per unit volume, D_l the diffusion coefficient of the lth component.

In lattice Boltzmann method the Eq. (23) is satisfied with calculations of proper distribution functions by solving the Lattice Boltzmann Equation.

To solve multicomponenet equations of chemical species in LBM, as same as temperature, in addition of f distribution function for flow, LBM utilizes multi distribution functions, g_l, for lth species concentration field. The g_l distribution function is as below (Delavar et al., 2010):

$$g_{lk}(\vec{x}+\vec{c}_k \Delta t, t+\Delta t) = g_{lk}(\vec{x},t) + \frac{\Delta t}{\tau_l}\left[g_{lk}^{eq}(\vec{x},t) - g_{lk}(\vec{x},t)\right] + \Delta t . \omega_k . S_l \qquad 25$$

where S_l is the source term of the lth species. Because that species concentration is scalar the corresponding equilibrium distribution functions are defined by (Mohammad, 2007; Mezrhab et al., 2006; Delavar et al., 2010):

$$g_{lk}^{eq} = \omega_k . C_l . \left[1 + \frac{\vec{c}_k . \vec{u}}{c_s^2}\right] \qquad 26$$

As same as temperature after computing the values of local distribution functions:

$$C_l = \sum_k g_{lk} \qquad 27$$

where sub-index i denotes the component of the Cartesian coordinates which implied summation for repeated indices.

6. CONSIDERING SOURCE TERM AND NO DIMENSIONAL NUMBERS

To use Eqs. (17) and (25) there is an important issue, which is existence of the LBM time step size, Δt. In order to correct modeling of source terms, this time step size must correlate to macroscopic time step size properly. This correlation attains if (Anwar and Sukop, 2009):

$$\left(\frac{Dt}{H^2}\right)_{real} = \left(\frac{Dt}{N^2}\right)_{LBM} \qquad 28$$

where H stands for characteristic length, and N is number of lattices used for H in LBM simulation, D stands for diffusion coefficient in equation 17 or 23, t is time. Therefore, in simulation of temperature or species concentration fields, the proper time step size must calculated and applied. A practical solution is to set $\Delta t = \Delta x = \Delta y$ as used before in (Peng et. al., 2003; Zhao et al.,2007; Chang et al., 2009, Delavar et al., 2010)

To achieve exact solutions the magnitude of concerning variable must changes between zero and one, so the dimensionless form of governing equations or proper variable change is necessary. To do this, diffusion coefficients, source terms and boundary condition will be change.

To ensure of correct simulation of flow, temperature and species equations, the dimensionless numbers should be equal in real and LBM. The important no dimensional numbers are Reynolds number, Prandtl number, Rayleigh number and Schmidt number.

To retain Reynolds number:

$$\text{Re} = \left(\frac{UH}{\upsilon}\right)_{real} = \left(\frac{UN}{\upsilon}\right)_{LBM} \qquad 29$$

where N is the number of grid used for meshing H as characteristic length in LBM simulation.

The next important dimensionless number is Prandtl number, which is the ratio of kinematic viscosity to thermal diffusivity, so for a given value of Prandtl number:

$$\text{Pr} = \left(\frac{\upsilon}{\alpha}\right)_{real} = \left(\frac{\upsilon}{\alpha}\right)_{LBM} \qquad 30$$

Another important dimensionless number is Schmidt number, which is defined as ratio of momentum diffusivity (viscosity) and mass diffusivity.

$$Sc = \frac{\upsilon}{D} \qquad 31$$

This number is used to characterize fluid flows in which there are simultaneous momentum and mass diffusion convection processes.

The relaxation times, τ, τ_g and τ_l, for flow, temperature and species in LB equations given in Eqs. (12), (17) and (25) can then be determined by:

$$\tau = v/\left(c_s^2 \Delta t\right) + 0.5 \qquad 32$$

$$\tau_g = \alpha/\left(c_s^2 \Delta t\right) + 0.5 \qquad 33$$

$$\tau_l = D_l/\left(c_s^2 \Delta t\right) + 0.5 \qquad 34$$

where v is the fluid kinematic viscosity, α stand for thermal diffusion coefficient and D_l is diffusion coefficient of species l.

7. FLOW SIMULATION IN POROUS MEDIA

The Darcy's equation has been used to study in porous medium.

$$\vec{u} = -\frac{K}{\mu}\vec{\nabla}P \tag{35}$$

For flows with low velocities, the Darcy's law can make a good prediction. However, for higher velocities it was shown that the theoretical prediction based on Darcy's law does not agree well with experimental data. Two modify this problem two notable modifications of the Darcy's law are proposed: first the Forchheimer's equation which is considered non-linear drag effect due to the solid matrix (Eqs. (36) and (37)), and second Brinkman's equation, which consider viscous stress introduced by the solid boundary (Eq. (38)). It should mention that for the case in which the Reynolds number or the Darcy number is large, the non-linear drag must be considered (Seta et al, 2006a, 2006b).

$$-\vec{\nabla}P = \frac{\mu}{K}\vec{u} + \frac{C\rho_f}{K^{1/2}}|\vec{u}|\vec{u} \tag{36}$$

$$C = \frac{1.75}{\sqrt{150\varepsilon^3}} \tag{37}$$

$$-\vec{\nabla}P = \frac{\mu}{K}\vec{u} - \mu_{eff}\nabla^2\vec{u} \tag{38}$$

Nield and Bejan combined these two equations and derived the Brinkman–Forchheimer equation, which includes the viscous and inertial terms by the local volume averaging technique (Nield and Bejan, 1992). This model successfully simulated flow in porous media in wide range of porosities, Rayleigh, Reynolds and Darcy numbers (Seta et al, 2006a, 2006b; Yan et al., 2008; Nield et al., 2005).

The Brinkman–Forchheimer equation is:

$$\frac{\partial \vec{u}}{\partial t} + (\vec{u}.\nabla)\left(\frac{\vec{u}}{\varepsilon}\right) = -\frac{1}{\rho}\nabla(\varepsilon p) + \upsilon_{eff}\nabla^2\vec{u} + \vec{F} \tag{39}$$

where ε is the porosity of the medium, and υ_{eff} the effective viscosity. $\vec{F}$ is the total body force and contains the viscous diffusion, the inertia due to the presence of porous medium, and an external force, which with the widely used Ergun's relation, can be written as (Seta et al, 2006a, 2006b; Ergun 1952):

$$\vec{F} = -\frac{\varepsilon \upsilon}{K}\vec{u} - \frac{1.75}{\sqrt{150\varepsilon K}}|\vec{u}|\vec{u} + \varepsilon\vec{G} \qquad (40)$$

where υ is the kinematic viscosity, K is the permeability, and $\vec{G}$ is the acceleration due to gravity. The permeability relates to Darcy number (Da), and the characteristic length (H) as:

$$K = DaH^2 \qquad (41)$$

For porous medium, the corresponding distribution functions for flow are as same as Eq. (11). But the equilibrium distribution functions are calculated by (Seta et al., 2006a, 2006b):

$$f_k^{eq} = w_k \rho \left[1 + \frac{\vec{c}_k \cdot \vec{u}}{c_s^2} + \frac{1}{2}\frac{(\vec{c}_k \cdot \vec{u})^2}{\varepsilon c_s^4} - \frac{1}{2}\frac{\vec{u}^2}{\varepsilon c_s^2} \right] \qquad (42)$$

It was shown that the best choice for the forcing term F_k to achieve correct equation of hydrodynamics is taking (Seta et al, 2006a, 2006b; Peng and Shu, 2003):

$$F_k = w_k \rho \left(1 - \frac{1}{2\tau_v}\right)\left[\frac{\vec{c}_k \cdot \vec{F}}{c_s^2} + \frac{(\vec{u} \cdot \vec{c}_k)(\vec{F} \cdot \vec{c}_k)}{\varepsilon c_s^4} - \frac{\vec{u} \cdot \vec{F}}{\varepsilon c_s^2}\right] \qquad (43)$$

The force term F_k defines the fluid velocity $\vec{u}$ as (Seta et al, 2006a, 2006b):

$$\vec{u} = \sum_k c_k F_k / \rho + \frac{\Delta t}{2}\vec{F} \qquad (44)$$

According the above equations, $\vec{F}$ relates to $\vec{u}$, so the Eq. (44) is nonlinear for the velocity. Guo and Zhao presented a temporal velocity $\vec{v}$ to solve this nonlinear problem as follows (Guo and Zhao, 2002):

$$\vec{u} = \frac{\vec{v}}{c_0 + \sqrt{c_0^2 + c_1|\vec{v}|}}, \quad \vec{v} = \sum_k c_k f_k / \rho + \frac{\Delta t}{2}\varepsilon\vec{G} \qquad (45)$$

$$c_0 = \frac{1}{2}\left(1 + \varepsilon\frac{\Delta t}{2}\frac{\upsilon}{K}\right), \quad c_1 = \varepsilon\frac{\Delta t}{2}\frac{1.75}{\sqrt{150\varepsilon^3 K}} \qquad (46)$$

To solve energy equation the overall thermal conductivity of porous medium should be identified which generally depend on porous solid structure and the fluid, which passes through it. The heat transfer via these solid structure and fluid occurs in parallel, and then the overall conductivity k_A is the weighted arithmetic mean of the conductivities k_s and k_f (Nield and Bejan, 2006):

$$k_A = (1-\varepsilon)k_s + \varepsilon k_f \qquad (47)$$

If the structure and orientation of the porous medium was such that the overall heat transfer takes place in series, then the overall conductivity k_B will be the weighted harmonic mean of k_s and k_f (Nield and Bejan, 2006):

$$\frac{1}{k_B} = \frac{(1-\varepsilon)}{k_s} + \frac{\varepsilon}{k_f} \qquad (48)$$

In general, k_A and k_B provide upper and lower bounds, respectively. For practical purposes, it is recommended to use k_m, the weighted geometric mean of k_s and k_f, defined by (Nield and Bejan, 2006):

$$k_m = k_s^{\varepsilon-1} k_f^{\varepsilon} \qquad (49)$$

The above equation is valid if k_s and k_f are not too different from each other, for other cases the effective thermal conductivity in porous media was calculated by (Jiang and Lu, 2006):

$$k_m = k_f \left[(1-\sqrt{1-\varepsilon}) + \frac{2\sqrt{1-\varepsilon}}{1-\sigma B} \left(\frac{(1-\sigma)B}{(1-\sigma B)^2} \ln\left(\frac{1}{\sigma B}\right) - \frac{B+1}{2} - \frac{B-1}{1-\sigma B} \right) \right]$$

$$B = 1.25 \left[\frac{1-\varepsilon}{\varepsilon} \right]^{10/9} \quad , \quad \sigma = \frac{k_f}{k_s} \qquad (50)$$

8. LATTICE MODELS

As mentioned above one of the major characteristics of LBM is limiting of particles movement in specific directions. The number of these directions and their orientation depends on used LBM model. DnQm is used to address LBM models (lattice arrangement), n identifies the dimension of model (1 for one dimensional and 2 for two dimensional), and m

identifies the number of streaming directions. In this section, following paragraphes address most used LBM models.

8.1. One Dimensional

Generally, two, one dimensional models are used for lattice structure, D1Q3 and D1Q5, Figure 2.

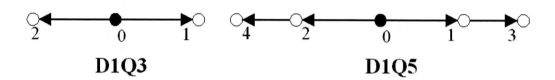

Figure 2. 1D lattice models.

D1Q3
In this model the lattice velocities and weighting factors are as below:

$$c_k = \begin{cases} 0, & k=0 \\ 1, & k=1 \\ -1, & k=2 \end{cases}$$

51

$$w_k = \begin{cases} 0, & k=0 \\ 1/2, & k=1 \\ 1/2, & k=2 \end{cases}$$

52

The weighting factor for f_0 is zero, so there is no need to consider the effect of it, so this model is reffered as D1Q2. Another possible weighting factors are as below:

$$w_k = \begin{cases} 4/6, & k=0 \\ 1/6, & k=1 \\ 1/6, & k=2 \end{cases}$$

53

This model is refered ad D1Q3 because that f_0 need to be considered.

D1Q5

For this scheme:

$$c_k = \begin{cases} 0, & k=0 \\ 2/6, & k=1,3 \\ -2/6, & k=2,4 \end{cases} \qquad 54$$

$$w_k = \begin{cases} 6/12, & k=0 \\ 2/12, & k=1,2 \\ 1/12, & k=3,4 \end{cases} \qquad 55$$

8.2. Two Dimensionl

D2Q5

This model is illustratd in Figure 3. For this model:

$$\vec{c}_k = \begin{cases} (0,0), & k=0 \\ (\pm 1,0), & k=1,3 \\ (0,\pm 1), & k=2,4 \end{cases} \qquad 56$$

$$w_k = \begin{cases} 6/12, & k=0 \\ 2/12, & k=1,2 \\ 1/12, & k=3,4 \end{cases} \qquad 57$$

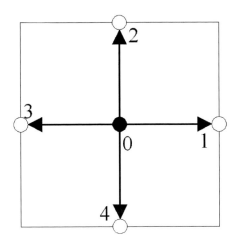

Figure 3. Lattice arrangement for 2-D models, D2Q5.

D2Q9

This model is illustrated in figure 4. The velocities and weighting factors are as:

$$\vec{c}_k = \begin{cases} (0,0), & k = 0 \\ (\pm 1,0),(0,\pm 1), & k = 1-4 \\ (\pm,\pm 1), & k = 5-8 \end{cases} \qquad 58$$

$$w_k = \begin{cases} 0, & k = 0 \\ 1/4, & k = 1-4 \end{cases} \qquad 59$$

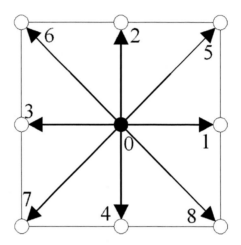

Figure 4. Lattice arrangement for 2-D models, D2Q9.

This model is very common in LBM simulation, which is use in example presented in this section.

8.3 Three Dimensional

D3Q15

This model is must used three dimensional LBM model, Figure 4. The velocity vectors and weighting factors are as below:

$$c_k = \begin{cases} (0,0,0), & k = 0; \\ (\pm 1,0,0),(0,\pm 1,0),(0,0,\pm 1), & k = 1,2,...,6; \\ (\pm 1,\pm 1,\pm 1) & k = 7,...,18. \end{cases} \qquad 60$$

$$w_k = \begin{cases} 2/9, k=0; \\ 1/9, k=1,2,..,6; \\ 1/72, k=7,..,14. \end{cases}$$

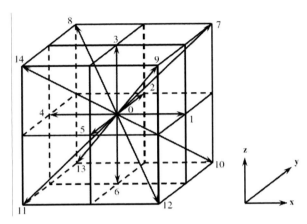

Figure 5. Lattice arrangement for 2-D models, D3Q15 (Yan and Zu, 2007).

9. BOUNDARY CONDITIONS

One of the important steps in LBM simulations like as other numerical investigations is proper setting of the boundary conditions. Implementation of boundary conditions in NS equations is straightforward but more attention is required in LBM. From the streaming process, the distribution functions out of the domain are known. The unknown distribution functions are those toward the domain. The dotted lines in Figure 6 are the unknown distribution functions, which needs to be determined (for D2Q9).

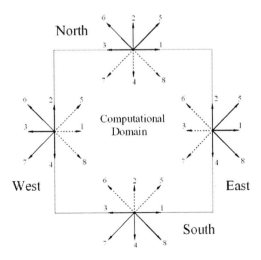

Figure 6. Domain boundaries and known (solid lines) and unknown (dotted lines) distribution functions.

9.1. Bounce Back Boundary Condition

For solid walls, which are assumed to be no slip, the bounce-back scheme will be applied. This scheme specifies the outgoing directions of the distribution functions as the reverse of the incoming directions at the boundary sites.

For example for flow field in the bottom boundary of the Figure 6 the following conditions will be used:

$$f_{2,n} = f_{4,n}, \quad f_{5,n} = f_{7,n} \quad f_{6,n} = f_{8,n} \tag{62}$$

where n is the lattice on the boundary.
For other boundaries:

North:

$$f_{4,n} = f_{2,n}, \quad f_{7,n} = f_{5,n} \quad f_{8,n} = f_{6,n} \tag{63}$$

East:

$$f_{6,n} = f_{4,n}, \quad f_{7,n} = f_{5,n} \quad f_{6,n} = f_{8,n} \tag{64}$$

West:

$$f_{4,n} = f_{6,n}, \quad f_{5,n} = f_{7,n} \quad f_{8,n} = f_{6,n} \tag{65}$$

9.2. Boundary Condition with Given Velocity

This boundary condition occurs in many practical applications, such as inlet port of the flow in the channel or pipe. This boundary condition for different boundaries (north, east, south or west) will satisfy by (Mohammed, 2007):

West:

$$\rho_W = \frac{1}{1-u_w}[f_0 + f_2 + f_4 + 2(f_3 + f_6 + f_7)] \tag{66}$$

$$f_1 = f_3 + \frac{2}{3}\rho_W u_W \tag{67}$$

$$f_5 = f_7 - \frac{1}{2}(f_2 - f_4) + \frac{1}{6}\rho_W u_W + \frac{1}{2}\rho_W v_W \qquad 69$$

$$f_8 = f_6 + \frac{1}{2}(f_2 - f_4) + \frac{1}{6}\rho_W u_W - \frac{1}{2}\rho_W v_W \qquad 69$$

East:

$$\rho_E = \frac{1}{1+u_E}[f_0 + f_2 + f_4 + 2(f_1 + f_5 + f_8) \qquad 70$$

$$f_3 = f_1 - \frac{2}{3}\rho_E u_E \qquad 71$$

$$f_6 = f_8 - \frac{1}{2}(f_2 - f_4) - \frac{1}{6}\rho_E u_E + \frac{1}{2}\rho_E v_E \qquad 72$$

$$f_7 = f_5 + \frac{1}{2}(f_2 - f_4) - \frac{1}{6}\rho_E u_E - \frac{1}{2}\rho_E v_E \qquad 73$$

North:

$$\rho_N = \frac{1}{1+v_N}[f_0 + f_1 + f_3 + 2(f_2 + f_5 + f_6) \qquad 74$$

$$f_4 = f_2 - \frac{2}{3}\rho_N v_N \qquad 75$$

$$f_7 = f_5 + \frac{1}{2}(f_1 - f_3) - \frac{1}{6}\rho_N v_N - \frac{1}{2}\rho_N u_N \qquad 76$$

$$f_8 = f_6 + \frac{1}{2}(f_3 - f_1) + \frac{1}{2}\rho_N u_N - \frac{1}{6}\rho_N v_N \qquad 77$$

South:

$$\rho_S = \frac{1}{1-v_S}[f_0 + f_1 + f_3 + 2(f_4 + f_7 + f_8) \qquad 78$$

$$f_2 = f_4 + \frac{2}{3}\rho_s v_s \qquad (79)$$

$$f_5 = f_7 + \frac{1}{2}(f_1 - f_3) + \frac{1}{6}\rho_s u_s + \frac{1}{2}\rho_s v_s \qquad (80)$$

$$f_6 = f_8 + \frac{1}{2}(f_1 - f_3) + \frac{1}{6}\rho_v V_s + \frac{1}{2}\rho_s v_s \qquad (81)$$

9.3. Open Boundary Condition

This condition assure for outlet ports by implementation of extrapolation. For example for east boundary:

$$f_{3,n} = 2f_{3,n-1} - f_{3,n-2} \qquad (82)$$

$$f_{6,n} = 2f_{6,n-1} - f_{6,n-2} \qquad (83)$$

$$f_{7,n} = 2f_{7,n-1} - f_{7,n-2} \qquad (84)$$

where n is the lattice on the boundary and $n-1$ and $n-2$, denote the lattices inside the cavity adjacent to the boundary. Another possible extrapolation, which can implement is (Seta e al., 2006b):

$$f_{3,n} = \frac{4}{3}f_{3,n-1} - \frac{1}{3}f_{3,n-2} \qquad (86)$$

$$f_{6,n} = \frac{4}{3}f_{6,n-1} - \frac{1}{3}f_{6,n-2} \qquad (86)$$

$$f_{7,n} = \frac{4}{3}f_{7,n-1} - \frac{1}{3}f_{7,n-2} \qquad (87)$$

9.4. Boundary Condition With Given Scalar

Furthermore, for the saclar field (such as temperature or species concentrations), the local scalar is defined as in Eq. (20) or (27). For boundaries with specified saclar, such as temperature or species concentrations in inlet ports ($C = C_{in}$), the unknown distribution functions are evaluates as (Delavar et al., 2010):

$$g_{1,n} = C_{in}(\omega_1 + \omega_3) - g_{3,n}$$
$$g_{5,n} = C_{in}(\omega_5 + \omega_7) - g_{7,n}$$
$$g_{8,n} = C_{in}(\omega_8 + \omega_6) - g_{6,n}$$
88

Forther information about different boundary conditions is presented in literature such as (Mohammad et al., 2007)

10. SOME EXAMPLES

10.1. Flow in Clear Channel

To be familiar the abilities of LBM, first it was used to silumate the flow between two infinite parallel plates, Poiseuille flow, Figure 7 illuastrates the configuration of the problem.

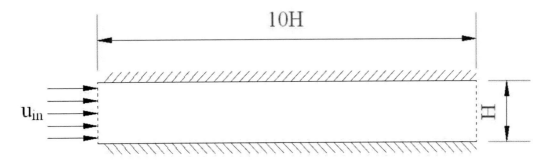

Figure 7. Flow in clear channel geometry.

For this 2D model, the D2Q9 was implemented. The LBM code is presented at appendix A.

This problem has an analytical solution, which is (Nield, 2004):

$$\frac{u}{u_{\max}} = 6y - 6y^2$$
89

The results have been compared well with analytical solution, figure 8. The flow field is shown in figure 9.

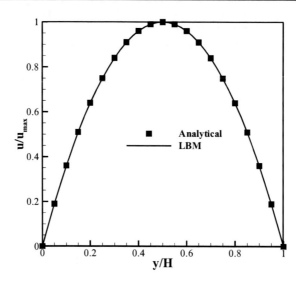

Figure 8. Comparison of velocity profile for clear channel flow between LBM and analytical solution.

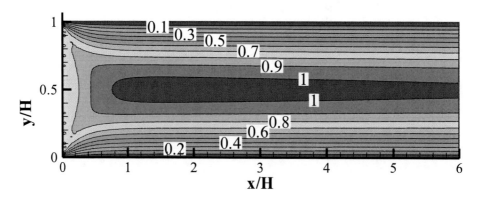

Figure 9. The velocity field in clear channel.

10.2. Free Convection in Square Cavity

The next axample is free convection in cavity. The initial stationary flow was heated from the left wall, $T_h = 1.0$, while the right wall was maintained at a constant low temperature, $T_c = 0.0$. The upper and bottom walls assigned adiabatic boundary conditions. As previous example, the simulations based on the D2Q9 model. The Prandtl number assumed to have a constant value of 0.71. For validation, the averaged Nusselt number calculated at different Rayleigh numbers. Table 1 shows the computed averaged Nusselt number in comparison with previous works (de Vahl Davis, 1983; and Kao et al., 2008). Results show a good agreement compared to the previous studies. Figure 10 illustrates the achieved isotherms and streamlines at $Ra = 10^6$ by LBM and CFD.

Table 1. Comparison of averaged Nusselt numbers computed at different Rayleigh numbers using different grids with results presented in (de Vahl Davis, 1983; Kao et al., 2008)

Ra		1,000	Diff.(%)	10,000	Diff.(%)	100,000	Diff.(%)	1,000,000	Diff.(%)
de Vahl Davis (1983)		1.118	---	2.243	---	4.519	---	8.825	---
Kao et al. (2008)		1.113	0.447	2.231	0.535	4.488	0.686	8.696	1.462
Present Study	110x110	1.130	1.064	2.276	1.455	4.584	1.438	8.851	0.299
	100x100	1.131	1.173	2.278	1.574	4.578	1.301	8.833	0.092
	80x80	1.134	1.468	2.285	1.885	4.581	1.377	8.770	0.627

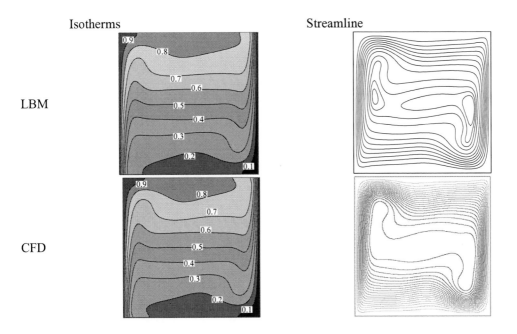

Figure 10. Comparison of isotherms and stream lines at $Ra = 10^6$, LBM and CFD The used LBM code is presented in appendix B.

10.3. Heat Generating Section in the Channel

To show modeling of heat source by LBM, in this example, flow in the channel passing over a heat generating section is simulated, Figure 11. In this simulation, the concerned fluid was air with $\Pr = 0.7$. The upper and bottom walls and inlet temperature were fixed at $T = 50°C$, the Reynolds number at inlet was set to 100. The heat generating rate in the heating section was set to $10^6 W.m^{-3}$. Heat generating sources take place in situations like as combustion and other chemical reacting flows. Figure 12 and 13 show the obtained isotherms and velocity field and vectors, respectively.

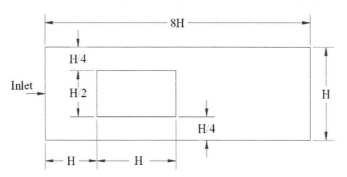

Figure 11. Flow in the channel passing over a heat generating section.

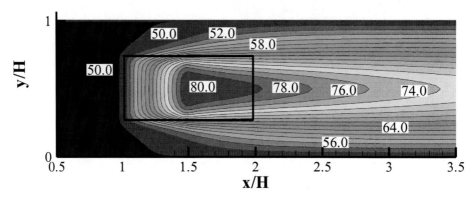

Figure 12. Isotherms for flow passing through heat generating section.

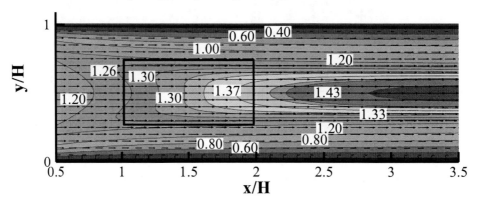

Figure 13. Flow field and velocity vectors for flow passing through heat generating section.

10.4. Discreet Heater on Vertical Wall of Cavity (Delavar Et Al. 2009)

The computational domain is a square cavity in which the left side and horizontal upper walls are isotherms. The heater (hot wall) is located at the right side wall of the cavity with length equal 0.4H and remained areas are adiabatic (Figure 14).The distance between heater and upper cold wall (S/H) was set equal to 0.4. The Rayleigh numbers change between 10^3 to 10^6.

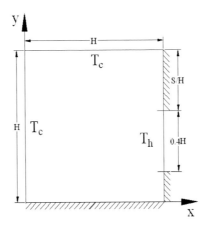

Figure 14. Schematic geometry of problem.

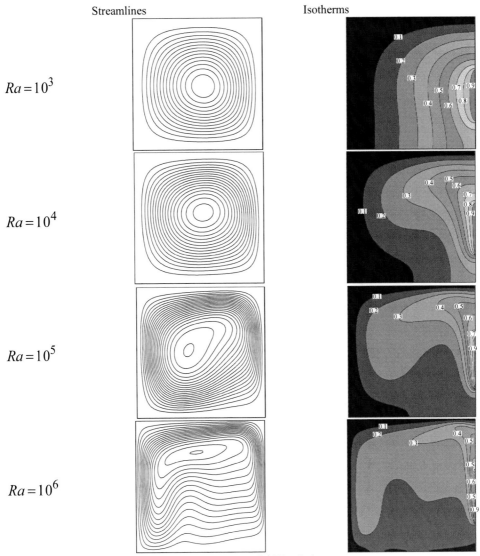

Figure 15. Streamlines and isotherm contours for $S/H = 0.4$.

10.5. Flow in Porous Channel

Flow in porous media take place in many important applications such as geothermal energy extraction, petroleum processing, catalytic and chemical particle beds, transpiration cooling, packed-bed regenerators, heat transfer enhancement, solid matrix or micro-porous heat exchangers, micro-thrusters, and many others. This example concern about flow in channel filled with porous media as hown in figure 16. In figure 17 velocity profiled acjhieved by LBM compare well with results obtained by Mahmud and Fraser (2005). The velocity fieldis illustrated in figure 18.

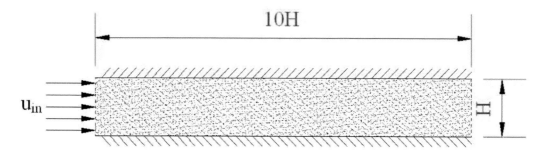

Figure 16. Channel filled with porous media.

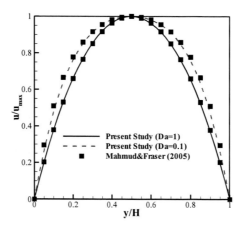

Figure 17. Velocity profile in channel filled with porous media obtained by LBM and Mahmud and Fraser (2005).

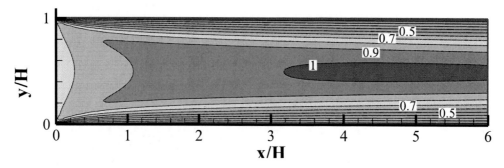

Figure 18. Velocity field in channel filled with porous media.

10.6. Free Convection in Porous Cavity

The computational domain is a square cavity in which the left and right side walls are isotherms, hot and cold walls, respectively. Upper and bottom walls are adiabatic. The cavity is filled with porous media. Tthe Rayleigh numbers change between 10^4 to 5×10^5. For two different porosities, 0.4 and 0.9 the simulations have been carried out.

Table 2. Comparison of averaged Nusselt numbers for free convection in porous cavity computed at different Rayleigh numbers using different grids with results presented in Seta et al. (2006a) and Nithiarasu et al. (1997), Da=0.01, Pr=1.0

Ra	Porosity	Nithiarasu et al. (1997)	Seta et al (2006a)	Present
10^4	0.4	1.408	1.362	1.391
	0.9	1.640	1.633	1.657
10^5	0.4	2.983	2.992	3.056
	0.9	3.910	3.902	4.001
5×10^5	0.4	4.990	4.923	5.087
	0.9	6.700	6.336	6.731

Isotherms   Streamlines

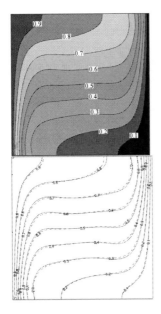 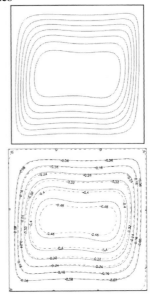

Figure 19. Isotherms and streamlines for free convection at $Ra = 10^5$ and , results obtained by LBM and Seta et al. (2006b) (LBM and FDM).

10.7. Porous Block in Channel

Flow passing over a porous block in the channel is the last example in this section. The configuration is as figure 20. Flow field and stream lines are shown in figure 21. For the porous media porosity and Darcy number was set equal to 0.5 and 0.001, respectively. The simulation has been carried out for Re =100

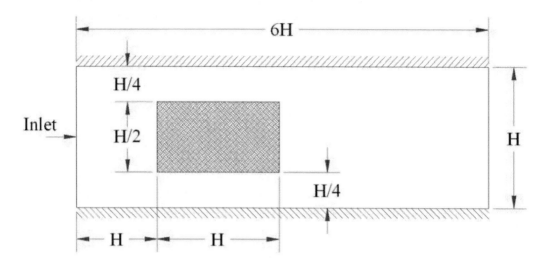

Figure 20. Model configuration.

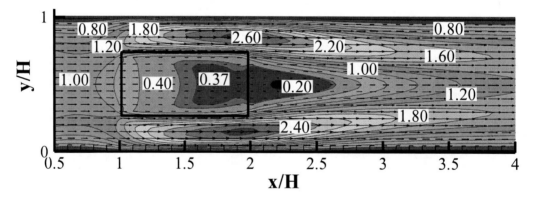

Figure 21. Velocity field and vectors for flow passing through porous block in channel.

REFERENCES

[1] S. Anwar, M.C. Sukop (2009) Regional scale transient groundwater flow modeling using Lattice Boltzmann methods, Computers and Mathematics with Applications, *Computers and Mathematics with Applications, Computers & Mathematics with Applications*, 58, pp:1015-1023

[2] C. Chang, C.H. Liu, C.A. Lin (2009) Boundary conditions for lattice Boltzmann simulations with complex geometry flows, *Computers and Mathematics with Applications, Computers & Mathematics with Applications*, 58, pp:940-949

[3] S. Chen and G. D. Doolen, (1998), Lattice Boltzmann method for fluid flows, *Ann. Rev. Fluid Mech.* 30, pp. 329-364.

[4] M.A. Delavar, M. Farhadi and K. Sedighi (2009) Effect of the heater location on heat transfer and entropy generation in the cavity using the lattice Boltzmann Methode, *Heat Transfer research*, Vol. 40, No. 6, pp: 521-536

[5] M. A. Delavar, M. Farhadi, K. Sedighi (2010) Numerical simulation of direct methanol fuel cells using lattice Boltzmann method, *Int. J. Hydrogen Energy*, Corrected proof

[6] M.A. Delavar, M. Farhadi,K. Sedighi (2010) Effect of discrete heater at the vertical wall of the cavity over the heat transfer and entropy generation using LBM, *Thermal Science*, in press. doi:10.2298/TSCI090422035A.

[7] G.D. Doolen, *Lattice Gas Method for Differential Equations*, (1989), Addison-Wesley Publishing Company

[8] G.D. Doolen, (1991), Lattice Gas Method: Theory, Applications and Hardware, *Physica D*, p. 47

[9] D.J. Evans, and G.P. Morris, (1983), Nonequilibrium molecular-dynamics simulation of coquette flow in two-dimensional fluids, *Phys. Rev. E*, 51(19), pp. 1776-1779.

[10] U. Frisch, B. Hasslacher and Y. Pomeau, (1986), Lattice-Gas automata for. the Navier-Stokes equation, *Phys. Rev. Lett.* 56. p.1505

[11] U. Frisch, D. d'Humieres, B. Hasslacher, P. Lallemand, Y. Pomeau and J. P. Rivet, (1987), Lattice Gas Hydrodynamics in Two and Three Dimensions, *Complex Systems* 1, p.649

[12] J. Goodfellow, (1991), *Molecular Dynamics*, Macmillan Press.

[13] J. Hardy, Yv. Pomeau and O. de Pazzis, (1973), Time Evolution of a Two-Dimensional Classical Lattice Syste, *Phys. Rev. Lett.* 31, pp. 276-279.

[14] P.-X. Jiang, X.-C. Lu , Numerical simulation of fluid flow and convection heat transfer in sintered porous plate channels, *International Journal of Heat and Mass Transfer* 49 (2006) 1685–1695

[15] L.-S. Luo, (2000), The lattice-gas and lattice Boltzmann methods: Past, Present, and Future, *Proceedings of the International Conference on Applied Computational Fluid Dynamics,* Beijing, China, pp. 52-83

[16] Mezrhab, M. Jami, C. Abid, M. Bouzidi, P. Lallemand (2006) Lattice-Boltzmann modelling of natural convection in an inclined square enclosure with partitions attached to its cold wall, *International Journal of Heat and Fluid Flow* 27, pp.456–465

[17] A.A. Mohamad (2007) Applied Lattice Boltzmann Method for Transport Phenomena, *Momentum, Heat and Mass Transfer*, The University of Calgary.

[18] D.A. Nield, A. Bejan (1992) *Convecton in porous media*, Second ed., Springer-Verlag, New York.

[19] D.A. Nield, A.V. Kuznetsov (2005) Thermally developing forced convection in a channel occupied by a porous medium saturated by a non-Newtonian fluid, *Int. J. Heat and Mass Transfer.* 48, pp:1214–1218

[20] D.A. Nield, A. Bejan (2006) *Convetion in porous media*, Springer Science+Business Media, Inc.

[21] P. Nithiarasu, K.N. Seetharamu, T. Sundararajan (1997) Natural convective heat transfer in a fluid saturated variable porosity medium, *Int. J. Heat Mass Transfer* 40, pp:3955–3967

[22] Y. Peng, C. Shu, Y.T. Chew (2003) A 3D incompressible thermal lattice Boltzmann model and its application to simulate natural convection in a cubic cavity, *Journal of Computational Physics* 193, pp260–274,

[23] R. Peyret, and T.D. Taylor, (1983), *Computational Methods for Fluid Flow*, Springer-Verlag, New York.
[24] T. Seta, E. Takegoshi, K. Okui (2006) Lattice Boltzmann simulation of natural convection in porous media, *Mathematics and Computers in Simulation* 72, pp:195–200
[25] T. Seta, E. Takegoshi, K. Kitano, K. Okui (2006) Thermal Lattice Boltzmann Model for Incompressible Flows through Porous Media, *J. Thermal Science And Technology*, Vol. 1, No. 2, pp:90-10
[26] S. Wolfram, (1986), Cellular Automaton Fluids. I. Basic Theory, *J. Stet. Phys*, 45, p. 471
[27] Y.Y. Yan, Y.Q. Zu (2007) A lattice Boltzmann method for incompressible two-phase flows on partial wetting surface with large density ratio, *Journal of Computational Physics* 227, pp. 763–775
[28] W.W. Yan, Y. Liu, Y.S. Xu, X.L. Yang (2008) Numerical simulation of air flow through a biofilter with heterogeneous porous media, *Bioresource Technology* 99, pp:2156–2161
[29] Y. Zhao, F. Qiu, Z. Fan, A. Kaufman (2007) Flow Simulation with Locally-Refined LBM, *In Proceedings of ACM SIGGRAPH Symposium on Interactive 3D Graphics and Games*, pp:181-188

APPENDIX A

LBM CODE for Simulation Flow in Clear Channel

```
! LBM code for flow simulation in clear channel

parameter (n=240,m=40)
real f(0:8,0:n,0:m),feq(0:8,0:n,0:m),visco(0:n,0:m),omega(0:n,0:m)
real rho(0:n,0:m),rhoo(0:n,0:m),uin(0:m)
real w(0:8),cx(0:8),cy(0:8)
real u(-1:n+1,-1:m+1),v(-1:n,-1:m+1)
integer i,j
cx(:)=(/0.0,1.0,0.0,-1.0,0.0,1.0,-1.0,-1.0,1.0/)
cy(:)=(/0.0,0.0,1.0,0.0,-1.0,1.0,1.0,-1.0,-1.0/)
w(:)=(/4./9.,1./9.,1./9.,1./9.,1./9.,1./36.,1./36.,1./36.,1./36./)
SV=0.0
dx=1.0
dy=dx
Re=50.0

!*******************************************!
! Setting initial values
!*******************************************!
do i=0,n
do j=0,m
rhoo(i,j)=6.0
visco(i,j)=0.02
u(i,j)=0.
```

```
v(i,j)=0.
end do
end do
do j=0,m
uin(j)=0.0
end do
! initial value at inlet
do j=1,m-1
uin(j)=Re*visco(0,j)/m !Re*visco/(channel height)
end do
do j=0,m
u(0,j)=uin(j)
v(0,j)=0.0
end do
!*********************************************!
! setting LBM solution parameters
!*********************************************!
do i=0,n
do j=0,m
omega(i,j)=1./(3.*visco(i,j)+0.5)
rho(i,j)=rhoo(i,j)
end do
end do
mstep=1
savenumber=0
kk=0
!*********************************************!
!main loop
!*********************************************!
1 do while (mstep==1)
call collesion(u,v,f,feq,rho,omega,w,cx,cy,n,m,visco)
call streaming(f,n,m)
call bouncon(f,n,m,uin)
call uvcalc(f,rho,u,v,cx,cy,n,m,visco)

! Showing some parameters to check solution in iterations
print*,kk,"**",u(n/2,m/2),"**",u(n-2,m/2),"**"

! maximum iteration criterion
if (kk==9999) mstep=2
! setting autosave period
if(savenumber==2000) call result(u,v,th,rho,uin,n,m,kk,savenumber);
kk=kk+1
savenumber=savenumber+1

END DO
!*********************************************!
! end of the main loop
!*********************************************!

call result(u,v,th,rho,uin,n,m,kk,savenumber)
stop
```

```
end
!*********************************************!
! end of the main program
!*********************************************!



!*********************************************!
! Subroutine of collesion for FLOW field
!*********************************************!
subroutine collesion(u,v,f,feq,rho,omega,w,cx,cy,n,m,visco)
real f(0:8,0:n,0:m),omega(0:n,0:m)
real feq(0:8,0:n,0:m),rho(0:n,0:m)
real w(0:8),cx(0:8),cy(0:8),visco(0:n,0:m)
real u(0:n,0:m),v(0:n,0:m)
do i=0,n
do j=0,m
t1=u(i,j)*u(i,j)+v(i,j)*v(i,j)
do k=0,8
t2=u(i,j)*cx(k)+v(i,j)*cy(k) !u.c(k)
feq(k,i,j)=rho(i,j)*w(k)*(1.0+3.0*t2+4.5*t2*t2-1.50*t1)
f(k,i,j)=omega(i,j)*feq(k,i,j)+(1.-omega(i,j))*f(k,i,j) !+force
end do
end do
end do
return
end


!*********************************************!
! Subroutine of streaming
!*********************************************!
subroutine streaming(f,n,m)
real f(0:8,0:n,0:m)
! streaming
 do j=0,m
 do i=n,1,-1 ! RIGHT TO LEFT
 f(1,i,j)=f(1,i-1,j)
 end do
 do i=0,n-1 ! LEFT TO RIGHT
 f(3,i,j)=f(3,i+1,j)
 end do
 end do
 do j=m,1,-1 ! TOP TO BOTTOM
 do i=0,n
 f(2,i,j)=f(2,i,j-1)
 end do
 do i=n,1,-1
 f(5,i,j)=f(5,i-1,j-1)
 end do
 do i=0,n-1
 f(6,i,j)=f(6,i+1,j-1)
 end do
 end do
```

```
do j=0,m-1 !BOTTOM TO TOP
do i=0,n
f(4,i,j)=f(4,i,j+1)
end do
do i=0,n-1
f(7,i,j)=f(7,i+1,j+1)
end do
do i=n,1,-1
f(8,i,j)=f(8,i-1,j+1)
end do
end do
return
end


!*********************************************!
! Subroutine of Boundary condition for flow field
!*********************************************!
subroutine bouncon(f,n,m,uin)
real f(0:8,0:n,0:m)
real uin(0:m+m)
! West boundary, V=Uin
do j=0,m
rhow=(f(0,0,j)+f(2,0,j)+f(4,0,j)+2.*(f(3,0,j)+f(6,0,j)+f(7,0,j)))/(1.-uin(j))
f(1,0,j)=f(3,0,j)+2.*rhow*uin(j)/3
f(5,0,j)=f(7,0,j)-(f(2,0,j)-f(4,0,j))/2.+rhow*uin(j)/6.
f(8,0,j)=f(6,0,j)+(f(2,0,j)-f(4,0,j))/2.+rhow*uin(j)/6.
! East Boundary, Open Boundary
f(3,n,j)=4.*f(3,n-1,j)/3-f(3,n-2,j)/3
f(6,n,j)=4.*f(6,n-1,j)/3-f(6,n-2,j)/3
f(7,n,j)=4.*f(7,n-1,j)/3-f(7,n-2,j)/3
end do
do i=0,n
! South Boundary
f(2,i,0)=f(4,i,0)
f(5,i,0)=f(7,i,0)
f(6,i,0)=f(8,i,0)
! North Boundary
f(4,i,m)=f(2,i,m)
f(8,i,m)=f(6,i,m)
f(7,i,m)=f(5,i,m)
end do
return
end


!*********************************************!
! Subroutine of velocity calculation
!*********************************************!
subroutine uvcalc(f,rho,u,v,cx,cy,n,m,visco)
real f(0:8,0:n,0:m),rho(0:n,0:m),u(0:n,0:m),v(0:n,0:m),cx(0:8),cy(0:8)
do j=0,m
do i=0,n
ssum=0.0
```

```
do k=0,8
ssum=ssum+f(k,i,j)
end do
rho(i,j)=ssum
end do
end do
do j=0,m
do i=0,n
usum=0.0
vsum=0.0
do k=0,8
usum=usum+f(k,i,j)*cx(k)
vsum=vsum+f(k,i,j)*cy(k)
end do
u(i,j)=usum/rho(i,j)
v(i,j)=vsum/rho(i,j)
end do
end do
return
end

!*******************************************!
! Subroutine of exporting results
!*******************************************!
subroutine result(u,v,th,rho,uin,n,m,kk,savenumber)
real u(0:n,0:m),v(0:n,0:m),rho(-1:n+1,-1:m+1)
real strf(0:n,0:m),uin(0:m)
CHARACTER FILOUT1*18
CHARACTER FILOUT2*18
CHARACTER FILOUT3*18
WRITE(FILOUT1,'(4HUVTS,I8,4H.lbm)')kk
WRITE(FILOUT2,'(4HUOUT,I8,4H.lbm)')kk


! Streamfunction Calculations
strf(0,0)=0.
do i=0,n
rhoav=0.5*(rho(i-1,0)+rho(i,0))
if(i.ne.0) strf(i,0)=strf(i-1,0)-rhoav*0.5*(v(i-1,0)+v(i,0))
do j=1,m
rhom=0.5*(rho(i,j)+rho(i,j-1))
strf(i,j)=strf(i,j-1)+rhom*0.5*(u(i,j-1)+u(i,j))
end do
end do
! Exporting velocity field
OPEN(2,FILE=FILOUT1)
write(2,*)"VARIABLES=X,Y,U,V,StreamF"
write(2,*)"ZONE ","I=",n+1,"J=",m+1,",","F=BLOCK"
do j=0,m
write(2,*)(i/float(m),i=0,n)
end do
do j=0,m
```

```
write(2,*)(j/float(m),i=0,n)
end do
do j=0,m
write(2,*)(u(i,j),i=0,n)
end do
do j=0,m
write(2,*)(v(i,j),i=0,n)
end do
do j=0,m
write(2,*)(strf(i,j),i=0,n)
end do


!
OPEN(3,FILE=FILOUT2)

! Exporting velocity profile near to outlet divided to inlet velocity
do j=0,m
write(3,*)(j/float(m),u(n-2,j)/u(0,m/2))
end do

savenumber=0.
return
end
```

APPENDIX B

LBM CODE for Free Convection in Cavity

```
! LBM code for flow and thermal fields in Free convection


parameter (n=90,m=90)
real f(0:8,0:n,0:m),feq(0:8,0:n,0:m),visco(0:n,0:m),omega(0:n,0:m)
real rho(0:n,0:m),rhoo(0:n,0:m),uin(0:m)
real g(0:8,0:n,0:m),geq(0:8,0:n,0:m),th(0:n,0:m)
real alpha(0:n,0:m),gbeta(0:n,0:m),omegat(0:n,0:m)
real w(0:8),cx(0:8),cy(0:8)
real u(-1:n+1,-1:m+1),v(-1:n,-1:m+1)
integer i,j
cx(:)=(/0.0,1.0,0.0,-1.0,0.0,1.0,-1.0,-1.0,1.0/)
cy(:)=(/0.0,0.0,1.0,0.0,-1.0,1.0,1.0,-1.0,-1.0/)
w(:)=(/4./9.,1./9.,1./9.,1./9.,1./9.,1./36.,1./36.,1./36.,1./36./)
ra=1.e6
SV=0.0
dx=1.0
dy=dx
tw1=1.0
tw2=0
thref=((tw1+tw2)/2)
```

```
dt=(tw1-tw2)
pr=.71
!********************************************!
! Setting initial values
!********************************************!
do i=0,n
do j=0,m
rhoo(i,j)=6.0
visco(i,j)=0.02
u(i,j)=0.
v(i,j)=0.
th(i,j)=0.0
end do
end do
do i=0,n
u(i,m)=0.0
v(i,m)=0.0
end do


!********************************************!
! setting LBM solution parameters
!********************************************!
do i=0,n
do j=0,m
alpha(i,j)=visco(i,j)/pr
rho(i,j)=rhoo(i,j)
end do
end do
do i=0,n
do j=0,m
omega(i,j)=1./(3.*visco(i,j)+0.5)
end do
end do
do i=0,n
do j=0,m
omegat(i,j)=1./(3.*alpha(i,j)+0.5)
end do
end do
do i=0,n
do j=0,m
gbeta(i,j)=ra*visco(i,j)*alpha(i,j)/(float(m*m*m)) ! Attention required
end do
end do

mstep=1 !
savenumber=0
kk=0

!********************************************!
!main loop
!********************************************!
!main loop
```

```
1 do while (mstep==1) !kk=1,mstep
call collesion(u,v,f,feq,rho,omega,w,cx,cy,n,m,th,gbeta,visco,thref)
call streaming(f,n,m)
call bouncon(f,n,m)
call uvcalc(f,rho,u,v,n,m,cx,cy)
! -----------------------------
!collesion for th
call colls(u,v,g,geq,th,omegat,w,cx,cy,n,m)
!streaming for th
call streaming(g,n,m)
call gbouncon(g,tw1,tw2,w,n,m)
call thcalcu(g,th,n,m)

! Showing some parameters to check solution in iterations
print*,kk,"**",th(n/2,m/2),"**",u(n/2,m/2),"**"

! maximum iteration criterion
if (kk==99999) mstep=2
! setting autosave period
if(savenumber==5000) call result(u,v,th,rho,n,m,kk,ra,pr,savenumber,tw1,tw2,dt,thref)

kk=kk+1
savenumber=savenumber+1
errmaxu=0.0
errmaxsc=0.0

END DO
!********************************************!
! end of the main loop
!********************************************!

call result(u,v,th,rho,n,m,kk,ra,pr,savenumber,tw1,tw2,dt,thref)
stop
end


!********************************************!
! Subroutine of collesion for FLOW field
!********************************************!
subroutine collesion(u,v,f,feq,rho,omega,w,cx,cy,n,m,th,gbeta,visco,thref)
real f(0:8,0:n,0:m),omega(0:n,0:m),gbeta(0:n,0:m)
real feq(0:8,0:n,0:m),rho(0:n,0:m),th(0:n,0:m)
real w(0:8),cx(0:8),cy(0:8),visco(0:n,0:m)
real u(0:n,0:m),v(0:n,0:m)
do i=0,n
do j=0,m
t1=u(i,j)*u(i,j)+v(i,j)*v(i,j)
do k=0,8
t2=u(i,j)*cx(k)+v(i,j)*cy(k) !u.c(k)
force=3.0*w(k)*gbeta(i,j)*((th(i,j)-thref)*cy(k)*rho(i,j))
if(i.eq.0.or.i.eq.n) force=0.0 !Attention required
if(j.eq.0.or.j.eq.m) force=0.0 !Attention required
feq(k,i,j)=rho(i,j)*w(k)*(1.0+3.0*t2+4.5*t2*t2-1.50*t1)
```

```
f(k,i,j)=omega(i,j)*feq(k,i,j)+(1.-omega(i,j))*f(k,i,j)+force
end do
end do
end do
return
end


!*********************************************!
! Subroutine of collesion for Thermal field
!*********************************************!
subroutine colls(u,v,g,geq,th,omegat,w,cx,cy,n,m)
real g(0:8,0:n,0:m),geq(0:8,0:n,0:m),th(0:n,0:m)
real w(0:8),cx(0:8),cy(0:8),omegat(0:n,0:m)
real u(0:n,0:m),v(0:n,0:m)
do i=0,n
do j=0,m
do k=0,8
geq(k,i,j)=th(i,j)*w(k)*(1.0+3.0*(u(i,j)*cx(k)+v(i,j)*cy(k)))
g(k,i,j)=omegat(i,j)*geq(k,i,j)+(1.0-omegat(i,j))*g(k,i,j)
end do
end do
end do
return
end


!*********************************************!
! Subroutine of streaming
!*********************************************!
subroutine streaming(f,n,m)
real f(0:8,0:n,0:m)
! streaming
 do j=0,m
 do i=n,1,-1 ! RIGHT TO LEFT
 f(1,i,j)=f(1,i-1,j)
 end do
 do i=0,n-1 ! LEFT TO RIGHT
 f(3,i,j)=f(3,i+1,j)
 end do
 end do
 do j=m,1,-1 ! TOP TO BOTTOM
 do i=0,n
 f(2,i,j)=f(2,i,j-1)
 end do
 do i=n,1,-1
 f(5,i,j)=f(5,i-1,j-1)
 end do
 do i=0,n-1
 f(6,i,j)=f(6,i+1,j-1)
 end do
 end do
 do j=0,m-1 !BOTTOM TO TOP
 do i=0,n
```

```fortran
f(4,i,j)=f(4,i,j+1)
end do
do i=0,n-1
f(7,i,j)=f(7,i+1,j+1)
end do
do i=n,1,-1
f(8,i,j)=f(8,i-1,j+1)
end do
end do
return
end


!*********************************************!
! Subroutine of Boundary condition for flow field
!*********************************************!
subroutine bouncon(f,n,m)
real f(0:8,0:n,0:m)!,feq(0:8,0:n,0:m)
real uin(0:m+m)
! West boundary, Bounce Back
do j=0,m
f(1,0,j)=f(3,0,j)
f(5,0,j)=f(7,0,j)
f(8,0,j)=f(6,0,j)
! East Boundary, Bounce Back
f(3,n,j)=f(1,n,j)
f(6,n,j)=f(8,n,j)
f(7,n,j)=f(5,n,j)
end do
do i=0,n
! South Boundary, Bounce Back
f(2,i,0)=f(4,i,0)
f(5,i,0)=f(7,i,0)
f(6,i,0)=f(8,i,0)
! North Boundary, Bounce Back
f(4,i,m)=f(2,i,m)
f(8,i,m)=f(6,i,m)
f(7,i,m)=f(5,i,m)
end do
return
end


!*********************************************!
! Subroutine of Boundary condition for Thermal field
!*********************************************!
subroutine gbouncon(g,tw1,tw2,w,n,m)
real g(0:8,0:n,0:m),geq(0:8,0:n,0:m)
real w(0:8),tw1,tw2

! Boundary Conditions
! West Boundary Condition, T=1
do j=0,m
g(1,0,j)=tw1*(w(1)+w(3))-g(3,0,j)
```

```
g(5,0,j)=tw1*(w(5)+w(7))-g(7,0,j)
g(8,0,j)=tw1*(w(8)+w(6))-g(6,0,j)
end do
! East Boundary Condition, T=0
do j=0,m
g(6,n,j)=tw2*(w(8)+w(6))-g(8,n,j)
g(3,n,j)=tw2*(w(1)+w(3))-g(1,n,j)
g(7,n,j)=tw2*(w(5)+w(7))-g(5,n,j)
end do
! Top Boundary Condition, Adiabatic
do i=0,n
do k=0,8
g(k,i,m)=g(k,i,m-1)
end do
end do
! Bottom Boundary Condition, Adiabatic
do i=0,n
do k=0,8
g(k,i,0)=g(k,i,1)
end do
end do
return
end


!*******************************************!
! Temperature calculation
!*******************************************!
subroutine thcalcu(g,th,n,m)
real g(0:8,0:n,0:m),th(0:n,0:m)
do j=0,m
do i=0,n
thsum=0.0
do k=0,8
thsum=thsum+g(k,i,j)
end do
th(i,j)=thsum
end do
end do
return
end


!*******************************************!
! Subroutine of velocity calculation
!*******************************************!
subroutine uvcalc(f,rho,u,v,n,m,cx,cy)
real f(0:8,0:n,0:m),rho(0:n,0:m),u(0:n,0:m),v(0:n,0:m)
real cx(0:8),cy(0:8)
do j=0,m
do i=0,n
uvsum=0.0
do k=0,8
uvsum=uvsum+f(k,i,j)
```

```
end do
rho(i,j)=uvsum
end do
end do
do j=0,m
do i=0,n
usum=0.0
vsum=0.0
do k=0,8
usum=usum+f(k,i,j)*cx(k)
vsum=vsum+f(k,i,j)*cy(k)
end do
u(i,j)=usum/rho(i,j)
v(i,j)=vsum/rho(i,j)
end do
end do
return
end

!*********************************************!
! Subroutine of exporting results
!*********************************************!
subroutine result(u,v,th,rho,n,m,kk,ra,pr,savenumber,tw1,tw2,dt,thref)
real u(0:n,0:m),v(0:n,0:m),rho(-1:n+1,-1:m+1),th(0:n,0:m)
real strf(0:n,0:m)
real tw1,tw2,dt,thref
real tt(0:n,0:m)
CHARACTER FILOUT1*18
CHARACTER FILOUT2*18
WRITE(FILOUT1,'(4HUVTS,I8,4H.lbm)')kk
WRITE(FILOUT2,'(4HNUAV,I8,4H.txt)')kk
! Streamfunction Calculations
strf(0,0)=0.
do i=0,n
rhoav=0.5*(rho(i-1,0)+rho(i,0))
if(i.ne.0) strf(i,0)=strf(i-1,0)-rhoav*0.5*(v(i-1,0)+v(i,0))
do j=1,m
rhom=0.5*(rho(i,j)+rho(i,j-1))
strf(i,j)=strf(i,j-1)+rhom*0.5*(u(i,j-1)+u(i,j))
end do
end do

! Exporting velocity and thermal fields
OPEN(2,FILE=FILOUT1)
write(2,*)"VARIABLES=X,Y,U,V,th,StreamF,rho"
write(2,*)"ZONE ","I=",n+1,"J=",m+1,",","F=BLOCK"
do j=0,m
write(2,*)(i/float(m),i=0,n)
end do
do j=0,m
write(2,*)(j/float(m),i=0,n)
end do
```

```
do j=0,m
write(2,*)(u(i,j),i=0,n)
end do
do j=0,m
write(2,*)(v(i,j),i=0,n)
end do
do j=0,m
write(2,*)(th(i,j),i=0,n)
end do
do j=0,m
write(2,*)(strf(i,j),i=0,n)
end do
do j=0,m
write(2,*)(rho(i,j),i=0,n)
end do

OPEN(3,FILE=FILOUT2)
! Nusselt Number Calculations
snul=0.0
snur=0.0
snuu=0.0
snud=0.0
do j=0,m
rnul=(th(0,j)-th(1,j))*float(m)
snul=snul+rnul
rnur=(th(n-1,j)-th(n,j))*float(m)
snur=snur+rnur
end do

do i=0,n
rnuu=(th(i,m)-th(i,m-1))*float(n)
snuu=snuu+rnuu
rnud=(th(i,0)-th(i,1))*float(n)
snud=snud+rnud
end do
avnl=snul/float(m)
avnr=snur/float(m)
avnu=snuu/float(n)
avnd=snud/float(n)
write(3,*)"(Ra=",ra,")**(Pr=",pr,")"
write(3,*)"(Ave Nu-Left=",avnl,")**(Ave Nu-right=",avnr,")**(Ave Nu=",(avnr+avnl)/2,")"
write(3,*)"(Ave Nu-Up=",avnu,")**(Ave Nu-Down=",avnd,")**(Ave Nu=",(avnu+avnd)/2,")"
write(3,*)"(T(hot)=",tw1,")**(T(cold)=",tw2,")**(T(ref)=",thref,")**(Delta-T=",dt,")"
write(3,*)"(Number of iteration=)",kk
savenumber=0.
return
end
```

In: Computational Fluid Dynamics
Editor: Alyssa D. Murphy

ISBN 978-1-61209-276-8
© 2011 Nova Science Publishers, Inc.

Chapter 18

MOVING MESH INTERFACE TRACKING

Shaoping Quan[*]
Large-Scale Complex Systems, Institute of High Performance Computing,
A*STAR, 1 Fusionopolis Way #16-16 Connexis, Singapore

ABSTRACT

In multiphase flow simulations, there are a number of approaches that can be used to capture the interfacial dynamics as well as the surrounding fluid flow. Front Tracking, Volume of Fluid, Immersed Boundary, Lattice Boltzmann, and Boundary Integral are only a few examples. The fluid properties are usually smoothed or smeared in the above methods. In this chapter, a newly developed method, Moving Mesh Interface Tracking (MMIT) is introduced. This method treats the interface as a surface mesh, and this surface mesh connects the interior volume element into a single mesh, and then the Navier-Stokes equations are solved on this single mesh. The interface mesh moves with the fluid velocity. Thus, this method has a zero-thickness interface and naturally conserves the total mass of each phase. The jumps in fluid properties and boundary conditions across the interface are directly implemented without smoothing or smearing of the fluid properties. The interface mesh moves in a Lagrangian fashion, while the interior nodes move by a smoothing approach. This motion usually does not guarantee good mesh quality, especially for large deformation. Therefore, mesh adaptations including 3-2 and 2-3 swapping, 4-4 flipping, edge bisection, and edge contraction, are implemented to achieve good mesh quality as well as to obtain computational efficiency. Mesh separation and mesh combination are introduced to handle topological transitions such as droplet pinch-off and interface merging.

This method has been validated against a number of theoretical predictions and experimental observations. The mesh adaptation schemes are capable of dealing with large deformation as well as achieving computation efficiency. The mesh separation and mesh combination are robust in simulations of droplet pinch-off and droplets colliding. The simulations also demonstrate the great potential of this method to investigate the detailed physics for small scales such as the thin film formed during the droplet near contact motion, and the necking thread in droplet pinch-off.

---

[*] Email: quansp@ihpc.a-star.edu.sg

1. INTRODUCTION

Numerical methods for simulating multiphase flows, such as droplets/bubbles in another fluid flow, have been significantly developed over the last several decades. The numerical approach for multiphase flows is a great alternative for experiments where the time scales and/or length scales are tiny, such as in spray and combustion. In general, most of these methods can be fitted into two categories.

In the first category, the governing equations for the two fluids are usually solved in a fixed background mesh or Cartesian grid, where the interface between the two fluids is captured or tracked by employing some functions or markers. One of the most popular methods in this category is the volume of fluid (VOF) method. In VOF, the Navier-Stokes equations are solved in a fixed Cartesian grid, and the interface is tracked using a volume fraction function (also called color function). VOF naturally conserves the total mass of each phase, as the color function is advected in a conserved fashion. However, the interface, which is constructed from the volume fraction either by simple line interface calculation (SLIC) or piecewise linear interface construction (PLIC), is usually not smooth, and does not guarantee the correct shape. Therefore, the calculation of the surface tension (interfacial curvature) might have errors. There is a great volume of literature on this method, a comprehensive review of VOF can be found in Scardovelli and Zaleski (1999). Another popular method is level set. Similar to VOF, the governing equations are solved on a fixed Cartesian grid. However, the interface is captured by a level set function. The interface is not smeared; however, the jumps in fluid properties and the jump in normal stresses are smoothed in an interfacial region whose width is normally several cells. The total mass of each phase is usually not conserved as the discrete level set function is not in a conserved form. Topological changes are usually handled automatically in VOF and level set. Detailed reviews of this method can be found in Osher and Fedkiw (2001). Using a linked list of elements or points to represent the interface, the front-tracking method is developed to solve multiphase flows. Similar to the above two methods, the fluid properties and the jump in pressure due to surface tension are smoothed in an interfacial region. However, as the interface is explicitly tracked, the surface tension can be calculated more accurately. The details of the method were described in the previous chapter, and can also be found in Unverdi and Tryggvason (1992). Lattice Boltzmann methods solve the discrete Boltzmann equations. The interface is usually captured by the density field. The next chapter will review this method in details.

The second category is moving mesh/grid approaches. The boundary integral method belongs to this category. A boundary integral formula is derived for fluid problems, and then the equation is solved on the boundary elements. The boundary elements evolve with the fluid velocity on the boundary, and thus the interface is zero-thickness. Boundary integral methods accurately and efficiently capture the interface dynamics; however, they are limited to viscous or potential flows. The newly developed moving mesh interface tracking method (MMIT) (Quan and Schmidt 2007; Quan et al. 2009) not only treats the interface as zero-thickness, but also applies to more general fluid regimes. MMIT has also been applied for simulations of free surface flows (Perot and Nallapati 2003; Dai and Schmidt 2005). In the MMIT method, the interface is represented as the surface triangular elements of the interior volume mesh. These triangular elements connect the interior meshes of the both phases into one single mesh. This single mesh is used to discretize the governing equations. As the interface is zero-thickness, the jumps in the fluid properties and jumps in the normal stresses are implemented directly without any smoothing. The interfacial nodes move with the fluid, and thus MMIT naturally conserves the total mass of each phase. However, there are challenges for this method. These challenges are: 1) as the mesh moves, the mesh deforms and the quality of mesh becomes worse, which results in possible discontinuation of the simulation; 2) extra efforts are needed to handle topological transitions such as droplet pinch-off and coalescence.

In this chapter, the governing equations and the numerical method to solve the equations for immiscible, incompressible multiphase fluid flows are discussed. The mesh adaptations which are employed locally to achieve good mesh quality and computing efficiency are described. To address the challenges with regards to topological changes, mesh separation and mesh combination are also addressed. Validations and applications of the method are also presented.

2. Governing Equations

2.1. Integral Form of Navier-Stokes Equations

In multiphase flows, the interface is moving with the fluid, so the interface elements are the natural control surfaces for the control volume on the interface. Therefore, the integral form of the Navier-Stokes equations for a moving and deforming control volume is used in the MMIT method, and the equations are

$$\frac{d}{dt}\iiint_{CV} dv = \iint_{CS} \mathbf{v} \cdot \mathbf{n}\, ds, \qquad (2.1)$$

$$\frac{d}{dt}\iiint_{CV} \rho dv + \iint_{CS} \rho(\mathbf{u} - \mathbf{v}) \cdot \mathbf{n} ds = 0, \qquad (2.2)$$

$$\frac{d}{dt}\iiint_{CV} \rho \mathbf{u} dv + \iint_{CS} \rho \mathbf{u}(\mathbf{u} - \mathbf{v}) \cdot \mathbf{n} ds = \iiint_{CV} \rho \mathbf{f} dv - \iint_{CS} p \mathbf{n} ds + \iint_{CS} \mu(\nabla \mathbf{u} + \nabla \mathbf{u}^T) \cdot \mathbf{n} ds \qquad (2.3)$$

where CV stands for a control volume and CS for the surfaces of the corresponding CV, $\mathbf{u}$ and $\mathbf{v}$ denote the fluid velocity and moving mesh velocity, respectively, $\mathbf{n}$ is the unit normal vector of CS pointing outward to the surrounding fluid, $\mathbf{f}$ denotes the body force per unit mass (usually gravitational forces), p is the pressure, ρ denotes the density and μ the viscosity, and the superscript T stands for the transpose. It should be noted that Eq. (2.1) is derived from Leibnitz's theorem and it tells that the volume change of a control volume should be balanced by the surface integral of moving mesh velocities. The only difference in the equations of the conservation of mass and linear momentum compared to the normally used ones lies in the convection terms, where a term regarding the moving mesh velocity is presented. In MMIT, **u** = **v** is required to conserve the total mass of each phase, so the convection terms in (2.2) and (2.3) are identically zero on the interface. Therefore, the jump in the fluid's density across the interface presents no numerical difficulties for the MMIT method. However, in MMIT, the moving mesh velocity does not necessary equal the fluid velocity for the region other than the interface.

2.2. Jump Conditions across the interface

It is usually the case that the properties of two fluids (such as the density and viscosity) are much different, for example, air bubbles in water or water droplets in air, the difference of the fluid properties is on the order of 100 to 1000. This huge difference across the zero-thickness interface results in jump and continuity conditions across the interface, and these conditions should be always satisfied in order to accurately solve the problem. These conditions can be derived from the governing equations by assuming that the neighboring control volumes on the interface are collapsing to the interface surface, and thus the volume is approaching zero, and in a limit, it becomes zero. The details can be found in (Quan and Schmidt 2006).

In the conservation of mass (Eq. 6.2.2), the only term remained is the second term, i.e. the surface integral term or the convection term. As the surface area is the same and not equal to zero, the net mass flux must be zero in order to satisfy the equation. Then

$$\rho_1(\mathbf{u}_1 - \mathbf{v}) \cdot \mathbf{n}_1 + \rho_2(\mathbf{u}_2 - \mathbf{v}) \cdot \mathbf{n}_2 = 0. \tag{2.4}$$

by considering the unit normal vectors of the two interfacial surfaces are opposite in direction, i.e. $\mathbf{n}_1 = -\mathbf{n}_2$, one has

$$\rho_1(u_{1n} - v_n) - \rho_2(u_{2n} - v_n) = 0, \tag{2.5}$$

where u_{1n} and u_{2n} are the normal component of the fluids' velocities, and v_n is denotes the normal velocity of the moving mesh. By considering the two fluids are immiscible, one finally has,

$$u_{1n} = u_{2n} = v_n, \tag{2.6}$$

as the two fluids are viscous, the no-slip condition must be enforced everywhere on the interface, i.e.

$$u_{1t} = u_{2t}, \tag{2.7}$$

where the subscript t stands for the tangential component.

In the conservation of linear momentum (Eq.2.3), the volume integral terms are zero as the two interfacial volumes collapse to the interface. The convection term is also zero because it was shown that **u** = **v** on the interface. Then this equation becomes

$$-\iint_{CS} p\mathbf{n}ds + \iint_{CS} \mu(\nabla\mathbf{u} + \nabla\mathbf{u}^T) \cdot \mathbf{n}ds = 0. \tag{2.8}$$

by taking the normal component of the above equation, one can have

$$-\iint_{CS} p\mathbf{n}\cdot\mathbf{n}ds + \iint_{CS} [\mu(\nabla\mathbf{u} + \nabla\mathbf{u}^T) \cdot \mathbf{n}]\cdot\mathbf{n}ds = 0. \tag{2.9}$$

then one obtains

$$[\![p]\!] = [\![2\mu(\nabla\mathbf{u} \cdot \mathbf{n}) \cdot \mathbf{n}]\!], \tag{2.10}$$

where $[\![\cdot]\!]$ denotes the difference across the interface. It should be noted that when the two interfacial cells are collapsing to the interface, there are two interfacial surfaces left with the normal vectors in opposite direction. By considering the Laplace's formula for the surface tension forces,

$$[\![p]\!] = -\sigma\left(\frac{1}{R_1} + \frac{1}{R_2}\right). \tag{2.11}$$

one finally gets,

$$[\![p]\!] = -\sigma\left(\frac{1}{R_1} + \frac{1}{R_2}\right) + [\![2\mu(\nabla\mathbf{u} \cdot \mathbf{n}) \cdot \mathbf{n}]\!], \tag{2.12}$$

which is the jump condition of the normal stresses across the interface. Here, σ is the surface tension coefficient, and R_1 and R_2 are the two principle radii of the surface.

If the tangential component of Eq. 2.8 is considered, then the continuity of the shear stresses across the interface is expressed as,

$$[\![\mu(\nabla\mathbf{u} + \nabla\mathbf{u}^T) \cdot \mathbf{n} \cdot \mathbf{t}]\!] = 0 \tag{2.13}$$

where *t* is the unit tangential vector.

3. Numerical Methods

3.1. Discretization of the Governing Equations

As the interface is moving and deforming, and the interface is an ideal control surface, the finite volume method is very suitable for multiphase flow problems. The discretization of the governing equations in a strongly conserved form (Eqs. 2.1, 2.2, and 2.3) for incompressible, immiscible two-phase flows is

$$\frac{V_c^{n+1} - V_c^n}{\Delta t} = \sum_{cell\ faces} U_f^{mesh}, \qquad (3.1)$$

$$\frac{\rho_c^{n+1} V_c^{n+1} - \rho_c^n V_c^n}{\Delta t} + \sum_{cell\ faces} \rho_f (U_f - U_f^{mesh}) = 0, \qquad (3.2)$$

which can be found in (Zhang et al. 2002; Dai et al. 2002; Perot and Nallapati 2003; Dai and Schmidt 2005; Quan and Schmidt 2007), and the detailed solution technique can also be found in the above mentioned literature. Here, V_c is the cell volume, U_{mesh} denotes the face flux due to the motion of the mesh, U_f stands for the face flux of the fluid, $\mathbf{u}_c$ is the cell centroid velocity, $\mathbf{u}_f$ is the velocity located on the cell surface center, $\mathbf{g}$ denotes the gravitational acceleration, $\mathbf{r}_c^{CG}$ stands for the position vector of the cell centroid, A_f is the surface area, and the subscripts f and c denote the surface and the cell. It should be noted, gravitation force is treated by employing

$$\nabla(\rho_c \mathbf{g} \cdot \mathbf{r}_c^{CG}) = \mathbf{g} \cdot \mathbf{r}_c^{CG} \nabla \rho_c + \rho_c \mathbf{g} \nabla \mathbf{r}_c^{CG} \qquad (3.4)$$

since $\nabla \mathbf{r}_c^{CG}$ is an identity matrix of size three, the above equation can be rewritten as

$$\rho_c \mathbf{g} = \nabla(\rho_c \mathbf{g} \cdot \mathbf{r}_c^{CG}) - \mathbf{g} \cdot \mathbf{r}_c^{CG} \nabla \rho_c \qquad (3.5)$$

3.2. Moving Mesh Scheme

In the moving mesh interface tracking method, the interface nodes move with the corresponding fluid velocity, i.e.

$$\frac{d\mathbf{x}_n}{dt} = \mathbf{u}_n \qquad (3.6)$$

where the node velocity $\mathbf{u}_n$ can be computed by the average of the surrounding cells' velocities, and $\mathbf{x}_n$ denotes the interface node position. The moving mesh scheme is displayed in Figure 3.1. Here, only a two-dimensional triangular surface for two consecutive time steps, namely n and n+1, is shown, and the thick line denotes the interface between phase 1 and phase 2 (with density and viscosity as ρ_1, μ_1 and ρ_2, μ_2). However, the interior nodes do not move in a Lagrangian fashion as the interfacial nodes, because the Lagrangian motion of both interior and interface nodes can cause highly distorted elements. Instead, the interior mesh moves in a way to achieve good mesh quality, like in a smoothing fashion, where the mesh velocity is not known a priori, thus the moving mesh face flux U_{mesh} needs to be calculated from the displacement of the neighboring nodes (Perot and Nallapati 2003)

$$U_f^{mesh} = \frac{\mathbf{x}_f^{n+1} - \mathbf{x}_f^n}{\Delta t} \cdot \left[\frac{1}{2}(\mathbf{n}_f^{n+1} A_f^{n+1} + \mathbf{n}_f^n A_f^n) - \frac{\Delta t^2}{12} \sum_{face\ edges} (\mathbf{v}_{n1} \times \mathbf{v}_{n2}) \right], \quad (3.7)$$

where $\mathbf{v}_{n1}$ and $\mathbf{v}_{n2}$ are the moving mesh velocities at nodes 1 and 2.

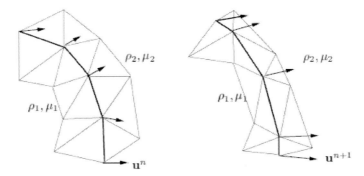

Figure 3.1. Schematic of moving mesh interface tracking scheme. The mesh and the interfacial velocities at time step n (left) and n+1 (right).

3.3. Shear Stress Calculation

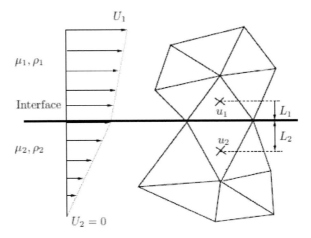

Figure 3.2. Neighboring cells of the interface. The velocity distribution is a two-layer Couette flow. For simplicity, Two-dimensional triangular elements are shown.

The shear stresses for the interior surfaces in multiphase flow simulations can be computed as the ones in single-phase flows, i.e. by doing the product of the velocity gradients and the viscosity on the surface. More specifically for a typical finite volume method, the velocity gradient is computed using a central difference scheme. However, the viscosity on the interface is not defined. The only condition for the shear stresses across the interface is the continuity requirement as in Eq. 2.13.

There are a number of ways to define the interface viscosity; the simplest one is averaging of the two fluids' viscosities. i.e. an effective viscosity on the interface is defined as

$$\mu_e^{aa} = \frac{\mu_1 + \mu_2}{2}. \tag{3.8}$$

another way is using harmonic mean, and an effective viscosity is expressed as

$$\mu_e^{hm} = \frac{2}{\frac{1}{\mu_1} + \frac{1}{\mu_2}}. \tag{3.9}$$

it is noticed that in the above two approaches; the dimensions of the neighboring cells are not considered. A geometric harmonic mean method is introduced by Quan and Schmidt (2007), and similar scheme was developed by Kang et al. (2000), however in a different scenario.

$$\mu_e^{gh} = \frac{1}{\frac{\delta_1}{\mu_1} + \frac{\delta_2}{\mu_2}}, \tag{3.10}$$

where δ_1 and δ_2 are the ratios between the distances from the two neighboring, interfacial cells' centroids to the total distances between the two cells' centroids, more specifically, they are defined as $\delta_1 = L_1/(L_1+L_2)$ and $\delta_2 = L_2/(L_1+L_2)$ as shown in Figure 3.2.

With the definition of interfacial viscosity, the shear stress on the interface is computed by

$$\tau = \mu_e \frac{u_1 - u_2}{L_1 + L_2}. \tag{3.11}$$

Another method for calculating the interfacial shear stress is based on the continuity condition for the shear stress. This method needs no averaging of the viscosities, but uses a least-squares fitting method to compute the shear stress on each side of the interface, and then takes the average of the two shear stresses as the interfacial shear stress. The averaging is used to reduce errors.

The accuracy and stability of the above methods were analyzed theoretically and numerically in the work of Quan and Schmidt (2007). The geometric harmonic mean and harmonic mean methods calculate the interfacial shear stress with higher fidelity. The error of averaging the viscosity is unacceptable for large viscosity ratios. Although, the least-squares fitting method is the most accurate one, it is not stable.

3.4. Jump in Normal Stresses

As shown in the previous section, there are jumps in normal stresses across the interface, which includes the jumps in both normal viscous stresses and the surface tension forces. The way of treating these jumps has been discussed in Quan and Schmidt (2007) and Dai and Schmidt (2005) in great detail, so only a summary is given here.

To calculate the jump in viscous normal stresses across the interface, the key is to obtain the velocity gradient in the face normal direction. To compute the gradient, a local coordinate system is set up with the face normal as one axis, one of the edge direction as another axis, the cross-product of these two vectors as the third one. Then, the neighboring cell-centroid velocities are projected to this coordinate system. Using a Taylor series expansion for a multivariable function, the normal velocity gradients of each phase can be calculated. Then the jump in viscous normal stresses across the interface is computed as

$$JUMP_{visc} = 2(\mu_1 D_1 - \mu_2 D_2), \tag{3.12}$$

where D_1 and D_2 are the velocity gradients in the face normal direction for phase 1 and 2, respectively.

As the interface is tracked by the interfacial surfaces, the curvature field is directly calculated using a least-squares parabola fitting method in a local coordinate system, which is similar to the local coordinate mentioned above. The parabola is fitted in a form of

$$z(x, y) = ax^2 + by^2 + cxy + dx + ey + f, \tag{3.13}$$

where z is the value in the face normal direction, and x and y are in the tangential directions. Using the surrounding node positions, the unknowns (a, b, c, d, e, f) can be computed. Then the curvature is calculated

$$k = \frac{1}{R_1} + \frac{1}{R_2} = -\frac{a(1+e^2) - 2cde + b(1+d^2)}{2(1+d^2+e^2)^{\frac{3}{2}}}. \tag{3.14}$$

4. MESH ADAPTATIONS

In the MMIT method, the interface nodes move with the fluid velocity, and the interior nodes moves in a way to achieve good mesh quality. However, as time progresses there are elements for which the quality becomes worse and the above motion cannot improve the quality any more. In such cases, mesh adaptations, in which the connectivity of the mesh needs to be changed, are necessary to restore good mesh quality. Also, in order to capture the details of fluid flow where the interfacial curvature variation is huge, local mesh refinement is needed. On the other hand, mesh coarsening is also necessary to remove small elements which do not help to resolve the physics and may limit the time step. In the previous work (Dai and Schmidt 2005; Quan and Schmidt 2007), a number of mesh adaptation schemes are introduced and are proved to be robust to handle large deformation and to achieve good mesh quality as well as obtain computing efficiency. Here only a summary of these schemes is presented.

4.1. Mesh Quality Metrics

A number of mesh quality criteria are employed to decide which element needs mesh adaptation, such as the dihedral angle of two neighboring triangle surfaces in a tetrahedron, the line angle between two edges, and also a quality metric proposed by Knupp (2003). The dihedral angle can be computed by the dot-product of the two face-unit-normal vectors, and the line angle is also calculated by the dot-product of the two edge directions. The quality metric for a triangle element by Knupp (2003) is defined as

$$Qt_s = \frac{\sqrt{3}\alpha}{\lambda_{11} + \lambda_{22} + \lambda_{12}}, \qquad (4.1)$$

where λ_{11} and λ_{22} stand for the lengths of the edges connected to a common node, and α is

$$\alpha = \sqrt{\lambda_{11}\lambda_{22} \sin^2 \vartheta}. \qquad (4.2)$$

here θ is the edge angle between the two edges of the triangular surfaces, and it can be seen that α is actually the area of the triangle surface. The quality metric for a tetrahedron is

$$Qt_v = \frac{3(\alpha\sqrt{2})^{2/3}}{\frac{3}{2}(\lambda_{11} + \lambda_{22} + \lambda_{33}) - (\lambda_{12} + \lambda_{13} + \lambda_{23})}, \qquad (4.3)$$

where α stands for the volume of the tetrahedron, λ_{11}, λ_{22}, and λ_{33} denote the square of the length of the three edges connected to a common node of the tetrahedron, while λ_{12}, λ_{23} and λ_{13} are the square of the length of the other three edges which do not share the common node. An optimization-based smoothing method using the above quality metrics is developed by Dai and Schmidt (2005).

4.2. 2-3/3-2 Swapping

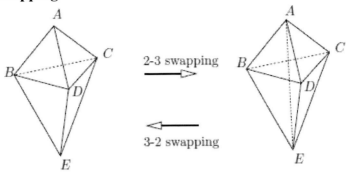

Figure 4.1 2-3 and 3-2 swapping.

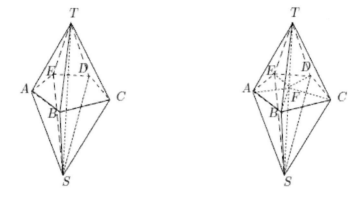

Figure 4.2. Edge bisection. Node F, the newly inserted node.

For interior edges, 2-3 or 3-2 swapping is applied to obtain better mesh quality, and the scheme is shown in Figure 4.1. For 2-3 swapping, initially, there are two tetrahedral sharing one common triangle surface, i.e. triangle *BCD*. Then node *A* and node *E* are connected and a new edge *AE* is created. The newly created three tetrahedra are *AEBD*, *AEDC*, and *AEBC*. Edge *AE* is the common edge for the three elements. On the other hand, for 3-2 swapping, Edge *AE* is deleted, and thus node *A* and *E* are disconnected. Finally, the number of the elements is reduced from 3 to 2. The criteria for applying these two kinds of swapping are based on mesh quality, if the new configuration improves mesh qualities, and then the swapping is performed. It should be noted that 2-3 swapping is only allowed for any edges (*BC*, *CD*, and *DB*) if the common surface is not convex, and the 3-2 swapping can only be applied to the common edge which has exactly three sharing tetrahedra. For the details readers are referred to (Dai and Schmidt 2005).

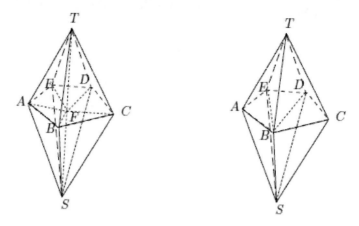

Figure 4.3. Edge contraction. Node F is to be deleted.

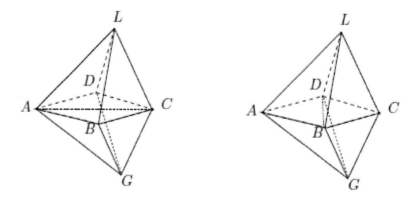

Figure 4.4. 4-4 flipping.

4.3. Edge Bisection and Contraction

The mesh adaptation scheme in our method is mainly edge based. Edge bisection is employed to refine the mesh where a high resolution is needed, while edge contraction is applied to edges where the mesh is too fine which may not be necessary for investigating the physics, but limit time steps. Figs. 4.2 and 4.3 show these two adaptive mesh approaches. To bisect a long edge, which is edge ST in the figure, a node F is inserted. Usually the node is inserted at the center of the edge. After the operation, five new edges are created, i.e. AF, BF, CF, DF, and EF, and five more tetrahedra are formed. Thus the mesh is refined. To contract an edge, which is edge BF in the figure, node F collapses to node B. Then, edge EF is removed, and two new edges BE and BD are created. The number of the tetrahedra is reduced from 10 to 6. It is obviously that the mesh becomes coarser. Dai and Schmidt (2005) and Quan (2006) have discussed these operations in detail. Quan and Schmidt (2007) discussed the cases where edge bisection cannot directly be applied. The basic idea is to apply edge

4.4. 4-4 Flipping

For multiphase flows, the above mentioned 2-3 and 3-2 swapping cannot be applied to the faces and edges on the interface, as these operations lead to artificial mixing of the two fluids, and also make the interface rough. However, for the interfacial surfaces which share one common edge, 4-4 flipping is applied to improve the surface mesh quality. A sketch of this flipping is displayed in Figure 4.4, where the triangles *ABC* and *ADC* are the interfacial surfaces. If the smallest edge angle is less than a certain criterion (15° is used in our simulations), and if after the operation, the smallest edge angle becomes larger, then 4-4 flipping is applied for the edge. It is noticed that after 4-4 flipping, two new triangle surfaces replace the old ones, and the face dihedral angle could be changed. In order to minimize this change, 4-4 flipping is allowed for the face dihedral angle for the initial two interfacial surfaces larger than 175°. From the figure, four initial tetrahedra *ABCL*, *ABCG*, *ACDL*, and *ACDG* are replaced by *ABDL*, *ABDG*, *BCDL*, and *BCDG*. The two new surfaces *ABD* and *BCD* replace the old ones; moreover, the surface mesh quality is greatly improved for this case. The details of this method can be found in (Quan and Schmidt 2007).

4.5. Generic Interfacial Edge Flipping

It is noticed from Figure 4.4 that 4-4 flip only applies to the interfacial edge which has exactly four sharing tetrahedra. However, it is not always the case that only 4 elements share the edge. Statistically, it is more likely that an interfacial edge has more than 4 surrounding elements. For such a scenario, a general edge swapping is introduced in Quan and Schmidt (2007). Figure 4.5 shows such an operation for an edge with 5 sharing tetrahedra, three on top and two at the bottom of the interfacial surfaces *ADB* and *ABC*. Edge *AB* is bisected by inserting a node *T*, and then the newly created edge *CT* is contracted to node *C*. Finally, the interfacial faces are *ACD* and *CDB* instead of *ABC* and *ABD*. Edge *CD* has four sharing tetrahedra (*ACDB*, *BCDG*, *ACDL*, and *BCDL*), and there are two other tetrahedra in the final configuration, i.e. *ACLM* and *CBLM*.

4.6. Mesh Smoothing

Mesh smoothing is also implemented to improve the mesh quality by relocating the node positions without reconnecting nodes or disconnecting edges, as shown in Figure 4.6 where Laplacian smoothing is employed (Dai and Schmidt 2005). This scheme relocates the position of each vertex to the centroid of a geometry formed by its neighboring nodes. For node *O*, Laplacian smoothing requires

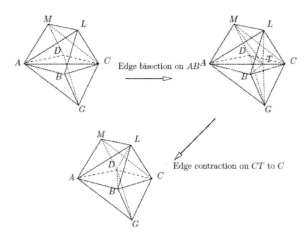

Figure 4.5. Generic edge flipping for edges which have more than 4 sharing tetrahedra.

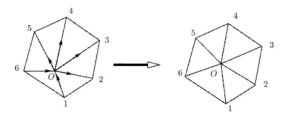

Figure 4.6. Mesh smoothing. A free vertex O, and its neighboring nodes labeled as $1 \ldots 6$. The left and right figures show the mesh before and after smoothing, respectively.

$$\mathbf{x}_o^{new} - \mathbf{x}_o^{old} = \sum_i k_i(\mathbf{x}_i - \mathbf{x}_o). \tag{4.4}$$

for an interface node, to conserve mass and the interface shape, the free vertex can only be moved in tangential direction, and thus it obeys

$$\mathbf{x}_o^{new} - \mathbf{x}_o^{old} = (\mathbf{I} - \mathbf{n}\mathbf{n}^T) \sum_i k_i(\mathbf{x}_i - \mathbf{x}_o). \tag{4.5}$$

Here, k_i is a spring constant, $\mathbf{n}$ denotes the surface normal, T stands for the transpose. It can be seen that these smoothing schemes try to uniformly distribute all the nodes without considering any quality metrics mentioned above, and thus the mesh quality is not guaranteed. Dai and Schmidt (2005) implemented a local optimization-based smoothing which was proposed by Freitag and Ollivier-Gooch et al. (1997). This method moves the free node to a position such that the minimum quality of the surrounding elements is maximized. Thus, this method always improves mesh quality. However, this method cannot be applied for interface smoothing, since it does not guarantee volume conservation. Therefore, in our simulations, the Laplacian smoothing is applied to the interface mesh smoothing, while the optimization-based smoothing is used for the interior mesh.

5. MESH COMBINATION AND SEPARATION

The interface is represented by the surface meshes in MMIT; therefore interface merging and breakup cannot be automatically handled as what in Level Set or Volume of Fluid methods. In order to deal with topological transitions in multiphase flows, mesh combination and mesh separation are developed to allow coalescence and pinch-off, respectively. Quan and Schmidt (2007) proposed a mesh separation scheme for specific cases where the axis of the neck region aligns with one of axe of the global coordinate. A more general mesh separation and a mesh combination were introduced by Quan and Schmidt (2009). Here a summary of these two schemes is given.

For interface breakup, there is always a neck region formed with the smallest radius of the region decreasing with time. So, in order to numerically simulate the pinch-off in two-phase flows, this neck region (see Figure 5.1) must be identified either by some numerical schemes or the phenomena of the problem. In Figure 5.1, a two-dimensional mesh, i.e. triangles, is shown for simplicity, and the interface is denoted by thick lines. Phase 1 is discretized by thin triangles, while phase 2 by thick ones. As the neck region, or numerically, conversion region, is assumed to be a rotation of some geometry; an axis of conversion is defined, and then the center of the region is identified as x_0. The cutting plane (or in 2D, it is actually a line) through x_0 and perpendicular to the axis of conversion is located. Then, the width (w_d) of the conversion region is identified. The cell in phase 1 of the neck region with the distance (d_c) from its centroid to the cutting plane forms a conversion region, and the distance is calculated by

$$d_c = \frac{|(\mathbf{x}_c - \mathbf{x}_0) \cdot (\mathbf{x}_a - \mathbf{x}_0)|}{|\mathbf{x}_a - \mathbf{x}_0|}. \tag{5.1}$$

The cells in the conversion region are transferred to the other phase (phase 2) by changing the fluid properties (ρ_1, μ_1) of the cells to the other fluid's properties (ρ_2, μ_2). After this conversion, two new interfaces are usually created, and the surfaces are rugged as shown in Figure 5.2 (a), (b) and (c). The simulation cannot be continued as the computed curvature field for these interfaces has huge errors. Thus, a smooth algorithm based on a projection approach is employed. The newly created interfacial nodes are moved in a projection way to the surface of a local hemisphere. The distance (δ_d, see Figure 5.1) for a new interfacial node ($\mathbf{x}_n$) moving to the projected position ($\mathbf{x}'_n$) is calculated by

$$\delta_d = d_n - d'_n, \tag{5.2}$$

where d_n and d'_n can be computed by Eq.5.1. It should be noted that if the node is moved to the new position in one step, there might be some invalid cells created, and thus the simulation is stopped. To avoid this situation, the node is moved in a number of sub-time steps (usually, we choose 50 for sub-iteration) and the node is moved uniformly by (δ_d/number of sub-iteration). In the sub-iteration, mesh adaptations, such as 2-3/3-2 swapping and mesh smoothing, are used to achieve good mesh quality. The interface after smoothing is shown in Figure5.2 (d). The simulation will be continued from this point.

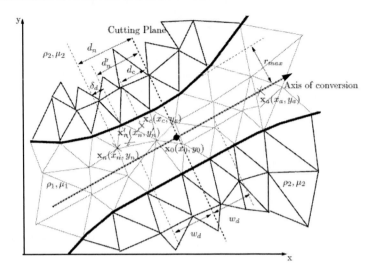

Figure 5.1. Two-dimensional schematic of mesh separation. Phase 1 with ρ1, μ1 is denoted by thin triangles, and phase 2 with ρ2, μ2 by thick triangles. The interface is displayed by thick lines. For colored figure, see (Quan and Schmidt 2009).

On the other hand, mesh combination is used to permit the merging of two interfaces. The basic idea of the mesh combination is similar to the one of the mesh separation, which is based on the conversion of the interior cell in the conversion region from one phase to another phase, and after the conversion, the newly created interfaces are smoothed by a projection approach. However, there are differences in the implementations. For the details of the mesh combination, readers are referred to Quan and Schmidt (2009).

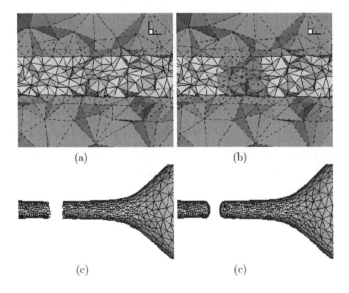

Figure 5.2. An example of mesh separation in 3D. One fluid is shaded green and the other fluid is shade gray. For colored figures, see (Quan and Schmidt 2007). (a) The necking region before the mesh separation; (b) The newly created interface, and the neighboring interior cells; (c) The newly created interface; (d) Interface after the projection.

6. VALIDATION AND APPLICATION

To validate the MMIT scheme, a number of multiphase flow cases were simulated and compared with experimental and/or theoretical solutions. The potential of MMIT in investigating underlying physics of multiphase flows is also demonstrated.

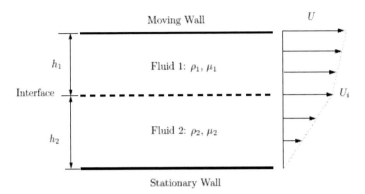

Figure 6.1. A sketch of a two-layer Couette flow.

6.1. Two-layer Couette Flow

Couette flow of a single fluid is one of the most common fluid flows which have a theoretical solution. A notable aspect of this flow is that the shear stress is a constant in the fluid field, and thus the velocity gradient if the viscosity of the fluid is a constant. This flow configuration can be obtained in reality by moving one of the plates of two parallel plates in which a fluid is contained. Two-layer Couette flow is an extension of the single-phase Couette flow, in which the fluid in between the two parallel plates consists of two immiscible liquids and the gravitational force is negligible. As the two fluids have different viscosities, although the shear in the fluid field is a constant, the velocity gradients in the two fluids are different. A jump in the velocity gradient occurs across the interface between the two phases. For a given velocity of the top wall, the two different fluids with viscosity of μ_1 and μ_2, and the distance between the two plates being $h=h_1+h_2$ as shown in Figure 6.1, the velocity at the interface can be computed by

$$U_i = \frac{\lambda}{\lambda + \delta} U, \qquad (6.1)$$

where $\lambda=\mu_1/\mu_2$, and $\delta= h_1/h_2$.

Thus, this two-layer Couette flow serves as an excellent test case to validate the numerical scheme. Here, a two-layer Couette flow in a three-dimensional box (as shown in Figure 6.2) is simulated with the viscosity ratio of $\lambda= 0.1$ and $\delta=1.0$, and matched densities because the density does not play any role for the steady state solutions. A periodic boundary condition is enforced for the left and right sides, and a moving wall with velocity of U is applied for the top wall, and the bottom side is a stationary wall. The initial velocity field is linear (see Figure 6.3 (a)), which satisfies the continuity equation, but does not satisfy the conservation of linear momentum. The simulation is run without mesh motion until a steady state solution is reached. The steady state velocity distribution is displayed in Figure 6.3 (b). The two phases are shaded in different colors for clarity. It can be seen that the jump of the velocity gradient across the interface is sharp. The simulated interfacial velocity is $0.11U$, and agrees well with the theoretical prediction of $0.091U$. Here the simulated interfacial velocity is the average of all the interfacial node velocities.

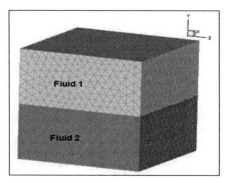

Figure 6.2. The simulation domain and mesh of a two-layer Couette flow.

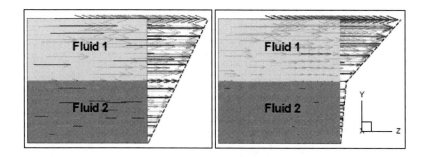

Figure 6.3. The initial velocity distribution (a) and the steady state velocity field (b) of a two-layer Couette flow.

6.2. Zero-gravity Droplet Oscillation

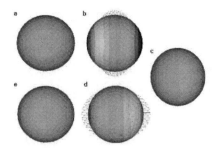

Figure 6.4. Shape evolution of the droplet oscillation with $\lambda = \eta = 0.01$. The time sequence from (a) to (e) are 0.0, 0.25, 0.75, 1.0, and 1.25 of the period T. The arrows are velocities on the surfaces. The shape is colored by the velocity magnitude.

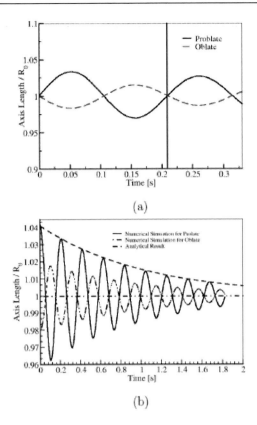

Figure 6.5. (a) Comparison of the computed period and the theoretical one of the droplet oscillation; (b) Comparison of the decay factor.

Droplet oscillation with negligible gravitational forces is another benchmark for numerical methods in modeling immiscible, multiphase flows. Here a three dimensional droplet oscillation is simulated with $\lambda = \mu_s/\mu_d = 0.01$, and $\eta = \rho_s/\rho_d = 0.01$. Here, the subscripts s and d stand for the surrounding and droplet phases, respectively. The Ohnesorge number ($Oh = \mu_d/\sqrt{\rho_d D_0 \sigma}$) based on the droplet properties is 0.013 with D_0 being the diameter of a volume equivalent sphere. The initial droplet is an ellipsoid with the major axis 5% longer than the radius of a sphere with an equivalent volume of the droplet. The droplet experiences a damped oscillation with the surface tension force as a restoring force. This validation has been reported by Quan and Schmidt (2007). Lamb (1945) had analytical solutions for this droplet oscillation. According to the Lamb theory, the period is

$$T = \frac{2\pi}{\sqrt{\frac{n(n+1)(n-1)(n+2)}{(n+1)\rho_d + n\rho_s}\frac{\sigma}{R_0^3}}}, \tag{6.2}$$

where ρ_d and ρ_s are the densities of the droplet and the suspending fluid, n is the oscillation mode number with n = 2 being the lowest mode for an incompressible drop, σ denotes the surface tension coefficient, and R_0 stands for the radius of volume-equivalent sphere; and the decay factor is

$$\tau = \frac{\rho_d R_0^2}{(n-1)(2n+1)\mu_d}. \tag{6.3}$$

Figure 6.4 shows the shape oscillation at t = 0.0, 0.25, 0.75, 1.0 and 1.25 of the period. The simulated period is 0.2108s and the decay factor is 1.03, compared well with the theoretical prediction of period of 0.2089 (Eq. 6.2) and decay factor of 1.034 (Eq. 6.3) by Lamb (1945). Figs.6.5 (a) and 6.5 (b) show the comparisons. It is noted that our numerical scheme is capable to simulate the droplet oscillation for around 9 periods and still very good accuracy is maintained. It is usually the case that for the majority of other methods, where the interface is not exactly zero-thickness, a parasitic current due to the non-perfect balance of the surface tension forces dominates for a nearly spherical liquid particle with a minimal velocity field caused by other forces. For the late stage of the simulation, the velocity field due to the initial resultant surface tension forces by the deformation is damped out and is almost approaching zero. However, our simulation still predicts the decay very well.

6.3 Single Drop in Shear Flow

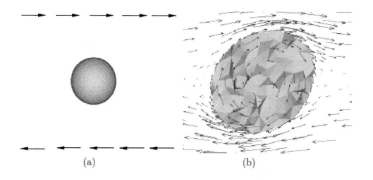

Figure 6.6. Droplet under shear. The viscosity ratio is λ =2.0, and the capillary number is Ca=0.05. (a) Initial state. (b) Steady state.

To test the accuracy of the shear stress calculation in a two-phase flow where the surface tension forces also present, a droplet in a linear shear flow is simulated and compared with Taylor's analytical result (Taylor 1932, 1934). Again, a three-dimensional box is employed for this simulation and a periodic boundary condition is imposed for the inlet and outlet. The droplet is located in the center of the domain. The upper wall and the lower wall are moving walls and with a velocity of U in opposite direction, as shown in Figure 6.6 (a). The fluid inside the simulation domain is initially at rest. It is believed that this configuration best mimics experiments. The shape of the droplet and the velocities of the steady state are displayed in Figure 6.6.6 (b). The simulated deformation factor,

$$D = \frac{L-B}{L+B} = Ca\frac{19+16\lambda}{16+16\lambda}, \tag{6.4}$$

is 0.076, while the Taylor theory gave 0.053, and his experimental result is 0.07. Here, Ca is the capillary number, L denotes the longest axis and B stands for the shortest axis of the droplet. The capillary number is defined as

$$Ca = \frac{\mu_d \gamma R_0}{\sigma}, \qquad (6.5)$$

where γ stands for the shear rate, and R_0 is the radius of the initial spherical droplet.

6.4. Ellipsoidal Drop Translating in another Fluid

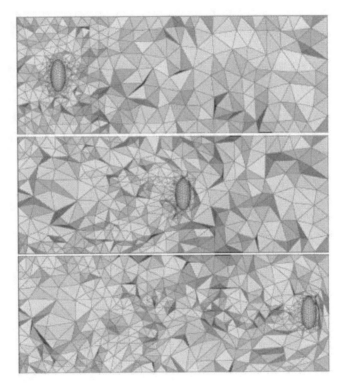

Figure 6.7. Ellipsoid translates in a fluid. Figures from the top to bottom are the initial mesh, the ellipsoid at the middle of the domain, and the ellipsoid near the right wall, respectively.

In the simulation of the above three cases, either the mesh is not moving, or the mesh movement is minimal. To test the adaptive mesh schemes mentioned previously, especially the mesh adaptation for interior meshes, an ellipsoidal which is translated in a box is employed. In this test case, the flow solver is turned off, and the ellipsoid is moving at a uniform velocity. Thus there is no deformation for the ellipsoid. The outer domain is a box. Therefore, the mesh adaptations, such as 4-4 flip, interface edge bisection and edge collapsing are also switched off for the testing. Figure 6.7 shows the evolution of the ellipsoid at the initial stage, the middle stage, and the last stage from the top to the bottom. The ellipsoid has

6.5. Relaxation and Pinch-off of an Elongated Droplet

An initially elongated droplet experiences relaxation, and possible breakup, and this process has been studied by Stone et al. (1986). In their experiments, a computer-controlled four-roll mill was used to continuously deform the droplet in another liquid, and finally the droplet became a ligament. Then, the suspending flow was stopped, the relaxation and pinch-off of the ligament was observed and investigated. In their work, an "end-pinching mechanism" was proposed. To further investigate this, Stone and Leal (1989) used a boundary integral method to study this problem. Their numerical results indicated that the two factors which determine the relaxation and breakup process are the viscosity ratio and the initial shape of the elongated droplet. They explained that the end-pinching was the competition between a surface tension driven flow near the end and a pressure-driven flow away from the center of the elongated droplet. They also found that the capillary-wave instability mechanism cannot be used to explain the relaxation and breakup process of a moderately elongated droplet in another fluid. However, it should be noted that the numerical simulations were limited to low Reynolds-number flow, where the inertia force is completely neglected. Based on the previous works by Stone and his co-workers, Qian and Law (1997) proposed a schematic of the end-pinching mechanism.

Recently, Tong and Wang (2007) simulated the time-dependent relaxation and pinch-off of a moderately elongated liquid droplet using a coupled level set and volume-of-fluid method. Interestingly, they found a "flaw" in the end-pinching mechanism by Stone and his co-workers, and a correction was proposed. Based on their simulations, they demonstrated that the lateral curvature of the neck region created a slightly concave surface, and thus decreased the surface tension forces, while in the previous work; there was no such concave surface at the neck region reported. It should be noted, that in the work by Stone and his co-workers, the viscosity ratio (droplet viscosity to suspending viscosity) for most cases is less than unity and also the flow is in Stokes regime, while the viscosity ratio is infinite and the flow has a finite Reynolds number.

Here, using the MMIT method, an initially elongated droplet immersed a quiescent fluid is investigated. Different viscosity ratios ($\lambda = \mu_d/\mu_s$) are studied, namely, 0.1 and 100. However, the density ratios ($\eta = \rho_d/\rho_s$) are fixed as 10, the initial shape of the ligament is the same for the two simulations. The outer domain is a cylinder with the wall boundary condition applied for all the outer walls. The length ratio (k) which relates the length of the ligament to the radius of the central cylinder part of the ligament is 20. The characteristic time based on the radius of the cylindrical section of the ligament (r_0) is $t_c = r_0\mu_d/\sigma$. The Ohnesorge number based on the ligament properties ($Oh_d = \mu_d/\sqrt{\rho_d r_0 \sigma}$) is 0.037, and $Oh_s = \mu_s/\sqrt{\rho_s r_0 \sigma}$ numbers based on the suspending fluid properties are 1.18 and 1.18 × 10^{-2}. An effective Reynolds number is defined as $Re = \sqrt{\rho_d \sigma r_0}/\mu_d$, which is the inverse of Oh_d. It should be noted that these two simulations have finite Reynolds numbers.

Figure 6.8 shows the shape evolution of the two cases with viscosity ratio of 0.1 and 100. The ligament is colored by the axial radius. Comparing these two cases, there are some differences. The two ends of $\lambda = 0.1$ are more like a prolate, while the ones of $\lambda = 100$ are more like an oblate (except the one at $t = 244.9$). The middle section of $\lambda = 0.1$ is much thinner than the one of $\lambda = 100$, and has less volume. It can also be observed that the shapes of the neck regions are different. However, it is difficult to see the details. To clearly compare the neck regions, Figure 6.9 displays an enlarged view of the neck regions. For $\lambda = 0.1$, the neck region has a cylinderical shape, while a cone shape appears for $\lambda = 100$ case. This difference in the neck shape was also observed by Burton et al. (2005) for the pinch-off of a water drop in air and an air bubble in water.

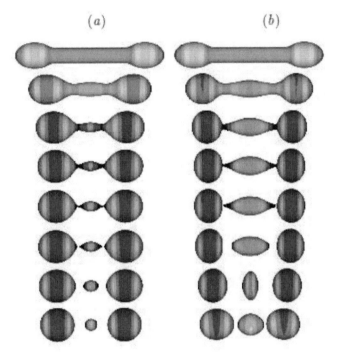

Figure 6.8. The effect of the viscosity ratio on the relaxation and breakup of an elongated droplet immersed in another viscous fluid. The droplets are shaded by the non-dimensional radius (r/r_0) in the radial direction. $\eta = 10$ and $Oh_d = 3.7 \times 10^{-2}$. (a) $\lambda = 0.1$, $Oh_s = 1.18$; t = 0.0, 251.3, 435.3, 489.6, 509.8, 511.2, 532.9, 613.7. (b) $\lambda = 100$, $Oh_s = 1.18 \times 10^{-3}$; t = 0.0, 108.0, 144.0, 157.0, 158.5, 165.6, 201.7, 244.9.

It is noted that a neck shape of a water droplet pinch-off in air was predicted in Tong and Wang's simulations (Tong and Wang 2007), while Stone and his coworkers observed a neck shape of an air bubble in water. This can be explained by the fact that Tong and Wang simulated a droplet in vacuum, while the viscosity ratio of Stone and his co-workers' research is mainly less than 1.0. So it is believed that the end-pinching mechanism by Stone and his co-workers is applied to the viscosity ratio less than $O(1)$, while Tong and Wang's applies to viscosity ratio larger than $O(1)$.

6.6. Droplets Coalescence

Droplet-droplet colliding is very common occurrence in a number of natural phenomena and engineering applications, such as raindrops, inkjet printings, fuel injection, etc. Because of its importance, it has been studied theoretically, numerically, and experimentally for decades. Here are only a few examples of the wide literature encountered, (Ashgriz and Poo 1990; Abid and Chesters 1994; Lafaurie et al. 1994; Qian and Law 1997; Guido and Simeone 1998; Eggers et al. 1999; Yang et al. 2001; Duchemin et al. 2003; Wu et al. 2004; Aarts et al. 2005; Pan and Suga 2005; Case and Nagel 2008; Yoon et al. 2007; Gotaas et al. 2007).

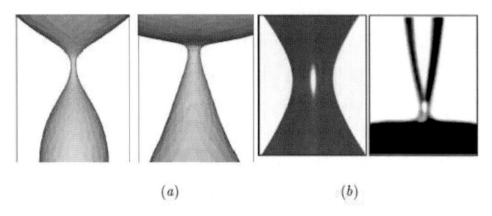

(a) (b)

Figure 6.9. Enlarged view of the neck shape showing the difference in the symmetry. On the left of (a) is the case in figure 18(a) with $\lambda = 0.1$. On the right of (a) is the case in figure 18(b) with $\lambda = 100$. (b) the experimental observation by Burton et al. (2005). On the left is an air bubble in water, and on the right is a water droplet in air.

Near Contact Motion

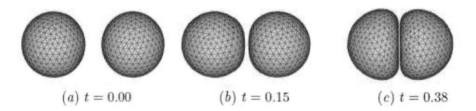

(a) $t = 0.00$ (b) $t = 0.15$ (c) $t = 0.38$

Figure 6.10. Droplets' shapes for the near contact motion. $Re = 20.0$ and $We = 8.0$. Time t is non-dimensionalized by $t_c = D/U_c$.

Since, in the MMIT method, the interface is explicitly tracked the surface mesh of the volume mesh for solving the Navier-Stokes equations, it has the potential to investigate the dynamics of the thin film which is usually created in the collision process of drops or bubbles. To demonstrate this capability, a near contact motion of two colliding droplets is simulated with $\lambda = 10.0$ and $\eta = 10.0$. The velocities (U_c) of the two-droplet centroids are the same in the magnitude but opposite in the direction. The details of the way to obtain a smoothed initial velocity field for this simulation can be found in (Quan et al. 2009), a similar approach had been reported in earlier work (Quan and Schmidt 2007). Here, a Weber number is defined as $We = \frac{\rho_d(2U_c)^2 D}{\sigma}$, where D is the diameter of the initial spherical droplet. A Reynolds number is $Re = \frac{\rho_d(2U_c)D}{\mu_d}$. For this simulation, $We = 8.0$ and $Re = 20.0$. The two drops are located in the center of a box, and wall boundary conditions are enforced on the sides of the box.

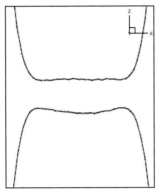

Figure 6.11. Details of the gap shape at $t = 0.38$ of the case of Figure 20. The enlarged view of the gap shape with the ratio of x-axis to z-axis distance scaled by 0.1.

6.10 shows the evolution of the shapes of the two droplets as they are approaching each other. It can be seen that there is a very thin film formed in between the two drops. From Figure 20 (c), it seems that the two sides of the thin film are parallel. However, a close view of the thin film region, as shown in Figure 6.11, reveals that there is a dimple in each side of the thin film. Similar shapes of the thin film were reported in previous works (Abid and Chesters 1994; Yoon et al. 2007; Manica, Klaseboer and Chan 2008). The smallest width of the thin film gap is around 1.5% of D, and the details of the evolution of the thin-film gap can be found in Quan et al. (2009). Figure 6.12 shows the interior mesh for the whole domain, it can be seen that the local mesh adaptations works very well.

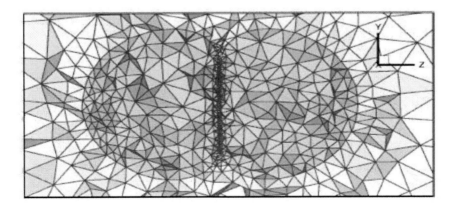

Figure 6.12. Mesh adaptation for the near contact motion of the two deformable droplets.

Off-center Collision

To test the capability of the mesh combination and mesh separation, an off-center droplet-pair collision is simulated with $Re = 80.0$ and $We = 32.0$. The Reynolds number and Weber number are defined the same way as in the previous section. In this simulation, the two droplets are moving at the same speed towards each other; however, the motion of the centroid at the initial stage is just parallel, not in-line as in the previous near contact motion of two droplets. The initial velocity field is obtained in the same way as the one in the previous section. The density ratio and viscosity ratio are 10.0 for this simulation.

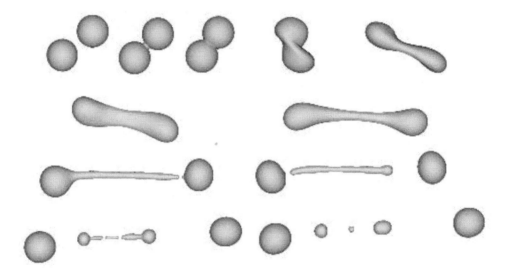

Figure 6.13. Snapshots of the droplets off-center collision. $Re = 80.0$ and $We = 32.0$. Time sequence non-dimensionalized by t_c from the top to the bottom and form the left to the right is 0.0, 0.511, 0.577, 0.820, 1.322, 1.702, 2.502, 3.745, 4.241, 5.139, and 5.463.

Figure 6.14. A similar experiment from Qian and Law (1997).

Figure 6.13 displays the evolution of the shape of the two droplet off-center collision. The time is non-dimensionalized by $t_c = U_c/D$. At $t=0.511$, i.e. the second image, the mesh combination scheme is applied to allow the coalescence to occur. The connecting bridge is smooth and symmetric for this case. It should be noted that for the off-center collision, the shape of the bridge might not be symmetric. However, the exact shape is not known based on the current knowledge. So for simplicity, an axi-symmetric shape is used for the simulation. Due to the large surface tension forces, the bridge opens up quickly (see $t = 0.577$). However, the surface tension force becomes smaller and smaller in the bridge region as the curvature is reducing. The momentum of the initial two droplets becomes dominant and moves the top and the bottom parts of the coalesced drop apart (as shown in t = 0.820, 1.322, and 1.702). Then, a liquid thread is formed in the center at t =2.502. The liquid thread breaks up and two primary droplets are created (see t = 3.745 and 4.241). Finally, the center liquid thread experiences relaxation due to the large surface tension forces at the two ends. A dump-bell shape is formed, and the liquid thread pinches off again at t = 5.139. Finally three satellite droplets are created. Figure6.14 shows a similar case by Qian and Law (1997). It can be seen that our numerical simulation captures most features of the droplet off-center collision.

6.7. Two Drops Impulsively Accelerated by a Gaseous Flow

Fluid-particle interactions are one of the most interesting research topics in the field and the study can be traced back to as early as 1752 by D'Alembert, a French mathematician. He proved that for incompressible, inviscid flow, the drag force is zero on a body moving with constant velocity through the fluid, which is in direct contradiction to measurements of substantial drag for bodies moving through fluids. This contradiction is well-known as D'Alembert's paradox. Viscosity for a real fluid is the key for the paradox. For a solid particle moving steadily in a fluid, it is found that the drag coefficient, C_D depends only on the Reynolds number based on the suspending fluid properties. However, the interaction between a liquid particle and the suspending flow is much more challenging as the liquid particle might be deforming or experience topological transitions as it is moves with the carrier fluid.

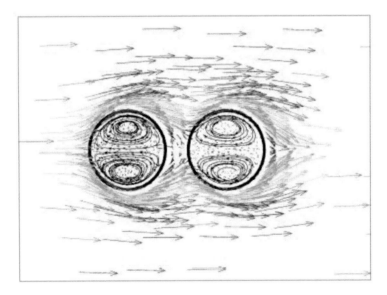

Figure 6.15. Smoothed initial velocity field for the droplet-pair impulsively accelerated by a gaseous flow. $Re_g^i = 40$ and $We_g^i = 40$, λ= 50, and η= 50. The initial distance between the two drops is $0.6r_0$.

Temkin and Kim (1980) studied the motion of a droplet accelerated by weak shock waves; they found that the drag coefficients for the droplets with the relative fluid velocity decreasing are larger than the steady drag at the same Reynolds numbers. Temkin and Mehta (1982) experimentally investigated the droplet drag in accelerating and decelerating flows, and they concluded that, for decelerating relative flows, the unsteady drag is always larger than the steady drag at the same Reynolds number; while for accelerating relative flows, the unsteady drag is smaller than the corresponding steady drag. Using the MMIT method, Quan and Schmidt (2006) studied the droplet dynamics when it impulsively accelerated by the surround gaseous flow. The Weber numbers based on the gas properties are in a range of 0.4 to 40.0. The simulations further confirm and also explain the observations by Temkin and his co-workers for the decelerating relative flows. Wadhwa et al. (2007) used a hybrid compressible-incompressible numerical method and simulated transient behavior of a decelerating drop in axisymmetric flows. They found that for small Weber numbers, i.e. small deformation, the drag coefficient is about the same for solid spheres and liquid droplets, while for the large deformation, the differences of the drag coefficient between solid spheres and droplets are increased. The above studies only focus on the interaction between a single droplet and the surrounding flow. However, it is more common in real applications, such as fuel injections, inkjet printings, that there are more than one droplet which interact with the surrounding flow and also other droplets. Temkin and Ecker (1989) experimentally studied droplet pair interactions in a weak shock-wave flow field with small deformation. The distance between the two droplets are in a range of 1.5 to 11 diameters of the droplet. They found that the upstream droplet is not affected by the downstream droplet, while the downstream one experiences significant reductions in drag coefficient. They also found that the drag reduction can be as much as 50% relative to the isolated droplet.

The MMIT method has been employed to simulate a single droplet impulsively accelerated by a gaseous flow and validated against experimental observations (Quan and Schmidt 2006; Temkin and Kim 1980; Temkin and Mehta 1982). To further investigate the droplet-droplet interaction and the interaction between drops and the surrounding flow, a droplet pair impulsively accelerated by a gaseous flow is simulated. The two equal-sized drops are aligned with the flow direction, and the distance between the two droplets in the flow direction is 0.6 r_0 with r_0 the radius of the initial spherical drop. The initial flow field is obtained in a similar way as the one used in a single droplet (Quan and Schmidt 2007). The outer domain is a box with moving wall boundary condition applied for the top and the bottom walls, and slip boundary condition for the two side walls. A free stream is enforced for the inlet and an outlet is applied for the exit. The box is large enough to minimize wall effects. An initial Reynolds number (Re_g^i) based on the gas properties $Re_g^i = \frac{\rho_g (2U_0)(2r_0)}{\mu_g}$ is 40, and an initial Weber number $We_g^i = \frac{\rho_g (2U_0)^2 (2r_0)}{\sigma}$ is also 40.0. Here, U_0 is the initial free stream velocity. The density ratio and viscosity ratio are 50, respectively.

Figure 6.15 shows the smoothed initial velocity field. The velocity is colored by its magnitude, and the two drops are denoted by the black thick solid lines. It can be seen that the fluid field inside the droplet has some differences. The second (downstream or rear) droplet is actually inside the wake of the first (upstream or front) drop. Figure 6.16 shows the evolution of the drop's shape. Initially, two droplets are spherical and the smallest distance between them is $0.6r_0$. As time progresses, the front droplet deforms to a nearly ellipsoidal shape (as shown in Figure 6.16 (a) – (c)), while the rear one behaves differently. The front part of the second droplet becomes more and more curved, while the rear part is becoming flattened. The smallest distance between the two drops changes slowly. The front part of the upstream droplet keeps flattening, while the rear part is flattened much faster, resulting a cap shape (as shown in Figure 6.16 (d)). The smallest distance is much smaller than the initial distance. As time progress, the distance between the two drops is becoming smaller and smaller, and a mushroom shape is formed by the two droplets. The rear part of the upstream droplet is actually dimpled (see Figure 6.16 (e) and (f)).

Figure 6.17 displays the velocity field near the two droplets at the last stage of Figure 6.16. The thick solid lines stand for the drops, and it can be observed that the rear part of the upstream droplet is dimpled and the front part of the downstream droplet is flattened as compared to the one in Figure 6.16 (d). From the left figure where the velocity field is shown with respect to the reference frame of the upstream drop centroid, it can be seen that the downstream droplet is totally inside of the large wake created by the flattened first droplet. The second droplet is actually sucked in by the wake of the first drop and thus the distance between the two drops is becoming smaller. From the right figure where the velocity is displayed with the reference frame located at to the centroid of the downstream droplet, it can be observed that the flow field in front of the downstream drop is not the free stream flow, but disturbed significantly by the upstream drop, and the first drop moves almost uniformly toward the second drop.

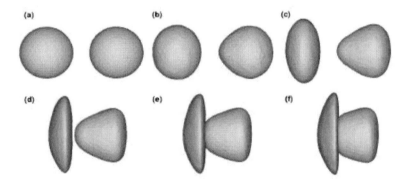

Figure 6.16. Evolution of the shape of the droplet-pair impulsively accelerated by a gaseous flow with the initial distance between the two drops being 0.6 time the radius of the droplet. Non-dimensional time for (a) to (e) is 0.0, 2.50, 7.50, 13.01, 14.51, and 15.51.

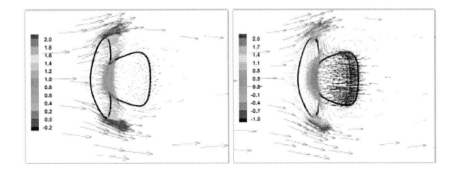

Figure 6.17. The velocity field near the two droplets in *xy*-plane. The left figure is at the reference frame of the upstream droplet centroid, while the right one is moving with the downstream droplet. The thick black solid lines stand for the of interfaces of the two drops. The velocity vectors are colored by their magnitude.

To further investigate the motion of the two drops, Figure 6.18 shows the z-axis velocity (w) and the position of the two droplet's centroid versus time. From the left figure, it can be seen that the rear droplet move much slower than the upstream droplet. However, it is noticed that for time larger than 6.5, the rear one accelerates much faster compared to the movement before this time, while the upstream drop accelerates slower. This is because after this time, the two droplets are very near, and the strong wake of the first droplet sucks the downstream droplet in, while the second droplet also disturbs the wake of the first drop, which leads to a slower motion of the first drop. The figure on the right demonstrates that the two droplets move nearer and nearer as time progresses and the distance is less than one radius (r_0).

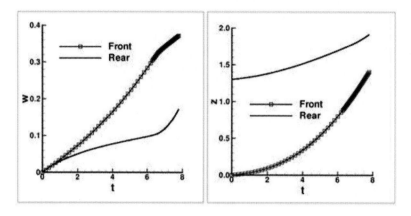

Figure 6.18. (a) Relative, instantaneous Reynolds number based on the surrounding fluid versus non-dimensional time for the front drop and rear drop in a same free stream. (b) The centroid z-axis position, and is non-dimensionalized by $2r_0$.

CONCLUSION AND FUTURE WORK

The moving mesh interface tracking method uses the surface meshes of the domain elements to represent the interface between the immiscible two-phase flows, and thus the interface is exact zero-thickness. The interfacial nodes move with the fluid velocity to conserve the total mass of each phase. This Lagrangian motion of the interface prevents the mixing of the two immiscible fluids, and also eases the difficulties in numerical treatments of large density differences. It should be noted that, compared to other methods where the interface (or the fluid properties) is (are) usually smeared (or smoothed); MMIT has the potential to investigate the physics of small scales, such as the thin film formed in two droplets near contact motion. The interior mesh moves in a smoothing fashion, as the Lagrangian motion is not necessary a good scheme to achieve good mesh quality. The smoothing includes local optimization-based smoothing, and also a spring-analogue global smoothing method. These two smoothing schemes help to achieve good mesh quality.

The interface moves in a Lagrangian mode, while the interior mesh moves in a smoothing fashion; as time progresses, the mesh quality in the domain will become too bad to continue the simulation, or to obtain reasonable resolutions. Therefore, it is one of the key issues in the MMIT method to employ mesh adaptation methods to obtain good mesh quality. A number of mesh adaptation schemes are implemented in the current code, which include 3-2 or 2-3 swapping, 4-4 flipping, edge bisection, edge contraction, etc. It should be noted that these mesh adaptations are applied locally to the region where the mesh quality is bad. This means that mesh adaptation in our method is locally applied, unlike other methods such as mesh redistribution, in which, once there is some mesh with bad quality, a new mesh is regenerated and also the solution for the new mesh is interpolated from the old mesh's. To apply the mesh adaptation locally, the errors due to the interpolation are minimized.

In two-phase flows, topological transitions such as interface breakup and/or merging are very common. To address these challenges, mesh combination and mesh separation are introduced for the MMIT method. These two schemes are based on the conversion of the liquid cells in the conversion region of one phase to another phase by reverting the liquid properties (density and viscosity) of the cells. The conversion of the liquid cells may introduce mass gain or mass loss in one phase; however, these errors can be controlled and minimized by allowing the mesh combination or mesh separation to be applied later such that the conversion region is much smaller.

The MMIT is robust in simulating various two-phase flows with high fidelity. The mesh adaptations are capable to obtain good mesh quality, to achieve computing efficiency, and to obtain good mesh resolution where it is necessary. The mesh combination and mesh separation is robust in dealing with topological changes such as interface merging and breaking. The MMIT has been tested against a number of two-phase flow cases and compared well against the available experimental and/or theoretical results.

In MMIT, the governing equations of incompressible flows are solved in a strongly-conserved form using tetrahedral as the basic mesh elements. This unstructured mesh method has the capabilities to simulate fluid flows with complex geometries. However, the geometry of the domain is rather simple for the cases presented above, and the wall effects are usually ignored. It is very interesting to investigate wall effects of the complex geometries on the fluid flows, or bubble/droplet dynamics, such as flows and blood cells in blood vessels, or bubbles/ droplets in micro-fluidics. As the interface is a real boundary in MMIT, it can prevent any dispersion of a solute from the liquid phase to the gas phase. Therefore, MMIT is an ideal method for studying the surfactant effect on bubble or droplet dynamics. MMIT can also be used to examine small-scale physics in the necking or merging process.

REFERENCES

Aarts, D.G.A.L., Lekkerkerker, H.N.W., Guo, H., Wegdam, G.H. & Bonn, D. (2005). Hydrodynamics of droplet coalescence. *Phys. Rev. Lett.*, 95:164503.

Abid, S. & Chesters, A.K. (1994). The drainage and rupture of partially-mobile films between colliding drops at constant approach velocity. *Int. J. Multiph. Flow*, 20:613–629.

Ashgriz, N. & Poo, J.Y. (1990). Coalescence and separation in binary collisions of liquid-drops. *J. Fluid Mech.*, 221:183–204.

Burton, J.C., Waldrep R., & Taborek, P. (2005). Scaling and instabilities in bubble pinch-off. *Phys. Rev. Lett.*, 94(18):184502.

Case, S.C. & Nagel, S.R. (2008). Coalescence in low-viscosity liquids. *Phys. Rev. Lett.*, 100:084503.

Dai, M.Z. & Schmidt, D.P. (2005). Adaptive tetrahedral meshing in free-surface flow. *J. Comput. Phys.*, 208:228–252.

Dai, M.Z., Wang, H.S., Perot, J.B. & Schmidt, D.P. (2002). Direct interface tracking of droplet deformation. *Atomization Spray*, 12:721–735.

Duchemin, L., Eggers, J. & Josserand, C. (2003). Inviscid coalescence of drops. *J. Fluid Mech.*, 487:167–178.

Eggers, J., Lister, J.R. & Stone, H.A. (1999). Coalescence of liquid drops. *J. Fluid Mech.*, 401:293–310.

Freitag, L.A. & Ollivier-Gooch, C. (1997). Tetrahedral mesh improvement using swapping and smoothing. *Int. J. Numer. Meth. Eng.*, 40:3979–4002.

Gotaas, C., Havelka, P., Jakobsen, H.A., Svendsen, H.F., Hase, M., Roth, N. & Weigand, B. (2007). Effect of viscosity on droplet-droplet collision outcome: Experimental study and numerical comparison. *Phys. Fluids*, 19:102106.

Guido, S. & Simeone, M. (1998). Binary collision of drops in simple shear flow by computer-assisted video optical microscopy. *J. Fluid Mech.*, 357:1–20.

Kang, M., Fedkiw, R. & Liu, X.-D. (2000). A boundary condition capturing method for multiphase incompressible flow. *J. Sci. Comput.*, 15:323–360.

Knupp, P. M. (2003). Algebraic mesh quality metrics for unstructured initial meshes. *Finite Elem. Anal. Des.*, 39:217–241.

Lafaurie, B., Nardone, C., Scardovelli, R., Zaleski, S. & Zanetti, G. (1994). Modelling merging and fragmentation in multiphase flows with SURFER. *J. Comput. Phys.*, 113:134–147.

Lamb, H. (1945). Hydrodynamics. Dover, New York.

Manica, R, Klaseboer, E. and Chan, D.Y.C. (2008). Dynamic Interactions between drops – a critical assessment. *Soft Matter* 4: 1613-1616.

Osher, S. & Fedkiw, R.P. (2001). Level set methods: An overview and some recent results. *J. Comput. Phys.*, 169:463–502.

Pan, Y. & Suga, K. (2005). Numerical simulation of binary liquid droplet collision. *Phys. Fluids*, 17:082105.

Perot, B. & Nallapati, R. (2003). A moving unstructured staggered mesh method for the simulation of incompressible free-surface flows. *J. Comput. Phys.*, 184: 192–214.

Qian, J. & Law, C.K. (1997). Regimes of coalescence and separation in droplet collision. *J. Fluid Mech.*, 331:59–80.

Quan, S.P. (2005). Simulating multiphase flows using an unstructured moving mesh interface tracking method. PhD thesis, University of Massachusetts Amherst, Amherst, MA, USA.

Quan, S.P. & Schmidt, D.P. (2006). Direct numerical study of a liquid droplet impulsively accelerated by gaseous flow. *Phys. Fluids*, 18:102103.

Quan, S.P. & Schmidt, D.P. (2007). A moving mesh interface tracking method for 3D incompressible two-phase flows. *J. Comput. Phys.*, 221:761–780.

Quan, S.P. & Schmidt, D.P. Computation of the interface shear stress for multiphase flow simulations, (Submitted).

Quan, S.P., Lou, J. & Schmidt, D. P. (2009). Modeling merging and breakup in the moving mesh interface tracking method for multiphase flow simulations. *J. Comput. Phys.*, 228, 2660-2675.

Scardovelli, R. & Zaleski, S. (1999). Direct numerical simulation of free-surface and interfacial flow. *Ann. Rev. Fluid Mech.*, 31:567–603.

Stone, H.A. & Leal, L.G. (1989). Relaxation and breakup of an initially extended drop in an otherwise quiescent fluid. *J. Fluid Mech.*, 198:399–427.

Stone, H.A., Bentley, B.J. & Leal, L.G. (1986). An experimental study of transient effects in the breakup of viscous drops. *J. Fluid Mech.*, 173:131–158.

Taylor, G.I. (1932). The viscosity of a fluid containing small drops of another fluid. *Proc. R. Soc. Lond. A*, 138:41–48.

Taylor, G.I. (1934). The formation of emulsions in definable fields of flow. *Proc. R. Soc. Lond. A*, 146:501–523.

Temkin, S. & Ecker, G.Z. (1989). Droplet pair interactions in a shock-wave flow field. *J. Fluid Mech.*, 202:467–497.

Temkin, S. & Kim, S.S. (1980). Droplet motion induced by weak shock waves. *J. Fluid Mech.*, 96:133–157.

Temkin, S. & Mehta, H.K. (1982). Droplet drag in an accelerating and decelerating flow. *J. Fluid Mech.*, 116:297–313.

Tong, A.Y. & Wang, Z.Y. (2007). Relaxation dynamics of a free elongated liquid ligament. *Phys. Fluids*, 19:092101.

Unverdi, S.O. & Tryggvason, G. (1992). A front tracking method for viscous, incompressible, multi-fluid flows. *J. Comput. Phys.*, 100:25–37.

Wadhwa, A.R., Magi, V., & Abraham J. (2007). Transient deformation and drag of decelerating drops in axisymmetric flows. *Phys. Fluids*, 19:113301.

Wu, M.M., Cubaud, T. & Ho, C.M. (2004). Scaling law in liquid drop coalescence driven by surface tension. *Phys. Fluids*, 16:L51–L54.

Yang, H., Park, C.C., Hu, Y.T. & Leal, L.G. (2001). The coalescence of two equalized drops in a two-dimensional linear flow. *Phys. Fluids*, 13:1087–1106.

Yoon, Y., Baldessari, F., Ceniceros, H.D. & Leal, L.G. (2007). Coalescence of two equalized deformable drops in an axisymmetric flow. *Phys. Fluids*, 19:102102.

Zhang, X., Schmidt, D. & Perot, B. (2002). Accuracy and conservation properties of a three-dimensional unstructured staggered mesh scheme for fluid dynamics. *J. Comput. Phys.*, 175:764–791.

INDEX

3

3D images, 298

A

absorption spectra, 345
acid, 3, 7, 14, 37, 421, 423, 476
acidic, 3
acoustics, 476, 512
activation energy, 27
adaptation, xvi, 43, 45, 54, 71, 86, 260, 593, 602, 604, 613, 617, 622
adaptations, xvi, 593, 595, 601, 607, 613, 617, 623
ADC, 605
adenine, 345
adhesion, 126, 185, 245, 255, 263
adhesions, 255
adsorption, xiii, 124, 302, 343, 348
aerospace, 238, 298
aggregation, 386
air temperature, 18, 141, 304, 308
albumin, 244
alters, 200
ambient air, 23, 142, 146, 285
ambient air temperature, 142, 146
American Heart Association, 132, 134
amplitude, 6, 189, 525
angiography, 132
anisotropy, 6, 38, 425
anticoagulant, 124
antigenicity, 245
aorta, 111, 112, 113, 114, 125, 126, 129
ARC, 340
architecture design, 265
Argentina, 511
Arrhenius equation, 306
arteries, 111, 112, 113, 244
artery, 111, 112, 114, 115, 129, 131, 259
artificial heart valve, ix, 110, 126
assessment, 257, 471, 624
asymmetry, ix, 137, 146, 170, 176, 437
atmospheric pressure, 31, 32, 272, 312
atoms, 346, 347
automata, 387, 388, 579
automate, 120
axisymmetric flow systems, viii, 41

B

bacteria, 299, 308, 310
base, xi, 16, 18, 23, 30, 31, 35, 36, 79, 206, 209, 215, 222, 227, 231, 236, 237, 395, 461, 539
batteries, 116
behaviors, xiii, 317, 416, 440
Beijing, 341, 385, 549, 579
Belgium, 179, 238, 250, 509
benchmarking, 197
bending, 29, 36, 246, 260, 348, 523, 543
benefits, ix, 110, 131, 247
benzene, 348
bias, 192, 520
bioassay, 400
biochemical processes, 437
biological activity, 244
biological processes, 304
biological responses, 255
biological systems, xi, 243, 245
biomechanics, 546
biomedical applications, 127
bioremediation, 508
biosensors, 299
birth rate, 218
bleeding, 134

blood, viii, ix, 107, 109, 110, 111, 113, 114, 116, 117, 118, 119, 121, 123, 124, 126, 128, 129, 130, 131, 132, 133, 134, 246, 259, 512, 623
blood clot, 124
blood flow, viii, 107, 109, 113, 114, 119, 121, 123, 124, 126, 131, 133, 134, 512
blood particles, 133
blood plasma, 123
blood vessels, 623
Boltzmann constant, 318
bone, xi, 244, 245, 246, 247, 248, 255, 257, 260, 261, 262, 263, 264
bone cells, 260, 261
bone form, 260
bone marrow, 261
bone mass, 246
bones, 246, 260
boundary surface, 435
boundary value problem, 297
brain, viii, 109, 111, 113, 114
building blocks, 345

C

C++, 480
CAD, 112, 125, 128, 298
calcium, 244, 254, 260, 263
calibration, 470, 508
candidates, 115
capillary, 8, 12, 21, 22, 81, 82, 95, 100, 108, 182, 465, 612, 613, 614
carbohydrates, 245
carbon, ix, 3, 4, 33, 97, 107, 110, 111, 125
carbon dioxide, ix, 97, 107, 110, 111
cardiac output, 116, 118
cardiopulmonary bypass, viii, 109, 110, 115, 121, 131, 132
cardiopulmonary bypass (CPB), viii, 109
cardiovascular function, 512
cardiovascular system, viii, 109, 111, 112, 128
carotid arteries, 112
cartilage, 262
case studies, xi, 244
case study, 246
casting, 189, 244, 512
castor oil, 102
catalyst, 3, 4, 8, 13, 22, 23, 25, 26, 27, 29, 31, 32, 33, 34, 35
catalytic activity, 3
catheter, 133
C-C, 348
CDC, 302
cell assembly, vii, 1, 6, 38

cell biology, 258
cell culture, 261
cell differentiation, 265
cell line, 261
cell membranes, 6
cell surface, 255, 256, 598
ceramic, 344, 345
cerebral blood flow, 111, 112, 113
challenges, ix, xii, 42, 45, 97, 100, 110, 113, 128, 130, 246, 255, 297, 401, 595, 623
charge density, 102, 281, 284, 285, 287, 401
chemical, ix, x, xiv, 3, 4, 5, 42, 94, 123, 179, 180, 245, 247, 314, 344, 348, 385, 396, 425, 437, 445, 449, 457, 459, 464, 472, 476, 477, 493, 494, 558, 559, 573, 576
chemical degradation, 4, 5
chemical industry, xiv, 385
chemical reactions, 42, 94, 396, 459
chimneys, 137
China, 205, 317, 385, 451, 461, 465, 467, 471, 549, 579
chondrocyte, 264
chromium, 3
chronic obstructive pulmonary disease, 131
circulation, 79, 85, 127, 207, 227, 231, 237, 438, 443, 447, 495, 509
classes, 72, 218, 386
classification, xiii, 131
closure, xiv, 130, 208, 224, 385, 426, 427, 495
clusters, 343
CMC, 310, 417
C-N, 348
CO_2, ix, 110, 111, 121, 123, 124, 129, 134
coal, 451
coal liquefaction, 451
coding, 197, 410
coefficient of variation, 461
collaboration, xi, 135, 243, 244
collagen, 244, 245, 248, 249, 250, 252, 253, 254, 257, 259, 260, 261, 262, 263, 265
collagen sponges, 262
collection electrodes (CEs), xi, 267
collisions, 219, 388, 403, 557, 623
color, 304, 400, 594
coma, viii, 109, 111
combined effect, 236, 238, 310
combustion, 42, 314, 459, 540, 573, 594
commercial, x, 122, 139, 164, 180, 181, 183, 191, 192, 196, 197, 249, 250, 251, 258, 298, 302, 311, 324, 334, 553
communities, xi, 243, 244
community, xii, 107, 297, 298
compaction, 252

compatibility, 116, 118
competition, 614
complex interactions, 208
complexity, xi, 42, 43, 72, 118, 127, 243, 245, 253, 394, 438, 460
compliance, xv, 511, 532, 533
complications, 110, 111, 124, 134, 512
composites, 245, 344, 345
composition, 2, 31, 248, 251, 252, 253
compressibility, 256, 466
compression, xiii, 244, 255, 256, 257, 317, 319
computation, xvi, 64, 68, 93, 107, 185, 186, 269, 296, 348, 398, 449, 470, 496, 501, 507, 515, 541, 593
Computational Fluid Dynamics (CFD), vii, x, xiv, 1, 8, 36, 39, 104, 109, 123, 132, 133, 180, 181, 182, 183, 188, 201, 208, 254, 259, 267, 298, 299, 312, 313, 315, 385, 387, 459, 468, 471, 476, 491, 494, 512, 553
computational grid, 53, 446, 496
computational modeling, 263, 345
computed tomography, 248
computer, 36, 55, 72, 183, 188, 189, 258, 259, 273, 298, 345, 387, 460, 494, 497, 498, 512, 553, 554, 614, 624
condensation, 5, 11, 22, 302, 318, 331, 332, 335, 336, 337, 338, 340
conditioning, 500
conduction, 34, 200, 282, 304, 308, 472, 552
conductivity, 2, 3, 4, 11, 12, 13, 15, 18, 23, 31, 33, 34, 35, 36, 98, 146, 301, 318, 334, 438, 472, 563
configuration, xii, xiv, xv, 6, 134, 164, 176, 194, 268, 279, 280, 293, 445, 454, 482, 493, 494, 497, 498, 502, 503, 504, 505, 506, 522, 546, 571, 577, 578, 603, 605, 609, 612
conjugate gradient method, 497
connectivity, 55, 57, 108, 601
consensus, 513
conservation, 10, 45, 46, 47, 48, 49, 51, 52, 61, 65, 69, 71, 99, 102, 107, 116, 269, 285, 298, 321, 332, 402, 410, 412, 415, 427, 431, 437, 449, 452, 466, 467, 471, 478, 479, 495, 548, 549, 552, 553, 558, 595, 596, 597, 606, 610, 625
constant rate, 301
constituents, 244, 245
construction, 61, 122, 343, 451, 496, 594
consumption, 29, 308, 446, 462, 470, 476, 505
contact time, 220, 454
containers, 310, 315
contour, 29, 30, 35, 88, 94, 108, 186, 187, 190, 212, 439, 440, 443, 448, 451, 460, 537, 538
contradiction, 121, 619

convergence, xv, 113, 185, 201, 437, 452, 496, 500, 502, 504, 507, 511, 529, 530, 531, 544
cooling, ix, 137, 138, 148, 174, 176, 177, 299, 322, 392, 552, 576
coronary artery bypass graft, 131
correlation, 99, 119, 138, 214, 215, 216, 217, 252, 270, 310, 419, 420, 426, 434, 438, 445, 457, 559
correlations, ix, x, 97, 137, 152, 153, 157, 158, 180, 181, 182, 222, 426, 438, 443, 446, 469
corrosion, 4, 464
cost, vii, viii, 1, 3, 36, 109, 121, 181, 257, 322, 324, 386, 398, 436, 451, 476, 505, 532
CPB, viii, 109, 111, 113, 121, 128, 131
CPU, 512, 539, 541
cracks, 36
critical value, 319
crust, 304, 305, 308
crystal growth, 459
crystalline, 138
crystallization, 303, 461
crystals, 303
CSF, 184, 191, 192, 194, 196, 197, 198
CT scan, 248, 254
cues, 245, 257
cultivation, xi, 243, 244, 245, 247, 255
culture, 247, 253, 261, 262, 263, 265
culture conditions, 247, 265
cures, 388
curvilinear grid, 43
cycles, vii, 1, 2, 4, 6, 7, 38, 504
cycling, 7
cytoplasm, 255, 256
cytoskeleton, 255, 256, 264

D

damage mechanisms, vii, 1, 6, 8
damping, 429, 460, 525, 527
data analysis, 186
data set, 128
data structure, 56
database, 185
death rate, 218, 219
decay, 477, 481, 611, 612
decomposition, 5, 496, 506
deconvolution, 473
decoupling, 247
deformability, 90
deformation, xvi, 15, 29, 30, 31, 32, 33, 34, 35, 42, 43, 71, 93, 100, 102, 103, 106, 108, 125, 199, 215, 245, 246, 252, 255, 257, 261, 321, 387, 415, 438, 495, 496, 533, 537, 542, 543, 544, 546, 593, 602, 612, 613, 620, 624, 625

degradation, vii, 1, 2, 4, 251
degradation mechanism, vii, 1, 4
dehydration, 6
Delta, 530, 531, 586
denaturation, 304, 312
deposition, 3, 134, 251, 261, 262, 290, 302, 314
deposition rate, 302, 314
depth, x, 205, 231, 245, 255, 258, 343, 462
derivatives, xv, 53, 65, 493, 497, 498, 507, 508, 524, 527
desorption, 218
destruction, 197, 311
detachment, 191, 195, 197, 200
detection, 345, 476
deviation, 81, 277, 401, 408, 410, 416, 473
DFT, xiii, 343, 345, 346
diaphragm, 322
diastole, 116, 125, 126, 130
dielectric constant, 284
differential equations, 127, 426, 496, 509, 526
differential scanning, 305
diffusion, 4, 8, 10, 11, 12, 14, 18, 21, 22, 27, 29, 33, 34, 36, 38, 43, 62, 69, 100, 101, 123, 189, 223, 247, 248, 282, 301, 386, 387, 396, 399, 413, 419, 421, 432, 435, 436, 460, 475, 476, 477, 478, 480, 481, 482, 519, 520, 521, 522, 532, 558, 559, 560, 561
diffusion process, 396, 399, 478
diffusion time, 399
diffusivities, 12
diffusivity, 14, 23, 29, 31, 124, 150, 282, 401, 413, 414, 453, 463, 478, 532, 560
direct numerical simulation, viii, 41, 42, 44, 104, 106, 122, 460
discharge electrodes (DEs), xi, 267
discontinuity, 42, 60, 62, 413
Discrete Geometric Conservation Law (DGCL), xv, 511, 513
discrete phase model (DPM), xii, 268, 299
discretization, 68, 185, 192, 197, 324, 387, 410, 421, 436, 446, 452, 465, 466, 496, 497, 502, 505, 533, 539, 553, 554, 598
diseases, 115, 388
dispersion, 4, 208, 214, 217, 218, 221, 232, 234, 240, 295, 300, 345, 426, 427, 430, 433, 434, 437, 447, 448, 449, 469, 623
displacement, 7, 29, 30, 31, 32, 33, 34, 35, 128, 209, 250, 252, 254, 255, 465, 466, 514, 522, 524, 525, 532, 544, 599
distribution function, xv, 71, 198, 218, 389, 390, 391, 392, 393, 394, 396, 401, 551, 554, 555, 556, 557, 558, 559, 562, 567, 568, 571
divergence, 10, 47, 63, 67, 69, 70, 183, 184, 281, 436

DNA, 476
donors, 246
draft, 454, 456, 470
drag reduction, x, 205, 206, 207, 208, 216, 232, 233, 235, 236, 237, 238, 239, 620
drainage, 623
drying, xii, 5, 23, 34, 244, 297, 299, 301, 302, 303, 311, 312, 313, 314, 315
DSC, 305
durability, vii, viii, ix, 1, 2, 4, 5, 6, 7, 37, 110, 115
dynamic viscosity, 119, 120, 126, 140, 281, 318, 319, 334, 463, 477, 480

E

ECM, 245
editors, 507, 508
egg, 308, 309, 314
electric charge, 102
electric field, xi, xii, 102, 103, 104, 106, 267, 268, 279, 280, 281, 284, 294
electrical conductivity, 5
electricity, 138
electrocatalyst, 4
electrode surface, 284
electrodes, xi, 3, 29, 33, 267, 269, 279, 287, 290, 293, 345
electrohydrodynamics, viii, 41, 97, 105
electrolyte, vii, 1, 2, 3, 5, 6, 14, 22, 23, 29, 31, 32, 37, 38, 401
electromagnetic waves, 304
electrons, 3, 4, 13, 27, 29, 33, 34
Electrostatic precipitators (ESPs), xi, 267
elongation, 201
embolism, 124
emission, 256, 295
emulsions, 108, 494, 625
endoscope, 446
energy, vii, xii, xv, 1, 2, 4, 8, 11, 12, 98, 116, 131, 138, 140, 141, 142, 177, 189, 198, 207, 213, 222, 270, 298, 299, 302, 308, 317, 318, 319, 321, 322, 331, 332, 392, 393, 394, 401, 430, 432, 434, 440, 445, 451, 452, 463, 469, 476, 493, 495, 509, 513, 515, 536, 537, 546, 563, 576
energy consumption, 451
energy transfer, 515, 546
engineering, xi, xiii, xiv, 42, 111, 113, 127, 130, 182, 243, 244, 245, 246, 247, 248, 254, 256, 258, 259, 261, 262, 263, 264, 265, 298, 312, 313, 385, 386, 457, 460, 476, 477, 490, 508, 512, 616
England, 315
entropy, 13, 27, 39, 331, 334, 579

Index

environment, xii, 2, 8, 34, 127, 246, 247, 255, 257, 317, 319, 322, 447
environmental conditions, 2, 4
environmental organizations, 298
enzymes, xii, 297, 299, 310
equilibrium, 142, 271, 273, 284, 331, 333, 388, 389, 390, 391, 392, 396, 402, 403, 464, 556, 557, 558, 559, 562
equipment, 98, 298, 303, 323, 476
erythrocytes, 123
eukaryotic cell, 264
Europe, 239, 240, 315
evaporation, 11, 18, 98, 99, 206, 301, 302, 304, 318, 319
evolution, 6, 7, 42, 92, 100, 101, 197, 208, 218, 231, 388, 393, 396, 401, 402, 416, 421, 459, 522, 525, 542, 543, 610, 613, 614, 617, 618, 620
exchange rate, ix, 110, 121
experimental condition, 75, 83, 201
experimental design, 297
exposure, ix, 110, 119, 126, 471
extracellular matrix, 244, 245, 247, 259, 261
extracorporeal blood pumps, viii, 109
extraction, 419, 421, 437, 467, 576
extrusion, 299

F

fabrication, ix, 4, 37, 179, 180, 194, 244
FEM, 246, 252, 255, 256, 479, 525, 539, 546, 549
femur, 258
fiber, 33, 121, 122, 123, 124, 129, 130, 134, 258, 322, 344, 447, 470
fiber bundles, 122
fiber membranes, 121, 122, 123
fibers, ix, 104, 110, 121, 122, 123, 124
fibrate, 104
fibroblasts, 260
fidelity, 600, 623
film formation, 194, 195
film thickness, x, 180, 182, 197, 220
films, 72, 192, 195, 345, 623
filtration, 397
financial, 201, 312, 340, 461, 490
financial support, 201, 312, 340, 461, 490
finite element method, xiii, 7, 259, 479, 547, 548, 549
fixed grid, viii, 41, 55, 56, 60, 61, 63, 67, 70, 71, 104, 192
flexibility, 104, 135, 291, 294, 308, 344, 386
flight, 260, 512
flocculation, 448
flooding, 440, 463

flow curves, 124, 126
flow field, x, xii, xiv, xv, 10, 43, 44, 46, 53, 54, 56, 63, 67, 68, 81, 99, 100, 101, 119, 120, 126, 133, 180, 185, 186, 205, 208, 233, 267, 279, 280, 299, 385, 410, 425, 431, 437, 438, 442, 447, 448, 450, 460, 468, 469, 470, 493, 494, 495, 496, 498, 502, 506, 541, 568, 571, 581, 586, 620, 621, 625
fluctuations, 57, 93, 270, 426
flue gas, xi, xii, 267, 268, 279
fluidized bed, 434, 469, 470
Fluid-Structure Interaction (FSI), xv
fluorescence, 191
foils, 345
food industry, 298, 311, 312, 314, 315, 319
food products, 299, 311, 314
formation, ix, 5, 37, 56, 94, 95, 96, 97, 105, 106, 107, 110, 116, 118, 123, 124, 125, 126, 130, 135, 181, 183, 186, 189, 190, 191, 193, 194, 197, 198, 200, 244, 247, 255, 257, 258, 262, 303, 304, 331, 332, 348, 401, 406, 407, 446, 459, 464, 465, 467, 625
formula, 12, 59, 101, 212, 270, 427, 437, 454, 594, 597
France, 240, 473
free energy, 39, 197
free surface, xiii, 106, 108, 192, 419, 448, 451, 453, 456, 594
freedom, 388, 519, 554
freezing, xii, 297, 303, 304, 311, 312
friction, x, 93, 205, 206, 207, 210, 211, 235, 236, 238, 241, 438, 446
front-tracking method, viii, 41, 104, 108, 594
fuel cell, vii, 1, 2, 3, 4, 5, 6, 7, 8, 10, 16, 17, 22, 23, 24, 27, 29, 30, 31, 32, 33, 34, 35, 36, 37, 38, 39, 579
fuel cell systems, vii, 1, 4, 8
fundamental needs, 494
fusion, 200

G

gait, 258
gas diffusion, 3, 4, 5, 6, 7, 8, 11, 12, 20, 22, 23, 27, 29, 33, 34, 35, 36, 396
gas–liquid two-phase flows, ix, 179
gel, xiii, 244, 302, 343, 344, 345
gene expression, 260
genes, 246, 247
genetic predisposition, xi, 243
geometrical parameters, 502, 503
Germany, 109, 315
glucose, 477, 503
glycerin, 84

glycosaminoglycans, 260
grading, 272
grants, 490
granules, 305
graphite, 6, 16
gravitation, 229, 598
gravitational force, ix, 118, 179, 181, 595, 609, 611
gravitational forces, ix, 118, 179, 181, 595, 611
gravity, 47, 189, 214, 228, 237, 300, 462, 562, 610
grid generation, 502
grid resolution, 192, 406
grids, xiv, 43, 55, 71, 194, 272, 385, 435, 446, 496, 497, 507, 547, 548, 549, 573, 577
groundwater, 508, 578
growth, xi, 7, 81, 189, 192, 200, 218, 243, 244, 245, 247, 251, 254, 255, 257, 261, 263, 303, 311, 331, 332, 466, 472
growth factor, 245, 263
guanine, 345

H

healing, 257, 264, 265
health, 297, 308
heart disease, 135
heart failure, 115
heart transplantation, 115
heart valves, ix, 110, 135
heat capacity, 300, 301, 319, 332
heat loss, 148
heat release, 332, 459
heat transfer, ix, xiv, 8, 12, 33, 98, 100, 137, 138, 139, 143, 144, 145, 146, 148, 149, 150, 151, 152, 153, 154, 156, 157, 158, 161, 162, 163, 164, 165, 166, 167, 168, 169, 170, 171, 172, 173, 174, 175, 176, 177, 178, 182, 200, 213, 240, 300, 301, 303, 304, 308, 310, 311, 312, 313, 314, 315, 332, 385, 396, 397, 398, 445, 451, 552, 554, 563, 576, 579
heating rate, 305
height, 18, 81, 93, 117, 139, 144, 148, 149, 150, 151, 153, 163, 167, 171, 175, 176, 234, 276, 277, 279, 395, 405, 406, 454, 477, 503, 581
hematocrit, 119
hemisphere, 607
hemoglobin, 123, 124
heterogeneity, 254, 258
hexane, 449
hip replacement, 258
histology, 260
homeostasis, 246, 260
homogeneity, 252, 254, 258, 448
hospitalization, 115
host, xi, 243, 244, 459

human, viii, ix, xi, 109, 110, 112, 116, 127, 130, 243, 244, 260, 261, 262, 264
human body, ix, xi, 110, 116, 127, 130, 243, 244
human neutrophils, 264
humidity, vii, 2, 4, 5, 6, 7, 8, 9, 15, 16, 18, 19, 29, 36, 37, 38, 299, 301, 314
hybrid, 42, 44, 119, 222, 386, 410, 436, 466, 548, 620
hydrogen, 2, 3, 4, 5, 10, 13, 14, 22, 23, 29, 31, 32, 305, 346
hydrogen bonds, 305
hydrogen peroxide, 5
hydroxyapatite, 244

I

IAM, 509
IMA, 508
image, 80, 133, 191, 248, 249, 250, 256, 258, 259, 618
images, 79, 112, 251, 260
immersion, 244, 322
impact assessment, 313
implants, xi, 243
improvements, 44, 68, 116, 121, 269, 290, 410
impurities, 2
in vitro, 132, 135, 262
in vivo, 132, 258, 260, 265
incidence, 115
indentation, 79
independence, 195
India, 297, 312, 343
individuals, 245, 258
induction, xi, 243, 244, 245, 253
industries, xi, xii, 102, 267, 297, 298, 448, 451, 494
industry, xii, 98, 100, 182, 247, 317, 443, 498
inequality, 504
inertia, 42, 44, 118, 282, 300, 400, 476, 477, 504, 561, 614
inertial effects, 82, 407
information exchange, 71
ingredients, 299, 314
initial state, 76
initiation, 7, 135, 476
inlet diffuser, xi, 267
inlet evase, xi, 267, 272, 291, 292, 293, 294
institutions, 126, 130
insulation, 2
insulin, 247
integration, vii, 1, 47, 63, 68, 71, 221, 415, 434, 435, 452, 467, 514, 522, 528, 533, 548
integrin, 261
integrity, 4, 305, 538

interaction process, 404
interfacial flow problems, viii, 41, 44, 52, 105
interfacial layer, 409
interference, 401, 465
interphase, xiv, 385, 387, 413, 414, 419, 421, 423, 426, 428, 429, 437, 438, 443, 461, 467
investment, 344
ions, 279, 284
Iran, 551
Iraq, 1
Ireland, 243
isotherms, 572, 573, 574, 577
Israel, 177, 472
issues, vii, xiii, xv, 1, 4, 5, 52, 247, 459, 511, 522, 544, 622
iteration, xv, 69, 70, 185, 186, 436, 500, 501, 511, 513, 544, 547, 581, 586, 607
iterative solution, 437

J

Japan, 134, 238, 240, 467, 548, 550
joints, 6

K

kerosene, 400, 401, 404, 405, 406, 407, 408, 409, 433
kidney, xi, 243, 244
kidney dialysis, xi, 243, 244
kidneys, xi, 243, 244
kinetic model, 305, 308, 315
kinetics, x, xv, xvi, 31, 32, 39, 179, 302, 315, 316, 463, 551, 552
knees, xi, 243, 244
Korea, 239

L

laminar, 113, 139, 140, 141, 177, 182, 186, 187, 188, 219, 240, 254, 420, 434, 443, 454, 461, 463, 476, 477, 480
large eddy simulation (LES), xiv, 385
lasers, 128
lattice Boltzmann method (LBM), x, 180, 386, 459
Lattice Gas Automata (LGA), xvi, 552
lattice parameters, 404
lattices, 387, 388, 405, 559, 570
laws, 45, 46, 51, 457, 490, 519, 548, 552, 553
leaching, 244

lead, vii, xiii, 1, 6, 8, 23, 27, 32, 34, 42, 53, 54, 57, 63, 65, 94, 100, 116, 119, 120, 124, 130, 170, 193, 200, 252, 254, 255, 257, 318, 325, 340, 387, 420, 437, 464, 478, 512, 605
leakage, 130, 208, 212, 222, 224, 227, 229, 238
Lebanon, 202
Left Ventricular Assist Devices (LVADs), viii, 109
life cycle, 538
life expectancy, 115
lifetime, xi, 4, 5, 37, 115, 117, 243, 244, 300, 434, 462
ligament, 614, 625
ligand, 248
light, xi, 94, 243, 246, 257, 387, 512
linear systems, 496
liquid interfaces, 343
liquid phase, xiii, 8, 10, 11, 12, 20, 22, 82, 96, 97, 182, 192, 207, 214, 221, 222, 234, 332, 427, 429, 440, 441, 442, 448, 452, 453, 463, 623
liquids, vii, xiii, 42, 87, 90, 95, 96, 105, 106, 107, 191, 240, 387, 416, 417, 437, 467, 494, 609, 623
local conditions, 8, 29, 303
localization, 135
lung function, 121
Luo, 203, 220, 231, 240, 435, 464, 465, 468, 470, 472, 507, 547, 553, 579

M

magnetic field, xvi, 111, 552
magnitude, 7, 13, 27, 34, 66, 117, 118, 152, 160, 164, 186, 195, 226, 229, 247, 252, 254, 255, 256, 260, 286, 428, 478, 521, 529, 559, 610, 616, 620, 621
majority, 119, 191, 207, 244, 252, 612
man, 491
management, 5, 32, 34, 134
manipulation, 131, 193
marrow, 261, 262, 263
mass loss, 623
material degradation, 6
materials, vii, 1, 2, 4, 5, 29, 37, 124, 185, 244, 246, 259, 299, 302, 303, 311, 313, 343, 344, 345, 464, 482, 494
matrix, 16, 219, 251, 261, 262, 497, 498, 499, 500, 501, 521, 541, 598
matter, 86, 118
Maxwell equations, 12
measurement, x, xv, 86, 121, 127, 129, 138, 179, 180, 208, 258, 260, 275, 277, 305, 307, 446, 449, 470, 493
measurements, viii, ix, 73, 86, 109, 110, 121, 125, 128, 129, 143, 223, 226, 277, 287, 299, 310, 311,

325, 335, 345, 437, 443, 447, 449, 450, 468, 469, 471, 619
mechanical degradation, 5
mechanical properties, 7, 37, 245, 260, 263, 265
mechanical stress, vii, 1, 5, 6, 7, 247, 248, 249, 255, 264
media, 122, 123, 178, 263, 552, 576, 578
medical, ix, 110, 115, 130, 244
medical science, 244
medication, ix, 110
medicine, xi, 127, 243, 244, 246, 476
melting, 305
membranes, 4, 5, 6, 7, 34, 37, 38, 105, 121, 345
mesenchymal stem cells, 261, 263
messages, 514
metabolism, 261
metallurgy, 319
meter, 185, 462
methanol, 345, 579
methodology, xv, 98, 126, 142, 192, 193, 194, 195, 197, 200, 259, 460, 473, 493, 509, 513, 522
methyl cellulose, 310
Mexico, 37
microchannels, ix, x, 179, 180, 181, 182, 187, 192, 193, 194, 200, 201, 386, 387, 400, 409, 464, 465, 480, 490
micrograms, ix, 179
micrometer, 180
microorganisms, 308
microscopy, 191, 251, 253
Microsoft, 187
microstructure, 105, 252
Middle East, 493
migration, 4, 208, 244, 245, 248, 252, 263
miniature, 476
missions, 310, 314
mixing, xii, xiii, xiv, xv, 86, 100, 182, 201, 212, 299, 317, 319, 324, 325, 326, 327, 332, 339, 340, 396, 446, 451, 457, 462, 465, 468, 470, 475, 476, 477, 478, 479, 480, 481, 482, 490, 493, 494, 495, 498, 505, 507, 605, 622
model system, 262
modelling, x, xi, xii, 5, 16, 38, 139, 165, 179, 180, 181, 182, 183, 184, 188, 189, 190, 191, 192, 193, 194, 195, 197, 199, 200, 201, 208, 237, 239, 240, 243, 244, 246, 247, 248, 253, 254, 255, 256, 257, 259, 267, 279, 280, 291, 294, 297, 302, 303, 311, 315, 468, 469, 579
modern society, 130
modifications, xii, xv, 55, 68, 268, 269, 310, 311, 315, 441, 511, 513, 544, 554, 561
modules, 174, 176, 177
modulus, 19, 140, 255, 265, 523, 540

moisture, 5, 7, 15, 302, 304, 314
moisture content, 5, 304
mold, xiii
mole, 10, 13, 14, 18, 558
molecular dynamics, 387, 553, 554
molecular mass, 319
molecular structure, 346
molecular weight, 13, 344
molecules, 14, 195, 247, 256, 305, 345, 477, 553, 554
morphology, x, 205, 208, 246, 250, 253, 254
motivation, 388
Moving Mesh Interface Tracking (MMIT), xvi, 593
MPI, 502, 514
MRI, ix, 110, 111, 112, 114, 128, 246, 248, 251
mRNA, 261
multidimensional, 240
multiphase materials, 106

N

nafion, 39
nanodevices, 343
nanometer scale, 344
nanometers, 195, 254
nanoparticles, 343
nanoscale materials, 345
nanostructures, 343, 345
natural gas, 121
natural science, 340
Navier–Stokes equations, viii, 41, 104, 279, 321
necrotic core, 247
nerve, 131
Netherlands, 473
neural function, viii, 109, 111
neutrophils, 264
nitrogen, 23, 348
no dimension, 560
nodes, xiii, xvi, 16, 54, 55, 56, 57, 59, 64, 65, 71, 125, 186, 188, 189, 195, 199, 200, 250, 252, 273, 386, 388, 389, 390, 391, 412, 437, 447, 515, 543, 544, 593, 595, 598, 599, 601, 605, 606, 607, 622
nonequilibrium, 332
non-isothermal computational fluid dynamics model, vii, 1, 8
Norway, 41
nozzle supersonic flow process, xiii, 318
nucleation, 200, 218, 303, 318, 331, 332
nuclei, 218
nucleus, 255, 256
null, 522, 523, 533, 536, 540
numerical analysis, 180, 459
numerical computations, 480

numerical tool, 73, 495
nutrients, 246, 247, 248, 309

O

obstacles, 127
oil, 73, 100, 102, 449, 450, 451, 463
one dimension, 189, 389, 563, 564
operating system, 16
operations, viii, ix, xii, 56, 57, 58, 59, 60, 72, 109, 111, 128, 179, 246, 297, 298, 299, 303, 311, 312, 494, 604, 605
optical microscopy, 624
optimization, vii, xiv, xv, 1, 36, 37, 120, 121, 133, 200, 259, 297, 315, 346, 452, 457, 476, 493, 494, 497, 498, 500, 501, 502, 504, 505, 507, 508, 509, 513, 550, 602, 606, 622
optimization method, 497, 507, 508
ordinary differential equations, 246
organ, xi, 243, 244, 245
organize, 261, 343
organs, 115, 124, 134
oscillation, 62, 86, 89, 189, 190, 192, 533, 610, 611, 612
osmosis, 38
osteocyte, 264
outlet convergent duct, xi, 267
outlet evase, xi, 267, 272, 291
oxidation, 2
oxygen, ix, 2, 3, 10, 13, 14, 22, 23, 29, 31, 32, 33, 34, 39, 110, 134, 247, 263, 265, 387, 399, 400, 457, 464, 471
oxygenators, viii, 109, 111, 121, 122, 123, 124, 134
ozonation, 471

P

parallel, xvi, 45, 104, 107, 138, 139, 144, 164, 177, 181, 247, 253, 279, 348, 405, 436, 459, 460, 500, 505, 506, 523, 549, 551, 563, 571, 609, 617, 618
parallel processing, xvi, 45, 552
parallelization, 496
partial differential equations, xvi, 182, 183, 246, 271, 284, 387, 435, 479, 480, 552
particle mass, 282, 284, 388, 446
partition, 419
pasteurization, xii, 297, 299, 308, 310, 313, 314
pathways, 34, 126, 245, 247
PDEs, 246, 480, 553
perfusion, 113, 131, 132, 253, 261, 262, 263, 264, 265
periodicity, 117

permeability, 11, 18, 22, 23, 122, 248, 251, 252, 253, 263, 270, 291, 562
petroleum, 42, 319, 576
pharmaceutical, 247, 299, 494
phase boundaries, 43
phenotype, 257
Philadelphia, 315, 472, 507
phosphate, 244, 254, 263
photons, 304
physical activity, 261
physical characteristics, 525
physical features, 269, 294
physical mechanisms, viii, 2, 8, 452
physical phenomena, 44, 208, 459
physical properties, 42, 305, 404, 417, 479, 494, 495
physicians, 110
physics, x, xvi, xvii, 11, 42, 45, 72, 73, 97, 102, 104, 127, 179, 180, 181, 191, 197, 200, 201, 229, 238, 459, 549, 551, 593, 602, 604, 609, 622, 623
pitch, 164, 165, 166, 171
plastic deformation, 5, 6, 7, 38
platelets, 126
platform, 257, 324, 334
platinum, 3, 4, 39
Poisson equation, 45, 63, 281
Poland, 475
polarization, 16, 17, 102
polymer, 2, 4, 5, 6, 7, 14, 23, 29, 37, 38, 261, 262, 263, 344, 345
polymer chains, 345
Polymer electrolyte membrane (PEM), vii, 1
polymer matrix, 345
polymerization, 437
polyurethane, 244
porosity, 11, 18, 31, 34, 35, 36, 122, 252, 253, 254, 255, 265, 291, 292, 293, 294, 561, 578, 579
porous media, 12, 122, 123, 124, 270, 466, 554, 561, 563, 576, 577, 578, 579, 580
power generation, 7, 8, 98
power plants, xi, 267
precipitation, 244, 279
prediction models, 5
preparation, 345
pressure gauge, 446, 470
pressure gradient, 12, 20, 22, 67, 70, 93, 209, 246, 449
probe, xiii, 42, 277, 343, 438, 447, 470
process control, 311
project, 312, 340, 348
proliferation, 245, 247, 261, 265
propagation, 250, 401, 404, 459, 472
prostheses, 124, 125, 134, 135
prosthesis, 110, 258

proteins, 123, 244, 310, 312
protons, 3, 8, 29, 31, 33
prototypes, viii, 109, 268
pumps, viii, 109, 111, 123, 128, 133, 452
pure water, 12, 86, 87

Q

quadratic curve, 499
quality control, 254
quantification, 249, 257
Queensland, 267

R

Rab, 151
radiation, xii, 37, 164, 165, 169, 170, 171, 173, 174, 177, 178, 297, 304, 305, 312, 472
radius, 81, 82, 85, 90, 192, 209, 285, 318, 333, 412, 415, 462, 498, 499, 500, 501, 504, 541, 542, 607, 611, 612, 613, 614, 615, 620, 621, 622
Raman spectra, xiii, 343, 345, 346
Raman spectral studies, xiii, 343
Raman spectroscopy, 345
REA, 302
reactants, 2, 3, 5, 8, 13, 23, 27, 34, 344
reaction rate, 306
reactions, 3, 8, 27, 32, 39, 247, 304, 464, 476
reading, 105
reagents, 476
reality, 110, 113, 311, 459, 609
reconstruction, 43, 108, 197, 248, 251, 252, 253, 257, 258, 522
recovery, 115, 178, 236
rectangular collection chamber, xi, 267
red blood cells, 118
redistribution, 247, 623
reference frame, xii, 45, 86, 282, 297, 387, 419, 428, 431, 432, 434, 435, 497, 522, 621
refractive index, 129
regeneration, 244, 259, 264
regression, 153, 180, 457
rehydration, 303
relative humidity, vii, 2, 5, 7, 8, 9, 15, 16, 36
relaxation, 7, 38, 185, 198, 386, 388, 389, 393, 396, 401, 402, 403, 404, 433, 436, 437, 459, 460, 463, 512, 556, 560, 614, 615, 619
relaxation times, 401, 404, 560
relevance, viii, 109, 124, 400
reliability, vii, x, 1, 7, 117, 180, 394
remodelling, 246, 247
renewable energy, 138

renormalization, 473
requirements, 2, 3, 22, 56, 67, 116, 117, 127, 248, 311
researchers, ix, x, 179, 180, 181, 182, 185, 194, 200, 208, 247, 255, 257, 269, 279, 298, 521, 537
residuals, 113, 185, 197
residues, 324, 334
resistance, 4, 6, 15, 24, 31, 33, 34, 116, 138, 154, 155, 159, 166, 167, 236, 291, 293, 294, 295, 419, 421, 423, 467, 504
resolution, 16, 43, 45, 54, 56, 60, 67, 71, 72, 81, 111, 123, 124, 126, 128, 190, 209, 223, 254, 263, 406, 410, 604, 623
resources, xii, 9, 120, 183, 267, 460
response, vii, 1, 5, 6, 7, 8, 16, 38, 128, 255, 256, 257, 258, 260, 261, 262, 434, 462, 525
response time, 434, 462
restoration, 500
restrictions, 117, 402, 498
restructuring, 56, 57, 60, 71
retail, 299
Reynolds-averaged Navier-Stokes method (RANS), xiv, 385
RFS, 345
rheology, 264
rings, 88
risk, 110, 111, 115, 116, 131, 246
root, 223, 270, 439, 524
root-mean-square, 270, 439
routes, 246
rubber, 322
rules, 114, 200, 388, 395, 535
ruthenium, 3

S

safety, 130, 297
saturation, 11, 12, 22, 32, 98, 130, 284, 301, 319, 331, 333
scaling, xiv, 443, 462, 475, 541
scanning electron microscopy, 256
scatter, 433
scattering, 178
SCO, 75
sea level, 537
sediment, 446
seed, 250
seeding, 250, 262
segregation, 447
self-assembly, 345
sensitivity, 208, 223, 237, 309, 454
sensors, 260
sequencing, 476

shade, 227, 608
shape, 29, 30, 35, 42, 46, 53, 57, 59, 60, 67, 73, 74, 75, 76, 78, 79, 80, 81, 82, 86, 92, 93, 101, 107, 108, 125, 131, 190, 192, 195, 196, 197, 198, 208, 214, 264, 283, 309, 310, 407, 415, 418, 419, 423, 429, 481, 482, 502, 538, 541, 594, 606, 610, 612, 614, 615, 616, 617, 618, 620, 621
shear, ix, 6, 12, 93, 104, 110, 113, 118, 119, 120, 123, 125, 126, 130, 182, 192, 193, 195, 197, 207, 209, 210, 212, 214, 219, 222, 237, 246, 247, 248, 250, 253, 254, 255, 257, 258, 262, 263, 264, 274, 416, 417, 418, 426, 440, 448, 453, 461, 463, 470, 512, 597, 600, 609, 612, 613, 624
shear rates, ix, 110, 113, 120, 123, 125, 130
sheep, 131
shock, xiii, xv, 43, 118, 317, 319, 325, 330, 386, 511, 513, 523, 534, 536, 538, 539, 540, 542, 543, 544, 549, 550, 619, 625
shock waves, 325, 386, 619, 625
shortage, 115
side chain, 245
signaling pathway, 261
signalling, 244
signals, 247, 345
silica, 302
silicon, 102, 138
silver, 345, 346, 347, 348
Singapore, 593
sintering, 4
skilled personnel, 183
skin, 137, 148, 236, 241, 245, 259, 515, 525
smoothing, xvi, 53, 61, 187, 549, 593, 595, 599, 602, 605, 606, 607, 622, 624
sodium, 345
Solar heated ventilation cavities, ix, 137
sol-gel, xiii, 343, 344
solid matrix, 3, 12, 34, 561, 576
solid phase, 434, 443, 444, 446, 449, 450, 451, 460, 463
solid surfaces, 191
species, 8, 10, 11, 12, 245, 401, 457, 460, 558, 559, 560, 570
specific heat, 11, 99, 141, 301, 319, 334
specific surface, 248
spectroscopy, 345
square lattice, 389
stability, viii, xv, 109, 116, 192, 390, 392, 394, 395, 401, 511, 512, 513, 521, 523, 526, 529, 532, 548, 550, 600
stabilization, 513, 517, 519, 520, 522, 523, 535, 539, 544, 548
starch, 304, 305, 306, 314, 315, 316
starch granules, 305

states, 51, 122, 221, 388, 514, 518, 526, 527
Steam-jet vacuum pump, xii, 317, 319, 320
steel, 6, 189
stem cell differentiation, 265
stent, 259
stimulus, 260
stochastic model, 473
storage, 138, 148, 308, 386
stress, 5, 6, 7, 16, 19, 29, 37, 47, 100, 102, 113, 115, 118, 119, 133, 192, 207, 209, 210, 212, 222, 237, 247, 248, 250, 252, 253, 254, 255, 258, 260, 262, 263, 264, 318, 321, 425, 427, 448, 460, 473, 495, 496, 504, 512, 514, 561, 600, 609, 612, 624
stress concentrators, 252
stretching, 261, 346, 348
stroke, 131, 132
stromal cells, 261
substitutes, xi, 243, 244, 246, 255, 258
substrate, 348
substrates, xiii, 343, 344, 345
suppression, 439
surface area, 3, 71, 189, 221, 283, 290, 300, 596, 598
surface energy, 189, 221, 248
surface properties, 165
surface tension, viii, ix, 12, 41, 42, 44, 51, 52, 53, 54, 55, 57, 60, 64, 65, 66, 67, 72, 73, 84, 87, 100, 101, 104, 105, 108, 179, 181, 184, 185, 189, 190, 191, 192, 194, 195, 196, 197, 198, 200, 209, 212, 217, 318, 387, 410, 415, 459, 463, 467, 594, 597, 601, 611, 612, 614, 618, 625
surfactant, viii, 41, 97, 100, 101, 102, 105, 107, 387, 623
surfactants, 42, 86, 87, 100, 101, 107, 195
surgical technique, 244
survival, ix, 110, 130
survival rate, 110
suspensions, 453, 465
swelling, vii, 1, 5, 6, 7, 15, 19, 29, 38, 305
Switzerland, 196, 508, 549
symmetry, 56, 76, 185, 188, 222, 249, 271, 280, 291, 346, 388, 419, 452, 482, 616
symptomatic treatment, 388
synchronization, 514
synthesis, ix, 179, 261, 345, 464

T

tanks, xiii, 425, 429, 431, 434, 435, 437, 438, 440, 443, 445, 446, 447, 448, 449, 450, 460, 469, 470, 473, 494, 495, 546
techniques, viii, x, xii, xiii, xiv, xv, xvi, 2, 37, 42, 43, 44, 97, 109, 111, 112, 113, 116, 124, 180, 182, 183, 194, 195, 197, 198, 200, 201, 205, 207, 208,

218, 244, 246, 247, 248, 317, 319, 437, 459, 493, 494, 496, 511, 512, 544, 552, 553
technological advances, 244, 258
technologies, xv, 42, 511
technology, 2, 5, 72, 81, 110, 115, 201, 238, 248, 315, 386, 451, 459
tension, 42, 44, 52, 53, 55, 64, 65, 66, 67, 73, 181, 184, 189, 190, 196, 197, 198, 200, 255, 400, 401, 403, 404, 405, 407, 408, 415, 464, 513, 594, 612, 618
tensions, 408, 409
testing, vii, 1, 4, 134, 613
thermal destruction, 310, 313
thermal energy, 98
thermal expansion, vii, 1, 5, 15, 19, 29, 150, 558
thermal properties, 146, 168, 472
thermal stability, 3
thermodynamic equilibrium, 11, 12
thermodynamics, 459, 552, 553
thinning, 190, 192, 193, 416, 417, 418
third dimension, 188, 189, 298
three-dimensional flow modeling, viii, 41, 104
three-dimensional model, 45, 188, 190, 191, 199
three-dimensionality, 181, 188
thrombosis, ix, 110, 111, 124, 135
thrombus, ix, 110, 116, 118, 125, 126, 130, 135
tibia, 260
tissue, xi, 111, 243, 244, 245, 246, 247, 248, 251, 253, 254, 255, 256, 257, 259, 260, 261, 262, 263, 264, 265
tissue perfusion, 111
titanium, 125, 244
topology, 54, 72, 90, 100, 513
torsion, 513
total energy, 318, 321, 518, 536, 537
trajectory, ix, 87, 88, 89, 90, 179, 180, 282, 290, 299, 300, 425, 448
transducer, 322
transformation, 304, 413, 414, 423, 515, 519, 521, 535
transformation processes, 304
transformations, xv, 304, 413, 511, 513, 517, 518, 519
translation, 102
transmission, 138, 246, 258, 539
transonic flow, xiii, 317, 319, 339, 340
transpiration, 576
transplant, viii, 109, 115
transplantation, 115
transport, vii, viii, x, xii, xv, xvi, 1, 2, 4, 5, 8, 11, 12, 15, 16, 19, 20, 22, 27, 31, 32, 33, 34, 36, 37, 38, 39, 41, 46, 47, 48, 49, 101, 105, 121, 122, 123, 124, 129, 180, 182, 188, 208, 218, 222, 223, 237, 265, 268, 269, 279, 282, 311, 314, 332, 333, 340, 386, 387, 396, 400, 420, 421, 423, 426, 435, 445, 460, 464, 477, 479, 480, 551, 552, 554, 555
transport processes, vii, 1, 22, 121
transportation, vii, 1, 4, 414
treatment, 34, 115, 142, 190, 197, 272, 289, 310, 419, 425, 434, 453
trial, 132, 298
turbulence, xii, xiv, 86, 93, 106, 126, 139, 142, 177, 207, 208, 214, 219, 220, 221, 222, 225, 226, 232, 237, 268, 269, 270, 272, 314, 318, 321, 322, 325, 385, 386, 425, 426, 428, 429, 433, 434, 437, 438, 439, 440, 443, 450, 452, 460, 463, 468, 469, 473, 477, 480, 494, 495, 496, 507, 509, 546
turbulent flows, xiv, 106, 385, 460, 473
Turkey, 493
turnover, 260
two-phase flows, viii, 41, 43, 104, 107, 108, 180, 387, 400, 428, 438, 449, 466, 467, 468, 580, 598, 607, 622, 623, 624

U

UK, 137, 251
ultrasound, 111
underwater vehicles, x, 205, 206, 238
United States (USA), 37, 115, 177, 259, 261, 262, 264, 295, 507, 624
universal gas constant, 14
User Defined Functions (UDFs), xii, 268
UV, 345

V

vacuum, xii, xiii, 317, 319, 320, 331, 337, 615
validation, viii, x, 41, 44, 81, 90, 102, 111, 118, 123, 124, 125, 126, 127, 128, 129, 130, 135, 143, 183, 193, 201, 205, 258, 259, 262, 307, 314, 322, 400, 416, 502, 522, 550, 572, 611
valve, 110, 124, 125, 126, 130, 134, 135
valvular heart disease, 134
vapor, 98, 99, 100, 206, 333, 334, 466
variables, ix, 55, 62, 68, 71, 73, 82, 113, 137, 157, 181, 183, 217, 218, 258, 271, 282, 284, 312, 388, 391, 400, 404, 410, 435, 452, 453, 494, 496, 498, 501, 502, 504, 505, 506, 508, 513, 518, 519, 527, 539, 552
variations, 2, 6, 84, 107, 147, 149, 150, 155, 159, 161, 231, 304, 308, 449, 504
vascular system, viii, 109, 110, 112, 127
vascular wall, 512

vector, 10, 47, 48, 49, 51, 55, 63, 64, 66, 71, 189, 196, 213, 214, 217, 411, 413, 415, 423, 424, 436, 441, 442, 443, 444, 447, 462, 478, 495, 498, 504, 514, 515, 517, 520, 555, 595, 597, 598
vehicles, x, 205, 206
ventilated cavitation, x, 205, 207, 224
ventilation, ix, xi, 137, 138, 139, 141, 143, 144, 145, 146, 148, 158, 171, 174, 175, 176, 177, 178, 206, 207, 212, 224, 236
vertebral artery, 112, 114
vessels, 111, 112, 113, 114, 129, 386, 425, 434, 468, 469, 470, 506, 507, 508
vibration, 227, 346, 348, 490
vision, 460
visualization, ix, 79, 110, 129, 186, 277, 496
Vitamin C, 310, 313
vitamins, 299, 310
volume-of-fluid (VOF), x, 180

W

wall temperature, 138
waste, 34, 248
waste heat, 34
water evaporation, 5
wells, 73
wettability, 401, 404, 408, 449, 465
wetting, 8, 27, 181, 192, 193, 194, 198, 199, 200, 201, 404, 465, 466, 580
wires, 279

Y

yield, xv, 5, 51, 93, 256, 435, 436, 551, 553
yolk, 308